MITOCHONDRIAL TRANSPLANTATION AND TRANSFER

TRANSLATIONAL AND APPLIED BIOENERGETICS SERIES

MITOCHONDRIAL TRANSPLANTATION AND TRANSFER

BIOLOGY, METHODS, APPLICATIONS, AND DISEASES

Series Editor

SERGEJ M. OSTOJIC

Department of Nutrition and Public Kristiansand, University of Agder, Kristiansand, Norway

Edited by

GOKHAN BURCIN KUBAT

Department of Mitochondria and Cellular Research, Gulhane Health Sciences Institute, University of Health Sciences, Ankara, Turkey

Gulhane Training and Research Hospital, University of Health Sciences, Ankara, Turkey

ONER ULGER

Department of Mitochondria and Cellular Research, Gulhane Health Sciences Institute, University of Health Sciences, Ankara, Turkey; Gulhane Training and Research Hospital, University of Health Sciences, Ankara, Turkey

SERDAR GUNAYDIN

Department of Cardiovascular Surgery, University of Health Sciences, Ankara City Hospital Campus, Ankara, Turkey

ELSEVIER

ACADEMIC PRESS

An imprint of Elsevier

Academic Press is an imprint of Elsevier
125 London Wall, London EC2Y 5AS, United Kingdom
525 B Street, Suite 1650, San Diego, CA 92101, United States
50 Hampshire Street, 5th Floor, Cambridge, MA 02139, United States

ISBN: 978-0-443-18858-9

For information on all Academic Press publications visit our website at https://www.elsevier.com/books-and-journals

Publisher: Mica H. Haley
Acquisitions Editor: Ali Afzal-Khan
Editorial Project Manager: Tracy I. Tufaga
Production Project Manager: Jayadivya Saiprasad
Cover Designer: Mark Rogers

Typeset by TNQ Technologies

Dedication

As the editors of this book, we dedicate it first and foremost to the people who lost their lives in the earthquake that occurred in Turkey on 6 February 2023.

Gokhan Burcin Kubat, the first editor of this book, dedicates the book to his parents Sait Kubat and Esma Kubat for their lifelong support and to Melike Celik for her continuous support in his journey in science, medicine, and mitochondria.

Oner Ulger, the second editor of this book, dedicates the book to his honorable wife Ceylan and children Eren and Beril. He would like to express gratitude for the support received during the writing of this book.

Serdar Gunaydin, the third editor of this book, dedicates the book to his dear wife, Berrin, and son, Dogac, for their continuous support.

Contents

6. The role of extracellular vesicles and gap junctions in inter-cellular mitochondrial transfer

Dayene Caldeira and Anna Krasnodembskaya

III

Mitochondrial transplantation from isolation process to application

7. Isolation process for mitochondrial transplantation

Romana Fato, Christian Bergamini, and Francesco Pallotti

8. Determination of mitochondrial number and function

Christian Bergamini, Romana Fato, and Francesco Pallotti

9. Storage of isolated mitochondria

Francesco Pallotti, Christian Bergamini, and Romana Fato

IV

Application methods and limitations of mitochondrial transplantation

10. Mitochondrial transplantation methods

Gokhan Burcin Kubat, Ibrahim Turkel, Berkay Ozerklig, Ozbeyen Atalay, Serpil Suvay, Meltem Tuncer, Omer Faruk Taner, and Oner Ulger

11. Mitochondrial replacement therapy

Duygu Dayanir

12. Application route of mitochondrial transplantation

Meghnad Ganesh Joshi, Mrunal N. Damle, and Rakesh Kumar Sharma

13. Limitations of mitochondrial transplantation

Leena Rajendra Chaudhari, Mrunal N. Damle, Rakesh Kumar Sharma, and Meghnad G. Joshi

14. The ethical issues of mitochondrial transplantation and transfer

John B. Appleby

V

Cardiovascular diseases and mitochondrial transplantation

15. Cardiovascular diseases and pathophysiology

Therese Jamison

16. Cardiovascular diseases and mitochondrial transplantation: Basic and preclinical studies and transitions to clinical applications

Andrei B. Borisov, Kevin McCusker, David McIntyre, and Jean D. Peduzzi-Nelson

VI

Neurological diseases and mitochondrial transplantation

17. Neurological diseases and pathophysiology

Rıza Sonkaya and Elvin Malikov

18. Preclinical studies

Pasquale Picone and Domenico Nuzzo

VII

Other diseases and mitochondrial transplantation

19. Hereditary mitochondrial diseases and mitochondrial transplantation

Aycan Celik and Sevim Karakas Celik

20. Mitochondrial transplantation in kidney injury

Patricia Crespo Braga, Anabela Rodrigues, Pedro Fontes Oliveira, and Marco G. Alves

21. Cancer and mitochondrial transplantation, and novel insights into its application in the cancer-like phenotype of pulmonary arterial hypertension

Olakunle Sanni, Carmen Payne, and Gerald Maarman

22. Diabetes and mitochondrial transplantation

Eli John Louwagie and Michelle Leigh Baack

23. Aging and mitochondrial transplantation

Mustafa Baran, Sevin Guney, Guler Ozturk, and Burcu Caykara Peran

24. Mitochondrial transfer in the immune compartment

Colwyn Ansel Headley

25. The past, present, and close future of mitochondria as a therapeutic agent

Romina Maya, Annalia Valdivieso, Paola Robayo, and Andrés Caicedo

Contributors

Deniz Akbay Department of Anesthesiology and Reanimation, American Hospital, Istanbul, Turkey

Marco G. Alves Institute of Biomedicine - iBiMED and Department of Medical Sciences, University of Aveiro, Aveiro, Portugal

John B. Appleby Lancaster Medical School, Lancaster University, Lancaster, United Kingdom

Ozbeyen Atalay Department of Physiology, Faculty of Medicine, Hacettepe University, Ankara, Turkey

Muhammed Mustafa Ayturk Department of Pathology, Gülhane Training and Research Hospital, Ankara, Turkey

Michelle Leigh Baack Sanford School of Medicine, University of South Dakota, Sioux Falls, SD, United States; Environmental Influences on Health and Disease Group, Sanford Research, Sioux Falls, SD, United States

Mustafa Baran Department of Physiology, Faculty of Medicine, Istanbul Medeniyet University, Istanbul, Turkey

Christian Bergamini Dipartimento di Farmacia e Biotecnologie, FABIT, Università Degli Studi di Bologna, Bologna, Italy

Andrei B. Borisov Department of Biomedical Engineering, Wayne State University, Detroit, MI, United States

Patricia Crespo Braga Institute of Biomedical Sciences Abel Salazar (ICBAS), University of Porto, Porto, Portugal; ITR-Laboratory for Integrative and Translational Research in Population Health, Porto, Portugal

Andrés Caicedo Universidad San Francisco de Quito USFQ, Colegio de Ciencias de la Salud, Escuela de Medicina, Quito, Ecuador; Universidad San Francisco de Quito USFQ, Instituto de Investigaciones en Biomedicina iBioMed, Quito, Ecuador; Mito-Act Research Consortium, Quito, Ecuador; Sistemas Médicos SIME, Universidad San Francisco de Quito USFQ, Quito, Ecuador; Universidad San Francisco de Quito USFQ, USFQ Space Front, Quito, Ecuador

Dayene Caldeira Wellcome-Wolfson Institute for Experimental Medicine, Queen's University of Belfast, Belfast, United Kingdom; Institute of Biophysics Carlos Chagas Filho, Federal University of Rio de Janeiro, Rio de Janeiro, Brazil

Burcu Caykara Peran Department of Physiology, Faculty of Medicine, Istanbul Medeniyet University, Istanbul, Turkey

Aycan Celik Department of Molecular Biology and Genetics, Faculty of Sciences, Zonguldak Bulent Ecevit University, Zonguldak, Turkey

Esin Akbay Cetin Department of Biology, Faculty of Science, Hacettepe University, Ankara, Turkey

Leena Rajendra Chaudhari Department of Stem Cell and Regenerative Medicine, D.Y. Patil Education Society (Deemed University), Kolhapur, Maharashtra, India

Mrunal N. Damle Stem Plus Biotech Pvt. Ltd., Research and Development, Sangli, Maharashtra, India; Department of Stem Cells and Regenerative Medicine, D.Y. Patil Education Society (Deemed University), Kolhapur, Maharashtra, India

Duygu Dayanir Department of Histology and Embryology, Faculty of Medicine, Gazi University, Ankara, Turkey

İlknur Melis Durasi Molecular Biology and Genetics Department, Faculty of Engineering and Natural Sciences, Istinye University, Istanbul, Turkey; Mdnutriacademy, Istanbul, Turkey

Tuba Edgünlü Department of Medical Biology, Faculty of Medicine, Mugla Sıtkı Koçman University, Mugla, Turkey

Romana Fato Dipartimento di Farmacia e Biotecnologie, FABIT, Università Degli Studi di Bologna, Bologna, Italy

Sevin Guney Department of Physiology, Faculty of Medicine, Gazi University, Ankara, Turkey

Colwyn Ansel Headley Stanford Cardiovascular Institute, Stanford University School of Medicine, Palo Alto, CA, United States

Therese Jamison Department of Nursing, Lawrence Technological University, Southfield, MI, United States

Meghnad G. Joshi Department of Stem Cell and Regenerative Medicine, D.Y. Patil Education Society (Deemed University), Kolhapur, Maharashtra, India; Stem Plus Biotech Pvt. Ltd., Research and Development, Sangli, Maharashtra, India

Sevim Karakas Celik Department of Medical Genetics, Faculty of Medicine, Zonguldak Bulent Ecevit University, Zonguldak, Turkey

Anna Krasnodembskaya Wellcome-Wolfson Institute for Experimental Medicine, Queen's University of Belfast, Belfast, United Kingdom

Gokhan Burcin Kubat Department of Mitochondria and Cellular Research, Gulhane Health Sciences Institute, University of Health Sciences, Ankara, Turkey; Gulhane Training and Research Hospital, University of Health Sciences, Ankara, Turkey

Eli John Louwagie Sanford School of Medicine, University of South Dakota, Sioux Falls, SD, United States

Gerald Maarman Centre for Cardio-Metabolic Research in Africa (CARMA), Division of Medical Physiology, Department of Biomedical Sciences, Faculty of Medicine & Health Sciences, Stellenbosch University, Stellenbosch, South Africa

Elvin Malikov Department of Neurology, Ganja Military Hospital, Ganja, Azerbaijan

Romina Maya Universidad San Francisco de Quito USFQ, Colegio de Ciencias de la Salud, Escuela de Medicina, Quito, Ecuador; Universidad San Francisco de Quito USFQ, Instituto de Investigaciones en Biomedicina iBioMed, Quito, Ecuador; Mito-Act Research Consortium, Quito, Ecuador

Kevin McCusker Lawrence Technological University, Southfield, MI, United States

David McIntyre Department of Product Development, Terumo Cardiovascular, Ann Arbor, MI, United States

Domenico Nuzzo Istituto per la Ricerca e l'Innovazione Biomedica, Consiglio Nazionale delle Ricerche, Palermo, Italy

Pedro Fontes Oliveira LAQV-REQUIMTE and Department of Chemistry, University of Aveiro, Aveiro, Portugal

Cilem Ozdemir Department of Medical Biology, Faculty of Medicine, Mugla Sıtkı Koçman University, Mugla, Turkey

Berkay Ozerklig Department of Exercise and Sport Sciences, Faculty of Sport Sciences, Hacettepe University, Ankara, Turkey

Guler Ozturk Department of Physiology, Faculty of Medicine, Istanbul Medeniyet University, Istanbul, Turkey

Francesco Pallotti Dipartimento di Medicina e Chirurgia, Università Degli Studi dell'Insubria, Varese, Italy; SSD Laboratorio Analisi-SMEL Specializzato in Citogenetica e Genetica Medica, ASST Sette Laghi-Ospedale di Circolo-Fondazione Macchi, Varese, Italy

Carmen Payne Centre for Cardio-Metabolic Research in Africa (CARMA), Division of Medical Physiology, Department of Biomedical Sciences, Faculty of Medicine & Health Sciences, Stellenbosch University, Stellenbosch, South Africa

Jean D. Peduzzi-Nelson Department of Neurosurgery, Department of Ophthalmology, Visual and Anatomical Sciences, School of Medicine, Wayne State University, Detroit, MI, United States

Pasquale Picone Istituto per la Ricerca e l'Innovazione Biomedica, Consiglio Nazionale delle Ricerche, Palermo, Italy

Paola Robayo Universidad San Francisco de Quito USFQ, Colegio de Ciencias de la Salud, Escuela de Medicina, Quito, Ecuador; Universidad San Francisco de Quito USFQ, Instituto de Investigaciones en Biomedicina iBioMed, Quito, Ecuador; Mito-Act Research Consortium, Quito, Ecuador

Anabela Rodrigues Hospital Center of Vila Nova de Gaia/Espinho, Vila Nova de Gaia, Portugal

Halit Yasin Sahin Department of Urology, Bilkent City Hospital, University of Health Sciences, Ankara, Turkey

Olakunle Sanni Centre for Cardio-Metabolic Research in Africa (CARMA), Division of Medical Physiology, Department of Biomedical Sciences, Faculty of Medicine & Health Sciences, Stellenbosch University, Stellenbosch, South Africa

Rakesh Kumar Sharma Department of Obstetrics and Gynecology, D.Y. Patil Medical College, Kolhapur, Maharashtra, India

Gulcin Simsek Department of Pathology, Gülhane Training and Research Hospital, Ankara, Turkey

Rıza Sonkaya Department of Neurology, Gülhane School of Medicine, University of Health Sciences, Ankara, Turkey

Serpil Suvay Department of Physiology, Faculty of Medicine, University of Health Sciences, Ankara, Turkey

Omer Faruk Taner Department of Plastic Surgery, Gulhane Training and Research Hospital, Ankara, Turkey

Meltem Tuncer Department of Physiology, Faculty of Medicine, Hacettepe University, Ankara, Turkey

Ibrahim Turkel Department of Exercise and Sport Sciences, Faculty of Sport Sciences, Hacettepe University, Ankara, Turkey

Oner Ulger Gulhane Training and Research Hospital, University of Health Sciences, Ankara, Turkey; Department of Mitochondria and Cellular Research, Gulhane Health Sciences Institute, University of Health Sciences, Ankara, Turkey

Annalia Valdivieso Universidad San Francisco de Quito USFQ, Colegio de Ciencias de la Salud, Escuela de Medicina, Quito, Ecuador; Universidad San Francisco de Quito USFQ, Instituto de Investigaciones en Biomedicina iBioMed, Quito, Ecuador; Mito-Act Research Consortium, Quito, Ecuador

Preface

The main aim of this book is to critically review a detailed overview of "mitochondrial transplantation" that aims to make organized information about transplantation and transfer of healthy mitochondria and to define future steps to improve new therapeutic applications in a variety of diseases. The importance of mitochondrial transplantation is emphasised in cases where mitochondrial dysfunction is observed in a group of diseases such as neurodegenerative diseases, ischaemia, cardiovascular diseases, toxicity and cancer. Mitochondrial transplantation would be innovative for the treatment of diseases in which conventional therapies are unsuccessful. This book will have detailed classical and literature information about the current applications of mitochondrial transplantation and the results of these studies.

This book contains 25 chapters, divided into 6 major sections. The first section discusses the role of mitochondria in cell metabolism under normal physiological conditions, mitochondrial dynamics, the mechanism of biogenesis, and mitochondrial DNA-heteroplasmy. It also covers the basic pathophysiology of diseases associated with mitochondrial dysfunction. The second section of this book focuses on the basic mechanisms of mitochondrial transfer under normal physiological conditions, tunneling nano tube (TNT), microvesicles, and gap junctions, highlighting their importance in mitochondrial transplantation. The third section describes the process of mitochondrial transplantation, from isolation to application. It explains the methods used to isolate mitochondria, the suitability of the number, and functions of the isolated mitochondria for the specific model being applied, and the storage of the isolated mitochondria until the application. Furthermore, this section presents the current methods of mitochondrial transplantation, discusses its potential applications, and explores the limitations and ethical considerations associated with this procedure. Sections four and five discuss the pathophysiology of cardiovascular and neurological diseases related to mitochondria, as well as the significance of mitochondrial transplantation in treating these diseases. The discussion includes results from both preclinical and clinical studies. Section six describes the various diseases and pathophysiological conditions in which mitochondrial transplantation has been or has the potential to be applied. It also presents the prospects of mitochondrial transplantation, which has shown promise as a treatment modality in recent years.

This book provides a comprehensive overview of the biochemical and physiological background of mitochondrial dysfunction, with chapters written by leading experts in the field of experimental and clinical molecular biology and medicine, and a wide range of mitochondrial transplantation interventions. Therefore, *Mitochondrial Transplantation and Transfer: Biology, Methods, Applications, and Diseases* might be a valuable resource for various audiences from both academic and industry sectors, including research and experimental medicine, biochemistry, and molecular biology, but also for graduate and undergraduate students, PhD students, physicians, and health practitioners.

June 2024

Acknowledgments

We remain very grateful to Sergej Ostojic, the editor of the series, and would like to thank him for his continued support.

We would like to thank the hardworking authors from all over the world, including the Americas, Asia, Africa, and Europe, who dedicated their time to writing and editing the chapters and preparing the rich scientific content.

We would like to thank Mrs. Tracy Tufaga from Elsevier for her prompt communication with the authors and professional preparation of content and technical material for this collection.

Acknowledgments

SECTION I

Introduction

CHAPTER

1

Mitochondrial transplantation, methods and mitochondrial dysfunction related diseases: An overview

Oner Ulger[1,2] *and Gokhan Burcin Kubat*[1,2]

[1]**Gulhane Training and Research Hospital, University of Health Sciences, Ankara, Turkey;**
[2]**Department of Mitochondria and Cellular Research, Gulhane Health Sciences Institute, University of Health Sciences, Ankara, Turkey**

Almost all eukaryotes have mitochondria, which are common intracellular organelles that participate in a broad range of cellular activities such as energy production, generation of reactive oxygen species (ROS), cell signaling regulation, cell death, and biosynthetic metabolism (Protasoni & Zeviani, 2021). Furthermore, mitochondria are essential to maintaining calcium homeostasis, regulating metabolism, and controlling body temperature (Vakifahmetoglu-Norberg et al., 2017; Von Bank et al., 2021). They are also a part of an interconnected system that communicates with other cellular compartments via cell signaling (Tait & Green, 2012).

A double-membrane system encircles mitochondria and divides the organelle into internal compartments. In this context, mitochondria contain a matrix, inner mitochondrial membrane (IMM), intermembrane space, and outer mitochondrial membrane (OMM) from the inside out (Cooper, 2000; McCarron et al., 2013). A mixture of proteins and phospholipids makes up the outer and IMM (Xian & Liou, 2021). Thanks to its porous structure, the OMM permits the free translocation of ions, small and uncharged molecules, while bigger molecules, such as proteins, must be imported by specific translocases (Kühlbrandt, 2015). The OMM proteins are key components in mediating continuous fission and fusion along with mitochondrial dynamics. These proteins are essential for mitophagy because they regulate mitochondrial integrity and function as mitophagy receptors for autophagosome

Mitochondrial Transplantation and Transfer
https://doi.org/10.1016/B978-0-443-18858-9.00001-9

encapsulation (Xian & Liou, 2021). The mitochondrial intermembrane space is an aqueous compartment that contains a diverse range of proteins with different and crucial roles including the exchange of molecules between the matrix and the cytosol, regulation of apoptosis and metabolism, prevention of ROS production, and control of mitochondrial morphogenesis (Edwards et al., 2020; Herrmann & Riemer, 2010). Cardiolipin, containing four fatty acids, makes up a significant component of the lipid bilayer of IMM (Bruce Alberts, 2002). The IMM has cristae which are formed by the folding inward of the membrane. The IMM mostly contains electron transport chain (ETC) complexes, proteins essential for the synthesis of iron sulfur groups, and protein translocation and membrane fusion apparatus (Klecker & Westermann, 2021). The proteome of IMM is an important component of mitochondrial activities and cellular metabolism including oxidative phosphorylation (OXPHOS) and mitochondrial biogenesis (Lee et al., 2017). The IMM protects the proton gradient that promotes OXPHOS and creates a physiological barrier between the cytosol and the matrix (Cooper, 2000). As a result, only a small subset of molecules are present in the matrix, whereas the cytosol and intermembrane space are chemically identical in terms of the small molecules they contain (Bruce Alberts, 2002). The matrix and the IMM that surround it are the main functional components of the mitochondria. The mitochondrial matrix's high pH generates the power necessary for ATP synthesis, which has a pH value between 7.9 and 8. The citric acid cycle is one of the biochemical mechanisms that occur in the matrix and it includes an abundance of protein so that each step requires a different enzyme. Moreover, organellar DNA replication, transcription, protein biosynthesis, and several enzymatic processes conduct in the mitochondrial matrix (Kühlbrandt, 2015).

The loss of maintaining the transmembrane potential of IMM, dysfunction of ETC, and decreased transport of metabolites into mitochondria are all examples of changes that result in a reduction in mitochondrial function (Nicolson, 2014). These result in disruption of the OXPHOS mechanisms of the ETC and adenosine triphosphate (ATP) production. Accordingly, mitochondrial disorders mostly harm tissues that demand high energy, such as the brain, muscles, and heart (Meyers et al., 2013).

Mitochondria are extremely dynamic organelles that coordinate fission and fusion to sustain their structure. Many cellular activities, including apoptosis and mitochondrial quality, etc. depend on their effective and fast morphological modifications (Tilokani et al., 2018). Mitochondrial dynamics control stem cell differentiation, proliferation, migration, and death through mesenchymal metabolism or epithelial mesenchymal transition (Seo et al., 2018). Besides, the metabolic programming of the mitochondrial dynamics, which includes aerobic glycolysis or fatty acid oxidation, organizes the immune cell development (Rambold & Pearce, 2018). Furthermore, the maintenance of mitochondrial DNA (mtDNA) depends heavily on mitochondrial dynamics. With the control of mtDNA, the fusion process is significant in the activation of OXPHOS. Three GTPases, Mitofusin 1, Mitofusin 2, and Optical Atrophy 1, work together to fuse mitochondria. Cellular physiology depends on the process of mitochondrial fission (Chan, 2012). The most crucial protein for fission, dynamin-related protein 1, is positioned on the OMM via Mff, Fis1, MiD49, and MiD51 receptor proteins (Atkins et al., 2016). Exogenous and endogenous mitochondria fuse after mitochondrial transplantation (Cowan et al., 2017).

Mitochondria have an extrachromosomal genome that is distinct and independent from the nuclear genome (Cowan et al., 2017). Since mitochondria carry their DNA (mtDNA), they are a unique organelle that is somewhat independent of the nucleus. Even though mitochondria have their mtDNA, mtDNA is insufficient to support all of these mitochondrial functions. There are ~1000 distinct proteins in mitochondria that are encoded in nuclear DNA (nDNA) (Kotrys & Szczesny, 2019). Thus, the mitochondrial and nuclear genomes control mitochondrial function.

One form of mtDNA is typically found in an organism, and it is virtually passed down through the maternal line. Homoplasmy refers to mitochondria that have just one type of mtDNA. But occasionally, the expression of various forms of mtDNA in a single cell or organism can be observed. This condition is known as heteroplasmy. To put it another way, heteroplasmy means the concurrence of mutant and healthy mtDNA (Tikochinski et al., 2020). Maternal transmission, paternal mtDNA leaking during fertilization, and mtDNA mutation over time can all result in heteroplasmy (Polovina et al., 2020). For each cell, various degrees of heteroplasmy can be found. During cell division, replicative segregation can alter the level of mtDNA heteroplasmy (Nissanka & Moraes, 2020). The ratio of mutant to normal mtDNA determines whether or not symptoms occur. Healthy mtDNA may substitute for the dysfunction driven by mtDNA mutations. The occurrence of symptoms, however, depends on whether changed variations surpass a certain threshold (Alston et al., 2017).

Since mtDNA repair mechanisms are less evolved than nDNA, and histone proteins do not preserve mtDNA like they do nDNA, the rate of mtDNA mutation is greater (Hahn & Zuryn, 2019). mtDNA mutations may be inherited or acquired. The distribution of inherited mutations in primary oocytes and tissues is unpredictable. Comparatively, sporadic acquired mutations ensure through aging, mtDNA conservation problems, and a variety of other disorders and develop in only mtDNA molecules (Lawless et al., 2020). Mutations in mitochondria-related genes can cause mitochondrial diseases primarily and lead to mitochondrial dysfunction and acquired or secondary diseases (Niyazov et al., 2016).

Several methods of cell communication have evolved as a result of the biological complexity of organisms. Intercellular signaling is crucial for organisms in their maintenance and growth. Through regulating intercellular signaling, the effects of mitochondria are not limited to the cell but also spread throughout the organism. Mitochondria can cross cell borders because of their extremely flexible and dynamic nature, which enables their movement across mammalian cells (Valenti et al., 2021). Mitochondria and mitochondrial genomes are mobile and may facilitate the communication between cells by playing a signaling function in intercellular communication. Mitochondria typically undergo morphological or positional alterations in response to stressors and can be transported across cells in addition to internal mitochondrial mobility. Intercellular transfer of mitochondria is involved in maintaining homeostasis (Liu et al., 2021). Recent studies found that the peripheral blood of humans contains undamaged and functioning mitochondria, free mtDNA, mitochondrial vesicles, and mitochondrial peptides (Al Amir Dache et al., 2020; Lindqvist et al., 2018; Miller et al., 2020; Picca et al., 2020). Mitochondria can control signals that travel via the blood to distant cells. As a result, the intravascular mitochondrial injection may be promising, but initially, the whole process must be clarified.

Tunneling nanotubes (TNTs), extracellular vesicles (EVs), gap junctions (GJs), and cell fusion/mitochondrial extrusion are some of the different mechanisms for the transfer of mitochondria. The primary cellular structures thought to mediate mitochondrial transport from cell to cell are TNTs. Many cargos, containing molecules (such as Ca^{2+} ions), macromolecules (such as lipid droplets and proteins), and organelles (such as mitochondria and lysosomes), can be transferred via TNTs (Vignais et al., 2017).

Almost all cells release a population of vesicles known as EVs. They can be separated into three primary groups based on their various biogenetic and functional characteristics (Woith et al., 2019). EVs play a role in intracellular and intercellular communication. Proteins, carbohydrates, lipids, and a variety of genetic elements can potentially be delivered to numerous cells in distant organs via EVs. The processes of EV mediated transfer of mitochondrial cargo have recently been explained, demonstrating the active packing of mitochondrial proteins into EVs. Recently, it was discovered that mitochondria can regulate interorganellar interaction by vesicles and inhibit mitochondrial damage (Amari & Germain, 2021). Mitovesicles (an EV subtype with mitochondrial origin) carrying several mitochondrial proteins were discovered as formerly unknown double membraned EVs in the brain (D'Acunzo et al., 2022). Moreover, mitochondria transferred via microvesicles have been reported to be internalized into the cell by endocytosis, phagocytosis, or micropinocytosis (Zhang, Tan, et al., 2021).

GJ provides chemical and electrical communication among all cell types. Metabolites, secondary messengers, mitochondria, etc. are all transported across GJs. Connexins are one of the most important proteins for GJs (Beyer & Berthoud, 2018). Connexins have a role in healthy tissue homeostasis during embryogenesis and organogenesis, with different connexin family members expressed at different stages and anatomic sites (Ngezahayo & Ruhe, 2021). Connexin-43 (Cx43) GJs may be involved in mitochondrial transport, according to evidence, but the findings on the role of Cx43 in mitochondrial transfer are unsettled (Ngezahayo & Ruhe, 2021).

It is possible to artificially transplant mitochondria, which are transferable organelles in vivo, thanks to the properties described in the prior paragraphs. Various methods and outcomes were obtained in the mitochondrial transplantation experiments. A healthy tissue or cell line can serve as a mitochondria source for autologous or heterologous mitochondrial transplantation/transfer. Cells from organs with high energy needs are theoretically better candidates for mitochondrial donation. With the help of commercial kits or a manual differential centrifuge technique, appropriate mitochondria isolation processes are carried out from these sources. To prevent harming the mitochondria, the mitochondrial isolation method should be carried out at +4°C hypothermic conditions throughout. In addition to purity, the preparation of isolated mitochondria needs to be assessed from a quantitative point of view by counting them, and from a qualitative and functional point of view by evaluating their capacity to generate ATP. For mitochondrial transplantation, the assessment of mitochondrial numbers is necessary. Hemocytometers, immunofluorescent labeling, particle counters, and the mitochondrial protein assay may be used to control the amount of isolated mitochondria (Kubat, Ulger et al., 2021).

Preservation of isolated mitochondria to be utilized in mitochondrial transplantation procedures is still not desirable. Several storage strategies have been thoroughly detailed (Liao et al., 2020). The purpose of these procedures was to retain the bioenergetic functioning of the preparations so that the total functionality of mitochondria could be quantified. As fresh mitochondria preparations are preferred for mitochondrial transplantation, these preservation techniques have not yet been used for this purpose. One of the biggest obstacles to the advancement of the mitochondrial transplantation process is still this challenge.

The majority of treatments for mitochondrial diseases focus on treating symptoms, even if therapeutic strategies to restore mitochondrial function are still being developed. By introducing exogenous mitochondria to target cells, the possibility of new cell basis therapies would rise. Coincubation, microinjection, photothermal nanoblade, BLAST (Biophotonic Laser Assisted Surgical Tool), magnetomitotransfer, penetrating peptides, MitoCeption, MitoPunch, fluidFM, EV mediated mitochondrial transfer and mitochondrial replacement therapy (MRT) are some of the mitochondrial transplantation methods (D'Amato et al., 2023; Ulger & Kubat, 2022).

Numerous studies have demonstrated the routes of administration of mitochondria, primarily direct injection, systemic injection, and intranasal administration. Direct injection into the tissue is desirable if there is localized injury since it allows for the quick transmission of many mitochondria to the affected area. With tissue/organ mitochondrial impairments, mitochondrial transplantation via injecting them into systemic circulation can be preferred. Recent studies have focused on the different ways of delivering mitochondria. For instance, intranasal administration of mitochondria was effective for Parkinson's disease by demonstrating mitochondrial internalization and biodistribution through the central nervous system (Chang et al., 2021). Notwithstanding certain shortcomings, each of these investigations is crucial for providing a new view and paving the way for further scientific research.

Transplanting healthy mitochondria into sick tissues may allow for major gains in the region where treatment is most challenging. Nonetheless, some issues to be overcome for the application of mitochondrial transplantation may be used to their highest extent. For instance, the Ca^{2+} concentration of a medium containing isolated free mitochondria negatively impacts the mitochondrial activity and ability to produce energy. Ca^{2+} allegedly builds up in the mitochondria, raises membrane permeability, and causes organelle edema. Another question is what happens to mitochondria when they enter the target cell like the number of mitochondria which can get inside the cell and the ability to preserve their integrity and functionality. The effects of mtDNA interacting with the nDNA as well as the reaction of the immune system to transplanted mitochondria must also be defined well (Oner Ulger, 2023). Moreover, procedures covering the administration route, timing, dose, and indication/contraindication must be performed with comprehensive research for the therapeutic use of mitochondrial transplantation.

There are several ethical concerns raised by the use of mitochondrial transplantation methods like MRT. The MRT method enables a mother who has inherited mtDNA disorders to have a child that shares her genetics but is free of those inherited mtDNA diseases. In order to prevent these diseases, the MRT approach combines the genetic material of the intended parents with healthy donor mtDNA. This procedure is the topic of significant ethical criticism since it uses genetic material that does not belong to the subject. The idea of having a genetically identical but protected from inherited mtDNA diseases child has been instrumental in the development of this method. Moreover, one of the ethical judgments on the subject is that there is no ethical reason not to allow the MRT even in the absence of a risk of hereditary disease. Despite these, other experts claim that it is not feasible to take the possible risks involved with using the MRT. There are now just two countries around the globe that have laws governing this matter. It has been applied in countries with no laws, though. The method can also be used to increase fertilization in addition to preventing mtDNA related diseases. Because the existing child will not be harmed after the MRT is confirmed safe, it is believed that using MRT for reproductive reasons is not ethically problematic. Ethical debates continue on many issues such as legal regulations, in which cases they can be used, whether the identity of the donor should be disclosed, concerns about baby design, or the use of sex selection for various reasons (Bredenoord & Appleby, 2017; Medicine, 2016; Pike, 2022).

The cardiovascular system, like the rest of the body, depends on mitochondria to maintain homeostasis and life. In addition to primer pathological situations like cardiovascular disease, aging, and external stress, mitochondrial dysfunction can also be brought on by abnormalities in the mitochondrial DNA or nDNA. Congestive heart failure, cardiomyopathies, uncompensated myocardial hypertrophy, atherosclerosis, and hypertension are caused by mitochondrial damage (Stamerra et al., 2022). Several types of circulatory insufficiencies are exacerbated by mutations in mitochondrial genes. Mutations may cause aberrant alterations in ATP synthesis, besides control of cell death mechanism and calcium transport. Delivering genes specifically to the mitochondria is a suggested method for treating mitochondria dysfunction brought on by mutations (Slone & Huang, 2020). Given that a single cell can include hundreds of mitochondria, it is difficult to achieve considerable efficiency in intramitochondrial gene transfer. By enhancing mitochondria respiration and survival, replenishing their numbers, and activating their metabolic activity, mitochondrial transplantation in this context would enable us to save the cardiovascular tissues affected by mitochondrial pathology and improve functional outcomes for patients. Clinical trials are now being conducted to treat cardiovascular diseases via mitochondrial transplantation (Emani et al., 2017). The cell to cell transfer of intact mitochondria is also a potential technique for restoring normal mitochondrial activity. Current research has demonstrated that different types of tissues experience the intercellular transfer and exchange of mitochondria both in healthy and unhealthy circumstances (Liu et al., 2021; Torralba et al., 2016).

A sizable set of incapacitating neurological conditions are categorized as neurodegenerative diseases. These diseases appear in various ways and have several underlying mechanisms, which represent the degeneration of neurons due to various conditions. Because mitochondria are assumed to be key in the formation of ROS, mitochondrial failure is thought to be the basis of several forms of neurodegenerative disorders (Wang et al., 2019). Thus, mitochondrial abnormalities are quite likely to affect neurons. Aside from that, neurons rely on OXPHOS to supply their energy necessities. Mitochondrial dysfunction disrupts energy metabolism, resulting in reduced ATP, increased ROS, and poor calcium buffering, all of which promote neuronal death in degenerative neurological disorders (Norat et al., 2020). Therefore, mitochondrial transplantation provides a unique and promising therapy that can result in higher ATP production, mtDNA replacement, and lower oxidative stress (McCully et al., 2016; Ulger et al., 2021). Many preclinical experiments on various neurological diseases have recently focused on mitochondrial transplantation/transfer. It has been discovered that transplanted mitochondria greatly improved motor function, reduced apoptosis, diminished ROS levels, and enhanced locomotor and sensory functioning (Shi et al., 2017; Zhang et al., 2019; Zhao et al., 2020). While this method is still in its infancy in terms of efficacy and security, more studies are required to completely determine its effects. Additionally, research into novel mitochondrial pharmaceutical formulations and delivery system strategies will be a critical next step in the treatment of numerous neurodegenerative diseases.

Several hereditary mitochondrial disorders with extremely diverse genetic etiologies and clinical prognosis are included in primary mitochondrial diseases (Muraresku et al., 2018). Primary mitochondrial diseases are genetic disorders that can be related to pathogenic mutations in ETC protein-coding genes (Adam, 2023). Human mitochondrial diseases such as MELAS syndrome, Leigh syndrome, depletion syndromes, Kearns-Sayre syndrome, and Pearson syndrome are triggered by mutations, deletions, or duplications in mtDNA or nDNA in genes connected to mitochondria (Orsucci et al., 2021). Healthy and abnormal cells can release mtDNA and mitochondria, which can trigger a variety of mechanisms. By modifying the endogenous mtDNA, it is possible to restore mtDNA content in human cells with damaged mitochondria. The consequences of mitochondrial transplantation on host mtDNA are not well investigated. For instance, according to one of these rare experiments, exogenous mtDNA has been found in mtDNA deficient cells long after mitochondrial transplantation (Orsucci et al., 2021). This result leads us to conclude that mtDNA transfer via mitochondrial transplantation may be beneficial. Mitochondrial transplantation improved mitochondrial functions in MELAS cells by reducing one of the most prevalent genetic mutations that cause mitochondrial disease (Lin et al., 2019). Research on patients with Pearson syndrome who received hematopoietic stem cells that are manipulated with wild type mitochondria via mitochondrial transplantation found improvements in energy and respiration capacities of lymphocytes, and patient quality of life (Jacoby et al., 2018).

Concerns regarding kidney illness have grown among the general public in recent years. Kidney injuries are identified and categorized as chronic kidney disease (CKD) and acute kidney disease (AKD). Acute kidney injury (AKI), which is one of a group of disorders known as AKD, is marked by a sudden reduction in excretory kidney function. An increased risk of CKD development has been associated with AKI (Kellum et al., 2021; Lameire et al., 2021). Many stressors, including ischemia-reperfusion injury and sepsis, may cause AKI. The beginning and development of AKI and CKD as well as the progression from AKI to CKD are accompanied by a disruption in mitochondria structure and functions (e.g., organelle swelling, loss of cristae, reduction in ATP synthesis, increase in ROS, activation of apoptosis) (Zhang, Agborbesong, & Li, 2021). Given that the kidney is a very metabolically active organ, it requires a lot of energy to carry out its functions, such as filtration and reabsorption (Bhatia et al., 2020). While there is currently no real treatment for kidney diseases like AKI or CKD, there is rising interest in therapeutic mitochondria targeted therapy, particularly mitochondrial transplantation, because mitochondria are a key factor for physiological and pathological processes (Breyer & Susztak, 2016; Jabbari et al., 2020). It has been revealed that mesenchymal stem cells (MSCs) are effective in preventing kidney damage and that EVs generated from stem-like cells may transfer mitochondria (Zou et al., 2018). Research on mitochondrial transplantation has revealed improvement in function, histologic structure, apoptotic markers, antioxidant capability, and tubular cell regeneration in kidney damage (Doulamis et al., 2020; Kubat, Kartal et al., 2021; Kubat, Ozler et al., 2021). These results suggest that mitochondrial transplantation/transfer may be recommended as a possible therapy for treating renal problems in the future since it looks to be a safe procedure with few risks and significant benefits in mitochondrial performance.

Nearly every tissue in the body is affected by the systemic metabolic condition known as diabetes. The mitochondrial malfunction has importance in terms of pathological conditions in diabetes. High-functioning mitochondria are essential for cellular respiration and ATP production in nearly all human body tissues. Excess ROS emerges following the ETC uncoupling that occurs in hyperglycemia and hyperlipidemia (Kaludercic & Di Lisa, 2020; Wang et al., 2022). Cellular signalings such as the insulin/AKT pathway and the peroxisome proliferator activated receptor gamma coactivator-1 alpha are interrelated to diabetes and have a broad impact on mitochondria and cells (Huang et al., 2018; Wu et al., 2016). Theoretically, it is possible to transplant or transfer healthy mitochondria into sick tissue, augment or replace unhealthy mitochondria, and facilitate healing. The transfer of mitochondria from MSC to beta cells was shown through EVs and MSC derived TNTs (Huang et al., 2018). Mitochondrial transplantation has been demonstrated to increase respiratory function and lessen stress induced apoptosis in cardiomyocytes damaged by pregestational diabetes (Louwagie et al., 2021). It was also shown that transplanting healthy mitochondria into the hippocampus of diabetic rats increased mitochondrial activity while decreasing neuron loss (Ma et al., 2020). Despite these encouraging findings in the setting of diabetes, there is no certainty that the reported experimental advantages will apply to clinical protocols, and more research is needed.

Aging is a very complicated process, and genomic instability, epigenetic changes, telomere shortening and proteostasis disruption, stem cell depletion, mitochondrial dysfunction, and changes in cellular communication contribute to aging. The mitochondrial hypothesis is one explanation for aging. The theory proposes that ROS generated by aerobic mitochondrial metabolism harm cells by reacting with a variety of macromolecules (Chistiakov et al., 2014). For the prevention of aging and its consequences, mitochondrial capacity needs to be increased by promoting mitochondrial biogenesis, protecting cardiolipin, expanding antioxidant capacity, and strengthening ETC activity. Examples include the repair of ETC function by coenzyme Q10, thiamine, and riboflavin, the restoration of glutathione levels and nitric oxide generation by cysteine donors, and the creation of nitric oxide by arginine (Avula et al., 2014). Recent developments in biochemical and molecular technologies have made it feasible to comprehend mitochondrial damage and malfunction, allowing the development of more efficient solutions, even though these methods are mostly symptomatic. Mitochondrial transfer/transplantation has developed as a cutting-edge strategy to get around the limits of these therapies (Avula et al., 2014). Since they attempted to enhance the number of healthy mitochondria in target organs, recent developments in mitochondrial transfer and transplantation research are hopeful. Mitochondrial transplantation may be one of the therapy options for aging and age-related disorders, even if these investigations are still in the experimental stage.

In addition to the known classical functions of mitochondria inside the cell, mitochondria have other new functions such as taking part in intercellular communication and immunoregulation. Researchers have been encouraged by the endosymbiotic source of mitochondria and the ability to be transferred through different ways to develop new methods. This leads to the improvement of mitochondrial transplantation methods based on the delivery of healthy mitochondria to damaged cells and tissues. These methods showed evidence of therapeutic benefits from in vitro to clinical studies. New findings on the physiology of extracellular mitochondria and the mechanisms behind their therapeutic potential are continuously being described. In this book, classical and new information about the role of mitochondria in physiological and pathophysiological mechanisms is presented as a foundation first. Then, all the aspects of mitochondrial transplantation processes from isolation to implementation are described with sample protocols. Then, current knowledge on the effects of mitochondrial transplantation in various pathological conditions and diseases is reported and projections are made by defining potential future effects.

References

Adam, M. P., Mirzaa, G. M., Pagon, R. A., et al. (2023). GeneReviews®. In M. P. Adam, G. M. Mirzaa, R. A. Pagon, et al. (Eds.), *Seattle*. University of Washington.

Al Amir Dache, Z., Otandault, A., Tanos, R., Pastor, B., Meddeb, R., Sanchez, C., … Thierry, A. R. (2020). Blood contains circulating cell-free respiratory competent mitochondria. *FASEB J, 34*(3), 3616–3630. https://doi.org/10.1096/fj.201901917RR

Alston, C. L., Rocha, M. C., Lax, N. Z., Turnbull, D. M., & Taylor, R. W. (2017). The genetics and pathology of mitochondrial disease. *The Journal of Pathology, 241*(2), 236–250. https://doi.org/10.1002/path.4809

Amari, L., & Germain, M. (2021). Mitochondrial extracellular vesicles - origins and roles. *Frontiers in Molecular Neuroscience, 14*, 767219. https://doi.org/10.3389/fnmol.2021.767219

Atkins, K., Dasgupta, A., Chen, K. H., Mewburn, J., & Archer, S. L. (2016). The role of Drp1 adaptor proteins MiD49 and MiD51 in mitochondrial fission: Implications for human disease. *Clinical Science, 130*(21), 1861–1874. https://doi.org/10.1042/cs20160030

Avula, S., Parikh, S., Demarest, S., Kurz, J., & Gropman, A. (2014). Treatment of mitochondrial disorders. *Current Treatment Options in Neurology, 16*(6), 292. https://doi.org/10.1007/s11940-014-0292-7

Beyer, E. C., & Berthoud, V. M. (2018). Gap junction gene and protein families: Connexins, innexins, and pannexins. *Biochimica et Biophysica Acta (BBA) - Biomembranes, 1860*(1), 5–8. https://doi.org/10.1016/j.bbamem.2017.05.016

Bhatia, D., Capili, A., & Choi, M. E. (2020). Mitochondrial dysfunction in kidney injury, inflammation, and disease: Potential therapeutic approaches. *Kidney Research and Clinical Practice, 39*(3), 244–258. https://doi.org/10.23876/j.krcp.20.082

Bredenoord, A. L., & Appleby, J. B. (2017). Mitochondrial replacement techniques: Remaining ethical challenges. *Cell Stem Cell, 21*(3), 301–304. https://doi.org/10.1016/j.stem.2017.08.009

Breyer, M. D., & Susztak, K. (2016). The next generation of therapeutics for chronic kidney disease. *Nature Reviews Drug Discovery, 15*(8), 568–588. https://doi.org/10.1038/nrd.2016.67

Bruce Alberts, A. J., Julian, L., Raff, M., Roberts, K., & Walter, P. (2002). *Molecular biology of the cell* (4th ed. ed.). New York: Garland Science.

Chan, D. C. (2012). Fusion and fission: Interlinked processes critical for mitochondrial health. *Annual Review of Genetics, 46*, 265–287. https://doi.org/10.1146/annurev-genet-110410-132529

Chang, J.-C., Chao, Y.-C., Chang, H.-S., Wu, Y.-L., Chang, H.-J., Lin, Y.-S., … Liu, C.-S. (2021). Intranasal delivery of mitochondria for treatment of Parkinson's Disease model rats lesioned with 6-hydroxydopamine. *Scientific Reports, 11*(1), 10597. https://doi.org/10.1038/s41598-021-90094-w

Chistiakov, D. A., Sobenin, I. A., Revin, V. V., Orekhov, A. N., & Bobryshev, Y. V. (2014). Mitochondrial aging and age-related dysfunction of mitochondria. *BioMed Research International, 2014*, 238463. https://doi.org/10.1155/2014/238463

Cooper, G. M. (2000). *The cell: A molecular approach* (2nd ed. ed.). Sunderland (MA): Sinauer Associates.

Cowan, D. B., Yao, R., Thedsanamoorthy, J. K., Zurakowski, D., Del Nido, P. J., & McCully, J. D. (2017). Transit and integration of extracellular mitochondria in human heart cells. *Scientific Reports, 7*(1), 17450. https://doi.org/10.1038/s41598-017-17813-0

D'Acunzo, P., Kim, Y., Ungania, J. M., Pérez-González, R., Goulbourne, C. N., & Levy, E. (2022). Isolation of mitochondria-derived mitovesicles and subpopulations of microvesicles and exosomes from brain tissues. *Nature Protocols, 17*(11), 2517–2549. https://doi.org/10.1038/s41596-022-00719-1

D'Amato, M., Morra, F., Di Meo, I., & Tiranti, V. (2023). Mitochondrial transplantation in mitochondrial medicine: Current challenges and future perspectives. *International Journal of Molecular Sciences, 24*(3). https://doi.org/10.3390/ijms24031969

Doulamis, I. P., Guariento, A., Duignan, T., Kido, T., Orfany, A., Saeed, M. Y., … McCully, J. D. (2020). Mitochondrial transplantation by intra-arterial injection for acute kidney injury. *American Journal of Physiology-Renal Physiology, 319*(3), F403–f413. https://doi.org/10.1152/ajprenal.00255.2020

Edwards, R., Gerlich, S., & Tokatlidis, K. (2020). The biogenesis of mitochondrial intermembrane space proteins. *Biological Chemistry, 401*(6–7), 737–747. https://doi.org/10.1515/hsz-2020-0114

Emani, S. M., Piekarski, B. L., Harrild, D., Del Nido, P. J., & McCully, J. D. (2017). Autologous mitochondrial transplantation for dysfunction after ischemia-reperfusion injury. *The Journal of Thoracic and Cardiovascular Surgery, 154*(1), 286–289. https://doi.org/10.1016/j.jtcvs.2017.02.018

Hahn, A., & Zuryn, S. (2019). Mitochondrial genome (mtDNA) mutations that generate reactive oxygen species. *Antioxidants (Basel), 8*(9), 392. https://doi.org/10.3390/antiox8090392

Herrmann, J. M., & Riemer, J. (2010). The intermembrane space of mitochondria. *Antioxid Redox Signal, 13*(9), 1341–1358. https://doi.org/10.1089/ars.2009.3063

Huang, X., Liu, G., Guo, J., & Su, Z. (2018). The PI3K/AKT pathway in obesity and type 2 diabetes. *International Journal of Biological Sciences, 14*(11), 1483–1496. https://doi.org/10.7150/ijbs.27173

Jabbari, H., Roushandeh, A. M., Rostami, M. K., Razavi-Toosi, M. T., Shokrgozar, M. A., Jahanian-Najafabadi, A., ... Roudkenar, M. H. (2020). Mitochondrial transplantation ameliorates ischemia/reperfusion-induced kidney injury in rat. *Biochimica et Biophysica Acta, Molecular Basis of Disease, 1866*(8), 165809. https://doi.org/10.1016/j.bbadis.2020.165809

Jacoby, E., Blumkin, M., Anikster, Y., Varda-Bloom, N., Pansheen, J., Bar Yoseph, O., ... Toren, A. (2018). First-in-human mitochondrial augmentation of hematopoietic stem cells in Pearson syndrome. *Blood, 132*, 1024. https://doi.org/10.1182/blood-2018-99-113773

Kühlbrandt, W. (2015). Structure and function of mitochondrial membrane protein complexes. *BMC Biology, 13*(1), 89. https://doi.org/10.1186/s12915-015-0201-x

Kaludercic, N., & Di Lisa, F. (2020). Mitochondrial ROS formation in the pathogenesis of diabetic cardiomyopathy. *Frontiers in Cardiovascular Medicine, 7*, 12. https://doi.org/10.3389/fcvm.2020.00012

Kellum, J. A., Romagnani, P., Ashuntantang, G., Ronco, C., Zarbock, A., & Anders, H.-J. (2021). Acute kidney injury. *Nature Reviews Disease Primers, 7*(1), 52. https://doi.org/10.1038/s41572-021-00284-z

Klecker, T., & Westermann, B. (2021). Pathways shaping the mitochondrial inner membrane. *Open Biology, 11*(12), 210238. https://doi.org/10.1098/rsob.210238

Kotrys, A. V., & Szczesny, R. J. (2019). Mitochondrial gene expression and beyond-novel aspects of cellular physiology. *Cells, 9*(1). https://doi.org/10.3390/cells9010017

Kubat, G. B., Kartal, Y., Atalay, O., Ulger, O., Ekinci, O., Celik, E., ... Budak, M. T. (2021). Investigation of the effect of isolated mitochondria transplantation on renal ischemia-reperfusion injury in rats. *Toxicology and Applied Pharmacology, 433*, 115780. https://doi.org/10.1016/j.taap.2021.115780

Kubat, G. B., Ozler, M., Ulger, O., Ekinci, O., Atalay, O., Celik, E., ... Budak, M. T. (2021). The effects of mesenchymal stem cell mitochondrial transplantation on doxorubicin-mediated nephrotoxicity in rats. *Journal of Biochemical and Molecular Toxicology, 35*(1), e22612. https://doi.org/10.1002/jbt.22612

Kubat, G. B., Ulger, O., & Akin, S. (2021). Requirements for successful mitochondrial transplantation. *Journal of Biochemical and Molecular Toxicology, 35*(11), e22898. https://doi.org/10.1002/jbt.22898

Lameire, N. H., Levin, A., Kellum, J. A., Cheung, M., Jadoul, M., Winkelmayer, W. C., & Stevens, P. E. (2021). Harmonizing acute and chronic kidney disease definition and classification: Report of a kidney disease: Improving global outcomes (KDIGO) consensus conference. *Kidney International, 100*(3), 516–526. https://doi.org/10.1016/j.kint.2021.06.028

Lawless, C., Greaves, L., Reeve, A. K., Turnbull, D. M., & Vincent, A. E. (2020). The rise and rise of mitochondrial DNA mutations. *Open Biology, 10*(5), 200061. https://doi.org/10.1098/rsob.200061

Lee, S.-Y., Kang, M.-G., Shin, S., Kwak, C., Kwon, T., Seo, J. K., ... Rhee, H.-W. (2017). Architecture mapping of the inner mitochondrial membrane proteome by chemical tools in live cells. *Journal of the American Chemical Society, 139*(10), 3651–3662. https://doi.org/10.1021/jacs.6b10418

Liao, P.-C., Bergamini, C., Fato, R., Pon, L. A., & Pallotti, F. (2020). Chapter 1 - isolation of mitochondria from cells and tissues. In L. A. Pon, & E. A. Schon (Eds.), *Methods in cell biology* (Vol 155, pp. 3–31). Academic Press.

Lin, T. K., Chen, S. D., Chuang, Y. C., Lan, M. Y., Chuang, J. H., Wang, P. W., ... Liou, C. W. (2019). Mitochondrial transfer of Wharton's jelly mesenchymal stem cells eliminates mutation burden and rescues mitochondrial bioenergetics in Rotenone-stressed MELAS fibroblasts. *Oxidative Medicine and Cellular Longevity*, 9537504. https://doi.org/10.1155/2019/9537504

Lindqvist, D., Wolkowitz, O. M., Picard, M., Ohlsson, L., Bersani, F. S., Fernström, J., ... Mellon, S. H. (2018). Circulating cell-free mitochondrial DNA, but not leukocyte mitochondrial DNA copy number, is elevated in major depressive disorder. *Neuropsychopharmacology, 43*(7), 1557–1564. https://doi.org/10.1038/s41386-017-0001-9

Liu, D., Gao, Y., Liu, J., Huang, Y., Yin, J., Feng, Y., ... Gao, J. (2021). Intercellular mitochondrial transfer as a means of tissue revitalization. *Signal Transduction and Targeted Therapy, 6*(1), 65. https://doi.org/10.1038/s41392-020-00440-z

Louwagie, E. J., Larsen, T. D., Wachal, A. L., Gandy, T. C. T., & Baack, M. L. (2021). Mitochondrial transfer improves cardiomyocyte bioenergetics and viability in male rats exposed to pregestational diabetes. *International Journal of Molecular Sciences, 22*(5). https://doi.org/10.3390/ijms22052382

Ma, H., Jiang, T., Tang, W., Ma, Z., Pu, K., Xu, F., ... Wang, Q. (2020). Transplantation of platelet-derived mitochondria alleviates cognitive impairment and mitochondrial dysfunction in db/db mice. *Clinical Science, 134*(16), 2161–2175. https://doi.org/10.1042/cs20200530

McCarron, J. G., Wilson, C., Sandison, M. E., Olson, M. L., Girkin, J. M., Saunter, C., & Chalmers, S. (2013). From structure to function: Mitochondrial morphology, motion and shaping in vascular smooth muscle. *Journal of Vascular Research, 50*(5), 357–371. https://doi.org/10.1159/000353883

McCully, J. D., Levitsky, S., Del Nido, P. J., & Cowan, D. B. (2016). Mitochondrial transplantation for therapeutic use. *Clinical and Translational Medicine, 5*(1), 16. https://doi.org/10.1186/s40169-016-0095-4

Committee on the Ethical and Social Policy Considerations of Novel Techniques for Prevention of Maternal Transmission of Mitochondrial DNA Diseases; Board on Health Sciences Policy; Institute of Medicine. (2016). *Mitochondrial replacement techniques: Ethical, social, and policy considerations*. In E. R. Claiborne A, & J. Kahn (Eds.). National Academies Press (US).

Meyers, D. E., Basha, H. I., & Koenig, M. K. (2013). Mitochondrial cardiomyopathy: Pathophysiology, diagnosis, and management. *Texas Heart Institute Journal, 40*(4), 385–394.

Miller, B., Kim, S. J., Kumagai, H., Mehta, H. H., Xiang, W., Liu, J., ... Cohen, P. (2020). Peptides derived from small mitochondrial open reading frames: Genomic, biological, and therapeutic implications. *Experimental Cell Research, 393*(2), 112056. https://doi.org/10.1016/j.yexcr.2020.112056

Muraresku, C. C., McCormick, E. M., & Falk, M. J. (2018). Mitochondrial disease: Advances in clinical diagnosis, management, therapeutic development, and preventative strategies. *Current Genetic Medicine Reports, 6*(2), 62–72. https://doi.org/10.1007/s40142-018-0138-9

Ngezahayo, A., & Ruhe, F. A. (2021). Connexins in the development and physiology of stem cells. *Tissue Barriers, 9*(4), 1949242. https://doi.org/10.1080/21688370.2021.1949242

Nicolson, G. L. (2014). Mitochondrial dysfunction and chronic disease: Treatment with natural supplements. *Integrative Medicine, 13*(4), 35–43.

Nissanka, N., & Moraes, C. T. (2020). Mitochondrial DNA heteroplasmy in disease and targeted nuclease-based therapeutic approaches. *EMBO Reports, 21*(3), e49612. https://doi.org/10.15252/embr.201949612

Niyazov, D. M., Kahler, S. G., & Frye, R. E. (2016). Primary mitochondrial disease and secondary mitochondrial dysfunction: Importance of distinction for diagnosis and treatment. *Molecular Syndromology, 7*(3), 122–137. https://doi.org/10.1159/000446586

Norat, P., Soldozy, S., Sokolowski, J. D., Gorick, C. M., Kumar, J. S., Chae, Y., ... Kalani, M. Y. S. (2020). Mitochondrial dysfunction in neurological disorders: Exploring mitochondrial transplantation. *Npj Regenerative Medicine, 5*(1), 22. https://doi.org/10.1038/s41536-020-00107-x

Oner Ulger, G. B. K. (2023). Role of mitochondria and mitochondrial transplantation in drug-induced toxic organ injury. Exploratory research and hypothesis. *Medicine*. https://doi.org/10.14218/ERHM.2022.00117, 000(000), 000–000.

Orsucci, D., Caldarazzo Ienco, E., Rossi, A., Siciliano, G., & Mancuso, M. (2021). Mitochondrial syndromes revisited. *Journal of Clinical Medicine, 10*(6). https://doi.org/10.3390/jcm10061249

Picca, A., Beli, R., Calvani, R., Coelho-Júnior, H. J., Landi, F., Bernabei, R., … Marzetti, E. (2020). Older adults with physical frailty and sarcopenia show increased levels of circulating small extracellular vesicles with a specific mitochondrial signature. *Cells, 9*(4). https://doi.org/10.3390/cells9040973

Pike, G. K. (2022). The benefits, risks and alternatives of mitochondrial replacement therapy — bringing proportionality into public policy debate. *Clinical Ethics, 17*(4), 368–376. https://doi.org/10.1177/14777509221091097

Polovina, E. S., Parakatselaki, M. E., & Ladoukakis, E. D. (2020). Paternal leakage of mitochondrial DNA and maternal inheritance of heteroplasmy in Drosophila hybrids. *Scientific Reports, 10*(1), 2599. https://doi.org/10.1038/s41598-020-59194-x

Protasoni, M., & Zeviani, M. (2021). Mitochondrial structure and bioenergetics in normal and disease conditions. *International Journal of Molecular Sciences, 22*(2). https://doi.org/10.3390/ijms22020586

Rambold, A. S., & Pearce, E. L. (2018). Mitochondrial dynamics at the interface of immune cell metabolism and function. *Trends in Immunology, 39*(1), 6–18. https://doi.org/10.1016/j.it.2017.08.006

Seo, B. J., Yoon, S. H., & Do, J. T. (2018). Mitochondrial dynamics in stem cells and differentiation. *International Journal of Molecular Sciences, 19*(12). https://doi.org/10.3390/ijms19123893

Shi, X., Zhao, M., Fu, C., & Fu, A. (2017). Intravenous administration of mitochondria for treating experimental Parkinson's disease. *Mitochondrion, 34*, 91–100. https://doi.org/10.1016/j.mito.2017.02.005

Slone, J., & Huang, T. (2020). The special considerations of gene therapy for mitochondrial diseases. *Npj Genomic Medicine, 5*, 7. https://doi.org/10.1038/s41525-020-0116-5

Stamerra, C. A., Di Giosia, P., Giorgini, P., Ferri, C., Sukhorukov, V. N., & Sahebkar, A. (2022). Mitochondrial dysfunction and cardiovascular disease: Pathophysiology and emerging therapies. *Oxidative Medicine and Cellular Longevity*, 9530007. https://doi.org/10.1155/2022/9530007, 2022.

Tait, S. W., & Green, D. R. (2012). Mitochondria and cell signalling. *Journal of Cell Science, 125*(Pt 4), 807–815. https://doi.org/10.1242/jcs.099234

Tikochinski, Y., Carreras, C., Tikochinski, G., & Vilaça, S. T. (2020). Population-specific signatures of intra-individual mitochondrial DNA heteroplasmy and their potential evolutionary advantages. *Scientific Reports, 10*(1), 211. https://doi.org/10.1038/s41598-019-56918-6

Tilokani, L., Nagashima, S., Paupe, V., & Prudent, J. (2018). Mitochondrial dynamics: Overview of molecular mechanisms. *Essays in Biochemistry, 62*(3), 341–360. https://doi.org/10.1042/ebc20170104

Torralba, D., Baixauli, F., & Sánchez-Madrid, F. (2016). Mitochondria know no boundaries: Mechanisms and functions of intercellular mitochondrial transfer. *Frontiers in Cell and Developmental Biology, 4*, 107. https://doi.org/10.3389/fcell.2016.00107

Ulger, O., Kubat, G. B., Cicek, Z., Celik, E., Atalay, O., Suvay, S., & Ozler, M. (2021). The effects of mitochondrial transplantation in acetaminophen-induced liver toxicity in rats. *Life Sciences, 279*, 119669. https://doi.org/10.1016/j.lfs.2021.119669

Ulger, O., & Kubat, G. B. (2022). Therapeutic applications of mitochondrial transplantation. *Biochimie, 195*, 1–15. https://doi.org/10.1016/j.biochi.2022.01.002

Vakifahmetoglu-Norberg, H., Ouchida, A. T., & Norberg, E. (2017). The role of mitochondria in metabolism and cell death. *Biochemical and Biophysical Research Communications, 482*(3), 426–431. https://doi.org/10.1016/j.bbrc.2016.11.088

Valenti, D., Vacca, R. A., Moro, L., & Atlante, A. (2021). Mitochondria can cross cell boundaries: An overview of the biological relevance, pathophysiological implications and therapeutic perspectives of intercellular mitochondrial transfer. *International Journal of Molecular Sciences, 22*(15). https://doi.org/10.3390/ijms22158312

Vignais, M. L., Caicedo, A., Brondello, J. M., & Jorgensen, C. (2017). Cell connections by tunneling nanotubes: Effects of mitochondrial trafficking on target cell metabolism, homeostasis, and response to therapy. *Stem Cells International*, 6917941. https://doi.org/10.1155/2017/6917941, 2017.

Von Bank, H., Hurtado-Thiele, M., Oshimura, N., & Simcox, J. (2021). Mitochondrial lipid signaling and adaptive thermogenesis. *Metabolites, 11*(2). https://doi.org/10.3390/metabo11020124

Wang, A., Lin, Y., Liang, B., Zhao, X., Qiu, M., Huang, H., … Kong, Y. (2022). Statins attenuate cholesterol-induced ROS via inhibiting NOX2/NOX4 and mitochondrial pathway in collecting ducts of the kidney. *BMC Nephrology, 23*(1), 184. https://doi.org/10.1186/s12882-022-02815-6

Wang, Y., Xu, E., Musich, P. R., & Lin, F. (2019). Mitochondrial dysfunction in neurodegenerative diseases and the potential countermeasure. *CNS Neuroscience and Therapeutics, 25*(7), 816–824. https://doi.org/10.1111/cns.13116

Woith, E., Fuhrmann, G., & Melzig, M. F. (2019). Extracellular vesicles-connecting Kingdoms. *International Journal of Molecular Sciences, 20*(22). https://doi.org/10.3390/ijms20225695

Wu, H., Deng, X., Shi, Y., Su, Y., Wei, J., & Duan, H. (2016). PGC-1α, glucose metabolism and type 2 diabetes mellitus. *Journal of Endocrinology, 229*(3), R99–r115. https://doi.org/10.1530/joe-16-0021

Xian, H., & Liou, Y.-C. (2021). Functions of outer mitochondrial membrane proteins: Mediating the crosstalk between mitochondrial dynamics and mitophagy. *Cell Death & Differentiation, 28*(3), 827–842. https://doi.org/10.1038/s41418-020-00657-z

Zhang, X., Agborbesong, E., & Li, X. (2021). The role of mitochondria in acute kidney injury and chronic kidney disease and its therapeutic potential. *International Journal of Molecular Sciences, 22*(20). https://doi.org/10.3390/ijms222011253

Zhang, Y., Tan, J., Miao, Y., & Zhang, Q. (2021). The effect of extracellular vesicles on the regulation of mitochondria under hypoxia. *Cell Death & Disease, 12*(4), 358. https://doi.org/10.1038/s41419-021-03640-9

Zhang, Z., Ma, Z., Yan, C., Pu, K., Wu, M., Bai, J., … Wang, Q. (2019). Muscle-derived autologous mitochondrial transplantation: A novel strategy for treating cerebral ischemic injury. *Behavioural Brain Research, 356*, 322–331. https://doi.org/10.1016/j.bbr.2018.09.005

Zhao, Z., Yu, Z., Hou, Y., Zhang, L., & Fu, A. (2020). Improvement of cognitive and motor performance with mitotherapy in aged mice. *International Journal of Biological Sciences, 16*(5), 849–858. https://doi.org/10.7150/ijbs.40886

Zou, X., Kwon, S. H., Jiang, K., Ferguson, C. M., Puranik, A. S., Zhu, X., & Lerman, L. O. (2018). Renal scattered tubular-like cells confer protective effects in the stenotic murine kidney mediated by release of extracellular vesicles. *Scientific Reports, 8*(1), 1263. https://doi.org/10.1038/s41598-018-19750-y

CHAPTER

2

Mitochondria and cell metabolism

Muhammed Mustafa Ayturk[1], *Halit Yasin Sahin*[2] *and Gulcin Simsek*[1]

[1]**Department of Pathology, Gülhane Training and Research Hospital, Ankara, Turkey;** [2]**Department of Urology, Bilkent City Hospital, University of Health Sciences, Ankara, Turkey**

Introduction

Mitochondria, located in the cytoplasm of all eukaryotic cells, are membrane dependent organelles. Mitochondria are thought to be like a bacteria, which form after endosymbiosis. Endosymbiosis is a symbiotic interaction in which one organism lives inside the other. Mitochondria consist of two separate and functionally distinct membranes: the outer mitochondrial membrane (OMM) and the inner mitochondrial membrane (IMM). These membranes comprise the intermembrane space (IMS) and matrix compartments (Kühlbrandt, 2015; Nunnari & Suomalainen, 2012). Those membranes are essential to produce energy. It is believed that only a small fraction of human mitochondria are derived from the original endosymbiont, although mammalian mitochondria have retained some bacterial characteristics (Gabaldón & Huynen, 2004).

Mitochondria contain a genetic structure that utilizes an individual DNA code in contrast to their eukaryotic hosts and their DNA encodes transfer RNAs, ribosomal RNAs, and some mitochondrial proteins (Gray et al., 1998). Mitochondrial DNA (mtDNA) is arranged into distinct nucleoids within the matrix (Nunnari & Suomalainen, 2012). It is believed that the inheritance of paternal mitochondria or mtDNA only maternally is inherited in all mammals, but some exceptional cases have been reported (Luo et al., 2018). The transfer of mitochondria from the father to the child is impeded by diverse physiological mechanisms. The mitochondria are in the tail of the spermatozoon and are eliminated from the oocyte after fertilization has taken place (Al Rawi et al., 2011; Sato & Sato, 2011). However, it has recently been demonstrated that human infants received mtDNA from both their fathers and mothers, resulting in mtDNA heteroplasmy (Luo et al., 2018).

Mitochondrial Transplantation and Transfer
https://doi.org/10.1016/B978-0-443-18858-9.00002-0

Mitochondrial structure

The development of technology over the last 30–40 years has enabled us to comprehend the structure of the mitochondria. Understanding mitochondrial compartments, dynamic mitochondrial networks, and mitochondrial DNA segregation has been made possible by various techniques. These include transmission electron microscopy, light microscopy, electron microscope tomography, high-resolution scanning electron microscopy, and three dimensional electron microscopy (Glancy, 2020; Perkins & Frey, 2000). Mitochondria have a unique double membrane structure that is different from that of other organelles. The OMM is a barrier between mitochondria and cytoplasm. The IMM physically divides the intermembrane space from the protein-dense central matrix. The IMM encompasses two differentiating features: the inner boundary membrane and the cristae. These regions merge at the crista junctions, which are also the main locations of the mitochondrial energy conversion process. These regions merge at the crista junctions, which are also the main locations of the mitochondrial energy conversion process. The cristae extend into the matrix to various depths (Perkins et al., 1997).

Pores on the OMM are formed by membrane proteins called porins, such as voltage dependent anion channels (VDACs) (Kühlbrandt, 2015). VDAC participates in mitochondria-associated apoptosis as well as metabolic exchange (Kühlbrandt, 2015). Specialized translocases (outer (TOM) and inner (TIM) membrane complexes) are used to transport larger molecules like proteins (Bayrhuber et al., 2008; Neupert & Herrmann, 2007). The OMM lacks membrane potential because of its porous structure (Llopis et al., 1998). Conversely, the IMM operates as a highly efficient diffusion barrier for every ion and molecule. Solely membrane transport proteins that are selective for individual ions or molecules allow the crossing of the IMM by ions. The ion selectivity of the IMM leads to the development of an electrochemical membrane potential of roughly 180 mV. The oxidative phosphorylation (OXPHOS) of a cluster of membrane protein complexes is responsible for generating the electrochemical gradient across the IMM, which serves as a means for synthesizing adenosine triphosphate (ATP) (Guo et al., 2018; Zorova et al., 2018).

The inner and outer membranes separate the mitochondria into three main functional regions. The area surrounded by the inner membrane is known as the mitochondrial matrix. With a pH of approximately 8, the matrix generates an electrochemical gradient that is vital for ATP production (Llopis et al., 1998). The mitochondrial matrix is the site of organellar DNA replication, transcription, protein synthesis, and several enzymatic processes. An example of the biochemical processes in the matrix is the tricarboxylic acid (TCA) cycle. The mitochondrial matrix is highly protein dense because each reaction is supported by a single enzyme (Kühlbrandt, 2015).

The IMS is the section between the inner and outer membranes of the mitochondria. This is the 20-nm space between the OMM and the inner boundary membrane, which is part of IMM (Schleyer & Neupert, 1985). The OMM, the intermembranous gap, and the IMM must be crossed by all matrix proteins that are imported into the mitochondria from the cytoplasm. The inner border membrane is filled with carrier proteins that move ions, ATP, adenosine diphosphate (ADP), and small metabolites from the cytoplasm to the matrix. Among the most prevalent proteins in the IMM is the ATP/ADP transporter (Pebay-Peyroula et al., 2003).

Cristae are deep invaginations of the matrix that are created by the IMM. The lumen of the crista is known to be the third compartment within the mitochondria. Cristae is the main site of energy conversion. Several factors contribute to this phenomenon, including the presence of ATP synthase in the cristae membranes and the abundance of the small, soluble electron transporter protein, cytochrome *c*, within the crista lumen (Von Ballmoos et al., 2009).

Powerhouse of the cell

Energy metabolism is performed by mitochondria and a portion of the free energy generated by the oxidation of food is transformed into ATP in mitochondria. Cells metabolize proteins, lipids, and carbohydrates in a variety of ways to produce energy. In mitochondria, electrons are oxidized and stored as NADH (reduced form of nicotinamide adenine dinucleotide) and $FADH_2$ (reduced form of flavine adenine dinucleotide). These transporters use the flow of electrons to push protons into the IMS. The electrons are deposited in the electron transport chain (ETC) (Sazanov, 2015). ETC is oxygen dependent, and when oxygen levels are low, the cytosol uses less effective anaerobic respiration to turn glycolytic products into energy.

Pyruvate is one of the most significant precursors in the oxidative energy synthesis of mitochondria. Pyruvate can be produced by a variety of processes, including lactate oxidation, alanine transamination, and as a by product of glycolysis (Gray et al., 2014). Glycolysis is the main source of pyruvate. Therefore, it is helpful to keep in mind glycolysis before moving on to how pyruvate is used in mitochondria to produce energy.

Glycolysis is a cytoplasmic enzyme activity that converts glucose into two pyruvic acids for energy. During the initial phase of glycolysis, hexokinase enzyme catalyzes the phosphorylation of glucose to prevent its efflux from the cell. Hexokinases need ATP usage for glucose phosphorylation in the cell. Following this reaction there is a ten step response sequence that includes five phases of preparation and the remaining five stages of payment (Kumari, 2018). During glycolysis, high-energy substances like 1,3-biphosphoglycerate and phosphoenolpyruvate are phosphorylated at the substrate level to create ATP (Walsh et al., 2018). NADH, which would subsequently be utilized in ETC, is also created during these processes. As a result, glycolysis produces ATP, NADH, and pyruvate and these are transported to the mitochondria. When exposed to hypoxia, mitochondrial respiration becomes impaired, which leads to cells producing lactate by sinking electrons onto pyruvate via lactate dehydrogenase (LDH). The lactate produced is converted to pyruvate in the presence of oxygen (Glancy et al., 2021). Lactate is the primary source of intermediates in the TCA in lung cancer patients according to a study (Faubert et al., 2017).

The function of the mitochondrial pyruvate carrier (MPC) is to coordinate the entry of pyruvate into the mitochondrial matrix (Gray et al., 2014). A study in yeast, Drosophila, and humans discovered MPC1 and MPC2, two proteins responsible for transporting pyruvate into the mitochondria (Bricker et al., 2012). The cytosolic pyruvate metabolism and TCA are therefore efficiently linked by the MPC. If the MPC fails to function properly, the TCA deteriorates and energy generation is interrupted (Brivet et al., 2003; Gray et al., 2014).

Once pyruvate passes through the MPC and enters the mitochondrial matrix, it can lead to several outcomes. The TCA cycle converts some pyruvate to carbon dioxide, facilitating ATP production through OXPHOS. In other instances, the PDH (pyruvate dehydrogenase) complex irreversibly transforms pyruvate and nicotinamide adenine dinucleotide (NAD+) into acetyl CoA, NADH, and carbon dioxide through a series of processes. Alternatively, pyruvate may combine with pyruvate carboxylase (PC) within the TCA (Gray et al., 2014; Patel & Korotchkina, 2006). In anaplerotic reactions, acetyl CoA, NADH, and carbon dioxide are replenished within the TCA cycle to maintain its products (Owen et al., 2002). The production of oxaloacetate in the TCA cycle occurs through its function as an acceptor for the acetyl CoA created by PDH (Cheng et al., 2011), and acetyl CoA regulates PC-specificity positively (Adina-Zada et al., 2012). Elevated levels of acetyl CoA indicate either that the cells' energy requirements have been met and pyruvate should be employed in gluconeogenesis, or that there is a shortage of oxaloacetate to utilize the acetyl CoA that is generated by PDH while experiencing energetic stress. PC is unquestionably impactful in guiding pyruvate carbon to the process of gluconeogenesis while also functioning as an anaplerotic shunt to uphold TCA carbon flux. The direction of overall carbon flux is determined by the joint regulation of PC and PDH (Gray et al., 2014).

Tricarboxylic acid cycle (Krebs cycle—citric acid cycle)

Pyruvate serves as the main carbon fuel input that propels the entire carbon flow in the TCA. The IMM facilitates pyruvate entry into the mitochondrial matrix via the MPC. Citrate or oxaloacetate, depending on the demand for oxaloacetate replenishment, are the forms of pyruvate carbon that enter the TCA in the matrix (Gray et al., 2014). The TCA enzymes emerged before the presence of oxygen on Earth, and over time, the TCA process has evolved into a major energy production mechanism for eukaryotic cells. To generate energy, one TCA cycle fundamentally functions to oxidize acetyl CoA into two CO_2 molecules. The conclusion of the TCA process produces ATP, 3 NADH, and 1 $FADH_2$, which are then transferred to ETC complexes I and II (NADH dehydrogenase and succinate dehydrogenase (SDH), respectively) to create ATP via OXPHOS (Martínez-Reyes & Chandel, 2020). At the beginning of the TCA cycle, acetyl CoA, a 2-carbon compound produced by the oxidation of fatty acids, amino acids, or pyruvate, combines with oxaloacetate, a 4-carbon compound, to form the 6-carbon compound citrate (Martínez-Reyes & Chandel, 2020).

Citrate is converted into its isomer, isocitrate, followed by two oxidative decarboxylations, where isocitrate is converted into the five-carbon α-ketoglutarate (α-KG), and then into the four-carbon succinyl-CoA. This process releases two molecules of CO_2 and produces two molecules of NADH. Succinyl-CoA is transformed into succinate, leading to the production of guanosine triphosphate (GTP) that can be turned into ATP. The four-carbon compound fumarate is formed when succinate is oxidized. In this process, $FADH_2$ is generated via the transfer of two hydrogen atoms into flavin adenine dinucleotide (FAD). The enzyme responsible for this process, SDH, is noteworthy for being a component of both the TCA and ETC. SDH serves as the only enzyme involved in both processes. Fumarate is converted to malate and then to oxaloacetate, which subsequently reacts with another acetyl CoA molecule to complete the TCA cycle. Consequently, TCA produces GTP, NADH, and $FADH_2$, contributing to energy production (Judge & Dodd, 2020; Martínez-Reyes & Chandel, 2020; Ryan et al., 2021).

The control of the TCA and its continual interaction with OXPHOS is crucial for maintaining cell stability. The metabolic flux of the TCA is controlled by several positive and negative allosteric regulators. NADH hinders all regulatory enzymes of TCA. The dysfunction of ETC leads to an accumulation of NADH, which ultimately results in the shutdown of TCA. As NADH produces ATP via ETC and OXPHOS, ATP allosterically inhibits PDH and isocitrate dehydrogenase (IDH). Therefore, when sufficient NADH and ATP are present, the cycle slows, leading to increased stimulation of TCA regulatory enzymes (Martínez-Reyes & Chandel, 2020).

Electron transport chain

Through successive enzymatic processes, TCA produces the reductants NADH and $FADH_2$, which are required for electron transfer to the mitochondrial respiratory chain also known as the ETC. As electrons flow via complexes in the IMM, a functioning ETC generates an ATP producing mitochondrial membrane potential ($\Delta\Psi$). This procedure, called OXPHOS, needs oxygen to function. The oxidative TCA is made possible by replenishing NAD+ and FAD by mitochondrial complex II in the ETC (Martínez-Reyes & Chandel, 2020).

The mitochondrial ETC accepts electron donations from NADH and $FADH_2$ produced by the TCA cycle at complexes I and II. Complex I contains a flavin mononucleotide (FMN) and seven iron sulfur (FeS) clusters, ranging from low to high potential, that are the initial cofactors transferring the electrons from NADH to ubiquinone (CoQ). Via the Q-cycle, the electron carrier CoQ is then converted to ubiquinol (QH_2). In this process, complex I pumps protons from the mitochondrial matrix into the IMS, with four protons being transported for every two electrons transferred from NADH (Nolfi-Donegan et al., 2020; Zhao et al., 2019). Complex II, also referred to as succinate dehydrogenase, serves as a link between metabolism and OXPHOS, as it is a constituent of both the TCA cycle and the ETC (Cecchini, 2003). As part of the TCA, complex II oxidizes succinate to fumarate (Sun et al., 2005). Similar to complex I, complex II is a further electron entry point that transfers succinate-derived electrons to CoQ via FeS clusters. In contrast to complex I, complex II does not involve the translocation of protons from the matrix to the IMS (Zhao et al., 2019). When complex III (coenzyme Q: cytochrome *c* reductase) enters the Q-cycle, electrons are transferred from it to cytochrome *c*. Upon binding of QH_2 to the cytoplasmic side of complex III, two protons are immediately released into the IMS. One electron is moved from QH_2 to the catalytically active and high potential iron–sulfur cluster in the Rieske center of complex III, while the other electron is given to cytochrome *b*. The next step involves moving an electron from cytochrome *b* to a second molecule of Q that is attached to the complex's matrix side to produce ubisemiquinone (Q-). Synchronous transmission of the electron from the 2Fe–2S cluster to cytochrome *c*1 results in its transfer and reduction to cytochrome *c* (Yang & Trumpower, 1986). Another QH_2 molecule attaches to the membrane side of complexes III–IV and undergoes the same oxidation process to convert Q- to QH_2, thereby concluding the Q-cycle and transporting two additional protons to the IMS (Trumpower, 2002). Following reduction, single electrons are transported from complex III to complex IV (cytochrome *c* oxidase, or COX) by the mobile electron transporter cytochrome *c*, where they form bonds with molecular oxygen and undergo reduction to water. Three core subunits make up complex IV:

flanking subunit II contains COX copper center CuA, the other redox-active metal center is heme a (Fea), a binuclear center made up of heme a3 (Fea3) and COX copper center CuB; and subunit III is involved in proton pumping (Kadenbach & Hüttemann, 2015; Konstantinov, 2012; Zhao et al., 2019). To transmit electrons from heme A to the binuclear core of complex IV subunit I, where O_2 binds and is reduced to H_2O, reduced cytochrome *c* can interact with subunit II of complex IV. Thus, eight protons are pumped from the matrix into complex IV; four of these protons are used to produce two water molecules, while the remaining four are transferred into the IMS (Nolfi-Donegan et al., 2020; Wikstrom, 1977). The electrochemical proton gradient known as $\Delta\Psi$ is created by the movement of 10 protons (H+) from the matrix into the IMS due to electron transport, with two from complex III and four from each of complex I and IV involved. $\Delta\Psi$ combines with proton concentration (pH) to generate a protonmotive force (Δp), which is crucial in the energy storage process of OXPHOS. This links the transfer of electrons (complexes I–IV) and the use of oxygen to the function of complex V (ATP synthase), whereby protons return to the matrix to reduce the proton gradient (Jonckheere et al., 2012; Nolfi-Donegan et al., 2020; Zhao et al., 2019). The extramembranous (also known as F1) and transmembranous (also known as F0) domains of complex V, a multi subunit complex, function by means of a rotating motor mechanism to generate ATP. Rotation is linked to proton transport from the IMS to F0 and results in the addition of a phosphate to ADP at specific places in F1 to form ATP (Dickson et al., 2006; Guo et al., 2018).

Biosynthesis of reactive oxygen species and mitochondria

Reactive oxygen species (ROS) are highly reactive molecules and free radicals derived from molecular oxygen. They function as signaling molecules and are involved in both physiological and pathological states, and they are essential in many different cellular activities (Fig. 2.1).

The production of ROS in mammalian cells occurs through several processes, such as ETC, oxidative enzymes, inflammation, and immune responses, and other cellular processes like exposure to environmental toxins, UV radiation, or certain drugs. The oxidation processes involved in the metabolism of fatty acids and amino acids are examples of typical metabolic reactions that can also result in the formation of ROS. Even though cells naturally produce ROS, oxidative stress can arise when there is an imbalance between the formation of ROS and the antioxidant defense mechanisms of the cells. Overproduction of ROS may harm lipids, proteins, and DNA in cells, causing cellular malfunction and a host of illnesses such as cancer, heart disease, neurological disorders, and aging (Juan et al., 2021).

Mammalian cells possess antioxidant defense systems, comprising enzymes such as superoxide dismutase (SOD), catalase, and glutathione peroxidase, to counteract ROS effects. These enzymes help neutralize and mitigate the harmful effects of ROS by converting them into less damaging substances.

Mitochondria can produce ROS, which are partially reduced metabolites of molecular oxygen during energy production and cellular respiration (Zorov et al., 2014). Therefore, mitochondria are both a center for ROS production and a target of these reactive molecules. These derivatives were only seen as damaging agents formerly but later, their roles in intracellular communication emerged.

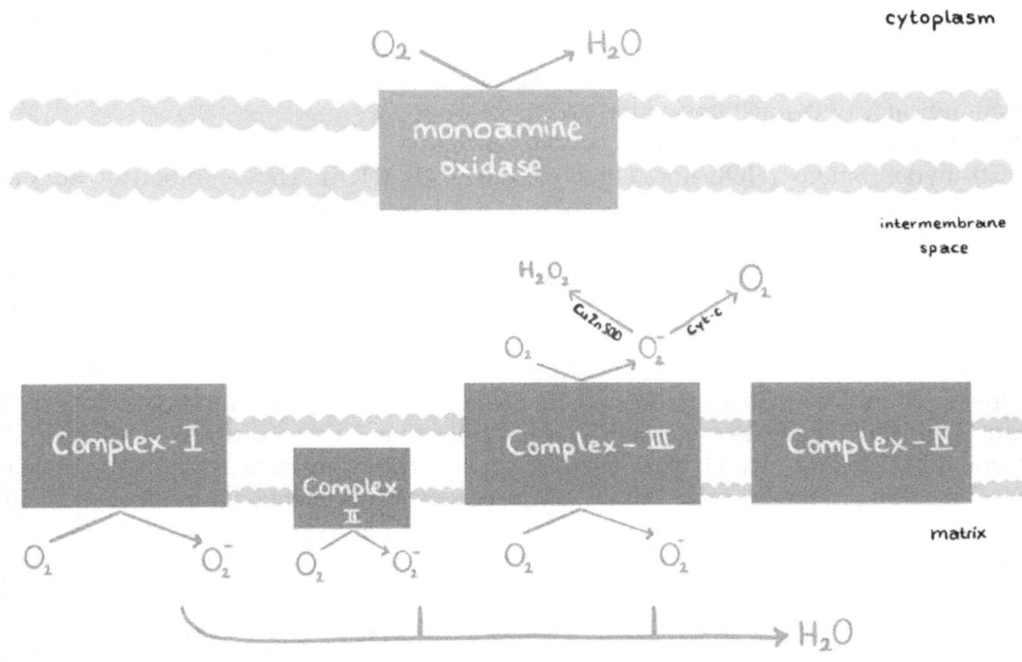

FIGURE 2.1 Production of reactive oxygen species with electron transport chain in mitochondria. *Created by Dilruba Aytürk.*

Several major sources of ROS produced by normal physiological processes have been identified. The most important center of these sources is assumed to be the mitochondrial ETC. In a process mediated by mitochondrial COX, almost 90% of the oxygen that a mammal consumes is converted to water. However, numerous redox centers in the mitochondrial ETC are capable of directly transferring an electron to oxygen (Mesa-Garcia et al., 2018). This reaction leads to the production of superoxide anion radical ($O_2^{\bullet -}$). $O_2^{\bullet -}$ can be converted to hydrogen peroxide (H_2O_2) either spontaneously or through enzymatic events ($2O_2^{\bullet -} + 2H^+$ à $H_2\ O_2 + O_2$).

H_2O_2 can also produce highly reactive hydroxyl (OH^-) ions:

$$\text{Fe (II)} + H_2O_2 \text{ à } OH^{\cdot} + OH^- + \text{FE (III)}$$

When an electron transfers to oxygen, ROS are produced at two different locations in the ETC: complex I and complex III. In vitro, $O_2^{\bullet -}$ is produced by these two mitochondrial sites from 1% to 2% of the oxygen molecules absorbed in normobaric and hyperbaric environments (Schottlender et al., 2021). Although it is believed that O_2-generated in complex I occurs inside the matrix, it is really released into the intermembrane space as well as the matrix when it is formed in complex III.

A free radical, a semiquinone anion species (SQ^-), which appears as an intermediate in the redox cycle of coenzyme Q10, is generated in the absence of the enzyme. Once formed, this anion can swiftly transfer its electrons to molecular oxygen. This aspect makes ROS production an indicator of the metabolic rate.

A more comprehensive definition of ROS encompasses compounds like hypochlorous (HOCl), hypobromous (HOBr), and hypoiodous acids (HOI). The definition of ROS also commonly includes peroxyl (ROO'), alkoxyl (RO'), semiquinone (SQ^-), and carbonate ($CO3^-$) radicals and organic hydroperoxides (ROOH).

Certain enzymes found in various cellular compartments, like peroxisomes and the endoplasmic reticulum, can generate ROS as part of their normal functions. Enzymes such as NADPH oxidases (NOX), xanthine oxidase, and cytochrome P450 can produce superoxide and other ROS. Oxidative enzymes themselves do not directly produce ROS. Instead, they are involved in reactions or processes that can lead to the generation of ROS as by products. NADPH oxidases are a family of enzymes found in various cell types, including immune cells. When activated, NOX enzymes transfer electrons from NADPH to molecular oxygen, resulting in the formation of $O_2^{\bullet-}$. This process is an intentional mechanism of ROS production for immune responses and defense against pathogens. Xanthine oxidase is an enzyme involved in purine metabolism. It can generate $O_2^{\bullet-}$ during the conversion of hypoxanthine to xanthine and further to uric acid. Cytochrome P450 enzymes are a large group of enzymes responsible for metabolizing various compounds, including drugs and xenobiotics. Some reactions catalyzed by specific cytochrome P450 enzymes generate ROS intermediates as part of the metabolic process. These ROS can include superoxide radicals and hydrogen peroxide (Patlevič et al., 2016).

Exposure to environmental toxins, such as pollutants, heavy metals, pesticides, and certain chemicals, can induce oxidative stress in cells. These toxic substances can generate ROS through different mechanisms, including direct interaction with cellular components or by disrupting cellular antioxidant defense systems. ROS production in response to environmental toxins can lead to oxidative damage and contribute to the development of various diseases. UV light, especially UVB and UVA, can cause skin cells to produce ROS. Skin molecular oxygen may absorb energy from UV light and produce ROS, such as singlet oxygen and superoxide radicals. ROS production by UV radiation contributes to oxidative stress, DNA damage, inflammation, and premature aging of the skin. Some medications, including certain antibiotics, chemotherapy drugs, and nonsteroidal antiinflammatory drugs (NSAIDs), can promote the generation of ROS. These drugs can interfere with cellular processes or interact with components like mitochondria, leading to increased ROS production. The excessive production of ROS by drugs can be damaging to cells and contribute to drug-induced toxicity (Ulger & Kubat, 2023).

High levels of ROS in cancer cells can arise from increased metabolic activity, mitochondrial dysfunction, peroxisome activity, elevated cellular receptor signaling, oncogene activity, increased oxidase activity, cyclooxygenase activity, lipoxygenase activity, and thymidine phosphorylase activity, or through cross-talk with infiltrating immune cells. Additionally, growth factors and cytokines can trigger various biological responses involving ROS (Bae et al., 2000). For instance, higher amounts of H_2O_2 and nitrite oxide are produced in tumor cells by interferon γ (IFNy) and tumor necrosis factor α (TNF-α) (Tiku et al., 1990). Furthermore, $O_2^{\bullet-}$ is formed by platelet-derived growth factor (PDGF), epidermal growth factor

(EGF), insulin, transforming growth factor β (TGFβ), interleukin-1 (IL-1), TNF-α, angiotensin, and lysophosphatidic acid (Meier et al., 1989). A direct correlation has been found between the increased functions of growth factors and increased $O_2^{\bullet-}$ production because of oncogenic mutation and the incidence of many cancers (Aunoble et al., 2000). Many chemotherapeutics are assumed to induce death in cancer cells by increasing the production of ROS in cancer cells, and multidrug therapies are thought to be one of the proposed mechanisms for tumor regression (Snezhkina et al., 2020).

Mitochondria and the surrounding cytosol are designed separately in terms of metabolic functions. The IMM is impermeable to NADH (Purvis & Lowenstein, 1961); for this reason, new definitions are needed such as malate–aspartate shuttle, a-glycerophosphate shuttle, citrate–malate shuttle, and fatty acid shuttle to explain the passage of NADH through the mitochondrial matrix. These shuttles are thought to serve to pass NADH into the mitochondrial matrix as well as produce the other side of the cycle, cytosolic NAD+. Cells use lactate oxidation and glycolysis to produce cytosolic NADH. Cytosolic NADH must be moved to the mitochondrial matrix to enter the ETC to maximize ATP synthesis. Similarly, cytosolic NAD+ must be replenished to sustain glycolytic flux and the conversion of lactate to pyruvate. On the other hand, these systems aid in the production of ROS. The main pathway for NADH transport, particularly in neurons, is the malate aspartate shuttle (Kauppinen et al., 1987).

Malate aspartate shuttle (Borst cycle)

The transfer of reducing equivalents, or electrons, across the mitochondrial membrane is facilitated by the malate–aspartate shuttle, a crucial process (Fig. 2.2). Facilitating the transfer of cytosolic NADH into the mitochondria allows for its utilization in the production of ATP through oxidative phosphorylation.

Glucose undergoes metabolism to pyruvate, which generates NADH within the cytosol during glycolysis. The cytosolic enzyme malate dehydrogenase (MDH) converts the excess cytosolic NADH into malate. Subsequently, the malate formed within the cytosol is transferred across the mitochondrial membrane via a malate-α-ketoglutarate antiporter. Inside the mitochondria, malate is converted to oxaloacetate by mitochondrial MDH. The oxidation of malate to oxaloacetate allows the reduced form of nicotinamide adenine dinucleotide, NADH, to be regenerated within the mitochondria. Aspartate is transported back to the cytosol where it can be converted back to oxaloacetate by the enzyme cytosolic aspartate aminotransferase. Aspartate is then produced from oxaloacetate by the enzyme aspartate aminotransferase. Aspartate is transported back to the cytosol where it can be converted back to oxaloacetate by the enzyme cytosolic aspartate aminotransferase. Oxaloacetate is subsequently converted to malate by cytosolic MDH, and the cycle continues.

Citrate–malate shuttle

The primary biological pathway that incorporates citrate is the citric acid cycle, which is also referred to as the Krebs cycle or TCA cycle. The citrate transporter (CIC) regulates the transfer of citrate and other metabolites from the mitochondrial matrix to the cytosol, thereby

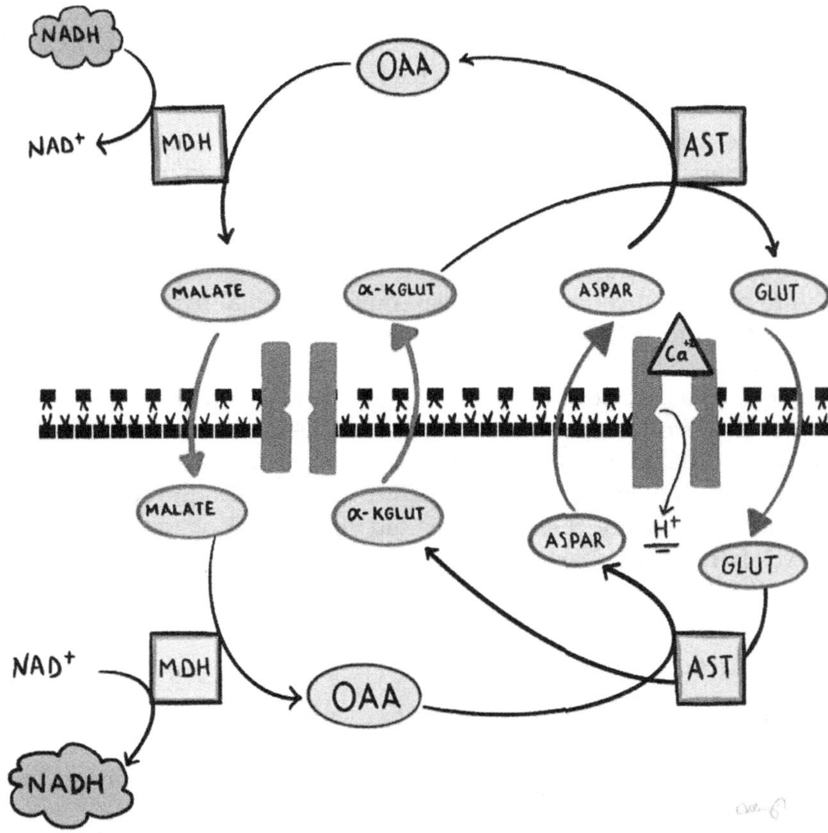

FIGURE 2.2 Malate–aspartate shuttle (Borst cycle). *Created by Dilruba Aytürk.*

linking intramitochondrial and extramitochondrial (cytosolic) reactions involving the transported substrates (Fig. 2.3).

Citrate synthase catalyzes the combination of acetyl CoA and oxaloacetate in the mitochondria matrix, forming the citrate molecule. CIC transports citrate out of the mitochondria, and ATP citrate lyase (CL), a cytosolic enzyme, breaks it down into acetyl CoA and oxaloacetate. Acetyl CoA is then directed toward lipogenesis (the biosynthesis of fatty acids and cholesterol), though its acetyl units can also be used for histone acetylation. Oxaloacetate can be converted to malate, which is transported to the mitochondria via citrate exchange. This process allows for the regeneration of cytosolic NAD+ by cytosolic MDH. Cytosolic NAD+ is essential for the continuous functioning of glycolysis. Additionally, the citrate molecule powerfully modulates the activity of several enzymes involved in cellular metabolism.

Malate can be transported to the mitochondrial matrix through NADH formation. Alternatively, it can be utilized as the substrate for malic enzyme (ME), which produces pyruvate and NADPH. Pyruvate then reenters the mitochondria through the pyruvate transporter (PYC). The mitochondrial enzyme PC converts the pyruvate molecule to oxaloacetate, which enhances the TCA's function. Pyruvate can also be converted to acetyl CoA by PDH (Zara et al., 2022).

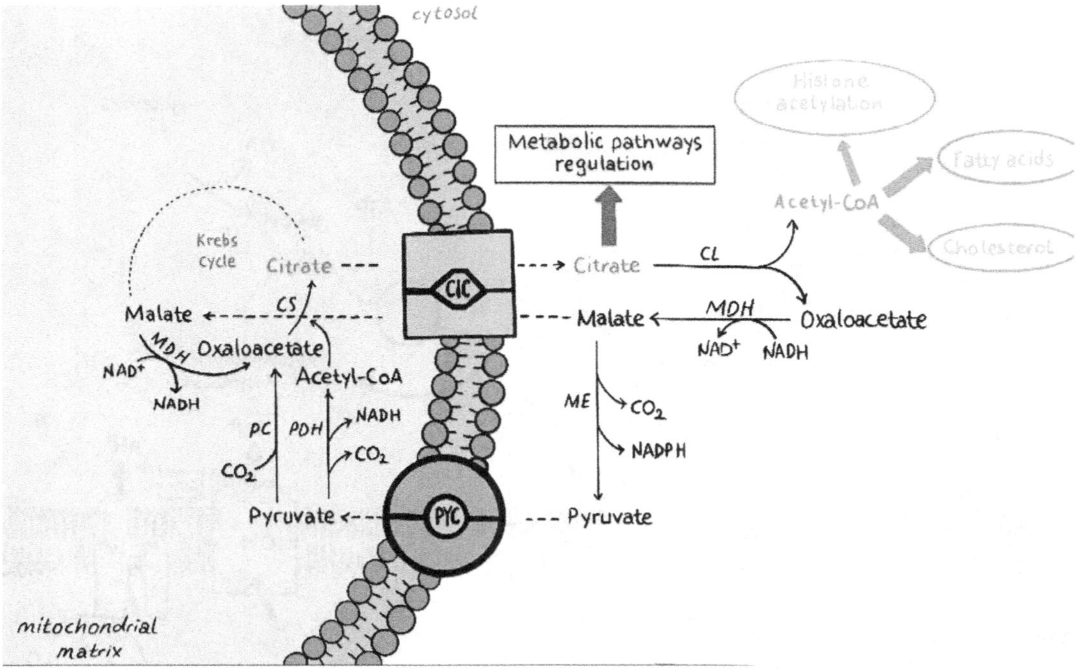

FIGURE 2.3 Citrate—malate shuttle. *Created by Dilruba Aytürk.*

α-Glycerophosphate shuttle

The mitochondrial matrix receives reducing equivalents from the cytosol by a process called the α-glycerophosphate shuttle, which is referred to as the glycerol-3-phosphate shuttle or the cytoplasmic NADH shuttle (Fig. 2.4). During glycolysis in the cytoplasm, reducing equivalents in the form of NADH are generated that must be transported into the mitochondria to undertake oxidative phosphorylation for ATP production. The α-glycerophosphate shuttle, unlike the malate aspartate shuttle, operates without requiring direct mitochondrial membrane transport to transfer NADH to the mitochondria. The process involves the use of two enzymes, namely glycerol-3-phosphate dehydrogenase (GPDH) and glycerol-3-phosphatase (G3Pase). During glycolysis, NADH produced in the cytoplasm donates its electrons to GPDH, an enzyme located in the cytosol. This action converts NADH to NAD+ and reduces dihydroxyacetone phosphate (DHAP) to α-glycerophosphate concurrently. A specific transporter transports α-glycerophosphate through the IMM and into the mitochondrial matrix. G3Pase, the glycerol-3-phosphate shuttle enzyme, then oxidizes α-glycerophosphate back to DHAP within the mitochondrial matrix. During this process of oxidation, electrons are transported to the ETC within the inner membrane of the mitochondria. The electrons obtained from α-glycerophosphate enter the ETC at the ubiquinone level and move along the respiratory chain, resulting in the generation of ATP through oxidative phosphorylation.

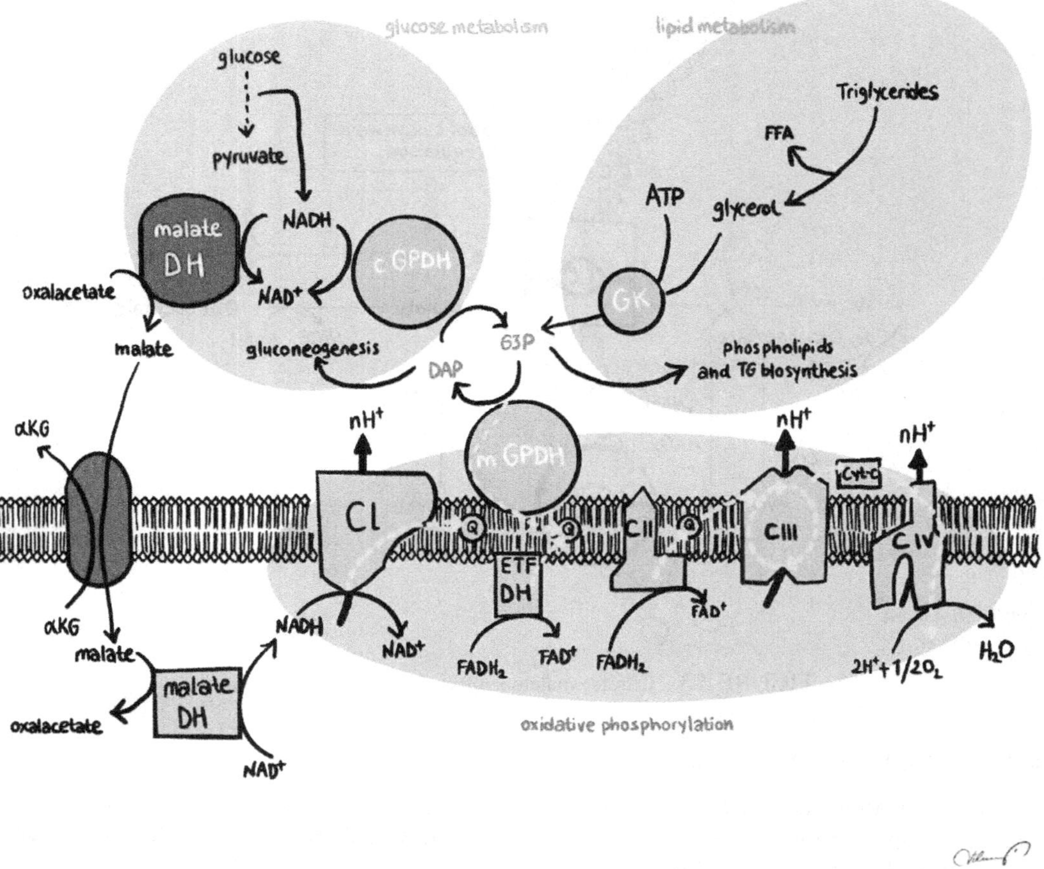

FIGURE 2.4 α-Glycerophosphate shuttle. *Created by Dilruba Aytürk.*

The α-glycerophosphate is regenerated by the GPDH enzyme found in the cytoplasm, thus allowing the transfer of extra reducing equivalents into the mitochondria and the continuation of the cycle (Mráček et al., 2013).

The α-glycerophosphate shuttle plays a crucial role in certain tissues, like the skeletal muscle and brain, by facilitating the transfer of cytoplasmic NADH into the mitochondria. This process efficiently produces ATP. Although fewer ATP molecules are generated when compared to the malate aspartate shuttle, the α-glycerophosphate shuttle is an indispensable mechanism for preserving energy balance and enabling cellular respiration. In essence, the α-glycerophosphate shuttle enables the transportation of reducing equivalents (NADH) from the cytoplasm to the mitochondrial matrix, which facilitates the production of ATP via oxidative phosphorylation. This process is made possible through the enzymatic action of GPDH and G3Pase, and it is imperative for cellular energy metabolism.

One carbon (1C) metabolism

A biochemical mechanism known as "one carbon metabolism" transfers one-carbon (C1) units for several vital cellular processes. These C1 units, usually in the form of methyl groups, are critical for the synthesis of several biomolecules, including nucleotides, amino acids, and lipids. Essential biological functions including DNA synthesis, amino acid metabolism, and cellular methylation pattern maintenance depend on one carbon metabolism (Ducker & Rabinowitz, 2017). The primary sources of one-carbon units are dietary nutrients, such as folate and choline, as well as endogenous metabolic intermediates, such as serine and glycine. The overall process of one-carbon metabolism involves a network of enzymatic reactions that regulate the transfer, utilization, and interconversion of C1 units (Ducker & Rabinowitz, 2017).

In addition to those normal cells, one-carbon metabolism is essential to elaborate detailed information in case of cancer. Cancer cells need intense anabolic activity and energy production. The cells engage in several anabolic pathways that are glucose and mitochondria dependent, resulting in the production of NADPH. NADPH provides synthesis for lipids, proteins, and nucleotides, while also decreasing equivalents in biosynthesis. Tumor development is promoted by anabolic pathways that are activated by oncogenic signaling and the loss of tumor suppressors. In fact, ROS generated by ETC is necessary for the proliferation, carcinogenesis, and proliferation of cancer cells due to mitochondrial metabolism. Areas at the center of a solid tumor can become hypoxic when rapidly proliferating tumor cells begin to become insufficiently available blood supplies. To activate the hypoxia inducible transcription factors (HIF) family of transcription factors and propel the expression of HIF target genes, including those involved in metabolism and angiogenesis, hypoxia also increases the generation of ROS within the mitochondria. More significantly, ROS must be maintained at a level that prevents cell death or senescence without enabling cancer cells to accumulate to levels that would compromise the redox balance required for cell proliferation and HIF activation. As a result, cancer cells have strictly controlled amounts of mitochondrial ROS.

The ETC in the mitochondria produces $O_2^{\bullet-}$ in response to hypoxia. $O_2^{\bullet-}$ is carried to the cytosol via the VDAC, where SOD1 converts it to H_2O_2. H_2O_2 is a signaling molecule that promotes the growth of cancer cells and HIF activation. SOD2 in the mitochondrial matrix may also convert $O_2^{\bullet-}$ to H_2O_2 and detoxify it into water. Due to their better intracellular cleaning capabilities, glutathione peroxidases (GPx) and peroxiredoxins (Prxs) participate in this water detoxification process to maintain the redox equilibrium. GPx and Prxs' capacity for scavenging is reliant on thioredoxin and glutathione. Additionally, glutathione and thioredoxin pools are refilled with NADPH.

Ammonia

The relationship between ammonia and mitochondria involves its production, detoxification, and utilization in various metabolic processes. Ammonia (NH3) can be produced through various cellular processes, including the breakdown of amino acids, deamination of glutamine, or the action of urease on urea. These processes occur in various cells and tissues throughout the body. Excess ammonia in the body is toxic, particularly to the central nervous system. The liver plays a crucial role in detoxifying ammonia by converting it into a less toxic form called urea through a process called the urea cycle. This cycle primarily occurs within the mitochondria of liver cells (hepatocytes). A part of the urea cycle, the mitochondrial ornithine cycle occurs predominantly within the liver mitochondria. Ammonia was detected to inhibit mitochondrial dehydrogenase activity, causing collapse, mitochondrial swelling, and a rise in ROS. Also, significant lipid peroxidation was detected in mitochondria exposed to ammonia, along with glutathione and ATP depletion (Niknahad et al., 2017). These mechanisms emphasize the importance of mitochondria in diseases such as ammonia accumulation and hepatic encephalopathy.

Hydrogen sulfide

Hydrogen sulfide can modulate mitochondrial function through both direct and indirect mechanisms. One important enzyme in the mitochondrial ETC, cytochrome oxidase, can be directly bound by it and inhibited in its function. Hydrogen sulfide inhibits electron transport by interfering with cytochrome oxidase, which in turn decreases ATP synthesis through oxidative phosphorylation. Hydrogen sulfide not only directly inhibits cytochrome oxidase but also modulates the activities of other respiratory enzyme complexes to control the ETC. For instance, it has been demonstrated to interact with cytochrome *c* oxidase complex IV, changing its activity and thereby impacting the bioenergetics of mitochondria (Fig. 2.5).

Hydrogen sulfide scavenges ROS and reactive nitrogen species, hence acting as a potent antioxidant. Under hazardous concentrations, hydrogen sulfide protects against cellular stress by promoting mitochondrial biogenesis and function. These factors have led to the association between some human diseases and increases in hydrogen sulfide and hydrogen sulfide-producing enzymes (Murphy et al., 2019).

The process of creating new mitochondria, known as mitochondrial biogenesis, has been demonstrated to be aided by hydrogen sulfur. It has the capability to activate signaling pathways similar to the peroxisome proliferator-activated receptor gamma coactivator 1 alpha (PGC-1α) pathway, resulting in increased expression of genes involved in mitochondrial biogenesis. This leads to an improvement in both the quantity and quality of mitochondria. Additionally, it serves as a signaling molecule implicated in numerous cellular processes. Its ability to stimulate protein kinases, such as protein kinase G (PKG) and Akt, yields downstream effects on mitochondrial function. Hydrogen sulfide's signaling properties can regulate energy metabolism, influence mitochondrial dynamics, and enhance cell survival by promoting cytoprotective mechanisms.

The control of cell death and mitochondrial membrane permeability is influenced by the opening and shutting of the mitochondrial permeability transition pore (MPTP). Hydrogen sulfide has been found to regulate the MPTP, possibly by preventing its opening or promoting its closure. This can have implications for protecting cells from apoptosis and maintaining mitochondrial integrity (Kent et al., 2021). The concentration dependence of hydrogen sulfide's effects is noteworthy, and the mechanisms responsible for its interactions with the mitochondria continue to be an active area of research. As more studies are conducted, our comprehension of hydrogen sulfide's impact on cellular processes and the mitochondria continues to improve.

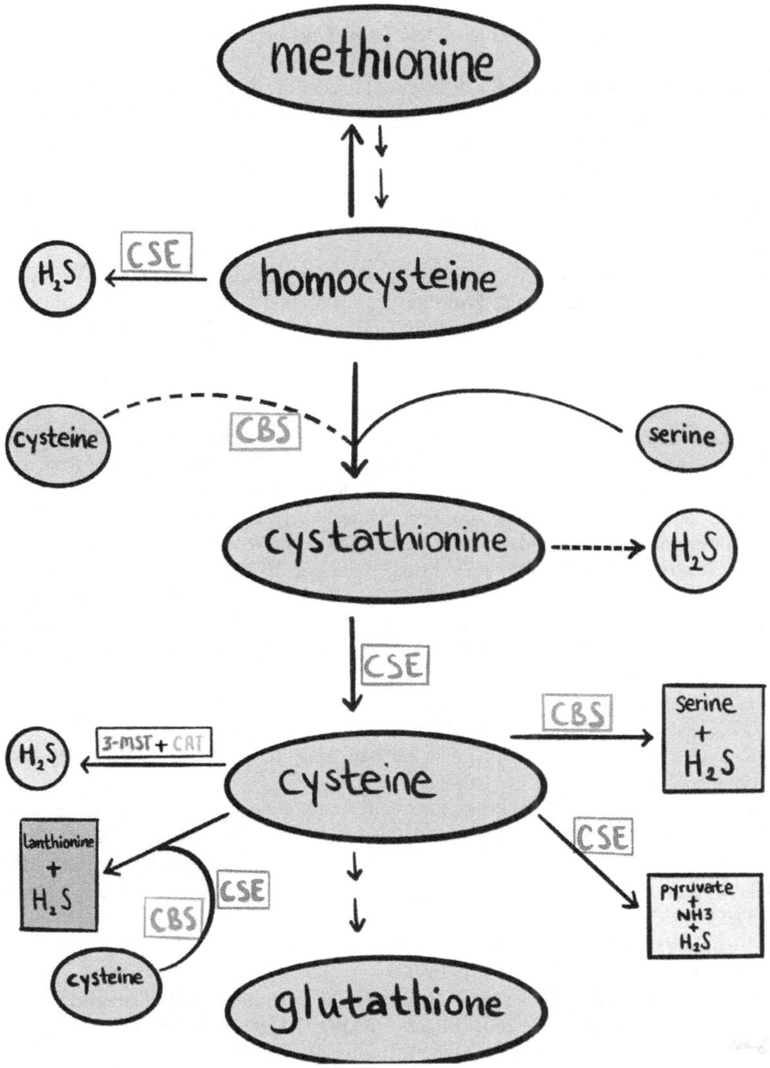

FIGURE 2.5 Glutathione and hydrogen sulfide. Glutathione is a tripeptide composed of three amino acids: cysteine, glycine, and glutamate. It is one of the most abundant antioxidants found within cells and plays a crucial role in maintaining cellular redox balance. *Created by Dilruba Aytürk.*

References

Adina-Zada, A., Zeczycki, T. N., & Attwood, P. V. (2012). Regulation of the structure and activity of pyruvate carboxylase by acetyl CoA. *Archives of Biochemistry and Biophysics, 519*(2), 118–130. https://doi.org/10.1016/j.abb.2011.11.015

Al Rawi, S., Louvet-Vallée, S., Djeddi, A., Sachse, M., Culetto, E., Hajjar, C., Boyd, L., Legouis, R., & Galy, V. (2011). Postfertilization autophagy of sperm organelles prevents paternal mitochondrial DNA transmission. *Science, 334*(6059), 1144–1147. https://doi.org/10.1126/science.1211878

Aunoble, B., Sanches, R., Didier, E., & Bignon, Y. J. (2000). Major oncogenes and tumor suppressor genes involved in epithelial ovarian cancer (review). *International Journal of Oncology, 16*(3), 567–576. https://doi.org/10.3892/ijo.16.3.567

Bae, Y. S., Sung, J. Y., Kim, O. S., Kim, Y. J., Hur, K. C., Kazlauskas, A., & Rhee, S. G. (2000). Platelet-derived growth factor-induced H_2O_2 production requires the activation of phosphatidylinositol 3-kinase. *Journal of Biological Chemistry, 275*(14), 10527–10531. https://doi.org/10.1074/jbc.275.14.10527

Bayrhuber, M., Meins, T., Habeck, M., Becker, S., Giller, K., Villinger, S., Vonrhein, C., Griesinger, C., Zweckstetter, M., & Zeth, K. (2008). Structure of the human voltage-dependent anion channel. *Proceedings of the National Academy of Sciences of the United States of America, 105*(40), 15370–15375. https://doi.org/10.1073/pnas.0808115105

Bricker, D. K., Taylor, E. B., Schell, J. C., Orsak, T., Boutron, A., Chen, Y. C., Cox, J. E., Cardon, C. M., Van Vranken, J. G., Dephoure, N., Redin, C., Boudina, S., Gygi, S. P., Brivet, M., Thummel, C. S., & Rutter, J. (2012). A mitochondrial pyruvate carrier required for pyruvate uptake in yeast, Drosophila, and humans. *Science, 336*(6090), 96–100. https://doi.org/10.1126/science.1218099

Brivet, M., Garcia-Cazorla, A., Lyonnet, S., Dumez, Y., Nassogne, M. C., Slama, A., Boutron, A., Touati, G., Legrand, A., & Saudubray, J. M. (2003). Impaired mitochondrial pyruvate importation in a patient and a fetus at risk. *Molecular Genetics and Metabolism, 78*(3), 186–192. https://doi.org/10.1016/S1096-7192(03)00016-7

Cecchini, G. (2003). Function and structure of complex II of the respiratory chain. *Annual Review of Biochemistry, 72*, 77–109. https://doi.org/10.1146/annurev.biochem.72.121801.161700

Cheng, T., Sudderth, J., Yang, C., Mullen, A. R., Jin, E. S., Matés, J. M., & DeBerardinis, R. J. (2011). Pyruvate carboxylase is required for glutamine-independent growth of tumor cells. *Proceedings of the National Academy of Sciences of the United States of America, 108*(21), 8674–8679. https://doi.org/10.1073/pnas.1016627108

Dickson, V. K., Silvester, J. A., Fearnley, I. M., Leslie, A. G. W., & Walker, J. E. (2006). On the structure of the stator of the mitochondrial ATP synthase. *EMBO Journal, 25*(12), 2911–2918. https://doi.org/10.1038/sj.emboj.7601177

Ducker, G. S., & Rabinowitz, J. D. (2017). One-carbon metabolism in health and disease. *Cell Metabolism, 25*(1), 27–42. https://doi.org/10.1016/j.cmet.2016.08.009

Faubert, B., Li, K. Y., Cai, L., Hensley, C. T., Kim, J., Zacharias, L. G., Yang, C., Do, Q. N., Doucette, S., Burguete, D., Li, H., Huet, G., Yuan, Q., Wigal, T., Butt, Y., Ni, M., Torrealba, J., Oliver, D., Lenkinski, R. E., & DeBerardinis, R. J. (2017). Lactate metabolism in human lung tumors. *Cell, 171*(2). https://doi.org/10.1016/j.cell.2017.09.019, 358-371.e9.

Gabaldón, T., & Huynen, M. A. (2004). Prediction of protein function and pathways in the genome era. *Cellular and Molecular Life Sciences, 61*(7–8), 930–944. https://doi.org/10.1007/s00018-003-3387-y

Glancy, B., Kane, D. A., Kavazis, A. N., Goodwin, M. L., Willis, W. T., & Gladden, L. B. (2021). Mitochondrial lactate metabolism: History and implications for exercise and disease. *Journal of Physiology, 599*(3), 863–888. https://doi.org/10.1113/JP278930

Glancy, B. (2020). Visualizing mitochondrial form and function within the cell. *Trends in Molecular Medicine, 26*(1), 58–70. https://doi.org/10.1016/j.molmed.2019.09.009

Gray, L. R., Tompkins, S. C., & Taylor, E. B. (2014). Regulation of pyruvate metabolism and human disease. *Cellular and Molecular Life Sciences, 71*(14), 2577–2604. https://doi.org/10.1007/s00018-013-1539-2

Gray, M. W., Lang, B. F., Cedergren, R., Golding, G. B., Lemieux, C., Sankoff, D., Turmel, M., Brossard, N., Delage, E., Littlejohn, T. G., Plante, I., Rioux, P., Saint-Louis, D., Zhu, Y., & Burger, G. (1998). Genome structure and gene content in protist mitochondrial DNAs. *Nucleic Acids Research, 26*(4), 865–878. https://doi.org/10.1093/nar/26.4.865

Guo, R., Gu, J., Zong, S., Wu, M., & Yang, M. (2018). Structure and mechanism of mitochondrial electron transport chain. *Biomedical Journal, 41*(1), 9–20. https://doi.org/10.1016/j.bj.2017.12.001

Jonckheere, A. I., Smeitink, J. A. M., & Rodenburg, R. J. T. (2012). Mitochondrial ATP synthase: Architecture, function and pathology. *Journal of Inherited Metabolic Disease, 35*(2), 211–225. https://doi.org/10.1007/s10545-011-9382-9

Juan, C. A., de la Lastra, J. M. P., Plou, F. J., & Pérez-Lebeña, E. (2021). The chemistry of reactive oxygen species (Ros) revisited: Outlining their role in biological macromolecules (dna, lipids and proteins) and induced pathologies. *International Journal of Molecular Sciences, 22*(9). https://doi.org/10.3390/ijms22094642

Judge, A., & Dodd, M. S. (2020). Metabolism. *Essays in Biochemistry, 64*(4), 607–647. https://doi.org/10.1042/EBC20190041

Kühlbrandt, W. (2015). Structure and function of mitochondrial membrane protein complexes. *BMC Biology, 13*(1). https://doi.org/10.1186/s12915-015-0201-x

Kadenbach, B., & Hüttemann, M. (2015). The subunit composition and function of mammalian cytochrome c oxidase. *Mitochondrion, 24*, 64–76. https://doi.org/10.1016/j.mito.2015.07.002

Kauppinen, R. A., Sihra, T. S., & Nicholls, D. G. (1987). Aminooxyacetic acid inhibits the malate-aspartate shuttle in isolated nerve terminals and prevents the mitochondria from utilizing glycolytic substrates. *Biochimica et Biophysica Acta, Molecular Cell Research, 930*(2), 173–178. https://doi.org/10.1016/0167-4889(87)90029-2

Kent, A. C., El Baradie, K. B. Y., & Hamrick, M. W. (2021). Targeting the mitochondrial permeability transition pore to prevent age-associated cell damage and neurodegeneration. *Oxidative Medicine and Cellular Longevity, 2021*, Article 6626484. https://doi.org/10.1155/2021/6626484, 2021.

Konstantinov, A. A. (2012). Cytochrome c oxidase: Intermediates of the catalytic cycle and their energy-coupled interconversion. *FEBS Letters, 586*(5), 630–639. https://doi.org/10.1016/j.febslet.2011.08.037

Kumari, A. (2018). Chapter 1 - Glycolysis. *Sweet biochemistry* (pp. 1–5). Academic Press.

Llopis, J., McCaffery, J. M., Miyawaki, A., Farquhar, M. G., & Tsien, R. Y. (1998). Measurement of cytosolic, mitochondrial, and Golgi pH in single living cells with green fluorescent proteins. *Proceedings of the National Academy of Sciences of the United States of America, 95*(12), 6803–6808. https://doi.org/10.1073/pnas.95.12.6803

Luo, S., Valencia, C. A., Zhang, J., Lee, N. C., Slone, J., Gui, B., Wang, X., Li, Z., Dell, S., Brown, J., Chen, S. M., Chien, Y. H., Hwu, W. L., Fan, P. C., Wong, L. J., Atwal, P. S., & Huang, T. (2018). Biparental inheritance of mitochondrial DNA in humans. *Proceedings of the National Academy of Sciences of the United States of America, 115*(51), 13039–13044. https://doi.org/10.1073/pnas.1810946115

Martínez-Reyes, I., & Chandel, N. S. (2020). Mitochondrial TCA cycle metabolites control physiology and disease. *Nature Communications, 11*(1). https://doi.org/10.1038/s41467-019-13668-3

Meier, B., Radeke, H. H., Selle, S., Younes, M., Sies, H., Resch, K., & Habermehl, G. G. (1989). Human fibroblasts release reactive oxygen species in response to interleukin-1 or tumour necrosis factor-α. *Biochemical Journal, 263*(2), 539–545. https://doi.org/10.1042/bj2630539

Mesa-Garcia, M. D., Plaza-Diaz, J., & Gomez-Llorente, C. (2018). Molecular basis of oxidative stress and inflammation. In *Obesity: Oxidative stress and dietary antioxidants* (pp. 41–62). Elsevier. https://doi.org/10.1016/B978-0-12-812504-5.00003-9

Mráček, T., Drahota, Z., & Houštěk, J. (2013). The function and the role of the mitochondrial glycerol-3-phosphate dehydrogenase in mammalian tissues. *Biochimica et Biophysica Acta, Bioenergetics, 1827*(3), 401–410. https://doi.org/10.1016/j.bbabio.2012.11.014

Murphy, B., Bhattacharya, R., & Mukherjee, P. (2019). Hydrogen sulfide signaling in mitochondria and disease. *Federation of American Societies for Experimental Biology Journal, 33*(12), 13098–13125. https://doi.org/10.1096/fj.201901304R

Neupert, W., & Herrmann, J. M. (2007). Translocation of proteins into mitochondria. *Annual Review of Biochemistry, 76*, 723–749. https://doi.org/10.1146/annurev.biochem.76.052705.163409

Niknahad, H., Jamshidzadeh, A., Heidari, R., Zarei, M., & Ommati, M. M. (2017). Ammonia-induced mitochondrial dysfunction and energy metabolism disturbances in isolated brain and liver mitochondria, and the effect of taurine administration: Relevance to hepatic encephalopathy treatment. *Clinical and Experimental Hepatology, 3*(3), 141–151. https://doi.org/10.5114/ceh.2017.68833

Nolfi-Donegan, D., Braganza, A., & Shiva, S. (2020). Mitochondrial electron transport chain: Oxidative phosphorylation, oxidant production, and methods of measurement. *Redox Biology, 37*, 101674. https://doi.org/10.1016/j.redox.2020.101674

Nunnari, J., & Suomalainen, A. (2012). Mitochondria: In sickness and in health. *Cell, 148*(6), 1145–1159. https://doi.org/10.1016/j.cell.2012.02.035

Owen, O. E., Kalhan, S. C., & Hanson, R. W. (2002). The key role of anaplerosis and cataplerosis for citric acid cycle function. *Journal of Biological Chemistry, 277*(34), 30409–30412. https://doi.org/10.1074/jbc.R200006200

Patel, M. S., & Korotchkina, L. G. (2006). Regulation of the pyruvate dehydrogenase complex. In *Biochemical society transactions* (Vol 34, pp. 217–222). https://doi.org/10.1042/BST20060217. Issue 2.

Patlevič, P., Vašková, J., Švorc, P., Vaško, L., & Švorc, P. (2016). Reactive oxygen species and antioxidant defense in human gastrointestinal diseases. *Integrative Medicine Research, 5*(4), 250–258. https://doi.org/10.1016/j.imr.2016.07.004

Pebay-Peyroula, E., Dahout-Gonzalez, C., Kahn, R., Trézéguet, V., Lauquin, G. J. M., & Brandolin, G. (2003). Structure of mitochondrial ADP/ATP carrier in complex with carboxyatractyloside. *Nature, 426*(6962), 39–44. https://doi.org/10.1038/nature02056

Perkins, G. A., & Frey, T. G. (2000). Recent structural insight into mitochondria gained by microscopy. *Micron, 31*(1), 97–111. https://doi.org/10.1016/S0968-4328(99)00065-7

Perkins, G., Renken, C., Martone, M. E., Young, S. J., Ellisman, M., & Frey, T. (1997). Electron tomography of neuronal mitochondria: Three-dimensional structure and organization of cristae and membrane contacts. *Journal of Structural Biology, 119*(3), 260–272. https://doi.org/10.1006/jsbi.1997.3885

Purvis, J. L., & Lowenstein, J. M. (1961). The relation between intra- and extramitochondrial pyridine nucleotides. *Journal of Biological Chemistry, 236*, 2794–2803.

Ryan, D. G., Frezza, C., & O'Neill, L. A. (2021). TCA cycle signalling and the evolution of eukaryotes. *Current Opinion in Biotechnology, 68*, 72–88. https://doi.org/10.1016/j.copbio.2020.09.014

Sato, M., & Sato, K. (2011). Degradation of paternal mitochondria by fertilization-triggered autophagy in *C. elegans* embryos. *Science, 334*(6059), 1141–1144. https://doi.org/10.1126/science.1210333

Sazanov, L. A. (2015). A giant molecular proton pump: Structure and mechanism of respiratory complex I. *Nature Reviews Molecular Cell Biology, 16*(6), 375–388. https://doi.org/10.1038/nrm3997

Schleyer, M., & Neupert, W. (1985). Transport of proteins into mitochondria: Translocational intermediates spanning contact sites between outer and inner membranes. *Cell, 43*(1), 339–350. https://doi.org/10.1016/0092-8674(85)90039-X

Schottlender, N., Gottfried, I., & Ashery, U. (2021). Hyperbaric oxygen treatment: Effects on mitochondrial function and oxidative stress. *Biomolecules, 11*(12). https://doi.org/10.3390/biom11121827

Snezhkina, A. V., Kudryavtseva, A. V., Kardymon, O. L., Savvateeva, M. V., Melnikova, N. V., Krasnov, G. S., & Dmitriev, A. A. (2020). ROS generation and antioxidant defense systems in normal and malignant cells. *Oxidative Medicine and Cellular Longevity*. https://doi.org/10.1155/2019/6175804, 2019.

Sun, F., Huo, X., Zhai, Y., Wang, A., Xu, J., Su, D., Bartlam, M., & Rao, Z. (2005). Crystal structure of mitochondrial respiratory membrane protein complex II. *Cell, 121*(7), 1043–1057. https://doi.org/10.1016/j.cell.2005.05.025

Tiku, M. L., Liesch, J. B., & Robertson, F. M. (1990). Production of hydrogen peroxide by rabbit articular chondrocytes. Enhancement by cytokines. *Journal of Immunology, 145*(2), 690–696.

Trumpower, B. L. (2002). A concerted, alternating sites mechanism of ubiquinol oxidation by the dimeric cytochrome bc1 complex. *Biochimica et Biophysica Acta, Bioenergetics, 1555*(1–3), 166–173. https://doi.org/10.1016/S0005-2728(02)00273-6

Ulger, O., & Kubat, G. B. (2023). Role of mitochondria and mitochondrial transplantation in drug-induced toxic organ injury. *Exploratory Research and Hypothesis in Medicine*. https://doi.org/10.14218/erhm.2022.00117, 000(000), 000–000.

Von Ballmoos, C., Wiedenmann, A., & Dimroth, P. (2009). Essentials for ATP synthesis by F1F0 ATP synthases. *Annual Review of Biochemistry, 78*, 649–672. https://doi.org/10.1146/annurev.biochem.78.081307.104803

Walsh, C. T., Tu, B. P., & Tang, Y. (2018). Eight kinetically stable but thermodynamically activated molecules that power cell metabolism. *Chemical Reviews, 118*(4), 1460–1494. https://doi.org/10.1021/acs.chemrev.7b00510

Wikstrom, M. K. F. (1977). Proton pump coupled to cytochrome c oxidase in mitochondria [24]. *Nature, 266*(5599), 271–273. https://doi.org/10.1038/266271a0

Yang, X. H., & Trumpower, B. L. (1986). Purification of a three-subunit ubiquinol-cytochrome c oxidoreductase complex from Paracoccus denitrificans. *Journal of Biological Chemistry, 261*(26), 12282–12289. https://doi.org/10.1016/s0021-9258(18)67236-9

Zara, V., Assalve, G., & Ferramosca, A. (2022). Multiple roles played by the mitochondrial citrate carrier in cellular metabolism and physiology. *Cellular and Molecular Life Sciences, 79*(8). https://doi.org/10.1007/s00018-022-04466-0

Zhao, R. Z., Jiang, S., Zhang, L., & Yu, Z. B. (2019). Mitochondrial electron transport chain, ROS generation and uncoupling (Review). *International Journal of Molecular Medicine, 44*(1), 3–15. https://doi.org/10.3892/ijmm.2019.4188

Zorov, D. B., Juhaszova, M., & Sollott, S. J. (2014). Mitochondrial reactive oxygen species (ROS) and ROS-induced ROS release. *Physiological Reviews, 94*(3), 909–950. https://doi.org/10.1152/physrev.00026.2013

Zorova, L. D., Popkov, V. A., Plotnikov, E. Y., Silachev, D. N., Pevzner, I. B., Jankauskas, S. S., Babenko, V. A., Zorov, S. D., Balakireva, A. V., Juhaszova, M., Sollott, S. J., & Zorov, D. B. (2018). Mitochondrial membrane potential. *Analytical Biochemistry, 552*, 50–59. https://doi.org/10.1016/j.ab.2017.07.009

CHAPTER

3

Mitochondrial dynamics and mitochondrial biogenesis

Esin Akbay Cetin[1] *and Deniz Akbay*[2]

[1]Department of Biology, Faculty of Science, Hacettepe University, Ankara, Turkey;
[2]Department of Anesthesiology and Reanimation, American Hospital, Istanbul, Turkey

Mitochondrial functions

Mitochondria were first described as granular cytoplasmic compartments in 1857 by Rudolph Albert von Kölliker, but their names were not given then. Benda (1898) renamed it mitochondrion from the Greek words mitos (meaning thread) and chondron (meaning grain). This name has generally been a name that describes the different morphologies in all cells where the organelle is found. In particular, the identification of mitochondrial function and biogenesis processes continued throughout the 20th century. In the 1990s, the synchronization of mitochondrial shape changes and fusion–fission events in living cells were visualized in detail by dynamically using fluorescent dyes and electron tomography analyses. In the same years, the first gene involved in mitochondrial fusion was identified in *Drosophila*. New genes that control mitochondrial fusion and fission have been identified during these years, most of them discovered in yeast.

Adenosine triphosphate (ATP) is produced by mitochondria through oxidative phosphorylation, so they are known as the powerhouse of eukaryotic cells. In addition to regulating calcium homeostasis and amino acid metabolism, they also control the citric acid and urea cycles, which determine a cell's life and death. The mitochondrial genome, or mitochondrial DNA (mtDNA), is packaged in nucleoid-like structures within the mitochondria. A human mtDNA molecule encodes 13 proteins, 2 ribosomal RNAs, and 22 transfer RNAs. In *Saccharomyces cerevisiae*, each mtDNA molecule encodes eight proteins.

A mitochondrion is an organelle with a double membrane system. The portion facing the cytosol is called the outer mitochondrial membrane (OMM) and the portion protruding into the mitochondrial matrix containing mtDNA is called the inner mitochondrial membrane (IMM). The partition delimited by the IMM and OMM is called the intermembrane space (IMS) (Liesa et al., 2009).

Mitochondrial Transplantation and Transfer
https://doi.org/10.1016/B978-0-443-18858-9.00003-2

Mitochondrial biogenesis

The reason mitochondria have so many copies of their circular DNA is that they are of bacterial origin. In light of this feature, mitochondria cannot regenerate, so new organelles are generated through a multistep process involving both fusion and fission. In order for mitochondrial adaptation to occur under cellular internal dynamics and/or intracellular stress conditions, the activation of advanced cellular mechanisms comes into action. This mechanism primarily occurs between the mitochondria and the nucleus. There are various mitochondrial stress signals and nuclear stress response pathways that are involved in this process. A mitochondrial response involves the release of metabolites like acetyl-CoA and "mito-stress" signals such as reactive oxygen species (ROS), providing retrograde signals to the nucleus that cause nuclear gene expression to change (da Cunha et al., 2015). In the nucleus, this information creates a response in the form of initiation of AMP-activated protein kinase (AMPK) signaling and activation of transcriptional programs in response to stress. Regulators involved in the transcriptional program include peroxisome proliferator–activated receptor γ coactivator 1-α (PGC-1α), cyclic adenosine monophosphate (cAMP)-response element binding protein (CREB), nuclear factor-κB, and p53 (Cagin & Enriquez, 2015). As a result of nuclear transcriptional responses, mtDNA is repaired or removed by fusion or fission, and glycolysis and lactate metabolism are switched on (Legaki et al., 2022). They are tightly regulated by nuclear and mitochondrial factors, and this entire process is referred to as mitochondrial biogenesis. In this process, most of the mitochondrial proteins are encoded by nuclear genes. Nuclear-encoded mitochondrial proteins constitute almost 90% of mitochondrial proteins, but their production requires the simultaneous transcription and translation of nuclear and mitochondrial genes. As an example of this mechanism, the formation of respiratory chain complexes appears to involve both nuclear and mitochondrial encoded products (Matsuda et al., 2010; McKenzie et al., 2009). There are also multiple aspects of mitochondrial biogenesis, including replication of mtDNA, transcription and translation of mtDNA-encoded genes, as well as loading of phospholipids and transfer of nuclear proteins into mitochondria (Ventura-Clapier et al., 2008; Zhu et al., 2013). Especially, nuclear-encoded proteins, after transcription and translation processes, are folded depending on their translocase by the outer membrane–mediated transport of mitochondria and they are divided into different compartments of the mitochondria according to their location (OMM, IMS, IMM or matrix). Control of this process can be attributed to AMPK, mechanical target of rapamycin, insulin-like signaling, and calcium- and nitric oxide–triggered signaling cascades (Fraga & Ventura, 2015; Schulz & Rehling, 2014). As mentioned above, mtDNA replication, transcription, and translation processes also have important contributions to mitochondrial biogenesis. It is believed that mitochondrial transcription occurs as a result of the presence of mitochondrial RNA polymerase and the transcription factors TFAM and TFB2M in mammals (Kanki et al., 2004; Yakubovskaya et al., 2014). Mainly this step is executed on the matrix-faceted side of the IMM. There are two contact sites between mitochondrial ribosomes and IMM in yeast. Two receptors exist, one of which contains 21S rRNA and one that fuses to the matrix-facing IMM bypassing AFG3. Translation of mtDNA-encoded transcripts is dependent on ribosomes, tRNA, and coactivators, as well as mitochondrial translation initiation factors mIF2 and mIF3 (homologous to *S. cerevisiae* AIM23) (Kuzmenko et al., 2016).

Mitochondrial biogenesis is a task of coordination and takes place by the joint functioning of both mitochondrial and nuclear genomes. As described above, nuclear genome factors play a significant role in controlling mitochondrial biogenesis. In the mitochondrial biogenesis regulatory network, nuclear respiration factors (NRF1 and NRF2) are essential components among the specific nuclear transcription factors that regulate mitochondrial gene expression. In vitro cell culture studies reveal that NRF1 regulates many target genes. These include the respiratory complex (cytochrome C oxidase subunit IV [COXIV], cytochrome *c*), the mediator responsible for mitochondrial protein transfer (TOMM34), heme proteins forming the detoxification response, and biosynthetic pathway (globin) and mtDNA can be counted (Biswas & Chan, 2010; Satoh et al., 2013; Yang et al., 2014). Like NRF1, NRF2 is deacetylated by SIRT7 and regulates mitochondrial transcription factors (TFAM, TFB1M, TFB2M) and respiratory chain components (COXIV subunits) (Bruni et al., 2010). A nuclear factor erythroid 2-like 2 (NFE2L2) transcription factor also contributes to mitochondrial biogenesis. Many studies have shown that NFE2L2 indirectly regulates NRF1 target genes by enhancing their expression (Piantadosi et al., 2008). However, NFE2L2 also targets different genes involved in mitochondrial processes and biosynthetic pathways (Taguchi et al., 2011). There are several genes involved in this process, including heme oxygenase-1, methylenetetrahydrofolate nicotinamide adenine dinucleotide phosphate-dependent dehydrogenase-2, and superoxide dismutase-1. This transcription factor, which is transferred to the nucleus in response to stress, is originally retained in the cytoplasm by interacting with Kelch like epichlorohydrin (ECH)-related protein 1 (Taguchi et al., 2011). NFE2L1, another member of the nuclear factor erythroid 2-like family, is known to be responsible for the regulation of detoxification mechanisms (Zhao et al., 2011). Since mitochondrial genes are among its potential targets, it is thought to be probably also involved in mitochondrial biogenesis. However, further research is required to clarify whether NFE2L1 is involved in mitochondrial-related functions.

In details, the emergence of knowledge regarding the transcriptional control of mitochondrial biogenesis has occurred with the identification of the transcriptional core regulator peroxisome proliferator–activated receptor γ (PPARγ) coactivator 1α (PGC-1α). PGC-1α is a member of a family of transcriptional coregulators, including the closely related PGC-1β and a more distant member, the PGC-1-associated coactivator (Andersson & Scarpulla, 2001; Dorn et al., 2015; Kressler et al., 2002; Lin et al., 2002). PGC-1 coactivators allow coordinate control of different pathways involved in mitochondrial biogenesis and perform their functions by interacting directly with target transcription factor effectors (Dorn et al., 2015; Vega et al., 2015). A specific LXXLL recognition site in the nucleus interacts directly with members of the nuclear receptor superfamily. This interaction leads to its connection with PPAR, estrogen-related receptor (ERR), and NRF-1, which is a transcription factor responsible for mitochondrial metabolism, respiration and biogenesis. TFAM and TFB2M, which mediate replication and transcription of the mitochondrial genome, are activated by NRF-1 (Dorn et al., 2015; Gleyzer et al., 2005). ERRα, on the other hand, regulates genes involved in almost every stage of mitochondrial energy transmission and ATP synthesis (Dorn et al., 2015; Huss et al., 2004). There are many transcription factors responsible for PGC-1α regulation. One of those that have been identified recently is necdin. Studies have shown that necdin stabilizes by inhibiting ubiquitination and facilitates mitochondrial biogenesis by regulating mitochondrial-specific genes (Hasegawa et al., 2016). Additionally, SIRT1 as well as transcription factors binding to IGHM enhancer 3-TFE3, the cAMP response element–binding protein CREB, and the forkhead transcription factor FoxO1 are also transcription factors that regulate PGC-1α other than necdin (Salma et al., 2015). Also, PGC-1β is upregulated by CREB and c-MYC (Chang & McDonnell, 2012; Herzig et al., 2001). Furthermore, mitofusin 1 (MFN1) and mitofusin 2 (MFN2) have been found to be direct targets of the ERRα/PGC-1α pathway (Dorn et al., 2015; Martin et al., 2014; Soriano et al., 2006). Therefore, this pathway has a role in coordinating mitochondrial biogenesis and dynamics. All these facts suggest that there are cross-regulatory circuits that coordinate mitochondrial dynamics, mitochondrial biogenesis, and also mitophagy. Coactivation of PGC-1α by ERR, PPAR, and other transcription factors not only plays a role in controlling biogenesis, but also regulates mitophagy and dynamic components. By dividing mitochondria, known as fission, smaller mitochondria are formed, encouraging the formation of new organelles. As opposed to this, fusion occurs when mitochondria communicate with other organelles, allowing them to grow longer and more interconnected. Several highly conserved guanosine triphosphatases and their interactors play important roles in these two mitochondrial dynamics. The fusion, division, and alteration of mitochondria's morphology have been reported as regulating a variety of processes, including local energy supply, interorganelle fusion, mitochondrial content mixing, mtDNA maintenance, mitochondrial distribution during mitosis, and mitochondrial segregation. Pathological conditions ranging from cancer to neurodegeneration have been associated with disruptions in this dynamic (Ploumi et al., 2017).

Mitochondrial dynamics

The mitochondria are dynamic organelles that undergo constant movement, division, and fusion. It has been shown that this process is essential to protect healthy mitochondria from negative processes occurring in the cell (Li et al., 2020). There is a constant transformation within mitochondria resulting in a homeostasis known as mitochondrial dynamics (Cormio et al., 2017; Di Pietro et al., 2017; Kim et al., 2017). This is especially possible due to mitochondrial quality control (MQC). The MQC system involves biogenesis, mitochondrial dynamics, and mitophagy (Chang et al., 2021). In addition to regulating mitochondria independently, mitochondrial fusions and fissions interact with other equilibrium equilibria in the MQC system, forming a network of regulation to maintain mitochondria's normal function. As a "monitoring controller," the MQC system maintains mitochondrial homeostasis by monitoring energy demands and metabolic activities within the cell (Simula et al., 2017). Using the MQC, mitochondria manage fission and fusion, biogenesis, and mitophagy, thus ensuring mitochondrial quantity, morphology, quality, and stability of the internal environment. Mitochondrial fission and fusion mainly refers to the process of remodeling, rebuilding, and recycling mitochondria (Fig. 3.1). In response to changes in the intracellular environment, these processes work together to coordinate mitochondrial morphology and create dynamic equilibrium. The area of the mitochondrial cristae expands as mitochondria fuse together. As a result, elongated mitochondria are able to produce more ATP in cells with higher metabolic demands (Khacho et al., 2014; Liu et al., 2023).

On the other hand, excessive ROS production occurs in aging cells and accordingly, the accumulation of damaged mitochondria in the cell increases. During mitochondrial fission, damaged mitochondria are first broken down into smaller pieces to prevent further ROS

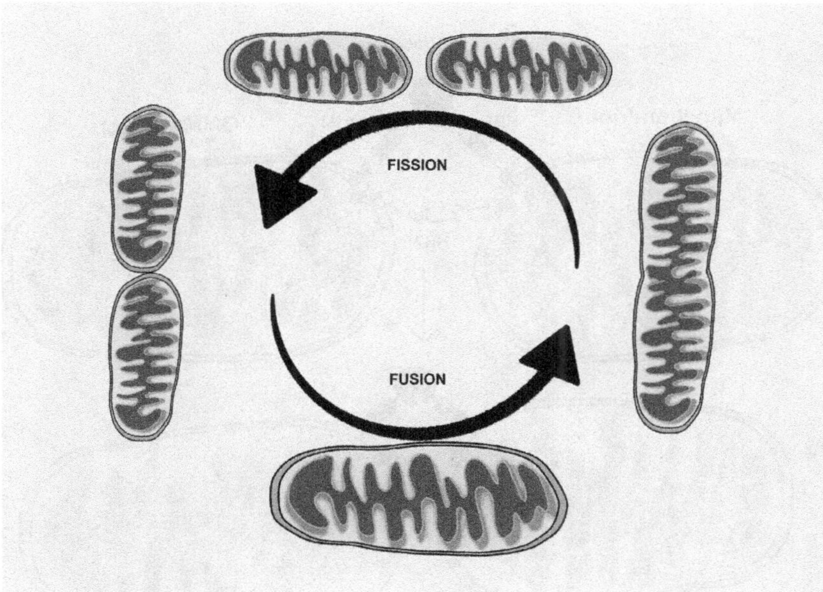

FIGURE 3.1 Typical mitochondrial dynamics diagram. A constant cycle of fusion and fission occurs in human mitochondria. *Diagram was drawn by Dr. Deniz Akbay.*

damage and then the environment is cleared by "mitophagy" in the next stage (Weir et al., 2017). The fine arrangement between fusion and fission keeps the mitochondrial content in the cell in dynamic equilibrium (Liu et al., 2023).

1. Fusion

Mitochondrial fusion is a three-step process of mitochondrial membranes (Fig. 3.2). The two mitochondria must first be connected in a trans state. It is then followed by the interlocking of the two membranes, which increases the contact area and decreases the distance between them (Brandt et al., 2016). Finally, fusion of the two OMMs induced by GTPase activity and conformational changes follow, resulting in fusion (Ishihara et al., 2004; Legros et al., 2002; Tilokani et al., 2018).

In detail, the first and second steps involve the fusion of the OMM. There are two different proteins (MFN1 and MFN2) in the mitofusin (MFNs) family involved in OMM fusion, which share approximately 80% sequence similarity with humans. It was discovered that *Drosophila melanogaster* has an ortholog of MFNs, the fuzzy onion (Fzo1) (Hales & Fuller, 1997). It has been preserved from yeast to human (Hermann et al., 1998; Santel & Fuller, 2001). In recent years, it has been reported that the process of mitochondrial fusion mechanism proposed by MFNs is especially based on the topologies of these proteins. As in yeast, MFNs contain two transmembrane domains separated by a short loop. The N-terminal region contains GTPase and coil-to-coil repeat 1 (HR1) domain, while the C-terminal region harbors HR2 domains. In

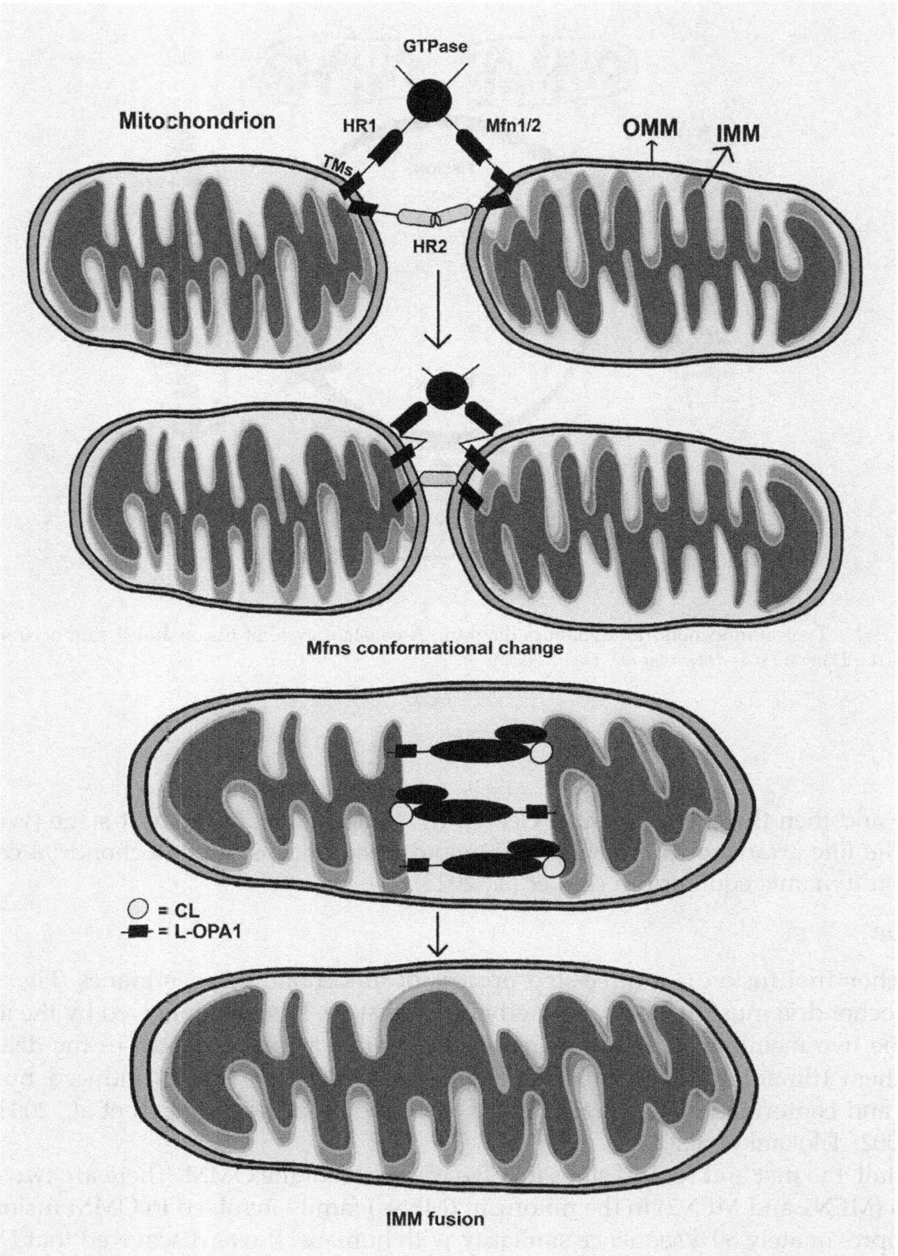

FIGURE 3.2 Schematic diagram of mitochondrial fusion steps. There are three major mitochondrial fusion proteins in mammals: MFN1, MFN2, and Opa1. Inner membrane processing is mediated by OPA1. Interactions between the HR2 and/or GTPase domains of Mfns tether the outer membranes of two opposing mitochondria. The binding of GTP causes Mfns to undergo conformational changes, resulting in mitochondrial docking and an increase in membrane contact sites. Additionally, oligomerization and power stroke are both governed by GTPase. This process is also influenced by the composition of the OMM in phospholipids. IMM fusion follows OMM fusion via OPA1 and CL. By interacting with CL on either side of the membrane, OPA1 tethers the two IMM, which fuse after the hydrolysis of GTP by OPA1. *Diagram was drawn by Dr. Deniz Akbay.*

previous studies, it has been assumed to connect to the OMM (Huang et al., 2011; Rojo et al., 2002).

Following the OMM fusion, the third stage of fusion occurs with the initiation of IMM fusion. GTPase protein family members OPA1 (optic atrophy 1) and specific lipid components of IMM mediate this step. The structure of OPA1 (Sesaki et al., 2003), which was first discovered in a yeast model (Mgm1p), has been conserved throughout evolution. It has been identified that OPA1 consists of eight splice variants. A membrane-associated metalloprotease (OMA1, YME1L) cleaves at the S1 and S2 sites of OPA1 to form short and soluble fragments. In order for fusion to occur, OPA1 cleavage by OMA1 and YME1L must be balanced (Ehses et al., 2009; Head et al., 2009). The OPA1 fragment contained two high molecular weight forms as well. Those are the long-form L-OPA1 and the short-form S-OPA1. In recent research, it has been shown that the L-OPA1 isoform alone is sufficient for fusion to occur. S-OPA1 has also been shown to induce membrane tubulation in liposome analysis and to induce mitochondrial fusion via YME1L-dependent S-OPA1 formation (Mishra et al., 2014).

A membrane lipid component called cardiolipin (CL) is also involved in the fusion process. In the remodeling and dynamics of membranes, this lipid plays a crucial role. In addition, CL is a mitochondria-specific negatively charged lipid that assembles and stabilizes large protein complexes such as OXPHOS complexes with mitochondrial contact site and cristae regulation system (MICOS) (Cogliati et al., 2016; Tilokani et al., 2018).

1. Fission

Mitochondrial fission consists of multiple steps that are organized by the large GTPase dynamin-related protein 1 (Drp1) (Fig. 3.3). Like other GTPases, Drp1 has evolved in a conserved manner. Yeast and *Caenorhabditis elegans* were the first organisms to describe it, and mammals have by now extensively studied its role in mitochondrial division (Bleazard et al., 1999; Labrousse et al., 1999; Smirnova et al., 2001).

After dynamically recruiting and oligomerizing into mitochondrial and peroxisomal membranes, this cytosolic protein stimulates membrane contraction through GTP (Kraus & Ryan, 2017). As a result of Drp1 forming a ring around mitochondria during division, the membrane becomes constricted (Fröhlich et al., 2013). In order to induce constriction in OMM, Drp1 lacks a pleckstrin homology domain that binds membrane phospholipids directly. A number of model organisms have been observed to recruit Drp1 orthologs to OMM through the membrane-bound protein Fis1 and two receptors Mdv1 and Caf4 (Griffin et al., 2005; Mozdy et al., 2000; Tieu et al., 2000). Nevertheless, these proteins and receptors are not found in mammals. Studies later found that mitochondrial fission factor (MFF) and mitochondrial dynamic proteins 49 and 51 (MiD49 and MiD51) acted as receptors for Drp1 in a manner similar to that found with this protein in mammals (Palmer et al., 2011). It is presumed that MFFs and MiDs play complementary roles in mitochondrial fission; MiDs take up GTP bound forms of Drp1 to facilitate oligomerization, while MFFs selectively take up oligomeric and active forms. It is necessary to make a final cut in order for the mitochondria to split. It was thought that Dnm2, which is initially involved in endocytic vesicle division, catalyzed this last step (Lee et al., 2016), but subsequent studies demonstrated that the endoplasmic reticulum (ER) is vital to the first step. Mitochondrial diameter is reduced from an average of 300–500 nm to about 150 nm by the ER to facilitate Drp1-oligomeric ring formation (Kameoka et al., 2018). During this stage, Drp1 and adapters MFF and MiD49/51 are also

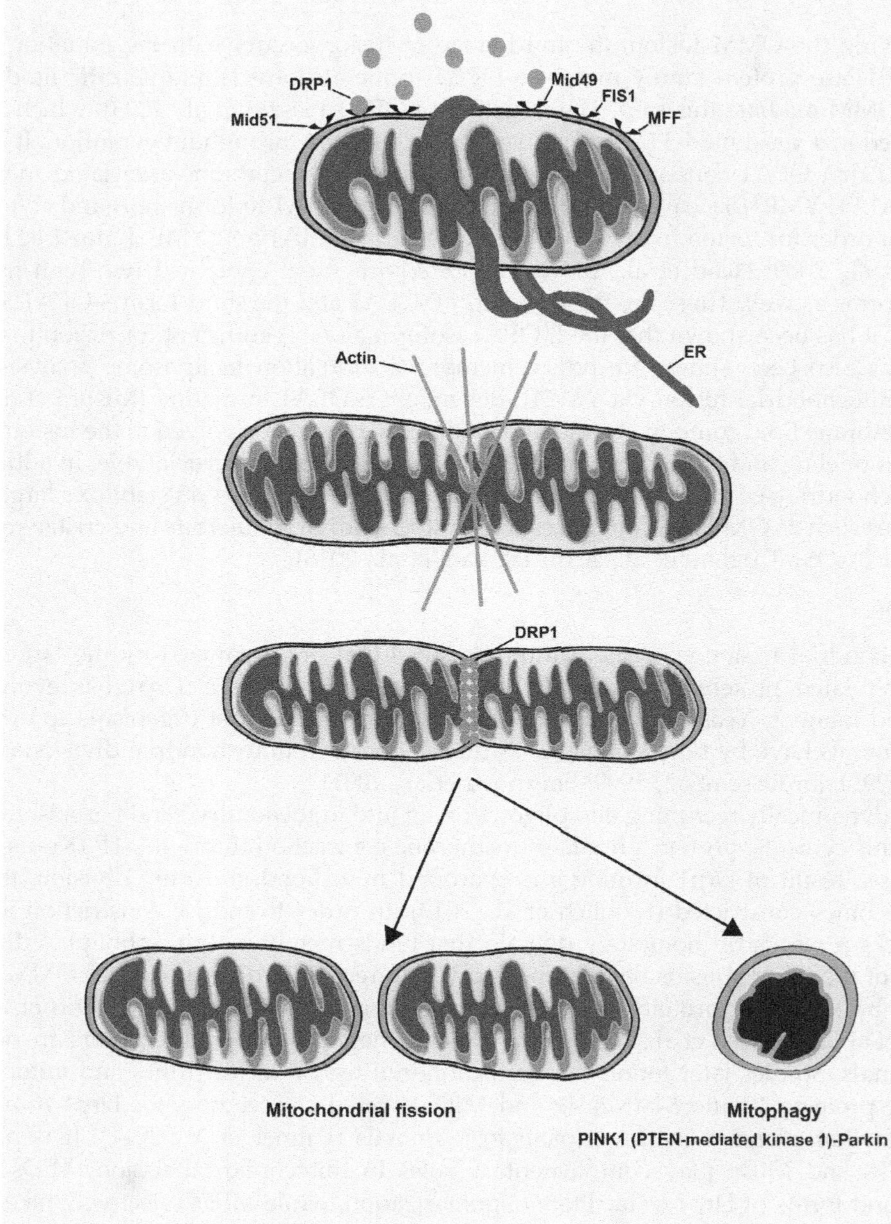

FIGURE 3.3 Schematic drawing showing the division of mitochondria. Cleavage results in either fission or mitophagy. The decision of this is completely determined by intracellular factors. ER tubules are responsible for marking the necessary regions for mitochondrial fission. During this process, Drp1, Fis1, Mid49, Mid51, and Mff receptors take them up into the mitochondria. Phosphorylation and ubiquitination play an important role in regulating the activity of Drp1. Mitochondrial division is performed by Drp1 and the preconstriction process is completed by actin filaments. This results in the formation of two mitochondria or one mitophagic cell. PINK1 (PTEN mediated kinase 1)-Parkin pathway regulates mitophagy. *Diagram was drawn by Dr. Deniz Akbay.*

inserted into mitochondrial-ER contact sites (Osellame et al., 2016; Yamashita et al., 2016). It is also important to establish actin assembly, which plays an important role during mitochondrial contraction. ER-bound reverse formin 2 (INF2) and mitochondrially linked formin-binding Spire1C are both actin-nucleating proteins. There is cooperation of these two proteins at the mitochondria/ER contact points (Manor et al., 2015). Cofilin, cortactin, Arp2/3, and septin 2, which regulate the actin cytoskeleton, have also been shown to be involved in mitochondrial fission (Li, Xu, et al., 2015; Pagliuso et al., 2016). As discussed so far, mitochondrial division is regulated by the simultaneous action of ER and actin.

Despite detailed information on the mechanisms that regulate OMM contraction in mitochondrial fission, little is known about how the IMM contracts or divides in this process. Recent studies have shown that narrowing of IMMs occurs at Ca^{2+} dependent and mitochondria-ER contact sites (Chakrabarti et al., 2018). As calcium enters the mitochondria, the membrane potential changes, activating OMA1. S-OPA1 accumulation impairs the MICOS complex's ability to stabilize OMM-IMM tethering, leading to IMM untethering and contraction (Cho et al., 2017).

This process is also mediated by phospholipids in mitochondrial membranes. A saturated lipid, phospholipids (PA), transfers directly from the ER to the mitochondria to be converted into CL (Vance, 2015). CL promotes liposome membrane tubulation by stimulating OPA1 assembly and GTPase activity, whereas PA increases MFN1/2-dependent OMM fusion (Ban et al., 2017). Previously, CL was described as having a major role to play in heterotypic interactions with OPA1, which promotes membrane fusion (Ban et al., 2017; Tilokani et al., 2018).

1. Mitophagy

Mitochondrial biogenesis and mitophagy are known to be opposing processes that coconstruct the mitochondrial cycle. In mitochondrial biogenesis, the nucleus directs the necessary proteins to the mitochondria as a mediator and supports the formation of new mitochondria, thus increasing the number of mitochondria. Contrary to this, mitophagy involves the self-specific destruction of damaged or aging mitochondria following fission. The balance between these two mitochondrial processes is crucial to maintain the necessary mitochondrial number stability and metabolic homeostasis in the cell. This balance is also important for the cells to recover from the stressful and damaged state against external factors (Moreira et al., 2017). In different studies, it has been shown that mitochondrial biogenesis is activated simultaneously with mitophagy (Palikaras & Tavernarakis, 2014; Zhu et al., 2013). It is possible that mitochondria require biogenesis to adapt to cellular changes in energy needs induced by mitophagy, which is regarded as an anabolic–catabolic equilibrium. There are many factors (aging, stress, etc.) that can negatively affect mitochondrial function, resulting in damage accumulation. Cells repair damage and save affected organelles as their first response to damage accumulations. It is notable, however, that mitochondrial ATP-dependent proteases detect misfolded and/or oxidized polypeptides and activate their proteolysis, while UPS selectively targets OMM proteins (Tatsuta & Langer, 2008). As a result, the mitochondrial unfolded protein response (UPRmt) is activated simultaneously, initiating a retrograde signaling cascade to the nucleus (Pellegrino et al., 2013). When mitochondrial dynamics are also involved, dilution and elimination of mitochondrial damage occurs through fission and fusion mechanisms (Tatsuta & Langer, 2008).

If we need to examine the mitophagic process in more detail, it is first necessary to discuss why mitophagy is important. Mitochondrial homeostasis is achieved by removal of old or damaged mitochondria and simultaneous activation of other essential mitochondrial dynamics. As it is known from cell physiology, the removal of damaged cells or organelles from the environment occurs by autophagy. This event has been termed "mitophagy" in mitochondria. The process of mitophagy, by contrast, involves the degrading of mitochondria by autophagosomes and transferring their load to the degrading lysosomes. Mitophagy targets mitochondria for degradation via receptor-mediated mechanisms. When mitophagy receptors are examined, we can divide them into three: receptors that bind ubiquitous mitochondrial proteins, receptors that are structurally localized in OMM, and finally receptor-like lipid molecules of mitochondrial membranes.

PINK1 (PTEN-mediated kinase 1)-Parkin is the first mitophagy pathway to be discussed, which has been studied primarily in mammalian cells. Damaged mitochondria are removed using this mitophagy pathway. Under normal conditions, PINK1 is actively recruited to mitochondria and undergoes rapid proteolytic degradation. As a result of mitochondrial dysfunction, PINK1 proteolysis becomes disturbed. Parkin is specifically attracted to damaged mitochondria by PINK1, which phosphorylates a number of substrates directly and indirectly. A mitochondrial outer membrane protein called Parkin acts as an E3 ubiquitin ligase after PINK1-mediated mitochondrial translocation, targeting the organelle for autophagy. Also, studies in mouse embryonic fibroblasts and cardiomyocytes related to this pathway have shown that MFN2 acts as a substrate for PINK1. Phosphorylated MFN2 has been shown to act as a Parkin receptor in the OMM of damaged mitochondria (Chen & Dorn, 2013). On the other hand, studies with organelles (like peroxisomes) lacking MFN2 have shown that MFN2 is not indispensable for Parkin uptake, and it can be emphasized that further studies on this subject are necessary (Lazarou et al., 2012). Once properly stabilized and activated in OMM, Parkin triggers engulfment by autophagosomes by ubiquitinating multiple OMM proteins that attract specific autophagy receptors (Nucleoporin 62/sequestosome 1, neighbor of BRCA 1 and optineurin) (Ploumi et al., 2017).

There are also other proteins involved in autophagy, such as Bcl-2/Adenovirus E1B interacting protein 3 (Bnip3), Bnip3-like/NIP3-like protein X (Bnip3L/Nix), Bcl-2-like 13 proteins, and protein 1 containing the FUN14 domain (FUNDC1) (Novak et al., 2010). There is a phosphorylation-regulated LIR motif on these receptors that mediates direct interaction with the LC3-II-tagged phagophore. The Bcl-2 (B-cell lymphoma 2) family contains Bnip3 and its homolog Bnip3L/Nix, which are proapoptotic proteins (Rikka et al., 2011). In addition to their proapoptotic roles, both proteins induce autophagy. Studies in cell culture have demonstrated that phagophores act as selective mitophagy receptors by binding directly to LC3 and GABA(A) receptor-associated proteins (Hanna et al., 2012). One of the aforementioned proteins, FUNDC1, also acts as an OMM receptor. In mammalian cells, this receptor is activated under hypoxic conditions (Liu et al., 2012; Wu et al., 2014). Additionally, this receptor interacts with both Drp1 and OPA1 factors. This information indicates that mitochondrial fission and fusion occur simultaneously during mitophagy (Chen et al., 2016; Ploumi et al., 2017).

Mitophagy is also controlled by lipids located in OMM in addition to protein receptors. CL is a dimeric phospholipid normally found in the IMM of eukaryotic cells that is transported to the OMM by interacting directly with damaged mitochondrial LC3 (LC3-II). Cl is transferred from the IMM to the OMM's IMS-facing side by nucleoside diphosphate kinase-D and mitochondrial creatine kinase (Li, Tsoi et al., 2015). Mitophagy occurs when phospholipid scramblase 3 allows CL to migrate to the OMM's cytoplasm to generate a "catch and eat me" signal. A sphingolipid called ceramide also interacts with LC3-II and contributes to mitophagy. Different types of cell carcinomas respond effectively to exogenous C18-ceramide or overexpression of ceramide synthase 1 (Ploumi et al., 2017; Sentelle et al., 2012).

Different cells and mitochondrial dynamics

Numerous biological processes require fission and fusion of mitochondria, as demonstrated by studies on mitochondrial dynamics. The expression levels and activities of proteins involved in fusion and fission regulate mitochondrial dynamics and mitochondrial shape changes in cells based on energy requirements, cell type, and metabolic resources (Anand et al., 2013; Chang & Blackstone, 2010; Leboucher et al., 2012; Lee et al., 2014). There are different shapes of mitochondria in different cell types, but each shape is suitable for the activity of the cell. Active metabolic cells have fused mitochondria, while dormant cells have fragmented mitochondria. Cells that require a lot of ATP have longer mitochondria with an organized cristae structure (Cogliati et al., 2013; Paumard et al., 2002). OXPHOS activity (Madan et al., 2022; Yao et al., 2019) and membrane potential are increased by mitochondrial fusion (Mitra et al., 2009), while ROS levels are reduced (Lee et al., 2014). Cells and tissues that are differentiated, such as skeletal muscles (Picard et al., 2013), heart (cardiomyocytes) (Pennanen et al., 2014; Wai et al., 2015), neuronal stem cells (Pennanen et al., 2014; Wai et al., 2015), and pancreatic cells (Twig et al., 2010) have reticular mitochondria, while stem cells and embryonic cells have fragmented mitochondria (Chen et al., 2012; Lees et al., 2017; Westermann, 2012). In addition to neuronal differentiation and synapse formation, mitochondrial fission is implicated in embryonic brain development (Ishihara et al., 2009; Wakabayashi et al., 2009) as well as differentiation of myoblasts (Kim et al., 2013; Madan et al., 2022).

When different cell/tissue and mitochondrial dynamics are considered, the brain stands out first. The high energy requirement of neurons, in particular, explains why mitochondria must have a well-ordered morphological adaptation and distribution in brain tissue. Several recent studies show that mitochondrial dynamics play a significant role both in mature neurons and during neural development, in particular during neurogenesis and the selection of neural stem cells (Chen et al., 2008; Khacho et al., 2017). As mitochondria take part in these processes, any disruption in mitochondrial dynamics can lead to neurodegenerative diseases. Studies have reported abnormal mitochondrial fragmentation and mitochondrial dysfunction, especially in neurodevelopmental disorders like tuberous sclerosis complex, Wolfram syndrome, and schizophrenia (Hjelm et al., 2015; Rajasekaran et al., 2015). It has not been clarified how mitochondrial dynamics contribute to neurodevelopmental disorders, but it is known that mitochondrial dynamics are an important factor. In recent studies on these uncertainties in the field, mitochondrial dynamics were found to play a crucial role during neurogenesis both in the adult brain and in embryonic development (Bartolome et al., 2018;

Gonçalves et al., 2016; Iwata & Vanderhaeghen, 2021). Neuronal survival and development are strongly regulated by mitochondrial dynamics in tissue homeostasis. Neural stem cells are differentiated through the cooperative activation of mitochondrial dynamics. Enhanced mitochondrial respiration plays a major role in neural stem cell differentiation. There should also be strong increases in mitochondrial mass and mtDNA copy number (Wang et al., 2010, 2011). A mitochondrial signaling program modulates a nuclear developmental program, which regulates the identity of stem cells (Khacho et al., 2016). During differentiation, changes in mitochondrial morphology are required to enable the activation of OXPHOS metabolism to produce the high amount of ATP required (Khacho et al., 2018; Shyh-Chang et al., 2013; Takubo et al., 2013).

Many cell types can easily adjust mitochondrial position and morphology based on metabolic conditions and energy demands. The mitochondrial network of active metabolic cardiomyocytes with numerous myofilaments and a rigid cytoskeleton is fragmented in adult heart cells (Wai et al., 2015). A spherical mitochondrion, on the other hand, is tightly packed between myofibrils, resulting in a rapid and efficient flow of energy (Tepp et al., 2011). As a result, mitochondria cannot move freely due to this positioning. Vendelin et al. (2005) found that mitochondrial fragmentation in adult cardiomyocytes remains dormant despite abundant expression of fusion–fission proteins. The physiology of the heart is influenced by mitochondrial dynamics proteins, which are hypodynamic in especially adult cardiomyocyte mitochondria. The findings suggest that adult cardiomyocytes are not as static as once thought (Eisner et al., 2017). Furthermore, mitochondria have been shown in different studies to play a role in the development of changes in the amount of nutrients and oxygen delivered to the heart after birth (Gong et al., 2015; Gottlieb & Bernstein, 2015). As the embryonic heart forms, oxygen requirements are low, so anaerobic glycolysis is the primary source of energy (Porter et al., 2011). It transforms from anaerobic glycolysis to oxidation of fatty acids during the process of heart development. The placenta matures during the embryonic process and increased blood flow promotes oxidative stress in the cells, resulting in ATP production using aerobic respiration (Burton, 2009). During the development of the heart and differentiation of cardiomyocytes, mitochondrial morphology and networks undergo continuous changes, which are closely related to mitochondrial function (Gong et al., 2015). Heart tissue in adult cardiomyocytes has been reported to still have some interconnected mitochondria (Huang et al., 2013). In adult cardiomyocytes, it was shown to transfer structurally constrained mitochondria contents by close contact, kissing, and nanotunnels instead of continuous mitochondrial fusion and fission (Huang et al., 2013). A mitochondrial dynamic has been indicated by Li et al. (2020) that maintains an imbalance in mitochondrial content among adjacent, touching mitochondria. "Kissing" was the term used to describe this event. Moreover, nanotunnels have been shown to allow neighboring or distant mitochondria to communicate by connecting through a thin tunnel. Nanotunneling is also a jumping event, unlike kissing, as confirmed in vivo. In a similar study of H9c2 cardiomyoblasts, the fusion and fission proteins undergo significant morphological changes, whereas the changes in adult cardiomyocytes are only slight. In cultured adult cardiomyocytes, kissing and contraction events have been demonstrated to dominate mitochondrial dynamics rather than common fission and fusion (Li et al., 2020). A mitochondrial fusion occurs with the lowest frequency (Qin et al., 2018), whereas the process from fusion to fission takes longer in other cell types (Li et al., 2020; Twig et al., 2008).

As part of lung homeostasis, mitochondria play a key role. Depending on their energy level, airway epithelial cells differ in mitochondrial number and intracellular organization. Furthermore, they regulate surfactant production, cellular aging, mucociliary function, and mucus secretion, as they do in alveolar macrophages (Aghapour et al., 2020). It is important that mitochondrial dynamics occur in an orderly manner during lung development. It has been extensively studied how mitochondrial dynamics may be altered in chronic respiratory diseases (asthma, chronic obstructive pulmonary disease, lung cancer, cystic fibrosis) (Mabalirajan et al., 2009; Malsin et al., 2018; Ryter et al., 2018). It is known that mitochondrial dysfunction prevents the recovery of the lung by changing the cell bioenergetics (Ten & Ratner, 2020). Environmental factors, pollutants, and allergens cause oxidative stress, with mitophagy activating the PINK/Parkin pathway, leading to mitochondrial dysfunction followed by cellular senescence. As a result, allergic airway inflammation and the development of asthma are triggered (Qiu et al., 2019; Sachdeva et al., 2019). Research has shown that CaMKII plays an important role in the production of ROS in allergic inflammation and correlates with asthma severity (Sebag et al., 2017). As per this information, allergic airway inflammation results from an axis involving allergens-ROS-oxidants-CaMKII-mitophagy (Zhang, Agborbesong et al., 2021). There is, however, a need for increased studies in the area of mitophagy and asthma as well as more detailed studies. There is also evidence that the ER–mitochondria interaction plays a role in a variety of pathophysiological processes, such as immunity and cell death (Marchi et al., 2014). A connection is made between ER stress and an increase in ROS production in this case as well. Proteins that are unfolded and mitochondria that are dysfunctional are responsible for ER stress (Delmotte & Sieck, 2020). The mechanism triggering the processes, however, remains unclear. It has been suggested that mitochondrial ROS (mtROS) formation may play an important role in supporting physiological signals of cellular stress as well as leading to mitochondrial translocation due to increased anatomical exposure of the lungs to the environment (Caldeira, Mesquita et al., 2021; van der Vliet et al., 2018). As such, excess mtROS formation triggers inflammatory responses through unbinding ETC, calcium imbalance, and disruption of mitochondrial-ER communication (Caldeira, Weiss et al., 2021; Liu et al., 2017).

One of the most important components of liver tissue is the hepatocyte, which plays an important role in regulating systemic metabolism. Different tissues have different mitochondrial density, often based on oxidative phosphorylation requirements. Mitochondria play a minimal role in maintaining homeostasis in tissue function by regulating metabolic networks. It is clear why mitochondria play such an important role in the metabolic processes of the liver because the pyruvate dehydrogenase complex is found within the mitochondrial matrix. Acetyl-CoA and CO_2 are formed by mitochondria when pyruvate, the final metabolite of aerobic glycolysis, is converted to acetyl-CoA. There are more mitochondria in these cells than any other type of cell. Different studies have examined mitochondrial dynamics in perfused livers and isolated primary rat and mouse hepatocytes (Das et al., 2012; Umbaugh et al., 2021). While hepatocyte mitochondria are either spherical or short tubular, they exhibit limited fusion activity and motility. Physiological conditions of hepatocytes maintain mitochondrial dynamics as in cardiac tissue (Ramachandran et al., 2021). The mitochondria of the liver also play an important role in insulin resistance. Researchers have found that insulin resistance is accompanied by changes in mitochondrial function, structure, and organization. According to a report, a decrease in ATP production correlates with an increase in mtROS production, which induces the oxidation of respiratory chain proteins (Degli Esposti et al., 2012).

Among the organs of the body, the kidney ranks second in terms of mitochondrial content and oxygen consumption after the heart. It is vital for tubular cells to reabsorb solutes, which is an ATP-dependent process. Therefore, the kidney contains a large number of mitochondria. As a result of disruptions in proton gradient regulation and membrane potential, mitochondrial dysfunction is triggered (decreased adenosine ATP, increased mtROS, inflammation) (Bhatia et al., 2020). Physiological processes requiring varying amounts of energy are managed by different segments of the nephrons in the kidneys. It is therefore known that mitochondria are distributed differently among them. A renal proximal tubular cell, for instance, has a high energy demand since it reabsorbs and secretes against chemical gradients. As a result of mitochondrial oxidative phosphorylation, most ATP is produced in proximal tubular cells, whereas podocytes, endothelial cells, and mesangial cells are capable of glycolysis (Zhang, Do et al., 2021).

During the reproductive process, mitochondrial biogenesis is crucial for ovarian function. An oocyte's mitochondrial biogenesis increases dramatically as it develops from an immature germinal vesicle to a mature oocyte. Fertilized eggs produce enough ATP to fuel fertilization, embryogenesis, and mitochondrial replication until implantation. During oogenesis, cAMP is essential for maintaining meiotic arrest and mitochondrial ATP production, which both depend on nuclear maturation (Chiang et al., 2020). Steroidogenesis is another important function of mitochondria in the ovary. During ovarian steroidogenesis, which is responsible for producing sex steroids, cholesterol is transferred to the IMM by the steroidogenic acute regulatory protein. Granulosa cells produce estrogen during folliculogenesis, which triggers the luteinizing hormone to emerge before ovulation. For pregnancy implantation, progesterone acts as a luteal support for the endometrium (Chiang et al., 2020). Ovaries also age along with humans. Increasing disruptions in mitochondrial homeostasis are the main reason for this. As a result of mtROS production and mtDNA damage, aging is accelerated (Wei & Lee, 2002). Researchers have demonstrated that increased mtROS leads to stagnant metabolism and structural changes associated with aging of the ovary in mice (Ben-Meir et al., 2015). Additionally, oocyte-specific mitochondrial fission protein Drp1 mutations have been linked to maturation defects and lower oocyte quality (Chiang et al., 2020).

In male reproductive systems, mitochondria provide energy for sperm motility. By microscopic examination, mitochondria in asthenozoospermia have irregularities and shorter middle segments, while mitochondria from normozoospermic samples are similar in length and width (Ramalho-Santos et al., 2009). Along with motility, enzymatic activities of the ETC complex contribute to sperm quality as well. During spermatogenesis, testicular mitochondria undergo marked changes in morphology, location, and energy metabolism. Three types of mitochondria are distinguished at different stages in this process, including "orthodox" mitochondria (in Sertoli cells, spermatogonia, preleptotene spermatocytes, and leptotene spermatocytes), "intermediate form" mitochondria (in zygothene spermatocytes), and "condensed" mitochondria (in pachytene and secondary spermatocytes and early spermatids) (De Martino et al., 1979). A number of factors such as Activin A released by Sertoli cells cause these changes in morphology (Meinhardt et al., 2000). Testicular mitochondria differ from mitochondria from other tissues in terms of their bioenergetic parameters. It has been shown that testosterone mitochondria require less oxygen to generate approximately the same electrical potential as other tissues, and they change in phosphorylation efficiency with younger animals having less efficient phosphorylation (Amaral et al., 2009; Ramalho-Santos et al., 2009).

The mucosal barrier function of intestinal epithelial cells is essential for digestion. Intestinal epithelial integrity depends on mitochondrial function, which is crucial for maintaining mucosal barrier function (Krzystek-Korpacka et al., 2020). Mitochondrial dysfunction can damage epithelium cells in the gastrointestinal tract. Consequently, intestinal inflammation is induced with decreased barrier function and increased permeability of epithelial cells. A reduction in Paneth cell number and/or damage is one of the most important mechanisms behind Crohn's disease (CD) causing intestinal pathology. Moreover, mitochondrial dysfunction increases ileal inflammation, especially in patients with CD, since mitochondrial endosomal antiproliferative proteins decrease as a result of mitochondrial dysfunction (Sui et al., 2022).

The mitochondria play an important functional role in all cells and tissues. Additionally, mitochondria are shaped and function differently depending on the tissue.

Conclusion

In mammalian biology, mitochondrial dynamics play an important role as a fundamental biological process. These dynamics determine mitochondrial morphology and involve a cascade of cyclical events important for cellular function (Madan et al., 2022; Xie et al., 2020). Further, it is well known that fusion and fission homeostasis are responsible for organisms' physiological and pathological repair processes. Researchers have examined the molecular mechanisms of the quality control of membrane remodeling events (Ng et al., 2021; Wang et al., 2022) considering the basic components of the fusion and fission mechanism, and based on these studies, detailed information on this subject has been obtained. However, in order to make further progress toward this goal, the details of these mechanisms need to be examined in more detail in terms of structural and biochemical aspects.

In addition to controlling mitochondrial shape, mitochondrial dynamics also plays a crucial role in mitochondrial function. New methods must be developed to understand how mitochondria interact with each other and why such interactions are important to the health of the overall mitochondrial population within a cell. Because mitochondrial damage can result in inflammatory responses in different ways by causing mitochondrial fusion and fission dysfunction, there is no doubt that detailed studies will lead to promising approaches to improve human health.

References

Aghapour, M., Remels, A. H. V., Pouwels, S. D., Bruder, D., Hiemstra, P. S., Cloonan, S. M., & Heijink, I. H. (2020). Mitochondria: At the crossroads of regulating lung epithelial cell function in chronic obstructive pulmonary disease. *American Journal of Physiology - Lung Cellular and Molecular Physiology, 318*(1), L149–L164. https://doi.org/10.1152/ajplung.00329.2019

Amaral, S., Mota, P. C., Lacerda, B., Alves, M., Pereira, M.d. L., Oliveira, P. J., & Ramalho-Santos, J. (2009). Testicular mitochondrial alterations in untreated streptozotocin-induced diabetic rats. *Mitochondrion, 9*(1), 41–50. https://doi.org/10.1016/j.mito.2008.11.005

Anand, R., Langer, T., & Baker, M. J. (2013). Proteolytic control of mitochondrial function and morphogenesis. *Biochimica et Biophysica Acta - Molecular Cell Research, 1833*(1), 195–204. https://doi.org/10.1016/j.bbamcr.2012.06.025

Andersson, U., & Scarpulla, R. C. (2001). PGC-1-related coactivator, a novel, serum-inducible coactivator of nuclear respiratory factor 1-dependent transcription in mammalian cells. *Molecular and Cellular Biology, 21*(11), 3738–3749. https://doi.org/10.1128/MCB.21.11.3738-3749.2001

Ban, T., Ishihara, T., Kohno, H., Saita, S., Ichimura, A., Maenaka, K., Oka, T., Mihara, K., & Ishihara, N. (2017). Molecular basis of selective mitochondrial fusion by heterotypic action between OPA1 and cardiolipin. *Nature Cell Biology, 19*(7), 856–863. https://doi.org/10.1038/ncb3560

Bartolome, F., De La Cueva, M., Pascual, C., Antequera, D., Fernandez, T., Gil, C., Martinez, A., & Carro, E. (2018). Amyloid β-induced impairments on mitochondrial dynamics, hippocampal neurogenesis, and memory are restored by phosphodiesterase 7 inhibition. *Alzheimer's Research & Therapy, 10*(1). https://doi.org/10.1186/s13195-018-0352-4

Ben-Meir, A., Burstein, E., Borrego-Alvarez, A., Chong, J., Wong, E., Yavorska, T., Naranian, T., Chi, M., Wang, Y., Bentov, Y., Alexis, J., Meriano, J., Sung, H. K., Gasser, D. L., Moley, K. H., Hekimi, S., Casper, R. F., & Jurisicova, A. (2015). Coenzyme Q10 restores oocyte mitochondrial function and fertility during reproductive aging. *Aging Cell, 14*(5), 887–895. https://doi.org/10.1111/acel.12368

Benda, C. (1898). Ueber die spermatogenese der vertebraten und höherer evertebraten, II. Theil: Die histiogenese der spermien. *Archiv für Anatomie Physiologie, 73*, 393–398.

Bhatia, D., Capili, A., & Choi, M. E. (2020). Mitochondrial dysfunction in kidney injury, inflammation, and disease: Potential therapeutic approaches. *Kidney Research and Clinical Practice, 39*(3), 244–258. https://doi.org/10.23876/j.krcp.20.082

Biswas, M., & Chan, J. Y. (2010). Role of Nrf1 in antioxidant response element-mediated gene expression and beyond. *Toxicology and Applied Pharmacology, 244*(1), 16–20. https://doi.org/10.1016/j.taap.2009.07.034

Bleazard, W., McCaffery, J. M., King, E. J., Bale, S., Mozdy, A., Tieu, Q., Nunnari, J., & Shaw, J. M. (1999). The dynamin-related GTPase Dnm1 regulates mitochondrial fission in yeast. *Nature Cell Biology, 1*(5), 298–304. https://doi.org/10.1038/13014

Brandt, T., Cavellini, L., Kühlbrandt, W., & Cohen, M. M. (2016). A mitofusin-dependent docking ring complex triggers mitochondrial fusion in vitro. *Elife, 5*. https://doi.org/10.7554/eLife.14618

Bruni, F., Polosa, P. L., Gadaleta, M. N., Cantatore, P., & Roberti, M. (2010). Nuclear respiratory factor 2 induces the expression of many but not all human proteins acting in mitochondrial DNA transcription and replication. *Journal of Biological Chemistry, 285*(6), 3939–3948. https://doi.org/10.1074/jbc.M109.044305

Burton, G. J. (2009). Oxygen, the Janus gas; its effects on human placental development and function. *Journal of Anatomy, 215*(1), 27–35. https://doi.org/10.1111/j.1469-7580.2008.00978.x

Cagin, U., & Enriquez, J. A. (2015). The complex crosstalk between mitochondria and the nucleus: What goes in between? *The International Journal of Biochemistry & Cell Biology, 63*, 10–15. https://doi.org/10.1016/j.biocel.2015.01.026

Caldeira, D.d. A. F., Mesquita, F. M., Pinheiro, F. G., Oliveira, D. F., Oliveira, L. F. S., Nascimento, J. H. M., Takiya, C. M., Maciel, L., & Zin, W. A. (2021). Acute exposure to C60 fullerene damages pulmonary mitochondrial function and mechanics. *Nanotoxicology, 15*(3), 352–365. https://doi.org/10.1080/17435390.2020.1863498

Caldeira, D.d. A. F., Weiss, D. J., Rocco, P. R. M., Silva, P. L., & Cruz, F. F. (2021). Mitochondria in focus: From function to therapeutic strategies in chronic lung diseases. *Frontiers in Immunology, 12*. https://doi.org/10.3389/fimmu.2021.782074

Chakrabarti, R., Ji, W. K., Stan, R. V., Sanz, J.d. J., Ryan, T. A., & Higgs, H. N. (2018). INF2-mediated actin polymerization at the ER stimulates mitochondrial calcium uptake, inner membrane constriction, and division. *Journal of Cell Biology, 217*(1), 251–268. https://doi.org/10.1083/jcb.201709111

Chang, C. R., & Blackstone, C. (2010). Dynamic regulation of mitochondrial fission through modification of the dynamin-related protein Drp1. *Annals of the New York Academy of Sciences, 1201*, 34–39. https://doi.org/10.1111/j.1749-6632.2010.05629.x. Blackwell Publishing Inc.

Chang, C. Y., & McDonnell, D. P. (2012). Molecular pathways: The metabolic regulator estrogen-related receptor α as a therapeutic target in cancer. *Clinical Cancer Research, 18*(22), 6089–6095. https://doi.org/10.1158/1078-0432.CCR-11-3221

Chang, X., Lochner, A., Wang, H. H., Wang, S., Zhu, H., Ren, J., & Zhou, H. (2021). Coronary microvascular injury in myocardial infarction: Perception and knowledge for mitochondrial quality control. *Theranostics, 11*(14), 6766–6785. https://doi.org/10.7150/thno.60143

Chen, C. T., Hsu, S. H., & Wei, Y. H. (2012). Mitochondrial bioenergetic function and metabolic plasticity in stem cell differentiation and cellular reprogramming. *Biochimica et Biophysica Acta - General Subjects, 1820*(5), 571–576. https://doi.org/10.1016/j.bbagen.2011.09.013

Chen, C. T., Shih, Y. R. V., Kuo, T. K., Lee, O. K., & Wei, Y. H. (2008). Coordinated changes of mitochondrial biogenesis and antioxidant enzymes during osteogenic differentiation of human mesenchymal stem cells. *Stem Cells, 26*(4), 960–968. https://doi.org/10.1634/stemcells.2007-0509

Chen, M., Chen, Z., Wang, Y., Tan, Z., Zhu, C., Li, Y., Han, Z., Chen, L., Gao, R., Liu, L., & Chen, Q. (2016). Mitophagy receptor FUNDC1 regulates mitochondrial dynamics and mitophagy. *Autophagy, 12*(4), 689–702. https://doi.org/10.1080/15548627.2016.1151580

Chen, Y., & Dorn, G. W. (2013). PINK1-phosphorylated mitofusin 2 is a parkin receptor for culling damaged mitochondria. *Science, 340*(6131), 471–475. https://doi.org/10.1126/science.1231031

Chiang, J. L., Shukla, P., Pagidas, K., Ahmed, N. S., Karri, S., Gunn, D. D., Hurd, W. W., & Singh, K. K. (2020). Mitochondria in ovarian aging and reproductive longevity. *Ageing Research Reviews, 63*. https://doi.org/10.1016/j.arr.2020.101168

Cho, B., Cho, H. M., Jo, Y., Kim, H. D., Song, M., Moon, C., Kim, H., Kim, K., Sesaki, H., Rhyu, I. J., Kim, H., & Sun, W. (2017). Constriction of the mitochondrial inner compartment is a priming event for mitochondrial division. *Nature Communications, 8*(1). https://doi.org/10.1038/ncomms15754

Cogliati, S., Enriquez, J. A., & Scorrano, L. (2016). Mitochondrial cristae: Where beauty meets functionality. *Trends in Biochemical Sciences, 41*(3), 261–273. https://doi.org/10.1016/j.tibs.2016.01.001

Cogliati, S., Frezza, C., Soriano, M. E., Varanita, T., Quintana-Cabrera, R., Corrado, M., Cipolat, S., Costa, V., Casarin, A., Gomes, L. C., Perales-Clemente, E., Salviati, L., Fernandez-Silva, P., Enriquez, J. A., & Scorrano, L. (2013). Mitochondrial cristae shape determines respiratory chain supercomplexes assembly and respiratory efficiency. *Cell, 155*(1), 160–171. https://doi.org/10.1016/j.cell.2013.08.032

Cormio, A., Musicco, C., Gasparre, G., Cormio, G., Pesce, V., Sardanelli, A. M., & Gadaleta, M. N. (2017). Increase in proteins involved in mitochondrial fission, mitophagy, proteolysis and antioxidant response in type I endometrial cancer as an adaptive response to respiratory complex I deficiency. *Biochemical and Biophysical Research Communications, 491*(1), 85–90. https://doi.org/10.1016/j.bbrc.2017.07.047

Das, S., Hajnóczky, N., Antony, A. N., Csordás, G., Gaspers, L. D., Clemens, D. L., Hoek, J. B., & Hajnóczky, G. (2012). Mitochondrial morphology and dynamics in hepatocytes from normal and ethanol-fed rats. *Pflugers Archiv European Journal of Physiology, 464*(1), 101–109. https://doi.org/10.1007/s00424-012-1100-4

Da Cunha, F. M., Torelli, N. Q., & Kowaltowski, A. J. (2015). Mitochondrial retrograde signaling: Triggers, pathways, and outcomes. *Oxidative Medicine and Cellular Longevity, 2015*. https://doi.org/10.1155/2015/482582

Degli Esposti, D., Hamelin, J., Bosselut, N., Saffroy, R., Sebagh, M., Pommier, A., Martel, C., & Lemoine, A. (2012). Mitochondrial roles and cytoprotection in chronic liver injury. *Biochemistry Research International*, 1–16. https://doi.org/10.1155/2012/387626

Delmotte, P., & Sieck, G. C. (2020). Endoplasmic reticulum stress and mitochondrial function in airway smooth muscle. *Frontiers in Cell and Developmental Biology, 7*. https://doi.org/10.3389/fcell.2019.00374

De Martino, C., Floridi, A., Marcante, M. L., Malorni, W., Barcellona, P. S., Bellocci, M., & Silvestrini, B. (1979). Morphological, histochemical and biochemical studies on germ cell mitochondria of normal rats. *Cell and Tissue Research, 196*(1), 1–22. https://doi.org/10.1007/BF00236345

Di Pietro, V., Lazzarino, G., Amorini, A. M., Signoretti, S., Hill, L. J., Porto, E., Tavazzi, B., Lazzarino, G., & Belli, A. (2017). Fusion or fission: The destiny of mitochondria in traumatic brain injury of different severities. *Scientific Reports, 7*(1). https://doi.org/10.1038/s41598-017-09587-2

Dorn, G. W., Vega, R. B., & Kelly, D. P. (2015). Mitochondrial biogenesis and dynamics in the developing and diseased heart. *Genes & Development, 29*(19), 1981–1991. https://doi.org/10.1101/gad.269894.115

Ehses, S., Raschke, I., Mancuso, G., Bernacchia, A., Geimer, S., Tondera, D., Martinou, J. C., Westermann, B., Rugarli, E. I., & Langer, T. (2009). Regulation of OPA1 processing and mitochondrial fusion by m-AAA protease isoenzymes and OMA1. *Journal of Cell Biology, 187*(7), 1023–1036. https://doi.org/10.1083/jcb.200906084

Eisner, V., Cupo, R. R., Gao, E., Csordás, G., Slovinsky, W. S., Paillard, M., Cheng, L., Ibetti, J., Chen, S. R. W., Chuprun, J. K., Hoek, J. B., Koch, W. J., & Hajnóczky, G. (2017). Mitochondrial fusion dynamics is robust in the heart and depends on calcium oscillations and contractile activity. *Proceedings of the National Academy of Sciences of the United States of America, 114*(5), E859–E868. https://doi.org/10.1073/pnas.1617288114

Fröhlich, C., Grabiger, S., Schwefel, D., Faelber, K., Rosenbaum, E., Mears, J., Rocks, O., & Daumke, O. (2013). Structural insights into oligomerization and mitochondrial remodelling of dynamin 1-like protein. *The EMBO Journal, 32*(9), 1280–1292. https://doi.org/10.1038/emboj.2013.74

Fraga, H., & Ventura, S. (2015). Influence of cytoplasmatic folding on mitochondrial import. *Current Medicinal Chemistry, 22*(19), 2349–2359. https://doi.org/10.2174/0929867322666150311153413

Gleyzer, N., Vercauteren, K., & Scarpulla, R. C. (2005). Control of mitochondrial transcription specificity factors (TFB1M and TFB2M) by nuclear respiratory factors (NRF-1 and NRF-2) and PGC-1 family coactivators. *Molecular and Cellular Biology, 25*(4), 1354–1366. https://doi.org/10.1128/MCB.25.4.1354-1366.2005

Gonçalves, J. T., Schafer, S. T., & Gage, F. H. (2016). Adult neurogenesis in the Hippocampus: From stem cells to behavior. *Cell, 167*(4), 897–914. https://doi.org/10.1016/j.cell.2016.10.021

Gong, G., Liu, X., Zhang, H., Sheu, S. S., & Wang, W. (2015). Mitochondrial flash as a novel biomarker of mitochondrial respiration in the heart. *American Journal of Physiology - Heart and Circulatory Physiology, 309*(7), H1166–H1177. https://doi.org/10.1152/ajpheart.00462.2015

Gottlieb, R. A., & Bernstein, D. (2015). Mitochondria shape cardiac metabolism. *Science, 350*(6265), 1162–1163. https://doi.org/10.1126/science.aad8222

Griffin, E. E., Graumann, J., & Chan, D. C. (2005). The WD40 protein Caf4p is a component of the mitochondrial fission machinery and recruits Dnm1p to mitochondria. *Journal of Cell Biology, 170*(2), 237–248. https://doi.org/10.1083/jcb.200503148

Hales, K. G., & Fuller, M. T. (1997). Developmentally regulated mitochondrial fusion mediated by a conserved, novel, predicted GTPase. *Cell, 90*(1), 121–129. https://doi.org/10.1016/S0092-8674(00)80319-0

Hanna, R. A., Quinsay, M. N., Orogo, A. M., Giang, K., Rikka, S., & Gustafsson, A. B. (2012). Microtubule-associated protein 1 light chain 3 (LC3) interacts with Bnip3 protein to selectively remove endoplasmic reticulum and mitochondria via autophagy. *Journal of Biological Chemistry, 287*(23), 19094–19104. https://doi.org/10.1074/jbc.M111.322933

Hasegawa, K., Yasuda, T., Shiraishi, C., Fujiwara, K., Przedborski, S., Mochizuki, H., & Yoshikawa, K. (2016). Promotion of mitochondrial biogenesis by necdin protects neurons against mitochondrial insults. *Nature Communications, 7*(1). https://doi.org/10.1038/ncomms10943

Head, B., Griparic, L., Amiri, M., Gandre-Babbe, S., & Van Der Bliek, A. M. (2009). Inducible proteolytic inactivation of OPA1 mediated by the OMA1 protease in mammalian cells. *Journal of Cell Biology, 187*(7), 959–966. https://doi.org/10.1083/jcb.200906083

Hermann, G. J., Thatcher, J. W., Mills, J. P., Hales, K. G., Fuller, M. T., Nunnari, J., & Shaw, J. M. (1998). Mitochondrial fusion in yeast requires the transmembrane GTPase Fzo1p. *Journal of Cell Biology, 143*(2), 359–373. https://doi.org/10.1083/jcb.143.2.359

Herzig, S., Long, F., Jhala, U. S., Hedrick, S., Quinn, R., Bauer, A., Rudolph, D., Schutz, G., Yoon, C., Puigserver, P., Spiegelman, B., & Montminy, M. (2001). CREB regulates hepatic gluconeogenesis through the coactivator PGC-1. *Nature, 413*(6852), 179–183. https://doi.org/10.1038/35093131

Hjelm, B. E., Rollins, B., Mamdani, F., Lauterborn, J. C., Kirov, G., Lynch, G., Gall, C. M., Sequeira, A., & Vawter, M. P. (2015). Evidence of mitochondrial dysfunction within the complex genetic etiology of schizophrenia. *Complex Psychiatry, 1*(4), 201–219. https://doi.org/10.1159/000441252

Huang, P., Galloway, C. A., & Yoon, Y. (2011). Control of mitochondrial morphology through differential interactions of mitochondrial fusion and fission proteins. *PLoS One, 6*(5). https://doi.org/10.1371/journal.pone.0020655

Huang, X., Sun, L., Ji, S., Zhao, T., Zhang, W., Xu, J., Zhang, J., Wang, Y., Wang, X., Franzini-Armstrong, C., Zheng, M., & Cheng, H. (2013). Kissing and nanotunneling mediate intermitochondrial communication in the heart. *Proceedings of the National Academy of Sciences, 110*(8), 2846–2851. https://doi.org/10.1073/pnas.1300741110

Huss, J. M., Pineda Torra, I., Staels, B., Giguère, V., & Kelly, D. P. (2004). Estrogen-related receptor α directs peroxisome proliferator-activated receptor α signaling in the transcriptional control of energy metabolism in cardiac and skeletal muscle. *Molecular and Cellular Biology, 24*(20), 9079–9091. https://doi.org/10.1128/MCB.24.20.9079-9091.2004

Ishihara, N., Eura, Y., & Mihara, K. (2004). Mitofusin 1 and 2 play distinct roles in mitochondrial fusion reactions via GTPase activity. *Journal of Cell Science, 117*(26), 6535–6546. https://doi.org/10.1242/jcs.01565

Ishihara, N., Nomura, M., Jofuku, A., Kato, H., Suzuki, S. O., Masuda, K., Otera, H., Nakanishi, Y., Nonaka, I., Goto, Y. I., Taguchi, N., Morinaga, H., Maeda, M., Takayanagi, R., Yokota, S., & Mihara, K. (2009). Mitochondrial fission factor Drp1 is essential for embryonic development and synapse formation in mice. *Nature Cell Biology, 11*(8), 958–966. https://doi.org/10.1038/ncb1907

Iwata, R., & Vanderhaeghen, P. (2021). Regulatory roles of mitochondria and metabolism in neurogenesis. *Current Opinion in Neurobiology, 69*, 231–240. https://doi.org/10.1016/j.conb.2021.05.003

Kameoka, S., Adachi, Y., Okamoto, K., Iijima, M., & Sesaki, H. (2018). Phosphatidic acid and cardiolipin coordinate mitochondrial dynamics. *Trends in Cell Biology, 28*(1), 67–76. https://doi.org/10.1016/j.tcb.2017.08.011

Kanki, T., Ohgaki, K., Gaspari, M., Gustafsson, C. M., Fukuoh, A., Sasaki, N., Hamasaki, N., & Kang, D. (2004). Architectural role of mitochondrial transcription factor a in maintenance of human mitochondrial DNA. *Molecular and Cellular Biology, 24*(22), 9823–9834. https://doi.org/10.1128/MCB.24.22.9823-9834.2004

Khacho, M., Clark, A., Svoboda, D. S., Azzi, J., MacLaurin, J. G., Meghaizel, C., Sesaki, H., Lagace, D. C., Germain, M., Harper, M. E., Park, D. S., & Slack, R. S. (2016). Mitochondrial dynamics impacts stem cell identity and fate decisions by regulating a nuclear transcriptional program. *Cell Stem Cell, 19*(2), 232–247. https://doi.org/10.1016/j.stem.2016.04.015

Khacho, M., Clark, A., Svoboda, D. S., MacLaurin, J. G., Lagace, D. C., Park, D. S., & Slack, R. S. (2017). Mitochondrial dysfunction underlies cognitive defects as a result of neural stem cell depletion and impaired neurogenesis. *Human Molecular Genetics, 26*(17), 3327–3341. https://doi.org/10.1093/hmg/ddx217

Khacho, M., & Slack, R. S. (2018). Mitochondrial dynamics in the regulation of neurogenesis: From development to the adult brain. *Developmental Dynamics, 247*(1), 47–53. https://doi.org/10.1002/dvdy.24538

Khacho, M., Tarabay, M., Patten, D., Khacho, P., MacLaurin, J. G., Guadagno, J., Bergeron, R., Cregan, S. P., Harper, M. E., Park, D. S., & Slack, R. S. (2014). Acidosis overrides oxygen deprivation to maintain mitochondrial function and cell survival. *Nature Communications, 5*. https://doi.org/10.1038/ncomms4550

Kim, B., Kim, J. S., Yoon, Y., Santiago, M. C., Brown, M. D., & Park, J. Y. (2013). Inhibition of Drp1-dependent mitochondrial division impairs myogenic differentiation. *American Journal of Physiology - Regulatory, Integrative and Comparative Physiology, 305*(8), R927–R938. https://doi.org/10.1152/ajpregu.00502.2012

Kim, J. C., Park, G. D., & Kim, S. H. (2017). Inhibition of oxidative stress by antioxidant supplementation does not limit muscle mitochondrial biogenesis or endurance capacity in rats. *Journal of Nutritional Science and Vitaminology, 63*(5), 277–283. https://doi.org/10.3177/jnsv.63.277

Kraus, F., & Ryan, M. T. (2017). The constriction and scission machineries involved in mitochondrial fission. *Journal of Cell Science, 130*(18), 2953–2960. https://doi.org/10.1242/jcs.199562

Kressler, D., Schreiber, S. N., Knutti, D., & Kralli, A. (2002). The PGC-1-related protein PERC is a selective coactivator of estrogen receptor α. *Journal of Biological Chemistry, 277*(16), 13918–13925. https://doi.org/10.1074/jbc.M201134200

Krzystek-Korpacka, M., Kempiński, R., Bromke, M. A., & Neubauer, K. (2020). Oxidative stress markers in inflammatory bowel diseases: Systematic review. *Diagnostics, 10*(8). https://doi.org/10.3390/diagnostics10080601

Kuzmenko, A., Derbikova, K., Salvatori, R., Tankov, S., Atkinson, G. C., Tenson, T., Ott, M., Kamenski, P., & Hauryliuk, V. (2016). Aim-less translation: Loss of *Saccharomyces cerevisiae* mitochondrial translation initiation factor mIF3/Aim23 leads to unbalanced protein synthesis. *Scientific Reports, 6*. https://doi.org/10.1038/srep18749

Labrousse, A. M., Zappaterra, M. D., Rube, D. A., & Van der Bliek, A. M. (1999). *C. elegans* dynamin-related protein DRP-1 controls severing of the mitochondrial outer membrane. *Molecular Cell, 4*(5), 815–826. https://doi.org/10.1016/S1097-2765(00)80391-3

Lazarou, M., Jin, S. M., Kane, L. A., & Youle, R. J. (2012). Role of PINK1 binding to the TOM complex and alternate intracellular membranes in recruitment and activation of the E3 ligase parkin. *Developmental Cell, 22*(2), 320–333. https://doi.org/10.1016/j.devcel.2011.12.014

Leboucher, G. P., Tsai, Y. C., Yang, M., Shaw, K. C., Zhou, M., Veenstra, T. D., Glickman, M. H., & Weissman, A. M. (2012). Stress-induced phosphorylation and proteasomal degradation of mitofusin 2 facilitates mitochondrial fragmentation and apoptosis. *Molecular Cell, 47*(4), 547–557. https://doi.org/10.1016/j.molcel.2012.05.041

Lee, J. E., Westrate, L. M., Wu, H., Page, C., & Voeltz, G. K. (2016). Multiple dynamin family members collaborate to drive mitochondrial division. *Nature, 540*(7631), 139–143. https://doi.org/10.1038/nature20555

Lee, J. Y., Kapur, M., Li, M., Choi, M. C., Choi, S., Kim, H. J., Kim, I., Lee, E., Taylor, J. P., & Yao, T. P. (2014). MFN1 deacetylation activates adaptive mitochondrial fusion and protects metabolically challenged mitochondria. *Journal of Cell Science, 127*(22), 4954–4963. https://doi.org/10.1242/jcs.157321

Lees, J. G., Gardner, D. K., & Harvey, A. J. (2017). Pluripotent stem cell metabolism and mitochondria: Beyond ATP. *Stem Cells International, 2017*. https://doi.org/10.1155/2017/2874283

Legaki, A. I., Moustakas, I. I., Sikorska, M., Papadopoulos, G., Velliou, R. I., & Chatzigeorgiou, A. (2022). Hepatocyte mitochondrial dynamics and bioenergetics in obesity-related non-alcoholic fatty liver disease. *Current Obesity Reports, 11*(3), 126–143. https://doi.org/10.1007/s13679-022-00473-1

Legros, F., Lombès, A., Frachon, P., & Rojo, M. (2002). Mitochondrial fusion in human cells is efficient, requires the inner membrane potential, and is mediated by mitofusins. *Molecular Biology of the Cell, 13*(12), 4343–4354. https://doi.org/10.1091/mbc.E02-06-0330

Li, A., Gao, M., Jiang, W., Qin, Y., & Gong, G. (2020). Mitochondrial dynamics in adult cardiomyocytes and heart diseases. *Frontiers in Cell and Developmental Biology, 8*. https://doi.org/10.3389/fcell.2020.584800

Li, S., Xu, S., Roelofs, B. A., Boyman, L., Lederer, W. J., Sesaki, H., & Karbowski, M. (2015). Transient assembly of F-actin on the outer mitochondrial membrane contributes to mitochondrial fission. *Journal of Cell Biology, 208*(1), 109–123. https://doi.org/10.1083/jcb.201404050

Li, X. X., Tsoi, B., Li, Y. F., Kurihara, H., & He, R. R. (2015). Cardiolipin and its different properties in mitophagy and apoptosis. *Journal of Histochemistry and Cytochemistry, 63*(5), 301–311. https://doi.org/10.1369/0022155415574818

Liesa, M., Palacín, M., & Zorzano, A. (2009). Mitochondrial dynamics in mammalian health and disease. *Physiological Reviews, 89*(3), 799–845. https://doi.org/10.1152/physrev.00030.2008

Lin, J., Puigserver, P., Donovan, J., Tarr, P., & Spiegelman, B. M. (2002). Peroxisome proliferator-activated receptor γ coactivator 1β (PGC-1β), a novel PGC-1-related transcription coactivator associated with host cell factor. *Journal of Biological Chemistry, 277*(3), 1645–1648. https://doi.org/10.1074/jbc.C100631200

Liu, A. R., Lv, Z., Yan, Z. W., Wu, X. Y., Yan, L. R., Sun, L. P., Yuan, Y., & Xu, Q. (2023). Association of mitochondrial homeostasis and dynamic balance with malignant biological behaviors of gastrointestinal cancer. *Journal of Translational Medicine, 21*(1). https://doi.org/10.1186/s12967-023-03878-1

Liu, L., Feng, D., Chen, G., Chen, M., Zheng, Q., Song, P., Ma, Q., Zhu, C., Wang, R., Qi, W., Huang, L., Xue, P., Li, B., Wang, X., Jin, H., Wang, J., Yang, F., Liu, P., Zhu, Y., & Chen, Q. (2012). Mitochondrial outer-membrane protein FUNDC1 mediates hypoxia-induced mitophagy in mammalian cells. *Nature Cell Biology, 14*(2), 177–185. https://doi.org/10.1038/ncb2422

Liu, X., & Chen, Z. (2017). The pathophysiological role of mitochondrial oxidative stress in lung diseases. *Journal of Translational Medicine, 15*(1). https://doi.org/10.1186/s12967-017-1306-5

Mabalirajan, U., Aich, J., Leishangthem, G. D., Sharma, S. K., Dinda, A. K., & Ghosh, B. (2009). Effects of vitamin E on mitochondrial dysfunction and asthma features in an experimental allergic murine model. *Journal of Applied Physiology, 107*(4), 1285–1292. https://doi.org/10.1152/japplphysiol.00459.2009

Madan, S., Uttekar, B., Chowdhary, S., & Rikhy, R. (2022). Mitochondria lead the way: Mitochondrial dynamics and function in cellular movements in development and disease. *Frontiers in Cell and Developmental Biology, 9*. https://doi.org/10.3389/fcell.2021.781933

Malsin, E. S., & Kamp, D. W. (2018). The mitochondria in lung fibrosis: Friend or foe? *Translational Research, 202*, 1–23. https://doi.org/10.1016/j.trsl.2018.05.005

Manor, U., Bartholomew, S., Golani, G., Christenson, E., Kozlov, M., Higgs, H., Spudich, J., & Lippincott-Schwartz, J. (2015). A mitochondria-anchored isoform of the actin-nucleating spire protein regulates mitochondrial division. *Elife, 4*. https://doi.org/10.7554/eLife.08828

Marchi, S., Patergnani, S., & Pinton, P. (2014). The endoplasmic reticulum-mitochondria connection: One touch, multiple functions. *Biochimica et Biophysica Acta - Bioenergetics, 1837*(4), 461–469. https://doi.org/10.1016/j.bbabio.2013.10.015

Martin, O. J., Lai, L., Soundarapandian, M. M., Leone, T. C., Zorzano, A., Keller, M. P., Attie, A. D., Muoio, D. M., & Kelly, D. P. (2014). A role for peroxisome proliferator-activated receptor γ coactivator-1 in the control of mitochondrial dynamics during postnatal cardiac growth. *Circulation Research, 114*(4), 626–636. https://doi.org/10.1161/CIRCRESAHA.114.302562

Matsuda, N., Sato, S., Shiba, K., Okatsu, K., Saisho, K., Gautier, C. A., Sou, Y. S., Saiki, S., Kawajiri, S., Sato, F., Kimura, M., Komatsu, M., Hattori, N., & Tanaka, K. (2010). PINK1 stabilized by mitochondrial depolarization recruits Parkin to damaged mitochondria and activates latent Parkin for mitophagy. *Journal of Cell Biology, 189*(2), 211–221. https://doi.org/10.1083/jcb.200910140

McKenzie, M., Lazarou, M., & Ryan, M. T. (2009). Chapter 18 analysis of respiratory chain complex assembly with radiolabeled nuclear- and mitochondrial-encoded subunits. *Methods in Enzymology, 456*, 321–339. https://doi.org/10.1016/S0076-6879(08)04418-2

Meinhardt, A., McFarlane, J. R., Seitz, J., & De Kretser, D. M. (2000). Activin maintains the condensed type of mitochondria in germ cells. *Molecular and Cellular Endocrinology, 168*(1–2), 111–117. https://doi.org/10.1016/S0303-7207(00)00308-7

Mishra, P., Carelli, V., Manfredi, G., & Chan, D. C. (2014). Proteolytic cleavage of Opa1 stimulates mitochondrial inner membrane fusion and couples fusion to oxidative phosphorylation. *Cell Metabolism, 19*(4), 630–641. https://doi.org/10.1016/j.cmet.2014.03.011

Mitra, K., Wunder, C., Roysam, B., Lin, G., & Lippincott-Schwartz, J. (2009). A hyperfused mitochondrial state achieved at G 1 –S regulates cyclin E buildup and entry into S phase. *Proceedings of the National Academy of Sciences, 106*(29), 11960–11965. https://doi.org/10.1073/pnas.0904875106

Moreira, O. C., Estébanez, B., Martínez-Florez, S., De Paz, J. A., Cuevas, M. J., & González-Gallego, J. (2017). Mitochondrial function and mitophagy in the elderly: Effects of exercise. *Oxidative Medicine and Cellular Longevity.* https://doi.org/10.1155/2017/2012798

Mozdy, A. D., McCaffery, J. M., & Shaw, J. M. (2000). Dnm1p Gtpase-mediated mitochondrial fission is a multi-step process requiring the novel integral membrane component Fis1p. *Journal of Cell Biology, 151*(2), 367–380. https://doi.org/10.1083/jcb.151.2.367

Ng, M. Y. W., Wai, T., & Simonsen, A. (2021). Quality control of the mitochondrion. *Developmental Cell, 56*(7), 881–905. https://doi.org/10.1016/j.devcel.2021.02.009

Novak, I., Kirkin, V., McEwan, D. G., Zhang, J., Wild, P., Rozenknop, A., Rogov, V., Löhr, F., Popovic, D., Occhipinti, A., Reichert, A. S., Terzic, J., Dötsch, V., Ney, P. A., & Dikic, I. (2010). Nix is a selective autophagy receptor for mitochondrial clearance. *EMBO Reports, 11*(1), 45–51. https://doi.org/10.1038/embor.2009.256

Osellame, L. D., Singh, A. P., Stroud, D. A., Palmer, C. S., Stojanovski, D., Ramachandran, R., & Ryan, M. T. (2016). Cooperative and independent roles of the Drp1 adaptors Mff, MiD49 and MiD51 in mitochondrial fission. *Journal of Cell Science, 129*(11), 2170–2181. https://doi.org/10.1242/jcs.185165

Pagliuso, A., Tham, T. N., Stevens, J. K., Lagache, T., Persson, R., Salles, A., Olivo-Marin, J. C., Oddos, S., Spang, A., Cossart, P., & Stavru, F. (2016). A role for septin 2 in Drp1-mediated mitochondrial fission. *EMBO Reports, 17*(6), 858–873. https://doi.org/10.15252/embr.201541612

Palikaras, K., & Tavernarakis, N. (2014). Mitochondrial homeostasis: The interplay between mitophagy and mitochondrial biogenesis. *Experimental Gerontology, 56*, 182–188. https://doi.org/10.1016/j.exger.2014.01.021

Palmer, C. S., Osellame, L. D., Laine, D., Koutsopoulos, O. S., Frazier, A. E., & Ryan, M. T. (2011). MiD49 and MiD51, new components of the mitochondrial fission machinery. *EMBO Reports, 12*(6), 565–573. https://doi.org/10.1038/embor.2011.54

Paumard, P., Vaillier, J., Coulary, B., Schaeffer, J., Soubannier, V., Mueller, D. M., Brèthes, D., di Rago, J.-P., & Velours, J. (2002). The ATP synthase is involved in generating mitochondrial cristae morphology. *The EMBO Journal, 21*(3), 221–230. https://doi.org/10.1093/emboj/21.3.221

Pellegrino, M. W., Nargund, A. M., & Haynes, C. M. (2013). Signaling the mitochondrial unfolded protein response. *Biochimica et Biophysica Acta - Molecular Cell Research, 1833*(2), 410–416. https://doi.org/10.1016/j.bbamcr.2012.02.019

Pennanen, C., Parra, V., López-Crisosto, C., Morales, P. E., del Campo, A., Gutierrez, T., Rivera-Mejías, P., Kuzmicic, J., Chiong, M., Zorzano, A., Rothermel, B. A., & Lavandero, S. (2014). Mitochondrial fission is required for cardiomyocyte hypertrophy mediated by a Ca2+-calcineurin signaling pathway. *Journal of Cell Science, 127*(12), 2659–2671. https://doi.org/10.1242/jcs.139394

Piantadosi, C. A., Carraway, M. S., Babiker, A., & Suliman, H. B. (2008). Heme oxygenase-1 regulates cardiac mitochondrial biogenesis via nrf2-mediated transcriptional control of nuclear respiratory factor-1. *Circulation Research, 103*(11), 1232–1240. https://doi.org/10.1161/01.RES.0000338597.71702.ad

Picard, M., White, K., & Turnbull, D. M. (2013). Mitochondrial morphology, topology, and membrane interactions in skeletal muscle: A quantitative three-dimensional electron microscopy study. *Journal of Applied Physiology, 114*(2), 161–171. https://doi.org/10.1152/japplphysiol.01096.2012

Ploumi, C., Daskalaki, I., & Tavernarakis, N. (2017). Mitochondrial biogenesis and clearance: A balancing act. *FEBS Journal, 284*(2), 183–195. https://doi.org/10.1111/febs.13820

Porter, G. A., Hom, J. R., Hoffman, D. L., Quintanilla, R. A., Bentley, K. L.d. M., & Sheu, S. S. (2011). Bioenergetics, mitochondria, and cardiac myocyte differentiation. *Progress in Pediatric Cardiology, 31*(2), 75–81. https://doi.org/10.1016/j.ppedcard.2011.02.002

Qin, Y., Gao, M., Li, A., Sun, J., Liu, B., & Gong, G. (2018). Mitoflash lights single mitochondrial dynamics events in mature cardiomyocytes. *Biochemical and Biophysical Research Communications, 503*(2), 729–736. https://doi.org/10.1016/j.bbrc.2018.06.068

Qiu, Y. N., Wang, G. H., Zhou, F., Hao, J. J., Tian, L., Guan, L. F., Geng, X. K., Ding, Y. C., Wu, H. W., & Zhang, K. Z. (2019). PM2.5 induces liver fibrosis via triggering ROS-mediated mitophagy. *Ecotoxicology and Environmental Safety, 167*, 178–187. https://doi.org/10.1016/j.ecoenv.2018.08.050

Rajasekaran, A., Venkatasubramanian, G., Berk, M., & Debnath, M. (2015). Mitochondrial dysfunction in schizophrenia: Pathways, mechanisms and implications. *Neuroscience & Biobehavioral Reviews, 48*, 10–21. https://doi.org/10.1016/j.neubiorev.2014.11.005

Ramachandran, A., Umbaugh, D. S., & Jaeschke, H. (2021). Mitochondrial dynamics in drug-induced liver injury. *Livers, 1*(3), 102–115. https://doi.org/10.3390/livers1030010

Ramalho-Santos, J., Varum, S., Amaral, S., Mota, P. C., Sousa, A. P., & Amaral, A. (2009). Mitochondrial functionality in reproduction: From gonads and gametes to embryos and embryonic stem cells. *Human Reproduction Update, 15*(5), 553–572. https://doi.org/10.1093/humupd/dmp016

Rikka, S., Quinsay, M. N., Thomas, R. L., Kubli, D. A., Zhang, X., Murphy, A. N., & Gustafsson, Å. B. (2011). Bnip3 impairs mitochondrial bioenergetics and stimulates mitochondrial turnover. *Cell Death & Differentiation, 18*(4), 721–731. https://doi.org/10.1038/cdd.2010.146

Rojo, M., Legros, F., Chateau, D., & Lombès, A. (2002). Membrane topology and mitochondrial targeting of mitofusins, ubiquitous mammalian homologs of the transmembrane GTPase Fzo. *Journal of Cell Science, 115*(8), 1663–1674.

Ryter, S. W., Rosas, I. O., Owen, C. A., Martinez, F. J., Choi, M. E., Lee, C. G., Elias, J. A., & Choi, A. M. K. (2018). Mitochondrial dysfunction as a pathogenic mediator of chronic obstructive pulmonary disease and idiopathic pulmonary fibrosis. *Annals of the American Thoracic Society, 15*, S266–S272. https://doi.org/10.1513/AnnalsATS.201808-585MG

Sachdeva, K., Do, D. C., Zhang, Y., Hu, X., Chen, J., & Gao, P. (2019). Environmental exposures and asthma development: Autophagy, mitophagy, and cellular senescence. *Frontiers in Immunology, 10*. https://doi.org/10.3389/fimmu.2019.02787

Salma, N., Song, J. S., Arany, Z., & Fisher, D. E. (2015). Transcription factor Tfe3 directly regulates Pgc-1alpha in muscle. *Journal of Cellular Physiology, 230*(10), 2330–2336. https://doi.org/10.1002/jcp.24978

Santel, A., & Fuller, M. T. (2001). Control of mitochondiral morphology by a human mitofusin. *Journal of Cell Science, 114*(5), 867–874.

Satoh, J., Kawana, N., & Yamamoto, Y. (2013). Pathway analysis of ChIP-Seq-based NRF1 target genes suggests a logical hypothesis of their involvement in the pathogenesis of neurodegenerative diseases. *Gene Regulation and Systems Biology, 7*(7). https://doi.org/10.4137/grsb.s13204. GRSB.S13204.

Schulz, C., & Rehling, P. (2014). Remodelling of the active presequence translocase drives motor-dependent mitochondrial protein translocation. *Nature Communications, 5*(1). https://doi.org/10.1038/ncomms5349

Sebag, S. C., Koval, O. M., Paschke, J. D., Winters, C. J., Jaffer, O. A., Dworski, R., Sutterwala, F. S., Anderson, M. E., & Grumbach, I. M. (2017). Mitochondrial CaMKII inhibition in airway epithelium protects against allergic asthma. *JCI Insight, 2*(3). https://doi.org/10.1172/jci.insight.88297

Sentelle, R. D., Senkal, C. E., Jiang, W., Ponnusamy, S., Gencer, S., Panneer Selvam, S., Ramshesh, V. K., Peterson, Y. K., Lemasters, J. J., Szulc, Z. M., Bielawski, J., & Ogretmen, B. (2012). Ceramide targets autophagosomes to mitochondria and induces lethal mitophagy. *Nature Chemical Biology, 8*(10), 831–838. https://doi.org/10.1038/nchembio.1059

Sesaki, H., Southard, S. M., Yaffe, M. P., & Jensen, R. E. (2003). Mgm1p, a dynamin-related GTPase, is essential for fusion of the mitochondrial outer membrane. *Molecular Biology of the Cell, 14*(6), 2342–2356. https://doi.org/10.1091/mbc.E02-12-0788

Shyh-Chang, N., Daley, G. Q., & Cantley, L. C. (2013). Stem cell metabolism in tissue development and aging. *Development, 140*(12), 2535–2547. https://doi.org/10.1242/dev.091777

Simula, L., Nazio, F., & Campello, S. (2017). The mitochondrial dynamics in cancer and immune-surveillance. *Seminars in Cancer Biology, 47*, 29–42. https://doi.org/10.1016/j.semcancer.2017.06.007

Smirnova, E., Griparic, L., Shurland, D. L., & Van der Bliek, A. M. (2001). Dynamin-related protein Drp1 is required for mitochondrial division in mammalian cells. *Molecular Biology of the Cell, 12*(8), 2245–2256. https://doi.org/10.1091/mbc.12.8.2245

Soriano, F. X., Liesa, M., Bach, D., Chan, D. C., Palacín, M., & Zorzano, A. (2006). Evidence for a mitochondrial regulatory pathway defined by peroxisome proliferator-activated receptor-γ coactivator-1α, estrogen-related receptor-α, and mitofusin 2. *Diabetes, 55*(6), 1783–1791. https://doi.org/10.2337/db05-0509

Sui, G. Y., Wang, F., Lee, J., & Roh, Y. S. (2022). Mitochondrial control in inflammatory gastrointestinal diseases. *International Journal of Molecular Sciences, 23*(23). https://doi.org/10.3390/ijms232314890

Taguchi, K., Motohashi, H., & Yamamoto, M. (2011). Molecular mechanisms of the Keap1-Nrf2 pathway in stress response and cancer evolution. *Genes to Cells, 16*(2), 123–140. https://doi.org/10.1111/j.1365-2443.2010.01473.x

Takubo, K., Nagamatsu, G., Kobayashi, C. I., Nakamura-Ishizu, A., Kobayashi, H., Ikeda, E., Goda, N., Rahimi, Y., Johnson, R. S., Soga, T., Hirao, A., Suematsu, M., & Suda, T. (2013). Regulation of glycolysis by Pdk functions as a metabolic checkpoint for cell cycle quiescence in hematopoietic stem cells. *Cell Stem Cell, 12*(1), 49–61. https://doi.org/10.1016/j.stem.2012.10.011

Tatsuta, T., & Langer, T. (2008). Quality control of mitochondria: Protection against neurodegeneration and ageing. *The EMBO Journal, 27*(2), 306–314. https://doi.org/10.1038/sj.emboj.7601972

Ten, V. S., & Ratner, V. (2020). Mitochondrial bioenergetics and pulmonary dysfunction: Current progress and future directions. *Paediatric Respiratory Reviews, 34*, 37–45. https://doi.org/10.1016/j.prrv.2019.04.001

Tepp, K., Shevchuk, I., Chekulayev, V., Timohhina, N., Kuznetsov, A. V., Guzun, R., Saks, V., & Kaambre, T. (2011). High efficiency of energy flux controls within mitochondrial interactosome in cardiac intracellular energetic units. *Biochimica et Biophysica Acta - Bioenergetics, 1807*(12), 1549–1561. https://doi.org/10.1016/j.bbabio.2011.08.005

Tieu, Q., & Nunnari, J. (2000). Mdv1p is a Wd repeat protein that interacts with the dynamin-related Gtpase, Dnm1p, to trigger mitochondrial division. *Journal of Cell Biology, 151*(2), 353–366. https://doi.org/10.1083/jcb.151.2.353

Tilokani, L., Nagashima, S., Paupe, V., & Prudent, J. (2018). Mitochondrial dynamics: Overview of molecular mechanisms. *Essays in Biochemistry, 62*(3), 341–360. https://doi.org/10.1042/EBC20170104

Twig, G., Elorza, A., Molina, A. J. A., Mohamed, H., Wikstrom, J. D., Walzer, G., Stiles, L., Haigh, S. E., Katz, S., Las, G., Alroy, J., Wu, M., Py, B. F., Yuan, J., Deeney, J. T., Corkey, B. E., & Shirihai, O. S. (2008). Fission and selective fusion govern mitochondrial segregation and elimination by autophagy. *EMBO Journal, 27*(2), 433–446. https://doi.org/10.1038/sj.emboj.7601963

Twig, G., Liu, X., Liesa, M., Wikstrom, J. D., Molina, A. J., Las, G., ... Shirihai, O. S. (2010). Biophysical properties of mitochondiral fusion events in pancreatic beta-cells and cardiac cells unravel potential control mechanisms of its selectivity. *American Journal of Physiology-Cell Physiology, 299*(2), C477–C487.

Umbaugh, D. S., Nguyen, N. T., Jaeschke, H., & Ramachandran, A. (2021). Mitochondrial membrane potential drives early change in mitochondrial morphology after acetaminophen exposure. *Toxicological Sciences, 180*(1), 186–195. https://doi.org/10.1093/toxsci/kfaa188

van der Vliet, A., Janssen-Heininger, Y. M. W., & Anathy, V. (2018). Oxidative stress in chronic lung disease: From mitochondrial dysfunction to dysregulated redox signaling. *Molecular Aspects of Medicine, 63*, 59–69. https://doi.org/10.1016/j.mam.2018.08.001

Vance, J. E. (2015). Phospholipid synthesis and transport in mammalian cells. *Traffic, 16*(1), 1–18. https://doi.org/10.1111/tra.12230

Vega, R. B., Horton, J. L., & Kelly, D. P. (2015). Maintaining ancient organelles: Mitochondrial biogenesis and maturation. *Circulation Research, 116*(11), 1820–1834. https://doi.org/10.1161/CIRCRESAHA.116.305420

Vendelin, M., Béraud, N., Guerrero, K., Andrienko, T., Kuznetsov, A. V., Olivares, J., Kay, L., & Saks, V. A. (2005). Mitochondrial regular arrangement in muscle cells: A \crystal- like\ pattern. *American Journal of Physiology - Cell Physiology, 288*(3), C757–C767. https://doi.org/10.1152/ajpcell.00281.2004

Ventura-Clapier, R., Garnier, A., & Veksler, V. (2008). Transcriptional control of mitochondrial biogenesis: The central role of PGC-1α. *Cardiovascular Research, 79*(2), 208–217. https://doi.org/10.1093/cvr/cvn098

Wai, T., García-Prieto, J., Baker, M. J., Merkwirth, C., Benit, P., Rustin, P., Rupérez, F. J., Barbas, C., Ibañez, B., & Langer, T. (2015). Imbalanced OPA1 processing and mitochondrial fragmentation cause heart failure in mice. *Science, 350*(6265). https://doi.org/10.1126/science.aad0116

Wakabayashi, J., Zhang, Z., Wakabayashi, N., Tamura, Y., Fukaya, M., Kensler, T. W., Iijima, M., & Sesaki, H. (2009). The dynamin-related GTPase Drp1 is required for embryonic and brain development in mice. *Journal of Cell Biology, 186*(6), 805–816. https://doi.org/10.1083/jcb.200903065

Wang, S., Tan, J., Miao, Y., & Zhang, Q. (2022). Mitochondrial dynamics, mitophagy, and mitochondria–endoplasmic reticulum contact sites crosstalk under hypoxia. *Frontiers in Cell and Developmental Biology, 10*. https://doi.org/10.3389/fcell.2022.848214

Wang, W., Esbensen, Y., Kunke, D., Suganthan, R., Rachek, L., Bjørås, M., & Eide, L. (2011). Mitochondrial DNA damage level determines neural stem cell differentiation fate. *Journal of Neuroscience, 31*(26), 9746–9751. https://doi.org/10.1523/JNEUROSCI.0852-11.2011

Wang, W., Osenbroch, P., Skinnes, R., Esbensen, Y., Bjørås, M., & Eide, L. (2010). Mitochondrial DNA integrity is essential for mitochondrial maturation during differentiation of neural stem cells. *Stem Cells, 28*(12), 2195–2204. https://doi.org/10.1002/stem.542

Wei, Y. H., & Lee, H. C. (2002). Oxidative stress, mitochondrial DNA mutation, and impairment of antioxidant enzymes in aging. *Experimental Biology and Medicine, 227*(9), 671–682. https://doi.org/10.1177/153537020222700901

Weir, H. J., Yao, P., Huynh, F. K., Escoubas, C. C., Goncalves, R. L., Burkewitz, K., Laboy, R., Hirschey, M. D., & Mair, W. B. (2017). Dietary restriction and AMPK increase lifespan via mitochondrial network and peroxisome remodeling. *Cell Metabolism, 26*(6), 884–896.e5. https://doi.org/10.1016/j.cmet.2017.09.024

Westermann, B. (2012). Bioenergetic role of mitochondrial fusion and fission. *Biochimica et Biophysica Acta - Bioenergetics, 1817*(10), 1833–1838. https://doi.org/10.1016/j.bbabio.2012.02.033

Wu, W., Tian, W., Hu, Z., Chen, G., Huang, L., Li, W., Zhang, X., Xue, P., Zhou, C., Liu, L., Zhu, Y., Zhang, X., Li, L., Zhang, L., Sui, S., Zhao, B., & Feng, D. (2014). ULK1 translocates to mitochondria and phosphorylates FUNDC1 to regulate mitophagy. *EMBO Reports, 15*(5), 566–575. https://doi.org/10.1002/embr.201438501

Xie, J. H., Li, Y. Y., & Jin, J. (2020). The essential functions of mitochondrial dynamics in immune cells. *Cellular and Molecular Immunology, 17*(7), 712–721. https://doi.org/10.1038/s41423-020-0480-1

Yakubovskaya, E., Guja, K. E., Eng, E. T., Choi, W. S., Mejia, E., Beglov, D., Lukin, M., Kozakov, D., & Garcia-Diaz, M. (2014). Organization of the human mitochondrial transcription initiation complex. *Nucleic Acids Research, 42*(6), 4100–4112. https://doi.org/10.1093/nar/gkt1360

Yamashita, S. I., Jin, X., Furukawa, K., Hamasaki, M., Nezu, A., Otera, H., Saigusa, T., Yoshimori, T., Sakai, Y., Mihara, K., & Kanki, T. (2016). Mitochondrial division occurs concurrently with autophagosome formation but independently of Drp1 during mitophagy. *Journal of Cell Biology, 215*(5), 649–665. https://doi.org/10.1083/jcb.201605093

Yang, Z. F., Drumea, K., Mott, S., Wang, J., & Rosmarin, A. G. (2014). GABP transcription factor (nuclear respiratory factor 2) is required for mitochondrial biogenesis. *Molecular and Cellular Biology, 34*(17), 3194–3201. https://doi.org/10.1128/MCB.00492-12

Yao, C. H., Wang, R., Wang, Y., Kung, C. P., Weber, J. D., & Patti, G. J. (2019). Mitochondrial fusion supports increased oxidative phosphorylation during cell proliferation. *Elife, 8*. https://doi.org/10.7554/eLife.41351

Zhang, X., Agborbesong, E., & Li, X. (2021). The role of mitochondria in acute kidney injury and chronic kidney disease and its therapeutic potential. *International Journal of Molecular Sciences, 22*(20), 11253. https://doi.org/10.3390/ijms222011253

Zhang, Y., Do, D. C., Hu, X., Wang, J., Zhao, Y., Mishra, S., Zhang, X., Wan, M., & Gao, P. (2021). CaMKII oxidation regulates cockroach allergen–induced mitophagy in asthma. *Journal of Allergy and Clinical Immunology, 147*(4), 1464–1477.e11. https://doi.org/10.1016/j.jaci.2020.08.033

Zhao, R., Hou, Y., Xue, P., Woods, C. G., Fu, J., Feng, B., Guan, D., Sun, G., Chan, J. Y., Waalkes, M. P., Andersen, M. E., & Pi, J. (2011). Long isoforms of NRF1 contribute to arsenic-induced antioxidant response in human keratinocytes. *Environmental Health Perspectives, 119*(1), 56–62. https://doi.org/10.1289/ehp.1002304

Zhu, J., Wang, K. Z. Q., & Chu, C. T. (2013). After the banquet: Mitochondrial biogenesis, mitophagy, and cell survival. *Autophagy, 9*(11), 1663–1676. https://doi.org/10.4161/auto.24135

CHAPTER

4

Mitochondrial DNA and heteroplasmy

İlknur Melis Durasi[1,2]

[1]Molecular Biology and Genetics Department, Faculty of Engineering and Natural Sciences, Istinye University, Istanbul, Turkey; [2]Mdnutriacademy, Istanbul, Turkey

Mitochondria are essential organelles of the cell that produce energy in the form of ATP by oxidative phosphorylation (OXPHOS). They are double-membraned organelles in the cytoplasm and produce endogenous reactive oxygen species (ROS) in addition to crucial roles they have on functional pathways of cells, such as cell apoptosis, calcium homeostasis, urea cycle, and TCA cycle (tricarboxylic acid cycle). These cellular processes are controlled by both genomic DNA and mitochondrial DNA (Hopper et al., 2006).

The view that mitochondria evolved from bacteria engulfed by ancestral cells can explain why it contains its own genome and its own biosynthetic machinery for making RNA and organelle proteins and suggests that mitochondria evolved from bacteria that were endocytosed more than 1 billion years ago. Mitochondria are the only organelle in the human cell with its own DNA, apart from the nucleus.

Human mtDNA contains around 1500 genes, 37 of which are encoded by mtDNA, while the rest are encoded by nuclear DNA (nDNA) (Anderson et al., 1981). Although mtDNA represents only 1% of cellular DNA, it is essential for the maintenance of cellular functions (Table 4.1). Human mitochondrial genome size is a small (~16.5 kb), circular, double-stranded, multicopy molecule. Fig. 4.1 shows the organization of the human mitochondrial genome. It contains two rRNA genes, 22 tRNA genes, which are needed for the translation of mitochondrial genes, and 13 protein-coding sequences in the electron transport chain (ETC) and the displacement loop (D-loop) (Bianchi et al., 2001; Smith et al., 2022). The D-loop is the control zone of the mtDNA consisting of promoter regions. But in addition, the vast majority of mitochondrial proteins like OXPHOS complexes (I—V) require nuclear-encoded proteins to be produced (Calvo & Mootha, 2010).

Mitochondrial Transplantation and Transfer
https://doi.org/10.1016/B978-0-443-18858-9.00004-4

TABLE 4.1 The top mtDNA Protein Synthesis Gene Mutations.

Syndromes	Locus	Disease	Allele	RNA	Ho	He	References
Cardiomyopathy							
Mitochondrial myopathy and cardiomyopathy	MTTL1	MMC	A3260G	tRNA Leu (UUR)	−	+	Sweeney, Brockington, et al. (1993), Zeviani et al. (1991)
Mitochondrial myopathy and cardiomyopathy	MTTL1	MMC	C3303T	tRNA Leu (UUR)	+	+	Silvestri et al. (1994)
Maternally inherited cardiomyopathy	MTTI	MICM	A4300G	tRNA Ile	+	+	Casali et al. (1995)
Deafness/sensorineural hearing loss							
Deafness	MTRNR1	DEAF	C1494T	12S rRNA	+	−	Zhao et al. (2004)
Deafness	MTRNR1	DEAF	A1555G	12S rRNA	+	−	Fischel-Ghodsian et al. (1993), Hutchin et al. (1993), Prezant et al. (1993)
Deafness/ sensorineural hearing loss	MTTS1 precursor	SNHL	A7445G	tRNA Ser (UCN) precursor	+	+	Reid et al. (1994)
Deafness/ sensorineural hearing loss	MTTS1	SNHL	T7511C	tRNA Ser (UCN)	+	+	Sue et al. (1999)
Diabetes mellitus							
Diabetes mellitus and deafness	MTTL1	DM/DMDF/MIDD/ SNHL/FSGS/ Cardiac + multiorgan dysfunction	A3243G	tRNA Leu (UUR)	−	+	Manouvrier et al. (1995), Massin et al. (1995), van den Ouweland et al. (1992)
Encephalomyopathy							
Encephalomyopathy, MELAS	MTTF	MELAS/MM and EXIT	G583A	tRNA Phe	−	+	Darin et al. (2006), Hanna et al. (1998)
Encephalomyopathy, ataxia, myoclonus, and deafness	MTTV	AMDF	G1606A	tRNA Val	−	+	Tiranti et al. (1998)
Encephalomyopathy, MELAS	MTTL1	MELAS/LS	A3243G	tRNA Leu (UUR)	−	+	Goto et al. (1990)
Encephalomyopathy, MELAS	MTTL1	MELAS	C3256T	tRNA Leu (UUR)	−	+	Moraes, Ciacci, Bonilla, et al. (1993), Sato et al. (1994)
Encephalomyopathy, MELAS	MTTL1	MELAS	T3271C	tRNA Leu (UUR)	−	+	Goto et al. (1991), Hayashi et al. (1993), Sakuta et al. (1993)

TABLE 4.1 The top mtDNA Protein Synthesis Gene Mutations.—cont'd

Syndromes	Locus	Disease	Allele	RNA	Ho	He	References
Encephalomyopathy, MELAS	MTTL1	MELAS/Myopathy/ Deafness + cognitive impairment	T3291C	tRNA Leu (UUR)	−	+	Goto (1995), Goto et al. (1994)
Encephalomyopathy, MELAS	MTTQ	MELAS/ encephalopathy	G4332A	tRNA Gln	−	+	Bataillard et al. (2001)
Encephalomyopathy, Leigh syndrome	MTTW	MILS	A5537insT	tRNA Trp	−	+	Santorelli et al. (1997)
Encephalomyopathy	MTTS1	PEM/AMDF/Motor neuron disease-like	C7472insC C7471CC (='7472insC')	tRNA Ser (UCN)	+	+	Jaksch, Hofmann, et al. (1998), Jaksch, Klopstock, et al. (1998), Schuelke et al. (1998), Tiranti et al. (1995)
Encephalomyopathy, MERRF	MTTK	MERRF	A8344G	tRNA Lys	−	+	Shoffner et al. (1990), Wallace et al. (1988)
Encephalomyopathy, MERRF	MTTK	MERRF	T8356C	tRNA Lys	−	+	Masucci et al. (1995), Silvestri et al. (1992), Zeviani et al. (1993)
Encephalomyopathy, MERRF	MTTK	MERRF/MICM? DEAF/Autism/LS/ Ataxia + Lipomas	G8363A	tRNA Lys	−	+	Ozawa et al. (1997), Santorelli et al. (1996)
Encephalomyopathy	MTTG	PEM	T10010C	tRNA Gly	−	+	Bidooki et al. (1997)
Encephalomyopathy, MERRF	MTTH	MERRF-MELAS/ cerebral edema	G12147A	tRNA His	−	+	Melone et al. (2004), Taylor et al. (2004)
MITOCHONDRIAL myopathy							
Mitochondrial myopathy	MTTL1	MM/CPEO	A3243G	tRNA Leu (UUR)	−	+	Jean-Francois et al. (1994), Moraes, Ciacci, Silverstri, et al. (1993)
Mitochondrial myopathy	MTTL1	MM	A3302G	tRNA Leu (UUR)	−	+	Bindoff et al. (1993), Maniura-Weber et al. (2006), Shoffner, Krawiecki, et al. (1993)
Mitochondrial myopathy, CPEO	MTTI	CPEO/MS	G4298A	tRNA Ile	−	+	Taylor et al. (1998)
Mitochondrial myopathy, CPEO	MTTI	CPEO	G4308A	tRNA Ile	−	+	Schaller et al. (2011), Souilem et al. (2011)
Mitochondrial myopathy	MTTA	Myopathy	G5650A	tRNA Ala	−	+	McFarland et al. (2008)
Mitochondrial myopathy, CPEO	MTTN	CPEO/MM	G5703A	tRNA Asn	−	+	Hao and Moraes (1997), Moraes, Ciacci, Bonilla, et al. (1993), Vives-Bauza et al. (2003)

(Continued)

TABLE 4.1 The top mtDNA Protein Synthesis Gene Mutations.—cont'd

Syndromes	Locus	Disease	Allele	RNA	Ho	He	References
Mitochondrial myopathy	MTTS1	MM/EXIT	G7497A	tRNA Ser (UCN)	+	+	Grafakou et al. (2003), Jaksch, Klopstock, et al. (1998), Mollers et al. (2005)
Mitochondrial myopathy, CPEO	MTTL2	CPEO/KSS	G12315A	tRNA Leu (CUN)	−	+	Fu et al. (1996)
Mitochondrial myopathy	MTTE	Reversible COX deficiency myopathy	T14674C	tRNA Glu	+	−	Horvath et al. (2009), Mimaki et al. (2010)
Mitochondrial myopathy	MTTE	MM + DM/ Encephalomyopathy	T14709C	tRNA Glu	+	+	Hanna et al. (1995), Hao et al., (1995)

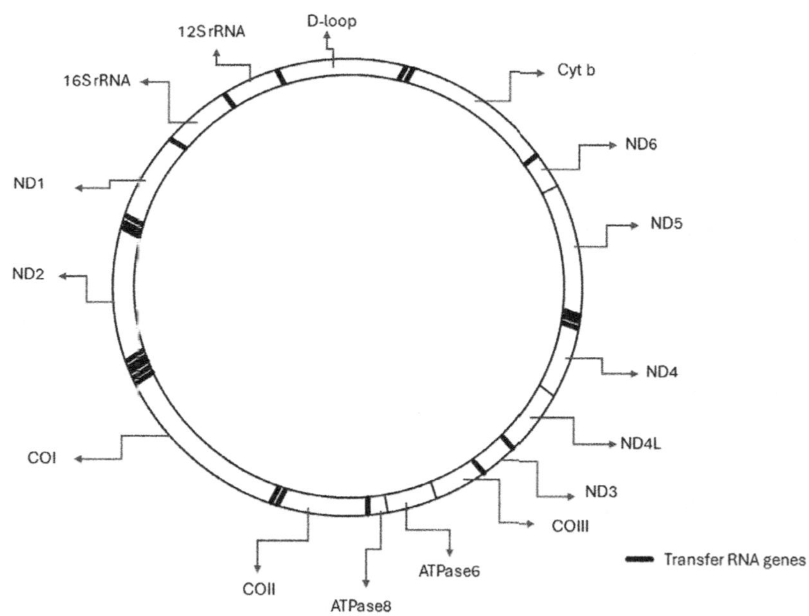

FIGURE 4.1 Structure of human mitochondrial DNA.

Replication, transcription, and translation of mtDNA

Understanding the inheritance system of mtDNA is important for revealing the underlying mechanism and the resulting effects of mutations of mtDNA. Mitochondrial replisome which is a large protein complex responsible for the mtDNA replication has several units as follows: (a) Heterodimeric mtDNA polymerase gamma within the replisome, (b) Twinkle protein which takes on the task of enzymatic activity of DNA helicase, (c) single-stranded binding protein. Through mitochondrial transcription factor, this protein complex with mtDNA forms structures called "nucleoids". There are two models represented for mtDNA replication: (a) strand-displacement-asynchronous model (Shadel & Clayton, 1997) and (b) symmetric strand-coupled replication model (Brown et al., 2005). You can find the details of the subject in the "Mitochondrial DNA Biogenesis" section.

With every cell generation, mitochondria must be doubled and distributed into each daughter cell. In addition, the rising need would produce new organelles in organelle turnover. Mitochondrial biosynthesis is a complex process as it has two separate genetic systems.

There are special genes that are responsible for encoding the proteins in mitochondria. Mitochondria use these proteins from the cytosol after cytosolic ribosome synthesis. The mitochondrial DNA encodes other proteins which are synthesized by mitochondrial ribosomes using amino acids encoded by mitochondrial mRNA. With the sequencing, it is understood that, unlike other genomes, nearly every nucleotide seems to be part of a coding sequence. In addition, although there are more than 30 tRNAs, only 22 are used to specify amino acids to produce mitochondrial proteins. tRNAs can recognize any of the four nucleotides in the third position, which permits protein synthesis with 22 tRNAs. And, one of the most surprising features of the mitochondrial genome is Variant Genetic Code. When the mitochondrial gene sequence and the amino acid sequences of the proteins are compared, it is observed that the sequences are different (Morgan-Richards et al., 2017). Mitochondria in living cells are dynamic, frequently dividing, fusing, and changing shape (Bereiter-Hahn et al., 1994). The number and shape of mitochondria vary dramatically in different cell types and can change in the same cell type under different conditions (Zahedi et al., 2018). GTPases that are placed on mitochondrial membranes are the regulators of this variability.

Inheritance of mitochondrial DNA

In contrast to the nuclear genome and unlike the mitochondrial genome in most organisms, the mtDNA in humans is maternally inherited and is transferred via oocytes, in other words, it is *uniparental* (Al Rawi et al., 2011; Giles et al., 1980). Nuclear DNA (nDNA) is diploid and conforms to Mendelian inheritance laws.

A fertilized human egg carries around 2000 copies of the human mitochondrial genome. A human carrying a deleterious mutation would generally not survive. Although a single cell has identical sequences of mtDNA which is described as homoplasmic, with somatic mutations heteroplasmy, different sequences of mtDNA can be observed (Chatterjee et al., 2006). In this case, oocytes carry both mutant and wildtype mitochondrial genomes and their offspring inherit the mixture of mtDNAs. The replication of mitochondrial DNA is not directly linked to the cell division cycle. Certain mitochondrial DNA is destroyed and replicated repeatedly and independently, as the cell divides; a process which is named as *relaxed replication* (Birky et al., 1994). During this replication process, the number of mitochondrial DNA carrying a single mutation may increase or may be repaired by the mtDNA repair mechanisms. Because mtDNA is more susceptible to oxidative disorders than cellular DNA, mtDNA repair mechanisms are not sufficient to avert high mutation rates (Lee et al., 2010). As a result of mitotic dissociation that takes place in a particular tissue, they become unhealthy or healthy according to the majority of mitochondrial DNA being healthy or defective. There are more than 1000 copies of mtDNA in a cell and its number differs according to tissue type and from person to person. Pathologies due to mitochondrial damage usually occur when heteroplasmy is above a certain level and this breakpoint also shows differences among tissues. All this means we can identify an inherited disease in humans caused by a mutation in mitochondrial DNA by diagnosing it in the succeeding generation from daughters, not sons, starting from the affected mother.

Mitochondrial homoplasmy and heteroplasmy

Two types of mtDNA populations can exist in a cell. Homoplasmy is the term used in genetics to describe the condition in which all the copies of mtDNA in a cell represent the same variant. It refers to a condition in which all mtDNA molecules within a cell or organism are genetically identical. In other words, there is no genetic variation among the mtDNA copies present. Homoplasmy is the default state of mtDNA and is typically observed in healthy individuals, where a single variant of mtDNA is uniformly present throughout all tissues. This uniformity ensures the efficient functioning of mitochondria and enables the synthesis of proteins required for cellular production. In contrast to homoplasmy, there are also cases where there are different types of mtDNAs in the cell which are mutated and named with the term heteroplasmy (Lee et al., 2010). This phenomenon arises due to genetic variation introduced during mtDNA replication or due to the coexistence of different mtDNA molecules from different origins. Heteroplasmy can manifest in a tissue-specific or random distribution, with varying proportions of different mtDNA variants across tissues and even within individual cells.

Implications of Heteroplasmy:

1. **Phenotypic variation:** Heteroplasmy can lead to phenotypic variation among individuals carrying the same mitochondrial mutation. The severity and manifestation of mitochondrial diseases are often influenced by the proportion of mutant mtDNA present. Higher levels of mutant mtDNA can result in more severe symptoms, while lower levels may be associated with milder phenotypes or even asymptomatic carriers.
2. **Maternal inheritance:** Heteroplasmy affects the transmission of mtDNA to subsequent generations. During gametogenesis, the proportion of mtDNA variants can change, resulting in a shift in heteroplasmy levels in offspring. This phenomenon, known as the bottleneck effect, contributes to the variability of mtDNA inheritance and can influence the likelihood of disease transmission.
3. **Tissue-specific heteroplasmy:** Heteroplasmy levels can vary among different tissues within an individual. Some tissues may have higher proportion of mutant mtDNA, leading to tissue-specific mitochondrial dysfunction. The variation in heteroplasmy among tissues contributes to the heterogeneity of mitochondrial diseases and poses challenges for diagnosis and treatment.
4. **The coexistence of maternal and paternal mtDNA:** While heteroplasmy is commonly associated with genetic variation introduced during mtDNA replication or the

coexistence of different mtDNA molecules, it has primarily been attributed to maternal mtDNA. However, recent evidence suggests that paternal mtDNA can also be detected in offspring, albeit at significantly lower levels compared to maternal mtDNA. The mechanism underlying the leakage of paternal mtDNA during fertilization is not yet fully understood. However, several hypotheses have been proposed to explain this intriguing phenomenon: (a) Paternal mtDNA in Sperm Cells: Traditionally, the paternal contribution to mitochondrial genetics was considered negligible due to the selective elimination of paternal mitochondria during fertilization. However, recent studies challenge this notion by demonstrating the presence of small amounts of paternal mtDNA in sperm cells. It is hypothesized that this paternal mtDNA may be transferred to the fertilized egg during the process of fertilization. (b) Entry and Elimination Mechanisms: Following the entry of paternal mtDNA into oocyte during fertilization, several mechanisms may contribute to its subsequent degradation or elimination. These mechanisms include the action of maternal enzymes that selectively degrade paternal mtDNA or the dilution of paternal mtDNA through the rapid replication and predominance of maternal mtDNA within developing embryo.

The leakage of paternal mtDNA results in low-level heteroplasmy in offspring. The proportion of paternal mtDNA is typically very small compared to maternal mtDNA, leading to minimal functional consequences. However, in rare cases, higher levels of paternal mtDNA may occur, potentially influencing mitochondrial function and manifesting as mitochondrial diseases or other phenotypic effects. The impact of paternal mtDNA on offspring health and development warrants further investigation. In addition to this, the presence of paternal mtDNA in offspring challenges the traditional understanding of mtDNA inheritance patterns and population genetics. Mitochondrial DNA has been a valuable tool for tracing maternal lineages and investigating population histories. The leakage of paternal mtDNA introduces additional genetic diversity, potentially complicating the analysis of mtDNA lineages and population studies. Understanding the extent and frequency of paternal mtDNA leakage can refine our understanding of human genetic variation and evolutionary processes. Heteroplasmy resulting from paternal mtDNA leakage can contribute to the manifestation and severity of mitochondrial diseases. The proportion of mutant mtDNA within the mitochondrial population determines the phenotypic expression of mitochondrial diseases. The presence of paternal mtDNA can introduce additional genetic variation, potentially influencing disease progression and clinical outcomes. Understanding the interplay between maternal and parental mtDNA variants in disease manifestation is crucial for accurate diagnosis and effective treatment strategies.

Dynamics of mitochondrial DNA inheritance

Random Segregation during Cell Division: During cell division, mitochondria distribute randomly into daughter cells, resulting in a mixture of mtDNA variants in each progeny. This process occurs in both mitotic cell division, where somatic cells divide, and meiotic cell division, which gives rise to gametes for sexual reproduction. The random segregation of mitochondria ensures that both daughter cells inherit a diverse population of mtDNA, leading to heteroplasmy. Several factors influence the random segregation of mtDNA during cell division: (a) Mitochondrial Network Dynamics: Mitochondria form a dynamic network within cells, continuously undergoing fusion and fission. The distribution of mtDNA during cell division depends on the balance between fusion and fission events. Unequal partitioning of mitochondria during fission can lead to uneven segregation of mtDNA variants in daughter cells. (b) Cytoskeletal Dynamics: The cytoskeleton, composed of microtubules and actin filaments, plays a role in the positioning and movement of mitochondria during cell division. It helps ensure proper distribution of mitochondria and their associated mtDNA during the division process. (c) Asymmetric Division: In certain cell types, such as stem cells or germ cells, asymmetric division occurs. Asymmetric division generates daughter cells with different fates, and this process can result in different proportions of mtDNA variants being passed on to the progeny cells.

Replicative Segregation: Replicative segregation refers to the distribution of mtDNA variants during mitochondrial replication. It occurs due to the replication of mtDNA molecules within individual mitochondria, leading to a separation of different mtDNA variants into daughter molecules. Unlike random segregation, replicative segregation can result in the homoplasmic inheritance of mtDNA variants, where a single variant dominates within a mitochondrial population. Three different genotypes can arise during cell divisions: normal homoplasmy, mutant homoplasmy, and heteroplasmy. The phenotype of the cell is determined by the ratio of mutant mtDNAs (Fig. 4.2). For example, the phenotype of a cell with heteroplasmic genotype may not show the genetic defect if the ratio of mutant mtDNAs is lower than normal mtDNAs (Craven et al., 2017).

mtDNA heteroplasmy level shows variability as a result of replicative segregation during cell division. Three different genotypes can arise during cell divisions: normal homoplasmy, mutant homoplasmy, and heteroplasmy. The phenotype of the cell is determined by the ratio of mutant mtDNAs. The phenotype of a cell with heteroplasmic genotype may not show the genetic defect if the ratio of mutant mtDNAs is lower than normal mtDNAs.

The exact mechanism of replicative segregation is still under investigation, but several hypotheses have been proposed: (a) Stochastic Replication: Stochastic replication suggests that the replication of mtDNA molecules is a random process, leading to the unequal partitioning of mtDNA variants into daughter molecules. This stochastic nature of replication can result in the preferential amplification of certain mtDNA variants over others. (b) Selective Replication Advantage: Selective replication advantage proposes that certain mtDNA variants have a replicative advantage, leading to their preferential amplification during replication. This advantage could be conferred by factors such as mutation rates, replication efficiency, or interactions with mitochondrial proteins involved in replication (Lawless et al., 2020).

The random segregation of mtDNA during cell division results in the presence of multiple mtDNA variants in different cells or tissues, leading to heteroplasmy. Heteroplasmy can influence the phenotypic variation observed in mitochondrial diseases, as the proportion of mutant mtDNA determines the severity and onset of symptoms. The interplay between random segregation and replicative segregation contributes to the inheritance patterns of mtDNA. While random segregation ensures the transmission of a diverse mtDNA population, replicative segregation can lead to the fixation of a single mtDNA variant within a lineage, resulting in homoplasmy. The dynamics of random segregation and replicative

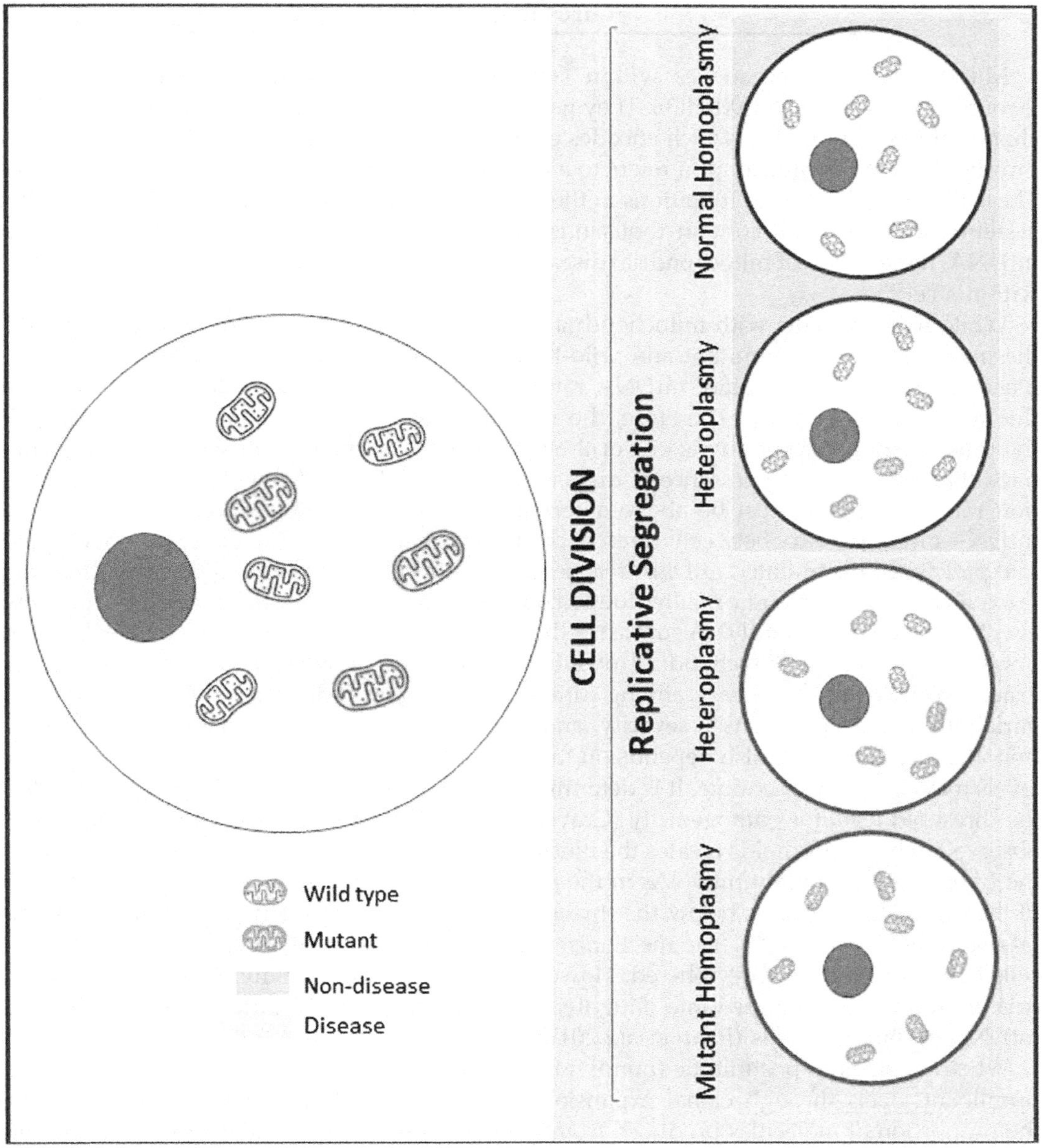

FIGURE 4.2 Mitochondrial homoplasmy and heteroplasmy.

segregation play a role in the evolution of mtDNA. The random segregation of mtDNA contributes to genetic diversity and allows for adaptive responses to changing environmental conditions. Replicative segregation, on the other hand, can result in the accumulation of beneficial mutations or the loss of deleterious mutations within mitochondrial lineages (Wallace & Chalkia, 2013).

Threshold effect

Mitochondria are organelles within cells that are responsible for generating energy through a process called OXPHOS. They have their own small circular DNA, known as mitochondrial DNA (mtDNA), which encodes essential components of the OXPHOS system. The threshold effect of mitochondria refers to a phenomenon observed in mitochondrial diseases. These diseases result from mutations in the mtDNA that impair the function of the OXPHOS system. Since cells can contain thousands of mitochondria, each with multiple copies of mtDNA, the severity of mitochondrial disease depends on the proportion of mutated mtDNA within a cell.

While some patients with mitochondrial disease are homoplasmic for mtDNA mutation, the majority have both mutant and wild-type mtDNA, meaning the majority have heteroplasmy. Although the mutant mtDNA ratio is high, some conditions do not cause disease due to low heteroplasmy. Therefore, the ratio of mutant mtDNA to wild-type mtDNA in the presence of heteroplasmy is one of the parameters associated with severe clinical symptoms that determine the presence of disease. This is called the "threshold effect". The mutation rate of mtDNA must be above a certain limit that the cell can tolerate. The mutant mtDNA ratio may vary between different organs and individuals. The threshold effect comes into play when the mutated mtDNA reaches a critical level within a cell. Below this threshold, the cell can still function normally due to compensation by the healthy mitochondria. However, once the mutated mtDNA surpasses the threshold, the cell's energy production becomes impaired, leading to mitochondrial dysfunction and the manifestation of symptoms. The exact threshold level can vary among different tissues and individuals, contributing to the variability in the age of onset, severity, and progression of mitochondrial diseases. In addition, the threshold level also depends on the particular mtDNA mutation, and heteroplasmy levels may also vary over time. It is determined that the mutant mtDNA level is 60%–80% of the threshold level for pathogenicity (Craven et al., 2017). The fact that pathogenicity can be observed at high thresholds creates the illusion that mutant mtDNAs are recessive. However, the frequencies of mutant mtDNAs in the population are substantial. Although about 1/200 of the population remains below the threshold, they carry potentially harmful mtDNA mutations (Elliott et al., 2008). The mechanisms that determine the different mutation levels in cells have not been fully deciphered. However, it is known that in dividing cells, mitochondria are randomly separated into daughter cells and there may be different levels of mutant mtDNA in daughter cells (Raap et al., 2012).

When it comes to postmitotic (nondividing) cells, mutant mtDNAs can be maintained at prominent levels through clonal expansion. In the model Elson presented in 2011, when the proportion of molecules produced from wild-type mtDNA in a cell decreases due to mutations, the cell responds by proliferation without selective action, by replicating the entire mitochondrial genome while simultaneously increasing the mutation load. Both the amount of wild-type and mutant alleles is increased (Elson et al., 2001).

The phenotypic threshold effect is closely related to the threshold effect of mitochondria. It refers to the point at which a person exhibits visible symptoms or a detectable phenotype due to the accumulation of mutated mtDNA above a critical threshold within a cell or tissue. This effect can be influenced by various factors, including the specific mutation, the type of tissue affected, the individual's overall health, and the presence of compensatory mechanisms. In some cases, the phenotypic threshold effect can result in a "threshold effect inheritance pattern," where the severity of the disease is variable and does not follow typical Mendelian inheritance. This is because the proportion of mutated mtDNA within cells can fluctuate across generations or even within an individual, leading to different family members or different tissues displaying varying degrees of disease severity. Understanding these threshold effects is crucial for diagnosing and managing mitochondrial diseases. Researchers are studying these effects to develop potential therapies aimed at reducing the proportion of mutated mtDNA or enhancing the function of healthy mitochondria within cells (Ghiselli & Milani, 2020; Rossignol et al., 2003).

In conclusion, both the threshold effect of mitochondria and the phenotypic threshold effect play significant roles in the manifestation and variability of mitochondrial diseases. These concepts highlight the complexity of mitochondrial genetics and the challenges associated with diagnosing and treating these disorders.

Mitochondrial diseases caused by the threshold effect of mitochondria are a subset of disorders resulting from mutations in the mitochondrial DNA (mtDNA). The severity and manifestation of these diseases depend on the proportion of mutated mtDNA within cells, as explained by the threshold effect. Here are a few examples influenced by this phenomenon highlighting the impact of the mitochondrial threshold effect on the development, progression, and variability of symptoms in mitochondrial diseases:

1. **Leber's hereditary optic neuropathy (LHON):** LHON is a mitochondrial disorder primarily affecting the optic nerve, leading to vision loss. It is caused by mutations in specific genes of the mtDNA that encode components of the ETC. The threshold effect plays a role in the development of symptoms. LHON typically manifests in young adults, as the proportion of mutated mtDNA in retinal ganglion cells crosses the threshold, causing energy production deficits and resulting in optic nerve damage (Nie et al., 2023).
2. **Mitochondrial encephalomyopathy, lactic acidosis, and stroke-like episodes (MELAS):** MELAS is a multisystem disorder characterized by a range of symptoms, including muscle weakness, lactic acidosis, and stroke-like episodes. It is caused by mutations in the mtDNA that affect various components of the ETC. The phenotypic threshold effect contributes to the variability in symptom severity and onset age. As the proportion of mutated mtDNA in affected tissues reaches a critical level, energy production becomes insufficient, leading to tissue damage (Shoop et al., 2023).
3. **Myoclonic epilepsy with ragged red fibers (MERRF):** MERRF is a rare mitochondrial disorder characterized by myoclonic seizures, muscle weakness, and ragged red fibers seen under a microscope in muscle biopsies. The condition is caused by mutations in the mtDNA that disrupt mitochondrial function. The threshold effect influences the age of onset and severity of symptoms. As the proportion of mutated mtDNA in muscle cells crosses the threshold, energy deficits and muscle dysfunction occur (Villanueva-Paz et al., 2019).
4. **Mitochondrial myopathy:** Mitochondrial myopathies are a group of disorders that primarily affect muscle tissue due to mitochondrial dysfunction. The phenotypic threshold effect can lead to variable symptom expression, including muscle weakness,

exercise intolerance, and muscle pain. The severity of the disease depends on the proportion of mutated mtDNA within muscle cells (Hardy et al., 2016).

5. **Kearns-sayre syndrome (KSS):** KSS is a rare disorder characterized by symptoms such as muscle weakness, progressive external ophthalmoplegia (PEO) (difficulty moving the eyes), and heart block. KSS is caused by large deletions or rearrangements of mtDNA. The threshold effect contributes to the clinical heterogeneity of KSS, as different tissues and individuals can exhibit varying degrees of symptoms based on the proportion of mutated mtDNA (Holloman et al., 2013; Turnbull et al., 2010).

Genetic bottleneck

Rapid shifts in heteroplasmy levels were first described in Holstein cows (Ashley et al., 1989; Olivo et al., 1983). It was found that with the limited transmission of mtDNA allele frequencies between generations a new allele could shift to full fixation, even at low levels. This is known as the mitochondrial bottleneck (Fig. 4.3). Similar mechanisms have subsequently been observed in mice, *Drosophila*, zebrafish, and human studies. The bottleneck hypothesis is based on population genetic theory (Cao et al., 2007; Otten et al., 2016; Solignac et al., 1984; Wai et al., 2008).

This hypothesis is explained with an interpretive example (Tang et al., 2022; Zhang et al., 2018). If a small number of molecules are to be taken several times from a pool of mtDNAs, the number of wild-type and mutant mtDNAs will vary each time. If the number of

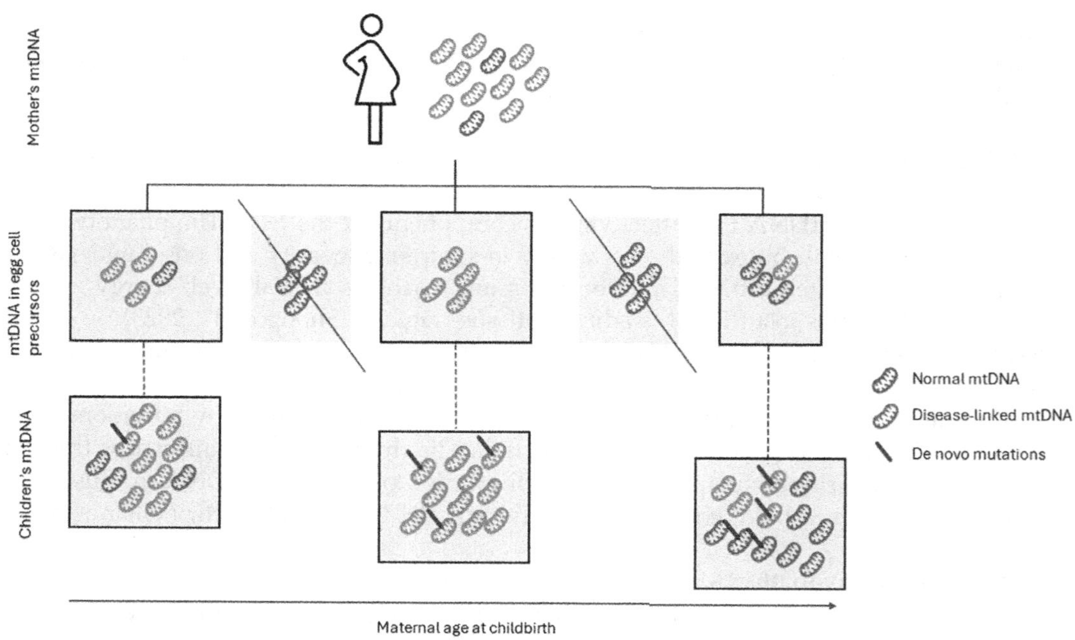

FIGURE 4.3 Mitochondrial DNA bottleneck.

molecules uptake increases, it will also change the ratio between mtDNA alleles. The larger the sample, the smaller the frequency difference between wild-type and mutant mtDNAs will be. This difference will be larger when the sample size is small. The source of mtDNA allele differences seen in generations was explained with this model. Although the "bottleneck" hypothesis explains the formation of heteroplasmy variability, it does not clarify by which cellular or molecular mechanism this process occurs (Johnston, 2019). There are several potential mechanisms explaining the bottleneck besides the physical restriction in the number of molecules during the maternal transmission of mtDNA. You can follow the illustrations of the proposed mechanisms, focal replication of mtDNA (Shoubridge & Wai, 2007; Wai et al., 2008), focal destruction of mtDNA (Diot et al., 2016), and the packaging of mtDNA (Cao et al., 2009) in Fig. 4.4. Although the mechanism of bottleneck is not fully known, for women carrying the pathogenic heteroplasmic mtDNA mutation, the bottleneck process enables their oocytes to have different heteroplasmy. This makes it possible for the most appropriate option to occur so that mitochondrial disease does not carry over to the offspring.

Heterogeneity

Mitochondrial and cellular functions like metabolic signaling, energy use, biosynthesis of iron–sulfur clusters, calcium homeostasis, triglyceride synthesis, and apoptosis are highly associated with the heterogeneity of mtDNA (Hopper et al., 2006; Stehling & Lill, 2013; Wang & Youle, 2009).

Mitochondrial heterogeneity is defined as the variation of a mitochondrial trait within the same single cell (das Neves, 2010). Mutations in different mitochondrial genomes can show similar phenotypic characteristics, though mutations in the same gene can result in different clinical phenotypes, even the ones that carry the same mtDNA mutations may show distinct traits. Interestingly, the mutation rate of mtDNA is significantly higher compared to nDNA. There are factors contributing to the higher mutation rate of mtDNA. Mutation rate refers to the frequency at which changes, or mutations, occur in DNA over a given time period. It is influenced by various factors, including DNA repair mechanisms, exposure to mutagens, and the fidelity of DNA replication processes. The mutation rate of mtDNA and nDNA differs significantly due to several unique characteristics of mtDNA.

Factors Contributing to Higher mtDNA Mutation Rate (Lax et al., 2011):

1. Lack of Protective Histones: Unlike nDNA, which is tightly bound to histone proteins that offer protection, mtDNA is not associated with histones. This lack of protection renders mtDNA more vulnerable to damage from environmental factors, ROS, and endogenous mutagens.
2. Close Proximity to Reactive Oxygen Species: Mitochondria are the primary producers of ROS during cellular respiration. The proximity of mtDNA to ROS generation increases the chances of oxidative damage to mtDNA molecules, leading to higher mutation rates.
3. Limited DNA Repair Mechanisms: Mitochondria have limited DNA repair mechanisms compared to the nucleus. While the nucleus possesses a sophisticated array of DNA

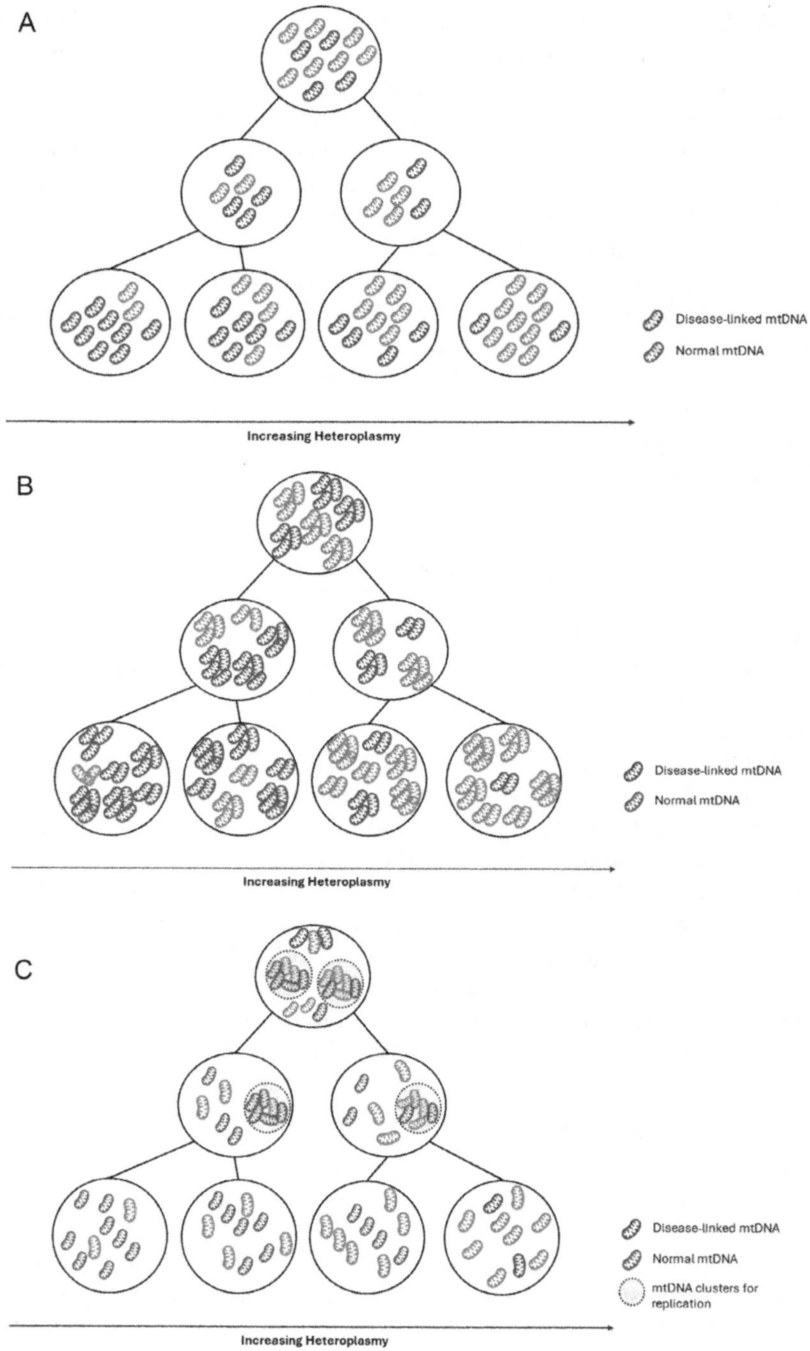

FIGURE 4.4 Alternative mechanisms of differing levels of heteroplasmy during genetic bottleneck. (A) Passive reduction of mtDNA, (B) mtDNA packaged into homoplasmic clusters, (C) focal mtDNA replication.

repair systems, mitochondria have fewer repair pathways and lower efficiency in correcting errors or damage to mtDNA.
4. Replication Process: The replication process of mtDNA contributes to its higher mutation rate. Unlike nDNA, mtDNA replication lacks proofreading mechanisms, which are specialized enzymes that correct errors during DNA synthesis. This absence of proofreading mechanisms in mtDNA replication increases the likelihood of errors and subsequent mutations.
5. With over 90% of the mitochondrial genome composed of coding sequences and the absence of introns (noncoding sequences within genes) along with minimal noncoding bases between genes, mutations in mtDNA can have substantial and direct impacts on cellular function.

Mitochondrial heterogeneity is not only seen in the development of pathogenic conditions but is actually a normal physiological process to a certain extent. Mitochondrial variance in the physiological state provides functional adaptation to particular groups of mitochondria within the cell; however, the physiological heterogeneity levels that are usual in the presence of chronic stress begin to reach levels that give birth to pathological results (Twig et al., 2008). We have the knowledge that a single cell has a population of mitochondria and it is important to understand the underlying mechanisms of mitochondrial heterogeneity. Genetic and nongenetic mitochondrial heterogeneity are the two sources that are not independent of each other (Aryaman et al., 2019).

Point mutations

mtDNA point mutations constitute an important cause of mitochondrial diseases with a population prevalence of approximately 1/200 and are usually inherited by the mother. Point mutations can occur in any mtDNA gene (protein genes, rRNA, and tRNA genes representing approximately 10% of the mitochondrial genome). Point mutations on genes encoding mitochondrial proteins that affect the function of the protein encoded by the gene on the respiratory chain complex can lead to various mitochondrial disorders, including NARP (Neuropathy, Ataxia, and Retinitis Pigmentosa) disease. NARP is primarily associated with specific point mutations in the ATP6 gene of mtDNA. The ATP6 gene encodes a subunit of the ATP synthase complex, which is a vital component of the mitochondrial ETC and OXPHOS system. In NARP disease, specific point mutations occur in the ATP6 gene, leading to dysfunction in the ATP synthase complex. These mutations typically involve substitutions of single nucleotide bases, resulting in the production of abnormal ATP6 protein subunits. This dysfunction disrupts the normal flow of electrons through the ETC, impairing the production of ATP and causing cellular energy deficits. Here are a few examples:

1. **m.8993T>G mutation:** This mutation involves a substitution of thymine (T) with guanine (G) at position 8993 in the mitochondrial genome. It is one of the most common mutations associated with NARP and also with another mitochondrial disorder called Leigh syndrome. The m.8993T>G mutation can lead to variable

symptoms, including neuropathy, ataxia, and retinitis pigmentosa (Balasubramaniam et al., 2017).
2. **m.8993T>C mutation:** This mutation involves a substitution of thymine (T), but with cytosine (C) at position 8993. This mutation can also result in NARP symptoms and is often considered in the context of mitochondrial disease diagnostics (Craig et al., 2007).
3. **m.9176T>C mutation:** In this mutation, thymine (T) is replaced by cytosine (C) at position 9176. The m.9176T>C mutation can lead to a range of symptoms, including those associated with NARP, as well as with other mitochondrial disorders (Balasubramaniam et al., 2017).
4. **m.9176T>G mutation:** In this mutation, thymine (T) is replaced by guanine (G) at position 9176. Similar to the m.9176T>C mutation, this mutation can contribute to NARP symptoms and other mitochondrial-related conditions (Khailany et al., 2020).

In addition, point mutations on mt-tRNA, which contribute to more than half of the mutations, reduce the number of mt-tRNAs involved in the mitochondrial translation process and may affect translation completion and protein synthesis. Mitochondrial protein synthesis, aminoacylation activity, three-dimensional structure, and folding of tRNA in mitochondria are the mentioned molecular examples of the mutations that affect the activity of tRNA genes (Pü et al., 2007; Ruiz-Pesini et al., 2007).

The two best-known diseases associated with mt-tRNA point mutations are mitochondrial encephalopathy with lactic acidosis and stroke-like episodes (MELAS), and MERRF. Different phenotypes show up with different point mutations. MELAS affects the organs: skeletal muscle, brain heart, endocrine pancreas, ears, and gut. The A3243G mutation on the tRNALeu (UUR) gene in a wobble position with 80% high prevalence causes MELAS (Goto et al., 1990). A8344G mutation on tRNALys causes MERRF is also located in a wobble position (Schon et al., 2012). The mechanism of why the wobble position is important and how it is used by the mitochondrial ribosome for correct coding is explained in (Kirino et al., 2004).

The deafness phenotype is aminoglycoside-induced and is passed on maternally in generations susceptible to its toxicity. The most common mutation in the mitochondrial rRNA genes is the A1555G mutation in the 12S rRNA gene. It changes the structure of the rRNA as in the 16S rRNA in the bacterial genome and interrupts the translation (Prezant et al., 1993). Pathogenic protein synthesis mutations cause multisystem diseases. Encephalopathy, mitochondrial myopathy, gastrointestinal diseases, diabetes, renal dysfunctions, and neurological complex diseases such as Alzheimer's and Parkinson's can be given as examples. Leigh syndrome is one of the mitochondrial syndromes that occur because of protein synthesis mutations. The most known mutations are T8993C and T8993G. Different pathogenic symptoms occur when different mutations affect mitochondrial functions. The frequency size of the mutations determines the type of symptoms: low mutation frequency causes the person to be a carrier of the mutation, moderate mutation frequency causes neurological susceptibility, and high frequency causes maternally inherited Leigh syndrome (Holt et al., 1990).

Deletion mutations

Single large-scale deletions are the most well-known disease-associated mutation type. Unlike point mutations that occur in mt-tRNA and protein-synthesizing genes, deletion mutations can occur in any region of the mitochondrial genome during the repair or replication process (Krishnan et al., 2008) and been associated with several pathologies like myopathies, cancer, pancytopenia, and more (Kato et al., 2011; Zhu et al., 2004). mtDNA deletion syndromes predominantly involve overlapping phenotypes that may evolve from one pathogenic syndrome to another over time (Rö et al., 1995; Sabella-Jimé et al., 2020). Although research studies have previously shown the relationship of large-scale deletion mutations with cancer pathologies (Glover et al., 2017), the correlation link has still not been fully clarified (Allegra et al., 2006; Fontana & Gahlon, 2020; Hopkins et al., 2017).

Well-studied KSS and Pearson Marrow-Pancreatic Syndrome (PMPS) are the pathologies, resulting from deletion mutations of mtDNA. When KSS and PMPS mitochondrial disease mechanisms are examined, it has been shown that fragile sites (FRA) in mtDNA also play a role in large-scale deletions (Richards & Macaulay, 2001). FRAs are large genomic regions that are susceptible to breaks and gaps on metaphase chromosomes upon replication stress are rearranged in many diseases (Blin et al., 2019). It causes the DNA to be unstable, and this also manifests itself in pathologies that develop due to mtDNA mutations. In cases where they fall within the scope of the transcribed gene regions, it can alter the replication process, causing R-loops to occur, resulting in fork stalling (Kaushal & Freudenreich, 2019). This prevents the completion of replication during mitosis phase of the cell. Structure-specific endonucleases are activated in this process. In PEO and KSS diseases, it has been determined that in mtDNA, which has alternative structures of G-quadruplex, triplex and cruciform, the motifs that provide G4 formation are located adjacent to the mitochondrial deletion regions (Bharti et al., 2014; Damas et al., 2012; Dong et al., 2014; Oliveira et al., 2013). The 9 bp deletion in the intergenic region of mtDNA is among the most prevalent mitochondrial deletion (Jin et al., 2012). A recent study reveals a novel fragility mechanism related to 9 bp deletion and presents a novel function of endonuclease-G on the fragile sites of mtDNA, especially on G-quadruplex structure (Dahal et al., 2022). Again, a novel study about revealing the underlying mechanism of mtDNA instability has found out FANC proteins provide the maintenance of mitochondrial genome stability on FRAs by tuning mitochondrial stress response (Fernandes et al., 2021).

KSS is a multisystem syndrome that develops before 20 years of age and is associated with pigmentary retinopathy and PEO. Any of the conditions of cardiac conduction disturbances, high CSF protein, and cerebellar ataxia are also observed (Gloria Pang et al., 2022). Pearson syndrome is manifested by bone marrow failure, sideroblastic anemia, and pancreatic insufficiency. This syndrome is variable and aggressive, so it is progressive. It can cause death in early infancy or childhood. If there is survival after bone marrow dysfunction, a transition to the manifestations of KSS can be made with spontaneous recovery (Gloria Pang et al., 2022). MtDNA deletion mutations reflect a specific cellular event associated with aging. They cause phenotypic changes that result in cell dysfunction, apoptosis, and cell loss leading to impaired tissue structure and function (Chinnery et al., 2002; Chomyn & Attardi, 2003; Li et al., 2021). Studies have shown that the frequency of mtDNA deletion strongly predicts the abundance of ETC-deficient cells. ETC-deficient cells are the indicators of apoptosis which leads to cell loss. This correlation makes the frequency of mtDNA deletions be used as a determinant factor of cell loss. In line with this information, analysis on human quadriceps muscle samples showed that the frequency of mtDNA deletion increases exponentially with age (Herbst et al., 2019, 2021).

Epigenetics

Epigenetic changes are factors that regulate gene expression but do not cause any changes in the primary DNA base sequence. While DNA methylation (5-mC) or hydroxymethylation (5-hmC) (Kriaucionis & Heintz, 2009; Tahiliani et al., 2009) affects cellular control of gene expression, histone modifications (on N-terminal tails of the nucleosome) function in transcriptional activity and accessibility of chromatin structure in the cell and noncoding RNAs (ncRNAs) are responsible from the regulation of gene expression (Lee, 2009; Nagano et al., 2008).

MtDNA can also be regulated under the influence of epigenetic mechanisms like cellular DNA, but the importance of models explaining how epigenetic mechanisms affect mtDNA has been recognized in recent years. There are three main epigenetic mechanisms that regulate the expression of genes in mtDNA: (1) DNA methylation, (2) ncRNAs, and (3) posttranslational modifications of mtDNA proteins.

Methylation

It is understood that mammalian mitochondria have mtDNA methyltransferase 1 (mtDNMT1) activity in both 5-mC and 5-hmC and the mitochondrial methylation location is mainly the noncoding region in the D-loop of mtDNA in CpC-, CpA-, and CpT sites (Liu et al., 2016). The very first studies showing that DNMT activity has an effect on mtDNA were shown in vertebrate studies (Vanyushin & Kirnos, 1976, 1977). There is a regulation mechanism of nuclear DNMT1 translocation to the matrix of mitochondria. DNMT1, DNMT3a, DNMT3b, and TET1 and TET2 enzymes are identified within the mitochondria (Chen et al., 2012; Chestnut et al., 2011). It can be stated that mtDNA demethylation occurs with OXPHOS in mitochondria in the presence of TET1 and TET2. At the upstream of the transcription start site of the gene, there exists the mitochondria targeting sequence. Changes in mtDNMT1 affect transcription in both strands of mtDNA. This is also an indication that the transcriptional regulation of mtDNA can be controlled by the epigenetic mechanism of the gene of nDNA (Shaughnessy et al., 2014). Another study, examining mitochondrial DNA methylation showed no increase in 5-mC levels in oocyte and early embryonic mice (Fan et al., 2019). A methylation study performed again on mice showed that

5-hmC mitochondrial methylation decreased in the frontal cortex of aged mice due to cell aging, while there was no change in the cerebellum (Dzitoyeva et al., 2012). This indicates mitochondrial methylation is involved in a cell-specific mechanism. A novel study also suggested that there might be a deamination of 5 mC in mtDNA via APOBEC3 in mitochondria, although the mechanism is still unknown (Wakae et al., 2018). Summing up, the next steps and mechanisms from mtDNA methylation products remain an area to be explored for further research.

Noncoding RNAs

ncRNAs are a large segment of the transcriptome that does not encode a protein but contain information and regulatory effect on the function of the cell. ncRNAs can be divided into two groups; long ncRNAs(lncRNAs) and short ncRNAs (small interfering RNAs—siRNA, microRNA—miRNA, piwi-interacting RNA—-piRNA). Although the ncRNAs used in mitochondria are mostly formed by nDNA, mitochondrial DNA also plays a role in this. On the contrary, lncRNAs formed in mitochondria can be effective in the cell nucleus as well.

All short ncRNAs (siRNA, miRNA, and piRNA) communicate with Argonaute family proteins whose members play a central role in the regulation of gene expression. They together with siRNA, miRNA, and piRNA form the RISC complex. The RISC complex suppresses gene expression by different mechanisms depending on which short ncRNA is active. siRNA and miRNA come together to form RNA interference (RNAi). RNAi suppresses expression by deconstructing mRNAs. These epigenetic pathway mechanisms can be passed between generations for three to five generations. Both nuclear and mitochondrial miRNAs act on mitochondrial mRNA, disrupting its expression. OXPHOS and regulation of the citric acid cycle, which are the basic biochemical mechanisms of mitochondria, are regulated especially by mitochondrial miRNAs, resulting in a decrease in ATP levels due to miRNA excess (Dong et al., 2017; Duarte et al., 2014; Ghildiyal & Zamore, 2009; Liberman et al., 2019).

Mitochondrial damage checkpoint (mitocheckpoint)

Mitochondria play a crucial role in cellular energy production and are involved in various metabolic processes. However, these organelles are vulnerable to damage due to the generation of ROS, accumulation of mutations, and impaired protein quality control. To safeguard cellular health and maintain mitochondrial integrity, cells have evolved a surveillance mechanism called the mitochondrial damage checkpoint. The checkpoint acts as a quality control mechanism to ensure the elimination of damaged mitochondria and the maintenance of a functional mitochondrial population.

The checkpoint is initiated by specialized proteins known as mitochondrial stress sensors. These sensors are strategically located within or associated with the mitochondria and can detect various types of mitochondrial damage, including ROS production, loss of membrane potential, and impaired protein import. Examples of mitochondrial stress sensors include PINK1 (PTEN-induced putative kinase 1) and Parkin, which play key roles in mitophagy, the selective degradation of damaged mitochondria (Xiong et al., 2018). Upon detection of mitochondrial damage, the stress sensors activate signaling pathways that coordinate the cellular response. The best-studied pathway is the PINK1/Parkin pathway, which leads to the recruitment of Parkin to damaged mitochondria, tagging them for subsequent degradation by autophagy. Other signaling pathways, such as the unfolded protein response and the mitonuclear communication pathway, are also involved in the response to mitochondrial stress (Gu et al., 2021). Autophagy, specifically mitophagy, is an essential component of the mitochondrial damage checkpoint. It involves the sequestration of damaged mitochondria into autophagosomes and their subsequent delivery to lysosomes for degradation. Proteins involved in the autophagy process, including LC3 (microtubule-associated protein 1A/1B-light chain 3) and p62, participate in the removal of damaged mitochondria and the maintenance of mitochondrial quality control (Iriondo et al., 2022).

Damages in mitochondria cause genetic and epigenetic changes in the nDNA and similar to the DNA damage checkpoint response, mitocheckpoint signaling pathways are hypothesized (Xiong et al., 2018). Mitocheckpoint adjusts the balance between apoptotic and nonapoptotic signals. It is involved in regulating the function of the mitochondria, repairing cells with damaged mitochondria, and compensating for the damage caused by mtDNA mutations to the cell (Minocherhomji et al., 2012; Singh, 2006). There is a communication between the nucleus and mitochondria, and this crosstalk between them can be used in repair by regulating the expression of mtDNA and cellular DNA through epigenetic mechanisms in response to damaged mitochondria function (Sharma et al., 2019). It is also suggested that also mtOXPHOS-induced cellular senescence may be a checkpoint mechanism for the development of cancer as cellular senescence is a critical defining feature of the instability of the nuclear genome (Dimri, 2005).

The primary role of the mitochondrial damage checkpoint is to preserve cellular health and maintain mitochondrial integrity. It serves several important functions: (a) removal of damaged mitochondria: The checkpoint facilitates the selective removal of damaged mitochondria through mitophagy. This process ensures the elimination of dysfunctional mitochondria that can lead to the accumulation of ROS, release of proapoptotic factors, and impaired energy production. (b) Preservation of cellular energetics: By eliminating damaged mitochondria, the checkpoint helps maintain optimal mitochondrial function and energy production. This ensures the availability of ATP for cellular processes and prevents metabolic disturbances that can affect cell viability and function. (c) Prevention of oxidative stress: Mitochondrial damage often results in increased ROS production, leading to oxidative stress and cellular damage. The checkpoint's activation of mitophagy and removal of damaged mitochondria help mitigate oxidative stress and protect cells from the detrimental effects of ROS. (d) Maintenance of cellular homeostasis: The mitochondrial damage checkpoint contributes to overall cellular homeostasis by coordinating mitochondrial quality control mechanisms. It ensures a balance between mitochondrial biogenesis, removal of damaged organelles, and the preservation of a functional mitochondrial network (Singh, 2004).

The mitochondrial damage checkpoint is a critical surveillance system that helps maintain cellular health and homeostasis by detecting and responding to mitochondrial stress and dysfunction. By activating mitophagy and coordinating cellular signaling pathways, the checkpoint facilitates the removal of damaged mitochondria, preserves energy production, and prevents oxidative stress. Understanding the mechanisms and regulation of the mitochondrial damage checkpoint has implications for various diseases, including neurodegenerative disorders and age-related mitochondrial dysfunction. Further research in this field will enhance our understanding of mitochondrial quality control mechanisms and open avenues for therapeutic interventions targeting mitochondrial damage and dysfunction (Chen et al., 2023).

Exploring mitochondrial genome—in silico interpretation of mtDNA

Understanding the relationship between mtDNA heteroplasmy and pathogenic mutations with cell function and diseases, as well as the mechanisms of their transmission through generations is crucial for the development of accurate diagnostic criteria. The field of mitochondrial bioinformatics may enable researchers to delve deeper into the mitochondrial genome and to uncover unexplored genetic mechanisms of diseases. In particular, there is a lack of bioinformatics and computational biology methods related to gene expression, tRNAs and epigenetic mechanisms, and the crosstalk of these mechanisms with each other.

In silico interpretation of mtDNA involves the analysis of mtDNA sequences using computational tools and algorithms. It aims to decipher the functional consequences of mtDNA variants, including their impact on mitochondrial protein function, structural changes, and potential associations with diseases. By combining computational approaches with available databases and bioinformatics resources, researchers can gain valuable insights into the functional implications of mtDNA variants.

Methodologies for in silico interpretation

1. **Sequence alignment and comparison:** In silico interpretation begins with aligning and comparing mtDNA sequences. This allows researchers to identify sequence variations, including single nucleotide polymorphisms, insertions, deletions, and structural changes. Tools such as BLAST and CLUSTALW facilitate sequence alignment and provide information on the conservation and uniqueness of specific mtDNA variants (McGinnis & Madden, 2004; Thompson et al., 1994).
2. **Functional prediction tools:** Several computational tools and algorithms are available to predict the functional consequences of mtDNA variants. These tools assess the impact of variants on protein structure, stability, and function. Examples include PolyPhen-2

(Adzhubei et al., 2013), SIFT (Sim et al., 2012), and PROVEAN (Choi & Chan, 2015), which predict the potential pathogenicity or functional significance of mtDNA variants based on evolutionary conservation and protein structure.
3. **Mitochondrial disease databases:** Databases dedicated to mitochondrial diseases, such as Mitomap (Brandon et al., 2005) and MitoCarta (Rath et al., 2021), provide curated information on known pathogenic mtDNA variants and their associated phenotypes. These databases serve as valuable resources for researchers and clinicians, allowing them to explore the clinical relevance of identified mtDNA variants and their potential implications for disease development.

Applications of in silico interpretation of mtDNA

1. **Disease gene discovery:** In silico interpretation aids in the identification of disease-causing mtDNA variants. By analyzing large-scale sequencing datasets and comparing them with reference sequences, researchers can pinpoint potentially pathogenic mtDNA variants that contribute to mitochondrial diseases, such as Leigh syndrome, MELAS syndrome, and Leber's hereditary optic neuropathy (LHON).
2. **Population genetics and evolution:** In silico interpretation provides insights into the genetic diversity and evolutionary history of human populations. By analyzing mtDNA sequences from different populations worldwide, researchers can explore the distribution of mtDNA haplogroups, identify population-specific variants, and reconstruct maternal ancestry.
3. **Personalized medicine:** In silico interpretation of mtDNA contributes to personalized medicine by predicting individual susceptibility to certain diseases and informing treatment strategies. By assessing the functional consequences of mtDNA variants, clinicians can tailor therapeutic approaches for patients with mitochondrial diseases and potentially identify individuals at risk for developing certain disorders.

mtDNA variant annotation

Variant annotation provides the functional information assigned to DNA variants from different sources. It is necessary for the understanding of sequences with unknown functions.

Although there are databases collecting mtDNA exome and genome sequence data, there is not enough information about the frequency of the variants in the population, as the genetic mechanisms of mtDNA affecting variant frequency are quite complex. Incomplete penetrance, level of heteroplasmy, and mitochondrial haplotypes make it difficult to determine the frequency of variance. To illustrate this, GenBank (Benson et al., 2008) provides information on how many times an mtDNA mutation has been reported. The MitoMap database contains this mutation information about variants, which has been found to be pathogenic and has been identified in patients and is also included in phylogenetic studies. However, what is missing here is that there is no definite value range for the variant threshold, which is a crucial parameter in the determination of the disease state.

In addition, clinical data about mtDNA variants is also very important. There are databases such as CLINVAR that combine mtDNA variants with clinical data. Databases that carry both nDNA and mtDNA information, such as OMIM and CLINVAR, are less functional than databases established especially for mtDNA. To illustrate again with an example, there are 178 variants of the MT−ND1 gene reported in CLINVAR (Landrum et al., 2018). When we look at the MitoMap database for the same gene, we observe that it contains around 957 variant information, and 96 of them are associated with confirmed clinical information (Table 4.2) (Brandon et al., 2005). Systemic updates are not made in every tool and database. Therefore, when using computational tools, they should be able to be used to examine pipelines and when necessary should combine different data banks in line with the analysis work (Table 4.3).

There are several online resources available for accessing information about mitochondrial DNA (mtDNA) variants. These resources provide comprehensive databases and tools for researchers and clinicians to explore and analyze mtDNA variants. Here are some prominent online resources for mtDNA variant information:

1. **Mitomap** (http://www.mitomap.org): Mitomap is a widely used and comprehensive database dedicated to human mitochondrial DNA variations. It provides information on known pathogenic and nonpathogenic mtDNA variants, including their frequencies in different populations, associated diseases, and functional annotations. Mitomap also offers various tools for variant analysis and visualization (Brandon et al., 2005).
2. **MitoCarta** (https://www.broadinstitute.org/mitocarta): MitoCarta is a database that catalogs mitochondrial proteins and their associated genes. While it primarily focuses on protein-coding genes, it also includes information on their corresponding mtDNA variants. This resource is particularly useful for investigating the relationship between mtDNA variants and mitochondrial protein function (Rath et al., 2021).
3. **mtDB** (http://www.mtdb.igp.uu.se): mtDB is a database that provides comprehensive information about mtDNA sequences and their variations. It includes data on mtDNA haplogroups, polymorphisms, disease-associated variants, and population-specific mtDNA profiles. mtDB offers search functionalities and tools for mtDNA variant analysis (Ingman & Gyllensten, 2006).
4. **PhyloTree** (https://www.phylotree.org): PhyloTree is a database that focuses on mtDNA haplogroup classification and phylogenetic relationships. It provides an updated and curated tree structure for mtDNA haplogroups, allowing users to navigate through different branches and explore the associated variants. PhyloTree aids in the classification and interpretation of mtDNA variants in an evolutionary context (van Oven & Kayser, 2009).
5. **mtDNA-Server** (https://mtdna-server.uibk.ac.at): mtDNA-Server is a web-based tool that offers comprehensive analysis and interpretation of mtDNA variants. It provides information on mtDNA haplogroups, predictions of functional effects for variants, and predictions of the pathogenicity of specific mutations. mtDNA-Server incorporates various algorithms and databases to assist researchers and clinicians in the analysis of mtDNA variants (Weissensteiner, Forer, et al., 2016).

These online resources offer valuable information and tools for researchers, clinicians, and individuals interested in studying and understanding mtDNA variants. They serve as valuable references for exploring the functional significance, population distribution, and clinical implications of mtDNA variants. Researchers and healthcare professionals can utilize these resources to enhance their understanding of mitochondrial genetics, disease associations, and evolutionary patterns related to mtDNA.

TABLE 4.2 Reported pathogenic protein synthesis mutations for various phenotypic presentations selected from Mitomap database.

Syndromes	Locus	Disease	Allele	RNA	References
Encephalomyopathy					
Encephalomyopathy, Leigh syndrome	MTTV	LS	C1624T	tRNA Val	McFarland et al. (2002)
Encephalomyopathy, Leigh syndrome	MTTV	Adult LS	G1644T	tRNA Val	Chalmers et al. (1997)
Encephalomyopathy Leigh syndrome	MTTW	MILS	A5537insT	tRNA Trp	Santorelli et al. (1997)
Encephalomyopathy MERRF	MTTF	MERRF	G611A	tRNA Phe	Mancuso et al. (2004)
Encephalomyopathy MERRF	MTTK	MERRF	G8361A	tRNA Lys	Rossmanith et al. (2003)
Encephalomyopathy MERRF	MTTK	MERRF/ MICM + DEAF/ Autism	G8363A	tRNA Lys	Ozawa et al. (1997), Santorelli et al. (1996)
Encephalomyopathy MERRF	MTTL1	MERRF/KSS overlap	G3255A	tRNA Leu (UUR)	Nishigaki et al. (2003)
Mitochondrial myopathy					
Mitochondrial myopathy	MTTF	MM	T582C	tRNA Phe	Moslemi et al. (2004)
Mitochondrial myopathy	MTTF	MM	T618C	tRNA Phe	Kleinle et al. (1998)
Mitochondrial myopathy	MTTF	EXIT and deafness	G622A	tRNA Phe	Deschauer et al. (2006)
Mitochondrial myopathy	MTTF	Ataxia, PEO, deafness	T642C	tRNA Phe	Valente et al. (2009)
Mitochondrial myopathy	MTTL1	MM	G3242A	tRNA Leu (UUR)	Kirino et al. (2005)
Mitochondrial myopathy	MTTL1	MM/CPEO	T3250C	tRNA Leu (UUR)	Goto et al. (1992), Grünewald et al. (1999)
Mitochondrial myopathy	MTTL1	MM	A3251G	tRNA Leu (UUR)	Sweeney, Bundey, et al. (1993)
Cardiomyopathy					
Maternally inherited hypertrophic cardiomyopathy	MTTI	MHCM	A4295G	tRNA Ile	Merante et al. (1996)
Maternally inherited cardiomyopathy	MTTI	MICM	A4300G	tRNA Ile	Casali et al. (1995)
Cardiomyopathy	MTTW	HCM, severe multisystem disorder	C5545T	tRNA Trp	Sacconi et al. (2008)

TABLE 4.2 Reported pathogenic protein synthesis mutations for various phenotypic presentations selected from Mitomap database.—cont'd

Syndromes	Locus	Disease	Allele	RNA	References
Cardiomyopathy	MTTK	Cardiomyopathy	A8348G	tRNA Lys	Terasaki et al. (2001)
Maternally inherited hypertrophic cardiomyopathy	MTTG	MHCM	T9997C	tRNA Gly	Merante et al. (1994)
Maternally inherited cardiomyopathy	MTTH	MICM	G12192A	tRNA His	Shin et al. (2000)
Cardiomyopathy	MTTL2	Dilated cardiomyopathy	T12297C	tRNA Leu (CUN)	Grasso et al. (2001)
Fatal infantile cardiomyopathy Plus (MELAS)	MTTI	FICP	A4269G	tRNA Ile	Hayashi et al. (1994), Taniike et al. (1992)
Fatal infantile cardiomyopathy Plus (MELAS)	MTTI	FICP	A4317G	tRNA Ile	Ito et al. (1992), Tanaka et al. (1990)
Deafness					
Deafness	MTTF	DEAF	A636G	tRNA Phe	Konings et al. (2008)
Deafness	MTRNR1	DEAF	A827G	12S rRNA	Li et al. (2005)
Deafness	MTRNR1	DEAF	T961C	12S rRNA	Guaran et al. (2013)
Deafness	MTRNR1	DEAF	T961delT + C(n)ins	12S rRNA	Bacino et al. (1995)
Deafness	MTRNR1	DEAF	T961insC	12S rRNA	Bacino et al. (1995)
DIABETES mellitus					
Diabetes mellitus	MTRNR1	DM	C1310T	12S	Tawata et al. (1998)
Diabetes mellitus	MTRNR1	DM	A1438G	12S	Tawata et al. (1998)
Diabetes mellitus	MTTL1	DM	T3264C	tRNALeu (UUR)	Suzuki et al. (1997)
Diabetes mellitus	MTTL1	DM	T3271C	tRNA Leu (UUR)	Tsukuda et al. (1997)
Diabetes mellitus; metabolic syndrome	MTTI	Metabolic syndrome and Hypomagnesemia	T4291C	tRNA Ile	Wilson et al. (2004)
Diabetes mellitus and deafness and cardiomyopathy	MTTK	DMDF/MERRF/HCM	A8296G	tRNA Lys	Arenas et al. (1999), Kameoka, Isotani, Tanaka, Azukari, (1998), Kameoka, Isotani, Tanaka, Kitaoka, (1998)
Diabetes mellitus and deafness	MTTS2	DMDF	C12258A	tRNA Ser (AGY)	Lynn et al. (1998)

(Continued)

TABLE 4.2 Reported pathogenic protein synthesis mutations for various phenotypic presentations selected from Mitomap database.—cont'd

Syndromes	Locus	Disease	Allele	RNA	References
Other					
Movement disorder	MTTV	Movement disorder	T1659C	tRNA Val	Blakely et al. (2004)
Alzheimer and Parkinson disease	MTRNR2	ADPD	G3196A	rRNA 16S	Shoffner, Brown, et al. (1993), Wallace et al. (1992)
Alzheimer and Parkinson disease; deafness and migraine	MTTQ	ADPD/Hearing loss and migraine	T4336C	tRNA Gln	Cortopassi and Hutchin (1994), Hutchin and Cortopassi (1995), Leroy and Norby (1994), Shoffner, Brown, et al. (1993)
Dementia and chorea	MTTW	DEMCHO	G5549A	tRNA Trp	Nelson et al. (1995)
Multi-organ failure	MTTN	Multi-organ failure	T5728C	tRNA Asn	Meulemans et al. (2006)

Prediction tools

Next-generation sequencing (NGS) technologies have revolutionized the field of genomics, enabling high-throughput sequencing of DNA samples. This includes the sequencing of mitochondrial DNA (mtDNA), which has important implications for understanding mitochondrial genetics and associated diseases. Computational prediction tools have emerged as invaluable resources for analyzing and interpreting mtDNA sequences generated by NGS platforms (Ahmad & Hasan, 2021).

Sequencing mtDNA with NGS introduces many novel, previously unknown variants into the pool. Computational prediction tools play a crucial role in analyzing large-scale mtDNA sequencing data generated by NGS platforms. These tools help researchers and clinicians interpret the functional consequences of mtDNA variants, identify disease-causing mutations, and gain insights into the mitochondrial genome's role in human health and disease. By combining computational approaches with experimental data and available databases, these tools facilitate the efficient analysis and interpretation of mtDNA sequences. Computational prediction tools attempt to calculate the function and significance of new variants by aligning them with conserved regions or structural features of other known variants. The prediction efficiency of pathogenic mtDNA variants varies among bioinformatic tools as the algorithm for prediction for each tool is different. There are four categories used for estimating pathogenicity: benign, medium, damaging, and no prediction (Bris et al., 2018).

Most prediction tools are designed according to nDNA variant features, and they do not work efficiently on mtDNA variant prediction. As in the case of mtDNA-focused data banks,

TABLE 4.3 Online resources for mtDNA variants.

tRNA and tRFs		
MINTbase 2.0	tRF data	Pliatsika et al. (2018)
MINTmap		Loher et al. (2017)
mitotRNAdb	tRNA data	Jü et al. (2009)
Gene expression		
MITOMAP	Population data, variability data	Brandon et al. (2005)
Protein expression		
HmtDB	Population data, variability data	Clima et al. (2017)
MitoCarta 2.0	Mitochondrial proteins	
MitoMiner 4.0	Mitochondrial proteins, variability data	
IMPI		
MitoPhAST		
Nuclear DNA		
MitoCarta 2.0		Calvo et al. (2016)
MSeqDR		Shen et al. (2016)
IMPI		Smith and Robinson (2019)
mtDNA		
MITOMASTER		Brandon et al. (2009)
mtDB—human	Population data	Ingman and Gyllensten (2006)
Mitochondrial genome database		Attimonelli et al. (2000)
ChloroMitoSSRDB 2.0		Sablok et al. (2015)
PhyloTree mt	Variability data, evolutionary data	van Oven and Kayser (2009)
MitoAge		Toren et al. (2016)
MitoPhAST		Tan et al. (2015)
MamMiBase		Vasconcelos et al. (2005)
H-mito		Shen-Gunther et al. (2023)
MitoSeek		Guo et al. (2013)
HAPLOFIND		Vianello et al. (2013)

(Continued)

TABLE 4.3 Online resources for mtDNA variants.—cont'd

HaploGrep2		Weissensteiner, Pacher, et al. (2016)
Phy-Mer		Navarro-Gomez et al. (2015)
MtDNA—Server		Weissensteiner, Forer, et al. (2016)
MITOMAP	Population data, variability data	Brandon et al. (2005)
MitoBreak	Variability data	Damas et al. (2014)
MSeqDR	Variability data	Shen et al. (2016)

bioinformatics tools that are created with dedicated methods for mtDNA variants and machine learning methods give more reliable results (Table 4.4). Widely used computational prediction tools for mtDNA sequences are:

1. **MitoTool:** MitoTool is a comprehensive computational framework that provides a range of functionalities for analyzing mtDNA sequences. It includes tools for alignment, variant calling, haplogroup classification, and prediction of functional consequences of mtDNA variants. MitoTool integrates various algorithms and databases to assist in the interpretation of mtDNA sequence data generated by NGS (Fan & Yao, 2011).
2. **MToolBox:** MToolBox is a user-friendly software package designed for the analysis of human mtDNA sequences. It incorporates modules for variant calling, haplogroup assignment, and pathogenicity prediction of mtDNA variants. MToolBox also provides annotations, such as allele frequencies and conservation scores, to aid in the interpretation of identified mtDNA variants (Calabrese et al., 2014).

TABLE 4.4 Online prediction tools.

Tool		References
APOGEE	Coding variant prediction	Bianco et al. (2023)
MToolbox	Coding variant prediction	Calabrese et al. (2014)
Mitimpact2	Coding variant prediction	Castellana et al. (2015)
Mitoclass.1	Coding variant prediction	Martí et al. (2017)
MITOTIP	tRNA variant prediction	Sonney et al. (2017)
PON-mt-tRNA	tRNA variant prediction	Niroula and Vihinen (2016)
Haplogrep2	Haplogroup prediction	Weissensteiner, Pacher, et al. (2016)

3. **HmtVar:** HmtVar is a database and prediction tool that focuses on mitochondrial variants associated with human diseases. It integrates information from public databases and literature to curate a comprehensive collection of disease-associated mtDNA variants. HmtVar offers functionalities for variant annotation, pathogenicity prediction, and visualization of variant distribution across different populations (Preste et al., 2019).
4. **PolyPhen-2:** PolyPhen-2 is a widely used tool for predicting the functional impact of amino acid substitutions in proteins, including those encoded by mtDNA. It employs a combination of sequence-based and structure-based features to assess the potential pathogenicity of mtDNA variants. PolyPhen-2 scores variants as benign, possibly damaging, or probably damaging, aiding in the identification of functional mtDNA variants (Adzhubei et al., 2013).
5. **SIFT:** SIFT (Sorting Intolerant from Tolerant) is another popular tool for predicting the impact of amino acid substitutions on protein function. It employs evolutionary conservation principles to determine the likelihood of a variant being damaging or tolerated. SIFT scores mtDNA variants as deleterious or tolerated, helping prioritize functional mtDNA variants for further investigation (Sim et al., 2012).

In silico interpretation of mtDNA sequences is a powerful approach that enables researchers to explore the functional implications of mtDNA variants. By employing computational tools, sequence alignment, functional prediction algorithms, and mitochondrial disease databases, researchers can uncover the genetic basis of mitochondrial diseases, unravel the evolutionary history of human populations, and contribute to personalized medicine. Further advancements in computational tools and bioinformatics resources will enhance our understanding of the mitochondrial genome and its role in human health and disease.

Conclusion

Heteroplasmy, characterized by the coexistence of multiple mtDNA variants, has been a subject of extensive research in mitochondrial genetics. Mitochondrial homoplasmy and heteroplasmy represent two distinct states of mtDNA variation, each carrying significant implications for human health and genetics. While homoplasmy is the default state, heteroplasmy introduces genetic diversity within mitochondrial populations, influencing disease manifestation, inheritance patterns, and tissue-specific effects. The leakage of paternal mtDNA during fertilization adds another layer of complexity to this phenomenon. While the mechanisms behind paternal mtDNA leakage are still being elucidated, low-level heteroplasmy resulting from paternal mtDNA can occur in offspring. This discovery challenges conventional assumptions about maternal inheritance and raises intriguing questions about the functional implications and population genetics of paternal mtDNA leakage. Advancements in analysis techniques have facilitated the exploration of homoplasmy and heteroplasmy, shedding light on their role in mitochondrial diseases and their potential as therapeutic targets. By unraveling the complexities of mitochondrial DNA variation, researchers continue to expand our understanding of human genetics and pave the way for improved diagnostics and treatments in mitochondrial disorders. Furthermore, computational tools have become indispensable for analyzing and interpreting mtDNA sequences generated by NGS technologies. These tools offer functionalities for variant calling, haplogroup classification, pathogenicity prediction, and visualization of mtDNA variants. By leveraging the power of computational approaches, researchers and clinicians can gain valuable insights into mitochondrial genetics, disease associations, and population diversity. Continued advancements in computational tools and integration with experimental data will further enhance our understanding of the functional implications of mtDNA variants and their role in human health and disease.

References

Adzhubei, I., Jordan, D. M., & Sunyaev, S. R. (January 2013). Predicting functional effect of human missense mutations using PolyPhen-2. *Current Protocols in Human Genetics*. https://doi.org/10.1002/0471142905.hg0720s76 (Chapter 7):Unit 7.20.

Ahmad, R., & Hasan, M. Y. (2021). Next-generation sequencing technology in the diagnosis of mitochondrial disorders. *International Journal of Health Sciences, 15*(1), 1–2.

Al Rawi, S., Louvet-Vallée, S., Djeddi, A., Sachse, M., Culetto, E., Hajjar, C., Boyd, L., Legouis, R., & Galy, V. (November 25, 2011). Postfertilization autophagy of sperm organelles prevents paternal mitochondrial DNA transmission. *Science, 334*(6059), 1144–1147. https://doi.org/10.1126/science.1211878. Epub 2011 Oct 27. PMID: 22033522.

Allegra, E., Garozzo, A., Lombardo, N., De Clemente, M., & Carey, T. E. (August 2006). Mutations and polymorphisms in mitochondrial DNA in head and neck cancer cell lines. *Acta Otorhinolaryngologica Italica, 26*(4), 185–190. PMID: 18236634; PMCID: PMC2639997.

Anderson, S., Bankier, A. T., Barrell, B. G., de Bruijn, M. H., Coulson, A. R., Drouin, J., Eperon, I. C., Nierlich, D. P., Roe, B. A., Sanger, F., Schreier, P. H., Smith, A. J., Staden, R., & Young, I. G. (April 9, 1981). Sequence and organization of the human mitochondrial genome. *Nature, 290*(5806), 457–465. https://doi.org/10.1038/290457a0. PMID: 7219534.

Arenas, J., Campos, Y., Bornstein, B., Ribacoba, R., Martin, M. A., Rubio, J. C., Santorelli, F. M., Zeviani, M., DiMauro, S., & Garesse, R. (1999). A double mutation (A8296G and G8363A) in the mitochondrial DNA tRNA (Lys) gene associated with myoclonus epilepsy with ragged-red fibers. *Neurology, 52*(2), 377–382.

Aryaman, J., Johnston, I.G., Jones, N.S. (2019). Mitochondrial heterogeneity. *Frontiers in Genetics* 9, 718. https://doi.org/10.3389/fgene.2018.00718. PMID: 30740126; PMCID: PMC6355694.

Ashley, M. V., Laipis, P. J., & Hauswirth, W. W. (September 25, 1989). Rapid segregation of heteroplasmic bovine mitochondria. *Nucleic Acids Research, 17*(18), 7325–7331. https://doi.org/10.1093/nar/17.18.7325. PMID: 2798094; PMCID: PMC334812.

Attimonelli, M., Altamura, N., Benne, R., Brennicke, A., Cooper, J. M., D'Elia, D., Montalvo, A., Pinto, B., De Robertis, M., Golik, P., Knoop, V., Lanave, C., Lazowska, J., Licciulli, F., Malladi, B. S., Memeo, F., Monnerot, M., Pasimeni, R., Pilbout, S., … Saccone, C. (January 1, 2000). MitBASE : A comprehensive and integrated mitochondrial DNA database. The present status. *Nucleic Acids Research, 28*(1), 148–152. https://doi.org/10.1093/nar/28.1.148. PMID: 10592207; PMCID: PMC102423.

Bacino, C., Prezant, T. R., Bu, X., Fournier, P., & Fischel-Ghodsian, N. (1995). Susceptibility mutations in the mitochondrial small ribosomal RNA gene in aminoglycoside induced deafness. *Pharmacogenetics, 5*(3), 165–172.

Balasubramaniam, S., Lewis, B., Mock, D. M., Said, H. M., Tarailo-Graovac, M., Mattman, A., van Karnebeek, C. D., Thorburn, D. R., Rodenburg, R. J., & Christodoulou, J. (2017). Leigh-like syndrome due to homoplasmic m.8993T>G variant with hypocitrullinemia and unusual biochemical features suggestive of multiple carboxylase deficiency (MCD). *JIMD Reports, 33*, 99–107. https://doi.org/10.1007/8904_2016_559. Erratum in: (2017) JIMD Reports, *33*, 111.

Bataillard, M., Chatzoglou, E., Rumbach, L., Sternberg, D., Tournade, A., Laforet, P., Jardel, C., Maisonobe, T., & Lombes, A. (2001). Atypical MELAS syndrome associated with a new mitochondrial tRNA glutamine point mutation. *Neurology, 56*(3), 405–407.

Benson, D. A., Karsch-Mizrachi, I., Lipman, D. J., Ostell, J., & Wheeler, D. L. (January 2008). GenBank. *Nucleic Acids Research, 36*(Database issue), D25–D30. https://doi.org/10.1093/nar/gkm929. Epub 2007 Dec 11. PMID: 18073190; PMCID: PMC2238942.

Bereiter-Hahn, J., & Vöth, M. (February 15, 1994). Dynamics of mitochondria in living cells: Shape changes, dislocations, fusion, and fission of mitochondria. *Microscopy Research and Technique, 27*(3), 198–219. https://doi.org/10.1002/jemt.1070270303. PMID: 8204911.

Bharti, S. K., Sommers, J. A., Zhou, J., Kaplan, D. L., Spelbrink, J. N., Mergny, J. L., & Brosh, R. M., Jr. (October 24, 2014). DNA sequences proximal to human mitochondrial DNA deletion breakpoints prevalent in human disease form G-quadruplexes, a class of DNA structures inefficiently unwound by the mitochondrial replicative Twinkle helicase. *Journal of Biological Chemistry, 289*(43), 29975–29993. https://doi.org/10.1074/jbc.M114.567073. Epub 2014 Sep 5. PMID: 25193669; PMCID: PMC4208006.

Bianchi, N. O., Bianchi, M. S., & Richard, S. M. (March 2001). Mitochondrial genome instability in human cancers. *Mutation Research, 488*(1), 9–23. https://doi.org/10.1016/s1383-5742(00)00063-6. PMID: 11223402.

Bianco, S. D., Parca, L., Petrizzelli, F., Biagini, T., Giovannetti, A., Liorni, N., Napoli, A., Carella, M., Procaccio, V., Lott, M. T., Zhang, S., Vescovi, A. L., Wallace, D. C., Caputo, V., & Mazza, T. (August 19, 2023). APOGEE 2: Multi-layer machine-learning model for the interpretable prediction of mitochondrial missense variants. *Nature Communications, 14*(1), 5058. https://doi.org/10.1038/s41467-023-40797-7. PMID: 37598215; PMCID: PMC10439926.

Bidooki, S. K., Johnson, M. A., Chrzanowska-Lightowlers, Z., Bindoff, L. A., & Lightowlers, R. N. (1997). Intracellular mitochondrial triplasmy in a patient with two heteroplasmic base changes. *The American Journal of Human Genetics, 60*(6), 1430–1438.

Bindoff, L. A., Howell, N., Poulton, J., McCullough, D. A., Morten, K. J., Lightowlers, R. N., Turnbull, D. M., & Weber, K. (1993). Abnormal RNA processing associated with a novel tRNA mutation in mitochondrial DNA. A potential disease mechanism. *Journal of Biological Chemistry, 268*(26), 19559–19564.

Birky, C. W. (September 1994). Relaxed and stringent genomes: Why cytoplasmic genes don't obey mendel's laws. *Journal of Heredity, 85*(5), 355–365.

Blakely, E. L., Poulton, J., Pike, M., Wojnarowska, F., Turnbull, D. M., McFarland, R., & Taylor, R. W. (2004). Childhood neurological presentation of a novel mitochondrial tRNA(Val) gene mutation. *Journal of the Neurological Sciences, 225*(1–2), 99–103.

Blin, M., Le Tallec, B., Nähse, V., Schmidt, M., Brossas, C., Millot, G. A., Prioleau, M. N., & Debatisse, M. (January 2019). Transcription-dependent regulation of replication dynamics modulates genome stability. *Nature Structural & Molecular Biology, 26*(1), 58–66. https://doi.org/10.1038/s41594-018-0170-1. Epub 2018 Dec 31. PMID: 30598553.

Brandon, M. C., Lott, M. T., Nguyen, K. C., Spolim, S., Navathe, S. B., Baldi, P., & Wallace, D. C. (January 1, 2005). MITOMAP: A human mitochondrial genome database–2004 update. *Nucleic Acids Research, 33*(Database issue), D611–D613. https://doi.org/10.1093/nar/gki079. PMID: 15608272; PMCID: PMC540033.

Brandon, M. C., Ruiz-Pesini, E., Mishmar, D., Procaccio, V., Lott, M. T., Nguyen, K. C., Spolim, S., Patil, U., Baldi, P., & Wallace, D. C. (January 2009). MITOMASTER: A bioinformatics tool for the analysis of mitochondrial DNA sequences. *Human Mutation, 30*(1), 1–6. https://doi.org/10.1002/humu.20801. PMID: 18566966; PMCID: PMC4292933.

Bris, C., Goudenege, D., Desquiret-Dumas, V., Charif, M., Colin, E., Bonneau, D., Amati-Bonneau, P., Lenaers, G., Reynier, P., & Procaccio, V. (December 11, 2018). Bioinformatics tools and databases to assess the pathogenicity of mitochondrial DNA variants in the field of next generation sequencing. *Frontiers in Genetics, 9*, 632. https://doi.org/10.3389/fgene.2018.00632. PMID: 30619459; PMCID: PMC6297213.

Brown, T. A., Cecconi, C., Tkachuk, A. N., Bustamante, C., & Clayton, D. A. (October 15, 2005). Replication of mitochondrial DNA occurs by strand displacement with alternative light-strand origins, not via a strand-coupled mechanism. *Genes & Development, 19*(20), 2466–2476. https://doi.org/10.1101/gad.1352105. PMID: 16230534; PMCID: PMC1257401.

Calabrese, C., Simone, D., Diroma, M. A., Santorsola, M., Guttà, C., Gasparre, G., Picardi, E., Pesole, G., & Attimonelli, M. (November 1, 2014). MToolBox: A highly automated pipeline for heteroplasmy annotation and prioritization analysis of human mitochondrial variants in high-throughput sequencing. *Bioinformatics, 30*(21), 3115–3117. https://doi.org/10.1093/bioinformatics/btu483. Epub 2014 Jul 14. PMID: 25028726; PMCID: PMC4201154.

Calvo, S. E., Clauser, K. R., & Mootha, V. K. (January 4, 2016). MitoCarta2.0: An updated inventory of mammalian mitochondrial proteins. *Nucleic Acids Research, 44*(D1), D1251–D1257. https://doi.org/10.1093/nar/gkv1003. Epub 2015 Oct 7. PMID: 26450961; PMCID: PMC4702768.

Calvo, S. E., & Mootha, V. K. (2010). The mitochondrial proteome and human disease. *Annual Review of Genomics and Human Genetics, 11*, 25–44. https://doi.org/10.1146/annurev-genom-082509-141720. PMID: 20690818; PMCID: PMC4397899.

Cao, L., Shitara, H., Horii, T., Nagao, Y., Imai, H., Abe, K., Hara, T., Hayashi, J., & Yonekawa, H. (March 2007). The mitochondrial bottleneck occurs without reduction of mtDNA content in female mouse germ cells. *Nature Genetics, 39*(3), 386–390. https://doi.org/10.1038/ng1970. Epub 2007 Feb 11. PMID: 17293866.

Cao, L., Shitara, H., Sugimoto, M., Hayashi, J., Abe, K., & Yonekawa, H. (December 2009). New evidence confirms that the mitochondrial bottleneck is generated without reduction of mitochondrial DNA content in early primordial germ cells of mice. *PLoS Genetics, 5*(12), e1000756. https://doi.org/10.1371/journal.pgen.1000756. Epub 2009 Dec 4. PMID: 19997484; PMCID: PMC2777314.

Casali, C., Santorelli, F. M., D'Amati, G., Bernucci, P., DeBiase, L., & DiMauro, S. (1995). A novel mtDNA point mutation in maternally inherited cardiomyopathy. *Biochemical and Biophysical Research Communications, 213*(2), 588–593.

Castellana, S., Rónai, J., & Mazza, T. (February 2015). MitImpact: An exhaustive collection of pre-computed pathogenicity predictions of human mitochondrial non-synonymous variants. *Human Mutation, 36*(2), E2413–E2422. https://doi.org/10.1002/humu.22720. Epub 2014 Dec 17. PMID: 25516408.

Chalmers, R. M., Lamont, P. J., Nelson, I., Ellison, D. W., Thomas, N. H., Harding, A. E., & Hammans, S. R. (1997). A mitochondrial DNA tRNA(Val) point mutation associated with adult-onset Leigh syndrome. *Neurology, 49*(2), 589–592.

Chatterjee, A., Mambo, E., & Sidransky, D. (August 7, 2006). Mitochondrial DNA mutations in human cancer. *Oncogene, 25*(34), 4663–4674. https://doi.org/10.1038/sj.onc.1209604. PMID: 16892080.

Chen, H., Dzitoyeva, S., & Manev, H. (September 5, 2012). Effect of valproic acid on mitochondrial epigenetics. *European Journal of Pharmacology, 690*(1–3), 51–59. https://doi.org/10.1016/j.ejphar.2012.06.019. Epub 2012 Jun 20. PMID: 22728245; PMCID: PMC3419440.

Chen, W., Zhao, H., & Li, Y. (2023). Mitochondrial dynamics in health and disease: Mechanisms and potential targets. *Signal Transduction and Targeted Therapy, 8*, 333. https://doi.org/10.1038/s41392-023-01547-9

Chestnut, B. A., Chang, Q., Price, A., Lesuisse, C., Wong, M., & Martin, L. J. (November 16, 2011). Epigenetic regulation of motor neuron cell death through DNA methylation. *Journal of Neuroscience, 31*(46), 16619–16636. https://doi.org/10.1523/JNEUROSCI.1639-11.2011. PMID: 22090490; PMCID: PMC3238138.

Chinnery, P. F., Samuels, D. C., Elson, J., & Turnbull, D. M. (October 26, 2002). Accumulation of mitochondrial DNA mutations in ageing, cancer, and mitochondrial disease: Is there a common mechanism? *Lancet, 360*(9342), 1323–1325. https://doi.org/10.1016/S0140-6736(02)11310-9. PMID: 12414225.

Choi, Y., & Chan, A. P. (August 15, 2015). PROVEAN web server: A tool to predict the functional effect of amino acid substitutions and indels. *Bioinformatics, 31*(16), 2745–2747. https://doi.org/10.1093/bioinformatics/btv195. Epub 2015 Apr 6. PMID: 25851949; PMCID: PMC4528627.

Chomyn, A., & Attardi, G. (May 9, 2003). MtDNA mutations in aging and apoptosis. *Biochemical and Biophysical Research Communications, 304*(3), 519–529. https://doi.org/10.1016/s0006-291x(03)00625-9. PMID: 12729587.

Clima, R., Preste, R., Calabrese, C., Diroma, M. A., Santorsola, M., Scioscia, G., Simone, D., Shen, L., Gasparre, G., & Attimonelli, M. (January 4, 2017). HmtDB 2016: Data update, a better performing query system and human mitochondrial DNA haplogroup predictor. *Nucleic Acids Research, 45*(D1), D698–D706. https://doi.org/10.1093/nar/gkw1066. Epub 2016 Nov 28. PMID: 27899581; PMCID: PMC5210550.

Cortopassi, G. A., & Hutchin, T. P. (1994). Germline inheritance of a rare mtDNA variant leads to greatly increased risk for Alzheimer's disease. (Abstract 857). *The American Journal of Human Genetics, 55*(Suppl. ment), A149.

Craig, K., Elliott, H. R., Keers, S. M., Lambert, C., Pyle, A., Graves, T. D., Woodward, C., Sweeney, M. G., Davis, M. B., Hanna, M. G., & Chinnery, P. F. (December 2007). Episodic ataxia and hemiplegia caused by the 8993T->C mitochondrial DNA mutation. *Journal of Medical Genetics, 44*(12), 797–799. https://doi.org/10.1136/jmg.2007.052902. PMID: 18055910; PMCID: PMC2652821.

Craven, L., Alston, C. L., Taylor, R. W., & Turnbull, D. M. (August 31, 2017). Recent advances in mitochondrial disease. *Annual Review of Genomics and Human Genetics, 18*, 257–275. https://doi.org/10.1146/annurev-genom-091416-035426. Epub 2017 Apr 17. PMID: 28415858.

Dahal, S., Siddiqua, H., Sharma, S., Babu, R. K., Rathore, D., Sharma, S., & Raghavan, S. C. (November 17, 2022). Unleashing a novel function of Endonuclease G in mitochondrial genome instability. *Elife, 11*, e69916. https://doi.org/10.7554/eLife.69916. PMID: 36394256; PMCID: PMC9711528.

Damas, J., Carneiro, J., Amorim, A., & Pereira, F. (January 2014). MitoBreak: The mitochondrial DNA breakpoints database. *Nucleic Acids Research, 42*(Database issue), D1261–D1268. https://doi.org/10.1093/nar/gkt982. Epub 2013 Oct 28. PMID: 24170808; PMCID: PMC3965124.

Damas, J., Carneiro, J., Gonçalves, J., Stewart, J. B., Samuels, D. C., Amorim, A., & Pereira, F. (September 2012). Mitochondrial DNA deletions are associated with non-B DNA conformations. *Nucleic Acids Research, 40*(16), 7606–7621. https://doi.org/10.1093/nar/gks500. Epub 2012 May 31. PMID: 22661583; PMCID: PMC3439893.

Darin, N., Kollberg, G., Moslemi, A. R., Tulinius, M., Holme, E., Gronlund, M. A., Andersson, S., & Oldfors, A. (2006). Mitochondrial myopathy with exercise intolerance and retinal dystrophy in a sporadic patient with a G583A mutation in the mt tRNA(phe) gene. *Neuromuscular Disorders, 16*(8), 504–506.

das Neves, R. P., Jones, N. S., Andreu, L., Gupta, R., Enver, T., & Iborra, F. J. (December 14, 2010). Connecting variability in global transcription rate to mitochondrial variability. *PLoS Biology, 8*(12), e1000560. https://doi.org/10.1371/journal.pbio.1000560. PMID: 21179497; PMCID: PMC3001896.

Deschauer, M., Swalwell, H., Strauss, M., Zierz, S., & Taylor, R. W. (2006). Novel mitochondrial transfer RNA(Phe) gene mutation associated with late-onset neuromuscular disease. *Archives of Neurology, 63*(6), 902–905.

Dimri, G. P. (June 2005). What has senescence got to do with cancer? *Cancer Cell, 7*(6), 505–512. https://doi.org/10.1016/j.ccr.2005.05.025. PMID: 15950900; PMCID: PMC1769521.

Diot, A., Dombi, E., Lodge, T., Liao, C., Morten, K., Carver, J., Wells, D., Child, T., Johnston, I. G., Williams, S., & Poulton, J. (August 15, 2016). Modulating mitochondrial quality in disease transmission: Towards enabling mitochondrial DNA disease carriers to have healthy children. *Biochemical Society Transactions, 44*(4), 1091–1100. https://doi.org/10.1042/BST20160095. PMID: 27528757; PMCID: PMC4984448.

Dong, D. W., Pereira, F., Barrett, S. P., Kolesar, J. E., Cao, K., Damas, J., Yatsunyk, L. A., Johnson, F. B., & Kaufman, B. A. (August 13, 2014). Association of G-quadruplex forming sequences with human mtDNA deletion breakpoints. *BMC Genomics, 15*(1), 677. https://doi.org/10.1186/1471-2164-15-677. PMID: 25124333; PMCID: PMC4153896.

Dong, Y., Yoshitomi, T., Hu, J. F., & Cui, J. (August 23, 2017). Long noncoding RNAs coordinate functions between mitochondria and the nucleus. *Epigenetics & Chromatin, 10*(1), 41. https://doi.org/10.1186/s13072-017-0149-x. PMID: 28835257; PMCID: PMC5569521.

Duarte, F. V., Palmeira, C. M., & Rolo, A. P. (September 26, 2014). The role of microRNAs in mitochondria: Small players acting wide. *Genes, 5*(4), 865–886. https://doi.org/10.3390/genes5040865. PMID: 25264560; PMCID: PMC4276918.

Dzitoyeva, S., Chen, H., & Manev, H. (December 2012). Effect of aging on 5-hydroxymethylcytosine in brain mitochondria. *Neurobiology of Aging, 33*(12), 2881–2891. https://doi.org/10.1016/j.neurobiolaging.2012.02.006. Epub 2012 Mar 22. PMID: 22445327; PMCID: PMC3462297.

Elliott, H. R., Samuels, D. C., Eden, J. A., Relton, C. L., & Chinnery, P. F. (August 2008). Pathogenic mitochondrial DNA mutations are common in the general population. *The American Journal of Human Genetics, 83*(2), 254–260. https://doi.org/10.1016/j.ajhg.2008.07.004. PMID: 18674747; PMCID: PMC2495064.

Elson, J. L., Samuels, D. C., Turnbull, D. M., & Chinnery, P. F. (March 2001). Random intracellular drift explains the clonal expansion of mitochondrial DNA mutations with age. *The American Journal of Human Genetics, 68*(3), 802–806. https://doi.org/10.1086/318801. Epub 2001 Feb 6. PMID: 11179029; PMCID: PMC1274494.

Fan, L., & Yao, Y. G. (March 2011). MitoTool: A web server for the analysis and retrieval of human mitochondrial DNA sequence variations. *Mitochondrion, 11*(2), 351–356. https://doi.org/10.1016/j.mito.2010.09.013. Epub 2010 Oct 7. PMID: 20933105.

Fan, L. H., Wang, Z. B., Li, Q. N., Meng, T. G., Dong, M. Z., Hou, Y., Ouyang, Y. C., Schatten, H., & Sun, Q. Y. (June 11, 2019). Absence of mitochondrial DNA methylation in mouse oocyte maturation, aging and early embryo development. *Biochemical and Biophysical Research Communications, 513*(4), 912–918. https://doi.org/10.1016/j.bbrc.2019.04.100. Epub 2019 Apr 17. PMID: 31005257.

Fernandes, P., Miotto, B., Saint-Ruf, C., Said, M., Barra, V., Nähse, V., Ravera, S., Cappelli, E., & Naim, V. (January 29, 2021). FANCD2 modulates the mitochondrial stress response to prevent common fragile site instability. *Communications Biology, 4*(1), 127. https://doi.org/10.1038/s42003-021-01647-8. PMID: 33514811; PMCID: PMC7846573.

Fischel-Ghodsian, N., Prezant, T. R., Bu, X., & Oztas, S. (1993). Mitochondrial ribosomal RNA gene mutation in a patient with sporadic aminoglycoside ototoxicity. *American Journal of Otolaryngology, 14*(6), 399–403.

Fontana, G. A., & Gahlon, H. L. (November 18, 2020). Mechanisms of replication and repair in mitochondrial DNA deletion formation. *Nucleic Acids Research, 48*(20), 11244–11258. https://doi.org/10.1093/nar/gkaa804. PMID: 33021629; PMCID: PMC7672454.

Fu, K., Hartlen, R., Johns, T., Genge, A., Karpati, G., & Shoubridge, E. A. (1996). A novel heteroplasmic tRNAleu(-CUN) mtDNA point mutation in a sporadic patient with mitochondrial encephalomyopathy segregates rapidly in skeletal muscle and suggests an approach to therapy. *Human Molecular Genetics, 5*(11), 1835–1840.

Ghildiyal, M., & Zamore, P. D. (February 2009). Small silencing RNAs: An expanding universe. *Nature Reviews Genetics, 10*(2), 94–108. https://doi.org/10.1038/nrg2504. PMID: 19148191; PMCID: PMC2724769.

Ghiselli, F., & Milani, L. (January 20, 2020). Linking the mitochondrial genotype to phenotype: A complex endeavour. *Philosophical Transactions of the Royal Society of London B Biological Sciences, 375*(1790), 20190169. https://doi.org/10.1098/rstb.2019.0169. Epub 2019 Dec 2. PMID: 31787041; PMCID: PMC6939375.

Giles, R. E., Blanc, H., Cann, H. M., & Wallace, D. C. (November 1980). Maternal inheritance of human mitochondrial DNA. *Proceedings of the National Academy of Sciences of the U S A, 77*(11), 6715–6719. https://doi.org/10.1073/pnas.77.11.6715. PMID: 6256757; PMCID: PMC350359.

Gloria Pang, S. W., Chih Lee, H. H., Ng Wing Kei, C., Yau, E. K. C., & Hui, J. (2022). Kearns-sayre syndrome minus: Two cases of identical large-scale mitochondrial DNA deletions with presentations outside the classical triad. *Case Rep Genet, 23*, 4153357. https://doi.org/10.1155/2022/4153357. PMID: 35502402; PMCID: PMC9056216.

Glover, T. W., Wilson, T. E., & Arlt, M. F. (July 25, 2017). Fragile sites in cancer: More than meets the eye. *Nature Reviews Cancer, 17*(8), 489–501. https://doi.org/10.1038/nrc.2017.52. PMID: 28740117; PMCID: PMC5546318.

Goto, Y. (1995). Clinical features of MELAS and mitochondrial DNA mutations. *Muscle & Nerve, 3*(12), S107–S112.

Goto, Y., Nonaka, I., & Horai, S. (December 13, 1990). A mutation in the tRNA(Leu)(UUR) gene associated with the MELAS subgroup of mitochondrial encephalomyopathies. *Nature, 348*(6302), 651–653. https://doi.org/10.1038/348651a0. PMID: 2102678.

Goto, Y., Nonaka, I., & Horai, S. (1991). A new mtDNA mutation associated with mitochondrial myopathy, encephalopathy, lactic acidosis and stroke-like episodes (MELAS). *Biochimica et Biophysica Acta, 1097*(3), 238–240.

Goto, Y., Tojo, M., Tohyama, J., Horai, S., & Nonaka, I. (1992). A novel point mutation in the mitochondrial tRNA-Leu(UUR) gene in a family with mitochondrial myopathy. *Annals of Neurology, 31*(6), 672–675.

Goto, Y., Tsugane, K., Tanabe, Y., Nonaka, I., & Horai, S. (1994). A new point mutation at nucelotide pair 3291 of the tRNALeu(UUR) gene in a patient with mitochondrial myopathy, encephalopathy, lactic acidosis, and stroke-like episodes (MELAS). *Biochemical and Biophysical Research Communications, 202*(3), 1624–1630.

Grünewald, T., Porschke, H., Goebel, H., Reichmann, H., & Seibel, P. (1999). Chronic progressive external ophthalmoplegia with a novel mitochondrial DNA deletion and a mutation in the tRNA (Leu(UUR)) gene. *Drug Development Research, 46*(1), 80–85.

Grafakou, O., Hol, F. A., Otfried Schwab, K., Siers, M. H., ter Laak, H., Trijbels, F., Ensenauer, R., Boelen, C., & Smeitink, J. (2003). Exercise intolerance, muscle pain and lactic acidaemia associated with a 7497G>A mutation in the tRNASer(UCN) gene. *Journal of Inherited Metabolic Disease, 26*(6), 593–600.

Grasso, M., Diegoli, M., Brega, A., Campana, C., Tavazzi, L., & Arbustini, E. (2001). The mitochondrial DNA mutation T12297C affects a highly conserved nucleotide of tRNA(Leu(CUN)) and is associated with dilated cardiomyopathy. *European Journal of Human Genetics, 9*(4), 311–315.

Gu, L. F., Chen, J. Q., Lin, Q. Y., & Yang, Y. Z. (July 26, 2021). Roles of mitochondrial unfolded protein response in mammalian stem cells. *World Journal of Stem Cells, 13*(7), 737–752. https://doi.org/10.4252/wjsc.v13.i7.737. PMID: 34367475; PMCID: PMC8316864.

Guaran, V., Astolfi, L., Castiglione, A., Simoni, E., Olivetto, E., Galasso, M., Trevisi, P., Busi, M., Volinia, S., & Martini, A. (October 2013). Association between idiopathic hearing loss and mitochondrial DNA mutations: A study on 169 hearing-impaired subjects. *International Journal of Molecular Medicine, 32*(4), 785–794. https://doi.org/10.3892/ijmm.2013.1470. Epub 2013 Aug 16. PMID: 23969527; PMCID: PMC3812239.

Guo, Y., Li, J., Li, C. I., Shyr, Y., & Samuels, D. C. (May 1, 2013). MitoSeek: Extracting mitochondria information and performing high-throughput mitochondria sequencing analysis. *Bioinformatics, 29*(9), 1210–1211. https://doi.org/10.1093/bioinformatics/btt118. Epub 2013 Mar 6. PMID: 23471301; PMCID: PMC4492415.

Hanna, M. G., Nelson, I. P., Morgan-Hughes, J. A., & Wood, N. W. (1998). MELAS: A new disease associated mitochondrial DNA mutation and evidence for further genetic heterogeneity. *Journal of Neurology Neurosurgery and Psychiatry, 65*(4), 512–517.

Hanna, M. G., Nelson, I., Sweeney, M. G., Cooper, J. M., Watkins, P. J., Morgan-Hughes, J. A., & Harding, A. E. (1995). Congenital encephalomyopathy and adult-onset myopathey and diabetes mellitus: Different phenotypic associations of a new heteroplasmic mtDNA tRNA glutamic acid mutation. *The American Journal of Human Genetics, 56*(5), 1026–1033.

Hao, H., Bonilla, E., Manfredi, G., DiMauro, S., & Moraes, C. T. (1995). Segregation patterns of a novel mutation in the mitochondrial tRNA glutamic acid gene associated with myopathy and diabetes mellitus. *The American Journal of Human Genetics, 56*(5), 1017–1025.

Hao, H., & Moraes, C. T. (1997). A disease-associated G5703A mutation in human mitochondrial DNA causes a conformational change and a marked decrease in steady-state levels of mitochondrial tRNA(Asn). *Molecular and Cellular Biology, 17*(12), 6831–6837.

Hardy, S. A., Blakely, E. L., Purvis, A. I., Rocha, M. C., Ahmed, S., Falkous, G., Poulton, J., Rose, M. R., O'Mahony, O., Bermingham, N., Dougan, C. F., Ng, Y. S., Horvath, R., Turnbull, D. M., Gorman, G. S., & Taylor, R. W. (June 23, 2016). Pathogenic mtDNA mutations causing mitochondrial myopathy: The need for muscle biopsy. *Neurol Genet., 2*(4), e82. https://doi.org/10.1212/NXG.0000000000000082. PMID: 27536729; PMCID: PMC4972142.

Hayashi, J., Ohta, S., Kagawa, Y., Takai, D., Miyabayashi, S., Tada, K., Fukushima, H., Inui, K., Okada, S., Goto, Y., & Nonaka, I. (1994). Functional and morphological abnormalities of mitochondria in human cells containing mitochondrial DNA with pathogenic point mutations in tRNA genes. *Journal of Biological Chemistry, 269*(29), 19060–19066.

Hayashi, J., Ohta, S., Takai, D., Miyabayashi, S., Sakuta, R., Goto, Y., & Nonaka, I. (1993). Accumulation of mtDNA with a mutation at position 3271 in tRNALeu(UUR) gene introduced from a MELAS patient to HeLa cells lacking mtDNA results in progressive inhibition of mitochondrial respiratory function. *Biochemical and Biophysical Research Communications, 197*(3), 1049–1055.

Herbst, A., Hoang, A. N., Woo, W., McKenzie, D., Aiken, J. M., Miller, R. A., Allison, D. B., Liu, N., & Wanagat, J. (September 2019). Mitochondrial DNA alterations in aged macrophage migration inhibitory factor-knockout mice. *Mechanism of Ageing and Development, 182*, 111126. https://doi.org/10.1016/j.mad.2019.111126. Epub 2019 Aug 2. PMID: 31381889; PMCID: PMC6718337.

Herbst, A., Lee, C. C., Vandiver, A. R., Aiken, J. M., McKenzie, D., Hoang, A., Allison, D., Liu, N., & Wanagat, J. (July 2021). Mitochondrial DNA deletion mutations increase exponentially with age in human skeletal muscle. *Aging-Clinical & Experimental Research, 33*(7), 1811–1820. https://doi.org/10.1007/s40520-020-01698-7. Epub 2020 Sep 23. PMID: 32965609; PMCID: PMC7985047.

Holloman, C. M., Wolfe, L. A., Gahl, W. A., & Boerkoel, C. F. (February 18, 2013). Kearns-Sayre syndrome presenting as isolated growth failure. *BMJ Case Reports*. https://doi.org/10.1136/bcr-2012-007272. PMID: 23420719; PMCID: PMC3604337.

Holt, I. J., Harding, A. E., Petty, R. K., & Morgan-Hughes, J. A. (March 1990). A new mitochondrial disease associated with mitochondrial DNA heteroplasmy. *The American Journal of Human Genetics, 46*(3), 428–433. PMID: 2137962; PMCID: PMC1683641.

Hopkins, J. F., Sabelnykova, V. Y., Weischenfeldt, J., Simon, R., Aguiar, J. A., Alkallas, R., Heisler, L. E., Zhang, J., Watson, J. D., Chua, M. L. K., Fraser, M., Favero, F., Lawerenz, C., Plass, C., Sauter, G., McPherson, J. D., van der Kwast, T., Korbel, J., Schlomm, T., Bristow, R. G., & Boutros, P. C. (September 22, 2017). Mitochondrial mutations drive prostate cancer aggression. *Nature Communications, 8*(1), 656. https://doi.org/10.1038/s41467-017-00377-y. PMID: 28939825; PMCID: PMC5610241.

Hopper, R. K., Carroll, S., Aponte, A. M., Johnson, D. T., French, S., Shen, R. F., Witzmann, F. A., Harris, R. A., & Balaban, R. S. (February 28, 2006). Mitochondrial matrix phosphoproteome: Effect of extra mitochondrial calcium. *Biochemistry, 45*(8), 2524–2536. https://doi.org/10.1021/bi052475e. PMID: 16489745; PMCID: PMC1415274.

Horvath, R., Kemp, J. P., Tuppen, H. A., Hudson, G., Oldfors, A., et al. (2009). Molecular basis of infantile reversible cytochrome c oxidase deficiency myopathy. *Brain, 132*(Pt 11), 3165–3174.

Hutchin, T., & Cortopassi, G. (1995). A mitochondrial DNA clone is associated with increased risk for Alzheimer disease. *Proceedings of the National Academy of Sciences of the United States of America, 92*(15), 6892–6895.

Hutchin, T., Haworth, I., Higashi, K., Fischel-Ghodsian, N., Stoneking, M., Saha, N., Arnos, C., & Cortopassi, G. (1993). A molecular basis for human hypersensitivity to aminoglycoside antibiotics. *Nucleic Acids Research, 21*(18), 4174–4179.

Ingman, M., & Gyllensten, U. (January 1, 2006). mtDB: Human Mitochondrial Genome Database, a resource for population genetics and medical sciences. *Nucleic Acids Research, 34*(Database issue), D749–D751. https://doi.org/10.1093/nar/gkj010. PMID: 16381973; PMCID: PMC1347373.

Iriondo, M. N., Etxaniz, A., Varela, Y. R., Ballesteros, U., Hervás, J. H., Montes, L. R., Goñi, F. M., & Alonso, A. (December 2022). LC3 subfamily in cardiolipin-mediated mitophagy: A comparison of the LC3A, LC3B and LC3C homologs. *Autophagy, 18*(12), 2985–3003. https://doi.org/10.1080/15548627.2022.2062111. Epub 2022 Apr 13. PMID: 35414338; PMCID: PMC9673933.

Ito, T., Hattori, K., Obayashi, T., Tanaka, M., Sugiyama, S., & Ozawa, T. (1992). Mitochondrial DNA mutations in cardiomyopathy. *Japanese Circulation Journal, 56*(10), 1045–1053.

Jühling, F., Mörl, M., Hartmann, R. K., Sprinzl, M., Stadler, P. F., & Pütz, J. (January 2009). tRNAdb 2009: Compilation of tRNA sequences and tRNA genes. *Nucleic Acids Research, 37*(Database issue), D159–D162. https://doi.org/10.1093/nar/gkn772. Epub 2008 Oct 28. PMID: 18957446; PMCID: PMC2686557.

Jaksch, M., Hofmann, S., Kleinle, S., Liechti-Gallati, S., Pongratz, D. E., Muller-Hocker, J., Jedele, K. B., Meitinger, T., & Gerbitz, K. D. (1998). A systematic mutation screen of 10 nuclear and 25 mitochondrial candidate genes in 21 patients with cytochrome c oxidase (COX) deficiency shows tRNA(Ser)(UCN) mutations in a subgroup with syndromal encephalopathy. *Journal of Medical Genetics, 35*(11), 895–900.

Jaksch, M., Klopstock, T., Kurlemann, G., Dorner, M., Hofmann, S., Kleinle, S., Hegemann, S., Weissert, M., Muller-Hocker, J., Pongratz, D., & Gerbitz, K. D. (1998). Progressive myoclonus epilepsy and mitochondrial myopathy associated with mutations in the tRNA(Ser(UCN)) gene. *Annals of Neurology, 44*(4), 635–640.

Jean-Francois, M. J., Lertrit, P., Berkovic, S. F., Crimmins, D., Morris, J., Marzuki, S., & Byrne, E. (1994). Heterogeneity in the phenotypic expression of the mutation in the mitochondrial tRNALeu(UUR) gene generally associated with the MELAS subset of mitochondrial encephalomyopathies. *Australian & New Zealand Journal of Medicine, 24*(2), 188–193.

Jin, Y., Yu, Q., Zhou, D., Chen, L., Huang, X., Xu, G., Huang, J., Gao, X., Gao, Y., & Shen, L. (May 2012). The mitochondrial DNA 9-bp deletion polymorphism is a risk factor for hepatocellular carcinoma in the Chinese population. *Genetic Testing and Molecular Biomarkers, 16*(5), 330–334. https://doi.org/10.1089/gtmb.2011.0208. Epub 2012 Jan 27. PMID: 22283196; PMCID: PMC3354583.

Johnston, I. G. (November 20, 2019). Varied mechanisms and models for the varying mitochondrial bottleneck. *Frontiers in Cell and Developmental Biology, 7*, 294. https://doi.org/10.3389/fcell.2019.00294. PMID: 31824946; PMCID: PMC6879659.

Kameoka, K., Isotani, H., Tanaka, K., Azukari, K., Fujimura, Y., Shiota, Y., Sasaki, E., Majima, M., Furukawa, K., Haginomori, S., Kitaoka, H., & Ohsawa, N. (1998). Novel mitochondrial DNA mutation in tRNA(Lys) (8296A–>G) associated with diabetes. *Biochemical and Biophysical Research Communications, 245*(2), 523–527.

Kameoka, K., Isotani, H., Tanaka, K., Kitaoka, H., & Ohsawa, N. (1998). Impaired insulin secretion in Japanese diabetic subjects with an A-to-G mutation at nucleotide 8296 of the mitochondrial DNA in tRNA(Lys). *Diabetes Care, 21*(11), 2034–2035.

Kato, M., Nakamura, M., Ichiba, M., Tomiyasu, A., Shimo, H., Higuchi, I., Ueno, S., & Sano, A. (April 2011). Mitochondrial DNA deletion mutations in patients with neuropsychiatric symptoms. *Neurosciences Research, 69*(4), 331–336. https://doi.org/10.1016/j.neures.2010.12.013. Epub 2010 Dec 24. PMID: 21185889.

Kaushal, S., & Freudenreich, C. H. (May 2019). The role of fork stalling and DNA structures in causing chromosome fragility. *Genes Chromosomes & Cancer, 58*(5), 270–283. https://doi.org/10.1002/gcc.22721. Epub 2019 Jan 29. PMID: 30536896; PMCID: PMC7083089.

Khailany, R., Gilani, N., Ozaslan, M., Safdar, M., Al-Shamari, I., Kanabe, B., Saadat, K., Homayounvash, J., Monfaredan, A., Al-Attar, M., & Arslan, A. (2020). Association of a point mutation (m.9176T>G) of the MT-ATP6 gene with Leigh syndrome: A case report. *Biomedical Research and Therapy, 7*(5), 3739–3743.

Kirino, Y., Yasukawa, T., Ohta, S., Akira, S., Ishihara, K., Watanabe, K., & Suzuki, T. (October 19, 2004). Codon-specific translational defect caused by a wobble modification deficiency in mutant tRNA from a human mitochondrial disease. *Proceedings of the National Academy of Sciences of the U S A, 101*(42), 15070–15075. https://doi.org/10.1073/pnas.0405173101. Epub 2004 Oct 11. PMID: 15477592; PMCID: PMC524061.

Kirino, Y., Goto, Y. I., Campos, Y., Arenas, J., & Suzuki, T. (2005). Specific correlation between the wobble modification deficiency in mutant tRNAs and the clinical features of a human mitochondrial disease. *Proceedings of the National Academy of Sciences of the United States of America, 102*(20), 7127–7132.

Kleinle, S., Schneider, V., Moosmann, P., Brandner, S., Krahenbuhl, S., & Liechti-Gallati, S. (1998). A novel mitochondrial tRNA(Phe) mutation inhibiting anticodon stem formation associated with a muscle disease. *Biochemical and Biophysical Research Communications, 247*(1), 112–115.

Konings, A., Van Camp, G., Goethals, A., Van Eyken, E., Vandevelde, A., Ben Azza, J., Peeters, N., Wuyts, W., Smeets, H., & Van Laer, L. (2008). Mutation analysis of mitochondrial DNA 12SrRNA and tRNASer(UCN) genes in non-syndromic hearing loss patients. *Mitochondrion, 8*(5–6), 377–382.

Kriaucionis, S., & Heintz, N. (May 15, 2009). The nuclear DNA base 5-hydroxymethylcytosine is present in Purkinje neurons and the brain. *Science, 324*(5929), 929–930. https://doi.org/10.1126/science.1169786. Epub 2009 Apr 16. PMID: 19372393; PMCID: PMC3263819.

Krishnan, K. J., Reeve, A. K., Samuels, D. C., Chinnery, P. F., Blackwood, J. K., Taylor, R. W., Wanrooij, S., Spelbrink, J. N., Lightowlers, R. N., & Turnbull, D. M. (March 2008). What causes mitochondrial DNA deletions in human cells? *Nature Genetics, 40*(3), 275–279. https://doi.org/10.1038/ng.f.94. PMID: 18305478.

Landrum, M. J., Lee, J. M., Benson, M., Brown, G. R., Chao, C., Chitipiralla, S., Gu, B., Hart, J., Hoffman, D., Jang, W., Karapetyan, K., Katz, K., Liu, C., Maddipatla, Z., Malheiro, A., McDaniel, K., Ovetsky, M., Riley, G., Zhou, G., … Maglott, D. R. (January 4, 2018). ClinVar: Improving access to variant interpretations and supporting evidence. *Nucleic Acids Research, 46*(D1), D1062–D1067. https://doi.org/10.1093/nar/gkx1153. PMID: 29165669; PMCID: PMC5753237.

Lawless, C., Greaves, L., Reeve, A. K., Turnbull, D. M., & Vincent, A. E. (May 2020). The rise and rise of mitochondrial DNA mutations. *Open Biol., 10*(5), 200061. https://doi.org/10.1098/rsob.200061. Epub 2020 May 20. PMID: 32428418; PMCID: PMC7276526.

Lax, N. Z., Turnbull, D. M., & Reeve, A. K. (December 2011). Mitochondrial mutations: Newly discovered players in neuronal degeneration. *The Neuroscientist, 17*(6), 645–658. https://doi.org/10.1177/1073858411385469. PMID: 22130639; PMCID: PMC3757997.

Lee, H. C., Chang, C. M., & Chi, C. W. (November 2010). Somatic mutations of mitochondrial DNA in aging and cancer progression. *Ageing Research Reviews, 9*(1), S47–S58. https://doi.org/10.1016/j.arr.2010.08.009. Epub 2010 Sep 8. PMID: 20816876.

Lee, J. T. (August 15, 2009). Lessons from X-chromosome inactivation: Long ncRNA as guides and tethers to the epigenome. *Genes & Development, 23*(16), 1831–1842. https://doi.org/10.1101/gad.1811209. PMID: 19684108; PMCID: PMC2725936.

Leroy, D., & Norby, S. (1994). A new human mtDNA polymorphism: tRNAGln/4336)(T-C). *Clinical Genetics, 45*(2), 109–110.

Li, Z., Zhang, Z., Ren, Y., Wang, Y., Fang, J., Yue, H., Ma, S., & Guan, F. (April 2021). Aging and age-related diseases: From mechanisms to therapeutic strategies. *Biogerontology, 22*(2), 165–187. https://doi.org/10.1007/s10522-021-09910-5. Epub 2021 Jan 27. PMID: 33502634; PMCID: PMC7838467.

Li, Z., Li, R., Chen, J., Liao, Z., Zhu, Y., Qian, Y., Xiong, S., Heman-Ackah, S., Wu, J., Choo, D. I., & Guan, M. X. (2005). Mutational analysis of the mitochondrial 12S rRNA gene in Chinese pediatric subjects with aminoglycoside-induced and non-syndromic hearing loss. *Human Genetics, 117*(1), 9–15.

Liberman, N., Wang, S. Y., & Greer, E. L. (December 2019). Transgenerational epigenetic inheritance: From phenomena to molecular mechanisms. *Current Opinion in Neurobiology, 59*, 189–206. https://doi.org/10.1016/j.conb.2019.09.012. Epub 2019 Oct 18. PMID: 31634674; PMCID: PMC6889819.

Liu, B., Du, Q., Chen, L., Fu, G., Li, S., Fu, L., Zhang, X., Ma, C., & Bin, C. (March 21, 2016). CpG methylation patterns of human mitochondrial DNA. *Scientific Reports, 6*, 23421. https://doi.org/10.1038/srep23421. PMID: 26996456; PMCID: PMC4800444.

Loher, P., Telonis, A. G., & Rigoutsos, I. (February 21, 2017). MINTmap: Fast and exhaustive profiling of nuclear and mitochondrial tRNA fragments from short RNA-seq data. *Scientific Reports, 7*, 41184. https://doi.org/10.1038/srep41184. PMID: 28220888; PMCID: PMC5318995.

Lynn, S., Wardell, T., Johnson, M. A., Chinnery, P. F., Daly, M. E., Walker, M., & Turnbull, D. M. (1998). Mitochondrial diabetes: Investigation and identification of a novel mutation. *Diabetes, 47*(11), 1800–1802.

Mancuso, M., Filosto, M., Mootha, V. K., Rocchi, A., Pistolesi, S., Murri, L., DiMauro, S., & Siciliano, G. (2004). A novel mitochondrial tRNAPhe mutation causes MERRF syndrome. *Neurology, 62*(11), 2119–2121.

Maniura-Weber, K., Helm, M., Engemann, K., Eckertz, S., Mollers, M., Schauen, M., Hayrapetyan, A., von Kleist-Retzow, J. C., Lightowlers, R. N., Bindoff, L. A., & Wiesner, R. J. (2006). Molecular dysfunction associated with the human mitochondrial 3302A>G mutation in the MTTL1 (mt-tRNALeu(UUR)) gene. *Nucleic Acids Research, 34*(22), 6404–6415.

Manouvrier, S., Rotig, A., Hannebique, G., Gheerbrandt, J. D., Royer-Legrain, G., Munnich, A., Parent, M., Grunfeld, J. P., Largilliere, C., Lombes, A., & Bonnefont, J. P. (1995). Point mutation of the mitochondrial tRNALeu gene (A 3243 G) in maternally inherited hypertrophic cardiomyopathy, diabetes mellitus, renal failure, and sensorineural deafness. *Journal of Medical Genetics, 32*(8), 654–656.

Martín-Navarro, A., Gaudioso-Simón, A., Álvarez-Jarreta, J., Montoya, J., Mayordomo, E., & Ruiz-Pesini, E. (March 7, 2017). Machine learning classifier for identification of damaging missense mutations exclusive to human mitochondrial DNA-encoded polypeptides. *BMC Bioinformatics, 18*(1), 158. https://doi.org/10.1186/s12859-017-1562-7. PMID: 28270093; PMCID: PMC5341421.

Massin, P., Guillausseau, P. J., Vialettes, B., Paquis, V., Orsini, F., Grimaldi, A. D., & Gaudric, A. (1995). Macular pattern dystrophy associated with a mutation of mitochondrial DNA. *American Journal of Ophthalmology, 120*(2), 247–248.

Masucci, J. P., Davidson, M., Koga, Y., Schon, E. A., & King, M. P. (1995). In vitro analysis of mutations causing myoclonus epilepsy with ragged-red fibers in the mitochondrial tRNALysgene: Two genotypes produce similar phenotypes. *Molecular and Cellular Biology, 15*(5), 2872–2881.

McFarland, R., Clark, K. M., Morris, A. A., Taylor, R. W., Macphail, S., Lightowlers, R. N., & Turnbull, D. M. (2002). Multiple neonatal deaths due to a homoplasmic mitochondrial DNA mutation. *Nature Genetics, 30*(2), 145–146.

McFarland, R., Swalwell, H., Blakely, E. L., He, L., Groen, E. J., Turnbull, D. M., Bushby, K. M., & Taylor, R. W. (2008). The m.5650G>A mitochondrial tRNA(Ala) mutation is pathogenic and causes a phenotype of pure myopathy. *Neuromuscular Disorders, 18*(1), 63–67.

McGinnis, S., & Madden, T. L. (July 1, 2004). BLAST: At the core of a powerful and diverse set of sequence analysis tools. *Nucleic Acids Research, 32*(Web Server issue), W20–W25. https://doi.org/10.1093/nar/gkh435. PMID: 15215342; PMCID: PMC441573.

Melone, M. A., Tessa, A., Petrini, S., Lus, G., Sampaolo, S., di Fede, G., Santorelli, F. M., & Cotrufo, R. (2004). Revelation of a new mitochondrial DNA mutation(G12147A) in a MELAS/MERFF phenotype. *Archives of Neurology, 61*(2), 269–272.

Merante, F., Myint, T., Tein, I., Benson, L., & Robinson, B. H. (1996). An additional mitochondrial tRNA(Ile) point mutation (A-to-G at nucleotide 4295) causing hypertrophic cardiomyopathy. *Human Mutation, 8*(3), 216–222.

Merante, F., Tein, I., Benson, L., & Robinson, B. H. (1994). Maternally inherited hypertrophic cardiomyopathy due to a novel T-to-C transition at nucleotide 9997 in the mitochondrial tRNA(glycine) gene. *The American Journal of Human Genetics, 55*(3), 437–446.

Meulemans, A., Seneca, S., Lagae, L., Lissens, W., De Paepe, B., Smet, J., Van Coster, R., & De Meirleir, L. (2006). A novel mitochondrial transfer RNA(Asn) mutation causing multiorgan failure. *Archives of Neurology, 63*(8), 1194–1198.

Mimaki, M., Hatakeyama, H., Komaki, H., Yokoyama, M., Arai, H., Kirino, Y., Suzuki, T., Nishino, I., Nonaka, I., & Goto, Y. (2010). Reversible infantile respiratory chain deficiency: A clinical and molecular study. *Annals of Neurology, 68*(6), 845–854.

Minocherhomji, S., Tollefsbol, T. O., & Singh, K. K. (April 2012). Mitochondrial regulation of epigenetics and its role in human diseases. *Epigenetics, 7*(4), 326–334. https://doi.org/10.4161/epi.19547. Epub 2012 Apr 1. PMID: 22419065; PMCID: PMC3368816.

Mollers, M., Maniura-Weber, K., Kiseljakovic, E., Bust, M., Hayrapetyan, A., Jaksch, M., Helm, M., Wiesner, R. J., & von Kleist-Retzow, J. C. (2005). A new mechanism for mtDNA pathogenesis: Impairment of post-transcriptional maturation leads to severe depletion of mitochondrial tRNASer(UCN) caused by T7512C and G7497A point mutations. *Nucleic Acids Research, 33*(17), 5647–5658.

Moraes, C. T., Ciacci, F., Bonilla, E., Jansen, C., Hirano, M., Rao, N., Lovelace, R. E., Rowland, L. P., Schon, E. A., & DiMauro, S. (1993). Two novel pathogenic mitochondrial DNA mutations affecting organelle number and protein synthesis. Is the tRNALeu(UUR) gene an etiologic hot spot? *The Journal of Clinical Investigation, 92*(6), 2906–2915.

Moraes, C. T., Ciacci, F., Silverstri, G., Shanske, S., Sciacco, M., Hirano, M., Schon, E. A., Bonilla, E., & DiMauro, S. (1993). Atypical clinical presentations associated with the MELAS mutation at position 3243 of human mitochondrial DNA. *Neuromuscular Disorders, 3*(1), 43–50.

Morgan-Richards, M., Bulgarella, M., Sivyer, L., Dowle, E. J., Hale, M., McKean, N. E., & Trewick, S. A. (November 29, 2017). Explaining large mitochondrial sequence differences within a population sample. *Royal Society Open Science, 4*(11), 170730. https://doi.org/10.1098/rsos.170730. PMID: 29291063; PMCID: PMC5717637.

Moslemi, A. R., Lindberg, C., Toft, J., Holme, E., Kollberg, G., & Oldfors, A. (2004). A novel mutation in the mitochondrial tRNA(Phe) gene associated with mitochondrial myopathy. *Neuromuscular Disorders, 14*(1), 46–50.

Nagano, T., Mitchell, J. A., Sanz, L. A., Pauler, F. M., Ferguson-Smith, A. C., Feil, R., & Fraser, P. (December 12, 2008). The Air noncoding RNA epigenetically silences transcription by targeting G9a to chromatin. *Science, 322*(5908), 1717–1720. https://doi.org/10.1126/science.1163802. Epub 2008 Nov 6. PMID: 18988810.

Navarro-Gomez, D., Leipzig, J., Shen, L., Lott, M., Stassen, A. P., Wallace, D. C., Wiggs, J. L., Falk, M. J., van Oven, M., & Gai, X. (April 15, 2015). Phy-mer: A novel alignment-free and reference-independent mitochondrial haplogroup classifier. *Bioinformatics, 31*(8), 1310–1312. https://doi.org/10.1093/bioinformatics/btu825. Epub 2014 Dec 12. PMID: 25505086; PMCID: PMC4393525.

Nelson, I., Hanna, M. G., Alsanjari, N., Scaravilli, F., Morgan-Hughes, J. A., & Harding, A. E. (1995). A new mitochondrial DNA mutation associated with progressive dementia and chorea: A clinical, pathological, and molecular genetic study. *Annals of Neurology, 37*(3), 400–403.

Nie, Z., Wang, C., Chen, J., Ji, Y., Zhang, H., Zhao, F., Zhou, X., & Guan, M. X. (January 6, 2023). Abnormal morphology and function in retinal ganglion cells derived from patients-specific iPSCs generated from individuals with Leber's hereditary optic neuropathy. *Human Molecular Genetics, 32*(2), 231–243. https://doi.org/10.1093/hmg/ddac190. PMID: 35947995; PMCID: PMC9840204.

Niroula, A., & Vihinen, M. (March 18, 2016). PON-mt-tRNA: A multifactorial probability-based method for classification of mitochondrial tRNA variations. *Nucleic Acids Research, 44*(5), 2020–2027. https://doi.org/10.1093/nar/gkw046. Epub 2016 Feb 3. PMID: 26843426; PMCID: PMC4797295.

Nishigaki, Y., Tadesse, S., Bonilla, E., Shungu, D., Hersh, S., Keats, B. J., Berlin, C. I., Goldberg, M. F., Vockley, J., DiMauro, S., & Hirano, M. (2003). A novel mitochondrial tRNA(Leu(UUR)) mutation in a patient with features of MERRF and Kearns-Sayre syndrome. *Neuromuscular Disorders, 13*(4), 334–340.

Oliveira, P. H., da Silva, C. L., & Cabral, J. M. (2013). An appraisal of human mitochondrial DNA instability: New insights into the role of non-canonical DNA structures and sequence motifs. *PLoS One, 8*(3), e59907. https://doi.org/10.1371/journal.pone.0059907. Epub 2013 Mar 29. PMID: 23555828; PMCID: PMC3612095.

Olivo, P. D., Van de Walle, M. J., Laipis, P. J., & Hauswirth, W. W. (November 24-30, 1983). Nucleotide sequence evidence for rapid genotypic shifts in the bovine mitochondrial DNA D-loop. *Nature, 306*(5941), 400–402. https://doi.org/10.1038/306400a0. PMID: 6646218.

Otten, A. B., Theunissen, T. E., Derhaag, J. G., Lambrichs, E. H., Boesten, I. B., Winandy, M., van Montfoort, A. P., Tarbashevich, K., Raz, E., Gerards, M., Vanoevelen, J. M., van den Bosch, B. J., Muller, M., & Smeets, H. J. (July 19, 2016). Differences in strength and timing of the mtDNA bottleneck between zebrafish germline and non-germline cells. *Cell Reports, 16*(3), 622–630. https://doi.org/10.1016/j.celrep.2016.06.023. Epub 2016 Jun 30. PMID: 27373161.

Ozawa, M., Nishino, I., Horai, S., Nonaka, I., & Goto, Y. I. (1997). Myoclonus epilepsy associated with ragged-red fibers: A G-to-A mutation at nucleotide pair 8363 in mitochondrial tRNA(lys) in two families. *Muscle & Nerve, 20*(3), 271–278.

Pütz, J., Dupuis, B., Sissler, M., & Florentz, C. (August 2007). Mamit-tRNA, a database of mammalian mitochondrial tRNA primary and secondary structures. *RNA, 13*(8), 1184–1190. https://doi.org/10.1261/rna.588407. Epub 2007 Jun 21. PMID: 17585048; PMCID: PMC1924894.

Pliatsika, V., Loher, P., Magee, R., Telonis, A. G., Londin, E., Shigematsu, M., Kirino, Y., & Rigoutsos, I. (January 4, 2018). MINTbase v2.0: A comprehensive database for tRNA-derived fragments that includes nuclear and mitochondrial fragments from all the cancer genome atlas projects. *Nucleic Acids Research, 46*(D1), D152–D159. https://doi.org/10.1093/nar/gkx1075. PMID: 29186503; PMCID: PMC5753276.

Preste, R., Vitale, O., Clima, R., Gasparre, G., & Attimonelli, M. (January 8, 2019). HmtVar: A new resource for human mitochondrial variations and pathogenicity data. *Nucleic Acids Research, 47*(D1), D1202–D1210. https://doi.org/10.1093/nar/gky1024. PMID: 30371888; PMCID: PMC6323908.

Prezant, T. R., Agapian, J. V., Bohlman, M. C., Bu, X., Oztas, S., Qiu, W. Q., Arnos, K. S., Cortopassi, G. A., Jaber, L., Rotter, J. I., Shohat, M., & Fischel-Ghodsian, N. (1993). Mitochondrial ribosomal RNA mutation associated with both antibiotic-induced and non-syndromic deafness. *Nature Genetics, 4*(3), 289–294.

Rötig, A., Bourgeron, T., Chretien, D., Rustin, P., & Munnich, A. (August 1995). Spectrum of mitochondrial DNA rearrangements in the Pearson marrow-pancreas syndrome. *Human Molecular Genetics, 4*(8), 1327–1330. https://doi.org/10.1093/hmg/4.8.1327. PMID: 7581370.

Raap, A. K., Jahangir Tafrechi, R. S., van de Rijke, F. M., Pyle, A., Wählby, C., Szuhai, K., Ravelli, R. B., de Coo, R. F., Rajasimha, H. K., Nilsson, M., Chinnery, P. F., Samuels, D. C., & Janssen, G. M. (2012). Non-random mtDNA segregation patterns indicate a metastable heteroplasmic segregation unit in m.3243A>G cybrid cells. *PLoS One, 7*(12), e52080. https://doi.org/10.1371/journal.pone.0052080. Epub 2012 Dec 18. PMID: 23272214; PMCID: PMC3525564.

Rath, S., Sharma, R., Gupta, R., Ast, T., Chan, C., Durham, T. J., Goodman, R. P., Grabarek, Z., Haas, M. E., Hung, W. H. W., Joshi, P. R., Jourdain, A. A., Kim, S. H., Kotrys, A. V., Lam, S. S., McCoy, J. G., Meisel, J. D., Miranda, M., Panda, A., … Mootha, V. K. (January 8, 2021). MitoCarta3.0: An updated mitochondrial proteome now with sub-organelle localization and pathway annotations. *Nucleic Acids Research, 49*(D1), D1541–D1547. https://doi.org/10.1093/nar/gkaa1011. PMID: 33174596; PMCID: PMC7778944.

Reid, F. M., Vernham, G. A., & Jacobs, H. T. (1994). A novel mitochondrial point mutation in a maternal pedigree with sensorineural deafness. *Human Mutation, 3*(3), 243–247.

Richards, M., & Macaulay, V. (June 2001). The mitochondrial gene tree comes of age. *The American Journal of Human Genetics, 68*(6), 1315–1320. https://doi.org/10.1086/320615. Epub 2001 May 10. PMID: 11349234; PMCID: PMC1226117.

Rossignol, R., Faustin, B., Rocher, C., Malgat, M., Mazat, J. P., & Letellier, T. (March 15, 2003). Mitochondrial threshold effects. *Biochemical Journal, 370*(Pt 3), 751–762. https://doi.org/10.1042/BJ20021594. PMID: 12467494; PMCID: PMC1223225.

Rossmanith, W., Raffelsberger, T., Roka, J., Kornek, B., Feucht, M., & Bittner, R. E. (2003). The expanding mutational spectrum of MERRF substitution G8361A in the mitochondrial tRNALys gene. *Annals of Neurology, 54*(6), 820\x96823.

Ruiz-Pesini, E., Lott, M. T., Procaccio, V., Poole, J. C., Brandon, M. C., Mishmar, D., Yi, C., Kreuziger, J., Baldi, P., & Wallace, D. C. (January 2007). An enhanced MITOMAP with a global mtDNA mutational phylogeny. *Nucleic Acids Research, 35*(Database issue), D823–D828. https://doi.org/10.1093/nar/gkl927. Epub 2006 Dec 18. PMID: 17178747; PMCID: PMC1781213.

Sabella-Jiménez, V., Otero-Herrera, C., Silvera-Redondo, C., & Garavito-Galofre, P. (November 2020). Mitochondrial DNA deletion and duplication in kearns-sayre syndrome (KSS) with initial presentation as Pearson marrow-pancreas syndrome (PMPS): Two case reports in barranquilla, Colombia. *Mol Genet Genomic Med, 8*(11), e1509. https://doi.org/10.1002/mgg3.1509. Epub 2020 Oct 8. PMID: 33030289; PMCID: PMC7667363.

Sablok, G., Padma Raju, G. V., Mudunuri, S. B., Prabha, R., Singh, D. P., Baev, V., Yahubyan, G., Ralph, P. J., & La Porta, N. (September 27, 2015). *ChloroMitoSSRDB 2.00: More genomes, more repeats, unifying SSRs search patterns and on-the-fly repeat detection. Database (Oxford)*. https://doi.org/10.1093/database/bav084. PMID: 26412851; PMCID: PMC4584093.

Sacconi, S., Salviati, L., Nishigaki, Y., Walker, W. F., Hernandez-Rosa, E., Trevisson, E., Delplace, S., Desnuelle, C., Shanske, S., Hirano, M., Schon, E. A., Bonilla, E., De Vivo, D. C., DiMauro, S., & Davidson, M. M. (2008). A functionally dominant mitochondrial DNA mutation. *Human Molecular Genetics, 17*(12), 1814–1820.

Sakuta, R., Goto, Y., Horai, S., & Nonaka, I. (1993). Mitochondrial DNA mutations at nucleotide positions 3243 and 3271 in mitochondrial myopathy, encephalopathy, lactic acidosis, and stroke- like episodes: A comparative study. *Journal of the Neurological Sciences, 115*(2), 158–160.

Santorelli, F. M., Mak, S. C., El-Schahawi, M., Casali, C., Shanske, S., Baram, T. Z., Madrid, R. E., & DiMauro, S. (1996). Maternally inherited cardiomyopathy and hearing loss associated with a novel mutation in the mitochondrial tRNALys gene (G8363A). *The American Journal of Human Genetics, 58*(5), 933–939.

Santorelli, F. M., Tanji, K., Sano, M., Shanske, S., El-Shahawi, M., Kranz-Eble, P., DiMauro, S., & De Vivo, D. C. (1997). Maternally inherited encephalopathy associated with a single-base insertion in the mitochondrial tRNATrp gene. *Annals of Neurology, 42*(2), 256–260.

Sato, W., Hayasaka, K., Shoji, Y., Takahashi, T., Takada, G., Saito, M., Fukawa, O., & Wachi, E. (1994). A mitochondrial tRNA(Leu)(UUR) mutation at 3256 associated with mitochondrial myopathy, encephalopathy, lactic acidosis, and stroke-like episodes (MELAS). *Biochemistry & Molecular Biology International, 33*(6), 1055–1061.

Schaller, A., Desetty, R., Hahn, D., Jackson, C. B., Nuoffer, J. M., Gallati, S., & Levinger, L. (2011). Impairment of mitochondrial tRNA(Ile) processing by a novel mutation associated with chronic progressive external ophthalmoplegia. *Mitochondrion, 11*(3), 488–496.

Schon, E. A., DiMauro, S., & Hirano, M. (December 2012). Human mitochondrial DNA: Roles of inherited and somatic mutations. *Nature Reviews Genetics, 13*(12), 878–890. https://doi.org/10.1038/nrg3275. PMID: 23154810; PMCID: PMC3959762.

Schuelke, M., Bakker, M., Stoltenburg, G., Sperner, J., & von Moers, A. (1998). Epilepsia partialis continua associated with a homoplasmic mitochondrial tRNA(Ser(UCN)) mutation. *Annals of Neurology, 44*(4), 700–704.

Shadel, G. S., & Clayton, D. A. (1997). Mitochondrial DNA maintenance in vertebrates. *Annual Review of Biochemistry, 66*, 409–435. https://doi.org/10.1146/annurev.biochem.66.1.409. PMID: 9242913.

Sharma, N., Pasala, M. S., Prakash, A., & Mitochondrial, D. N. A. (October 2019). Epigenetics and environment. *Environmental and Molecular Mutagenesis, 60*(8), 668–682. https://doi.org/10.1002/em.22319. Epub 2019 Aug 6. PMID: 31335990; PMCID: PMC6941438.

Shaughnessy, D. T., McAllister, K., Worth, L., Haugen, A. C., Meyer, J. N., Domann, F. E., Van Houten, B., Mostoslavsky, R., Bultman, S. J., Baccarelli, A. A., Begley, T. J., Sobol, R. W., Hirschey, M. D., Ideker, T.,

Santos, J. H., Copeland, W. C., Tice, R. R., Balshaw, D. M., & Tyson, F. L. (December 2014). Mitochondria, energetics, epigenetics, and cellular responses to stress. *Environmental Health Perspectives, 122*(12), 1271–1278. https://doi.org/10.1289/ehp.1408418. Epub 2014 Aug 15. PMID: 25127496; PMCID: PMC4256704.

Shen, L., Diroma, M. A., Gonzalez, M., Navarro-Gomez, D., Leipzig, J., Lott, M. T., van Oven, M., Wallace, D. C., Muraresku, C. C., Zolkipli-Cunningham, Z., Chinnery, P. F., Attimonelli, M., Zuchner, S., Falk, M. J., & Gai, X. (June 2016). MSeqDR: A centralized knowledge repository and bioinformatics web resource to facilitate genomic investigations in mitochondrial disease. *Human Mutation, 37*(6), 540–548. https://doi.org/10.1002/humu.22974. Epub 2016 Mar 21. PMID: 26919060; PMCID: PMC4846568.

Shen-Gunther, J., Gunther, R. S., Cai, H., & Wang, Y. (August 31, 2023). A customized human mitochondrial DNA database (hMITO DB v1.0) for rapid sequence analysis, haplotyping and geo-mapping. *International Journal of Molecular Sciences, 24*(17), 13505. https://doi.org/10.3390/ijms241713505. PMID: 37686313; PMCID: PMC10488239.

Shin, W. S., Tanaka, M., Suzuki, J., Hemmi, C., & Toyo-Oka, T. (2000). A novel homoplasmic mutation in mtDNA with a single evolutionary origin as a risk factor for cardiomyopathy. *The American Journal of Human Genetics, 67*(6), 1617–1620.

Shoffner, J. M., Brown, M. D., Torroni, A., Lott, M. T., Cabell, M. R., Mirra, S. S., Beal, M. F., Yang, C., Gearing, M., Salvo, R., Watts, R. L., Juncos, J. L., Hansen, L. A., Crain, B. J., Fayad, M., Reckord, C. L., & Wallace, D. C. (1993). Mitochondrial DNA variants observed in Alzheimer disease and Parkinson disease patients. *Genomics, 17*(1), 171–184.

Shoffner, J. M., Krawiecki, N., Cabell, M. F., Torroni, A., & Wallace, D. C. (1993). A novel tRNALeu(UUR) mutation in childhood mitochondrial myopathy. Poster 949. *The American Journal of Human Genetics, 53*(Suppl. l), 287.

Shoffner, J. M., Lott, M. T., Lezza, A. M., Seibel, P., Ballinger, S. W., & Wallace, D. C. (1990). Myoclonic epilepsy and ragged-red fiber disease (MERRF) is associated with a mitochondrial DNA tRNALys mutation. *Cell, 61*(6), 931–937.

Shoop, W. K., Lape, J., Trum, M., Powell, A., Sevigny, E., Mischler, A., Bacman, S. R., Fontanesi, F., Smith, J., Jantz, D., Gorsuch, C. L., & Moraes, C. T. (December 2023). Efficient elimination of MELAS-associated m.3243G mutant mitochondrial DNA by an engineered mitoARCUS nuclease. *Nature Metabolism, 5*(12), 2169–2183. https://doi.org/10.1038/s42255-023-00932-6. Epub 2023 Nov 30. PMID: 38036771; PMCID: PMC10730414.

Shoubridge, E. A., & Wai, T. (2007). Mitochondrial DNA and the mammalian oocyte. *Current Topics in Developmental Biology, 77*, 87–111. https://doi.org/10.1016/S0070-2153(06)77004-1. PMID: 17222701.

Silvestri, G., Moraes, C. T., Shanske, S., Oh, S. J., & DiMauro, S. (1992). A new mtDNA mutation in the tRNALys gene associated with myoclonic epilepsy and ragged red fibers (MERRF). *The American Journal of Human Genetics, 51*(6), 1213–1217.

Silvestri, G., Santorelli, F. M., Shanske, S., Whitley, C. B., Schimmenti, L. A., Smith, S. A., & DiMauro, S. (1994). A new mtDNA mutation in the tRNALeu(UUR) gene associated with maternally inherited cardiomyopathy. *Human Mutation, 3*(1), 37–43.

Sim, N. L., Kumar, P., Hu, J., Henikoff, S., Schneider, G., & Ng, P. C. (July 2012). SIFT web server: Predicting effects of amino acid substitutions on proteins. *Nucleic Acids Research, 40*(Web Server issue), W452–W457. https://doi.org/10.1093/nar/gks539. Epub 2012 Jun 11. PMID: 22689647; PMCID: PMC3394338.

Singh, K. K. (November 2004). Mitochondria damage checkpoint in apoptosis and genome stability. *FEMS Yeast Research, 5*(2), 127–132. https://doi.org/10.1016/j.femsyr.2004.04.008. PMID: 15489195.

Singh, K. K. (May 2006). Mitochondria damage checkpoint, aging, and cancer. *Annals of the New York Academy of Sciences, 1067*, 182–190. https://doi.org/10.1196/annals.1354.022. PMID: 16803984.

Smith, A. C., & Robinson, A. J. (January 8, 2019). MitoMiner v4.0: An updated database of mitochondrial localization evidence, phenotypes and diseases. *Nucleic Acids Research, 47*(D1), D1225–D1228. https://doi.org/10.1093/nar/gky1072. PMID: 30398659; PMCID: PMC6323904.

Smith, A. L. M., Whitehall, J. C., & Greaves, L. C. (September 2022). Mitochondrial DNA mutations in ageing and cancer. *Molecular Oncology, 16*(18), 3276–3294. https://doi.org/10.1002/1878-0261.13291. Epub 2022 Jul 28. PMID: 35842901; PMCID: PMC9490137.

Solignac, M., Génermont, J., Monnerot, M., et al. (1984). Genetics of mitochondria in Drosophila: mtDNA inheritance in heteroplasmic strains of *D. mauritiana*. *Molecular and General Genetics, 197*, 183–188. https://doi.org/10.1007/BF00330961

Sonney, S., Leipzig, J., Lott, M. T., Zhang, S., Procaccio, V., Wallace, D. C., & Sondheimer, N. (December 11, 2017). Predicting the pathogenicity of novel variants in mitochondrial tRNA with MitoTIP. *PLoS Computational Biology, 13*(12), e1005867. https://doi.org/10.1371/journal.pcbi.1005867. PMID: 29227991; PMCID: PMC5739504.

Souilem, S., Chebel, S., Mancuso, M., Petrozzi, L., Siciliano, G., FrihAyed, M., Hentati, F., & Amouri, R. (2011). A novel mitochondrial tRNA(Ile) point mutation associated with chronic progressive external ophthalmoplegia and hyperCKemia. *Journal of the Neurological Sciences, 300*(1–2), 187–190.

Stehling, O., & Lill, R. (August 1, 2013). The role of mitochondria in cellular iron-sulfur protein biogenesis: Mechanisms, connected processes, and diseases. *Cold Spring Harb Perspect Biology, 5*(8), a011312. https://doi.org/10.1101/cshperspect.a011312. PMID: 23906713; PMCID: PMC3721283.

Sue, C. M., Tanji, K., Hadjigeorgiou, G., Andreu, A. L., Nishino, I., Krishna, S., Bruno, C., Hirano, M., Shanske, S., Bonilla, E., Fischel-Ghodsian, N., DiMauro, S., & Friedman, R. (1999). Maternally inherited hearing loss in a large kindred with a novel T7511C mutation in the mitochondrial DNA tRNA(Ser(UCN)) gene. *Neurology, 52*(9), 1905–1908.

Suzuki, Y., Suzuki, S., Hinokio, Y., Chiba, M., Atsumi, Y., Hosokawa, K., Shimada, A., Asahina, T., & Matsuoka, K. (1997). Diabetes associated with a novel 3264 mitochondrial tRNA(Leu)(UUR) mutation. *Diabetes Care, 20*(7), 1138–1140.

Sweeney, M. G., Brockington, M., Weston, M. J., Morgan-Hughes, J. A., & Harding, A. E. (1993). Mitochondrial DNA transfer RNA mutation Leu(UUR) A-G 3260: A second family with myopathy and cardiomyopathy. *Quarterly Journal of Medicine, 86*(7), 435–438.

Sweeney, M. G., Bundey, S., Brockington, M., Poulton, K. R., Winer, J. B., & Harding, A. E. (1993b). Mitochondrial myopathy associated with sudden death in young adults and a novel mutation in the mitochondrial DNA leucine transfer RNA(UUR) gene. *Quarterly Journal of Medicine, 86*(11), 709–713.

Tahiliani, M., Koh, K. P., Shen, Y., Pastor, W. A., Bandukwala, H., Brudno, Y., Agarwal, S., Iyer, L. M., Liu, D. R., Aravind, L., & Rao, A. (May 15, 2009). Conversion of 5-methylcytosine to 5-hydroxymethylcytosine in mammalian DNA by MLL partner TET1. *Science, 324*(5929), 930–935. https://doi.org/10.1126/science.1170116. Epub 2009 Apr 16. PMID: 19372391; PMCID: PMC2715015.

Tan, M. H., Gan, H. M., Schultz, M. B., & Austin, C. M. (April 2015). MitoPhAST, a new automated mitogenomic phylogeny tool in the post-genomic era with a case study of 89 decapod mitogenomes including eight new freshwater crayfish mitogenomes. *Molecular Phylogenetics and Evolution, 85*, 180–188. https://doi.org/10.1016/j.ympev.2015.02.009. Erratum in: (June 2015). Molecular Phylogenetics and Evolution, 87(1), 118. PMID: 25721538.

Tanaka, M., Ino, H., Ohno, K., Hattori, K., Sato, W., Ozawa, T., Tanaka, T., & Itoyama, S. (1990). Mitochondrial mutation in fatal infantile cardiomyopathy. *Lancet, 336*(8728), 1452.

Tang, Z., Lu, Z., Chen, B., Zhang, W., Chang, H. Y., Hu, Z., & Xu, J. (May 3, 2022). A genetic bottleneck of mitochondrial DNA during human Lymphocyte development. *Molecular Biology and Evolution, 39*(5), msac090. https://doi.org/10.1093/molbev/msac090. PMID: 35482398; PMCID: PMC9113143.

Taniike, M., Fukushima, H., Yanagihara, I., Tsukamoto, H., Tanaka, J., Fujimura, H., Nagai, T., Sano, T., Yamaoka, K., Inui, K., & Okada, S. (1992). Mitochondrial tRNAIle mutation in fatal cardiomyopathy. *Biochemical and Biophysical Research Communications, 186*(1), 47–53.

Tawata, M., Ohtaka, M., Iwase, E., Ikegishi, Y., Aida, K., & Onaya, T. (1998). New mitochondrial DNA homoplasmic mutations associated with Japanese patients with type 2 diabetes. *Diabetes, 47*(2), 276–277.

Taylor, R. W., Chinnery, P. F., Bates, M. J., Jackson, M. J., Johnson, M. A., Andrews, R. M., & Turnbull, D. M. (1998). A novel mitochondrial DNA point mutation in the tRNA(Ile) gene: Studies in a patient presenting with chronic progressive external ophthalmoplegia and multiple sclerosis. *Biochemical and Biophysical Research Communications, 243*(1), 47–51.

Taylor, R. W., Schaefer, A. M., McDonnell, M. T., Petty, R. K., Thomas, A. M., Blakely, E. L., Hayes, C. M., McFarland, R., & Turnbull, D. M. (2004). Catastrophic presentation of mitochondrial disease due to a mutation in the tRNA(His) gene. *Neurology, 62*(8), 1420–1423.

Terasaki, F., Tanaka, M., Kawamura, K., Kanzaki, Y., Okabe, M., Hayashi, T., Shimomura, H., Ito, T., Suwa, M., Gong, J. S., Zhang, J., & Kitaura, Y. (2001). A case of cardiomyopathy showing progression from the hypertrophic to the dilated form: Association of Mt8348A->G mutation in the mitochondrial tRNA(Lys) gene with severe ultrastructural alterations of mitochondria in cardiomyocytes. *Japanese Circulation Journal, 65*(7), 691–694.

Thompson, J. D., Higgins, D. G., & Gibson, T. J. (November 11, 1994). CLUSTAL W: Improving the sensitivity of progressive multiple sequence alignment through sequence weighting, position-specific gap penalties and weight matrix choice. *Nucleic Acids Research, 22*(22), 4673–4680. https://doi.org/10.1093/nar/22.22.4673. PMID: 7984417; PMCID: PMC308517.

Tiranti, V., Chariot, P., Carella, F., Toscano, A., Soliveri, P., Girlanda, P., Carrara, F., Fratta, G. M., Reid, F. M., Mariotti, C., & Zeviani, M. (1995). Maternally inherited hearing loss, ataxia and myoclonus associated with a novel point mutation in mitochondrial tRNASer(UCN) gene. *Human Molecular Genetics, 4*(8), 1421–1427.

Tiranti, V., D'Agruma, L., Pareyson, D., Mora, M., Carrara, F., Zelante, L., Gasparini, P., & Zeviani, M. (1998). A novel mutation in the mitochondrial tRNA(Val) gene associated with a complex neurological presentation. *Annals of Neurology, 43*(1), 98–101.

Toren, D., Barzilay, T., Tacutu, R., Lehmann, G., Muradian, K. K., & Fraifeld, V. E. (January 4, 2016). MitoAge: A database for comparative analysis of mitochondrial DNA, with a special focus on animal longevity. *Nucleic Acids Research, 44*(D1), D1262–D1265. https://doi.org/10.1093/nar/gkv1187. Epub 2015 Nov 20. PMID: 26590258; PMCID: PMC4702847.

Tsukuda, K., Suzuki, Y., Kameoka, K., Osawa, N., Goto, Y., Katagiri, H., Asano, T., Yazaki, Y., & Oka, Y. (1997). Screening of patients with maternally transmitted diabetes for mitochondrial gene mutations in the tRNA [Leu(UUR)] region. *Diabetic Medicine, 14*(12), 1032–1037.

Turnbull, H. E., Lax, N. Z., Diodato, D., Ansorge, O., & Turnbull, D. M. (January 2010). The mitochondrial brain: From mitochondrial genome to neurodegeneration. *Biochimica et Biophysica Acta, 1802*(1), 111–121. https://doi.org/10.1016/j.bbadis.2009.07.010. Epub 2009 Aug 6. PMID: 19647794; PMCID: PMC2795853.

Twig, G., Elorza, A., Molina, A. J., Mohamed, H., Wikstrom, J. D., Walzer, G., Stiles, L., Haigh, S. E., Katz, S., Las, G., Alroy, J., Wu, M., Py, B. F., Yuan, J., Deeney, J. T., Corkey, B. E., & Shirihai, O. S. (January 23, 2008). Fission and selective fusion govern mitochondrial segregation and elimination by autophagy. *EMBO Journal, 27*(2), 433–446. https://doi.org/10.1038/sj.emboj.7601963. Epub 2008 Jan 17. PMID: 18200046; PMCID: PMC2234339.

Valente, L., Piga, D., Lamantea, E., Carrara, F., Uziel, G., Cudia, P., Zani, A., Farina, L., Morandi, L., Mora, M., Spinazzola, A., Zeviani, M., & Tiranti, V. (2009). Identification of novel mutations in five patients with mitochondrial encephalomyopathy. *Biochimica et Biophysica Acta, 1787*(5), 491–501.

van den Ouweland, J. M., Lemkes, H. H. P., Ruitenbeek, W., Sandkjujl, L. A., deVijlder, M. F., Struyvenberg, P. A. A., van de Kamp, J. J. P., & Maassen, J. A. (1992). Mutation in mitochondrial tRNALeu(UUR) gene in a large pedigree with maternally transmitted type II diabetes mellitus and deafness. *Nature Genetics, 1*(5), 368–371.

van Oven, M., & Kayser, M. (February 2009). Updated comprehensive phylogenetic tree of global human mitochondrial DNA variation. *Human Mutation, 30*(2), E386–E394. https://doi.org/10.1002/humu.20921. PMID: 18853457.

Vanyushin, B. F., & Kirnos, M. D. (March 18, 1977). Structure of animal mitochondrial DNA (base composition, pyrimidine clusters, character of methylation). *Biochimica et Biophysica Acta, 475*(2), 323–336. https://doi.org/10.1016/0005-2787(77)90023-5. PMID: 843531.

Vanyushin, B. F., & Kirnos, M. D. (1976). Structure of animal mitochondrial DNA: Nucleotide composition, pyrimidine clusters, and methylation character. *Molecular Biology, 10*(4), 715–724. PMID: 1023050.

Vasconcelos, A. T., Guimarães, A. C., Castelletti, C. H., Caruso, C. S., Ribeiro, C., Yokaichiya, F., Armôa, G. R., Pereira Gda, S., da Silva, I. T., Schrago, C. G., Fernandes, A. L., da Silveira, A. R., Carneiro, A. G., Carvalho, B. M., Viana, C. J., Gramkow, D., Lima, F. J., Corrêa, L. G., Mudado Mde, A., ... Russo, C. A. (May 15, 2005). MamMiBase: A mitochondrial genome database for mammalian phylogenetic studies. *Bioinformatics, 21*(10), 2566–2567. https://doi.org/10.1093/bioinformatics/bti326. Epub 2005 Feb 15. PMID: 15713730.

Vianello, D., Sevini, F., Castellani, G., Lomartire, L., Capri, M., & Franceschi, C. (September 2013). HAPLOFIND: A new method for high-throughput mtDNA haplogroup assignment. *Human Mutation, 34*(9), 1189–1194. https://doi.org/10.1002/humu.22356. Epub 2013 Jun 12. PMID: 23696374.

Villanueva-Paz, M., Povea-Cabello, S., Villalón-García, I., Suárez-Rivero, J. M., Álvarez-Córdoba, M., de la Mata, M., Talaverón-Rey, M., Jackson, S., & Sánchez-Alcázar, J. A. (May 2019). Pathophysiological characterization of MERRF patient-specific induced neurons generated by direct reprogramming. *Biochimica et Biophysica Acta (BBA) - Molecular Cell Research, 1866*(5), 861–881. https://doi.org/10.1016/j.bbamcr.2019.02.010. Epub 2019 Feb 22. PMID: 30797798.

Vives-Bauza, C., Del Toro, M., Solano, A., Montoya, J., Andreu, A. L., & Roig, M. (2003). Genotype-phenotype correlation in the 5703G>A mutation in the tRNA(ASN) gene of mitochondrial DNA. *Journal of Inherited Metabolic Disease, 26*(5), 507–508.

Wai, T., Teoli, D., & Shoubridge, E. A. (December 2008). The mitochondrial DNA genetic bottleneck results from replication of a subpopulation of genomes. *Nature Genetics, 40*(12), 1484–1488. https://doi.org/10.1038/ng.258. PMID: 19029901.

Wakae, K., Nishiyama, T., Kondo, S., Izuka, T., Que, L., Chen, C., Kase, K., Kitamura, K., Mohiuddin, M., Wang, Z., Ahasan, M. M., Nakamura, M., Fujiwara, H., Yoshizaki, T., Hosomochi, K., Tajima, A., Nakahara, T., Kiyono, T., & Muramatsu, M. (June 27, 2018). Keratinocyte differentiation induces APOBEC3A, 3B, and mitochondrial DNA hypermutation. *Scientific Reports, 8*(1), 9745. https://doi.org/10.1038/s41598-018-27930-z. PMID: 29950685; PMCID: PMC6021414.

Wallace, D. C., & Chalkia, D. (November 1, 2013). Mitochondrial DNA genetics and the heteroplasmy conundrum in evolution and disease. *Cold Spring Harbor Perspectives in Biology, 5*(11), a021220. https://doi.org/10.1101/cshperspect.a021220. PMID: 24186072; PMCID: PMC3809581.

Wallace, D. C., Shoffner, J. M., Brown, M. D., Torroni, A., Lott, M. T., & Cabell, M. (1992). Mitochondrial DNA mutations associated with Alzheimer's and Parkinson's disease. *The American Journal of Human Genetics, 51*, A30.

Wallace, D. C., Zheng, X., Lott, M. T., Shoffner, J. M., Hodge, J. A., Kelley, R. I., Epstein, C. M., & Hopkins, L. C. (1988). Familial mitochondrial encephalomyopathy (MERRF): Genetic, pathophysiological, and biochemical characterization of a mitochondrial DNA disease. *Cell, 55*(4), 601–610.

Wang, C., & Youle, R. J. (2009). The role of mitochondria in apoptosis*. *Annual Review of Genetics, 43*, 95–118. https://doi.org/10.1146/annurev-genet-102108-134850. PMID: 19659442; PMCID: PMC4762029.

Weissensteiner, H., Forer, L., Fuchsberger, C., Schöpf, B., Kloss-Brandstätter, A., Specht, G., Kronenberg, F., & Schönherr, S. (2016). mtDNA-Server: next-generation sequencing data analysis of human mitochondrial DNA in the cloud. *Nucleic Acids Research, 44*(W1), W64–W69. https://doi.org/10.1093/nar/gkw247. Epub 2016 Apr 15. PMID: 27084948; PMCID: PMC4987870.

Weissensteiner, H., Pacher, D., Kloss-Brandstätter, A., Forer, L., Specht, G., Bandelt, H. J., Kronenberg, F., Salas, A., & Schönherr, S. (2016a). HaploGrep 2: Mitochondrial haplogroup classification in the era of high-throughput sequencing. *Nucleic Acids Research, 44*(W1), W58–W63. https://doi.org/10.1093/nar/gkw233. Epub 2016 Apr 15. PMID: 27084951; PMCID: PMC4987869.

Wilson, F. H., Hariri, A., Farhi, A., Zhao, H., Petersen, K. F., Toka, H. R., Nelson-Williams, C., Raja, K. M., Kashgarian, M., Shulman, G. I., Scheinman, S. J., & Lifton, R. P. (2004). A cluster of metabolic defects caused by mutation in a mitochondrial tRNA. *Science, 306*(5699), 1190–1194.

Xiong, W., Hua, J., Liu, Z., Cai, W., Bai, Y., Zhan, Q., Lai, W., Zeng, Q., Ren, H., & Xu, D. (September 1, 2018). PTEN induced putative kinase 1 (PINK1) alleviates angiotensin II-induced cardiac injury by ameliorating mitochondrial dysfunction. *International Journal of Cardiology, 266*, 198–205. https://doi.org/10.1016/j.ijcard.2018.03.054. PMID: 29887448.

Zahedi, A., On, V., Phandthong, R., Chaili, A., Remark, G., Bhanu, B., & Talbot, P. (November 5, 2018). Deep analysis of mitochondria and cell health using machine learning. *Scientific Reports, 8*(1), 16354. https://doi.org/10.1038/s41598-018-34455-y. PMID: 30397207; PMCID: PMC6218515.

Zeviani, M., Gellera, C., Antozzi, C., Rimoldi, M., Morandi, L., Villani, F., Tiranti, V., & DiDonato, S. (1991). Maternally inherited myopathy and cardiomyopathy: Association with mutation in mitochondrial DNA tRNA-Leu(UUR). *Lancet, 338*(8760), 143–147.

Zeviani, M., Muntoni, F., Savarese, N., Serra, G., Tiranti, V., Carrara, F., Mariotti, C., & DiDonato, S. (1993). A MERRF/MELAS overlap syndrome associated with a new point mutation in the mitochondrial DNA tRNALys gene. *European Journal of Human Genetics, 1*(1), 80–87.

Zhang, H., Burr, S. P., & Chinnery, P. F. (July 20, 2018). The mitochondrial DNA genetic bottleneck: Inheritance and beyond. *Essays in Biochemistry, 62*(3), 225–234. https://doi.org/10.1042/EBC20170096. PMID: 29880721.

Zhao, H., Li, R., Wang, Q., Yan, Q., Deng, J. H., Han, D., Bai, Y., Young, W. Y., & Guan, M. X. (2004). Maternally inherited aminoglycoside-induced and nonsyndromic deafness is associated with the novel C1494T mutation in the mitochondrial 12S rRNA gene in a large Chinese family. *The American Journal of Human Genetics, 74*(1), 139–152.

Zhu, W., Qin, W., & Sauter, E. R. (2004). Large-scale mitochondrial DNA deletion mutations and nuclear genome instability in human breast cancer. *Cancer Detection and Prevention, 28*(2), 119–126. https://doi.org/10.1016/j.cdp.2003.11.008. PMID: 15068836.

Further reading

Kassauei, K., Habbe, N., Mullendore, M. E., Karikari, C. A., Maitra, A., & Feldmann, G. (2006). Mitochondrial DNA mutations in pancreatic cancer. *International Journal of Gastrointestinal Cancer, 37*(2–3), 57–64. https://doi.org/10.1007/s12029-007-0008-2. PMID: 17827523.

SECTION II

Physiological mitochondrial transfer

CHAPTER

5

Tunneling nanotubes

Tuba Edgünlü and Cilem Ozdemir

Department of Medical Biology, Faculty of Medicine, Mugla Sıtkı Koçman University, Mugla, Turkey

Tunneling nanotubes

As a result of the biological complexity of organisms, different mechanisms have evolved for communication between cells. In multicellular organisms, intercellular communication is of great importance for the survival and development of organisms. Because in multicellular organisms, the communication and interaction between cells in complex cell communities ensure the development and protection of tissues, organs, and organisms (Zhu et al., 2021). One of the common mechanisms in cell communication is that soluble molecules bind to target cell receptors, triggering paracrine or endocrine signaling. Other mechanisms are immunological and neurological synapses and extracellular vesicles. In addition, there are mechanisms in which direct physical interaction takes place (Gerdes et al., 2007). For example, in higher plants, cells are connected to each other by plasmodesmata structures, while in fungi, intercellular communication occurs through their septal pores. Cell-to-cell communication in animals occurs in short and long-range routes: (1) gap junctions and (2) tunneling nanotubes (TNTs) (Bloemendal & Kück, 2013).

TNTs are long and thin channels with a tubular structure of 50–500 nm in diameter. TNTs that can bind two or more cells over long distances (for example, in macrophages they can extend up to 200 μm) contain F-actin and adhesion proteins, usually at one end (Abounit & Zurzolo, 2012; Dupont et al., 2018). The ability to slip across the substrate and interconnect cells at distances of several cell diameters distinguishes TNTs from filopodia and cinema structures (Han & Wang, 2021). They have been defined as the internet of cells thanks to their multilevel transport capabilities over long distances (Baker, 2017; Han & Wang, 2021).

Mitochondrial Transplantation and Transfer
https://doi.org/10.1016/B978-0-443-18858-9.00005-6

TNTs morphology and structure

TNTs were first described in 2004, by Rustom et al. discovered in PC12 cells (Rustom et al., 2004). Subsequent studies have shown that these connections can occur de novo between animal cells and that various organelles, membrane components, and cellular components can be transferred between cells with these connections (Gerdes et al., 2007). It has some features that distinguish TNTs from other cellular protrusions. These features: Contain F-actin, connect at least 2 cells together, and do not touch the substrate. The tip of the protrusion from the donor cell communicates with the target cell by a receptor-ligand or synapse-like mechanism, thus forming closed-ended TNTs. The cytoplasm of the 2 cells can fuse, thus forming open-ended TNTs (Dupont et al., 2018). TNTs directly transfer cellular molecules or signals such as calcium ions, lysosomes, lipid droplets, endoplasmic reticulum components, miRNAs, and mitochondria between nonadjacent cells (Kimura et al., 2012; Abounit & Zurzolo, 2012; Ady et al., 2016; Haimovich et al., 2021) (Fig. 5.1).

TNTs are classified morphologically according to their size and the cytoskeletal elements they contain since protein markers specific to TNTs are not yet explained (Austefjord et al., 2014).

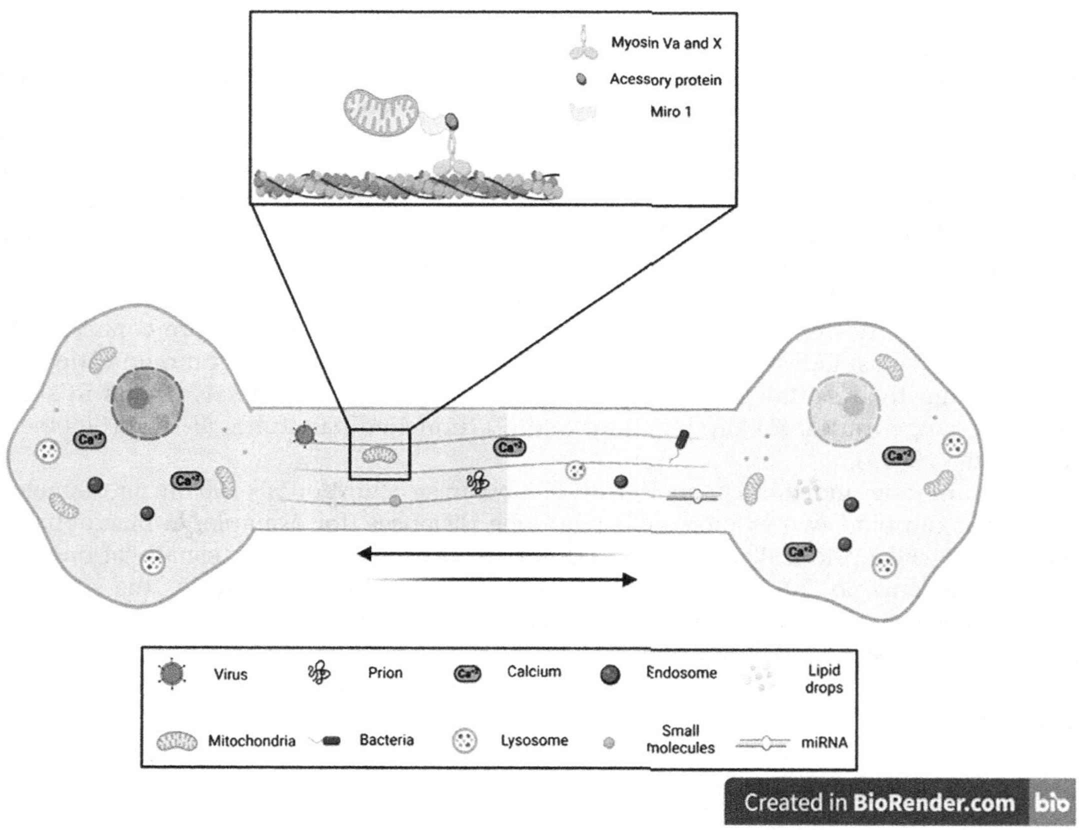

FIGURE 5.1 Representation of molecules transported along TNTs.

When TNTs are identified in vitro, they differ greatly from other cell protrusions because they are flat and bridge-like structures connecting cells. Different lengths have been reported for different cell types. For example, the average length of TNTs that bind T cells is 22 μm, while in PC12 cells this length is much less (Rustom et al., 2004; Sowinski et al., 2008). The diameters of TNTs, which consist of membranous filaments of varying lengths, also vary (Gerdes et al., 2013). The diameter of TNTs has been defined between 50 and 200 nm in PC12 cells and 180–380 nm in T cells (Rustom et al., 2004; Sowinski et al., 2008). The length and diameter vary according to the cell of origin of TNT and the skeletal proteins (Austefjord et al., 2014). Their different cytoskeletal proteins provide them to vary greatly in diameter. Actin-only TNTs are short and thin (<0.7 μm) those containing both actin and microtubules are thicker (>0.7 μm). In vitro characterization studies have shown that TNTs are composed of highly organized, uniform parallel F-actin-positive bundles (Ljubojevic et al., 2021). Although TNTs are mostly composed of F-actin, TNTs containing microtubules are also available. Microtubule-containing TNTs have been found in neurons, astrocytes, HUVEC cells, and some immune cells (Önfelt et al., 2006; Wang et al., 2012; Mineo et al., 2012). Although the mechanism of formation of microtubule-containing TNTs is unknown, it has been shown that end-binding protein 3 moves with microtubules in PC12 cells (Wang & Gerdes, 2015). Based on the cytoskeletal differences, TNTs are divided into three types: actin-based TNTs, microtubule-based TNTs, and actin- and microtubule-based TNTs (Zhu et al., 2021).

TNTs containing actin and microtubules use motor proteins to transport various cargos. TNTs containing F-actin can use myosin motor proteins, while TNTs containing microtubules can use dynein–kinesin motor proteins for cargo transport. The fact that TNTs contain different motor proteins determines their formation mechanism and the substances being transported between 2 cells (Gittes et al., 1993; Han & Wang, 2021). In in vivo studies, TNTs are defined as thin filament structures with actin structures that allow cargo transport between at least 2 cells. It is possible for more than one TNT to come together. TNTs formed in vivo do not require cytoplasmic continuity between cells to transport molecules and organelles. TNTs in the mouse retina were found to have an open end that continues with the cytoplasm and the other end closed end that connects to the gap junction (Alarcon-Martinez et al., 2020). These TNTs are called closed-ended TNTs and are also found in T cells (Sowinski et al., 2008).

Biogenesis and formation of the TNT

As in other cellular protrusions, TNTs require cytoskeletal reorganization to ensure intercellular connectivity, disrupt the cell membrane, remodel the cell, and protect the resulting protrusions. TNTs are dynamic structures and can form rapidly (Ariazi et al., 2017; Han & Wang, 2021). Leukocyte-specific transcript 1 (LST1), one of the Ras family proteins, interacts with Ras-related protein A (RalA). Then, RalGPS2 and M-sec interact with RalA and LST1 to form the exocyst complex. Formation of the complex induces TNT formation (Hase et al., 2009; Schiller et al., 2013). LST1 recruits filamin, the actin cross-linking protein involved in filopodia formation, into the exocyst complex. LST1 thus enables the formation of both thin and thick TNTs

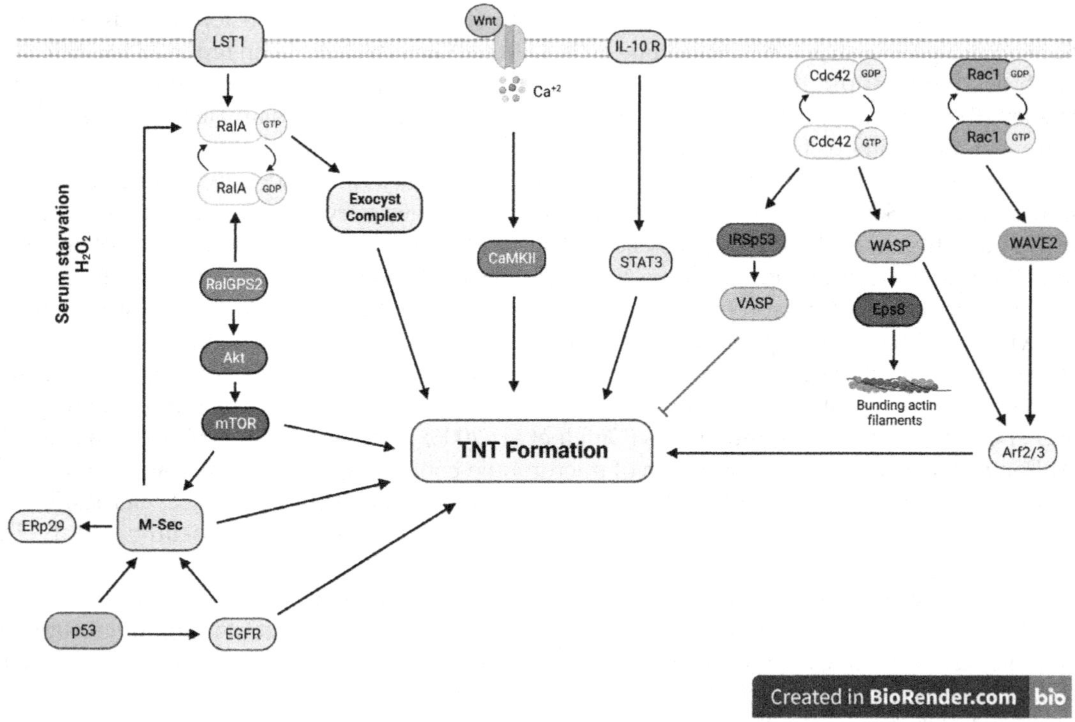

FIGURE 5.2 Molecular mechanisms of TNT formation.

(Mittal et al., 2019). M-sec, another myeloid cell-specific protein, which is one of the oldest markers discovered in TNT formation mechanisms, induces the formation of thin TNTs. Actin polymerization starting with M-sec is one of the important steps in TNT formation (Hase et al., 2009). ERp29, an endoplasmic reticulum chaperone, contributes to TNT formation by stabilizing M-sec (Pergu et al., 2019). During TNT biogenesis among macrophages, actin-associated protein 2/3 (Arp 2/3) is activated. In contrast, the complex of VASP/CDC42/IRSp53 induces filopodia formation and inhibits TNT formation (Keller, 2020). Similarly, it promotes TNT formation by activating Eps8 actin (Delage et al., 2016) (Fig. 5.2).

When TNT formation begins, actin polymerization is mediated by various Rho-GTPase proteins. CDC42 and Rac1 are RhoGTPases and activate WAVE2 and WASP, respectively. It has been reported that different GTPases are used to transport different molecules. For example, GTPase encoded by the Miro gene is involved in the mitochondrial transfer, while Rho-GTPase, the product of the Rab8 gene, is involved in the transport of transferrin protein (Burtey et al., 2015; Las & Shirihai, 2014). Myosin X (Myo 10) is functionally involved in de novo TNT formation. Inhibition of Myo 10 expression resulted in a decrease in TNT formation (Gousset et al., 2013). TNT formation can also be promoted by different signaling pathways. Ca^{2+}/calmodulin-dependent protein kinase II participates in TNT-mediated vesicle transport and TNT formation in CAD cells (Vargas et al., 2019). Also, IL-10/STAT3 and FAK/MMP-2 signaling pathways are involved in TNT-mediated signaling (Sáenz-de-Santa-María et al.,

2017; Souriant et al., 2019) (Fig. 5.2). The role of recipient cells in TNT biogenesis is undeniable. Recipient cells express phosphatidylserine and CSF-1 to direct the formed TNTs (Mittal et al., 2019).

Two mechanisms have been proposed for TNT formation. The first of these is the actin-driven de novo mechanism. Accordingly, TNT is formed as a result of the fusion of the filopodia-like protrusion from the donor cell with the neighboring cell. The protrusion from the donor cell membrane is elongated by actin polymerization and elongation is initiated by Rho-GTPases (Dupont et al., 2018; Matejka & Reindl, 2019). Studies have shown that actin inhibitors inhibit TNT formation and cargo transport. These results support F-actin–driven TNT biogenesis (Delage et al., 2016; Dilsizoglu et al., 2019; Hanna et al., 2017). These actin-driven structures are called closed-ended TNTs. In closed-ended TNTs, intercellular communication can occur via receptors and ligands, as well as by endocytosis/exocytosis (Fig. 5.3). Once a connection is established between two cells, the same cell can establish a connection with other cells. This creates a network of cells (Zhu et al., 2021).

The other mechanism of TNT formation is the cell dislodgement mechanism. TNT is formed when two cells in close contact separate from each other. It is seen in cancer cells, macrophages, natural killer (NK) cells, neural cells, and T cells. Studies have reported that with this mechanism, cells must remain in contact for at least 4 minutes for TNT formation (Sowinski et al., 2008). TNT stabilization and initiation of cell-to-cell contact are mediated by receptor–ligand interactions and adhesion molecules (such as N-cadherin and β-catenin) (Kimura et al., 2013). TNT biogenesis can occur in some cells by either of the two mechanisms

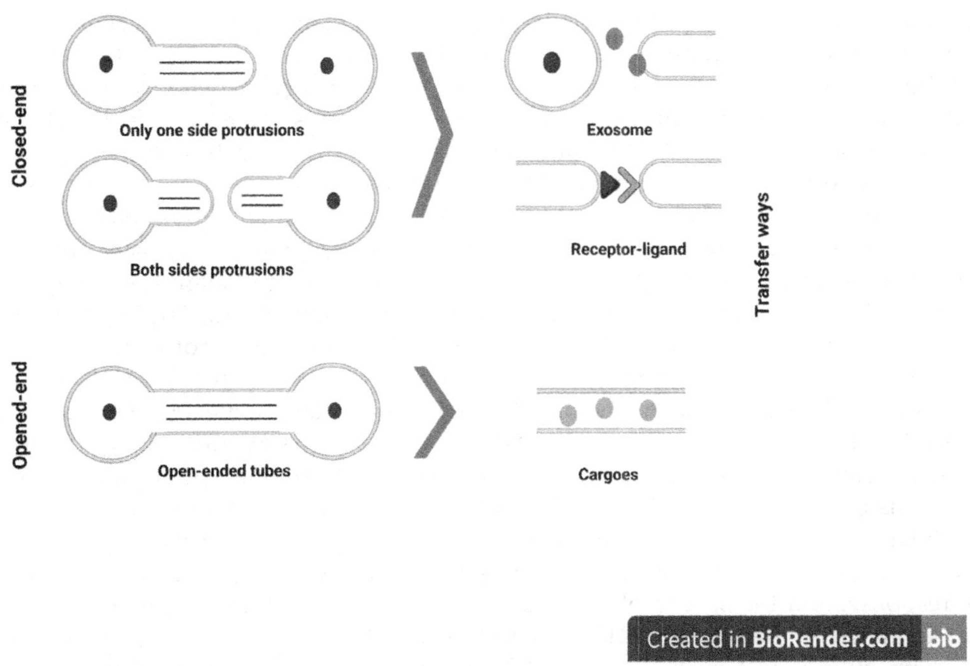

FIGURE 5.3 Models of TNTs formation.

described, or by cell dislodgement alone. Therefore, the mechanisms of TNT formation may be influenced by intracellular signaling pathways, paracrine signals, and cell type (Bukoreshtliev et al., 2009; Driscoll et al., 2022).

TNT modulation

TNT formation is sensitive to extracellular environmental factors such as neurodegenerative oligomerization, infection, inflammation, trauma, hydrogen peroxide (H_2O_2), radiation, and oxidative stress (Dilna et al., 2021; Eugenin et al., 2009; Ratzinger et al., 2011; Zhu et al., 2005). In addition, TNT formation is increased in cells undergoing apoptosis and the apoptotic process is prevented by transferring intact mitochondria (Wang & Gerdes, 2015). In addition to the increase in TNT formation, the stress state increases the transport rate of substances. The increase in TNT number in various cellular stress situations is not the same for all cells. TNT is formed in the stressed cell in neurons and astrocytes, and substance transport is supported. Stressed endothelial cells increase routing signals and increase TNT formation in nonstressed cells (Marzo et al., 2012; Matejka & Reindl, 2019). Promoting the formation of TNTs under cellular stress suggests that TNTs help cells adapt to changing conditions.

TNT-mediated transfer

As described earlier, TNTs formed in NK cells and among some subset cells of macrophages contain both F-actin and microtubules (Chauveau et al., 2010; Önfelt et al., 2006). Vesicular transport via structures described as thick TNTs (containing both F-actin and microtubules and greater than 0.7 μm in diameter) is energy dependent, indicating the involvement of microtubule motor proteins such as kinesins for transport. In addition, it has been revealed that thin TNTs (containing only actin) allow bacteria to navigate on the TNT surface due to ATP (Önfelt et al., 2006). In vitro studies have proven the existence of TNT-like structures that bind macrophages (Kadiu & Gendelman, 2011a, 2011b; Xu et al., 2009). Structures containing F-actin and microtubules were shown to transfer endocytosed HIV-1 particles between macrophages in endosomes and ER and Golgi-derived vesicles by an actomyosin-dependent mechanism (Kadiu & Gendelman, 2011a, 2011b). It is unclear whether microtubules are involved in the transfer. Therefore, the motor protein required for transport may differ according to the cell membrane contained in TNT. It is also unclear whether microtubule and F-actin–mediated transport occur simultaneously (Abounit & Zurzolo, 2012). TNT formation has been observed in many cell types in vitro. However, the transfer mechanisms of TNTs and whether the cargoes are cell-type specific have not been clarified. Also, it is unknown whether different TNT formation mechanisms lead to different intercellular junctions. There is increasing evidence to suggest that TNTs represent a new pathway for intercellular communication in events such as cell migration, development, immune response, and the spread of pathogens (Panasiuk et al., 2018; Valdinocci et al., 2021; Vedula et al., 2014; Zhu et al., 2021). In recent years, TNTs' use of electrical connections as a new connection path has shown that TNTs are involved in more physiological processes than previously thought (Wang et al., 2010).

A detailed explanation of the mechanisms underlying TNT biogenesis and cargo transfer will provide further insight into the cellular and organismal level function of TNTs caused by the physiological events in which they participate. The mechanisms elucidated can be opportunities for the prevention of many diseases when considered therapeutic targets. For this purpose, modeling of TNT formation and transfer mechanism with in silico modeling approaches can facilitate the prediction of biological processes caused by TNTs (Abounit & Zurzolo, 2012).

Nucleic acid transfer

Intracellular cargoes (nucleic acid, organelle, lipid, protein, etc.) are transported bidirectionally with TNTs. TNTs serve for the intercellular transfer of nucleic acids, including DNA, mRNA, noncoding RNA, mitochondrial DNA (mtDNA), and viral RNA. The transfer of nucleic acid cargoes between interconnected cells via TNTs can alter gene and protein expression in the recipient cell, revealing various cellular activities (Haimovich et al., 2021; Su & Igyártó, 2019) (Fig. 5.4). TNT-mediated transfer of nucleic acids differs from other intercellular transport mechanisms in that it is both directed and bidirectional (Driscoll et al., 2022).

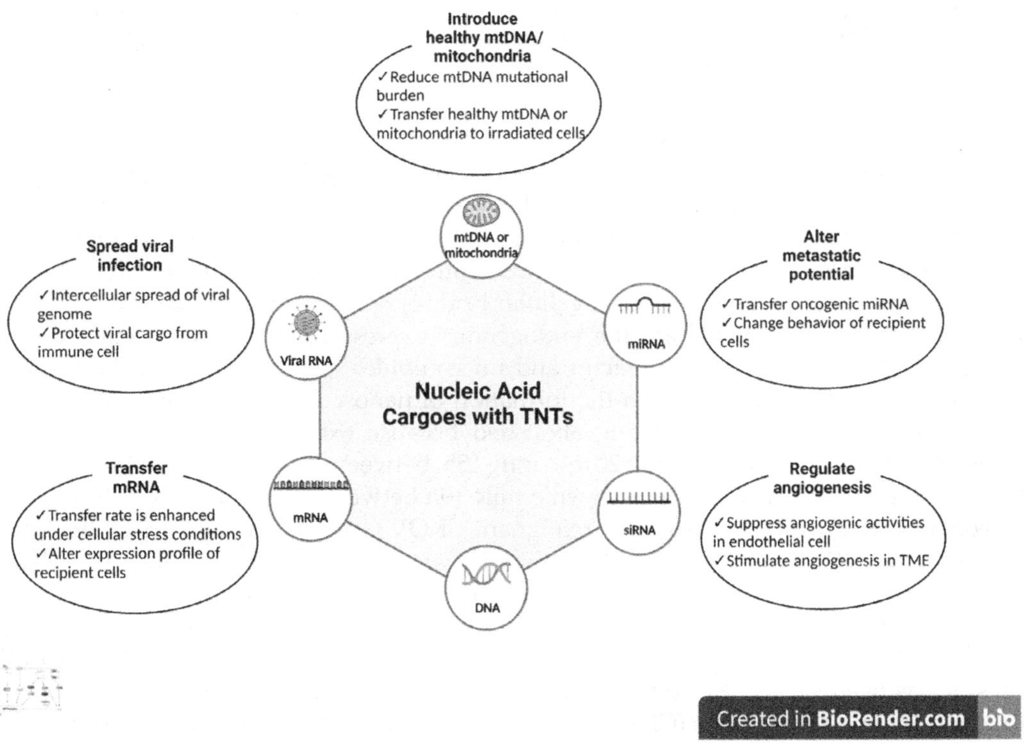

FIGURE 5.4 Effects of nucleic acid cargoes in recipient cells.

DNA transferring

Membranous tunneling tube formations were found among laryngeal squamous cell carcinoma. The identity of DAPI-stained vesicle-like structures within tubular structures was not elucidated but suggested that nuclear DNA may be displaced (Antanavičiūtė et al., 2014). DNA transfer is carried out via nondirected manners such as the release/uptake or extracellular vesicles of cell-free DNA and may result in nonspecific cell responses. Contrary to these pathways, the directed transfer of DNA by TNTs can regulate cell behavior (Driscoll et al., 2022).

mRNA transferring

Transfer of mRNAs via TNTs has been demonstrated between keratinocytes and Langerhans cells. Imaging studies have shown that transfer occurs by cell contact. Cellular contract suggests that TNTs act as channels and are responsible for the transfer of mRNAs (Su & Igyártó, 2019). In situ hybridization studies showed that a modified β-actin mRNA transcript carrying a 24-repeat sequence in the 3′UTR region can be transferred from donor murine embryonic fibroblasts to recipient cells expressing WT β-actin via TNTs. The stress state of the recipient cells determines mRNA transfer. Exposure to protein misfolding, oxidative stress, or serum starvation accelerated mRNA transfer, while heat shock decreases mRNA uptake (Haimovich et al., 2017). Although the transfer of mRNAs via extracellular vesicles is possible, TNT-mediated mRNA transfer may support a more physiological mechanism of cell-to-cell RNA transfer due to the limited size of the vesicles (Batagov & Kurochkin, 2013; Driscoll et al., 2022; Hung & Leonard, 2016).

Noncoding RNA transferring

TNT-mediated miRNA transfer is frequently observed in cancer cells. In cancer cells, miRNAs can elicit tumorigenic responses. Studies have shown that TNTs alter the microenvironment to promote tumor growth. For example, in breast cancer cells, cells with high metastatic potential had more cell connections than those with low metastatic potential. Transfer of miR-132 to endothelial cells by nano-sized cellular bridges between endothelial cells and metastatic breast cancer cells modulate the endogenous expression of a downstream target. The nanoscale structures contained F-actin and microtubules like TNTs. miR-132 expression in endothelium was decreased when the formation of nanoscale junctions was inhibited. miR-132 expression was not completely abolished because extracellular vesicles continued to transfer miR-132 (Connor et al., 2015). miR-155 between high-grade bladder cancer and low-grade bladder cancer cells, oncogenic miR-19a between osteosarcoma cells and stromal osteoblasts, and miR-199a between malignant SKOV3 and nonmalignant ISOE cells were transferred unidirectionally via TNTs (Driscoll et al., 2022; Lu et al., 2019). When cell contact occurs between vascular smooth muscle cells (SMCs) and endothelial cells, TGF-β secreted from endothelial cells enables SMCs to differentiate. Between differentiated SMC and endothelium, TNTs allowed the unidirectional transfer of mature miR-145/143 into endothelial cells. It has been shown that transferred miRNAs inhibited the angiogenic and proliferative functions of endothelial cells (Clement et al., 2015). Also, it has been determined that siRNAs can also be transferred with TNTs (Antanavičiūtė et al., 2014).

Viral RNA transferring

TNTs, which are highly effective links in the spread of viral infections, facilitate the spread of viral infections by increasing the surface area for the spread of viruses, transferring viruses from infected cells to uninfected cells, and bypassing cellular responses/antiviral defenses. TNTs can facilitate the intercellular transfer of viruses without the death of the host cell. TNTs deliver intact virions, viral proteins, and viral genomes to uninfected cells (Driscoll et al., 2022; Guo et al., 2016).

Nanotubes consisting of F-actin and myosin IIA were observed among MARC-145 cells infected with the porcine reproductive and respiratory virus (Guo et al., 2016). The effects of viral RNA transferred by TNTs are not yet known. A positive-sense viral genome was detected in naïve cells when PR8-influenza virus-infected and naïve A549 adenocarcinoma cells were cocultured. Considering the existence of heterotypic TNTs, it was concluded that viral RNA transfer occurs with TNTs (Kumar et al., 2017).

Cellular junctions consisting of F-actin between human metapneumovirus–infected and uninfected lung epithelial cells enabled the unidirectional transfer of viral load. Some of the fluorescent-label viral RNA appeared as large punctate dots, suggesting that structures resembling intact viral nucleocapsids could also be transferred via TNTs (El Najjar et al., 2016).

Mitochondrial transfer

The formation and transfer mechanisms of TNTs in intercellular communication are unclear; it is known that TNTs perform organelle transfer in pathological and stress conditions. Mitochondria are damaged by inflammation, injury, genotoxic agents, oxidative stress, and aging. Intercellular transfer of mitochondria/mitochondrial genomes can alter biological function in the recipient cell. Mitochondrial transfer plays both pathogenic and reparative roles (Mittal et al., 2019). Mitochondria or mtDNA transferred from MSCs via TNT to damaged organs in some respiratory, cardiovascular, neuronal, and immune system-related diseases have been reported to have a restorative effect (Cselenyák et al., 2010; Babenkoet al., 2015; Jackson et al., 2016; Pilling et al., 2006) (Table 5.1).

Studies have shown that donor cells must sense cellular damage for mitochondria transfer via TNT. MSCs promoted survival by transferring mitochondria to endothelial cells or cardiomyocytes exposed to ischemia-perfusion with TNTs (Han et al., 2016; Li et al., 2014). TNT formation in human umbilical cord endothelial cells exposed to ischemia–perfusion depends on MSCs detecting phosphatidylserine residues on the stressed cell surface. Although phosphatidylserine is usually found on the part of the cell membrane facing the cytoplasm, in the case of apoptosis, it is exhibited on the surface of the plasma membrane. Masking of phosphatidylserine on the cell surface reduces TNT formation. Phosphatidylserine on the surface of stressed cells is detected by MSCs and induces TNT formation. Transfer of mitochondria from MSCs to stressed human umbilical cord endothelial cells restored aerobic respiration. Aerobic respiration was not restored when stressed cells were cultured with MSCs containing dysfunctional mitochondria (Li et al., 2014; Mohammadalipour et al., 2020).

TABLE 5.1 Summary of mitochondrial transfer from donor cells to recipient cells.

Donor cell	Recipient cell	Transport mechanism	Cargo	Consequence of transportation	References
iPSC-MSC BMSC	Bronchial cells (cigarette Smoke-induced lung damage)	TNTs	Mitochondria	Increase of bioenergetics and ATP production in cell Reduce of cigarette smoke induced alveolar damage	Li et al. (2014)
hBMSCs mBMSCs	Alveolar epithelial cells	EVs and Cx-43 mediated TNTs	Mitochondria	Increase of ATP production Improve of mitochondrial respiratory function	Islam et al. (2012)
MMSCs[9]	PC12, astrocytes	TNTs	Mitochondria	Improvement of neurological functions Restore of the bioenergetics Stimulate of recipient cell proliferation	Babenko et al. (2018)
MMSCs	Rat cortical neuron, astroglial cells	TNTs (Miro-1 mediated)	Mitochondria	Neuroprotection	Babenko et al. (2015)
WJMSCs	Fibroblasts (rotenone-stressed)	TNTs	Mitochondria	Eliminate of mutation burden Improve of mitochondrial and cellular functions (protein translation of respiratory complexes, mitochondrial membrane potential, mitochondrial morphology and bioenergetics, mitochondrion-dependent viability, cell proliferation and apoptotic resistance, etc.)	Lin et al. (2019)
mBMSC	Epithelial cell (allergen-induced asthma)	TNTs (Miro-1 mediated)	Mitochondria	Rescue of epithelial injury Decrease in bronchial epithelial apoptosis and inflammation	Ahmad et al. (2014)
BMMSCs	Bronchial epithelial cells	EVs and TNTs	Mitochondria	Repair of wounded lung epithelial cells	Sinclair et al. (2016)

BMSCs, bone marrow stromal cells; *EVs*, extracellular vesicles; *hBMSCs*, human bone marrow-derived mesenchymal stem cells; *iPSC-MSCs*, induced pluripotent stem cell–derived mesenchymal stem cells; *mBMSCs*, mouse bone marrow–derived mesenchymal stem cells; *MMSCs*, mesenchymal multipotent stromal cells; *TNTs*, tunneling nanotubes; *WJMSCs*, Wharton's Jelly mesenchymal stem cells.

Wang et al. suggested that in hippocampal neurons and astrocytes, hydrogen peroxide or serum depletion leads to p53 activation and TNT formation trigger (Wang et al., 2011). Also, the studies with p53 knockout mice and p53-deficient cell lines suggested that TNT formation may occur in p53-independent ways and the existence of different stress-sensing pathways (Andresen et al., 2013). In light of the results obtained, cellular stress is required for organelle transfer (Torralba et al., 2016). The transfer of functional mitochondria is related to the degree of recipient cell damage. The mechanisms by which a healthy cell detects damage to the stressed cell and restores the cell's function rather than allowing the cell to be removed by apoptosis are noteworthy. These mechanisms will be of interest, especially when considering nonreplaceable cell lines such as adult neurons (Cho et al., 2012; Wang & Gerdes, 2015). In this context, the transfer of mitochondria from astrocytes to neurons showed a neuroprotective effect after stroke (Hayakawa et al., 2016; Torralba et al., 2016).

Transfer of mitochondria by TNTs is mediated by mitochondrial Rho GTPase Miro1 motor-adapter protein complexes. Overexpression of Miro1 promoted survival in stressed epithelial cells from ROS accumulation. The reparative effect of MSCs by mitochondria transfer has been shown in mice models of asthma and lung injury. MSCs overexpressing Miro1 decreased proinflammatory cytokines, airway hyperresponsiveness, and restored ATP levels, and the MSCs did not exhibit changes in their secretory properties or immunomodulatory paracrine signaling. In addition, the knockdown of Miro1 prevented the mitochondrial transfer of MSCs (Ahmad et al., 2014; Torralba et al., 2016). Mitochondria were transferred from MSCs to astrocytes when stress was induced in astrocytes by ROS. Also, intravenous administration of Miro1-overexpressing MSCs to stroke-mice models significantly improved their neurological function (Babenko et al., 2018) (Fig. 5.1).

Another proposed mechanism for mitochondria transfer is gap junctions. Connexin 43 (Cx43) encoded by GJA1 is a transmembrane protein that forms hemichannels. These hemichannels allow the transfer of metabolites, ions, miRNAs, peptides, and second messengers to the extracellular environment or between cells (Ribeiro-Rodrigues et al., 2017). Besides, Cx43 mediates TNT formation for mitochondrial transfer by using gap junctions. In the lipopolysaccharide (LPS)-induced lung injury model, Cx43 expression in both injured epithelial alveolar epithelial cells and donor bone marrow MSCs has been shown to be required for mitochondrial transfer via TNTs. TNT formation starts from cells in which Cx43 was expressed at high levels. When Cx43 was knocked out, TNT formation was inhibited and damaged epithelial cells did not survive (Islam et al., 2012). The relationship between TNT formation and Cx43 has also been shown in lung injury. Overexpression of Cx43 in induced pluripotent stem cells reduced cytokines and inflammatory cells by restoring mitochondrial membrane potential (Yao et al., 2015, 2018).

The transfer of mitochondria with TNTs has been known for a long time and this transfer plays an active role in modifying cellular responses. Studies have suggested evidence that mtDNA is transferred via TNTs. The mtDNA in healthy cells was labeled with ethynyl-2′-deoxyuridine and was detected in UV-irritated cells after coculture (Wang & Gerdes, 2015). TNT-mediated unidirectional mitochondria transfer was observed between Wharton's-jelly-derived MSCs (WJMSCs) and patient-derived fibroblasts carrying a point mutation in mtDNA (mt3243AΔG). Mitochondria derived from WJMSC reduced the mutation load in recipient fibroblast cells. These results showed that the transfer of mitochondria with TNTs enables the transfer of donor cell mtDNA to recipient cells (Lin et al., 2019).

Roles of TNTs in disease and healthy

TNTs in normal cells

TNTs are involved in a number of physiological processes as well as in the pathogenesis of multiple diseases (Fig. 5.5). The intercellular transfer of Ca^{2+} is a well-studied physiological event regulated by TNTs (Zhang et al., 2020). Long-range Ca^{2+} signaling between cells controls physiological activities such as synaptic plasticity and gene expression (Abounit & Zurzolo, 2012). Ca^{2+} influx is also inhibited when TNT formation is limited by inhibiting M-Sec expression. TNTs have spontaneous and inositol triphosphate (IP3)–induced Ca^{2+} signals. Depolarization signals transmitted via TNTs enable the activation of low voltage-gated calcium channels. Intercellular diffusion of Ca^{2+} has been suggested to involve IP3-evoked Ca^{2+} signals within TNTs (Wang et al., 2010). TNTs are found in B cells, T cells, dendritic cells, NK cells, and macrophages. Processed/unprocessed antigens can be transferred via TNTs. Antigen transfer by a TNT network in dendritic cells can be induced by CD40L and CD4 T cells. Induction contributes to enhancing the specific T-cell response to bacterial and viral antigens (Zaccard et al., 2016). TNTs in human NK cells connect with distant cells and allow the disintegration of distant target cells. In addition, IL-2, IL-15, IL-18, and IL-12 cytokines induce TNT formation in inhuman NK cells (Chauveau et al., 2010). Briefly, TNTs mediate the transmission of signals necessary for the immune response. TNT formation has been associated with osteoclastogenesis. When the formation of TNTs in osteoclast cells was inhibited by M-sec gene suppression, it suppressed osteoclastogenesis (Takahashi et al., 2013). Cell–cell interaction is very important during embryonic development. The

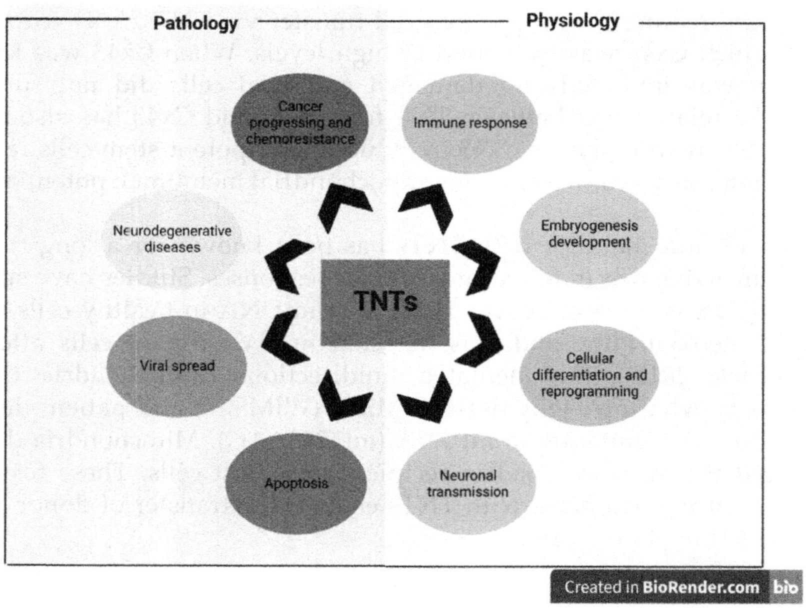

FIGURE 5.5 Tunneling nanotubes (TNTs) in normal physiology and pathology.

contribution of TNTs to development has begun to be elucidated with the identification of TNT-like structures in sea urchin embryonic cells (Miller et al., 1995). Following this study, the presence of TNT-like structures was demonstrated between inner cell mass cells and trophectoderm cells in mouse blastocysts (Salas-Vidal and Lomelı, 2004), between cranial neural crest cells in chicken embryos (Teddy & Kulesa, 2004) between epiblast cells in the blastula in zebrafish embryos (Caneparo et al., 2011), and between nonneuronal ectoderm cells in mouse embryos (Pyrgaki et al., 2010).

Cancer

The ability of cancer cells to communicate with each other results in more aggressive cancer phenotypes and resistance to treatment. TNTs contribute to tumorigenesis by facilitating long or short-distance communication. TNTs are formed in response to various cellular stresses and the cancer microenvironment induces TNT formation (Matejka & Reindl, 2019; Sosa et al., 2013). Many signaling pathways function dysregulated in cancer (Kretschmer et al., 2019; Ranzinger et al., 2011). p53, K-Ras, and PI3K/Akt/mTOR pathways are at the top of the irregular pathways and these pathways have severe roles in TNT formation (Dubois et al., 2018; Wang et al., 2011). During TNT formation, cellular signals in these pathways are involved in actin polymerization and can activate M-Sec (Ohno et al., 2010). Cancer cells can stimulate stromal cells to increase signals to trigger their proliferation. Cancer cell survival is supported by increasing cytokines and signals sent by TNTs from stromal cells. In this respect, TNTs appear to serve an oncogenic microenvironment (Polak et al., 2015). TNTs have been shown to exist for different cancers both in vitro and in vivo (Venkatesh & Lou, 2019; Osswald et al., 2019; Thayanithy et al., 2014). In studies with cancer cell lines suggest TNTs may contribute to metastasis and proliferation by enabling the communication and transfer of molecules between cancer cells. This homocellular transfer could alter apoptotic signals, metabolism, and chemotherapeutic response, although the specific mechanisms require further investigation (Lou et al., 2012). TNTs can transfer information about various mutations between cancer cells and increase cancer-specific phenotypes in recipient cells. For example, colon cancer cells have increased the invasiveness of the recipient cells by spreading the oncogenic information of the K-RAS mutation through TNTs. TNTs have many effects on cancer development (Desir et al., 2019). For example, TNTs permit cancer cells to reset the normal neighbor cells to render them more favorable to the generation of a tumor niche (Allegra et al., 2022).

miRNAs that have activities such as oxidative stress, neural development, DNA repair, immune response, and tumorigenesis are single-stranded RNAs. miRNAs are regulators of gene expression and contribute to tumorigenesis through TNTs (Allegra et al., 2022). Transfer of miRNAs via TNTs between tumor cells and surrounding stromal cells has been demonstrated in ovarian cancer (Thayanithy et al., 2014). Also, it has been reported that miRNAs transferred by TNTs activate the mTOR signaling pathway and stimulate cell proliferation (Huang & He, 2011; Lu et al., 2019). miRNA profiles in recipient cells may vary with TNTs. TNT-mediated communication between metastatic cancer cells and endothelial cells has been shown to alter the miRNA profile. The study showed that TNTs alter endothelial cells and form a pathway for the transfer of oncogenic factors (Connor et al., 2015).

Apoptosis is one of the critical cellular activities in the removal of cancerous cells. Apoptotic and antiapoptotic signals can also be transmitted by TNTs. In T cells, it has been shown that TNTs increase with the stimulation of the immune system, and apoptotic signals are carried by TNTs. With TNTs, phagocytic signals were transferred from apoptotic cells to living cells (Bittins & Wang, 2017).

Angiogenesis is defined as the formation of new vessels in order to provide the necessary factors for tumor growth and is one of the general characteristics of cancers. Endothelial cells and pericytes are known to induce angiogenesis through TNTs. The presence of TNT-like structures in the cerebral cortex and glioblastoma, where vascularization is intense, draws attention (Errede et al., 2018). Long TNTs can connect distant vessels for tumor growth. In contrast, short TNTs bind adjacent sprouting vessels (Astanina et al., 2015).

The effect of TNTs on energy metabolism in cancer cells is undeniable. Cells use mitochondria to produce energy, and increased mitochondrial activity has been shown in some cancers (Ashton et al., 2018; Saavedra, 2007). Mitochondria transfer via TNTs affects chemoresistance. It can promote chemoresistance through the reduction of ROS, prevention of apoptosis, and rescue of cells from dysfunctional mitochondria (Pasquier et al., 2013). Another effect of endothelial cells on cancer is to induce the transfer of mitochondria from mesenchymal cells via TNTs under chemotherapy stress. In this way, damaged endothelial cells get rid of chemotherapy stress, and proliferation, migration, and invasion can be promoted again by reducing the stress on endothelial cells (Feng et al., 2019).

Motor proteins are responsible for the transport of mitochondria via TNTs (Allegra et al., 2022). In TNTs, myosin Va and myosin X act together with mitochondria. In some donor cells, the small GTPase Miro1 provides mitochondria transfer (Gousset et al., 2013; Tardivel et al., 2016; Zhang et al., 2016) (Fig. 5.1). Decreased Miro1 expression in mesenchymal cells inhibited mitochondria transport (Ahmad et al., 2014). Transfer of mitochondria by TNTs alters energy metabolism in recipient cells. Due to the increased production of ATP and OXPHOS, the development of cancer cells is triggered (Dong et al., 2017; Tan et al., 2015).

Communication between cancer cells affects also resistance to treatment. In addition to promoting tumorigenesis, TNT-like structures, and miRNA, specific factors and mitochondria transfer render cancer cells resistant to treatment. Chemotherapy and radiotherapy, which are effective methods used in cancer treatment, induce TNT formation by increasing ROS production (Matejka & Reindl, 2019).

TNTs facilitate communication between cancer cells by facilitating the transfer of soluble molecules. This is perhaps the main suspect of chemoresistance. TNT biogenesis increased chemoresistance and cytokine secretion signals in B cell precursor acute lymphoblastic leukemia and acute myeloid leukemia (AML) cells (Omsland et al., 2017; Parker et al., 2017). Nuclear factor-kB (NF-kB) and tumor necrosis factor-alpha (TNF-α), which are responsible for the control of the M-sec promoter, induce TNT biogenesis in AML cells (Guzman et al., 2001). Accordingly, the increase in NF-kB and TNF-α leads to an increase in cytokine release and stress signals and the development of drug resistance (Mittal et al., 2019). Daunorubicin and cytarabine are chemotherapeutic agents for AML. Cytarabine inhibited TNT formation by inhibiting the NF-kB pathway in AML cells. Although daunorubicin is known to activate the NF-kB pathway, it reduced TNT formation in AML cells (Omsland et al., 2017). In light of these findings, it is possible to conclude that inhibition of the NF-kB pathway reduces chemotherapy resistance by reducing TNT biogenesis. Mitochondria transfer promotes metabolism and aerobic respiration in recipient cells to overcome treatment stress. It has been shown that endothelial cells and mesenchymal cells transfer mitochondria to cancer cells, triggering resistance to doxorubicin in recipient cells (Pasquier et al., 2013). Leukemic cells grafted in murine bone marrow have been shown to acquire mitochondria from stromal cells, which may affect cancer metabolism, chemoresistance, and cell proliferation. Inhibition of this transfer increased sensitivity to chemotherapy (Pinto et al., 2020). Increasing the production of ROS by chemotherapy increases the transfer of mitochondria. Inhibition of mitochondria transfer from mesenchymal cells to leukemic cells by cell contact increased apoptosis (Marlein et al., 2019). One of the important factors in chemotherapy resistance is the transfer of miRNAs. miR-199a is differentially expressed in chemosensitive cells and chemoresistant cells. This suggests that the recipient cells gain chemoresistance as a result of the transfer (Pinto et al., 2020).

On the other hand, the removal of autophagosomes, damaged organelles, ROS, or drug agents from the donor cell with TNTs is the cell's self-defense mechanism (Pinto et al., 2020; Sáenz-de-Santa-María et al., 2017). In a study of pancreatic and ovarian cancers, the transport of soluble drugs by TNT-like protrusions was observed (Desir et al., 2018). The drug is transferred from chemoresistance cells to chemosensitive cells via TNT-like protrusions and treatment-resistant cells occur. Although this situation is important for cancer treatment, there is no quantitative data supporting the relationship between the mechanisms (Pinto et al., 2020). Inhibition of TNT formation should be considered in the development of treatment strategies.

Neurodegenerative disease

Neurodegenerative diseases occur due to the accumulation of pathological proteins in the intracellular or extracellular areas of the central nervous system. Prions, amyloid-beta (Aβ) plaques, tau neurofibrils, alpha-synuclein (α-syn), and mutant Huntington (mHTT) protein aggregates cause common neurodegenerative diseases such as Parkinson's disease (PD), Alzheimer's disease (AD), and Huntington's disease (HD) (Vaquer-Alicea & Diamond, 2019). Considering the pathology of neurodegenerative diseases, protein aggregates are effectively transferred from cell to cell via exosome, secretion, transsynaptic transmission, and TNTs.

Cellular prion proteins (PrPCs) are infectious protein particles associated with glycosylphosphatidylinositol (GPI) and as a result of structural changes in PrPCs, pathogenic PrPCs (PrPSc) are formed (Engelke et al., 2018; Spagnolli et al., 2019). Although the mechanisms of PrPSc transfer have not yet been elucidated, an increased number of TNT and filopodia protrusions have been found in cells with PrPC (Miyazawa et al., 2010). The increase in the biogenesis and number of TNTs for PrPSc transport is due to cholesterol and ER stress (Zhu et al., 2015). Although the mechanisms of cell-to-cell transfer of PrPSc have not yet been elucidated, it is clear that TNTs that provide cell-to-cell contact play an important role. TNTs transfer intercellular PrPSc between neurons and from the periphery to the central nervous system via neuroimmune interactions (Gousset et al., 2009). There are some predictions about the transmission mechanisms of PrPSc in the development of transmissible spongiform encephalopathies. The basis of the first mechanism is the connection of PrPs with GPI. GPI linkages are a means of transporting prion proteins across the surface of TNTs. It was shown that PrPSc transferred across the cell membrane surface between ScGT1 cells in virtue of thin, micrometer-long strings sheltered from endocytosis (Rouvinski et al., 2014). According to the second proposed mechanism, PrPSc needs specific vesicles and organelles to be transported in TNTs. Studies with CAD cells support this mechanism. Transport of PrPSc in CAD cells via TNTs can be achieved by vesicles (Gousset et al., 2009). There is evidence to support that prions are transported by specific organelles. Cell-specific endosomes and endolysosomal vesicles can be used to transport PrPSc via TNT (Marzo et al., 2012; Victoria et al., 2016; Zhu et al., 2015). Moreover, PrPSc can be transferred by TNTs in the recycling compartments (ERC), lysosome, and early endosome where transformation takes place (Wang et al., 2021).

Prion-like protein aggregates such as tau and Aβ are at the basis of the pathogenesis of AD. TNTs are an important alternative pathway for cell-to-cell transfer of both tau neurofibrils and Aβ aggregates. Along with exogenous aggregates, endogenous aggregates are also transferred between neurons with thin TNTs (Wang et al., 2021). After exogenous tau is formed endogenously in CAD cells, it can be transferred from CAD cells to astrocytes and neurons by TNTs in ex vivo (Chastagner et al., 2020). There are studies that aim to prevent TNT formation and tau transfer as a treatment strategy in AD. CK-666 is an Arp2/3 inhibitor and inhibited the actin regulation effect of Arp2/3 and prevented TNT formation and tau transfer (Sartori-Rupp et al., 2019). Based on these results, inhibition of TNT biogenesis may be an important target as it will prevent the spread of pathological proteins. It has been suggested that oligomeric and fibrillar tau induce TNT biogenesis in primary neurons and neuronal CAD cells, thereby facilitating tau transfer (Abounit, Wu, et al., 2016b; Peggion et al., 2014). In addition, studies have suggested that tau can bind with actin. The proline-rich domain of tau and the microtubule-binding domain bind with F-actin, allowing F-actin to be bundled. Therefore, tau aggregates are assumed to be transported by other cellular extensions, including TNT, as they are associated with actin (Elie et al., 2015; Wang et al., 2021). Aβ, which has a cytotoxic effect on neurons, also increases neurotoxicity by spreading with TNTs, just like tau neurofibrillary. Under stress conditions, TNTs can promote cell-to-cell diffusion of Aβ aggregates. It has been suggested that extracellular Aβ plaques of rat hippocampal neurons and astrocytes can be transported unidirectionally by TNTs under certain stress conditions (Wang et al., 2011). The transport rate of Aβ aggregates is faster than that of organelles. The protrusions for the transport of Aβ aggregates by TNTs develop from stressed cells to nonstressed cells. Thus, it suggests that support is provided to ensure the survival of the stressed cell (Wang et al., 2021; Wang & Gerdes, 2015).

PD, in which α-syn aggregates form Lewy bodies, is the second most common neurodegenerative disease. The formation, aggregation, and transfer of α-syn aggregates promote PH progression (Spillantini et al., 1998). Clarification of transferring mechanisms is very important as it will bring new treatment strategies. α-syn is transported with TNTs between human neural progenitor cells, primary neurons, and astrocytes (Rostami et al., 2017; Grudina et al., 2019; Abounit, Bousset, et al., 2016a). Considering the intercellular transfer efficiency of various cellular components, cellular contact increases the transfer efficiency. It has been shown that fibrillar α-syn are transferred by TNTs within lysosomal vesicles in neuronal-like cells and primary neurons. Cells to which fibrillar α-syn were transferred showed a significant increase in the level of ROS, and this increase promoted TNT formation, resulting in more fibrillar α-syn transfer (Abounit, Bousset, et al., 2016a). Astrocytes play significant roles in α-syn transfer and PD progression. The accumulation of α-syn resulted in lysosomal–autophagosomal dysfunction and mitochondria dysfunction. With the increased stress created by aggregates in astrocytes, a defense mechanism may be triggered by transferring α-syn fibrils to healthy astrocytes to reduce stress. In addition, healthy astrocyte cells can transfer functional mitochondria to the stressed cell where the aggregates are located. Inhibition of actin polymerization can inhibit α-syn transfer via TNTs. In this context, it has been shown that α-syn/MHC-II aggregates of antigen-presenting astrocytes are transferred with TNTs. This shows that TNTs contribute to the spread of inflammation as well as pathological protein aggregates (Rostami et al., 2020). The transfer of α-syn aggregates between the brain and the periphery using the blood–brain barrier is another mechanism in the pathogenesis of PD. It has been shown that α-syn transfer between pericytes in the blood–brain barrier occurs through a network of TNTs (Sui et al., 2014; Wang et al., 2021). TNTs can be considered as effective pharmacological targets for inhibiting the transport of prion-like α-syn fibrils in the brain.

mHTT protein aggregates encoded by the mutant HTT gene contribute to HD progression by causing neurotoxicity. As in other neurodegenerative diseases, the transfer mechanisms of mHTT proteins are a target for HD. Studies have shown that mHTT contributes to the early stage of HD by spreading through TNTs in neuronal cells. Accumulation of mHTT aggregates in neuronal cells facilitates TNT formation and also increases mHTT transfer (Costanzo et al., 2013). GTPase/SUMO E3-like protein Rhes, a highly enriched in the striatum protein, dissolves mHTT aggregates and has been associated with behavioral disorders related to HD (Subramaniam et al., 2009; Swarnkar et al., 2015). An important role of TNT-like Rhes tunnels in the development of HD has recently been reported. It has been suggested that the SUMO E3 ligase domain of the Rhes protein facilitates the formation of TNT-like cellular protrusions in striatal neuronal cells (Sharma & Subramaniam, 2019). mHTT interacts with the phospholipid bilayer that forms the cell membrane. This interaction appears to be sufficient to explain the transport of mHTT aggregates across the TNT membrane (Wang et al., 2021).

TNTs play important roles in the transport of PrP and PrP-like proteins (fibrillar α-syn, Aβ, mHTT, etc.) in the brain. Although the transport is rapid, the transfer rate varies according to the size of the transported proteins. The number of TNTs formed for transport depends on cell motility and the difference in the number of interacting cells. Considering the studies, understanding the mechanisms underlying TNT biogenesis may be an important strategy for the treatment of neurodegenerative diseases.

Viral infections

Viruses evade host cell immunity by using TNTs to transfer their genomes to naive cells. The transport of viruses with TNTs provides some advantages in the transfer of viral particles. (1) It is an advantageous way in terms of energy compared to conventional processes. (2) Viruses travel on the cell surface using actin filaments and attachment to the target cell is facilitated. (3) TNTs increase the surface area; thus, the docking of viruses increases (4) TNTs protect viral components from proteases, immune cells, drugs, and phagocytic cells. (5) Transport by TNTs is faster than extracellular transport (Ljubojevic et al., 2021; Mittal et al., 2019; Tiwari et al., 2021). The rapid actin polymerization in TNTs makes TNT movement much more efficient in transporting viral particles (Önfelt et al., 2004). They form a pathway not only for the transfer of viral components but also for virions to surf. Thus, the propagation time between cells is shortened. This mechanism explains the long-distance spread of Herpes viruses between fibroblasts and epithelial cells via TNTs (Panasiuk et al., 2018). Cell-to-cell spread of viruses by TNTs protects viruses from circulating immune cells and inhibits virus–cell interaction for host defense (Tiwari et al., 2021). Understanding TNT mechanisms that effectively protect viruses from pharmaceutical targets and immune cells is an important strategy in the treatment of viral infections.

HIV infection in macrophages increased TNTs and HIV particles were found in TNTs during infection (Eugenin et al., 2009). It has been observed that virions travel extracellularly and intracellularly by endocytic transfer. This suggests that HIV uses a mechanism that promotes the formation of TNTs to spread (Sowinski et al., 2008). The mechanism by which HIV induces TNT formation has been described in a study with monocyte-derived macrophages. It can be said that M-sec and Nef proteins play a leading role in induction. Nef interacts with the Rho GTPase Ral A and the exocyst complex described in TNT biogenesis. It has been observed that there is no TNT induction in Nef deficiency in HIV. In addition, HIV infection was reduced when a molecule that inhibits TNT formation via M-sec was administered. This reduction was not observed in Nef-deficient mutant HIV-1 or the M-Sec-deficient $CD4^+$ T cell line (Hashimoto et al., 2016). In light of the results, it can be said that the M-Sec-Nef signaling pathway is critical in terms of pathogenesis for HIV transmission. In addition, the Nef protein controls intercellular communication by inducing TNT biogenesis via MyoX. Although the roles of Nef in TNT biogenesis are unclear, it has been accepted that the proteins of the exocyst complex act as effectors for Nef-mediated TNT formation (Mukerji et al., 2012; Tiwari et al., 2021).

TNTs express connexion-43 at their ends, enabling viral replication. Thus, the virus spreads between HIV-infected and noninfected macrophages (Okafo et al., 2017). Initial host infection and HIV reactivation refer to low viral load states that require cell-free virus infection. Therefore, TNT biogenesis is associated with initial host infection and HIV reactivation in the HIV life cycle (Mittal et al., 2019; Okafo et al., 2017). A study reported a decrease in TNT formation when the Gap junction connections at the end of cell-to-cell TNT connections were inhibited. This inhibition decreased intercellular spread and viral replication (Okafo et al., 2017). HIV uses TNTs to transfer Nef from infected macrophages to B cells. This transfer inhibits immunoglobulin A and IgG class switching and helps HIV escape host immunity (Mukerji et al., 2012; Xu et al., 2009).

Transfer of influenza proteins by TNTs has also been reported in the pathogenesis of influenza (Roberts et al., 2015). The influenza virus promoted TNT formation in macrophages but not in T cells. Transfer of viral proteins by TNTs is sufficient for viral replication and infection of uninfected cells (Kumar et al., 2017). Herpes viruses use gap junctions, endocytic pathways and TNTs to spread from cell to cell. It has been shown in a study in bovine herpes viruses that not only Herpes-1 but also organelles and envelope glycoprotein gE, tegument protein US3, and capsid protein VP26 proteins are transferred with TNTs (Panasiuk et al., 2018).

Conclusion

In summary, TNTs are a form of connection between cells at a close or far distance. TNTs have heterogeneous composition and morphology, myosin X, M-sec, LST1, and p53 are major molecules in TNT formation, and biogenesis is triggered under various stress situations (UV, serum starvation, oxidative stress, etc.). This connection contributes to many biological processes such as tissue repair, the spread of pathogens, drug resistance, signal transduction, apoptosis, cell differentiation, cancer development, and neurodegenerative diseases. It is necessary to elucidate the effects of TNT structure, which is thought to be especially effective in mitochondrial transfer, on cellular communication and regulation processes with advanced research methods. Cell-to-cell transfer of mitochondria causes remarkable cellular effects in target cells because mitochondria play critical roles in cellular functions such as lipid and amino acid biosynthesis, ATP production, and apoptosis.

Mitochondrial transfer promotes healing in slowly proliferating or nonregenerating injured cells such as neurons and cardiomyocytes, while transferring survival signals from drug-susceptible cancer cells to nonsusceptible cells may render cancer cells resistant to apoptosis and therapy. In such a scenario, existing strategies will need to be developed to find means to inhibit treatment resistance and induce tissue repair.

Therefore, cell-to-cell transfer of mitochondria via TNTs appears to be a double-edged sword. A better understanding of this cellular communication system, which can be a potential therapeutic target in different pathological conditions, will increase our chances of developing effective therapeutic approaches. TNTs might be a mechanism for cells to respond to harmful signals and transfer cellular substances or energy to another cell under stress.

References

Abounit, S., Bousset, L., Loria, F., Zhu, S., de Chaumont, F., Pieri, L., Olivo-Marin, J. C., Melki, R., & Zurzolo, C. (2016). Tunneling nanotubes spread fibrillar α-synuclein by intercellular trafficking of lysosomes. *EMBO Journal, 35*(19), 2120–2138. https://doi.org/10.15252/embj.201593411

Abounit, S., Wu, J. W., Duff, K., Victoria, G. S., & Zurzolo, C. (2016). Tunneling nanotubes: A possible highway in the spreading of tau and other prion-like proteins in neurodegenerative diseases. *Prion, 10*(5), 344–351. https://doi.org/10.1080/19336896.2016.1223003

Abounit, S., & Zurzolo, C. (2012). Wiring through tunneling nanotubes – From electrical signals to organelle transfer. *Journal of Cell Science, 125*(5), 1089–1098. https://doi.org/10.1242/jcs.083279

Ady, J., Thayanithy, V., Mojica, K., Wong, P., Carson, J., Rao, P., Fong, Y., & Lou, E. (2016). Tunneling nanotubes: An alternate route for propagation of the bystander effect following oncolytic viral infection. *Molecular Therapy – Oncolytics, 3*, 16029. https://doi.org/10.1038/mto.2016.29

Ahmad, T., Mukherjee, S., Pattnaik, B., Kumar, M., Singh, S., Rehman, R., Tiwari, B. K., Jha, K. A., Barhanpurkar, A. P., Wani, M. R., Roy, S. S., Mabalirajan, U., Ghosh, B., & Agrawal, A. (2014). Miro1 regulates intercellular mitochondrial transport and enhances mesenchymal stem cell rescue efficacy. *EMBO Journal, 33*(9), 994–1010. https://doi.org/10.1002/embj.201386030

Alarcon-Martinez, L., Villafranca-Baughman, D., Quintero, H., Kacerovsky, J. B., Dotigny, F., Murai, K. K., Prat, A., Drapeau, P., & Di Polo, A. (2020). Interpericyte tunnelling nanotubes regulate neurovascular coupling. *Nature, 585*(7823), 91–95. https://doi.org/10.1038/s41586-020-2589-x

Allegra, A., Di Gioacchino, M., Cancemi, G., Casciaro, M., Petrarca, C., Musolino, C., & Gangemi, S. (2022). Specialized intercellular communications via tunnelling nanotubes in acute and chronic leukemia. *Cancers, 14*(3), 659. https://doi.org/10.3390/cancers14030659

Andresen, V., Wang, X., Ghimire, S., Omsland, M., Gjertsen, B. T., & Gerdes, H. H. (2013). Tunneling nanotube (TNT) formation is independent of p53 expression. *Cell Death & Differentiation, 20*(8), 1124. https://doi.org/10.1038/cdd.2013.61

Antanavičiūtė, I., Rysevaitė, K., Liutkevičius, V., Marandykina, A., Rimkutė, L., Sveikatienė, R., Uloza, V., Skeberdis, V. A., & Scemes, E. (2014). Long-distance communication between laryngeal carcinoma cells. *PLoS One, 9*(6), e99196. https://doi.org/10.1371/journal.pone.0099196

Ariazi, J., Benowitz, A., De Biasi, V., Den Boer, M. L., Cherqui, S., Cui, H., Douillet, N., Eugenin, E. A., Favre, D., Goodman, S., Gousset, K., Hanein, D., Israel, D. I., Kimura, S., Kirkpatrick, R. B., Kuhn, N., Jeong, C., Lou, E., Mailliard, R., … Zurzolo, C. (2017). Tunneling nanotubes and gap junctions—their role in long-range intercellular communication during development, health, and disease conditions. *Frontiers in Molecular Neuroscience, 10*. https://doi.org/10.3389/fnmol.2017.00333

Ashton, T. M., Gillies McKenna, W., Kunz-Schughart, L. A., & Higgins, G. S. (2018). Oxidative phosphorylation as an emerging target in cancer therapy. *Clinical Cancer Research, 24*(11), 2482—2490. https://doi.org/10.1158/1078-0432.CCR-17-3070

Astanina, K., Koch, M., Jüngst, C., Zumbusch, A., & Kiemer, A. K. (2015). Lipid droplets as a novel cargo of tunnelling nanotubes in endothelial cells. *Scientific Reports, 5*. https://doi.org/10.1038/srep11453

Austefjord, M. W., Gerdes, H. H., & Wang, X. (2014). Tunneling nanotubes: Diversity in morphology and structure. *Communicative & Integrative Biology, 7*(2). https://doi.org/10.4161/cib.27934

Babenko, V. A., Silachev, D. N., Popkov, V. A., Zorova, L. D., Pevzner, I. B., Plotnikov, E. Y., Sukhikh, G. T., & Zorov, D. B. (2018). Miro1 enhances mitochondria transfer from multipotent mesenchymal stem cells (MMSC) to neural cells and improves the efficacy of cell recovery. *Molecules, 23*(3). https://doi.org/10.3390/molecules23030687

Babenko, V. A., Silachev, D. N., Zorova, L. D., Pevzner, I. B., Khutornenko, A. A., Plotnikov, E. Y., Sukhikh, G. T., & Zorov, D. B. (2015). Improving the post-stroke therapeutic potency of mesenchymal multipotent stromal cells by cocultivation with cortical neurons: The role of crosstalk between cells. *Stem Cells Translational Medicine, 4*(9), 1011—1020. https://doi.org/10.5966/sctm.2015-0010

Baker, M. (2017). How the Internet of cells has biologists buzzing. *Nature, 549*(7672), 322—324. https://doi.org/10.1038/549322a

Batagov, A. O., & Kurochkin, I. V. (2013). Exosomes secreted by human cells transport largely mRNA fragments that are enriched in the 3′-untranslated regions. *Biology Direct, 8*(1). https://doi.org/10.1186/1745-6150-8-12

Bittins, M., & Wang, X. (2017). TNT-induced phagocytosis: Tunneling nanotubes mediate the transfer of pro-phagocytic signals from apoptotic to viable cells. *Journal of Cellular Physiology, 232*(9), 2271—2279. https://doi.org/10.1002/jcp.25584

Bloemendal, S., & Kück, U. (2013). Cell-to-cell communication in plants, animals, and fungi: A comparative review. *Naturwissenschaften, 100*(1), 3—19. https://doi.org/10.1007/s00114-012-0988-z

Bukoreshtliev, N. V., Wang, X., Hodneland, E., Gurke, S., Barroso, J. F. V., & Gerdes, H. H. (2009). Selective block of tunneling nanotube (TNT) formation inhibits intercellular organelle transfer between PC12 cells. *FEBS Letters, 583*(9), 1481—1488. https://doi.org/10.1016/j.febslet.2009.03.065

Burtey, A., Wagner, M., Hodneland, E., Skaftnesmo, K. O., Schoelermann, J., Mondragon, I. R., Espedal, H., Golebiewska, A., Niclou, S. P., Bjerkvig, R., Kögel, T., & Gerdes, H. H. (2015). Intercellular transfer of transferrin receptor by a contact-, Rab8-dependent mechanism involving tunneling nanotubes. *Federation of American Societies for Experimental Biology Journal, 29*(11), 4695—4712. https://doi.org/10.1096/fj.14-268615

Caneparo, L., Pantazis, P., Dempsey, W., Fraser, S. E., & Laudet, V. (2011). Intercellular bridges in vertebrate gastrulation. *PLoS One, 6*(5), e20230. https://doi.org/10.1371/journal.pone.0020230

Chastagner, P., Loria, F., Vargas, J. Y., Tois, J., I Diamond, M., Okafo, G., Brou, C., & Zurzolo, C. (2020). Fate and propagation of endogenously formed Tau aggregates in neuronal cells. *EMBO Molecular Medicine, 12*(12). https://doi.org/10.15252/emmm.202012025

Chauveau, A., Aucher, A., Eissmann, P., Vivier, E., & Davis, D. M. (2010). Membrane nanotubes facilitate long-distance interactions between natural killer cells and target cells. *Proceedings of the National Academy of Sciences, 107*(12), 5545—5550. https://doi.org/10.1073/pnas.0910074107

Cho, Y. M., Kim, J. H., Kim, M., Park, S. J., Koh, S. H., Ahn, H. S., Kang, G. H., Lee, J. B., Park, K. S., & Lee, H. K. (2012). Mesenchymal stem cells transfer mitochondria to the cells with virtually no mitochondrial function but not with pathogenic mtDNA mutations. *PLoS One, 7*(3). https://doi.org/10.1371/journal.pone.0032778

Climent, M., Quintavalle, M., Miragoli, M., Chen, J., Condorelli, G., & Elia, L. (2015). TGFβ triggers miR-143/145 transfer from smooth muscle cells to endothelial cells, thereby modulating vessel stabilization. *Circulation Research, 116*(11), 1753–1764. https://doi.org/10.1161/CIRCRESAHA.116.305178

Connor, Y., Tekleab, S., Nandakumar, S., Walls, C., Tekleab, Y., Husain, A., Gadish, O., Sabbisetti, V., Kaushik, S., Sehrawat, S., Kulkarni, A., Dvorak, H., Zetter, B., Edelman, E. R., & Sengupta, S. (2015). Physical nanoscale conduit-mediated communication between tumour cells and the endothelium modulates endothelial phenotype. *Nature Communications, 6*(1). https://doi.org/10.1038/ncomms9671

Costanzo, M., Abounit, S., Marzo, L., Danckaert, A., Chamoun, Z., Roux, P., & Zurzolo, C. (2013). Transfer of polyglutamine aggregates in neuronal cells occurs in tunneling nanotubes. *Journal of Cell Science, 126*(16), 3678–3685. https://doi.org/10.1242/jcs.126086

Cselenyák, A., Pankotai, E., Horváth, E. M., Kiss, L., & Lacza, Z. (2010). Mesenchymal stem cells rescue cardiomyoblasts from cell death in an in vitro ischemia model via direct cell-to-cell connections. *BMC Cell Biology, 11*. https://doi.org/10.1186/1471-2121-11-29

Delage, E., Cervantes, D. C., Pénard, E., Schmitt, C., Syan, S., Disanza, A., Scita, G., & Zurzolo, C. (2016). Differential identity of Filopodia and Tunneling Nanotubes revealed by the opposite functions of actin regulatory complexes. *Scientific Reports, 6*. https://doi.org/10.1038/srep39632

Desir, S., O'Hare, P., Vogel, R. I., Sperduto, W., Sarkari, A., Dickson, E. L., Wong, P., Nelson, A. C., Fong, Y., Steer, C. J., Subramanian, S., & Lou, E. (2018). Chemotherapy-induced tunneling nanotubes mediate intercellular drug efflux in pancreatic cancer. *Scientific Reports, 8*(1). https://doi.org/10.1038/s41598-018-27649-x

Desir, W., Turbyville, C., Shetty, C., Zhai, R., Manova-Todorova, Starr, N., Steer, S., & Lou. (2019). Intercellular transfer of oncogenic KRAS via tunneling nanotubes introduces intracellular mutational heterogeneity in colon cancer cells. *Cancers, 11*(7), 892. https://doi.org/10.3390/cancers11070892

Dilna, A., Deepak, K. V., Damodaran, N., Kielkopf, C. S., Kagedal, K., Ollinger, K., & Nath, S. (2021). Amyloid-β induced membrane damage instigates tunneling nanotube-like conduits by p21-activated kinase dependent actin remodulation. *Biochimica et Biophysica Acta – Molecular Basis of Disease, 1867*(12), 166246. https://doi.org/10.1016/j.bbadis.2021.166246

Dilsizoglu Senol, A., Pepe, A., Grudina, C., Sassoon, N., Reiko, U., Bousset, L., Melki, R., Piel, J., Gugger, M., & Zurzolo, C. (2019). Effect of tolytoxin on tunneling nanotube formation and function. *Scientific Reports, 9*(1). https://doi.org/10.1038/s41598-019-42161-6

Dong, L. F., Kovarova, J., Bajzikova, M., Bezawork-Geleta, A., Svec, D., Endaya, B., Sachaphibulkij, K., Coelho, A. R., Sebkova, N., Ruzickova, A., Tan, A. S., Kluckova, K., Judasova, K., Zamecnikova, K., Rychtarcikova, Z., Gopalan, V., Andera, L., Sobol, M., Yan, B., … Neuzil, J. (2017). Horizontal transfer of whole mitochondria restores tumorigenic potential in mitochondrial DNA-deficient cancer cells. *Elife, 6*. https://doi.org/10.7554/eLife.22187

Driscoll, J., Gondaliya, P., & Patel, T. (2022). Tunneling nanotube-mediated communication: A mechanism of intercellular nucleic acid transfer. *International Journal of Molecular Sciences, 23*(10), 5487. https://doi.org/10.3390/ijms23105487

Dubois, F., Jean-Jacques, B., Roberge, H., Bénard, M., Galas, L., Schapman, D., Elie, N., Goux, D., Keller, M., Maille, E., Bergot, E., Zalcman, G., & Levallet, G. (2018). A role for RASSF1A in tunneling nanotube formation between cells through GEFH1/Rab11 pathway control. *Cell Communication and Signaling, 16*(1). https://doi.org/10.1186/s12964-018-0276-4

Dupont, M., Souriant, S., Lugo-Villarino, G., Maridonneau-Parini, I., & Vérollet, C. (2018). Tunneling nanotubes: Intimate communication between myeloid cells. *Frontiers in Immunology, 9*. https://doi.org/10.3389/fimmu.2018.00043

El Najjar, F., Cifuentes-Muñoz, N., Chen, J., Zhu, H., Buchholz, U. J., Moncman, C. L., & Dutch, R. E. (2016). Human metapneumovirus induces reorganization of the actin cytoskeleton for direct cell-to-cell spread. *PLoS Pathogens, 12*(9). https://doi.org/10.1371/journal.ppat.1005922

Elie, A., Prezel, E., Guérin, C., Denarier, E., Ramirez-Rios, S., Serre, L., Andrieux, A., Fourest-Lieuvin, A., Blanchoin, L., & Arnal, I. (2015). Tau co-organizes dynamic microtubule and actin networks. *Scientific Reports, 5*(1). https://doi.org/10.1038/srep09964

Engelke, A. D., Gonsberg, A., Thapa, S., Jung, S., Ulbrich, S., Seidel, R., Basu, S., Multhaup, G., Baier, M., Engelhard, M., Schätzl, H. M., Winklhofer, K. F., & Tatzelt, J. (2018). Dimerization of the cellular prion protein inhibits propagation of scrapie prions. *Journal of Biological Chemistry, 293*(21), 8020–8031. https://doi.org/10.1074/jbc.ra117.000990

Errede, M., Mangieri, D., Longo, G., Girolamo, F., De Trizio, I., Vimercati, A., Serio, G., Frei, K., Perris, R., & Virgintino, D. (2018). Tunneling nanotubes evoke pericyte/endothelial communication during normal and tumoral angiogenesis. *Fluids and Barriers of the CNS, 15*(1). https://doi.org/10.1186/s12987-018-0114-5

Eugenin, E. A., Gaskill, P. J., & Berman, J. W. (2009). Tunneling nanotubes (TNT) are induced by HIV-infection of macrophages: A potential mechanism for intercellular HIV trafficking. *Cellular Immunology, 254*(2), 142–148. https://doi.org/10.1016/j.cellimm.2008.08.005

Feng, Y., Zhu, R., Shen, J., Wu, J., Lu, W., Zhang, J., Zhang, J., & Liu, K. (2019). Human bone marrow mesenchymal stem cells rescue endothelial cells experiencing chemotherapy stress by mitochondrial transfer via tunneling nanotubes. *Stem Cells and Development, 28*(10), 674–682. https://doi.org/10.1089/scd.2018.0248

Gerdes, H. H., Bukoreshtliev, N. V., & Barroso, J. F. V. (2007). Tunneling nanotubes: A new route for the exchange of components between animal cells. *FEBS Letters, 581*(11), 2194–2201. https://doi.org/10.1016/j.febslet.2007.03.071

Gerdes, H. H., Rustom, A., & Wang, X. (2013). Tunneling nanotubes, an emerging intercellular communication route in development. *Mechanisms of Development, 130*(6–8), 381–387. https://doi.org/10.1016/j.mod.2012.11.006

Gittes, F., Mickey, B., Nettleton, J., & Howard, J. (1993). Flexural rigidity of microtubules and actin filaments measured from thermal fluctuations in shape. *Journal of Cell Biology, 120*(4), 923–934. https://doi.org/10.1083/jcb.120.4.923

Gousset, K., Marzo, L., Commere, P. H., & Zurzolo, C. (2013). Myo10 is a key regulator of TNT formation in neuronal cells. *Journal of Cell Science, 126*(19), 4424–4435. https://doi.org/10.1242/jcs.129239

Gousset, K., Schiff, E., Langevin, C., Marijanovic, Z., Caputo, A., Browman, D. T., Chenouard, N., de Chaumont, F., Martino, A., Enninga, J., Olivo-Marin, J. C., Männel, D., & Zurzolo, C. (2009). Prions hijack tunnelling nanotubes for intercellular spread. *Nature Cell Biology, 11*(3), 328–336. https://doi.org/10.1038/ncb1841

Grudina, C., Kouroupi, G., Nonaka, T., Hasegawa, M., Matsas, R., & Zurzolo, C. (2019). Human NPCs can degrade α–syn fibrils and transfer them preferentially in a cell contact-dependent manner possibly through TNT-like structures. *Neurobiology of Disease, 132*, 104609. https://doi.org/10.1016/j.nbd.2019.104609

Guo, R., Katz, B. B., Tomich, J. M., Gallagher, T., & Fang, Y. (2016). Porcine reproductive and respiratory syndrome virus utilizes nanotubes for intercellular spread. *Journal of Virology, 90*(10), 5163–5175. https://doi.org/10.1128/JVI.00036-16

Guzman, M. L., Neering, S. J., Upchurch, D., Grimes, B., Howard, D. S., Rizzieri, D. A., Luger, S. M., & Jordan, C. T. (2001). Nuclear factor-κB is constitutively activated in primitive human acute myelogenous leukemia cells. *Blood, 98*(8), 2301–2307. https://doi.org/10.1182/blood.V98.8.2301

Haimovich, G., Dasgupta, S., & Gerst, J. E. (2021). RNA transfer through tunneling nanotubes. *Biochemical Society Transactions, 49*(1), 145–160. https://doi.org/10.1042/BST20200113

Haimovich, G., Ecker, C. M., Dunagin, M. C., Eggan, E., Raj, A., Gerst, J. E., & Singer, R. H. (2017). Intercellular mRNA trafficking via membrane nanotube-like extensions in mammalian cells. *Proceedings of the National Academy of Sciences of the United States of America, 114*(46), E9873–E9882. https://doi.org/10.1073/pnas.1706365114

Han, H., Hu, J., Yan, Q., Zhu, J., Zhu, Z., Chen, Y., Sun, J., & Zhang, R. (2016). Bone marrow-derived mesenchymal stem cells rescue injured H9c2 cells via transferring intact mitochondria through tunneling nanotubes in an in vitro simulated ischemia/reperfusion model. *Molecular Medicine Reports, 13*(2), 1517–1524. https://doi.org/10.3892/mmr.2015.4726

Han, X., & Wang, X. (2021). Opportunities and challenges in tunneling nanotubes research: How far from clinical application? *International Journal of Molecular Sciences, 22*(5), 2306. https://doi.org/10.3390/ijms22052306

Hanna, S. J., McCoy-Simandle, K., Miskolci, V., Guo, P., Cammer, M., Hodgson, L., & Cox, D. (2017). The role of Rho-GTPases and actin polymerization during macrophage tunneling nanotube biogenesis. *Scientific Reports, 7*(1). https://doi.org/10.1038/s41598-017-08950-7

Hase, K., Kimura, S., Takatsu, H., Ohmae, M., Kawano, S., Kitamura, H., Ito, M., Watarai, H., Hazelett, C. C., Yeaman, C., & Ohno, H. (2009). M-Sec promotes membrane nanotube formation by interacting with Ral and the exocyst complex. *Nature Cell Biology, 11*(12), 1427–1432. https://doi.org/10.1038/ncb1990

Hashimoto, M., Bhuyan, F., Hiyoshi, M., Noyori, O., Nasser, H., Miyazaki, M., Saito, T., Kondoh, Y., Osada, H., Kimura, S., Hase, K., Ohno, H., & Suzu, S. (2016). Potential role of the formation of tunneling nanotubes in HIV-1 spread in macrophages. *Journal of Immunology, 196*(4), 1832–1841. https://doi.org/10.4049/jimmunol.1500845

Hayakawa, K., Esposito, E., Wang, X., Terasaki, Y., Liu, Y., Xing, C., Ji, X., & Lo, E. H. (2016). Transfer of mitochondria from astrocytes to neurons after stroke. *Nature, 535*(7613), 551–555. https://doi.org/10.1038/nature18928

Huang, S., & He, X. (2011). The role of microRNAs in liver cancer progression. *British Journal of Cancer, 104*(2), 235–240. https://doi.org/10.1038/sj.bjc.6606010

Hung, M. E., & Leonard, J. N. (2016). A platform for actively loading cargo RNA to elucidate limiting steps in EV-mediated delivery. *Journal of Extracellular Vesicles, 5*(1). https://doi.org/10.3402/jev.v5.31027

Islam, M. N., Das, S. R., Emin, M. T., Wei, M., Sun, L., Westphalen, K., Rowlands, D. J., Quadri, S. K., Bhattacharya, S., & Bhattacharya, J. (2012). Mitochondrial transfer from bone-marrow–derived stromal cells to pulmonary alveoli protects against acute lung injury. *Nature Medicine, 18*(5), 759–765. https://doi.org/10.1038/nm.2736

Jackson, M. V., Morrison, T. J., Doherty, D. F., McAuley, D. F., Matthay, M. A., Kissenpfennig, A., O'Kane, C. M., & Krasnodembskaya, A. D. (2016). Mitochondrial transfer via tunneling nanotubes is an important mechanism by which mesenchymal stem cells enhance macrophage phagocytosis in the in vitro and in vivo models of ARDS. *Stem Cells, 34*(8), 2210–2223. https://doi.org/10.1002/stem.2372

Kadiu, I., & Gendelman, H. E. (2011a). Human immunodeficiency virus type 1 endocytic trafficking through macrophage bridging conduits facilitates spread of infection. *Journal of Neuroimmune Pharmacology, 6*(4), 658–675. https://doi.org/10.1007/s11481-011-9298-z

Kadiu, I., & Gendelman, H. E. (2011b). Macrophage bridging conduit trafficking of HIV-1 through the endoplasmic reticulum and Golgi network. *Journal of Proteome Research, 10*(7), 3225–3238. https://doi.org/10.1021/pr200262q

Keller, K. (2020). Tunneling nanotubes and actin cytoskeleton dynamics in glaucoma. *Neural Regeneration Research, 15*(11), 2031–2032. https://doi.org/10.4103/1673-5374.282254

Kimura, S., Hase, K., & Ohno, H. (2013). The molecular basis of induction and formation of tunneling nanotubes. *Cell and Tissue Research, 352*(1), 67–76. https://doi.org/10.1007/s00441-012-1518-1

Kimura, S., Hase, K., & Ohno, H. (2012). Tunneling nanotubes: Emerging view of their molecular components and formation mechanisms. *Experimental Cell Research, 318*(14), 1699–1706. https://doi.org/10.1016/j.yexcr.2012.05.013

Kretschmer, A., Zhang, F., Somasekharan, S. P., Tse, C., Leachman, L., Gleave, A., Li, B., Asmaro, I., Huang, T., Kotula, L., Sorensen, P. H., & Gleave, M. E. (2019). Stress-induced tunneling nanotubes support treatment adaptation in prostate cancer. *Scientific Reports, 9*(1). https://doi.org/10.1038/s41598-019-44346-5

Kumar, A., Kim, J. H., Ranjan, P., Metcalfe, M. G., Cao, W., Mishina, M., Gangappa, S., Guo, Z., Boyden, E. S., Zaki, S., York, I., García-Sastre, A., Shaw, M., & Sambhara, S. (2017). Influenza virus exploits tunneling nanotubes for cell-to-cell spread. *Scientific Reports, 7*. https://doi.org/10.1038/srep40360

Las, G., & Shirihai, O. S. (2014). Miro1: New wheels for transferring mitochondria. *The EMBO Journal, 33*(9), 939–941. https://doi.org/10.1002/embj.201488441

Li, X., Zhang, Y., Yeung, S. C., Liang, Y., Liang, X., Ding, Y., Ip, M. S. M., Tse, H. F., Mak, J. C. W., & Lian, Q. (2014). Mitochondrial transfer of induced pluripotent stem cell-derived mesenchymal stem cells to airway epithelial cells attenuates cigarette smoke-induced damage. *American Journal of Respiratory Cell and Molecular Biology, 51*(3), 455–465. https://doi.org/10.1165/rcmb.2013-0529OC

Lin, T.-K., Chen, S.-D., Chuang, Y.-C., Lan, M.-Y., Chuang, J.-H., Wang, P.-W., Hsu, T.-Y., Wang, F.-S., Tsai, M.-H., Huang, S.-T., Wang, X.-W., Tsai, P.-C., Lin, H.-Y., & Liou, C.-W. (2019). Mitochondrial transfer of Wharton's jelly mesenchymal stem cells eliminates mutation burden and rescues mitochondrial bioenergetics in Rotenone-stressed MELAS fibroblasts. *Oxidative Medicine and Cellular Longevity, 2019*, 1–17. https://doi.org/10.1155/2019/9537504

Ljubojevic, N., Henderson, J. M., & Zurzolo, C. (2021). The ways of actin: Why tunneling nanotubes are unique cell protrusions. *Trends in Cell Biology, 31*(2), 130–142. https://doi.org/10.1016/j.tcb.2020.11.008

Lou, E., Fujisawa, S., Barlas, A., Romin, Y., Manova-Todorova, K., Moore, M. A. S., & Subramanian, S. (2012). Tunneling nanotubes: A new paradigm for studying intercellular communication and therapeutics in cancer. *Communicative & Integrative Biology, 5*(4), 399–403. https://doi.org/10.4161/cib.20569

Lu, J. J., Yang, W. M., Li, F., Zhu, W., & Chen, Z. (2019). Tunneling nanotubes mediated microRNA-155 intercellular transportation promotes bladder cancer cells' invasive and proliferative capacity. *International Journal of Nanomedicine, 14*, 9731–9743. https://doi.org/10.2147/IJN.S217277

Marlein, C. R., Piddock, R. E., Mistry, J. J., Zaitseva, L., Hellmich, C., Horton, R. H., Zhou, Z., Auger, M. J., Bowles, K. M., & Rushworth, S. A. (2019). CD38-driven mitochondrial trafficking promotes bioenergetic plasticity in multiple myeloma. *Cancer Research, 79*(9), 2285–2297. https://doi.org/10.1158/0008-5472.CAN-18-0773

Marzo, L., Gousset, K., & Zurzolo, C. (2012). Multifaceted roles of tunneling nanotubes in intercellular communication. *Frontiers in Physiology, 3*. https://doi.org/10.3389/fphys.2012.00072

Matejka, N., & Reindl, J. (2019). Perspectives of cellular communication through tunneling nanotubes in cancer cells and the connection to radiation effects. *Radiation Oncology, 14*(1). https://doi.org/10.1186/s13014-019-1416-8

Miller, J., Fraser, S. E., & McClay, D. (1995). Dynamics of thin filopodia during sea urchin gastrulation. *Development, 121*(8), 2501–2511.

Mineo, M., Garfield, S. H., Taverna, S., Flugy, A., De Leo, G., Alessandro, R., & Kohn, E. C. (2012). Exosomes released by K562 chronic myeloid leukemia cells promote angiogenesis in a src-dependent fashion. *Angiogenesis, 15*(1), 33–45. https://doi.org/10.1007/s10456-011-9241-1

Mittal, R., Karhu, E., Wang, J. S., Delgadó, S., Zukerman, R., Mittal, J., & Jhaveri, V. M. (2019). Cell communication by tunneling nanotubes: Implications in disease and therapeutic applications. *Journal of Cellular Physiology, 234*(2), 1130–1146. https://doi.org/10.1002/jcp.27072

Miyazawa, K., Emmerling, K., & Manuelidis, L. (2010). Proliferative arrest of neural cells induces prion protein synthesis, nanotube formation, and cell-to-cell contacts. *Journal of Cellular Biochemistry, 111*(1), 239–247. https://doi.org/10.1002/jcb.22723

Mohammadalipour, A., Dumbali, S. P., & Wenzel, P. L. (2020). Mitochondrial transfer and regulators of mesenchymal stromal cell function and therapeutic efficacy. *Frontiers in Cell and Developmental Biology, 8*. https://doi.org/10.3389/fcell.2020.603292

Mukerji, J., Olivieri, K. C., Misra, V., Agopian, K. A., & Gabuzda, D. (2012). Proteomic analysis of HIV-1 Nef cellular binding partners reveals a role for exocyst complex proteins in mediating enhancement of intercellular nanotube formation. *Retrovirology, 9*. https://doi.org/10.1186/1742-4690-9-33

Ohno, H., Hase, K., & Kimura, S. (2010). M-Sec: Emerging secrets of tunneling nanotube formation. *Communicative & Integrative Biology, 3*(3), 231–233. https://doi.org/10.4161/cib.3.3.11242

Okafo, G., Prevedel, L., & Eugenin, E. (2017). Tunneling nanotubes (TNT) mediate long-range gap junctional communication: Implications for HIV cell to cell spread. *Scientific Reports, 7*(1). https://doi.org/10.1038/s41598-017-16600-1

Omsland, M., Bruserud, Ø., Gjertsen, B. T., & Andresen, V. (2017). Tunneling nanotube (TNT) formation is downregulated by cytarabine and NF-κB inhibition in acute myeloid leukemia (AML). *Oncotarget, 8*(5), 7946–7963. https://doi.org/10.18632/oncotarget.13853

Önfelt, B., Nedvetzki, S., Benninger, R. K. P., Purbhoo, M. A., Sowinski, S., Hume, A. N., Seabra, M. C., Neil, M. A. A., French, P. M. W., & Davis, D. M. (2006). Structurally distinct membrane nanotubes between human macrophages support long-distance vesicular traffic or surfing of bacteria. *Journal of Immunology, 177*(12), 8476–8483. https://doi.org/10.4049/jimmunol.177.12.8476

Önfelt, B., Nedvetzki, S., Yanagi, K., & Davis, D. M. (2004). Cutting edge: Membrane nanotubes connect immune cells. *Journal of Immunology, 173*(3), 1511–1513. https://doi.org/10.4049/jimmunol.173.3.1511

Osswald, M., Jung, E., Wick, W., & Winkler, F. (2019). Tunneling nanotube-like structures in brain tumors. *Cancer Reports, 2*(6). https://doi.org/10.1002/cnr2.1181

Panasiuk, M., Rychłowski, M., Derewońko, N., & Bieńkowska-Szewczyk, K. (2018). Tunneling nanotubes as a novel route of cellto-cell spread of herpesviruses. *Journal of Virology, 92*(10). https://doi.org/10.1128/JVI.000090-18

Parker, I., Evans, K. T., Ellefsen, K., Lawson, D. A., & Smith, I. F. (2017). Lattice light sheet imaging of membrane nanotubes between human breast cancer cells in culture and in brain metastases. *Scientific Reports, 7*(1). https://doi.org/10.1038/s41598-017-11223-y

Pasquier, J., Guerrouahen, B. S., Al Thawadi, H., Ghiabi, P., Maleki, M., Abu-Kaoud, N., Jacob, A., Mirshahi, M., Galas, L., Rafii, S., Le Foll, F., & Rafii, A. (2013). Preferential transfer of mitochondria from endothelial to cancer cells through tunneling nanotubes modulates chemoresistance. *Journal of Translational Medicine, 11*(1). https://doi.org/10.1186/1479-5876-11-94

Peggion, C., Sorgato, M. C., & Bertoli, A. (2014). Prions and prion-like pathogens in neurodegenerative disorders. *Pathogens, 3*(1), 149–163. https://doi.org/10.3390/pathogens3010149

Pergu, R., Dagar, S., Kumar, H., Kumar, R., Bhattacharya, J., & Mylavarapu, S. V. S. (2019). The chaperone ERp29 is required for tunneling nanotube formation by stabilizing MSec. *Journal of Biological Chemistry, 294*(18), 7177–7193. https://doi.org/10.1074/jbc.RA118.005659

Pilling, A. D., Horiuchi, D., Lively, C. M., & Saxton, W. M. (2006). Kinesin-1 and dynein are the primary motors for fast transport of mitochondria in Drosophila motor axons. *Molecular Biology of the Cell, 17*(4), 2057–2068. https://doi.org/10.1091/mbc.E05-06-0526

Pinto, G., Brou, C., & Zurzolo, C. (2020). Tunneling nanotubes: The fuel of tumor progression? *Trends in Cancer, 6*(10), 874–888. https://doi.org/10.1016/j.trecan.2020.04.012

Polak, R., De Rooij, B., Pieters, R., & Den Boer, M. L. (2015). B-cell precursor acute lymphoblastic leukemia cells use tunneling nanotubes to orchestrate their microenvironment. *Blood, 126*(21), 2404–2414. https://doi.org/10.1182/blood-2015-03-634238

Pyrgaki, C., Trainor, P., Hadjantonakis, A. K., & Niswander, L. (2010). Dynamic imaging of mammalian neural tube closure. *Developmental Biology, 344*(2), 941–947. https://doi.org/10.1016/j.ydbio.2010.06.010

Ranzinger, J., Rustom, A., Abel, M., Leyh, J., Kihm, L., Witkowski, M., Scheurich, P., Zeier, M., Schwenger, V., & Bereswill, S. (2011). Nanotube action between human mesothelial cells reveals novel aspects of inflammatory responses. *PLoS One, 6*(12), e29537. https://doi.org/10.1371/journal.pone.0029537

Ribeiro-Rodrigues, T. M., Martins-Marques, T., Morel, S., Kwak, B. R., & Girão, H. (2017). Role of connexin 43 in different forms of intercellular communication – gap junctions, extracellular vesicles and tunnelling nanotubes. *Journal of Cell Science, 130*(21), 3619–3630. https://doi.org/10.1242/jcs.200667

Roberts, K. L., Manicassamy, B., & Lamb, R. A. (2015). Influenza a virus uses intercellular connections to spread to neighboring cells. *Journal of Virology, 89*(3), 1537–1549. https://doi.org/10.1128/JVI.03306-14

Rostami, J., Fotaki, G., Sirois, J., Mzezewa, R., Bergström, J., Essand, M., Healy, L., & Erlandsson, A. (2020). Astrocytes have the capacity to act as antigen-presenting cells in the Parkinson's disease brain. *Journal of Neuroinflammation, 17*(1). https://doi.org/10.1186/s12974-020-01776-7

Rostami, J., Holmqvist, S., Lindström, V., Sigvardson, J., Westermark, G. T., Ingelsson, M., Bergström, J., Roybon, L., & Erlandsson, A. (2017). Human astrocytes transfer aggregated alpha-synuclein via tunneling nanotubes. *Journal of Neuroscience, 37*(49), 11835–11853. https://doi.org/10.1523/JNEUROSCI.0983-17.2017

Rouvinski, A., Karniely, S., Kounin, M., Moussa, S., Goldberg, M. D., Warburg, G., Lyakhovetsky, R., Papy-Garcia, D., Kutzsche, J., Korth, C., Carlson, G. A., Godsave, S. F., Peters, P. J., Luhr, K., Kristensson, K., & Taraboulos, A. (2014). Live imaging of prions reveals nascent PrPSc in cell surface, raft-associated amyloid strings and webs. *Journal of Cell Biology, 204*(3), 423–441. https://doi.org/10.1083/jcb.201308028

Rustom, A., Saffrich, R., Markovic, I., Walther, P., & Gerdes, H. H. (2004). Nanotubular highways for intercellular organelle transport. *Science, 303*(5660), 1007–1010. https://doi.org/10.1126/science.1093133

Sáenz-de-Santa-María, I., Bernardo-Castiñeira, C., Enciso, E., García-Moreno, I., Chiara, J. L., Suarez, C., & Chiara, M. D. (2017). Control of long-distance cell-to-cell communication and autophagosome transfer in squamous cell carcinoma via tunneling nanotubes. *Oncotarget, 8*(13), 20939–20960. https://doi.org/10.18632/oncotarget.15467

Saavedra, E. (2007). Energy metabolism in tumor cells. *FEBS Journal, 274*(6).

Salas-Vidal, E., & Lomelı, H. (2004). Imaging filopodia dynamics in the mouse blastocyst. *Developmental Biology, 265*(1), 75–89. https://doi.org/10.1016/j.ydbio.2003.09.012

Sartori-Rupp, A., Cordero Cervantes, D., Pepe, A., Gousset, K., Delage, E., Corroyer-Dulmont, S., Schmitt, C., Krijnse-Locker, J., & Zurzolo, C. (2019). Correlative cryo-electron microscopy reveals the structure of TNTs in neuronal cells. *Nature Communications, 10*(1). https://doi.org/10.1038/s41467-018-08178-7

Schiller, C., Diakopoulos, K. N., Rohwedder, I., Kremmer, E., Toerne, C.v., Ueffing, M., Weidle, U. H., Ohno, H., & Weiss, E. H. (2013). LST1 promotes the assembly of a molecular machinery responsible for tunneling nanotube formation. *Journal of Cell Science, 126*(3), 767–777. https://doi.org/10.1242/jcs.114033

Sharma, M., & Subramaniam, S. (2019). Rhes travels from cell to cell and transports Huntington disease protein via TNT-like protrusion. *Journal of Cell Biology, 218*(6), 1972–1993. https://doi.org/10.1083/JCB.201807068

Sinclair, K. A., Yerkovich, S. T., Hopkins, P. M. A., & Chambers, D. C. (2016). Characterization of intercellular communication and mitochondrial donation by mesenchymal stromal cells derived from the human lung. *Stem Cell Research & Therapy, 7*(1). https://doi.org/10.1186/s13287-016-0354-8

Sosa, V., Moliné, T., Somoza, R., Paciucci, R., Kondoh, H., & Lleonart, M. E. (2013). Oxidative stress and cancer: An overview. *Ageing Research Reviews, 12*(1), 376–390. https://doi.org/10.1016/j.arr.2012.10.004

Souriant, S., Balboa, L., Dupont, M., Pingris, K., Kviatcovsky, D., Cougoule, C., Lastrucci, C., Bah, A., Gasser, R., Poincloux, R., Raynaud-Messina, B., Al Saati, T., Inwentarz, S., Poggi, S., Moraña, E. J., González-Montaner, P., Corti, M., Lagane, B., Vergne, I., … Vérollet, C. (2019). Tuberculosis exacerbates HIV-1 infection through IL-10/STAT3-dependent tunneling nanotube formation in macrophages. *Cell Reports, 26*(13), 3586–3599.e7. https://doi.org/10.1016/j.celrep.2019.02.091

Sowinski, S., Jolly, C., Berninghausen, O., Purbhoo, M. A., Chauveau, A., Köhler, K., Oddos, S., Eissmann, P., Brodsky, F. M., Hopkins, C., Önfelt, B., Sattentau, Q., & Davis, D. M. (2008). Membrane nanotubes physically connect T cells over long distances presenting a novel route for HIV-1 transmission. *Nature Cell Biology, 10*(2), 211–219. https://doi.org/10.1038/ncb1682

Spagnolli, G., Rigoli, M., Orioli, S., Sevillano, A. M., Faccioli, P., Wille, H., Biasini, E., & Requena, J. R. (2019). Full atomistic model of prion structure and conversion. *PLoS Pathogens, 15*(7). https://doi.org/10.1371/journal.ppat.1007864

Spillantini, M. G., Crowther, R. A., Jakes, R., Hasegawa, M., & Goedert, M. (1998). α-Synuclein in filamentous inclusions of Lewy bodies from Parkinson's disease and dementia with Lewy bodies. *Proceedings of the National Academy of Sciences, 95*(11), 6469–6473. https://doi.org/10.1073/pnas.95.11.6469

Su, Q., & Igyártó, B. Z. (2019). Keratinocytes share gene expression fingerprint with epidermal Langerhans cells via mRNA transfer. *Journal of Investigative Dermatology, 139*(11), 2313–2323.e8. https://doi.org/10.1016/j.jid.2019.05.006

Subramaniam, S., Sixt, K. M., Barrow, R., & Snyder, S. H. (2009). Rhes, a striatal specific protein, mediates mutant-huntingtin cytotoxicity. *Science, 324*(5932), 1327–1330. https://doi.org/10.1126/science.1172871

Sui, Y. T., Bullock, K. M., Erickson, M. A., Zhang, J., & Banks, W. A. (2014). Alpha synuclein is transported into and out of the brain by the blood–brain barrier. *Peptides, 62*, 197–202. https://doi.org/10.1016/j.peptides.2014.09.018

Swarnkar, S., Chen, Y., Pryor, W. M., Shahani, N., Page, D. T., & Subramaniam, S. (2015). Ectopic expression of the striatal-enriched GTPase Rhes elicits cerebellar degeneration and an ataxia phenotype in Huntington's disease. *Neurobiology of Disease, 82*, 66–77. https://doi.org/10.1016/j.nbd.2015.05.011

Takahashi, A., Kukita, A., Li, Y. J., Zhang, J. Q., Nomiyama, H., Yamaza, T., Ayukawa, Y., Koyano, K., & Kukita, T. (2013). Tunneling nanotube formation is essential for the regulation of osteoclastogenesis. *Journal of Cellular Biochemistry, 114*(6), 1238–1247. https://doi.org/10.1002/jcb.24433

Tan, A. S., Baty, J. W., Dong, L. F., Bezawork-Geleta, A., Endaya, B., Goodwin, J., Bajzikova, M., Kovarova, J., Peterka, M., Yan, B., Pesdar, E. A., Sobol, M., Filimonenko, A., Stuart, S., Vondrusova, M., Kluckova, K., Sachaphibulkij, K., Rohlena, J., Hozak, P., … Berridge, M. V. (2015). Mitochondrial genome acquisition restores respiratory function and tumorigenic potential of cancer cells without mitochondrial DNA. *Cell Metabolism, 21*(1), 81–94. https://doi.org/10.1016/j.cmet.2014.12.003

Tardivel, M., Bégard, S., Bousset, L., Dujardin, S., Coens, A., Melki, R., Buée, L., & Colin, M. (2016). Tunneling nanotube (TNT)-mediated neuron-to neuron transfer of pathological Tau protein assemblies. *Acta Neuropathologica Communications, 4*(1), 117. https://doi.org/10.1186/s40478-016-0386-4

Teddy, J. M., & Kulesa, P. M. (2004). *In vivo evidence for short-and long-range cell communication in cranial neural crest cells.*

Thayanithy, V., Babatunde, V., Dickson, E. L., Wong, P., Oh, S., Ke, X., Barlas, A., Fujisawa, S., Romin, Y., Moreira, A. L., Downey, R. J., Steer, C. J., Subramanian, S., Manova-Todorova, K., Moore, M. A. S., & Lou, E. (2014). Tumor exosomes induce tunneling nanotubes in lipid raft-enriched regions of human mesothelioma cells. *Experimental Cell Research, 323*(1), 178–188. https://doi.org/10.1016/j.yexcr.2014.01.014

Tiwari, V., Koganti, R., Russell, G., Sharma, A., & Shukla, D. (2021). Role of tunneling nanotubes in viral infection, neurodegenerative disease, and cancer. *Frontiers in Immunology, 12*. https://doi.org/10.3389/fimmu.2021.680891

Torralba, D., Baixauli, F., & Sánchez-Madrid, F. (2016). Mitochondria know no boundaries: Mechanisms and functions of intercellular mitochondrial transfer. *Frontiers in Cell and Developmental Biology, 4*. https://doi.org/10.3389/fcell.2016.00107

Valdinocci, D., Kovarova, J., Neuzil, J., & Pountney, D. L. (2021). Alpha-synuclein aggregates associated with mitochondria in tunnelling nanotubes. *Neurotoxicity Research, 39*(2), 429–443. https://doi.org/10.1007/s12640-020-00285-y

Vaquer-Alicea, J., & Diamond, M. I. (2019). Propagation of protein aggregation in neurodegenerative diseases. *Annual Review of Biochemistry, 88*, 785–810. https://doi.org/10.1146/annurev-biochem-061516-045049

Vargas, J. Y., Loria, F., Wu, Y. J., Córdova, G., Nonaka, T., Bellow, S., Syan, S., Hasegawa, M., van Woerden, G. M., Trollet, C., & Zurzolo, C. (2019). The Wnt/Ca2+ pathway is involved in interneuronal communication mediated by tunneling nanotubes. *EMBO Journal, 38*(23). https://doi.org/10.15252/embj.2018101230

Vedula, S. R. K., Hirata, H., Nai, M. H., Brugués, A., Toyama, Y., Trepat, X., Lim, C. T., & Ladoux, B. (2014). Epithelial bridges maintain tissue integrity during collective cell migration. *Nature Materials, 13*(1), 87–96. https://doi.org/10.1038/nmat3814

Venkatesh, V. S., & Lou, E. (2019). Tunneling nanotubes: A bridge for heterogeneity in glioblastoma and a new therapeutic target? *Cancer Reports, 2*(6). https://doi.org/10.1002/cnr2.1185

Victoria, G. S., Arkhipenko, A., Zhu, S., Syan, S., & Zurzolo, C. (2016). Astrocyte-to-neuron intercellular prion transfer is mediated by cell-cell contact. *Scientific Reports, 6*. https://doi.org/10.1038/srep20762

Wang, X., Bukoreshtliev, N. V., & Gerdes, H. H. (2012). Developing neurons form transient nanotubes facilitating electrical coupling and calcium signaling with distant astrocytes. *PLoS One, 7*(10). https://doi.org/10.1371/journal.pone.0047429

Wang, Y., Cui, J., Sun, X., & Zhang, Y. (2011). Tunneling-nanotube development in astrocytes depends on p53 activation. *Cell Death & Differentiation, 18*(4), 732–742. https://doi.org/10.1038/cdd.2010.147

Wang, X., & Gerdes, H. H. (2015). Transfer of mitochondria via tunneling nanotubes rescues apoptotic PC12 cells. *Cell Death & Differentiation, 22*(7), 1181–1191. https://doi.org/10.1038/cdd.2014.211

Wang, X. T., Sun, H., Chen, N. H., & Yuan, Y. H. (2021). Tunneling nanotubes: A novel pharmacological target for neurodegenerative diseases? *Pharmacological Research, 170*. https://doi.org/10.1016/j.phrs.2021.105541

Wang, X., Veruki, M. L., Bukoreshtliev, N. V., Hartveit, E., & Gerdes, H.-H. (2010). Animal cells connected by nanotubes can be electrically coupled through interposed gap-junction channels. *Proceedings of the National Academy of Sciences, 107*(40), 17194–17199. https://doi.org/10.1073/pnas.1006785107

Xu, W., Santini, P. A., Sullivan, J. S., He, B., Shan, M., Ball, S. C., Dyer, W. B., Ketas, T. J., Chadburn, A., Cohen-Gould, L., Knowles, D. M., Chiu, A., Sanders, R. W., Chen, K., & Cerutti, A. (2009). HIV-1 evades virus-specific IgG2 and IgA responses by targeting systemic and intestinal B cells via long-range intercellular conduits. *Nature Immunology, 10*(9), 1008–1017. https://doi.org/10.1038/ni.1753

Yao, Y., Fan, X. L., Jiang, D., Zhang, Y., Li, X., Xu, Z. B., Fang, S. B., Chiu, S., Tse, H. F., Lian, Q., & Fu, Q. L. (2018). Connexin 43-mediated mitochondrial transfer of iPSC-MSCs alleviates asthma inflammation. *Stem Cell Reports, 11*(5), 1120–1135. https://doi.org/10.1016/j.stemcr.2018.09.012

Yao, Y., Zeng, Q. X., Deng, X. Q., Tang, G. N., Guo, J. B., Sun, Y. Q., Ru, K., Rizzo, A. N., Shi, J. B., & Fu, Q. L. (2015). Connexin 43 upregulation in mouse lungs during ovalbumin-induced asthma. *PLoS One, 10*(12). https://doi.org/10.1371/journal.pone.0144106

Zaccard, C. R., Rinaldo, C. R., & Mailliard, R. B. (2016). Linked in: Immunologic membrane nanotube networks. *Journal of Leukocyte Biology, 100*(1), 81–94. https://doi.org/10.1189/jlb.4VMR0915-395R

Zhang, S., Kazanietz, M. G., & Cooke, M. (2020). Rho GTPases and the emerging role of tunneling nanotubes in physiology and disease. *American Journal of Physiology – Cell Physiology, 319*(5), C877–C884. https://doi.org/10.1152/ajpcell.00351.2020

Zhang, Y., Yu, Z., Jiang, D., Liang, X., Liao, S., Zhang, Z., Yue, W., Li, X., Chiu, S. M., Chai, Y. H., Liang, Y., Chow, Y., Han, S., Xu, A., Tse, H. F., & Lian, Q. (2016). iPSC-MSCs with high intrinsic MIRO1 and sensitivity to TNF-α yield efficacious mitochondrial transfer to rescue anthracycline-induced cardiomyopathy. *Stem Cell Reports, 7*(4), 749–763. https://doi.org/10.1016/j.stemcr.2016.08.009

Zhu, C., Shi, Y., & You, J. (2021). Immune cell connection by tunneling nanotubes: The impact of intercellular cross-talk on the immune response and its therapeutic applications. *Molecular Pharmaceutics, 18*(3), 772–786. https://doi.org/10.1021/acs.molpharmaceut.0c01248

Zhu, D., Tan, K. S., Zhang, X., Sun, A. Y., Sun, G. Y., & Lee, J. C. M. (2005). Hydrogen peroxide alters membrane and cytoskeleton properties and increases intercellular connections in astrocytes. *Journal of Cell Science, 118*(16), 3695–3703. https://doi.org/10.1242/jcs.02507

Zhu, S., Victoria, G. S., Marzo, L., Ghosh, R., & Zurzolo, C. (2015). Prion aggregates transfer through tunneling nanotubes in endocytic vesicles. *Prion, 9*(2), 125–135. https://doi.org/10.1080/19336896.2015.1025189

CHAPTER

6

The role of extracellular vesicles and gap junctions in inter-cellular mitochondrial transfer

Dayene Caldeira[1,2] *and Anna Krasnodembskaya*[1]

[1]Wellcome-Wolfson Institute for Experimental Medicine, Queen's University of Belfast, Belfast, United Kingdom; [2]Institute of Biophysics Carlos Chagas Filho, Federal University of Rio de Janeiro, Rio de Janeiro, Brazil

Introduction

Mitochondria are at the center of many vital processes from ATP production to cell cycle regulation (Torralba et al., 2016). Damaged mitochondria and their products, especially mitochondrial DNA (mtDNA) and oxidized components, can act as endogenous danger signals, also known as damage-associated molecular patterns, leading to immune activation, production of reactive oxygen species (ROS), and adverse effects on cells, tissues, and organs (Park et al., 2011). Thus, cells have several mitochondrial quality controls promoting the degradation of damaged mitochondrial components, including mitophagy and mitochondria-derived vesicles (MDVs) (Amari & Germain, 2021). Failure of these quality control mechanisms can culminate in a diversity of pathologies from cardiovascular diseases to cancer (Johnson et al., 2021; Singel et al., 2019; Srinivasan et al., 2017). However, extracellular mitochondrial content was also found under physiological conditions, either free (such as naked mitochondria) or encapsulated in vesicles. Mitochondrial donation is an efficient physiological mechanism to replace dysfunctional mitochondria and attenuate disease severity (Paliwal et al., 2018). Intercellular mitochondrial transfer can occur via tunneling nanotubes and extracellular vesicles (EVs), while some evidence suggests the involvement of the gap junctions (Islam et al., 2012). Recently, mechanisms of EV-mediated transfer of mitochondrial cargo have been clarified showing an active and selective packaging of mitochondrial proteins into EVs resulting in the blocking of mitochondrial damage and prevention of the release of proinflammatory oxidized mitochondrial content (Todkar et al., 2021).

Mitochondrial Transplantation and Transfer
https://doi.org/10.1016/B978-0-443-18858-9.00006-8

This emerging field highlights mitochondria-targeted drugs as promising therapeutic approaches for numerous pathologies. Furthermore, the unexpected aspects of mitochondrial biology have been uncovered recently: mitochondria-derived vesicles (MDVs) or mitovesicles, as small vesicles generated as carriers of mitochondrial cargo to other intracellular organelles and also present in the extracellular space, besides other types of vesicles as ARMMs (arrestin domain-containing protein 1 (ARRDC1)–mediated microvesicles). This chapter focuses on the mechanisms of mitochondrial transfer through EVs and gap junctions and the implications of such transfer to the functionality of recipient cells.

Extracellular vesicles

Ability to secrete EVs is universal to all cells, a process conserved throughout evolution (Deatherage et al., 2012). EVs are implicated in the intra- and intercellular communication. EVs are released under both physiological and pathological conditions, and their cargo is unique because it depends on the physiological state of the cell of origin. The membrane-limited space of EVs contains proteins, sugars, lipids, and a wide variety of genetic materials, with the potential to deliver information to multiple cells in remote tissues (Henderson & Azorsa, 2012). Once internalized via endocytosis or membrane fusion, EVs release their contents into the recipient cells. Although EVs are the general term used to describe all secreted membrane vesicles, they are heterogeneous in their function, content, and biogenesis pathways. Based on this latter factor, EVs are broadly divided into two main categories: exosomes and microvesicles (MVs) (Cocucci & Meldolesi, 2015). Both are able to package bioactive molecules and deliver them to other cells altering recipient cell function and metabolism; this property increases interest in the EV potential to act as biomarkers of diseases and therapeutic targets (Amari & Germain, 2021). Additionally, other types of vesicles involved in mitochondrial cargo delivery will be addressed in the discussion. Their distinct size, content, and biogenesis pathways are summarized in Figs. 6.1 and 6.2.

EV biogenesis and content

Exosomes

Biogenesis

Exosomes are small vesicles with a diameter of 30–100 nm, biogenesis of exosomes starts in multivesicular bodies (MVBs), which are endosomal structures involved in the sorting, recycling, storage, transportation, and release of proteins. MVBs internalize limiting membranes, thereby forming intraluminal vesicles (ILVs) that begin to accumulate in the early stage and continue to accumulate in the late endosomal stage. During this process, a cargo of proteins and lipids is sorted into these vesicles. Late endosomes eventually fuse with the lysosomes to degrade ILVs or move toward the plasma membrane. After fusion with the plasma membrane, MVBs release their ILVs (now exosomes) into the extracellular compartment via exocytosis (Doyle & Wang, 2019).

The formation of MVBs and exosomes can be mediated by endosomal sorting complex required for transport (ESCRT)-dependent and -independent pathways. In an ESCRT-dependent manner, MVBs are formed through the ubiquitination of the target proteins. These

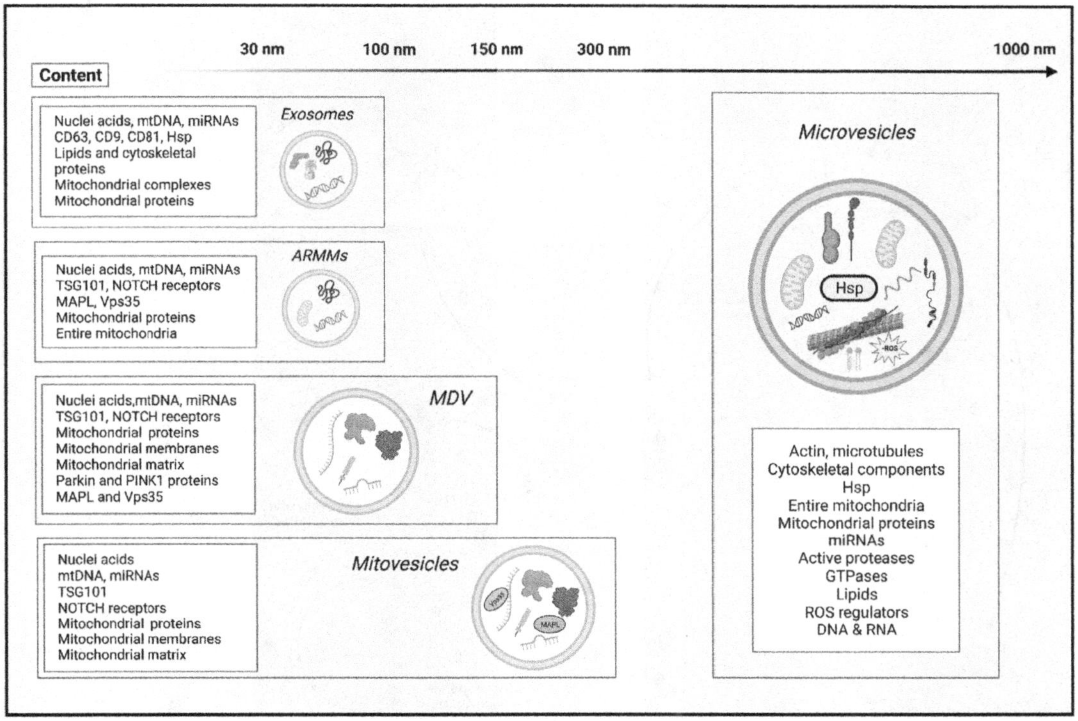

FIGURE 6.1 Subtypes of extracellular vesicles involved in mitochondrial cargo transfer. Exosomes are the smaller vesicles, with a size range of 30–150 nm, originating in late endosomes. Similar in size are ARMMs (ARRDC1-mediated microvesicles), which originated on the plasma membrane. Mitochondria-derived vesicles (MDVs) are small vesicles with a size range of 70–150 nm derived intracellularly from mitochondria. MDV are analogous to mitovesicles (50–300 nm) but with important distinctions in their mitochondrial cargo. Lastly, microvesicles (MV) are large EVs ranging in size from 50 to 1000 nm and generated by the direct budding of the plasma membrane, resulting in shedding from the cell. *Created with BioRender.com.*

ubiquitin tags are recognized by the ESCRT complex. Simultaneously, phosphatidylinositol-3-phosphate (PI3P), a phospholipid found in early and late endosomes, recruits the ESCRT-0 complex to early endosomes. PI3P, through interactions with the subunit Hrs (hepatocyte growth factor–regulated tyrosine kinase substrate) of ESCRT-0, promotes cargo organization, captures ubiquitinated proteins, and enriches MVBs compartments (Gurung et al., 2021). Then, the ESCRT machinery acts in a stepwise manner: ESCRT-1 interacts with ESCRT-II promoting inward budding around clusters of ubiquitinated proteins, and recruiting the ESCRT-III that performs budding and fission of these ILVs (futures exosomes) into the MVB lumen (Van Niel et al., 2018). Cargo clustering and membrane budding of exosomes can also occur through an ESCRT-independent mechanism. In this pathway, exosome formation is regulated by tetraspanins, a membrane integral protein family. Tetraspanins control exosome release and can influence exosome biogenesis and composition, regulating cargo sorting for release or lysosomal degradation of exosomes (Gurung et al., 2021).

MVBs transport and fusion with the plasma membrane require cytoskeletal and microtubule networks associated with motor proteins (dynein, kinesins, and myosins) and molecular switches (small GTPases) (Van Niel et al., 2018). Members of the Rab GTPase family and

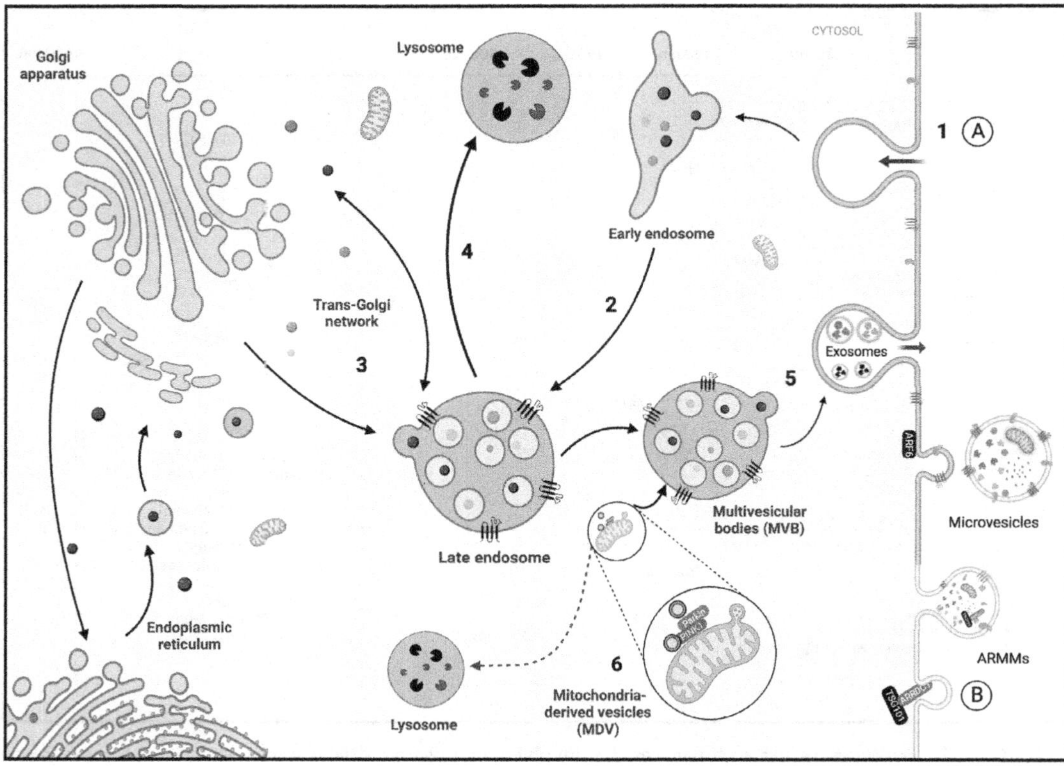

FIGURE 6.2 Pathways of EV biogenesis. (A) Exosomes formation firstly requires the budding of inner membrane forming early endosomes (1), that mature into late endosomes (or multivesicular bodies, MVBs) and accumulate intraluminal vesicles (ILVs) containing cargoes of proteins, lipids, and other components (2). During this maturation process, a cargo of proteins, lipids, and other components are sorted from the trans-Golgi network and possibly from the cytosol to MVBs, a process that requires ESCRT-complex or tetraspanins proteins (3). MVBs fuse with lysosomes (4) or avoid degradation and fuse to the plasma membrane for exosome secretion (5). (B) The mitochondrial content can either be incorporated into MVBs, released further as exosomes, or have their oxidized content targeted by Parkin-PINK1 complex to lysosomal degradation via mitochondria-derived vesicles (MDVs) (6). Microvesicles are generated by the direct budding of the plasma membrane, a process that requires actomyosin rearrangement regulated by ARF6. Whenever this release is specifically driven by the interaction of TSG101 and ARRDC1, these microvesicles can be classified as ARMMs (arrestin domain-containing protein 1 (ARRDC1)–mediated microvesicles). *Created with BioRender.com.*

RAL1 (Ras-related GTPases homolog) interact with SNARE proteins and take part in transporting vesicles between intracellular compartments, as well as in the transport of MVBs to the plasma membrane for exosome secretion (Van Niel et al., 2018). MVBs avoid lysosomal degradation and release exosomes through yet partially unclear mechanisms. However, lysosome inhibitors (such as bafilomycin A) increased EV secretion as measured by nanoparticle tracking analysis (Miao et al., 2015). Similarly, mutations in the Tau protein that induce lysosome inhibition increased the expression of proteins associated with EV secretion, and mutations in VPS4, required for MVB maturation and fusion with the lysosome, enriched EV associated α-synuclein (Hasegawa et al., 2011). Therefore, lysosomal dysfunction or overload to dispose of misfolded proteins increases EV secretion as a compensatory mechanism (Eitan et al., 2016).

Biological content

The biogenesis of exosomes helps to understand their content. Exosomes can contain accessory proteins involved in the ESCRT dependent mechanism (Alix, TSG101, HSC70, and HSP90β) often termed "exosomal marker proteins" (Doyle & Wang, 2019). In the ESCRT independent mechanism, proteins of the tetraspanin family (CD63, CD9, and CD81) are often used as exosome biomarkers (Doyle & Wang, 2019). Together with heat shock proteins (Hsp) and cytoskeletal components, all these are common components of all exosomes, while the presence of major histocompatibility complex (MHC) Class I and II are specific to the donor cell type. Exosomes also carry nucleic acids, including DNA sequences, and various RNA species. Exosomal miRNAs are the most abundant RNA species in the exosomes and have a strong potential to modulate the function of recipient cells (Jurj et al., 2020; Phinney et al., 2015; Su et al., 2022). The biomarker potential of EV-associated miRNAs is being extensively investigated in multiple diseases, including cardiovascular diseases, cancer, and infectious diseases (Hu et al., 2012). Complex lipids, such as ceramides and sphingomyelin, are also components of exosomes; these lipids are also involved in exosome secretion, cargo sorting, and signaling. Exosomal cargoes also can be delivered from the trans Golgi network and possibly from the cytosol (Gurung et al., 2021), and the impairment or depletion of these network regulators or transport might generally impact the targeting of some cargoes to the EVs (Zimmermann et al., 2005).

Exosomes can contain various mitochondrial proteins and other cargoes, such as cardiolipin and mtDNA (Liu et al., 2020). Proteomic analysis showed that mitochondrial cargo represents up to 10% of the total secreted proteins (Choi et al., 2013). Interestingly, Brushi et al. report "mitochondria like" exosomes in urinary samples from healthy humans, enriched by aerobic metabolism proteins and active in oxidative phosphorylation (OXPHOS), consuming oxygen coupled to ATP synthesis although electron microscopy analysis excluded the presence of intact mitochondria (Bruschi et al., 2016). Also, the package of mitochondrial cargo into exosomes was associated with the formation of mitochondria-derived vesicles (MDVs) (Todkar et al., 2021), discussed later in this chapter.

Microvesicles

Biogenesis

Microvesicles (MVs), also referred to as ectosomes, are large vesicles ranging in size from 50 to 1000 nm in diameter. However, the mechanism of MV formation is not completely understood. It is known that MVs are generated by direct budding of the plasma membrane, resulting in shedding from the cell (Clancy et al., 2021). Initially, lipids and membrane-associated proteins are clustered in discrete membrane microdomains, and the plasma membrane undergoes molecular rearrangement following the release of Ca^{2+} from the endoplasmic reticulum (Van Niel et al., 2018). Furthermore, the plasma membrane integrates signaling processes, such as changes in lipid and protein components, with the coordination of actomyosin contractile machinery partially regulated by ADP ribosylation factor 6 (ARF6) (Clancy et al., 2021). During MVs shedding, ARF6 (a member of the Ras GTPase family) activates phospholipase D and recruits extracellular signal-regulated kinase (ERK) to the plasma membrane. Subsequently, ERK in the plasma membrane triggers phosphorylation of myosin light chain kinase, which in turn phosphorylates the myosin light chain (MLC), facilitating actomyosin based contractility. Phosphorylated MLCs are enriched at the "necks" of vesicles, facilitating fission and releasing vesicles into the extracellular space (Clancy et al., 2021). Additionally, the formation of nascent MVs specifically requires cholesterol, as its depletion results in the loss of MV release (Clancy et al., 2021). Furthermore, a Rho GTPase dependent pathway seems to be required in MV formation, activating RHO-associated protein kinase, Lim kinase, which phosphorylates cofilin, an intracellular actin modulating protein. The following elongation of actin and formation of "actin-rings" results in the MV maturation, but also in the malignant transformation of cancer cells (Li et al., 2012).

Biological content

Based on their plasma membrane origin, MVs contain mainly cytosolic and plasma membrane associated proteins, especially proteins known to cluster at the membrane surface, such as tetraspanins, and cytoskeleton components, such as actin and microtubules (Doyle & Wang, 2019). The proteomic profile of MVs includes Hsp, integrins (ITGs), and proteins containing posttranslational modifications such as glycosylation and phosphorylation (Doyle & Wang, 2019). Identified cargoes include a range of ITG receptors, active proteases, multiple small GTPases, miRNA processing machinery, multiple forms of nucleic acids (both DNA and RNA), active lipids, ROS regulators, and, importantly, intact mitochondria (Clancy et al., 2021). However, the composition of MVs depends largely on the cell type from which they originate. Additionally, specific markers for distinguishing MVs from exosomes are still lacking (Jeppesen et al., 2019).

ARMMS

ARMMs (ARRDC1-mediated MVs) are EVs similar in size to exosomes but capable of direct budding from the plasma membrane. Supporting the nonexosomal nature of ARMMs, general exosomal markers such as CD9 and CD81 were not enriched in ARMMs, similar to significant amount of other proteins that are frequently found in exosomes (Wang & Lu, 2017). Direct budding from the plasma membrane is driven by a specific interaction of TSG101 with the tetrapeptide PSAP motif of an accessory protein, ARRDC1. This interaction results in the relocation of TSG101 from endosomes to the plasma membrane and mediates the release of MVs that contain TSG101, ARRDC1, and other cellular proteins (Nabhan et al., 2012). These vesicles are also enriched by NOTCH receptors (Wang & Lu, 2017). Mitochondrial transfer through ARMMs has been demonstrated as a mechanism by which MSCs mediate transmitophagy, delivering depolarized mitochondria to be degraded by macrophages to improve survival (Phinney et al., 2015).

Noncanonical vesicles mediating mitochondrial transfer

Mitochondria derived vesicles

Mitochondria are the central point to provide cells with energy and also to ensure efficient intracellular signaling. Therefore, they are physically connected to other cellular compartments through membrane contact sites and tethering molecules, which are essential components of this interaction. However, it was recently shown that mitochondria can mediate interorganellar communication through vesicles. Mitochondria-derived vesicles (MDVs) are single or double membrane small vesicles that differ morphologically and by size from intracellular mitochondria and canonical EV subtypes. MDVs can be defined as (1) a uniform structure (70–100 nm) (2) formed from the lateral segregation of mitochondrial membrane (3) in a dynamin-related protein 1 independent manner and (4) that contain multiple mitochondrial cargoes. MDVs can usually be limited to the outer membrane or can also include inner membrane and matrix content (Soubannier, Rippstein, et al., 2012).

An elegant study by Todkar et al. (2021) showed that the inclusion of the mitochondrial content into exosomes required MDVs. MDVs reach the endolysosomal system and incorporate into MVBs and subsequently into exosomes. EV preparations from unstimulated mouse embryonic fibroblasts were enriched for mitochondrial proteins and mtDNA. The inhibition of MDVs formation by either knocking down sorting nexin 9 or deleting optic atrophy-1 (OPA1) inhibited the inclusion of mitochondrial content into EVs, indicating that MDVs play a role in the EV cargo selection. Interestingly, activation of the ROS-induced mitophagy pathway targeted mitochondria via Parkinson's disease associated protein (Parkin) and phosphatase and tensin homolog induced kinase 1 (PINK1) to lysosomes for degradation and blocked the inclusion of mitochondrial proteins into EVs (Todkar et al., 2021).

MDVs have two distinct fates. MDVs selectively enriched by damaged and oxidized proteins are delivered to lysosomes for degradation, as an early response to oxidative stress (Soubannier, McLelland, et al., 2012). Neuspiel et al. (2008) showed that a subpopulation of MDVs, containing outer-membrane mitochondria-anchored protein ligase (MAPL) excludes another outer-membrane marker, translocase of outer mitochondrial membrane 20 (TOM20). This MAPL-MDV fuse with peroxisomes for degradation, while TOM20-MDVs do not, evidencing that the intracellular transport of MDVs depends on their cargo (Neuspiel et al., 2008). These pathways seem to be an elegant mechanism of mitochondria quality control utilized by the cell to preserve homeostasis, dispose of defective mitochondrial components and avoid the clearance of entire organelles under stress conditions (McLelland et al., 2014; Soubannier, Rippstein, et al., 2012). However, MDV generation is independent of mitochondrial depolarization, autophagy signaling, and mitochondrial fission, which further supports the hypothesis that MDVs also may act as a means to dispose of mildly oxidized mitochondria or mitochondria cargo and perhaps recycle them (Picca et al., 2020).

MDVs are actively involved in the pathway for mitochondrial antigen presentation (MitAP) in inflammatory conditions. During MitAP, MDV fuse with the endosomal compartments where mitochondrial antigens can be loaded on MHC class I of antigen-presenting cells (APCs), activating T cells. These data provide a link between mitochondrial dynamics and the autoimmune mechanisms in the etiology of Parkinson disease (Matheoud et al., 2016). Also, exogenous MDVs showed significant protective effects in an ischemic/hypoxic myocardial injury model, inhibiting hypoxia induced cardiomyocyte apoptosis and mitigating hypoxic damage (Li et al., 2020). Therefore, MDVs can be considered an efficient mechanism of mitochondrial cargo delivery.

Mitovesicles

Mitovesicles are another type of EVs derived intracellularly from mitochondria, with a size range of 50–300 nm (D'Acunzo et al., 2021). These vesicles share important features and are distinct from MDVs. While these vesicles also bud from the mitochondrial surface and may include the mitochondrial outer and inner membranes, as well as components of the mitochondrial matrix, other MDV characteristics such as the presence of MAPL and vacuolar protein sorting-associated protein 35 were not found in the MVs preparations. Several subsets of MDVs with specific functions may exist. At the same time, mitovesicles cannot be considered exosomes because the mechanisms responsible for their biogenesis from the mitochondria and their secretion into the extracellular space remain unclear. Compared with intracellular mitochondria, they are up to 10-fold smaller, have a narrower intermembrane space, and lack several mitochondrial structures, including cristae and proteins constitutively found in mitochondria, such as TOM20 (D'Acunzo et al., 2021). Mitovesicles were first identified in the brain as a previously unknown population of double membraned EVs containing multiple mitochondrial proteins. It was shown that levels and cargo of these vesicles are altered in the brains of patients with Down syndrome; however, the role of these vesicles in the pathophysiological process still needs to be elucidated (D'Acunzo et al., 2021).

EV uptake

Once released into the extracellular space, EVs can be taken up by cells nearby as well as reach remote tissues via systemic circulation, EV uptake can affect the functional properties of the recipient cells (Van Niel et al., 2018). EV internalization normally requires docking at the plasma membrane of target cells (anchoring), followed by the activation of surface receptors leading to vesicle internalization via endocytosis or membrane fusion (Van Niel et al., 2018). Cell-specific uptake depends on cell type. Despite the different contents and sizes, the principles of uptake are likely to be shared, as shown in Fig. 6.3.

Protein interactions

The EV uptake involves specific protein−protein interactions, some of which have been reported in the literature. On the surface, EV membranes express tetraspanins (CD9, CD63, and CD81), ITGs, cell adhesion molecules, and receptors for growth factor (Lötvall et al., 2014). Tetraspanins (Tspan), which have numerous functions involved in cell adhesion and motility, are highly abundant on the EV surface (Charrin et al., 2014). Treatment with anti-CD81 and CD9 reduces the uptake of EVs by dendritic cells (DCs) (Morelli et al., 2004) while overexpressing of Tspan8 contributes to EV uptake by rat aortic endothelial cells (Nazarenko et al., 2010). These data suggest that tetraspanins play an important role in EV internalization. ITGs and immunoglobulins are also involved in the uptake of EVs. Blocking of CD11a and its ligand ICAM-1, as well as ITGs αv (CD51) and β3 (CD61), can reduce DC uptake of EVs (Morelli et al., 2004). At the same time, DC−derived EVs were internalized via TCR-MHC and LFA-1−ICAM-1 interaction in $CD4^+$ cells (Hao et al., 2007), highlighting the role of protein−protein interactions in the uptake of vesicles by the immune cells.

Interestingly, exosomes can play a role in antigen presentation by APCs (Barros et al., 2018). During antigen presentation, APC cells, such as DCs, macrophages, and B-cells, expose peptide antigens bound to MHC class I or class II molecules, activating $CD8^+$ cytotoxic T lymphocytes and natural killer cells or $CD4^+$ helper T cells, respectively. Exosomes derived from mature DCs can transfer peptide MHC class II complexes to deficient DCs, which expose them to the cell surface allowing the activation of $CD4^+$ T cells (Segura et al., 2005; Théry et al., 2002). B cells also present the capacity to elicit T helper response through

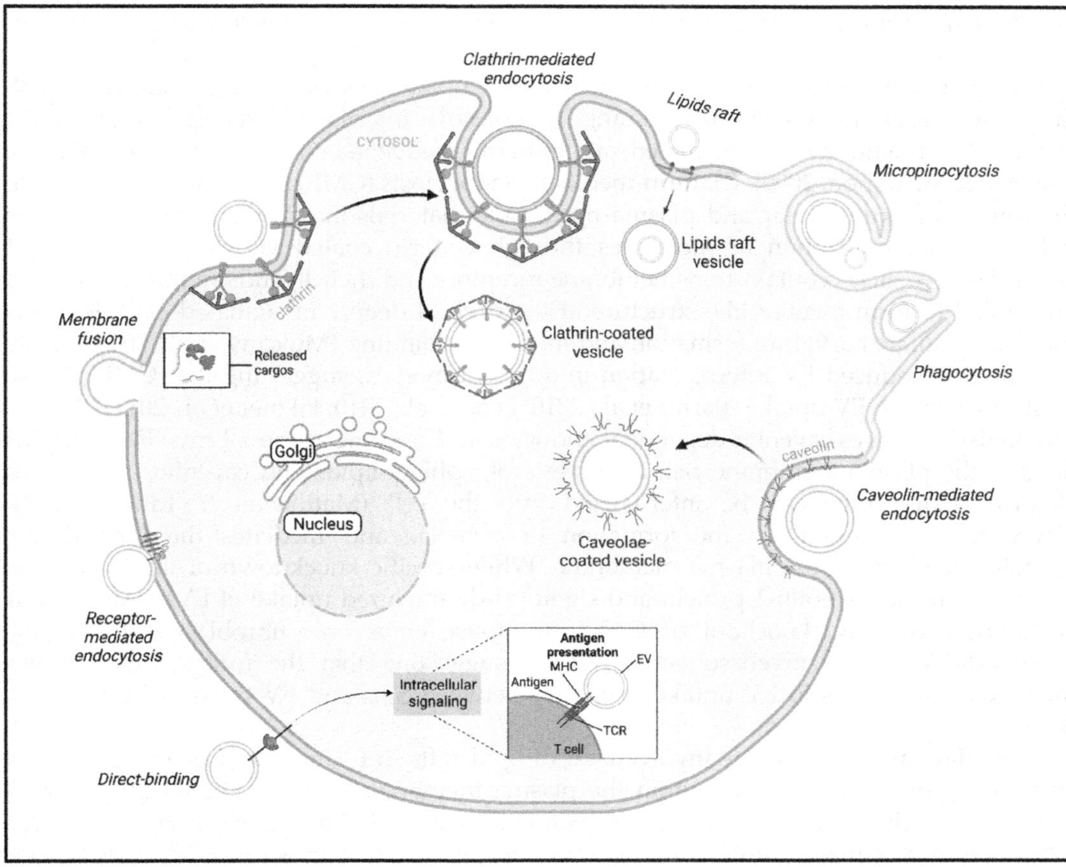

FIGURE 6.3 Mechanisms of the EV uptake by recipient cells. Once anchored into the plasma membrane of target cells through protein interactions, EVs can directly fuse with the recipient cell membrane or be engulfed by recipient cells through multiple pathways of endocytosis, including clathrin-mediated, caveolae-mediated, lipid raft-mediated, and receptor-mediated. Other cell-specific uptake mechanisms are utilized by immune cells to internalize EVs by immune cells, such as phagocytosis and macropinocytosis. Also, EVs can act as an antigen presenter interacting with T cells. *Created with BioRender.com.*

exosomes containing MHC class II molecules associated with peptides (Raposo et al., 1996). DC-derived exosomes have the capacity to activate $CD8^+$ T cells in a DC-independent manner via peptide MHC class I complex interaction (Hsu et al., 2003).

Endocytosis

Once attached to recipient cells, EVs can directly fuse with the recipient cell membrane or be engulfed by recipient cells through endocytosis pathways. The direct fusion of the EV membrane with the cell plasma membrane has been observed in studies using dequenching methods (Montecalvo et al., 2012; Parolini et al., 2009). The bilayer membranes come into direct contact, first forming a hemifusion of the outer leaflets, then expanding to the hemifusion diaphragm bilayer from which a fusion pore opens, ultimately forming a consistent structure for the release of EV content (Mulcahy et al., 2014).

Endocytosis is a well-recognized pathway for the uptake of EVs and mitochondria by cells. Endocytosis is an umbrella term for a range of molecular internalization pathways, including clathrin dependent, caveolin-mediated, and lipid raft mediated endocytosis (Chen et al., 2021; Doherty & McMahon, 2009). Clathrin-mediated endocytosis (CME) is a common pathway for the uptake of extracellular and plasma membrane materials in all eukaryotic cells and involves the internalization of molecules through clathrin coated vesicles (Mettlen et al., 2018). These vesicles contain transmembrane receptors and their ligands require the GTPase Dynamin2 to form a collar-like structure at the neck of deeply invaginated clathrin-coated pits and facilitate curvature formation and membrane binding (Mulcahy et al., 2014). Inhibition of CME reduced EV internalization in different models, suggesting that CME plays at least some part in EV uptake (Barrès et al., 2010; Feng et al., 2010; Fitzner et al., 2011). Another mechanism involves caveolin-dependent endocytosis. Caveolae are small cave-like invaginations in the plasma membrane rich in cholesterol, sphingolipids and caveolins, which, like clathrin coated pits, can be internalized into the cell (Matthaeus & Taraska, 2021). Caveolin-1 is required for the formation of caveolae and mediates the formation of caveolin rich rafts in the plasma membrane. While specific knockdown of the CAV1 gene leads to reduced caveolin-1 protein and significantly impaired uptake of EVs (Nanbo et al., 2013), paradoxically, knockout of CAV1 in mouse embryonic fibroblast cells leads to increased EV uptake (Svensson et al., 2013), suggesting that the function of caveolae-mediated endocytosis in EV uptake may vary between cells and EV types (Mulcahy et al., 2014).

Some data also support the involvement of lipid rafts in EV uptake. Lipid rafts or membrane rafts are microdomains within the plasma membrane with an altered phospholipid composition. They are rich in protein receptors and sphingolipids, such as sphingomyelin, which act as organizing centers for the assembly of signaling molecules. Lipid rafts affect membrane fluidity and mediate membrane protein trafficking (Ripa et al., 2021). In particular, lipids rafts have been tested for their roles in EV uptake (Ripa et al., 2021). Inhibitor agents for cholesterol and glycosphingolipids reduce EV uptake by disrupting lipid raft mediated endocytosis in recipient cells (Svensson et al., 2013), suggesting the possibility of EV internalization through this pathway. However, the scale and precise mechanisms of this route in cells remain unclear.

Phagocytosis and macropinocytosis

Phagocytosis is a receptor mediated process utilized by macrophages that involve the internalization of opsonized particulate matter, including bacteria and fragments of apoptotic cells. This mechanism includes the progressive formation of invaginations surrounding the material destined for internalization, which is usually employed for larger particles. The role of phagocytosis in EVs uptake was further substantiated by the inhibition of PI3K kinase, which led to a dose-dependent reduction in EV uptake (Feng et al., 2010). In addition, DCs can phagocytose EVs (Montecalvo et al., 2012), showing that phagocytosis is implicated in the EVs uptake. A mechanism similar to phagocytosis but with no direct contact with the internalized material is micropinocytosis. Uptake via micropinocytosis involves the formation of invaginated plasma membrane ruffles. The protrusion of these ruffled extensions allows EVs to be internalized as a result of the fusion of protrusions with themselves or back with the plasma membrane (Kwok et al., 2021). Micropinocytosis depends on the activity of the Na^+/H^+ exchanger, rac1 GTPase, and cholesterol (Kwok et al., 2021). Inhibition of the Na^+/H^+ exchanger and rac1 GTPase resulted in significantly reduced EV uptake by microglia (Fitzner et al., 2011). However, the results of micropinocytosis inhibition are inconsistent (Christianson et al., 2013) and suggest that micropinocytosis is either a minor pathway used by cells to internalize EVs or a mechanism used in specific cell types.

Effects of EV-mediated mitochondrial transfer

EVs are a powerful vehicle for mitochondrial delivery (Cai et al., 2020). In addition to keeping the mitochondria stable and maintaining their morphological and functional integrity encapsulated in an outer bilayer membrane, EVs can also deliver mitochondrial cargo such as mtDNA, cardiolipin, and other proteins (Ikeda et al., 2021). Classically, the engulfment of damaged or unnecessary mitochondria triggers a multistep process involving the endolysosomal pathway and autophagosome (Liu et al., 2022; Youle & Narendra, 2011). However, most exogenous mitochondria can escape this degradation axis and fuse with the endogenous mitochondrial network in cardiomyocytes and fibroblasts via the proteins mitofusin 1 (Mfn1) and 2 (Mfn2) and OPA1 (Cowan et al., 2017; Yao et al., 2020). The mechanism behind this exogenous and endogenous mitochondrial integration and pathways involved in the escape from degradation remains unclear. The most consistent effects of mitochondrial transfer in target cells are shown in Fig. 6.4.

Immune response

Previous studies have demonstrated the immunoregulatory role of EVs (Robbins & Morelli, 2014). Although their small size should preclude the incorporation of intact mitochondria, exosomes reportedly contain molecules commonly associated with mitochondria, including mitochondrial lipids (cardiolipin), mtDNA, and mitochondrial proteins (Wang et al., 2020). MSCs package and transfer partially depolarized mitochondria and mitochondrial content to macrophages via MVs as a prosurvival mechanism in response to oxidative stress, thereby increasing macrophage bioenergetics. In an elegant study by Phinney et al. it was demonstrated that this effect was possible due to MSCs release of MVs and exosomes with mtDNA and microRNA content. MVs containing mitochondria expressed TSG101 (an ESCRT-I-associated protein) and ARRDC, and were extruded from cells in ARMMs. These vesicles were taken up by macrophages, while exosome microRNAs facilitated macrophage tolerance to mitochondrial donation by regulating TLRs and cytokine expression (Phinney et al., 2015). Similarly, MSCs promoted an antiinflammatory and highly phagocytic macrophage phenotype in the preclinical models of ARDS, at least partially induced by EVs carrying mitochondria; notably this effect on macrophage polarization was associated with a metabolic shift from glycolysis to OXPHOS (Morrison et al., 2017). Moreover, pneumolysin-stimulated alveolar epithelial cells in an acute lung injury (ALI) model released MVs containing mitochondrial cargo that can be taken up by neutrophils and were able to suppress the neutrophil production of ROS (Letsiou et al., 2021).

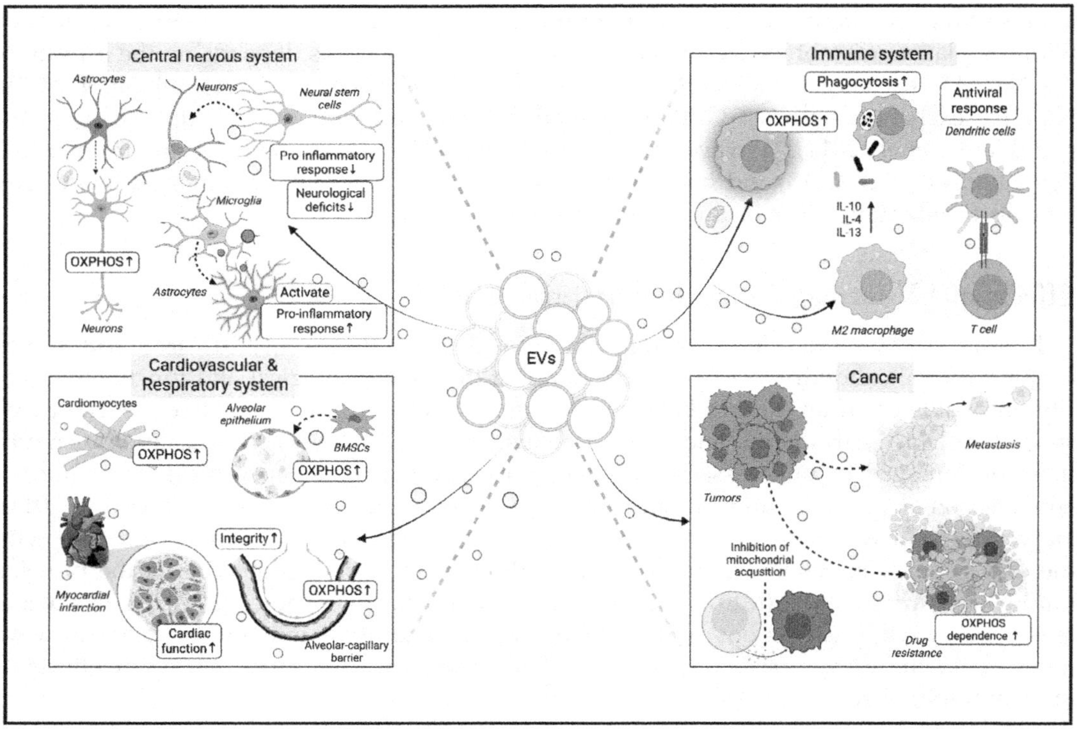

FIGURE 6.4 Effects of the EV-mediated mitochondrial transfer. Mitochondrial cargo transfer through EVs has been shown to be effective in modulating the immune system as well as restore functionality of the damaged tissues, especially in diseases where mitochondrial dysfunction contributes to pathogenesis (e.g., acute lung injury, cardiac infarction, and neurodegenerative diseases). Tumors use EVs for communication and progression, requiring functional mitochondria from adjacent healthy cells. Therefore, inhibition of the acquisition of functional mitochondria can be a target for cancer advancement and drug resistance. *Created with BioRender.com.*

Indeed, EVs' modulation of immune responses in some cells redefines the classic concepts usually attributed to cytokines. KSHV virus-infected cells release EVs to mediate early antiviral defenses to favor cell survival. These EVs contain not only viral content but also mtDNA as a triggering factor for an innate immune antiviral response (Jeon et al., 2019). Interestingly, immune cells are not acting as EV recipient cells but also can release EVs with mitochondrial content. Immune interactions between T cells and APCs play an important role in T cell activation (Gaudino & Kumar, 2019). Torralba et al. (2018) demonstrated that signals are transmitted from T cells to DCs through EVs that contain genomic and mtDNA in retrograde feedback. The mitochondrial content promotes an antiviral response with the induction of interferon (IFN)-related stimulated genes (Torralba et al., 2018).

Nevertheless, the literature shows controversial effects of EVs mitochondrial cargo on the immune system. MVs derived from LPS-activated THP1 are enriched in mitochondrial content and demonstrate a proinflammatory capacity to activate endothelial cells. Together with free mitochondria, these MV induced type 1 IFN and tumor necrosis factor responses, which were significantly reduced after immunodepletion of free mitochondria, indicating that this fraction may be the main factor responsible for the immune response (Puhm et al., 2019). Although the transfer of healthy mitochondria appears to be effective in the restoration of functional properties of the injured cells, mitochondrial proteins are known to participate in the inflammatory responses (Weinberg et al., 2015).

Central nervous system

Mitochondrial dysfunction is a feature of several inflammatory and degenerative disorders of the central nervous system (CNS). The transfer of functional mitochondria via EVs is a mechanism used by some CNS cells to modulate the physiology and metabolism of the target cells. Rat cortical astrocytes produce EVs containing functional mitochondria to support neurons subjected to oxygen-glucose deprivation. Mitochondrial cargo from these astrocytes conferred neuroprotection, restoring bioenergetics and viability in vitro and as a treatment after ischemic stroke in vivo (Hayakawa et al., 2016). Similarly, neural stem cell (NSC) encapsulated mitochondria are taken up by both somatic cells and mononuclear phagocytes, inhibiting the activation of proinflammatory responses. When delivered to mice with induced experimental autoimmune encephalomyelitis, both NSCs and EVs significantly ameliorated neurological deficits (Peruzzotti-Jametti et al., 2021). Mitochondria were also transferred by MVs from the human brain endothelial cell (BEC) line to recipient BECs and neurons in mice acute brain cortical and hippocampal slices. These MVs enhance ATP levels during oxygen-glucose deprivation, an in vitro model of cerebral ischemia (D'Souza et al., 2021). Otherwise, microglia disseminate neuroinflammation triggering the activation of naïve astrocytes in great part by the release of fragmented and dysfunctional microglial mitochondria to the neuronal milieu (Joshi et al., 2019).

Until now, mitovesicles have been linked to neurodegenerative diseases. Mitovesicles were augmented, and with composition altered in the brains of Ts2 mice and individuals with Down syndrome (DS), characterized by mitochondrial dysfunction, compared with age-matched diploid controls (D'Acunzo et al., 2021). Mitovesicle secretion in neurodegenerative diseases with mitochondrial dysfunction may contribute to the disease or act to eliminate detrimental mitochondrial material from the cell and mitigate oxidative stress; interestingly, it was shown that DS cells are deficient in mitophagy (Bordi et al., 2019). However, further studies are needed to elucidate the role of the mitovesicles in neurodegenerative diseases.

Cardiovascular and respiratory system

Abnormal morphology and mitochondrial dysfunction are the important mechanisms underlying the pathogenesis of cardiovascular diseases. EVs containing functional mitochondria restore intracellular bioenergetics and contractile properties in induced pluripotent stem cell-derived cardiomyocytes (iCMs) injured by hypoxia. Mitochondria enriched fraction of EVs also improved post-MI cardiac function in mice (Ikeda et al., 2021). Additionally, in a cardiomyopathy model using iCMs treated with doxorubicin (DOX), these cells were able to take up large MSC-derived vesicles enriched with mitochondria, which mitigated DOX injury and improved mitochondrial function. This result suggests a mechanism by which MSCs may improve cardiovascular performance in anthracycline-induced cardiomyopathy (O'Brien et al., 2021). In addition, exogenous MDVs showed a protective effect against hypoxia-induced H9c2 cardiomyocyte apoptosis by transferring Bcl-2 proteins, related to mitochondrial pathway of cell death and involved in antiapoptotic role due to control of cytochrome *c* release (Li et al., 2020; Skommer et al., 2010). Proteomic findings revealed that under oxidative stress by exposure to the anthracycline DOX, a well known inducer of superoxide production in mitochondria, H9c2 cardiac myoblasts increased oxidized mitochondrial proteins in MDVs, which are partly released in EVs as a mitochondria quality control mechanism (Vasam et al., 2021). Cardiomyocytes can also release mitochondria. Recently, a new vesiculated fragment containing mitochondria called exospheres was uncovered. These are large membrane vesicles (diameter of 3.5 μm) ejected by healthy cardiomyocytes as a noncanonical route to eliminate defective mitochondria and other cargo (Nicolás-Ávila et al., 2020). Macrophages process these exospheres containing mitochondria, and failure of these mechanisms results in activation of the inflammasome and autophagy arrest, compromising mitochondrial function and ultimately cardiac function (Nicolás-Ávila et al., 2020). However, these special large membrane vesicles have not been characterized as EVs and we do not discuss them here.

Remarkably, the mitochondrial transfer has been extensively studied in lung diseases. It was shown that bone marrow-derived mesenchymal stromal cells transferred their mitochondria to the alveolar epithelium after alveolar attachment via MVs and nanotubes, exhibiting a protective response by restituting alveolar bioenergetics in the in vivo ALI model (Islam et al., 2012). Investigating the ability of MSC-EVs to modulate alveolar-capillary barrier integrity in preclinical models of ARDS, Silva et al. (2021) showed the critical role of mitochondrial transfer via EVs improving mitochondrial dysfunction and the pulmonary epithelial endothelial barrier integrity. These effects were mitochondria dependent as EVs generated from a rhodamine-6G treated cells (such treatment results in mitochondria depletion of MSCs, however, does not impair their capacity to secrete EVs) were not effective (Silva et al., 2021).

On the other hand, EV-mediated mitochondrial transfer contributed to disease pathology. Exosomes from bronchoalveolar lavage fluid were isolated and characterized from asthmatic patients. These exosomes were enriched with functional mitochondria compared to healthy controls, and integrated with the mitochondrial network of recipient T lymphocytes, increasing ROS generation (Hough et al., 2018).

Cancer

Tumors are complex systems composed of multiple cells that constantly communicate with their microenvironment. EVs are important mediators of paracrine signaling required for cancer cell communication and progression (Dominiak et al., 2020; Martinez-Outschoorn et al., 2011). Significant mitochondrial abnormalities have been observed throughout cancer development and are considered to be involved in the formation and progression of tumors. Therefore, treatments targeting tumor mitochondria and the transfer of healthy mitochondria were considered. mtDNA carrying enhanced pathogenic mutations derived from metastatic tumor cells was shown to be transferred to low metastatic tumor and stromal cells via exosomes in vitro and in the tumor microenvironment, probably as a mechanism to enhance metastasis (Takenaga et al., 2021). Similarly, activation of metabotropic glutamate receptor 3 in breast cancer cells activates Rab27-dependent release of EVs containing mtDNA, which can transfer invasive characteristics to "recipient" tumor cells, disseminating proinvasive microenvironments during mammary carcinoma progression (Rabas et al., 2021). In addition, EVs enriched with mitochondrial proteins are elevated in the plasma of patients with melanoma, ovarian, and breast cancer (Jang et al., 2019). mtDNA copy numbers are increased in some types of carcinomas, such as esophageal, malignant head and neck, ovarian, and prostate (Sun et al., 2018). Increased mtDNA copy number in colorectal cancer (CRC) cells was associated with promotion of the microsatellite-stabilized CRC survival and metastasis caused by elevated mitochondrial OXPHOS, indicating a possible mechanism for cancer progression (Sun et al., 2018). Therefore, mitochondrial transfer may enhance tumor viability by increasing mtDNA copy number and OXPHOS. Hence, from the perspective of therapy, the essence of mitochondrial transfer is to enhance the overall mitochondria fitness of the recipient cells. In cancer, this appears to be the main challenge. Increased evidence shows that mitochondrial transfer is a survival mechanism of tumors, and inhibiting the acquisition of functional mitochondria from adjacent healthy cells can be used as an approach to stop cancer progression and drug resistance (Liu et al., 2021).

Gap junctions

Since Spees et al. (2006) first described mitochondrial transfer from hMSCs to A549 cells with defective mtDNA, the acquisition of functional mitochondria through intercellular transfer has been extensively investigated (Spees et al., 2006). Cellular mechanisms by which dysfunctional mitochondria trigger organelle exchange in response to damage signals remain unclear. However, several mechanisms have been proposed and gap junctions seem to play a role in intercellular mitochondrial transfer, especially in the lung (Torralba et al., 2016).

Gap junctions are clusters of intercellular channels involved in cell to cell signaling. These highly specialized and regulated membrane structures are essential for physiological events, through the diffusion of ions, metabolites, secondary messengers, and other small molecules, in addition to allowing electrical and biochemical coupling between cells (Goodenough et al., 2009; Meşe et al., 2007). Gap junctions are found in all cells of the human body and are encoded by at least 21 members of the connexin gene family. They can form channels with the same connexin isoform (homomeric), or hemichannels with different connexin isoforms (heteromeric). These channels have four transmembrane domains, with both the N-terminal and C-terminal domains residing in the cytosol (Bonacquisti & Nguyen, 2019). The role of connexins in healthy tissue homeostasis has been described in embryogenesis and organogenesis, with different connexin family members expressed at various stages and anatomic locations (Laird & Lampe, 2018). Connexins are also associated with cardiovascular and neurovascular diseases, and act as pathological pores during inflammation (Leybaert et al., 2017).

Connexin-43

The most extensively studied and well-characterized connexin is connexin-43 (Cx43). Cx43 is oligomerized into connexons probably in the trans Golgi network after its exit from the endoplasmic reticulum, and subsequently trafficked to its destination in the plasma membrane through vesicles (Leithe et al., 2018). Aberrant connexins, including Cx43, are continuously degraded via endoplasmic reticulum quality control mechanisms. Moreover, older Cx43 internalized from gap junctions could be enzymatically digested in lysosomes or recycled to the plasma membrane (Falk et al., 2014).

Evidence suggests that Cx43 is involved in mitochondrial transfer. Upregulation of the GJA1-20K gene promotes Cx43 overexpression in murine astrocytes, triggering mitochondrial transmission to damaged neurons. This mechanism improves protection and recovery in a traumatic brain injury model (Ren et al., 2022). The use of Gap26, a specific Cx43 hemichannel blocker, to achieve Cx43 knockdown completely blocks astrocyte-neuron mitochondrial transmission (Ren et al., 2022). In the murine LPS-induced lung injury model Cx43 expression level was markedly increased in the plasma membranes of alveolar epithelial cells after LPS exposure, facilitating attachment of mouse BMSCs (mBMSCs) to the alveolar epithelium, the release of MVs, and nanotube formation to donate mitochondria to alveolar cells (Islam et al., 2012). Confirming this hypothesis, Cx43-mutated mBMSCs demonstrated the attenuation of mitochondrial transfer coupled with reduced therapeutic effect in the lung, resulting from the failure of attachment between BMSCs and alveoli. The transfer of functional mitochondria from mBMSCs to alveolar epithelial cells occurs via both TNTs and MVs in a Cx43-dependent manner, supported by Ca^{2+} exchange between cells through Cx43-GJCs (Islam et al., 2012). These findings were supported by Yao et al. who demonstrated that mitochondrial transfer from iPSC-MSC to epithelial cells alleviated asthmatic inflammation in the mouse model of asthma, with Cx43 overexpression or silencing playing a critical role in TNT formation and mitochondrial transfer efficiency between these cells (Yao et al., 2018).

However, the results regarding the importance of Cx43 in mitochondrial transfer are controversial. Sinclair et al. (2016) found a lack of impact of Cx43 inhibition on mitochondrial transfer from different types of MSC to epithelial cells, while the mitochondrial transfer was not observed when inhibitors of TNT formation and MVs were used (Sinclair et al., 2016). Interestingly, although Cx43 was shown to be important for mitochondrial transfer through TNT in the murine model of asthma (Yao et al., 2018); however, silencing of Cx43 in iPSC-MSCs only partially abrogated their protective effects, suggesting that other mechanisms may be involved (Yao et al., 2018). Taken together, these results suggest that the role of Cx43 in the mitochondrial transfer is predominantly associated with the regulation of TNT formation and can change depending on the cell activation state. One of the proposed mechanisms suggests that Cx43 cooperates with adhesion factors such as N-cadherin to stabilize the formation of the TNTs between communicating cells (Ribeiro-Rodrigues et al., 2017).

Concluding remarks

Mitochondrial transfer plays an important role in normal physiological processes and has a broad biomedical significance. Entire mitochondria or mitochondrial cargo delivery encapsulated in different subtypes of EVs has been reported in several disease models, driving pathogenesis or promoting restoration of function in the target cells. Mitochondrial transfer opens new directions for the treatment of multiple conditions underpinned by mitochondrial dysfunction. Transplantation of healthy mitochondria or mitochondrial content via EVs to recipient cells is a feasible therapeutic approach to restore mitochondrial function and has shown good preclinical results. However, the mechanisms that regulate EVs release as well as EV mitochondrial content are still not sufficiently investigated.

References

Amari, L., & Germain, M. (2021). Mitochondrial extracellular vesicles – origins and roles. *Frontiers in Molecular Neuroscience, 14*. https://doi.org/10.3389/fnmol.2021.767219, 16625099.

Barrès, C., Blanc, L., Bette-Bobillo, P., André, S., Mamoun, R., Gabius, H. J., & Vidal, M. (2010). Galectin-5 is bound onto the surface of rat reticulocyte exosomes and modulates vesicle uptake by macrophages. *Blood, 115*(3), 696–705. https://doi.org/10.1182/blood-2009-07-231449, 15280020.

Barros, F. M., Carneiro, F., Machado, J. C., & Melo, S. A. (2018). Exosomes and immune response in cancer: Friends or foes? *Frontiers in Immunology, 9*. https://doi.org/10.3389/fimmu.2018.00730, 16643224.

Bonacquisti, E. E., & Nguyen, J. (2019). Connexin 43 (Cx43) in cancer: Implications for therapeutic approaches via gap junctions. *Cancer Letters, 442*, 439–444. https://doi.org/10.1016/j.canlet.2018.10.043, 18727980.

Bordi, M., Darji, S., Sato, Y., Mellén, M., Berg, M. J., Kumar, A., Jiang, Y., & Nixon, R. A. (2019). mTOR hyperactivation in Down Syndrome underlies deficits in autophagy induction, autophagosome formation, and mitophagy. *Cell Death and Disease, 10*(8). https://doi.org/10.1038/s41419-019-1752-5, 20414889.

Bruschi, M., Santucci, L., Ravera, S., Candiano, G., Bartolucci, M., Calzia, D., Lavarello, C., Inglese, E., Ramenghi, L. A., Petretto, A., Ghiggeri, G. M., & Panfoli, I. (2016). Human urinary exosome proteome unveils its aerobic respiratory ability. *Journal of Proteomics, 136*, 25–34. https://doi.org/10.1016/j.jprot.2016.02.001, 18767737.

Cai, J., Wu, J., Wang, J., Li, Y., Hu, X., Luo, S., & Xiang, D. (2020). Extracellular vesicles derived from different sources of mesenchymal stem cells: Therapeutic effects and translational potential. *Cell and Bioscience, 10*(1). https://doi.org/10.1186/s13578-020-00427-x, 20453701.

Charrin, S., Jouannet, S., Boucheix, C., & Rubinstein, E. (2014). Tetraspanins at a glance. *Journal of Cell Science, 127*(17), 3641–3648. https://doi.org/10.1242/jcs.154906, 14779137.

Chen, J., Zhong, J., Wang, L.-lin, & Chen, Y.-ying (2021). Mitochondrial transfer in cardiovascular disease: From mechanisms to therapeutic implications. *Frontiers in Cardiovascular Medicine, 8*. https://doi.org/10.3389/fcvm.2021.771298, 2297-055X.

Choi, D. S., Kim, D. K., Kim, Y. K., & Gho, Y. S. (2013). Proteomics, transcriptomics and lipidomics of exosomes and ectosomes. *Proteomics, 13*(10–11), 1554–1571. https://doi.org/10.1002/pmic.201200329, 16159861.

Christianson, H. C., Svensson, K. J., Van Kuppevelt, T. H., Li, J. P., & Belting, M. (2013). Cancer cell exosomes depend on cell-surface heparan sulfate proteoglycans for their internalization and functional activity. *Proceedings of the National Academy of Sciences of the United States of America, 110*(43), 17380–17385. https://doi.org/10.1073/pnas.1304266110, 10916490.

Clancy, J. W., Schmidtmann, M., & D'Souza-Schorey, C. (2021). The ins and outs of microvesicles. *FASEB BioAdvances, 3*(6), 399–406. https://doi.org/10.1096/fba.2020-00127, 25739832.

Cocucci, E., & Meldolesi, J. (2015). Ectosomes and exosomes: Shedding the confusion between extracellular vesicles. *Trends in Cell Biology, 25*(6), 364–372. https://doi.org/10.1016/j.tcb.2015.01.004, 18793088.

Cowan, D. B., Yao, R., Thedsanamoorthy, J. K., Zurakowski, D., Del Nido, P. J., & McCully, J. D. (2017). Transit and integration of extracellular mitochondria in human heart cells. *Scientific Reports, 7*(1). https://doi.org/10.1038/s41598-017-17813-0, 20452322.

D'Acunzo, P., Pérez-González, R., Kim, Y., Hargash, T., Miller, C., Alldred, M. J., Erdjument-Bromage, H., Penikalapati, S. C., Pawlik, M., Saito, M., Saito, M., Ginsberg, S. D., Neubert, T. A., Goulbourne, C. N., & Levy, E. (2021). Mitovesicles are a novel population of extracellular vesicles of mitochondrial origin altered in Down syndrome. *Science Advances, 7*(7). https://doi.org/10.1126/sciadv.abe5085, 2375-2548.

D'Souza, A., Burch, A., Dave, K. M., Sreeram, A., Reynolds, M. J., Dobbins, D. X., Kamte, Y. S., Zhao, W., Sabatelle, C., Joy, G. M., Soman, V., Chandran, U. R., Shiva, S. S., Quillinan, N., Herson, P. S., & Manickam, D. S. (2021). Microvesicles transfer mitochondria and increase mitochondrial function in brain endothelial cells. *Journal of Controlled Release, 338*, 505–526. https://doi.org/10.1016/j.jconrel.2021.08.038, 18734995.

Deatherage, B. L., Cookson, B. T., & Andrews-Polymenis, H. L. (2012). Membrane vesicle release in bacteria, eukaryotes, and archaea: A conserved yet underappreciated aspect of microbial life. *Infection and Immunity, 80*(6), 1948–1957. https://doi.org/10.1128/iai.06014-11, 0019-9567.

Doherty, G. J., & McMahon, H. T. (2009). Mechanisms of endocytosis. *Annual Review of Biochemistry, 78*, 857–902. https://doi.org/10.1146/annurev.biochem.78.081307.110540, 00664154.

Dominiak, A., Chełstowska, B., Olejarz, W., & Nowicka, G. (2020). Communication in the cancer microenvironment as a target for therapeutic interventions. *Cancers, 12*(5). https://doi.org/10.3390/cancers12051232, 20726694.

Doyle, L. M., & Wang, M. Z. (2019). Overview of extracellular vesicles, their origin, composition, purpose, and methods for exosome isolation and analysis. *Cells, 8*(7). https://doi.org/10.3390/cells8070727, 20734409.

Eitan, E., Suire, C., Zhang, S., & Mattson, M. P. (2016). Impact of lysosome status on extracellular vesicle content and release. *Ageing Research Reviews, 32*, 65–74. https://doi.org/10.1016/j.arr.2016.05.001, 18729649.

Falk, M. M., Kells, R. M., & Berthoud, V. M. (2014). Degradation of connexins and gap junctions. *FEBS Letters, 588*(8), 1221–1229. https://doi.org/10.1016/j.febslet.2014.01.031, 18733468.

Feng, D., Zhao, W. L., Ye, Y. Y., Bai, X. C., Liu, R. Q., Chang, L. F., Zhou, Q., & Sui, S. F. (2010). Cellular internalization of exosomes occurs through phagocytosis. *Traffic, 11*(5), 675–687. https://doi.org/10.1111/j.1600-0854.2010.01041.x, 16000854.

Fitzner, D., Schnaars, M., Van Rossum, D., Krishnamoorthy, G., Dibaj, P., Bakhti, M., Regen, T., Hanisch, U. K., & Simons, M. (2011). Selective transfer of exosomes from oligodendrocytes to microglia by macropinocytosis. *Journal of Cell Science, 124*(3), 447–458. https://doi.org/10.1242/jcs.074088, 14779137.

Gaudino, S. J., & Kumar, P. (2019). Cross-talk between antigen presenting cells and T cells impacts intestinal homeostasis, bacterial infections, and tumorigenesis. *Frontiers in Immunology, 10*. https://doi.org/10.3389/fimmu.2019.00360, 16643224.

Goodenough, D. A., Paul, D. L., Gap, & Junctions. (2009). *Cold spring harbour Perspective Biology* (Vol. 1).

Gurung, S., Perocheau, D., Touramanidou, L., & Baruteau, J. (2021). The exosome journey: From biogenesis to uptake and intracellular signalling. *Cell Communication and Signaling, 19*(1). https://doi.org/10.1186/s12964-021-00730-1, 1478811X.

Hao, S., Bai, O., Li, F., Yuan, J., Laferte, S., & Xiang, J. (2007). Mature dendritic cells pulsed with exosomes stimulate efficient cytotoxic T-lymphocyte responses and antitumour immunity. *Immunology, 120*(1), 90–102. https://doi.org/10.1111/j.1365-2567.2006.02483.x, 13652567.

Hasegawa, T., Konno, M., Baba, T., Sugeno, N., Kikuchi, A., Kobayashi, M., Miura, E., Tanaka, N., Tamai, K., Furukawa, K., Arai, H., Mori, F., Wakabayashi, K., Aoki, M., Itoyama, Y., Takeda, A., & Smith, W. (2011). The AAA-ATPase VPS4 regulates extracellular secretion and lysosomal targeting of α-synuclein. *PLoS One, 6*(12), e29460. https://doi.org/10.1371/journal.pone.0029460, 1932-6203.

Hayakawa, K., Esposito, E., Wang, X., Terasaki, Y., Liu, Y., Xing, C., Ji, X., & Lo, E. H. (2016). Transfer of mitochondria from astrocytes to neurons after stroke. *Nature, 535*(7613), 551–555. https://doi.org/10.1038/nature18928, 14764687.

Henderson, M. C., & Azorsa, D. O. (2012). The genomic and proteomic content of cancer cell-derived exosomes. *Frontiers in Oncology, 2*. https://doi.org/10.3389/fonc.2012.00038, 2234943X.

Hough, K. P., Trevor, J. L., Strenkowski, J. G., Wang, Y., Chacko, B. K., Tousif, S., Chanda, D., Steele, C., Antony, V. B., Dokland, T., Ouyang, X., Zhang, J., Duncan, S. R., Thannickal, V. J., Darley-Usmar, V. M., & Deshane, J. S. (2018). Exosomal transfer of mitochondria from airway myeloid-derived regulatory cells to T cells. *Redox Biology, 18*, 54–64. https://doi.org/10.1016/j.redox.2018.06.009, 22132317.

Hsu, D. H., Paz, P., Villaflor, G., Rivas, A., Mehta-Damani, A., Angevin, E., Zitvogel, L., & Le Pecq, J. B. (2003). Exosomes as a tumor vaccine: Enhancing potency through direct loading of antigenic peptides. *Journal of Immunotherapy, 26*(5), 440–450. https://doi.org/10.1097/00002371-200309000-00007, 15374513.

Hu, G., Drescher, K. M., & Chen, X. M. (2012). Exosomal miRNAs: Biological properties and therapeutic potential. *Frontiers in Genetics, 3*. https://doi.org/10.3389/fgene.2012.00056, 16648021.

Ikeda, G., Santoso, M. R., Tada, Y., Li, A. M., Vaskova, E., Jung, J. H., O'Brien, C., Egan, E., Ye, J., & Yang, P. C. (2021). Mitochondria-rich extracellular vesicles from autologous stem cell–derived cardiomyocytes restore energetics of ischemic myocardium. *Journal of the American College of Cardiology, 77*(8), 1073–1088. https://doi.org/10.1016/j.jacc.2020.12.060, 15583597.

Islam, M. N., Das, S. R., Emin, M. T., Wei, M., Sun, L., Westphalen, K., Rowlands, D. J., Quadri, S. K., Bhattacharya, S., & Bhattacharya, J. (2012). Mitochondrial transfer from bone-marrow-derived stromal cells to pulmonary alveoli protects against acute lung injury. *Nature Medicine, 18*(5), 759–765. https://doi.org/10.1038/nm.2736, 1546170X.

Jang, S. C., Crescitelli, R., Cvjetkovic, A., Belgrano, V., Olofsson Bagge, R., Sundfeldt, K., Ochiya, T., Kalluri, R., & Lötvall, J. (2019). Mitochondrial protein enriched extracellular vesicles discovered in human melanoma tissues can be detected in patient plasma. *Journal of Extracellular Vesicles, 8*(1), 1635420. https://doi.org/10.1080/20013078.2019.1635420, 2001-3078.

Jeon, H., Lee, J., Lee, S., Kang, S. K., Park, S. J., Yoo, S. M., & Lee, M. S. (2019). Extracellular vesicles from KSHV-infected cells stimulate antiviral immune response through mitochondrial DNA. *Frontiers in Immunology, 10*. https://doi.org/10.3389/fimmu.2019.00876, 16643224.

Jeppesen, D. K., Fenix, A. M., Franklin, J. L., Higginbotham, J. N., Zhang, Q., Zimmerman, L. J., Liebler, D. C., Ping, J., Liu, Q., Evans, R., Fissell, W. H., Patton, J. G., Rome, L. H., Burnette, D. T., & Coffey, R. J. (2019). Reassessment of exosome composition. *Cell, 177*(2), 428–445. https://doi.org/10.1016/j.cell.2019.02.029, 10974172.

Johnson, J., Mercado-Ayon, E., Mercado-Ayon, Y., Dong, Y. N., Halawani, S., Ngaba, L., & Lynch, D. R. (2021). Mitochondrial dysfunction in the development and progression of neurodegenerative diseases. *Archives of Biochemistry and Biophysics, 702*, 108698. https://doi.org/10.1016/j.abb.2020.108698, 00039861.

Joshi, A. U., Minhas, P. S., Liddelow, S. A., Haileselassie, B., Andreasson, K. I., Dorn, G. W., & Mochly-Rosen, D. (2019). Fragmented mitochondria released from microglia trigger A1 astrocytic response and propagate inflammatory neurodegeneration. *Nature Neuroscience, 22*(10), 1635–1648. https://doi.org/10.1038/s41593-019-0486-0, 15461726.

Jurj, A., Zanoaga, O., Braicu, C., Lazar, V., Tomuleasa, C., Irimie, A., & Berindan-neagoe, I. (2020). A comprehensive picture of extracellular vesicles and their contents. Molecular transfer to cancer cells. *Cancers, 12*(2). https://doi.org/10.3390/cancers12020298, 20726694.

Kwok, Z. H., Wang, C., & Jin, Y. (2021). Extracellular vesicle transportation and uptake by recipient cells: A critical process to regulate human diseases. *Processes, 9*(2), 1–16. https://doi.org/10.3390/pr9020273, 22279717.

Lötvall, J., Hill, A. F., Hochberg, F., Buzás, E. I., Di Vizio, D., Gardiner, C., Gho, Y. S., Kurochkin, I. V., Mathivanan, S., Quesenberry, P., Sahoo, S., Tahara, H., Wauben, M. H., Witwer, K. W., & Théry, C. (2014). Minimal experimental requirements for definition of extracellular vesicles and their functions: A position statement from the International Society for extracellular vesicles. *Journal of Extracellular Vesicles, 3*(1), 26913. https://doi.org/10.3402/jev.v3.26913, 2001-3078.

Laird, D. W., & Lampe, P. D. (2018). Therapeutic strategies targeting connexins. *Nature Reviews Drug Discovery, 17*(12), 905–921. https://doi.org/10.1038/nrd.2018.138, 14741784.

Leithe, E., Mesnil, M., & Aasen, T. (2018). The connexin 43 C-terminus: A tail of many tales. *Biochimica et Biophysica Acta (BBA) - Biomembranes, 1860*(1), 48–64. https://doi.org/10.1016/j.bbamem.2017.05.008, 18792642.

Letsiou, E., Teixeira Alves, L. G., Fatykhova, D., Felten, M., Mitchell, T. J., Müller-Redetzky, H. C., Hocke, A. C., & Witzenrath, M. (2021). Microvesicles released from pneumolysin-stimulated lung epithelial cells carry mitochondrial cargo and suppress neutrophil oxidative burst. *Scientific Reports, 11*(1). https://doi.org/10.1038/s41598-021-88897-y, 20452322.

Leybaert, L., Lampe, P. D., Dhein, S., Kwak, B. R., Ferdinandy, P., Beyer, E. C., Laird, D. W., Naus, C. C., Green, C. R., & Schulz, R. (2017). Connexins in cardiovascular and neurovascular health and disease: Pharmacological implications. *Pharmacological Reviews, 69*(4), 396–478. https://doi.org/10.1124/pr.115.012062, 15210081.

Li, B., Antonyak, M. A., Zhang, J., & Cerione, R. A. (2012). RhoA triggers a specific signaling pathway that generates transforming microvesicles in cancer cells. *Oncogene, 31*(45), 4740–4749. https://doi.org/10.1038/onc.2011.636, 14765594.

Li, B., Zhao, H., Wu, Y., Zhu, Y., Zhang, J., Yang, G., Yan, Q., Li, J., Li, T., & Liu, L. (2020). Mitochondrial-derived vesicles protect cardiomyocytes against hypoxic damage. *Frontiers in Cell and Developmental Biology, 8*. https://doi.org/10.3389/fcell.2020.00214, 2296634X.

Liu, D., Dong, Z., Wang, J., Tao, Y., Sun, X., & Yao, X. (2020). The existence and function of mitochondrial component in extracellular vesicles. *Mitochondrion, 54*, 122–127. https://doi.org/10.1016/j.mito.2020.08.005, 18728278.

Liu, D., Gao, Y., Liu, J., Huang, Y., Yin, J., Feng, Y., Shi, L., Meloni, B. P., Zhang, C., Zheng, M., & Gao, J. (2021). Intercellular mitochondrial transfer as a means of tissue revitalization. *Signal Transduction and Targeted Therapy, 6*(1). https://doi.org/10.1038/s41392-020-00440-z, 20593635.

Liu, Z., Sun, Y., Qi, Z., Cao, L., & Ding, S. (2022). Mitochondrial transfer/transplantation: An emerging therapeutic approach for multiple diseases. *Cell and Bioscience, 12*(1). https://doi.org/10.1186/s13578-022-00805-7, 20453701.

Martinez-Outschoorn, U. E., Pavlides, S., Sotgia, F., & Lisanti, M. P. (2011). Mitochondrial biogenesis drives tumor cell proliferation. *American Journal of Pathology, 178*(5), 1949–1952. https://doi.org/10.1016/j.ajpath.2011.03.002, 15252191.

Matheoud, D., Sugiura, A., Bellemare-Pelletier, A., Laplante, A., Rondeau, C., Chemali, M., Fazel, A., Bergeron, J. J., Trudeau, L. E., Burelle, Y., Gagnon, E., McBride, H. M., & Desjardins, M. (2016). Parkinson's disease-related proteins PINK1 and Parkin repress mitochondrial antigen presentation. *Cell, 166*(2), 314–327. https://doi.org/10.1016/j.cell.2016.05.039, 10974172.

Matthaeus, C., & Taraska, J. W. (2021). Energy and dynamics of caveolae trafficking. *Frontiers in Cell and Developmental Biology, 8*. https://doi.org/10.3389/fcell.2020.614472, 2296634X.

McLelland, G. L., Soubannier, V., Chen, C. X., McBride, H. M., & Fon, E. A. (2014). Parkin and PINK1 function in a vesicular trafficking pathway regulating mitochondrial quality control. *EMBO Journal, 33*(4), 282–295. https://doi.org/10.1002/embj.201385902, 14602075.

Meşe, G., Richard, G., & White, T. W. (2007). Gap junctions: Basic structure and function. *Journal of Investigative Dermatology, 127*(11), 2516–2524. https://doi.org/10.1038/sj.jid.5700770, 15231747.

Mettlen, M., Chen, P. H., Srinivasan, S., Danuser, G., & Schmid, S. L. (2018). Regulation of clathrin-mediated endocytosis. *Annual Review of Biochemistry, 87*, 871–896. https://doi.org/10.1146/annurev-biochem-062917-012644, 15454509.

Miao, Y., Li, G., Zhang, X., Xu, H., & Abraham, S. N. (2015). A TRP channel senses lysosome neutralization by pathogens to trigger their expulsion. *Cell, 161*(6), 1306–1319. https://doi.org/10.1016/j.cell.2015.05.009, 10974172.

Montecalvo, A., Larregina, A. T., Shufesky, W. J., Stolz, D. B., Sullivan, M. L. G., Karlsson, J. M., Baty, C. J., Gibson, G. A., Erdos, G., Wang, Z., Milosevic, J., Tkacheva, O. A., Divito, S. J., Jordan, R., Lyons-Weiler, J., Watkins, S. C., & Morelli, A. E. (2012). Mechanism of transfer of functional microRNAs between mouse dendritic cells via exosomes. *Blood, 119*(3), 756–766. https://doi.org/10.1182/blood-2011-02-338004, 15280020.

Morelli, A. E., Larregina, A. T., Shufesky, W. J., Sullivan, M. L. G., Stolz, D. B., Papworth, G. D., Zahorchak, A. F., Logar, A. J., Wang, Z., Watkins, S. C., Falo, L. D., & Thomson, A. W. (2004). Endocytosis, intracellular sorting, and processing of exosomes by dendritic cells. *Blood, 104*(10), 3257–3266. https://doi.org/10.1182/blood-2004-03-0824, 00064971.

Morrison, T. J., Jackson, M. V., Cunningham, E. K., Kissenpfennig, A., McAuley, D. F., O'Kane, C. M., & Krasnodembskaya, A. D. (2017). Mesenchymal stromal cells modulate macrophages in clinically relevant lung injury models by extracellular vesicle mitochondrial transfer. *American Journal of Respiratory and Critical Care Medicine, 196*(10), 1275–1286. https://doi.org/10.1164/rccm.201701-0170OC, 15354970.

Mulcahy, L. A., Pink, R. C., & Carter, D. R. F. (2014). Routes and mechanisms of extracellular vesicle uptake. *Journal of Extracellular Vesicles, 3*(1). https://doi.org/10.3402/jev.v3.24641, 20013078.

Nabhan, J. F., Hu, R., Oh, R. S., Cohen, S. N., & Lu, Q. (2012). Formation and release of arrestin domain-containing protein 1-mediated microvesicles (ARMMs) at plasma membrane by recruitment of TSG101 protein. *Proceedings of the National Academy of Sciences of the United States of America, 109*(11), 4146–4151. https://doi.org/10.1073/pnas.1200448109, 10916490.

Nanbo, A., Kawanishi, E., Yoshida, R., & Yoshiyama, H. (2013). Exosomes derived from epstein-barr virus-infected cells are internalized via caveola-dependent endocytosis and promote phenotypic modulation in target cells. *Journal of Virology, 87*(18), 10334–10347. https://doi.org/10.1128/JVI.01310-13, 10985514.

Nazarenko, I., Rana, S., Baumann, A., McAlear, J., Hellwig, A., Trendelenburg, M., Lochnit, G., Preissner, K. T., & Zöller, M. (2010). Cell surface tetraspanin Tspan8 contributes to molecular pathways of exosome-induced endothelial cell activation. *Cancer Research, 70*(4), 1668–1678. https://doi.org/10.1158/0008-5472.CAN-09-2470, 15387445.

Neuspiel, M., Schauss, A. C., Braschi, E., Zunino, R., Rippstein, P., Rachubinski, R. A., Andrade-Navarro, M. A., & McBride, H. M. (2008). Cargo-selected transport from the mitochondria to peroxisomes is mediated by vesicular carriers. *Current Biology, 18*(2), 102–108. https://doi.org/10.1016/j.cub.2007.12.038, 09609822.

Nicolás-Ávila, J. A., Lechuga-Vieco, A. V., Esteban-Martínez, L., Sánchez-Díaz, M., Díaz-García, E., Santiago, D. J., Rubio-Ponce, A., Li, J. L. Y., Balachander, A., Quintana, J. A., Martínez-de-Mena, R., Castejón-Vega, B., Pun-García, A., Través, P. G., Bonzón-Kulichenko, E., García-Marqués, F., Cussó, L., A-González, N., González-Guerra, A., … Hidalgo, A. (2020). A network of macrophages supports mitochondrial homeostasis in the heart. *Cell, 183*(1), 94–109. https://doi.org/10.1016/j.cell.2020.08.031, 10974172.

O'Brien, C. G., Ozen, M. O., Ikeda, G., Vaskova, E., Jung, J. H., Bayardo, N., Santoso, M. R., Shi, L., Wahlquist, C., Jiang, Z., Jung, Y., Zeng, Y., Egan, E., Sinclair, R., Gee, A., Witteles, R., Mercola, M., Svensson, K. J., Demirci, U., & Yang, P. C. (2021). Mitochondria-rich extracellular vesicles rescue patient-specific cardiomyocytes from doxorubicin injury: Insights into the SENECA trial. *Journal of the American College of Cardiology: CardioOncology, 3*(3), 428–440. https://doi.org/10.1016/j.jaccao.2021.05.006, 26660873.

Paliwal, S., Chaudhuri, R., Agrawal, A., & Mohanty, S. (2018). Regenerative abilities of mesenchymal stem cells through mitochondrial transfer. *Journal of Biomedical Science, 25*(1). https://doi.org/10.1186/s12929-018-0429-1, 14230127.

Park, J., Lee, J., Choi, C., & Pandey, S. (2011). Mitochondrial network determines intracellular ROS dynamics and sensitivity to oxidative stress through switching inter-mitochondrial messengers. *PLoS One, 6*(8), e23211. https://doi.org/10.1371/journal.pone.0023211, 1932-6203.

Parolini, I., Federici, C., Raggi, C., Lugini, L., Palleschi, S., De Milito, A., Coscia, C., Iessi, E., Logozzi, M., Molinari, A., Colone, M., Tatti, M., Sargiacomo, M., & Fais, S. (2009). Microenvironmental pH is a key factor for exosome traffic in tumor cells. *Journal of Biological Chemistry, 284*(49), 34211–34222. https://doi.org/10.1074/jbc.M109.041152, 1083351X.

Peruzzotti-Jametti, L., Bernstock, J. D., Willis, C. M., Manferrari, G., Rogall, R., Fernandez-Vizarra, E., Williamson, J. C., Braga, A., van den Bosch, A., Leonardi, T., Krzak, G., Kittel, Á., Benincá, C., Vicario, N., Tan, S., Bastos, C., Bicci, I., Iraci, N., Smith, J. A., … Storey, K. G. (2021). Neural stem cells traffic functional mitochondria via extracellular vesicles. *PLoS Biology, 19*(4), e3001166. https://doi.org/10.1371/journal.pbio.3001166, 1545-7885.

Phinney, D. G., Di Giuseppe, M., Njah, J., Sala, E., Shiva, S., St Croix, C. M., Stolz, D. B., Watkins, S. C., Di, Y. P., Leikauf, G. D., Kolls, J., Riches, D. W. H., Deiuliis, G., Kaminski, N., Boregowda, S. V., McKenna, D. H., & Ortiz, L. A. (2015). Mesenchymal stem cells use extracellular vesicles to outsource mitophagy and shuttle microRNAs. *Nature Communications, 6*. https://doi.org/10.1038/ncomms9472, 20411723.

Picca, A., Guerra, F., Calvani, R., Coelho-Junior, H. J., Bossola, M., Landi, F., Bernabei, R., Bucci, C., & Marzetti, E. (2020). Generation and release of mitochondrial-derived vesicles in health, aging and disease. *Journal of Clinical Medicine, 9*(5). https://doi.org/10.3390/jcm9051440, 20770383.

Puhm, F., Afonyushkin, T., Resch, U., Obermayer, G., Rohde, M., Penz, T., Schuster, M., Wagner, G., Rendeiro, A. F., Melki, I., Kaun, C., Wojta, J., Bock, C., Jilma, B., Mackman, N., Boilard, E., & Binder, C. J. (2019). Mitochondria are a subset of extracellular vesicles released by activated monocytes and induce type I IFN and TNF responses in endothelial cells. *Circulation Research, 125*(1), 43–52. https://doi.org/10.1161/CIRCRESAHA.118.314601, 15244571.

Rabas, N., Palmer, S., Mitchell, L., Ismail, S., Gohlke, A., Riley, J. S., Tait, S. W. G., Gammage, P., Soares, L. L., Macpherson, I. R., & Norman, J. C. (2021). PINK1 drives production of mtDNA-containing extracellular vesicles to promote invasiveness. *Journal of Cell Biology, 220*(12). https://doi.org/10.1083/jcb.202006049, 0021-9525.

Raposo, G., Nijman, H. W., Stoorvogel, W., Leijendekker, R., Harding, C. V., Melief, C. J. M., & Geuze, H. J. (1996). B lymphocytes secrete antigen-presenting vesicles. *Journal of Experimental Medicine, 183*(3), 1161–1172. https://doi.org/10.1084/jem.183.3.1161, 00221007.

Ren, D., Zheng, P., Zou, S., Gong, Y., Wang, Y., Duan, J., Deng, J., Chen, H., Feng, J., Zhong, C., & Chen, W. (2022). GJA1-20K enhances mitochondria transfer from astrocytes to neurons via Cx43-TnTs after traumatic brain injury. *Cellular and Molecular Neurobiology, 42*(6), 1887–1895. https://doi.org/10.1007/s10571-021-01070-x, 15736830.

Ribeiro-Rodrigues, T. M., Martins-Marques, T., Morel, S., Kwak, B. R., & Girão, H. (2017). Role of connexin 43 in different forms of intercellular communication - gap junctions, extracellular vesicles and tunnelling nanotubes. *Journal of Cell Science, 130*(21), 3619–3630. https://doi.org/10.1242/jcs.200667, 14779137.

Ripa, I., Andreu, S., López-Guerrero, J. A., & Bello-Morales, R. (2021). Membrane rafts: Portals for viral entry. *Frontiers in Microbiology, 12*. https://doi.org/10.3389/fmicb.2021.631274, 1664302X.

Robbins, P. D., & Morelli, A. E. (2014). Regulation of immune responses by extracellular vesicles. *Nature Reviews Immunology, 14*(3), 195–208. https://doi.org/10.1038/nri3622, 14741741.

Segura, E., Nicco, C., Lombard, B., Véron, P., Raposo, G., Batteux, F., Amigorena, S., & Théry, C. (2005). ICAM-1 on exosomes from mature dendritic cells is critical for efficient naive T-cell priming. *Blood, 106*(1), 216–223. https://doi.org/10.1182/blood-2005-01-0220, 00064971.

Silva, J. D., Su, Y., Calfee, C. S., Delucchi, K. L., Weiss, D., McAuley, D. F., O'Kane, C., & Krasnodembskaya, A. D. (2021). Mesenchymal stromal cell extracellular vesicles rescue mitochondrial dysfunction and improve barrier integrity in clinically relevant models of ARDS. *European Respiratory Journal, 58*(1). https://doi.org/10.1183/13993003.02978-2020, 13993003.

Sinclair, K. A., Yerkovich, S. T., Hopkins, P. M. A., & Chambers, D. C. (2016). Characterization of intercellular communication and mitochondrial donation by mesenchymal stromal cells derived from the human lung. *Stem Cell Research and Therapy, 7*(1). https://doi.org/10.1186/s13287-016-0354-8, 17576512.

Singel, K. L., Grzankowski, K. S., Khan, A. N. M. N. H., Grimm, M. J., D'Auria, A. C., Morrell, K., Eng, K. H., Hylander, B., Mayor, P. C., Emmons, T. R., Lénárt, N., Fekete, R., Környei, Z., Muthukrishnan, U., Gilthorpe, J. D., Urban, C. F., Itagaki, K., Hauser, C. J., Leifer, C., … Segal, B. H. (2019). Mitochondrial DNA in the tumour microenvironment activates neutrophils and is associated with worse outcomes in patients with advanced epithelial ovarian cancer. *British Journal of Cancer, 120*(2), 207–217. https://doi.org/10.1038/s41416-018-0339-8, 15321827.

Skommer, J., Brittain, T., & Raychaudhuri, S. (2010). Bcl-2 inhibits apoptosis by increasing the time-to-death and intrinsic cell-to-cell variations in the mitochondrial pathway of cell death. *Apoptosis, 15*(10), 1223–1233. https://doi.org/10.1007/s10495-010-0515-7, 1573675X.

Soubannier, V., McLelland, G. L., Zunino, R., Braschi, E., Rippstein, P., Fon, E. A., & McBride, H. M. (2012). A vesicular transport pathway shuttles cargo from mitochondria to lysosomes. *Current Biology, 22*(2), 135–141. https://doi.org/10.1016/j.cub.2011.11.057, 09609822.

Soubannier, V., Rippstein, P., Kaufman, B. A., Shoubridge, E. A., & McBride, H. M. (2012). Reconstitution of mitochondria derived vesicle formation demonstrates selective enrichment of oxidized cargo. *PLoS One, 7*(12). https://doi.org/10.1371/journal.pone.0052830, 19326203.

Spees, J. L., Olson, S. D., Whitney, M. J., & Prockop, D. J. (2006). Mitochondrial transfer between cells can rescue aerobic respiration. *Proceedings of the National Academy of Sciences of the United States of America, 103*(5), 1283–1288. https://doi.org/10.1073/pnas.0510511103, 00278424.

Srinivasan, S., Guha, M., Kashina, A., & Avadhani, N. G. (2017). Mitochondrial dysfunction and mitochondrial dynamics-The cancer connection. *Biochimica et Biophysica Acta - Bioenergetics, 1858*(8), 602–614. https://doi.org/10.1016/j.bbabio.2017.01.004, 18792650.

Su, Y., Silva, J. D., Doherty, D., Simpson, D. A., Weiss, D. J., Rolandsson-Enes, S., McAuley, D. F., O'Kane, C. M., Brazil, D. P., & Krasnodembskaya, A. D. (2022). Mesenchymal stromal cells-derived extracellular vesicles reprogramme macrophages in ARDS models through the miR-181a-5p-PTEN-pSTAT5-SOCS1 axis. *Thorax*. https://doi.org/10.1136/thoraxjnl-2021-218194, 14683296.

Sun, X., Zhan, L., Chen, Y., Wang, G., He, L., Wang, Q., Zhou, F., Yang, F., Wu, J., Wu, Y., Xing, J., He, X., & Huang, Q. (2018). Increased mtDNA copy number promotes cancer progression by enhancing mitochondrial oxidative phosphorylation in microsatellite-stable colorectal cancer. *Signal Transduction and Targeted Therapy, 3*(1). https://doi.org/10.1038/s41392-018-0011-z, 20593635.

Svensson, K. J., Christianson, H. C., Wittrup, A., Bourseau-Guilmain, E., Lindqvist, E., Svensson, L. M., Mörgelin, M., & Belting, M. (2013). Exosome uptake depends on ERK1/2-heat shock protein 27 signaling and lipid raft-mediated endocytosis negatively regulated by caveolin-1. *Journal of Biological Chemistry, 288*(24), 17713–17724. https://doi.org/10.1074/jbc.M112.445403, 1083351X.

Takenaga, K., Koshikawa, N., & Nagase, H. (2021). Intercellular transfer of mitochondrial DNA carrying metastasis-enhancing pathogenic mutations from high- to low-metastatic tumor cells and stromal cells via extracellular vesicles. *BMC Molecular and Cell Biology, 22*(1). https://doi.org/10.1186/s12860-021-00391-5, 26618850.

Théry, C., Duban, L., Segura, E., Væron, P., Lantz, O., & Amigorena, S. (2002). Indirect activation of naïve CD4+ T cells by dendritic cell-derived exosomes. *Nature Immunology, 3*(12), 1156–1162. https://doi.org/10.1038/ni854

Todkar, K., Chikhi, L., Desjardins, V., El-Mortada, F., Pépin, G., & Germain, M. (2021). Selective packaging of mitochondrial proteins into extracellular vesicles prevents the release of mitochondrial DAMPs. *Nature Communications, 12*(1). https://doi.org/10.1038/s41467-021-21984-w, 20411723.

Torralba, D., Baixauli, F., Sánchez-Madrid, F., & Mitochondria. (2016). Know No boundaries: Mechanisms and functions of intercellular mitochondrial transfer. *Frontiers in Cell and Developmental Biology, 0*.

Torralba, D., Baixauli, F., Villarroya-Beltri, C., Fernández-Delgado, I., Latorre-Pellicer, A., Acín-Pérez, R., Martín-Cófreces, N. B., Jaso-Tamame, Á. L., Iborra, S., Jorge, I., González-Aseguinolaza, G., Garaude, J., Vicente-Manzanares, M., Enríquez, J. A., Mittelbrunn, M., & Sánchez-Madrid, F. (2018). Priming of dendritic cells by DNA-containing extracellular vesicles from activated T cells through antigen-driven contacts. *Nature Communications, 9*(1). https://doi.org/10.1038/s41467-018-05077-9, 20411723.

Van Niel, G., D'Angelo, G., & Raposo, G. (2018). Shedding light on the cell biology of extracellular vesicles. *Nature Reviews Molecular Cell Biology, 19*(4), 213–228. https://doi.org/10.1038/nrm.2017.125, 14710080.

Vasam, G., Nadeau, R., Cadete, V. J. J., Lavallée-Adam, M., Menzies, K. J., & Burelle, Y. (2021). Proteomics characterization of mitochondrial-derived vesicles under oxidative stress. *Federation of American Societies for Experimental Biology Journal, 35*(4). https://doi.org/10.1096/fj.202002151R, 15306860.

Wang, Q., & Lu, Q. (2017). Plasma membrane-derived extracellular microvesicles mediate non-canonical intercellular NOTCH signaling. *Nature Communications, 8*(1). https://doi.org/10.1038/s41467-017-00767-2, 20411723.

Wang, X., Weidling, I., Koppel, S., Menta, B., Perez Ortiz, J., Kalani, A., Wilkins, H. M., & Swerdlow, R. H. (2020). Detection of mitochondria-pertinent components in exosomes. *Mitochondrion, 55*, 100–110. https://doi.org/10.1016/j.mito.2020.09.006, 18728278.

Weinberg, S. E., Sena, L. A., & Chandel, N. S. (2015). Mitochondria in the regulation of innate and adaptive immunity. *Immunity, 42*(3), 406–417. https://doi.org/10.1016/j.immuni.2015.02.002, 10974180.

Yao, B., Wang, R., Wang, Y., Zhang, Y., Hu, T., Song, W., Li, Z., Huang, S., & Fu, X. (2020). Biochemical and structural cues of 3D-printed matrix synergistically direct MSC differentiation for functional sweat gland regeneration. *Science Advances, 6*(10). https://doi.org/10.1126/sciadv.aaz1094, 23752548.

Yao, Y., Fan, X. L., Jiang, D., Zhang, Y., Li, X., Xu, Z. B., Fang, S. B., Chiu, S., Tse, H. F., Lian, Q., & Fu, Q. L. (2018). Connexin 43-mediated mitochondrial transfer of iPSC-MSCs alleviates asthma inflammation. *Stem Cell Reports, 11*(5), 1120–1135. https://doi.org/10.1016/j.stemcr.2018.09.012, 22136711.

Youle, R. J., & Narendra, D. P. (2011). Mechanisms of mitophagy. *Nature Reviews Molecular Cell Biology, 12*(1), 9–14. https://doi.org/10.1038/nrm3028, 14710080.

Zimmermann, P., Zhang, Z., Degeest, G., Mortier, E., Leenaerts, I., Coomans, C., Schulz, J., N'Kuli, F., Courtoy, P. J., & David, G. (2005). Syndecan recyling is controlled by syntenin-PIP2 interaction and Arf6. *Developmental Cell, 9*(3), 377–388. https://doi.org/10.1016/j.devcel.2005.07.011

SECTION III

Mitochondrial transplantation from isolation process to application

CHAPTER

7

Isolation process for mitochondrial transplantation

Romana Fato[3], Christian Bergamini[3] and Francesco Pallotti[1,2]

[1]Dipartimento di Medicina e Chirurgia, Università Degli Studi dell'Insubria, Varese, Italy; [2]SSD Laboratorio Analisi-SMEL Specializzato in Citogenetica e Genetica Medica, ASST Sette Laghi-Ospedale di Circolo-Fondazione Macchi, Varese, Italy; [3]Dipartimento di Farmacia e Biotecnologie, FABIT, Università Degli Studi di Bologna, Bologna, Italy

Introduction

Mitochondrial diseases are a group of rare genetic disorders caused by mutations in the mitochondrial DNA (mtDNA) that affect principally tissues with a large energy demand (Kanungo et al., 2018) leading to a wide range of diseases, including neurological disorders, muscular disorders, metabolic disorders, and even cancer (Taylor & Turnbull, 2005).

The involvement of mitochondria in several pathologies has generated the term "Mitochondrial medicine" (Di Mauro et al., 2006) that involves a variety of approaches in order to achieve new diagnostic tools and therapies to treat mitochondrial diseases. Unfortunately, there are no specific therapies for mitochondrial diseases but treatments can help to manage symptoms and improve quality of life. Treatments may include antioxidant supplementation to reduce the oxidative stress and counteract the energy decline (Apostolova & Victor, 2015; Enns, 2014). Mitochondrial Replacement Therapy, also Known as Mitochondrial Transplantation, is a technique that was used to prevent the transmission of mtDNA mutations from a mother to her offspring. Ooplasmic transfer was a relatively successful technique aiming to enhance oocyte quality (Barritt et al., 2001; Darbandi et al., 2017); it involves the transfer of 5%–15% of donor cytoplasm into the oocyte of the patient (Cohen et al., 1998). This technique aims to improve the poor-quality oocyte by efficient transfer of essential cellular components including mitochondria and other important organelles and molecules. Among all these components, mitochondria were considered to play a pivotal role in promoting the quality of the oocyte required to successfully form of a healthy zygote (Darbandi et al., 2017). Perez et al. reported a protective effect of microinjected mitochondria on the occurrence of spontaneous apoptosis in murine oocytes (Perez et al., 2000). Nevertheless, the

https://doi.org/10.1016/B978-0-443-18858-9.00007-X

introduction of a foreign mtDNA can cause conflicts between the nuclear DNA (nDNA) and the donor mtDNA increasing the heteroplasmy which raised concerns regarding the safety of this technique (Brenner et al., 2000). Mitochondrial transplantation has been recently explored for the treatment of ischemia reperfusion injury in cardiomyocytes. Blitzer et al. described a protective effect of autologous mitochondrial transplantation in a porcine model of ischemia and reperfusion (Blitzer et al., 2020); mitochondria derived from blood cells were transplanted into hematopoietic stem cells (Almannai et al., 2020), and lung injury, in a mouse model of ischemia and reperfusion, was reduced by mitochondrial transplantation (Moskowitzova et al., 2020). Although mitochondrial transplantation was successfully tested in animal and human models, the clinical application of this technique requires further validation to ensure its safety and efficacy. Mitochondria have a their own DNA which codify for only few mitochondrial components; for this reason, a correct cross-talk between nDNA and mtDNA is very important for the cells viability. The problem of nuclear–mitochondrial incompatibility was pointed out by Beale and Knowles (Beale & Knowles, 1976). Moreover, mitochondrial transplantation requires coupled mitochondria with a high degree of purity and since it is well known that mitochondria stay coupled only for few hours after isolation the choice of an appropriate isolation method and the time spent for the isolation may have a strong impact on the success of the whole procedure.

Mitochondria structure and function

Mitochondria are essential organelle present in all eukaryotic cell; their proper function ensures to the cell the right energy supply and exert a central role in the cell metabolism.

The central core of the mitochondria is the OXPHOS machinery which is a complex system to drive electrons, delivered by nutrient oxidation, to molecular oxygen in order to retrieve free energy to build ATP. The electron transfer is divided into discrete steps to maximize the free energy recovery with a maximal stoichiometry of 3 ATP for an NADH oxidation and 2 ATP for the oxidation of the higher potential electrons delivered by FADH2. The rate of ATP production depends on several factors such as the presence of components of the electron transfer chain (ETC), the amount of reducing substrate and oxygen, as well as the presence of ADP and Pi: the substrates for the mitochondrial ATPase. Following the Mitchell's chemiosmotic hypothesis, the electron transfer to molecular oxygen causes a proton accumulation into the intermembrane space that is known as the protonmotive force used by the ATPase to synthetize ATP. The OXPHOS is not the only metabolic pathway present into the mitochondria: the Krebs cycle fuels electrons to the OXPHOS oxidizing the acetyl-CoA derived from glucose, fatty acid, and amino acid catabolism. Moreover, all the Krebs cycle intermediate may be used as building blocks for anabolic pathways.

For this reason, mitochondria are well known as the "powerhouse" of all the eukaryotic cells: they are involved in energy production and are able to connect the cellular metabolism to the fate of the cells. They are bacteria-resembling subcellular particles whose origin is traced back to bacteria engulfed by larger cells. Like bacteria, mitochondria have their own DNA and ribosomes. During evolution, the major part of the bacterial genome was englobed into the nDNA; nevertheless, they retained part of their original genome that encode for 37 genes: 13 for polypeptide components of the oxidative phosphorylation machinery; 22 for tRNAs and 2 for ribosomal RNAs required for the transcription and translation processes within the organelle.

Mitochondria contains about other 1500 different proteins that are encoded by nDNA, translated in the cytosol and transported into the mitochondria through the TOM/TIM translocase system (Robin et al., 2012).

The 13 polypeptides encoded by the mtDNA are all hydrophobic subunits of the OXPHOS machinery and in particular seven subunits belong to complex I; one subunit for complex III; three subunits for complex IV; and two subunits for ATPase. This complex organization requires a highly regulated cross-talk between the nuclear and the mitochondrial genome.

In 1982 Clark and Shay demonstrated that purified mitochondria isolated from cells resistant to the antibiotics chloramphenicol and efrapeptin were able to transfer antibiotics resistance to sensitive cells (Clark & Shay, 1982). However, this finding remained underestimated for several years.

An important breakthrough in the study of the interrelationship between nuclear and mtDNA occurred in 1990 when Attardi's group introduced the use of cells devoided of mtDNA (Rho zero cells) (King & Attardi, 1989) which were then used in fusion experiments with denucleated cells. The cell lines thus obtained called "cytoplasmic hybrids" have contributed to clarify the functional consequences of mtDNA mutations.

For a long time, mitochondria were considered small organelle with a beans-like shape. Indeed, they undergo frequent morphology reorganization mediated by fusion and fission processes. The fusion and fission processes are responsible for the generation of the so-called "mitochondrial network" in which the organization depends on the functional state of the cell and it is essential to maintain healthy mitochondria.

Mitochondrial injuries are deleterious for cell survival and can be driven by different factors that can directly affect nuclear or mitochondrial genes codifying for mitochondrial components. Moreover, environmental factors and aging can alter mitochondrial functions through increased production and release of reactive oxygen species (ROS). Mitochondria are considered the major source of ROS in the cells and are also the first target of the damaging effect of ROS. In particular, mtDNA damage affects the OXPHOS machine leading to an increased ROS production triggering a vicious circle which brings to mitochondrial destruction, energy decline, and cell death (Lenaz et al., 1999, 2010; Ozawa, 1997).

Due to their central role, mitochondrial dysfunction causes several diseases as neurodegenerative diseases, diabetes, cancer, strokes and aging. Administration of antioxidant is the most common therapeutic strategy which aims to maintain general patient health, but it is not able to repair the mitochondrial damage.

From a general point of view, the mitochondrial diseases can be divided into two types: primary and secondary dysfunction. Primary mitochondrial diseases depend on mutation to a nuclear or mitochondrial gene encoding for a mitochondrial component.

Primary mitochondrial diseases are assumed to be rare and encompass a heterogeneous spectrum of clinical phenotypes. In contrast, the secondary mitochondrial diseases are the consequences of pathological states that can cause a mitochondrial damage like ischemia/reperfusion injury or metabolic syndromes, cancer, and diabetes.

The energy decline induced by ETC dysfunction is rapid. It was shown that in ischemic heart the ATP synthesis is strongly reduced and the ATP stores are depleted in 6 min. The energy decline is prolonged for 60–180 min after reperfusion (Tsukube et al., 1997).

Mitochondrial isolation

A critical step in mitochondrial transplantation is related to the isolation of healthy and pure mitochondria. The traditional methods developed for mitochondrial isolation are based on tissue disaggregation and homogenization followed by differential centrifugations (DCs). Mitochondria isolation procedure can alter morphology and function of these organelles. Isolated mitochondria are fragmented, with a more spheroid shape, altered respiratory capacity, higher sensitivity to Ca^{2+} mPTP opening and higher levels of ROS production; based on these observations, Picard et al. concluded that isolated mitochondria are more fragile with morphological functional characteristic that resembles more to the characteristic of mitochondria present in stressed cells than that one of healthy mitochondria (Picard et al., 2011).

Kesner et al. demonstrated that the integrity of the outer membrane was essential for mitochondrial transplantation into receiving cells (Kesner & Saada-Reich, 2016). They demonstrated that digitonin-treated mitochondria were not incorporated in the receiving cells suggesting that specific proteins located on the outer mitochondrial membrane could mediate the mitochondrial transfer. Moreover, recipient cells should have specific cell membrane proteins that may be required for the mitochondria–cell interaction. These results suggest that the way of incorporation of mitochondria is different from normal endocytosis and can be a biological phenomenon.

In a recent paper, using UV radiation to damage mitochondria, Hu et al. showed that the structural integrity of mitochondria for transplantation was responsible for the successfully treatment of childhood heart diseases (Hu et al., 2022). On the other hand, the molecules present in the mitochondria (such as , ATP, mDNA, or GSH etc.) alone were not able to restore the cellular functionality.

Actually, there is no way to evaluate the number of healthy or damaged mitochondria present in a mitochondrial preparation for transplantation. One way to assess mitochondrial function is the respiratory control ratio (RCR, state 3/state 4 respiratory rate). However, the RCR value differs from the mitochondria source. Katrangi et al. (2007) determined an RCR of 5.8 ± 0.56 in mitochondria from mouse liver and an RCR of 1.94 ± 0.26 for mitochondria isolated from the mammary cancer cell line 4T1 mouse. The authors have successfully used these mitochondrial preparations to reinsert mitochondria into human ρ0 cells.

Natural mitochondrial transfer from one cell to another contributes to maintain the cellular homeostasis. Thomas showed that extracellular vesicle–mediated mitochondrial transfer from mesenchymal stem cells to human macrophage enhanced their phagocytic capacity reducing inflammation and lung injury (Morrison et al., 2017).

A second way for mitochondrial transfer between cell deal with intracellular structures like tunneling nanotubes (TNTs) (Gousset et al., 2013; Sanchez & Brusco, 2016; Torralba et al., 2016).

In the last two decades, many experimental assays were carried out to improve technique for mitochondrial transplantation. The problems related to mitochondrial isolation for transplantation can be divided into: (1) the mitochondrial source; (2) the isolation method; and (3) the methods for transplantation.

Mitochondrial source

The first clinical application of isolated mitochondrial transplantation was related to cardioprotection in ischemia reperfusion injury in pediatric patients. This involved mitochondrial autotransplantation in which healthy autologous mitochondria derived from nonischemic skeletal muscle were transplanted into the injured myocardium (Emani et al., 2017). In patients with mitochondrial myopathies, autologous mitochondria would not be suitable for transplantation, in these cases, mitochondria can be harvested from another individual of the same species. Ramirez-Barbieri et al. showed that allogenic transplantation is safe and without alloreactivity (Ramirez-Barbieri et al., 2019).

Subsequently, several studies have shown that transplantation of isolated mitochondria from different sources, was able to restore damaged tissues, organs, or cells (Ali et al., 2020; Blitzer et al., 2020; Guariento et al., 2020; Kubat et al., 2021; Pourmohammadi-Bejarpasi et al., 2020; Ulger & Kubat, 2022; Ulger et al., 2021).

Moskowitzova et al. showed that mitochondrial transplantation prolonged the cold ischemia time for heart transplantation in murine model (Moskowitzova et al., 2018). In this case, mitochondria delivery was done in two times: first delivery was before cold ischemia and the second one was just before reperfusion; by this experimental strategy, the authors demonstrated that mitochondrial transplantation reduces heart tissue injury after prolonged cold ischemia and improve transplantation outcomes.

Isolation methods

There are different methods for mitochondria isolation depending on the mitochondrial source; the most common methods are based on DC (Bensley & Hoerr, 1934), density gradient high-speed ultracentrifugation (Sims, 1990), and magnetic cell separation (Hubbard et al., 2019).

Several protocols based on DC have been reported that include homogenization of source tissue or cells followed by DC steps. Briefly, the cell pellet was resuspended in cold isotonic buffer containing sucrose 0.25 M (in alternative mannitol 0.22 M and sucrose 0.07 M), Tris 2 mM, EDTA 1 mM, Hepes 20 mM pH 7.2 and undertaken to mechanical homogenization by 10 gently passes of pestle. The homogenate is centrifuged (750 × g for 10 min) to remove nucleus and unbroken cells. Successively, the supernatant is centrifugated (10,000 × g for 15 min) and then the mitochondrial pellet is resuspended in few microliters of extraction buffer (Liao et al., 2020).

The DC technique allows to obtain high yield of mitochondria with a lower purity compared with density gradient centrifugation techniques. The most recent separation technique named "fractionated mitochondrial magnetic separation" leads to high yield and purity mitochondrial preparations. The technique described by Hubbard et al. is based on the labeling of mitochondria with anti-TOM 22 magnetically tagged antibody, and then the lysate of the treated cell is trapped in a chromatography column and eluted (Hubbard et al., 2019).

The main problem with these methods is the isolation time: the whole procedure takes between 90 to 120 min, which is a too long for the use of mitochondria to rescue cardiomyocytes following ischemia and reperfusion injury. Commercial isolation kits can shorten the isolation time.

To solve this problem, McCully et al. described a rapid method for isolating mitochondria from tissue biopsies by replacing the centrifugation steps with filtrations through nylon mesh filters (Preble et al., 2014). This method shortens the time of mitochondria isolation to less than 30 min.

In the method of Mc Cully et al., 0.1 g of the autologous tissue sample was homogenated with an automated homogenizer which ensures the achievement of a homogeneous and reproducible homogenate solution. The homogenate was then subjected to a short digestion with Subtilisin A and the free organelles were separated by filtration; finally, the mitochondria were concentrated by centrifugation. By this method, the authors obtained about 1×10^9 mitochondria from 0.1 g of initial tissue mass (McCully et al., 2017).

Table 7.1 summarizes some of the most common isolation methods used with respect the mitochondrial source and the animal model used for mitochondrial transplantation.

Transplantation methods

Viable mitochondria can be delivered from one cell to another by different ways: microvesicle transport or by intercellular structures such as TNTs (Hayakawa et al., 2016; Islam et al., 2012; Spees et al., 2006). The mitochondrial transfer along TNT is mediated by some protein like MIRO 1 (Ahmad et al., 2014) which facilitates the mitochondrial transfer. The role of exchanging material between cells or the directionality of the exchange is still under investigations. Koyanagi et al. showed evidence of unidirectionally exchange of mitochondria from human endothelial progenitor cells to neonatal undifferentiated cardiomyocytes to sustain their maturation (Koyanagi et al., 2005). In 2012, Islam et al. used bone marrow–derived stem cells to supply healthy mitochondria to alveolar epithelial cells in a mouse model of *E. coli* LPS-induced acute lung injury (Islam et al., 2012).

Due to the central role of mitochondria in the energy balance of the cells, several processes such as proliferation, migration, differentiation and stress resistance are dependent on their correct function and integrity. Damaged mitochondria are degradated by a specific process called mitophagy; an erroneous discharge of nonfunctional mitochondria is deleterious for the cells' survival. Artificial mitochondria transfer led to improved bioenergetics with increased levels of ATP, reduction of ROS production, and stimulated oxygen consumption rate. Initially, mitochondrial transfer was achieved in vivo by single or repeated injections into the damaged area, and then mitochondria were delivered by intravenous injection while

TABLE 7.1 Mitochondrial isolation methods related to pathological model and bibliography.

Isolation methods	Pathological model	Bibliography
Differential centrifugation	Ischemia/reperfusion in animal models	(Ko et al., 2020)
		(Sun et al., 2015)
		(McCully et al., 2009)
		(Huang et al., 2016)
		(Pourmohammadi-Bejarpasi et al., 2020)
		(Ko et al., 2020)
		(Lin et al., 2013)
Different filters	Animal ischemia	(Blitzer et al., 2020)
Centrifugation,		(Shin et al., 2019)
<30 min		(Cowan et al., 2016)
		(Moskowitzova et al., 2020)
		(Doulamis, 2020)
		(Doulamis et al., 2020)
Commercial	Animal ischemia	(Fang et al., 2021)
Isolation kit,		
<60 min		

From Kubat, G. B., Ulger, O. & Akin, S. (2021). Requirements for successful mitochondrial transplantation. Journal of Biochemical and Molecular Toxicology, *35(11), e22898. https://doi.org/10.1002/jbt.22898*

mitochondrial transfer into in vitro cell culture by coincubation of mitochondria resuspended in respiration buffer allows cells to acquire functional mitochondria.

Discussion

Mitochondria are considered the powerhouse of all the eukaryotic cells. They are the main site for ATP synthesis through oxidative phosphorylation and fulfill several important functions. Mitochondria are involved in calcium homeostasis, contribute to cell metabolism, and provide intermediates for several biosynthetic pathways. in addition to their bioenergetic and metabolic role, mitochondria are also the main source of ROS whose production is increased by respiratory chain dysfunction. The primary targets of increased ROS production are the mitochondrial structures including mtDNA. Cumulative mtDNA mutations trigger a vicious circle which leads to a severe ATP depletion, mitochondrial destruction and cell death. There are no curative treatments for mitochondrial dysfunction: the treatment with antioxidant can only reduce the damage. Recently, mitochondrial transplantation has attracted great interest in translational medicine. In particular, the administration of competent mitochondria can recover the lost mitochondrial function in the recipient cells. The first clinical application of mitochondrial transplantation was for cardioprotection in patients with ischemia and reperfusion injury. The mitochondria used for transplantation were taken from the patient's tissue (autologous); but further studies in animal and in vitro models have shown that mitochondria can be taken from the tissues of different individuals of the same species (allogenic) or different species (heterologous). Regardless of their origin, the mitochondrial preparation used for cell transplantation must consist of an intact mitochondrial population with a high degree of purity highlighting the crucial role of the isolation procedure. In addition , the success of the mitochondrial isolation process depends on various factors, including the type of tissue used, the quality of the starting material, and the expertise of the operator. Mitochondria are also sensitive to changes in temperature, pH, and oxygen levels, so careful attention must be paid to these factors throughout the isolation process to ensure the viability and function of the organelles. Particular attention should be paid to the presence of calcium ions in the isolation buffer. The addition of ions chelators such as EGTA or EDTA is essential to avoid calcium uptake into the mitochondrial matrix that will trigger depolarization of the inner mitochondrial membrane followed by swelling of the mitochondria due to the opening of mPTP. Since one of the pathways used for the transfer of mitochondria to damaged organs involves the release of these organelles into the recipient's bloodstream, the problem related to the presence of calcium ions is the main challenge for their survival. The ability of isolated mitochondria to survive in the high calcium ions concentration in the blood is still an open question. Bertero et al. showed that mitochondria were severely damaged at a calcium concentration of 1mM (Bertero et al., 2020), while Mc Cully et al. showed that mitochondria could survive at a calcium concentration of 1.8 mM (McCully et al., 2020).

Another open question concerns the mechanism by which mitochondria are taken up into the receiving cells and how the few mitochondria taken up can supply the cells with sufficient ATP.

The therapeutic use of mitochondrial transplantation is still far from being a common clinical practice. Experiments in vitro and in animal models confirm the benefits of mitochondrial transplantation for the treatment of pathologies due to mitochondrial dysfunction. however, further studies needs to be conducted to assess the safety of this procedure before clinical application.

Isolation of intact mitochondria from cultured cells

Before you begin

1. Cell culture medium recommended for your cell line. The widest used medium is Dulbecco's modified Eagle's medium supplemented with 10% (v/v) fetal bovine serum, 0.1 mM minimal essential medium nonessential amino acids, and 2 mM L-glutamine, penicillin–streptomycin 50 U/mL and 50 mg/mL, respectively.
2. Seed 10^6–10^7 cells in a 500 cm^2 dish 2 or 3 days before performing the experiment. It is important to reach almost 100% of confluence before to start the isolation procedure.

3. Prepare the extraction buffer (0.25-M sucrose, 20-mM HEPES-KOH, pH 7.5, 10-mM KCl, 1.5-mM $MgCl_2$, 1-mM EDTA, 1-mM EGTA, 1-mM dithiothreitol, 0.1-mM PMSF) the same day of the isolation procedure to avoid contamination and check the pH of the buffer at 25°C before the use.

Materials and equipment

Cell pellet derived from culture cells (approx. 5×10^7–1×10^8 cells).

Extraction buffer (0.25-M sucrose, 20-mM HEPES-KOH, pH 7.5, 10-mM KCl, 1.5-mM MgCl2, 1-mM EDTA, 1-mM EGTA, 1-mM dithiothreitol, 0.1-mM PMSF).

Phase contrast microscope.

Refrigerated bench centrifuge.

Teflon-glass or glass-glass Potter-Elvehjem homogenizer.

Glassware and buffer must be precooled in an ice-bath. The extraction procedure must be performed at 0–4°C to minimize the activation of damaging phospholipases and proteases.

Step-by-step method details

Timing: 90–120 min

Remove the medium from the cells and wash with PBS. Remove the PBS and detach the cells with trypsin (Trypsin-EDTA 0.25% solution).

Centrifuge the cells (300 × g for 3 min) and resuspend the pellet in PBS. Centrifuge again as above.

Discard the supernatant and resuspend the pellet in five volumes of extraction buffer. Transfer the cell suspension to a glass homogenizer and homogenize with 20 strokes of Teflon pestle. Pay attention to keep the tube containing the cell suspension in an ice bath.

Transfer the homogenate in a Falcon tube and centrifuge at 750 × g for 10 min at 4°C. Collect the supernatant. Resuspend the pellet in five volumes of the extraction buffer and centrifuge at 750 × g for 10 min.

Pool the supernatants obtained from the two centrifugation steps and transfer it in a glass centrifuge tube. Centrifuge the pooled supernatants at 10.000 × g for 15 min at 4°C.

Resuspend the pellet, which represents a crude mitochondrial fraction, in 200/300 μL of cold extraction buffer (4°C) and centrifuge at 10.000 × g for 15 min. Discard the supernatant and resuspend the pelleted mitochondria in a small volume of extraction buffer.

Note

Concentrated resuspensions of mitochondria can be stored longer without loss of activity than dilute solutions.

The isolation procedure reported is derived from the method described by Yang et al. (1997) with minor modifications. The mitochondrial yield depends on the cell type used. Mitochondrial quantification can be achieved by protein determination by biuret method (Gornall et al., 1949).

Prepare the extraction buffer (0.25-M sucrose, 20-mM HEPES-KOH, pH 7.5, 10-mM KCl, 1.5-mM $MgCl_2$, 1-mM EDTA, 1-mM EGTA, 1-mM dithiothreitol, 0.1-mM PMSF) the same day of the isolation procedure to avoid contamination and check the pH of the buffer at 25°C before the use.

Expected outcomes

The mitochondrial isolation procedure described above allows intact and functional mitochondria to be obtained. Structural integrity can be verified by polarographic determination of oxygen consumption rate by supplying the respiratory chain with NADH-dependent substrates (such as the malate/glutamate mixture) in the presence (respiration state III) and absence (respiration state IV) of ADP.

The respiratory control ratio (RCR = ratio of state III/state IV respiration rate) is indicative of the quality of the mitochondrial preparation. An RCR value of 5 or more indicates good mitochondrial preparation.

Quantification and statistical analysis

This method allows a crude fraction of mitochondria that can be used for polarographic determinations, spectrophotometric determinations, and protein analysis, such as Western blotting, ELISA, and proteomic research. The yield depends on the type of cells used (e.g., HeLa cells are rich in mitochondria) and can range from 7 to 50 μg of mitochondrial protein per 10^6 cells. For experiments requiring a high degree of purity (e.g., in vitro protein synthesis and proteomic studies), separation methods such as gradient centrifugation should be used (see Related Techniques section).

The purity and the enrichment of mitochondria from whole cells can be assessed by Western blot analyses, measuring cytosolic (e.g., actin), nuclear (e.g., lamin A/C), and mitochondrial (e.g., OXPHOS subunits and porin) marker proteins. The integrity of the mitochondrial fraction can be tested by quantitation of an intermembrane space protein (e.g., cytochrome c) or a matrix protein (e.g., cyclophilin D) by immunoblot or by measuring ADP stimulation of oxygen consumption by polarography. Alternatively, measuring mitochondrial Ca^{2+} buffering capacity is a rapid and efficient method to assess the integrity of isolated mitochondria.

Advantages

Time saving: the entire protocol takes about 90 min.

Cultured cells are a standardized model and are easily obtained.

Limitations

The integrity of mitochondria must be checked carefully after isolation.

All the experiments must be performed within 3 h from isolation to obtain intact mitochondria.

The mitochondria isolation protocol should be adapted according to the cell type.

Alternative methods/procedures

An alternative method that achieves a preparation of viable, respiratory-competent mitochondria in less than 30 min and with a yield of approximately 2×10^{10} mitochondria per mg of mammalian tissue was described by Preble et al.

In this protocol, sample homogenization is achieved by using a commercial tissue dissociator and treating the sample with subtilisin A. Then, the homogenate is filtered through nylon filters instead of time-consuming DC steps. Because, even when kept on ice, the isolated mitochondria show a decrease in functional activity over time; the main advantage of this mitochondrial isolation protocol is that it allows the isolation of a large amount of intact, contamination-free mitochondria in a short time.

References

Ahmad, T., Mukherjee, S., Pattnaik, B., et al. (2014). Miro1 regulates intercellular mitochondrial transport and enhances mesenchymal stem cell rescue efficacy. *The EMBO Journal, 33*(9), 994–1010.

Ali, P. P., Kenney, M. C., & Kheradvar, A. (2020). Bioenergetics consequences of mitochondrial transplantation in cardiomyocytes. *Journal of American Heart Association, 9*(7), e014501 [PubMed: 32200731].

Almannai, M., El-Hattab, A. W., Ali, M., Soler-Alfonso, C., & Scaglia, F. (2020). Clinical trials in mitochondrial disorders, an update. *Molecular Genetics and Metabolism, 131*(1–2), 1–13.

Apostolova, N., & Victor, V. M. (March 2015). Molecular strategies for targeting antioxidants to mitochondria: Therapeutic implications. *Antioxidants and Redox Signaling, 686*, 729. https://doi.org/10.1089/ars.2014.5952

Barritt, J. A., Brenner, C. A., Malter, H. E., & Cohen, J. (March 2001). Mitochondria in human offspring derived from ooplasmic transplantation. *Human Reproduction, 16*(3), 513–516. https://doi.org/10.1093/humrep/16.3.513

Beale, G. H., & Knowles, J. K. C. (1976). Interspecies transfer of mitochondria in Paramecium aurelia. *MGG Molecular & General GeneticsVolume, 143*(Issue 2), 197–201 January.

Bensley, R. R., & Hoerr, N. L. (1934). Studies on cell structure by the freezing-drying method. VI. The preparation and properties of mitochondria. *The Anatomical Record, 60*, 449–455. https://doi.org/10.1002/ar.1090600408

Bertero, E., O'Rourke, B., & Maack, C. (2020). Mitochondria do not survive calcium overload during transplantation. *Circulation Research, 126*(6), 784–786 [PubMed: 32078444].

Blitzer, D., Guariento, A., Doulamis, I. P., Shin, B., Moskowitzova, K., Barbieri, G. R., Orfany, A., Del Nido, P. J., & McCully, J. D. (2020). *The Annals of Thoracic Surgery, 109*(3), 711. https://doi.org/10.1016/j.athoracsur.2019.06.07

Brenner, C. A., Barritt, J. A., Willadsen, S., & Cohen, J. (2000). Mitochondrial DNA heteroplasmy after human ooplasmic transplantation. *Fertility and Sterility, 74*(3), 573–578.

Clark, M. A., & Shay, J. W. (1982). Mitochondrial transformation of mammalian cells. *Nature, 295*(5850), 605–607.

Cohen, J., Scott, R., Alikani, M., Schimmel, T., Munné, S., Levron, J., et al. (1998). Ooplasmic transfer in mature human oocytes. *Molecular Human Reproduction, 4*(3), 269–280.

Cowan, D. B., Yao, R., Akurathi, V., Snay, E. R., Thedsanamoorthy, J. K., Zurakowski, D., Ericsson, M., Friehs, I., Wu, Y., Levitsky, S., Del Nido, P. J., Packard, A. B., & McCully, J. D. (2016). Intracoronary delivery of mitochondria to the ischemic heart for cardioprotection. *PLoS One, 11*(8), e0160889. https://doi.org/10.1371/journal.-pone.0160889. PMID: 27500955, PMCID: PMC4976938.

Darbandi, S., Darbandi, M., Khorram Khorshid, H. R., Sadeghi, M. R., Agarwal, A., Sengupta, P., Al-Hasani, S., & Akhondi, M. M. (2017). *Reproductive Biology and Endocrinology, 15*. Article number: 77.

Di, M. S., Hirano, M., & Schon, E. A. (Eds.). (2006). *Mitochondrial medicine*. London: Informa Healthcare.

Doulamis, P., Guariento, A., Duignan, T., Orfany, A., Kido, T., Zurakowski, D., Del Nido, P. J., McCully, J. D., & Cardiothorac, E. J. (2020). *Surgery, 57*(5), 836. https://doi.org/10.1093/ejcts/ezz326

Emani, S. M., Piekarski, B. L., Harrild, D., Del Nido, P. J., & McCully, J. D. (2017). Autologous mitochondrial transplantation for dysfunction after ischemia-reperfusion injury. *The Journal of Thoracic and Cardiovascular Surgery, 154*(1), 286–289. https://doi.org/10.1016/j.jtcvs.2017.02.018. Epub 2017 Feb 15, PMID: 28283239.

Enns, G. M. (2014). Treatment of mitochondrial disorders: Antioxidants and beyond. *Journal of Child Neurology, 29*(9), 1235–1240. https://doi.org/10.1177/0883073814538509

Fang, S. Y., Roan, J. N., Lee, J. S., Chiu, M. H., Lin, M. W., Liu, C. C., & Lam, C. F. (2021). Transplantation of viable mitochondria attenuates neurologic injury after spinal cord ischemia. *Journal of Thoracic and Cardiovascular Surgery, 161*(5), e337–e347. https://doi.org/10.1016/j.jtcvs.2019.10.151. Epub 2019 Nov 14.

Gornall, A. G., & Bardawill, C. J. (1949). Determination of serum proteins by means of the biuret reaction. *Journal of Biological Chemistry, 177*(2), 751–766. https://doi.org/10.1016/S0021-9258(18)57021-6. ISSN 0021-9258.

Gousset, K., Marzo, L., Commere, P.-H., & Zurzolo, C. (2013). Myo10 is a key regulator of TNT formation in neuronal cells. *Journal of Cell Science, 126*(Part 19), 4424–4435.

Guariento, A., Doulamis, I. P., Duignan, T., Kido, T., Regan, W. L., Saeed, M. Y., Hoganson, D. M., Emani, S. M., Fynn-Thompson, F., Matte, G. S., Del Nido, P. J., & McCully, J. D. (2020). *The Journal of Heart and Lung Transplantation, 39*, 1279. https://doi.org/10.1016/j.healun.2020.06.023

Hayakawa, K., Esposito, E., Wang, X., et al. (2016). Transfer of mito chondria from astrocytes to neurons after stroke. *Nature, 535*(7613), 551–555.

Hu, S. S., Li, R. Y., Cao, X. H., Liu, J. J., Wang, Z. H., Li, Z., Yang, M. L., Liu, J. W., Hu, L. M., Lin, C. J., Liu, J., & Wang, C. M. (September 2022). *Journal of Photochemistry and Photobiology B, 234*, 112534. https://doi.org/10.1016/j.jphotobiol.2022.112534. Epub 2022 Jul 23. PMID: 35905626.

Huang, P. J., Kuo, C. C., Lee, H. C., Shen, C. I., Cheng, F. C., Wu, S. F., Chang, J. C., Pan, H. C., Lin, S. Z., Liu, C. S., & Su, H. L. (2016). Transferring xenogenic mitochondria provides neural protection against ischemic stress in ischemic rat brains. *Cell Transplantation, 25*(5), 913–927. https://doi.org/10.3727/096368915X689785. Epub 2015 Nov 6. PMID: 26555763.

Hubbard, W. B., Harwood, C. L., Prajapati, P., Springer, J. E., Saatman, K. E., & Sullivan, P. G. (2019). Fractionated mitochondrial magnetic separation for isolation of synaptic mitochondria from brain tissue. *Scientific Reports, 9*, 9656. https://doi.org/10.1038/s41598-019-45568-3

Islam, M. N., Das, S. R., Emin, M. T., et al. (2012). Mitochondrial transfer from bone-marrow-derived stromal cells to pulmonary alveoli protects against acute lung injury. *Nature Medicine, 18*(5), 759–765.

Kanungo, S., Morton, J., Neelakantan, M., Ching, K., Saeedian, J., & Goldstein, A. (2018). Mitochondrial disorders. *Annals of Translational Medicine, 6*(24), 475. https://doi.org/10.21037/atm.2018.12.13

Katrangi, E., D'Souza, G., Boddapati, S. V., Kulawiec, M., Singh, K. K., Bigger, B., & Weissig, V. (2007). *Rejuvenation Research, 10*(4). https://doi.org/10.1089/rej.2007.0575

Kesner, E. E., Saada-Reich, A., & Lorberboum-Galski, H. (2016). *Scientific Reports, 6*, 26057. https://doi.org/10.1038/srep26057

King, M., & Attardi, G. (1989). Human cells lacking mtDNA: Repopulation with exogenous mitochondria by complementation. *Science, 246*(4929), 500–503. https://doi.org/10.1126/science.2814477

Ko, S. F., Chen, Y. L., Sung, P. H., Chiang, J. Y., Chu, Y. C., Huang, C. C., Huang, C. R., & Yip, H. K. (2020). *Journal of Cellular and Molecular Medicine, 24*(17), 10088. https://doi.org/10.1111/jcmm.15617

Koyanagi, M., Brandes, R. P., Haendeler, J., Zeiher, A. M., & Dimmeler, S. (2005). Cell-to-cell connection of endothelial progenitor cells with cardiac myocytes by nanotubes: A novel mechanism for cell fate changes? *Circulation Research, 96*(10), 1039–1041.

Kubat, G. B., Ozler, M., Ulger, O., Ekinci, O., Atalay, O., Celik, E., Safali, M., & Budak, M. T. (2021). *Journal of Biochemical and Molecular Toxicology, 35*(1), e22612. https://doi.org/10.1002/jbt.22612

Lenaz, G., Baracca, A., Barbero, G., Bergamini, C., Dalmonte, M. E., Del, S. M., Faccioli, M., Falasca, A., Fato, R., Genova, M. L., Sgarbi, G., & Solaini, G. (2010). Mitochondrial respiratory chain super-complex I–III in physiology and pathology. *BBA – Bioenergetics, 1797*(6–7), 633–640.

Lenaz, G., Bovina, C., Formiggini, G., & Parenti Castelli, G. (1999). Mitochondria, oxidative stress, and antioxidant defences. *Acta Biochimica Polonica, 46*, 1–21.

Liao, P.-C., Bergamini, C., Fato, R., & Pon, L. A. (2020). *Francesco Pallotti Methods Cell Biology, 155*, 3–31. https://doi.org/10.1016/bs.mcb.2019.10.002

Lin, H.-C., Liu, S.-Y., Lai, H.-S., & Lai, I.-R. (2013). Isolated mitochondria infusion mitigates ischemia-reperfusion injury of the liver in rats. *Shock, 39*(3), 304–310. https://doi.org/10.1097/SHK.0b013e318283035f

McCully, J. D., Cowan, D. B., Emani, S. M., & Del Nido, P. J. (2017). Mitochondrial transplantation: From animal models to clinical use in humans. *Mitochondrion, 34*, 127–134. https://doi.org/10.1016/j.mito.2017.03.004

McCully, J. D., Cowan, D. B., Pacak, C. A., Toumpoulis, I. K., Dayalan, H., & Levitsky, S. (2009). Injection of isolated mitochondria during early reperfusion for cardioprotection. *American Journal of Physiology-Heart and Circulatory Physiology, 296*(1), H94–H105. https://doi.org/10.1152/ajpheart.00567.2008. Epub 2008 Oct 31. PMID: 18978192, PMCID: PMC2637784.

McCully, J. D., Emani, S. M., & Del Nido, P. J. (2020). Letter by McCully et al regarding article, "mitochondria do not survive calcium overload". *Circulation Research, 126*(8), e56–e57 [PubMed: 32271682].

Moskowitzova, K., Orfany, A., Liu, K., Ramirez-Barbieri, G., Thedsanamoorthy, J. K., Yao, R., Guariento, A., Doulamis, I. P., Blitzer, D., Shin, B., Snay, E. R., Inkster, J. A. H., Iken, K., Packard, A. B., Cowan, D. B., Visner, G. A., Del Nido, P. J., & McCully, J. D. (2020). Mitochondrial transplantation enhances murine lung viability and recovery after ischemia-reperfusion injury. *The American Journal of Physiology-Lung Cellular and*

Molecular Physiology, 318(1), L78–L88. https://doi.org/10.1152/ajplung.00221.2019. Epub 2019 Nov 6. PMID: 31693391, PMCID: PMC6985877.

Moskowitzova, K., Shin, B., Liu, K., Ramirez-Barbieri, G., Guariento, A., Blitzer, D., Thedsanamoorthy Bs, J. K., Rouan Yao, B. A., Snay Cnmt, E. R., Inkster, J. A. H., Orfany, A., Zurakowski, D., Cowan, D. B., Packard, A. B., Visner Do, G. A., del Nido, P. J., & McCully, J. D. (2018). Mitochondrial transplantation prolongs cold ischemia time in murine heart transplantation. *The Journal of Heart and Lung Transplantation*. https://doi.org/10.1016/j.healun.2018.09.025

Morrison, T. J., Jackson, M. V., Cunningham, E. K., Kissenpfennig, A., McAuley, D. F., O'Kane, C. M., & Krasnodembskaya, A. D. (2017). Mesenchymal stromal cells modulate macrophages in clinically relevant lung injury models by extracellular vesicle mitochondrial transfer. *American Journal of Respiratory and Critical Care Medicine, 196*(10), 1275–1286. https://doi.org/10.1164/rccm.201701-0170OC. PMID: 28598224, PMCID: PMC5694830.

Moskowitzova, K., Orfany, A., Liu, K., Ramirez-Barbieri, G., Thedsanamoorthy, J. K., Yao, R., Guariento, A., Doulamis, S. P., Blitzer, D., Shin, B., Snay, E. R., Inkster, J. A. H., Iken, K., Packard, A. B., Cowan, D. B., Visner, G. A., del, N. P. J., & McCull, J. D. (2020). Mitochondrial transplantation enhances murine lung viability and recovery after ischemia-reperfusion injury. *American Journal of Physiology - Lung Cellular and Molecular Physiology*, L78–L88.

Ozawa, T. (1997). Genetic and functional changes in mitochondria associated with aging. *Physiological Reviews, 77*, 425–464.

Perez, G. I., Trbovich, A. M., Gosden, R. G., & Tilly, J. L. (2000). *Mitochondria and the death of oocytes Nature* (Vol. 403, p. 500).

Picard, M., Taivassalo, T., & Ritchieetal, D. (2011). Mitochondrial structure and function are disrupted by standard isolation methods. *PLoS One, 6*(3), e18317.

Pourmohammadi-Bejarpasi, Z., Roushandeh, A. M., Saberi, A., Rostami, M. K., Toosi, S., Jahanian-Najafabadi, A., Tomita, K., Kuwahara, Y., Sato, T., & Roudkenar, M. H. (2020). *Brain Research Bulletin, 165*, 70. https://doi.org/10.1016/j.brainresbull.2020.09.018

Preble, J. M., Pacak, C. A., Kondo, H., MacKay, A. A., Cowan, D. B., & McCully, J. D. (September 2014). *Journal of Visualized Experiments, 91*, e51682. Page 1 of 6.

Ramirez-Barbieri, G., Moskowitzova, K., Shin, B., Blitzer, D., Orfany, A., Guariento, A., Iken, K., Friehs, I., Zurakowski, D., del Nido, P. J., & McCully, J. D. (May 2019). *Alloreactivity and allorecognition of syngeneic and allogeneic mitochondria Mitochondrion* (Vol. 46, pp. 103–115).

Smith, R. A. J., Hartley, R. C., Cochemé, H. M., & Murphy, M. P. (June 2012). *Trends in Pharmacological Sciences, 33*(6). https://doi.org/10.1016/j.tips.2012.03.010

Sanchez, V., & Brusco, A. (2016). Mitochondrial transfer by intercellular nanotubes. In R. J. Gelpi, A. Boveris, & J. J. Poderoso (Eds.), *Biochemistry of oxidative stress* (pp. 95–108). Springer International Publishing. http://link.springer.com/chapter/10.1007/978-3-319-45865-6_7.

Shin, B., Saeed, M. Y., Esch, J. J., Guariento, A., Blitzer, D., Moskowitzova, K., Ramirez-Barbieri, G., Orfany, A., Thedsanamoorthy, J. K., Cowan, D. B., Inkster, J. A., Snay, E. R., Staffa, S. J., Packard, A. B., Zurakowski, D., Del Nido, P. J., & McCully, J. D. (2019). *JACC: Basic to Translational Science, 4*(8), 871. https://doi.org/10.1016/j.jacbts.2019.08.007

Sims, N. R. (1990). Rapid isolation of metabolically active mitochondria from rat brain and subregions using Percoll density gradient centrifugation. *Journal of Neurochemistry, 55*, 698–707. https://doi.org/10.1111/j.1471-4159.1990.tb04189.x

Spees, J. L., Olson, S. D., Whitney, M. J., & Prockop, D. J. (2006). Mitochondrial transfer between cells can rescue aerobic respiration. *Proceedings of the National Academy of Sci ences of the United States of America, 103*(5), 1283–1288.

Sun, C. K., Lee, F. Y., Kao, Y. H., Chiang, H. J., Sung, P. H., Tsai, T. H., Lin, Y. C., Leu, S., Wu, Y. C., Lu, H. I., Chen, Y. L., Chung, S. Y., Su, H. L., & Yip, H. K. (2015). *Journal of Pineal Research, 58*(2), 137. https://doi.org/10.1111/jpi.12199

Taylor, R., & Turnbull, D. (2005). Mitochondrial DNA mutations in human disease. *Nature Reviews Genetics, 6*, 389–402. https://doi.org/10.1038/nrg1606

Torralba, D., Baixauli, F., & Sánchez-Madrid, F. (2016). Mitochondria know no boundaries: Mechanisms and functions of intercellular mitochondrial transfer. *Frontiers in Cell and Developmental Biology, 4*, 107. https://doi.org/10.3389/fcell.2016.00107. PMID: 27734015, PMCID: PMC5039171.

Tsukube, T., McCully, J. D., Metz, R. M., Cook, C. U., & Levitsky, S. (1997). Ameliorationofischemiccalciumoverloadcorrelateswithhighenergyphosphatesinthesenescentmyocardium. *American Journal of Physiology - Heart and Circulatory Physiology, 273*, H418–H427.

Ulger, O., Kubat, G. B., Cicek, Z., Celik, E., Atalay, O., Suvay, S., & Ozler, M. (2021). *Life Sciences, 279*, 119669. https://doi.org/10.1016/j.lfs.2021.119669

Ulger, O., & Kubat, G. B. (2022). *Biochimie, 195*, 1–15. https://doi.org/10.1016/j.biochi.2022.01.002

Yang, J., Liu, X., Bhalla, K., Kim, C. N., Ibrado, A. M., Cai, J., Peng, T. I., Jones, D. P., & Wang, X. (1997). Prevention of apoptosis by Bcl-2: Release of cytochrome c from mitochondria blocked. *Science, 275*(5303), 1129–1132. https://doi.org/10.1126/science.275.5303.1129. PMID: 9027314.

Further reading

O'Kane, C. M., & Krasnodembskaya, Anna D. (November 15, 2017). *American Journal of Respiratory and Critical Care Medicine, 196*(10), 1275–1286. https://doi.org/10.1164/rccm.201701-0170OC. (Accessed 9 June 2017)

Shi, X. M. Z., Fu, C., & Fu, A. (2017). *Mitochondrion, 34*, 91. https://doi.org/10.1016/j.mito.2017.02.005

Morrison, T. J., Jackson, M. V., Cunningham, E. K., Kissenpfennig, A., McAuley, D. F., Torralba, D., Baixauli, F., & Sánchez-Madrid, F. (2016). Mitochondria know no boundaries: Mechanisms and functions of intercellular mitochondrial transfer. *Frontiers in Cell and Developmental Biology, 4*, 107.

CHAPTER

8

Determination of mitochondrial number and function

Christian Bergamini[1], *Romana Fato*[1] *and Francesco Pallotti*[2,3]

[1]Dipartimento di Farmacia e Biotecnologie, FABIT, Università Degli Studi di Bologna, Bologna, Italy; [2]Dipartimento di Medicina e Chirurgia, Università Degli Studi dell'Insubria, Varese, Italy; [3]SSD Laboratorio Analisi-SMEL Specializzato in Citogenetica e Genetica Medica, ASST Sette Laghi-Ospedale di Circolo-Fondazione Macchi, Varese, Italy

Introduction

Mitochondria for transplantation are isolated from cells or tissues by various techniques that yield mitochondria preparation with varying degrees of purity and quality. The purity of isolated mitochondria must be carefully evaluated, as contaminants such as microsomal, nuclear, or cytosolic components are usually present and may cause adverse effects. For example, nuclear DNA contamination can induce immune or autoimmune responses following mitochondrial transplantation (McCully et al., 2017; D'Amato et al., 2023; Zhang & Miao, 2022). In addition to purity, isolated mitochondria must be evaluated from a quantitative point of view by determining their number and integrity. To this end, various techniques are employed, ranging from colorimetric methods for protein determination, spectrophotometric methods, haemocytometers, particle counters, specific fluorescent probes, and transmission electron microscopy (TEM). Functionally, mitochondrial preparations can be assessed for oxygen consumption by polarographic techniques and for their ability to produce energy in the form of ATP by luminometric assays . In addition, mitochondrial functionality can be assessed by measuring the transmembrane potential with fluorescent probes and by evaluating the specific activities of the respiratory complexes.

Mitochondrial number and integrity

The number of mitochondria can be quickly determined by correlating the number with protein content using several common colorimetric methods: Bicinchoninic acid assay,

Mitochondrial Transplantation and Transfer
https://doi.org/10.1016/B978-0-443-18858-9.00008-1

Bradford, Biuret, and micro-Lowry methods (Azimzadeh et al., 2016; Becker et al., 1996; Chang et al., 2013; Huang et al., 2016; Jabbari et al., 2020; Jiang et al., 2020; Ko et al., 2020; Lee et al., 2021; McCully et al., 2014).

These colorimetric techniques have the advantage of being fast and inexpensive, but they do not provide information about the functionality of isolated mitochondria or the possible presence of contaminants.

Alternatively, the number of isolated mitochondria can be determined by using haemocytometry (Louwagie et al., 2021; McCully et al., 2009). This technique is fast and inexpensive; however, it is highly observer dependent, and the estimated number of mitochondria can be inconsistent. Preble JM et al. (McCully et al., 2014; Preble et al., 2014) report that the use of the haemocytometer underestimates the number of mitochondria by one to three times and recommend that all estimattes be made by a single operator. The same authors recommend the use of particle counters, such as the Beckman Multisizer 4e Coulter Counter, for optimal estimation of mitochondrial number. The use of particle counters offers multiple advantages over haemocytometers, as it also allows the purity of the preparation to be assessed at the same time, and it is an automated system that does not depend on the operator (Bahr et al., 1966; Preble et al., 2014; Rhyner, 2010).

Staining with fluorescent probes allows the determination of mitochondrial number and function by exploiting the ability of specific cationic fluorescent probes to accumulate in the mitochondrial matrix in a transmembrane potential-dependent manner. The mitochondrial transmembrane potential is a key indicator of mitochondrial activity and the driving force behind ATP production, because it reflects the process of electron transport and oxidative phosphorylation.

Several probes are commercially available for this purpose: MitoSpy Orange CMTMRos (BioLegends, USA) (Ulger et al., 2021), MitoTracker Red CMXRos (Chang et al., 2013; Masuzawa et al., 2013; Norat et al., 2020), and MitoTracker Orange CMTMRos (Thermo Fisher Scientific, USA) (Ene et al., 2022; McCully, Cowan, et al., 2007). Fluorescent probes that specifically stain mitochondria independent of mitochondrial potential can be used as counterstaining, including MitoTracker Green FM (MTG), MitoFluor Green (MFG), and nonyl acridine orange.

In a recent paper by McCully JD et al. (McCully et al., 2009), viable isolated mitochondrial number was determined by labeling an aliquot (10 μL) of isolated mitochondria with MitoTracker Orange CMTMRos (5 μmol/L). Aliquots of labeled mitochondria were spotted onto slides and counted using a spinning disk confocal microscope with a 63× C-apochromat objective (1.2 W Korr/0.17 NA, Zeiss). Mitochondria were counterstained with the mitochondria-specific dye MitoFluor Green (MFG, Invitrogen). Appropriate wavelengths were chosen to measure autofluorescence and background fluorescence using unstained cells and tissue. Briefly, 1 μL of labeled mitochondria was placed on a microscope slide and covered. Mitochondrial number was determined at low (×10) magnification covering the full specimen area, using MetaMorph Imaging Analysis software.Fluorescence techniques have the advantage of simultaneously correlating the number and functional status of isolated mitochondria; however, aggregation and overlapping mitochondria negatively affect the accuracy of the determination.

The use of fluorescent mitochondrial labels is not without its caveats (Gollihue et al., 2017). 10-nonyl acridine orange, MFG, and MTG are dyes whose fluorescence intensity changes with membrane potential (Keij et al., 2000) and it has been reported that the MitoTracker dyes can inhibit mitochondrial respiration; in particular, Buckman JF et al. (Buckman et al., 2001) reported that upon mitochondrial damage, such as uncoupling using FCCP, MitoTracker dyes were released into the cell cytoplasm, indicating that these dyes are not irreversibly bound to the mitochondria.

Moreover, MTG FM is reported to be cytotoxic in Hela cells even at low concentrations of 250 nM (Han et al., 2013), and MitoTracker Red CMXRos is toxic to human 143B osteosarcoma cells (Minamikawa et al., 1999). CMXRos is a photosensitizer that causes chemical damage when subjected to laser scanning, such as used in confocal imaging. When possible, transgenic labeling of mitochondria in vitro provides a stable alternative to labeling with more photosensitive MitoTracker dyes (Rizzuto et al., 1996; Shitara et al., 2001).

During the isolation and transplantation process, mitochondria are exposed to the extracellular environment, with changes in temperature and ionic composition. These parameters can alter mitochondrial structural integrity, adversely affecting the ability to produce energy.

TEM is an essential technique for visualizing mitochondrial ultrastructure. Using this technique, it is possible to determine that isolated mitochondria exhibit preserved cristae and intact outer and inner mitochondrial membranes (Collins et al., 2021; Cowan et al., 2017; Patel et al., 2009). It is also a valuable method for identifying contaminants in the preparation and for assessing mitochondrial purity (Collins et al., 2021; Farah et al., 2019; Kaza et al., 2017; Zhang & Miao, 2022). However, the use of TEM is not suitable for rapid assessment of the preparation.

Gollihue JL et al. (Gollihue et al., 2017) report the use of TEM to verify the structural integrity of tGFP-labeled mitochondria for intraspinal transplantation. After 10,000 rcf centrifugation, the mitochondrial pellet was fixed in 3% glutaraldehyde in sodium cacodylate buffer overnight at 4°C. The pellet was then rinsed in 0.2M sodium cacodylate buffer 3 times for 5 min each at 4°C. The pellet was then placed in 2% osmium tetroxide for 1 h at 4°C, followed by washing in distilled water 5 times for 5 min each at 4°C. The pellet was then placed in 2% uranyl acetate overnight at 4°C. The sample was rinsed in distilled water 5 times for 5 min each at room temperature, followed by sequential ethyl alcohol dehydrations in 30%, 50%, 70%, and 90% EtOH for 10 min each at room temperature. This was followed by three washes in 100% EtOH for 15 min each wash at room temperature. The pellet was then washed in resin (23.1% ERL 4221 (3,4-Epoxycyclohexylmethyl-3′,4′-epoxycyclohexane carboxylate) Electron Microscopy Sciences, Hatfield, PA), 18.5% DER (diglycidyl ether of polypropylene glycol, Electron Microscopy Sciences, Hatfield, PA), 57.7% NSA (nonenyl succinic anhydride, Electron Microscopy Sciences, Hatfield, PA), and 0.69% DMAE (dimethylaminoethanol, Electron Microscopy Sciences, Hatfield, PA) 3 times, each for 1 h at room temperature with gentle spinning. The pellet was then cut into smaller sections, placed in fresh resin in molded blocks, and incubated at 16°C for 48 h. The blocks were then trimmed and cut at 90 nm thickness using Ultracut UCT (Leica Microsystems, Buffalo Grove, IL). The sections were placed onto copper grids and allowed to dry at room temp, covered to protect from dust accumulation, overnight. The mounted grids were then processed using lead citrate. Briefly, each grid was placed on top of a drop of 0.5% aqueous lead citrate at room temperature in a Petri dish containing sodium hydroxide pellets to absorb carbon dioxide. The grids were left on the droplet for 3 min, and then moved to a new droplet of distilled water, in turn "washing" the grid. The grid was then moved to a fresh droplet of water every 5 min, for a total of five washes. The grid was then dabbed with a Kimwipe to remove the excess water and allowed to dry. Grids were then imaged at 60 kV on a Philips TECNAI 12 BT transmission electron microscope. The images captured represent the inner portion of the mitochondria pellet.

Another method to determine the integrity of isolated mitochondria is to spectrophotometrically follow their swelling after exposure to Ca^{2+}. Ca^{2+}-induced swelling is a marker of mitochondrial intactness after isolation (Gunter & Pfeiffer, 1990) and was used by Shi et al. (2017) to determine the mitochondrial condition of isolated mitochondria from HepG2 cells, which were then transplanted into the human neuroblastoma cell line SH-SY5Y or injected intravenously into Parkinson's disease model mice. Briefly, fresh prepared mitochondria (0.5 mg/mL) were incubated in iso-osmotic buffer (140 mmol/L KCl, 10 mmol/L Tris-Cl, pH 7.4) for 60 min; then 10 mmol/L Pi(K), 10 mmol/L succinic acid and 6.5 mmol/L $CaCl_2$ were added at room temperature to initiate mitochondrial swelling. The absorbance (520 nm) was recorded for 10 min following $CaCl_2$ addition.

Similarly, Brenner-Lavie et al. (2008) measured spectrophotometrically the mitochondrial swelling as a marker of integrity in isolated mitochondria from SH-SY5Y cells. The decrease in absorbance at 540 nm was monitored in a standard swelling buffer (215 mM mannitol, 50 mM sucrose, 10 mM KCl, 3 mM KH_2PO_4, 0.5 mM $MgCl_2$, 10 mM HEPES pH 7.4, 5 mM succinate, 2 μM rotenone, and 1 μg/mL oligomycin). Ca^{2+} (100 μM) was added 2 min after the addition of intact mitochondria.

Mitochondrial function

The ability of isolated mitochondria to sustain the process of oxidative phosphorylation can be examined through polarographic techniques. In particular, oxygen consumption rate (OCR) can be assessed in the presence of substrates capable of donating electrons at the level of Complex I (glutamate-malate) or Complex II (succinate), in the absence (respiratory state 4) and in the presence (respiratory state 3) of adenosine diphosphate (ADP), which is converted by the ATPase to ATP exploiting the transmembrane potential. The use of specific inhibitors such as oligomycin, which inhibits the ATP synthase activity and allows the proton leak determination, or uncouplers such as FCCP, which is capable of dissipating the proton gradient and thus maximizing the mitochondrial oxygen consumption, allows the functional characterization of isolated mitochondria. Respiration parameters such as the respiratory control index (given by the ratio of state 3 to state 4 respiration) or proton leak assessment provide a functional assessment of the isolated mitochondria to be transplanted.

OCR measurements in isolated mitochondria can be examined by polarographic techniques; Clark electrode measurements are the best and most reliable means of assessing mitochondrial function. Clark type electrodes include YSI oxygraph (McCully, Rousou, et al., 2007; Rousou et al., 2004) (Yellow Springs Instruments Co, Yellow Springs, OH), Oxytherm (Hansatech Instruments Ltd., Norfolk, UK), Strathkelvin oxygraph (Robicsek et al., 2018) (Strathkelvin 782 Oxygen System, Scotland), and Oxygraph-2K (Wu et al., 2018) (OROBOROS instruments Corp, Innsbruck, AUT). Alternatively, XF Analyzer can be used to determine the oxygen consumption togheter with extracellular acidification (Gupta et al., 2016; Louwagie et al., 2021; Zhang et al., 2021).

Katrangi et al. (2007) reported that xenogenic transfer of murine mitochondria into human cells lacking functional mitochondria can functionally restore respiration. The OCR determination in isolated mitochondria was performed using the Oxytherm System from Hansatech Instruments (Norfolk, United Kingdom). Briefly, isolated mitochondria were suspended in respiration medium containing glutamate/malate (20 mM HEPES-Tris (pH 7.0), 125 mM KCl, 2 mM K_2HPO_4, 1 mM $MgCl_2$, 0.01 mM EGTA, 1% BSA, 5 mM glutamate, 5 mM malate). The Oxytherm chamber was filled with the respiration medium supplemented with 0.2 mM ADP. State 3 respiration was measured immediately after the addition of mitochondria into the chamber. After a few minutes, oligomycin (1.25 μg/mL, final) was added to allow measurement of state 4 respiration. The respiratory control ratio was calculated according to RCR (state 3 rate)/(state 4 rate).

Gollihue et al. (2017) optimized techniques in vitro to prepare suspensions of isolated mitochondria for transplantation in vivo, with the aim to test the potential therapeutic effects of mitochondrial transplantation following spinal cord injury and other animal models of neurotrauma. To assay mitochondrial functionality after isolation, mitochondria were assayed for OCRs using the Seahorse Bioscience XFe Flux Analyzer. Isolated mitochondria were applied into cell plate cartridges that were then centrifuged to concentrate mitochondria to the bottom of each well. Then, 5 mM pyruvate/2.5 mM malate/1 mM ADP, 1μg/mL oligomycin, 3 μM carbonyl cyanide p-trifluoromethoxyphenylhydrazone (FCCP), and 100 nM rotenone/10 mM succinate were sequentially added to each well (Patel et al., 2014).

The advantage of using microplates over traditional Clarke electrodes is that only small amounts of sample are needed for metabolic analysis. Traditionally, experiments utilizing a Clark-type oxygen electrode require between 50 and 100 μg of mitochondrial protein per assay (Pandya et al., 2009).

The ability of isolated mitochondria to phosphorylate ADP to ATP can be considered a parameter of functionality and can be determined through colorimetric (Doulamis et al., 2020) (ATP Assay Kit (Colorimetric/Fluorometric) (ab83355)) or luminometric (Cowan et al., 2016; Louwagie et al., 2021; Pacak et al., 2015; Preble et al., 2014) (ATPlite Luminescence ATP Detection Assay System, PerkinElmer) techniques using commercially available kits. Determination of ATP synthesis should be carried out in the presence of ADP (1 mM); in the case of mitochondria labeled with fluorescent probes, care should be taken since it has been reported that some dyes can interfere with mitochondrial function (Pacak et al., 2015; Poot et al., 1996).

Alternatively, JC-1 dye can be used as an indicator of mitochondrial membrane potential for mitochondrial viability. JC-1 is a fluorescent ratiometric dye which undergoes potential-dependent accumulation in mitochondria, indicated by a fluorescence emission shift from the green monomer at lower dye concentrations to red aggregates which are formed as the dye concentration increases. Jabbari H. et al. (Jabbari et al., 2020) assessed the transmembrane potential in mitochondria isolated from rat muscle for autotransplantation in injured kidney cells using the JC-1 mitochondrial membrane potential Assay Kit (Cayman, USA). According to the manufacture's instruction, the mitochondria were incubated with 100 μL of the JC-1 staining solution/mL of respiration buffer at 37 C for 30 min in dark. Then, the stained mitochondria were observed under a fluorescent microscope and digital photographs were captured for aggregate (red), and monomer (green) forms of JC-1.

Sun et al. (2019) assayed the intactness of the isolated mitochondria by measuring the mitochondrial membrane potential ($\Delta\Psi m$) with JC-1. Briefly, after incubation with 10 μM JC-1 (Sigma–Aldrich) staining solution at 37°C, mitochondria were washed with JC-1 staining buffer two times and then analyzed using a flow cytometry (FlowSight, Amnis).

In addition, to further confirm that mitochondrial function has not been compromised by the isolation or storage process, the single activities of individual respiratory chain complexes and ATPase can be assessed spectrophotometrically as extensively reported in Birch-Machin et al. (1994), Frazier et al. (2020), and Spinazzi et al. (2012).

Conclusions

Several studies show that there are no marked differences in either cellular uptake or protective effects of mitochondria isolated from different tissues (Masuzawa et al., 2013; McCully et al., 2009). On the other hand, the effectiveness of mitochondrial transplantation is highly dependent on the quantity and functionality of the isolated organelles, which enable their uptake, internalization, and efficacy. It has been reported that non-viable or structurally compromised mitochondria, mitochondrial fractions, or administration of exogenous ATP or ADP do not provide cytoprotection (Masuzawa et al., 2013; McCully et al., 2014; Roushandeh et al., 2019). Therefore, it is of paramount importance to use experimental protocols that can assess not only the number of isolated mitochondria, but also their structural integrity taking into account various bioenergetic parameters (oxygen consumption, transmembrane potential, ability to synthesize ATP, etc.). Since mitochondrial dysfunction plays a crucial role in the pathogenesis of many diseases, mitochondrial transplantation may not only serve as a new therapeutic strategy, but also be applicable to the treatment of many diseases where conventional therapies have so far proved unsuccessful.

Flow Cytometry analysis of transmembrane potential in isolated mitochondria stained with JC-1 dye

Before you begin

Timing: 1 h

- JC-1 stock solution: Prepare JC-1 (5,5′,6,6′-Tetrachloro-1,1′,3,3′-tetraethyl-imidacarbocyanine iodide, 5,5′,6,6′-Tetrachloro-1,1′,3,3′-tetraethylbenzimidazolocarbocyanine iodide) stock solution in DMSO at 1 mg/mL concentration vortexing vigorously. Store the JC-1 Stock Solution at −20°C in working aliquots.

Note: Stock solutions of JC-1 are stable at −20°C, but degrade after several thawing and refreezing cycles.

- JC-1 working solution: prepare the JC-1 working solution (0.2 mg/mL) by diluting the JC-1 Stock Solution fivefold with DMSO. Centrifuge the working solution at 13,000 × g for 3 min to sediment insoluble particles. Keep the JC-1 Working Solution at room temperature.

Note: JC-1 is extremely sensitive to light and must be shielded from direct light at all stages of its preparation.

- Incubation buffer: 100 mM NaCl; 10 mM sodium potassium phosphate buffer (pH 7.4); 10 mM Tris–HCl buffer (pH 7.4); 10 mM $MgCl_2$.

Materials and equipment

- Flow cytometer equipped with a 488-nm excitation light source and band-pass filters centered around 525 and 590 nm
- 0.25 mM carbonyl cyanide m-(trifluoromethoxy)phenylhydrazone (FCCP) in DMSO
- 1 mg/mL JC-1 stock solution in dimethyl sulfoxide (DMSO)
- Incubation buffer (IB) reagents.

Step-by-step method details

Timing: 1 h

- Incubate the mitochondrial suspensions in IM (incubation medium) with JC-1 at a concentration of 1.5 μg/mg protein in 500 μL final volume for 1 min at room temperature and in the dark.
- Perform cytofluorimetric analysis immediately after JC-1 staining with a flow cytometer equipped with a 488 argon laser. For fluorescence detection, use a band-pass filter centered around 525 nm (green fluorescence, FL1) and a band-pass filter centered around 590 nm (orange fluorescence, FL2). Set the PMT value of the detector in FL1 near 310 V and in FL2 near 259V. Set FL1–FL2 compensation near 2.0%–4%, and FL2–FL1 compensation near 8.0%–9.5%. Acquire a minimum of 20,000 events per sample in list mode. Analyze data using logarithmic signal amplification for the green and orange channels.

Optional steps

- To measure the physical parameters of isolated mitochondria, analyze the forward scatter, which is related to dimensions, and side angle scatter, which is related to density. Forward and side scatter should be set at 10 (log 1).
- To achieve depolarization, add to the mitochondrial suspension 0.5 μM FCCP (carbonyl cyanide 4- (trifluoromethoxy) phenylhydrazone) after JC-1 incubation.

Expected outcomes

The fluorescence emission of the JC-1 dye changes depending on mitochondrial $\Delta\Psi$, due to the reversible formation of JC-1 aggregates upon membrane polarization, which causes shifts in emitted light from ~530 nm (i.e., emission of JC-1 monomeric form) to ~590 nm (i.e., emission of JC-1 aggregates) when excited at 488 nm. The color of the dye changes reversibly from green to greenish-orange as the mitochondrial membrane becomes more polarized. FCCP addition completely dissipates mitochondrial potential, leaving only green fluorescence emission.

Quantification and statistical analysis

Acquire fluorescence data for experimental samples in list mode using a log scale for the fluorescence channels. Acquire a minimum of 20,000 events per sample and use a flow rate no higher than 250 to 300 cells/sec.

Advantages

Flow cytometry (FCM) evaluates only the fluorescence associated with single particles, utilizing small amounts of biological samples and probe, avoiding quenching and light-scattering variations as well as toxic effects.

FCM allows the quantification of morphological and functional characteristics of mitochondria.

Measurements made using JC-1 can be both qualitative, based on the shift in fluorescence emission, and quantitative, based on the intensity of fluorescence emission.

Limitations

The fluorescence emission of JC-1 covers the green, yellow, and part of the red wavelengths of the light spectrum, so attention should be paid when using JC-1 in combination with other fluorochromes.

Optimization and troubleshooting

The flow cytometer setting described in the protocol is the one generally used in the literature. Because each instrument has a different sensitivity, a different setting may be needed.

Since JC-1 dye is very sensitive to alterations of temperature and pH, all reagents must be kept at 25°C and carefully checked for pH (7.4) during the experiment.

Due to the light sensitivity of JC-1 Staining procedure must be carried under no direct intense light and incubation in the dark.

FCCP-treated mitochondria should be used to confirming the sensitivity of JC-1 dye to changes in membrane potential.

In the absence of fluorescence emission, the main causes could be structurally altered mitochondria during isolation; low uptake of the probe into the organelle due to degradation; or sub-optimal concentration.

If the fluorescence signal is excessive, adjust the PMT voltage or decrease the probe concentration. Check the pH of the solution containing the sample.

Safety considerations and standards

DMSO enhances the uptake of dye through the skin. Wearing nitrile gloves and eye protection during all stages of sample handling is strongly recommended.

References

Azimzadeh, P., et al. (2016). Comparison of three methods for mitochondria isolation from the human liver cell line (HepG2). *Gastroenterology and Hepatology from Bed to Bench, 9*, 105–113.

Bahr, G. F., Herbener, G. H., & Glas, U. (1966). A method for the counting of mitochondria. *Experimental Cell Research, 41*, 99–108.

Becker, J. M., Caldwell, G. A., & Zachgo, E. A. (1996). Exercise 13—Protein assays. In J. M. Becker, G. A. Caldwell, & E. A. Zachgo (Eds.), *Biotechnology* (2nd ed., pp. 119–124). Academic Press. https://doi.org/10.1016/B978-012084562-0/50069-2

Birch-Machin, M. A., Briggs, H. L., Saborido, A. A., Bindoff, L. A., & Turnbull, D. M. (1994). An evaluation of the measurement of the activities of complexes I-IV in the respiratory chain of human skeletal muscle mitochondria. *Biochemical Medicine and Metabolic Biology, 51*, 35–42.

Brenner-Lavie, H., et al. (2008). Dopamine modulates mitochondrial function in viable SH-SY5Y cells possibly via its interaction with complex I: Relevance to dopamine pathology in schizophrenia. *Biochimica et Biophysica Acta (BBA)—Bioenergetics, 1777*, 173–185.

Buckman, J. F., et al. (2001). MitoTracker labeling in primary neuronal and astrocytic cultures: Influence of mitochondrial membrane potential and oxidants. *Journal of Neuroscience Methods, 104*, 165–176.

Chang, J.-C., et al. (2013). Functional recovery of human cells harbouring the mitochondrial DNA mutation MERRF A8344G via peptide-mediated mitochondrial delivery. *Nippon Suisan Gakkaishi, 21*, 160–173.

Collins, H. E., et al. (2021). Mitochondrial morphology and Mitophagy in heart diseases: Qualitative and quantitative Analyses using transmission electron microscopy. *Frontiers in Aging, 2*.

Cowan, D. B., et al. (2016). Intracoronary delivery of mitochondria to the ischemic heart for cardioprotection. *PLoS One, 11*, e0160889.

Cowan, D. B., et al. (2017). Transit and integration of extracellular mitochondria in human heart cells. *Scientific Reports, 7*, 17450.

D'Amato, M., Morra, F., Di Meo, I., & Tiranti, V. (2023). Mitochondrial transplantation in mitochondrial medicine: Current Challenges and future perspectives. *International Journal of Molecular Sciences, 24*, 1969.

Doulamis, I. P., et al. (2020). Mitochondrial transplantation for myocardial protection in diabetic hearts. *European Journal of Cardio-Thoracic Surgery, 57*, 836–845.

Ene, H. M., Karry, R., Farfara, D., & Ben-Shachar, D. (2022). Mitochondria play an essential role in the trajectory of adolescent neurodevelopment and behavior in adulthood: Evidence from a schizophrenia rat model. *Molecular Psychiatry*, 1–12. https://doi.org/10.1038/s41380-022-01865-4

Farah, N. K., Liu, X., Wu, C. H., & Wu, G. Y. (2019). An improved method for preparation of uniform and functional mitochondria from fresh liver. *Journal of Clinical and Translational Hepatology, 7*, 46–50.

Frazier, A. E., Vincent, A. E., Turnbull, D. M., Thorburn, D. R., & Taylor, R. W. (2020). Assessment of mitochondrial respiratory chain enzymes in cells and tissues. *Methods in Cell Biology, 155*, 121–156.

Gollihue, J. L., et al. (2017). Optimization of mitochondrial isolation techniques for intraspinal transplantation procedures. *Journal of Neuroscience Methods, 287*, 1–12.

Gunter, T. E., & Pfeiffer, D. R. (1990). Mechanisms by which mitochondria transport calcium. *American Journal of Physiology, 258*, C755–C786.

Gupta, N., Wu, C. H., & Wu, G. Y. (2016). Targeted transplantation of mitochondria to hepatocytes. *Hepatic Medicine, 8*, 115–134.

Han, J., Han, M. S., & Tung, C.-H. (2013). A non-toxic fluorogenic dye for mitochondria labeling. *Biochimica et Biophysica Acta, 1830*, 5130–5135.

Huang, P.-J., Kuo, C.-C., Lee, H.-C., Shen, C.-I., Cheng, F.-C., Wu, S.-F., Chang, J.-C., Pan, H.-C., Lin, S.-Z., Liu, C.-S., & Su, H.-L. (2016). Transferring xenogenic mitochondria provides neural protection against ischemic stress in ischemic rat brains. *Cell Transplant, 25*(5), 913–927. https://doi.org/10.3727/096368915X689785. Epub 2015 Nov 6.

Jabbari, H., et al. (2020). Mitochondrial transplantation ameliorates ischemia/reperfusion-induced kidney injury in rat. *Biochimica et Biophysica Acta—Molecular Basis of Disease, 1866*, 165809.

Jiang, X.-P., Baucom, C. C., & Elliott, R. L. (2020). Mitochondria dynamically transplant into cells in vitro and in mice and rescue aerobic respiration of mitochondrial DNA-depleted motor Neuron NSC-34. *Journal of Biomedical Science and Engineering, 13*, 203–221.

Katrangi, E., et al. (2007). Xenogenic transfer of isolated murine mitochondria into human rho0 cells can improve respiratory function. *Rejuvenation Research, 10*, 561–570.

Kaza, A. K., et al. (2017). Myocardial rescue with autologous mitochondrial transplantation in a porcine model of ischemia/reperfusion. *The Journal of Thoracic and Cardiovascular Surgery, 153*, 934–943.

Keij, J. F., Bell-Prince, C., & Steinkamp, J. A. (2000). Staining of mitochondrial membranes with 10-nonyl acridine orange, MitoFluor Green, and MitoTracker Green is affected by mitochondrial membrane potential altering drugs. *Cytometry, 39*, 203–210.

Ko, S.-F., et al. (2020). Hepatic 31P-magnetic resonance spectroscopy identified the impact of melatonin-pretreated mitochondria in acute liver ischaemia-reperfusion injury. *Journal of Cellular and Molecular Medicine, 24*, 10088–10099.

Lee, J. M., et al. (2021). Mitochondrial transplantation modulates inflammation and apoptosis, alleviating Tendinopathy both in vivo and in vitro. *Antioxidants, 10*, 696.

Louwagie, E. J., Larsen, T. D., Wachal, A. L., Gandy, T. C. T., & Baack, M. L. (2021). Mitochondrial transfer improves cardiomyocyte bioenergetics and viability in male rats exposed to pregestational diabetes. *International Journal of Molecular Sciences, 22*, 2382.

Masuzawa, A., et al. (2013). Transplantation of autologously derived mitochondria protects the heart from ischemia-reperfusion injury. *American Journal of Physiology—Heart and Circulatory Physiology, 304*, H966–H982.

McCully, J. D., Cowan, D. B., Emani, S. M., & del Nido, P. J. (2017). Mitochondrial transplantation: From animal models to clinical use in humans. *Mitochondrion, 34*, 127–134.

McCully, J. D., Cowan, D. B., Pacak, C. A., & Levitsky, S. (2007). Abstract 2272: Mitochondrial transplantation for cardioprotection. *Circulation, 116*. II_496-II_496.

McCully, J. D., et al. (2009). Injection of isolated mitochondria during early reperfusion for cardioprotection. *American Journal of Physiology—Heart and Circulatory Physiology, 296*, H94–H105.

McCully, J. D., Rousou, A. J., Parker, R. A., & Levitsky, S. (2007). Age- and gender-related differences in mitochondrial oxygen consumption and calcium with cardioplegia and diazoxide. *The Annals of Thoracic Surgery, 83*, 1102–1109.

McCully, J., Preble, J., Kondo, H., Levitsky, S., & James, D. (2014). Quality control parameters for mitochondria transplant. In *Cardiac tissue*.

Minamikawa, T., et al. (1999). Chloromethyl-X-rosamine (MitoTracker Red) photosensitises mitochondria and induces apoptosis in intact human cells. *Journal of Cell Science, 112*(Pt 14), 2419–2430.

Norat, P., et al. (2020). Mitochondrial dysfunction in neurological disorders: Exploring mitochondrial transplantation. *Npj Regenerative Medicine, 5*, 1–9.

Pacak, C. A., et al. (2015). Actin-dependent mitochondrial internalization in cardiomyocytes: Evidence for rescue of mitochondrial function. *Biology Open, 4*, 622–626.

Pandya, J. D., Pauly, J. R., & Sullivan, P. G. (2009). The optimal dosage and window of opportunity to maintain mitochondrial homeostasis following traumatic brain injury using the uncoupler FCCP. *Experimental Neurology, 218*, 381–389.

Patel, S. P., et al. (2014). N-acetylcysteine amide preserves mitochondrial bioenergetics and improves functional recovery following spinal trauma. *Experimental Neurology, 257*, 95–105.

Patel, S. P., Gamboa, J. L., McMullen, C. A., Rabchevsky, A., & Andrade, F. H. (2009). Lower respiratory capacity in extraocular muscle mitochondria: Evidence for intrinsic differences in mitochondrial composition and function. *Investigative Ophthalmology and Visual Science, 50*, 180–186.

Poot, M., et al. (1996). Analysis of mitochondrial morphology and function with novel fixable fluorescent stains. *Journal of Histochemistry and Cytochemistry, 44*, 1363–1372.

Preble, J. M., et al. (2014). Rapid isolation and purification of mitochondria for transplantation by tissue dissociation and differential filtration. *Journal of Visualized Experiments: JoVE*, e51682. https://doi.org/10.3791/51682

Rhyner, M. N. (2010). The coulter principle for analysis of subvisible particles in protein formulations. *The AAPS Journal, 13*, 54–58.

Rizzuto, R., et al. (1996). Double labelling of subcellular structures with organelle-targeted GFP mutants in vivo. *Current Biology, 6*, 183–188.

Robicsek, O., et al. (2018). Isolated mitochondria transfer improves neuronal differentiation of schizophrenia-derived induced pluripotent stem cells and rescues deficits in a rat model of the disorder. *Schizophrenia Bulletin, 44*, 432–442.

Roushandeh, A. M., Kuwahara, Y., & Roudkenar, M. H. (2019). Mitochondrial transplantation as a potential and novel master key for treatment of various incurable diseases. *Cytotechnology, 71*, 647–663.

Rousou, A. J., Ericsson, M., Federman, M., Levitsky, S., & McCully, J. D. (2004). Opening of mitochondrial KATP channels enhances cardioprotection through the modulation of mitochondrial matrix volume, calcium accumulation, and respiration. *American Journal of Physiology—Heart and Circulatory Physiology, 287*, H1967–H1976.

Shi, X., Zhao, M., Fu, C., & Fu, A. (2017). Intravenous administration of mitochondria for treating experimental Parkinson's disease. *Mitochondrion, 34*, 91–100.

Shitara, H., et al. (2001). Non-invasive visualization of sperm mitochondria behavior in transgenic mice with introduced green fluorescent protein (GFP). *FEBS Letters, 500*, 7–11.

Spinazzi, M., Casarin, A., Pertegato, V., Salviati, L., & Angelini, C. (2012). Assessment of mitochondrial respiratory chain enzymatic activities on tissues and cultured cells. *Nature Protocols, 7*, 1235–1246.

Sun, C., et al. (2019). Endocytosis-mediated mitochondrial transplantation: Transferring normal human astrocytic mitochondria into glioma cells rescues aerobic respiration and enhances radiosensitivity. *Theranostics, 9*, 3595–3607.

Ulger, O., et al. (2021). The effects of mitochondrial transplantation in acetaminophen-induced liver toxicity in rats. *Life Sciences, 279*, 119669.

Wu, S., et al. (2018). Polymer functionalization of isolated mitochondria for cellular transplantation and metabolic Phenotype alteration. *Advanced Science, 5*, 1700530.

Zhang, T., & Miao, C. (2022). Mitochondrial transplantation as a promising therapy for mitochondrial diseases. *Acta Pharmaceutica Sinica B*. https://doi.org/10.1016/j.apsb.2022.10.008

Zhang, Z., et al. (2021). Muscle-derived mitochondrial transplantation Reduces inflammation, enhances bacterial clearance, and improves survival in Sepsis. *Shock, 56*, 108.

CHAPTER

9

Storage of isolated mitochondria

Francesco Pallotti[1,2], *Christian Bergamini*[3] *and Romana Fato*[3]

[1]Dipartimento di Medicina e Chirurgia, Università Degli Studi dell'Insubria, Varese, Italy; [2]SSD Laboratorio Analisi-SMEL Specializzato in Citogenetica e Genetica Medica, ASST Sette Laghi-Ospedale di Circolo-Fondazione Macchi, Varese, Italy; [3]Dipartimento di Farmacia e Biotecnologie, FABIT, Università Degli Studi di Bologna, Bologna, Italy

Introduction

The current strategies for mitochondrial transplantation consist in mitochondria isolation and purification from either normal tissues or cells and then transplanting them into the target tissue. After isolation, mitochondria should be properly stored by using different methods, but it is essential a rapid operation. Therefore, methods for mitochondrial storage are devoted to maintain mitochondria viable for a prolonged period of time. Refrigeration or freezing alter mitochondrial membranes; damage to inner membrane impairs energy production from oxidative phosphorylation, ATP synthesis, transport of molecules by the inner membrane, biomolecular synthesis of amino acids, and decreased membrane potential along with the pH gradient. Impairment of outer membrane leads to a loss of integrity associated to an increase of permeability resulting in release of proteinaceous material (i.e., cytochrome *c*) from the intermembrane space. The final result of such modifications is apoptosis due to caspase activation (Bock et al., 2019).

Therefore, storage conditions of isolated mitochondria should preserve inner and outer membrane integrity, thus maintaining mitochondrial function during refrigeration or freezing. Here, we present some methods related to cold storage and cryopreservation of mitochondria performed to preserve mitochondrial stability during long term storage. It has been suggested that isolated mitochondria should be frozen rapidly in liquid nitrogen and stored in a liquid-nitrogen refrigerator at −196°C, being mitochondria one of the lists of biological materials that can be successfully preserved for prolonged periods (Fleischer, 1979).

One of the most important points to evaluate in pretransplantation tests is the isolation of mitochondria: it must be rapid and also the purification step to be preferred is the differential filtration rather than differential centrifugation and density gradient. The conservation step is

Mitochondrial Transplantation and Transfer
https://doi.org/10.1016/B978-0-443-18858-9.00009-3

also limiting for transplantation if a storable preparation is considered instead of a per-time preparation. Therefore, it is of great importance to establish a reliable method to preserve mitochondria for a long period of time. At the present time, it is recommended that once isolated, regardless of the techniques used, mitochondria can be used immediately for transfer and internalization.

Cold storage of mitochondria

It is well known that the isolated mitochondria activity is reduced after more than 1-h storage in ice (McCully et al., 2017). Repetitive centrifugation steps usually increase the isolation time of mitochondria and reduce mitochondrial viability. Several previous reports have shown that the bioenergetic function of mitochondria is decreased to <10%–15% of normal after freezing isolated mitochondria, irrespectively the use of preservatives (Olson et al., 1967; Wechsler, 1961).

Gnaiger and colleagues (Gnaiger et al., 2000) used different buffers to test the viability of mitochondria after isolation from rat heart mitochondria. Two different buffers were prepared, Mito Med S, an isotonic sucrose medium, and Mito Med P, a specific medium for preservation of mitochondria.

Properties of this medium regards ROS scavenging (Sucrose), membrane stabilizer (Taurine, BSA) and antioxidant (Taurine, Histidine, BSA), lipophilic/lipidic antioxidants (Vitamin E, Glutathion). Respiration rates of mitochondria from both buffers were assayed in mitochondrial suspensions at concentration of 7–10 mg protein/mL, stored on ice and used for respirometry within 1 h on day 0 and at various intervals up to 4 days. Respiration of rat heart mitochondria showed that cold storage in sucrose medium causes an immediate loss of respiratory capacity within 24 h, with significantly lower respiration at all times compared with mitochondria in Mito Med P; the decline in state 3 respiration was 52% after 24 h of cold storage in sucrose medium, whereas this defect was reduced to 18% using MitoMed P. However, preservation of mitochondria in Mito Med P for more than 48 h caused decreased respiratory capacity by impairing multiple mitochondrial functions. The cytochrome *c* release, as the earliest event in cold damage of preserved mitochondria, and the complex II electron transport impairment were the most sensitive targets of mitochondrial injury in cold storage.

Another interesting method to minimize mitochondrial damage and improve bioenergetics to maintain good graft function is to preserve organs to be transplanted using different media, such as Eurocollins (EU) (Collins & Halasz, 1976) and University of Wisconsin (UW) (Belzer & Southard, 1988); in rat livers stored in EU and UW it has been shown a significant reduction of the activity of mitochondrial electron transport chain (ETC) complexes (Jassem et al., 2006), thus increasing the mitochondrial oxidative stress that should be prevented by effective preservation. The EU solution imitates the composition and concentration of electrolytes found within cells and contains phosphate as a buffering agent along with glucose to maintain a hypertonic osmolarity of 406 mOsmol. The UW's solution for preserving organs depends on maintaining osmotic concentration via inert substrates and a colloidal carrier (hydroxyethyl starch) instead of glucose; furthermore it includes antioxidants such as adenosine and glutathione. Livers were either transplanted immediately or stored at 4°C for 7 or 24 h before transplantation; all livers were removed 30 min after graft reperfusion for analysis. Mitochondria were then isolated and their respiratory chain activities were assayed. The level of cytochrome within the mitochondria remained steady for 30 min following the reperfusion in livers that were preserved in UW solution. Conversely, all livers preserved in EU experienced notable depletion of cytochrome *c* following the reperfusion. This may result from the increased ability of UW solution to maintain the integrity of the mitochondrial membrane and reduce the frequency of permeability transition events, attributed to the inclusion of Mg in the UW solution as a stabilizer of the membrane and colloids that counteract cellular and mitochondrial swelling. After reperfusion, the activities of complexes III and IV remained stable for up to 7 hours in the UW solution due to improved preservation of mitochondrial antioxidants, such as glutathione. However, with preservation extending to 24 h, the activities significantly decreased. The activity of complex II did not vary across all experimental conditions, suggesting a potential resilience to ischemic and reperfusion damage. It is worth noting that both EU and UW solutions have constraints in mitigating oxidative harm post-cold storage and reperfusion.

These studies have shown that optimization of cold storage conditions must consider the use of antioxidants and colloids to maintain the structure and the function of mitochondria.

Cryopreservation of isolated mitochondria

Cryopreservation of mitochondria by using preservation solution has been reported. Greiff and Myers (1961) reported assays in rat liver mitochondria after freezing in dimethyl sulfoxide (DMSO) and glycerol at 10% and 30% solutions at −65°C for 18 (DMSO) and 15 (glycerol) days. The oxidative phosphorylation capacity of the frozen mitochondria did not change in comparison to the nonfrozen mitochondria, even when the mitochondria were stored for 18 days at −65°C with 10% DMSO or for 15 days at −65°C with 10% glycerol. On the other hand, when frozen isolated mitochondria were stored in 30% glycerol, the oxidative phosphorylation capacity of the mitochondria was inactivated, whereas the capacity of mitochondria stored in 30% DMSO was maintained. Thus, DMSO is less toxic than glycerol.

Nukala and colleagues (Nukala et al., 2006) analyzed mitochondrial activity from rat cerebral cortex after frozen storage at −80°C for 1 week in DMSO 10%. Electron microscopy sections showed intactness of inner and outer mitochondrial membranes (OMMs). A significant decrease in the activity of freshly isolated brain mitochondria washed with DMSO was reported in all respiration states. This decrease indicates that DMSO can affect mitochondrial bioenergetics. The activity was further reduced in DMSO-cryopreserved mitochondria, despite their significantly higher activity in all states compared to mitochondria frozen without DMSO. Although DMSO lowers respiration, it does preserve mitochondrial bioenergetics. The decrease in cytochrome *c* levels observed with the addition of DMSO alone was further reduced upon cryopreservation. However, the reduction in respiratory activity of cryopreserved mitochondria is not caused by rupture of mitochondrial membranes. The use of DMSO is a more effective method for mitochondrial preservation than glycerol, especially in muscle fibers, because DMSO is a more penetrating reagent that may cause greater dehydration, thus minimizing freezing injury (Kuznetsov et al., 2003).

Yamaguchi et al. (2007) tested the effect of cryopreservation in mouse liver mitochondrial preparations. After isolation, the mitochondrial pellet was resuspended in a small amount of trehalose buffer (300 mM trehalose, 10 mM HEPES-KOH pH 7.7, 1 mM EGTA, 1 mM EDTA, 0.1% BSA) at a concentration of 50 mg protein/mL. For freezing, 5 mL samples of mitochondria were pipetted in 0,6 mL-Eppendorf tubes placed on dry ice, without touching the tube wall with the pipette tip. After mitochondria were frozen in a sucrose-mannitol buffer, 95% cytochrome *c* was lost within the initial 15 min of being incubated at 37°C in the same buffer. Trehalose-frozen mitochondria retained nearly all cytochrome *c* for a minimum of 45 min, regardless of whether they were stored at 22 or 37°C. Additionally, trehalose-frozen mitochondria maintained cytochrome *c* levels for at least 2–3 h at 37°C, and more than 95% of the total cytochrome *c* was retained during a 45-minute incubation at 37°C. Regarding apoptosis, trehalose-frozen mitochondria released cytochrome *c* in response to cleaved Bid at rates comparable to those of freshly isolated mitochondria. ATP synthesis and respiratory capacity in trehalose-frozen mitochondria were reduced to approximately 30% of the normal level. Taken together these data confirm that apoptosis can be maintained in trehalose-freeze mitochondria, even though there is a considerable reduction in bioenergetic production capacity.

It is not possible to maintain mitochondrial energy production at equivalent levels as in normal conditions during cryopreservation. while mitochondrial membrane activity is maintained. Improvement in cryopreservation techniques needs efforts in maintaining normal bioenergetic production capacity.

In a brief technical note, assessment for a method aimed to cryopreserve isolated placental mitochondria has been described by Fischer et al. (Fisher et al., 2019). After isolating mitochondria subpopulation by differential centrifugation method, two different samples were submitted to cryopreservation, Cyto-Mito, from cytotrophoblast, and Syncytio-Mito, from syncytiotrophoblast, and resuspended in an isolation media containing 250 mM sucrose, 0.5 mM Na_2EDTA, and 10 mM Tris, pH 7.4. Cryopreservation of both fractions was carried out at −80°C; storage at this temperature was maintained up to 5 months and cryopreserved isolates were prewarmed for 5 min at 37°C prior to the evaluation of mitochondrial respiration. Respiration and maximum respiratory capacity did not differ significantly between cryopreserved mitochondrial fractions and fresh isolates. The respiratory rates from the cryopreserved samples were recovered at 83%–93% of those from freshly isolated mitochondria. Evaluation of the outer mitochondrial membrane (OMM) using exogenous cytochrome *c* revealed that cryopreservation isolates had only a 2% increase in respiration compared to fresh isolates. Therefore, cryopreservation did not appear to significantly affect mitochondrial outer membrane integrity. Furthermore, the inner mitochondrial membrane (IMM) was protected and functional.

Critical points

The maintenance of the integrity of both the IMM and the OMM is essential for proper organelle function and its impact on target cells (Casares et al., 2019). The OMM contains high levels of cholesterol, making it more permeable to oxidative stressors such as ultraviolet rays and hypoxia (Redza-Dutordoir & Averill-Bates, 2016). These stressors may cause irreversible damage to mitochondrial DNA and activate proapoptotic proteins (Reubold & Eschenburg, 2012). Cells receiving damaged mitochondria with a damaged OMM either fail to uptake them or suffer cell death (Kitani et al., 2014). The IMM is similar in chemical structure to bacterial membranes and contains large amounts of cardiolipin (Li et al., 2015). The IMM is crucial to the complexes of ETC and ATP synthesis (Pacak et al., 2015). Disruption of the IMM leads to modifications in cristae dynamics, which subsequently disrupt mitochondria fusion and ATP production (Chan, 2012).

Methods

Cold storage

Method from Gnaiger et al. (2000)

Source: Hearts of adult male rats.

Starting material: Cardiac mitochondria and muscle fibers.

Permeabilized fibers: Washed in ice-cold respiration medium (refer to Table I) and kept on ice in the same medium for 1–6 h until respirometric measurements were taken (Table 9.1).

Mitochondria: Isolated from left ventricles in isotonic sucrose medium (Table 9.2).

After last centrifugation, mitochondrial pellets must be resuspended in respiration medium (Table 9.1), or mitochondrial preservation solution (Mito Med P) (Table 9.3).

Storage: Mitochondrial suspension (7–10 mg protein/mL) kept on ice and utilized within 1 h, and at intervals up to 4 days.

Assays performed: High-resolution respirometry.

Problems: Short-term storage in ice; cytochrome *c* release, and Complex II damage.

Method of organ preservation in solutions (Collins & Halasz, 1976)

Source: Livers from male inbred Wistar rats.

Starting material: Liver mitochondria.

Organ conservation: After orthotopic transplantation, livers were removed from attachments, flushed with 5 mL of heparinized cold UW solution (Table 9.4) (Quader et al., 2020) or Eurocollins (EU) solution (Table 9.5) (Hirata et al., 2013).

Procedure in brief:

TABLE 9.1 Respiration medium for permeabilized fibers.

Sucrose 110 mM EGTA 0.5 mM (0.46 mM free EGTA) $MgCl_2$ 3.0 mM (2.1 mM free Mg^{2+}) KCl 80 mM (total K^+) K-lactobionate 60 mM KH_2PO_4 10 mM Taurine 20 mM HEPES 20 mM BSA 1 g/L glutamate 10 mM (for complex I respiration) malate 5 mM (for complex I respiration) Mg-ATP 1 mM (for complex I respiration) Leupeptin 1.0 mMpH 7.1 (30°C) osmolarity 320 mosmol/L ionic strength 97 mM

Gnaiger, E., Kuznetsov, A.V., Schneeberger, S., Seiler, R., Brandacher, G., Steurer, W., Margreiter, R. (2000). Mitochondria in the cold. In: "Life in the cold" (Eds.) Heldmaier, G., Klingenspor, M., Springer Berlin Heidelberg New York, pp. 431–442. https://doi.org/10.1007/978-3-662-04162-8_45.

TABLE 9.2 Isotonic Sucrose Medium (310 mosmol/L) for heart isolation.

Sucrose 310 mM EDTA 0.2 mM HEPES 10 mM pH 7.4 (with KOH at 25°C)

Gnaiger, E., Kuznetsov, A.V., Schneeberger, S., Seiler, R., Brandacher, G., Steurer, W., Margreiter, R. (2000). Mitochondria in the cold. In: "Life in the cold" (Eds.) Heldmaier, G., Klingenspor, M., Springer Berlin Heidelberg New York, pp. 431–442. https://doi.org/10.1007/978-3-662-04162-8_45.

TABLE 9.3 Mito Med P composition.

Sucrose 110 mM EGTA 0.5 mM (0.46 mM free EGTA)$MgCl_2$ 3.0 mM (2.1 mM free Mg^{2+}) KCl 80 mM (total K^+) K-lactobionate 60 mM KH_2PO_4 10 mM taurine 20 mM histidine 20 mM HEPES 20 mM BSA 1 g/L vitamin E 20 glutathione 3.0 mM glutamate 2.0 mM malate 2.0 mM Mg-ATP 2.0 mM leupeptine 1.0 mM pH 7.1 (30*C) osmolarity 340 mosmol/L ionic strength 108 mM

Gnaiger, E., Kuznetsov, A.V., Schneeberger, S., Seiler, R., Brandacher, G., Steurer, W., Margreiter, R. (2000). Mitochondria in the cold. In: "Life in the cold" (Eds.) Heldmaier, G., Klingenspor, M., Springer Berlin Heidelberg New York, pp. 431–442. https://doi.org/10.1007/978-3-662-04162-8_45.

TABLE 9.4 UW solution.

Adenine (free base) 5 mmol/L calcium chloride (dihydrate) 0.5 mmol/L glucose 10 mmol/L glutathione (reduced) 3 mmol/L HEPES (free acid) 10 mmol/L hydroxyethyl starch 50 g/L magnesium gluconate (anhydrous) 5 mmol/L mannitol 30 mmol/L potassium phosphate (monobasic) 25 mmol/L ribose, D(−) 5 mmol/L sodium gluconate 80 mmol/L sodium hydroxide 0.7 g/L sterile water for injection to 1 L final volume

From Quader, M., Torrado, J. F., Mangino, M. J., Toldo, S. (2020). Temperature and flow rate limit the optimal ex-vivo perfusion of the heart—An experimental study. Journal of Cardiothoracic Surgery, 15(1):180. https://doi.org/10.1186/s13019-020-01223-x.

TABLE 9.5 Eurocollins solution (EU).

Potassium 115 mmol/L sodium 10 mmol/L phosphate 100 mmol/L chloride 15 mmol/L bicarbonate 10 mmol/L dextrose 198 mmol/L osmolarity 406 mOsm/L

From Hirata, E. S., Pereira, R. I., Filho, G. A., Braga Ade, F. (2013). Hyperkalemia by Euro-Collins solution in anesthesia for renal transplantation: A case report. Brazilian Journal of Anesthesiology, 63(5):429–432. https://doi.org/10.1016/j.bjan.2012.08.004.

I. The livers can be immediately either transplanted or stored at 4°C for 7 h or 24 h before transplantation.
II. The livers should be removed 30 min after the reperfusion of the graft.
III. Mitochondria are isolated from portions of liver (10–12 g), removed and homogenized 1:10 W/V in ice cold buffer (Sucrose 75 mM, Mannitol 225 mM, EDTA 1 mM, HEPES 5 mM, and BSA 0.5 mg/mL).
IV. The homogenate is centrifuged for 10 min at 600 g at 4°C.
V. Supernatant is then centrifuged for 15 min at 12,000 g at 4°C, and the obtained pellet is washed two times and resuspended in 1-mL ice-cold homogenization buffer.

Assays performed: Respiratory chain activities, cytochrome contents, coenzyme Q_9, CoQ_9H_2 and vitamin E contents, hydroperoxide contents, mitochondrial fatty acid oxidation.

Problems: Both preservation solutions have limitations in preventing oxidative injuries after cold storage and reperfusion; mitochondrial impairment is higher in EC-stored livers, even if UW has not to be considered the best preservation medium.

Cryopreservation

Method from Nukala et al. (2006)

Source: Brain mitochondria from male Sprague–Dawley rats.
Starting material: Cortical mitochondria.
Procedure in brief:

I. After decapitation of the rats (~250 g), brains should be promptly removed.
II. The cortices are extracted and placed in an all-glass Dounce homogenizer with five times the volume of isolation buffer (Mannitol 215 mM, Sucrose 75 mM, BSA 0.1%, HEPES 20 mM, EGTA 1 mM and pH is adjusted to 7.2 with KOH).
III. After homogenization, mitochondria are isolated by differential centrifugation: centrifuge homogenate at 1300 g for 3 min in an Eppendorf microcentrifuge at 4°C and the supernatant is transferred to new tubes.
IV. Resuspend the loose pellet in isolation buffer with EGTA and centrifuge again at 1300 g for 3 min.
V. Transfer the resulting supernatant to new microcentrifuge tubes filled up completely with isolation buffer with EGTA and centrifuge again at 13,000 g for 10 min.
VI. Discard the supernatant and resuspend the pellet in 500 μL of isolation buffer with EGTA.
VII. Aliquot the samples for cryopreservation.

Cryopreservation: Add DMSO to the mitochondria preparation (10%, v/v), and place in Nalgene freezing container and allow to cool to −80°C at a uniform rate of ~−1°C/min. The sample can be stored at −80°C for up to 6 months for later use.

To revive the mitochondria, briefly thaw frozen samples on ice. The sample tubes should be topped off with isolation buffer with EGTA and centrifuged at 13,000 g for 10 min at 4°C. Resuspend the pellets in 500 μL of isolation buffer with or without EGTA, topped off and centrifuged at 10,000 g for 10 min at 4°C. Resuspend the obtained pellets in appropriate volume.

Assays performed: Electron microscopy, respiration studies, ELISA for cytochrome *c*, Western blotting.

Problems: Mitochondrial membranes and cristae structure are found to be preserved, while mitochondrial respiratory capacity is reduced (80% compared with unfrozen samples).

Method from Yamaguchi et al. (2007)

Source: Mouse liver.
Starting material: Mitochondria isolated form mouse liver.

Procedure in brief:

I. Homogenate mouse liver mitochondria by Dounce homogenization in AT buffer (Trehalose 300 mM, HEPES-KOH pH 7.7 10 mM, KCl 10mM, EGTA 1 mM, EDTA 1 mM, and BSA 0.1%).
II. Centrifuge the homogenate at 600 g for 10 min.
III. Centrifuge the supernatant at 3500 g for 15 min, and resuspend the resulting pellet in 10 mL of isolation buffer and centrifuge at 1500 g for 5 min.
IV. Centrifuge the mitochondrial supernatant at 5500 g for 10 min, and repeat the last two steps.
V. Resuspend the mitochondrial pellet in a small volume of trehalose buffer at 50 mg of protein/mL.
VI. For freezing, 5-mL samples of mitochondrial suspension should pipetted into 0.6-mL Eppendorf tubes placed on the bed of dry ice. Avoid contact of the pipet tips with the tube wall, as this may cause immediate freezing of the mitochondria inside the tip.
VII. The tubes can be immediately thrown into liquid nitrogen and then stored at −80°C. For defrosting, tubes can be held between the fingers for rapid thawing.

Assays performed: Cytochrome *c* release assay, ATP bioluminescence assay, calcium-induced mitochondrial swelling, flow cytometric analysis of isolated liver mitochondria, mitochondrial protein import assay, electron microscopy.

Problems: Reduction of oxidative phosphorylation and ATP production. The inducibility of apoptosis and outer and inner membranes are preserved.

Method from Fisher et al. (2019)

Source: Human placentae.

Starting material: Mitochondria from syncytiotrophoblast and trophoblast.

Collect placentae from healthy term pregnancies, postvaginal birth, and villous tissue as previously described (Burton et al., 2014). Isolation of mitochondrial subpopulations is obtained by differential centrifugation method from Martinez et al. (Martínez et al., 1997). Homogenize samples in a glass Dounce homogenizer with 1 mL of chilled isolation media (Sucrose 250 mM, Na_2 EDTA 0.5 mM, Tris 10 mM, pH 7.4), centrifuge at 1500 g for 10 min at 4°C; collect the supernatant and centrifuged at 4000 g for 15 min at 4°C to produce a pellet of enriched mitochondria (Fraction 1 or Cyto-Mito).

Centrifuge the collected supernatant at 12,000 g for 15 min at 4°C to produce a second pellet of enriched mitochondria (Fraction 2 or Syncytio-Mito). The isolation media is designed to allow for cryopreservation of extracted samples, both fractions are snap frozen and stored at −80°C. Revive cryopreserved isolates by warming to 37°C for 5 min. Isolates can be revived from storage at−80°C up to 5 months later.

Assays performed: Transmission electron microscopy, real-time respirometry and ATP measurements, flow cytometry measurement of membrane potential.

Problems: Presented only a limited set of data concerning key functional parameters.

Conclusions

Storage of isolated mitochondria for use in mitochondria transplantation techniques is still not to be considered optimal (Yamada et al., 2020). Several storage techniques have been comprehensively described since the 1960s. The goal of these methods was to preserve the bioenergetic functionality of the preparations in order to quantify the total functionality of mitochondria. To date, these methods have not yet been used for mitochondria transplantation, preferring the use of fresh mitochondria preparations for transplantation at the cellular level. This topic remains one of the main challenges regarding the development of mitotransplantation technique.

Cryopreservation of mitochondria isolated from rat heart (Nukala et al., 2006)

Before you begin

Timing: 2.5 h.

- Prepare isolation buffer medium:
  - 215 mM mannitol,
  - 75 mM sucrose,
  - 0.1% BSA,
  - 20 mM HEPES,
  - 1 mM EGTA,
  - pH is adjusted to 7.2 with KOH
- Sacrifice adult Sprague–Dawley rats,
- Dissect cortices, pooling up cortices from three to four rats from the same strain
- Perform mitochondrial isolation of crude mitochondria using isolation buffer
- Double centrifugation method is sufficient for a good yield of crude mitochondria
- Resuspend the final pellet in 500 μL of isolation buffer

Note: This method allows the isolation crude mitochondrial fraction from brain, containing both synaptic and nonsynaptic mitochondria.

Materials and equipment

- Mannitol (M4697, Sigma–Aldrich)
- Sucrose (S9378, Sigma–Aldrich)
- Bovine serum albumin (BSA) (A7906, Sigma–Aldrich)
- EGTA (324626, Millipore)
- HEPES potassium salt (H0527, Sigma–Aldrich)
- Potassium phosphate monobasic anhydrous (KH_2PO_4) (795488, Sigma–Aldrich)
- DMSO (D2650, Sigma–Aldrich)
- Nalgene Mr. Frosty, freezing container (C1562, Sigma–Aldrich)

Step-by-step method details

Cryopreservation step.
Timing: 10 m

- Add DMSO (10% v/v) in the resuspended pellet
- Place the sample in a Nalgene Mr. Frosty, freezing container
- Fill the container with room-temperature 100% isopropyl alcohol to the indicated fill line and discard the foam insert and tube holder.
- Insert your sample tubes into the tube holder and place the Mr. Frosty container in −80°C freezer for a minimum of 4 h.
- Remove frozen tubes and place in long-term storage freezer, at −180°C.
- Sample can be stored at −180°C up to 6 months for later use.

Reanimation of brain mitochondria preparation.
Timing: 30 m

- Fill up the tubes with isolation buffer
- Centrifuge at 13,000 g for 10 min at 4°C
- Resuspend the pellet in 500 μL of isolation buffer by filling up the tube
- Centrifuge at 10,000 g for 10 min at 4°C
- Resuspend the pellet in 500–700 μL in isolation buffer

Optional steps

- Most methods are designed to isolate distinct populations of mitochondria from rat brain: a) nonsynaptic mitochondria, called "free mitochondria" (FM); b) synaptosomal mitochondria (synaptic), subdivided in two fractions based on sedimentation properties, heavy (HM) and light (LM)
- If different brain populations are needed, Percoll density gradient centrifugation is necessary for further purification before or after the reanimation step.

Expected outcomes

- DMSO preserves the structure of cryopreserved mitochondrial preparation; inner and outer membranes, together with tight cristae, are intact.
- Electron transfer system is well coupled to oxidative phosphorylation.
- Respiratory activity is maintained in all states, despite the negative effect of brief incubation of mitochondria with DMSO in noncryopreserved mitochondrial preparations (lowering respiration)

Advantages

- Cryopreservation with DMSO of mitochondrial preparations allows the mitochondria to be structurally and biochemically suitable for further manipulations

Limitations

- This method has been used to characterize brain mitochondrial populations by completely ensuring mitochondrial properties close to their living tissue activities.
- Some limits have been noted in these preparations, restricting the possible applications in mito transplantation, such as a reduced respiratory activity (80%) compared to fresh mitochondrial preparations; this reduction can be due to a reduction of cytochrome *c* content observed in cryopreserved mitochondria (40% lower than fresh mitochondrial preparations).

Safety considerations and standards

DMSO is toxic; therefore, the use of eyeshields, gloves, and respiration filter is suggested. For all the reagents described above, the use of eyeshields and gloves is necessary.

References

Belzer, F. O., & Southard, J. H. (April 1988). Principles of solid-organ preservation by cold storage. *Transplantation, 45*(4), 673–676.

Bock, F. J., & Tait, S. W. G. (2019). Mitochondria as multifaceted regulators of cell death. *Nature Reviews Molecular Cell Biology, 21*, 85–100.

Burton, G. J., Sebire, N. J., Myatt, L., Tannetta, D., Wang, Y. L., Sadovsky, Y., Staff, A. C., & Redman, C. W. (January 2014). Optimising sample collection for placental research. *Placenta, 35*(1), 9–22.

Casares, D., Escribá, P. V., & Rosselló, C. A. (2019). Membrane lipid composition: Effect on membrane and organelle structure, function and compartmentalization and therapeutic avenues. *International Journal of Molecular Sciences, 20*(9), 2167.

Chan, D. C. (2012). Fusion and fission: Interlinked processes critical for mitochondrial health. *Annual Review of Genetics, 46*, 265–287.

Collins, G. M., & Halasz, N. A. (April 1976). Forty-eight hour ice storage of kidneys: Importance of cation content. *Surgery, 79*(4), 432–435.

Fisher, J., McKeating, D., Pennell, E., Cuffe, J., Holland, O., & Perkins, A. (July 2019). Mitochondrial isolation, cryopreservation and preliminary biochemical characterisation from placental cytotrophoblast and syncytiotrophoblast. *Placenta, 82*, 1–4.

Fleischer, S. (1979). Long-term storage of mitochondria to preserve energy-linked functions. *Methods in Enzymology, 55*, 28–32.

Gnaiger, E., Kuznetsov, A. V., Schneeberger, S., Seiler, R., Brandacher, G., Steurer, W., & Margreiter, R. (2000). Mitochondria in the cold. In G. Heldmaier, & M. Klingenspor (Eds.), *Life in the cold* (pp. 431–442). New York: Springer Berlin Heidelberg.

Greiff, D., & Myers, M. (1961). Effect of dimethyl sulphoxide on the cryo-tolerance of mitochondria. *Nature, 190*, 1202–1204.

Hirata, E. S., Pereira, R. I., Filho, G. A., & Braga Ade, F. (2013 September–October). Hyperkalemia by Euro-Collins solution in anesthesia for renal transplantation: A case report. *Brazilian Journal of Anesthesiology, 63*(5), 429–432.

Jassem, W., Armeni, T., Quiles, J. L., Bompadre, S., Principato, G., & Battino, M. (2006). Protection of mitochondria during cold storage of liver and following transplantation: Comparison of the two solutions, University of Wisconsin and Eurocollins. *Journal of Bioenergetics and Biomembranes, 38*, 49–55.

Kitani, T., Kami, D., Matoba, S., & Gojo, S. (August 2014). Internalization of isolated functional mitochondria: Involvement of macropinocytosis. *Journal of Cellular and Molecular Medicine, 18*(8), 1694–1703.

Kuznetsov, A. V., Kunz, W. S., Saks, V., Usson, Y., Mazat, J. P., Letellier, T., Gellerich, F. N., & Margreiter, R. (August 15, 2003). Cryopreservation of mitochondria and mitochondrial function in cardiac and skeletal muscle fibers. *Analytical Biochemistry, 319*(2), 296–303.

Li, X. X., Tsoi, B., Li, Y. F., Kurihara, H., & He, R. R. (May 2015). Cardiolipin and its different properties in mitophagy and apoptosis. *Journal of Histochemistry and Cytochemistry, 63*(5), 301–311.

Martínez, F., Kiriakidou, M., & Strauss, J. F., 3rd (May 1997). Structural and functional changes in mitochondria associated with trophoblast differentiation: Methods to isolate enriched preparations of syncytiotrophoblast mitochondria. *Endocrinology, 138*(5), 2172–2183.

McCully, J. D., Cowan, D. B., Emani, S. M., & Del Nido, P. J. (2017). Mitochondrial transplantation: From animal model to clinical use in humans. *Mitochondrion, 34*, 127–134.

Nukala, V. N., Singh, I. N., Davis, L. M., & Sullivan, P. G. (2006). Cryopreservation of brain mitochondria: A novel methodology for functional studies. *Journal of Neuroscience Methods, 152*, 48–54.

Olson, M. S., & Von Korff, R. W. (1967). Changes in endogenous substrates of isolated rabbit heart mitochondria during storage. *Journal of Biological Chemistry, 242*, 325–332.

Pacak, C. A., Preble, J. M., Kondo, H., Seibel, P., Levitsky, S., Del Nido, P. J., Cowan, D. B., & McCully, J. D. (April 10, 2015). Actin-dependent mitochondrial internalization in cardiomyocytes: Evidence for rescue of mitochondrial function. *Biology Open, 4*(5), 622–626.

Quader, M., Torrado, J. F., Mangino, M. J., & Toldo, S. (July 22, 2020). Temperature and flow rate limit the optimal ex-vivo perfusion of the heart—An experimental study. *Journal of Cardiothoracic Surgery, 15*(1), 180.

Redza-Dutordoir, M., & Averill-Bates, D. A. (December 2016). Activation of apoptosis signalling pathways by reactive oxygen species. *Biochimica et Biophysica Acta, 1863*(12), 2977–2992.

Reubold, T. F., & Eschenburg, S. (July 2012). A molecular view on signal transduction by the apoptosome. *Cellular Signalling, 24*(7), 1420–1425.

Wechsler, M. B. (1961). Studies on oxidative phosphorylation and ATPase activity of fresh and frozen brain mitochondria. *Archives of Biochemistry and Biophysics, 95*, 494–498.

Yamada, Y., Ito, M., Arai, M., Hibino, M., Tsujioka, T., & Harashima, H. (September 2, 2020). Challenges in promoting mitochondrial transplantation therapy. *International Journal of Molecular Sciences, 21*(17), 6365.

Yamaguchi, R., Andreyev, A., Murphy, A. N., Perkins, G. A., Ellisman, M. H., & Newmeyer, D. D. (2007). Mitochondria frozen with trehalose retain a number of biological functions and preserve outer membrane integrity. *Cell Death & Differentiation, 14*, 616–624.

SECTION IV

Application methods and limitations of mitochondrial transplantation

CHAPTER

10

Mitochondrial transplantation methods

Gokhan Burcin Kubat[1,2], *Ibrahim Turkel*[3], *Berkay Ozerklig*[3], *Ozbeyen Atalay*[4], *Serpil Suvay*[5], *Meltem Tuncer*[4], *Omer Faruk Taner*[6] *and Oner Ulger*[1,2]

[1]Department of Mitochondria and Cellular Research, Gulhane Health Sciences Institute, University of Health Sciences, Ankara, Turkey; [2]Gulhane Training and Research Hospital, University of Health Sciences, Ankara, Turkey; [3]Department of Exercise and Sport Sciences, Faculty of Sport Sciences, Hacettepe University, Ankara, Turkey; [4]Department of Physiology, Faculty of Medicine, Hacettepe University, Ankara, Turkey; [5]Department of Physiology, Faculty of Medicine, University of Health Sciences, Ankara, Turkey; [6]Department of Plastic Surgery, Gulhane Training and Research Hospital, Ankara, Turkey

Coincubation

While there are several techniques for transferring mitochondria, they may be of limited use for experiments involving multiple cells. Coincubation, a method of mitochondrial transfer, provides a simpler approach to mimic the natural transfer of mitochondria to recipient cells. Although coincubation allows mitochondria to be delivered to many cells in a single experiment, the efficiency of transferring mitochondria is still lacking. In addition, the transfer of mitochondria to the recipient cells by simple coincubation may not be equally efficient in different cell types due to the cellular uptake of exogenous mitochondria, and some cells may be more receptive to mitochondria than others. It is also crucial to maintain the integrity of the mitochondria once they are isolated and exposed to the extracellular environment. By overcoming the limitations of the coincubation method and improving efficiency, further studies are needed to better understand the cellular mechanisms of transplanted mitochondria, which may open new avenues for exploring diseases and developing novel therapies. Mitochondrial transplantation has been proposed as a potential therapeutic strategy for the

https://doi.org/10.1016/B978-0-443-18858-9.00010-X

treatment of diseases associated with mitochondrial dysfunction or damage to the mitochondrial DNA (mtDNA) (D'Amato et al., 2023; McCully et al., 2022). Recent studies have shown promising results using isolated mitochondria from a variety of sources, such as cultured stem cells or autologous tissues, for the treatment of damaged tissues, organs, or cells (D'Amato et al., 2023; Ulger & Kubat, 2022). The idea of mitochondrial transplantation was generated for the first time in 1982 by Clark and Shay (Clark & Shay, 1982). They isolated mitochondria from human fibroblasts that were affected by mtDNA mutations by using antibiotics chloramphenicol and efrapeptin. They then transferred the isolated mitochondria into antibiotic-sensitive, dysfunctional recipient cells, producing viable mammalian cells. The study demonstrated that isolated mitochondria could enter host cells through endocytosis mechanisms and that mtDNA from donor cells can be integrated into recipient cells. This leads to functional changes, including improved antibiotic resistance and increased viability of the recipient cells (Clark & Shay, 1982; Kubat et al., 2021). Collectively, these results demonstrate that mitochondrial transplantation can be achieved in large numbers of cells using the coincubation method. It has been reported by several research groups that mammalian cells can take up exogenous isolated mitochondria by coincubation method (Pour et al., 2020; Elliott et al., 2012; Katrangi et al., 2007). Katrangi and colleagues were the first to report the presence of mitochondria in recipient cells stained with a specific dye using fluorescence microscopy (Katrangi et al., 2007). In this study, the mitochondria isolated from mouse liver tissue were examined for their integrity by electron microscopy, and their functionality was determined by the respiration analysis. Later, the isolated mitochondria were coincubated with MDA-MB-231 and MCF-7 cell lines, and the results showed that the exogenous mitochondria were taken up by the recipient cells and the transfer of mouse mitochondria restored cellular respiration (Katrangi et al., 2007). The key finding of this study was that exogenous mitochondria can improve cellular functions, providing a potential therapeutic approach for diseases associated with mitochondrial dysfunction. Elliott et al. investigated the effect of exogenous mitochondria isolated from normal MCF-12A cells on cancer lines (Elliott et al., 2012). The study showed that the isolated mitochondria were able to enter malignant MCF-7, MDA-MB-231, and NCI/ADR-Res cells, but were not able to enter normal MCF-12A cells. Furthermore, there was a dose-dependent suppression of cancer cell proliferation and an increase in cancer cell sensitivity to chemotherapy (Elliott et al., 2012). Kesner and colleagues determined that the uptake of exogenous mitochondria by recipient cells occurs by macropinocytosis, based on their investigation using inhibitors of micropinocytosis (Kesner et al., 2016). Furthermore, exogenous mitochondria were taken up by HEK 293 cells within 10 min of coincubation (Kesner et al., 2016). To summarize, there are several issues to be considered for mitochondrial transplantation by coincubation, including donor mitochondrial viability and metabolic activity, recipient cell membrane characteristics, and recipient cell phenotype. However, further research is needed to understand whether additional physical and chemical agents can be used to maximize the efficiency of the coincubation method and to induce the desired effects in the recipient cells. The transfer of mitochondria from one cell to another under normal or stressful conditions is of critical importance for the maintenance of cellular homeostasis in complex organisms. Various intercellular structures like tunneling nanotubes (TNTs) and microvesicles (MVs) can mediate mitochondrial transport from one cell to another (Caicedo et al., 2017; Sinha et al., 2016; Sluijter et al., 2014). Since its first application, mitochondrial transplantation/transfer has been the subject of extensive research and the coincubation technique has been an important method for experimental research. While this technique allows the simultaneous transplantation of numerous mitochondria into multiple cells, it cannot be applied in vivo. In addition, cells may have different mitochondrial uptake capabilities depending on their physiological state. Despite the development of new techniques which add additional steps to the coincubation to enhance mitochondrial uptake (Caicedo et al., 2015; Liu et al., 2022), further approaches are needed to maximize coincubation efficiency.

Microinjection

The use of microinjection for the transfer of substances into cells has been well established in various fields of cell biology, including genetically modified animals and fertilization experiments. In addition, this method offers some advantages in terms of delivery the dose and timing of delivery, high efficiency, and less damage to the cell. However, the microinjection method for mitochondrial transplantation is limited to the delivery of isolated mitochondria to larger sample sizes. The use of appropriate mitochondrial injection techniques is crucial in improving or restoring cellular homeostasis, particularly for women who are at risk of age-related changes in the reproductive system, as mitochondria play a vital role in oocyte quality and development. Microinjection is a unique method for the delivery of substances, such as proteins, peptides, and nucleic acids directly into cells. This method involves physical penetration of the cell membrane with a small injection needle, allowing for precise delivery of the substance to the desired location performed under direct visual control using a microscope. Microinjection also delivers with precise volume and timing into either the cytosol or the nucleus, which are difficult to control in many cases with electroporation, carrier peptide delivery, transfection, or infection (Zhang & Yu, 2008a, 2008b). Furthermore, injecting substances into the cells is less stressful than chemical transfection and viral infection, resulting in a lower rate of cell death than is often seen with liposome-mediated transfection or viral infection (Zhang & Yu, 2008a, 2008b). In addition, the transduction efficiency of the cells that survive the injection is close to 100%, which makes microinjection an effective and reliable method for the delivery of substances into cells (Jinturkar et al., 2011; Zhang & Yu, 2008a, 2008b). Although microinjection is a simple method, it can be challenging to deliver substances to target cells. Since only one cell can be injected at a time, this technique limits the number of transfected cells in an experiment. On the other hand, cytoplasmic injection of substances such as DNA appears to be less effective due to the possibility that DNA may be degraded by cytoplasmic nuclease enzymes (Zhang & Yu, 2008a, 2008b). Microinjection is a widespread method for delivering genes, genetic material, and molecules into isolated cells. It is used in a variety of fields, such as transgenic animal technology and fertilization experiments (Jaffe & Terasaki, 2004; Liu et al., 2013). In addition, the microinjection technique is also used for the delivery of exogenous mitochondria into the target cells that have nonfunctional and mutated mitochondria (Irwin et al., 1999; King & Attardi, 1988; Mobarak et al., 2019; Pinkert et al., 1997). King et al. were the first to describe the process of direct injection of mitochondria into a target cell (King & Attardi, 1988). The authors treated the mtDNA with ethidium bromide to reduce its presence and then transplanted isolated mitochondria from Chloramphenicol antibiotic resistant cells into the damaged cells.

This resulted in the development of antibiotic resistance in the cultured cells, as well as the detection of donor mtDNA in the recipient cells (King & Attardi, 1988). Improving the cellular health of damaged cells by replacing their dysfunctional mitochondria with functional ones is one of the main goals of mitochondrial transplantation. In vivo studies have shown that the introduction of young mitochondria into aged oocytes by microinjection improves oocyte fertility and prevents the decline in oocyte quality (Li et al., 2010; Liu et al., 2022; Mobarak et al., 2019). Ebert and colleagues performed an experiment where mitochondria isolated from the liver or testes of *Mus molossinus* were microinjected into the fertilized eggs of CD1 mice (Ebert et al., 1989). The zygotes were viable and underwent normal development, resulting in the birth of the offspring. However, no exogenous mtDNA was detected in the fetal samples or the brain, liver, heart, testis, or ovary of the mature offspring (Ebert et al., 1989). This suggests that the microinjected mitochondria were not selectively replicated during embryonic development. Microinjection is a physical delivery method used to introduce substances such as proteins, peptides, nucleic acids, and mitochondria into target cells (Ebert et al., 1989; Irwin et al., 1999; Jinturkar et al., 2011; Ulger & Kubat, 2022). Due to its high efficiency, minimal cell damage, and human-controlled nature, it is a time-consuming technique that delivers substances one cell at a time (Jinturkar et al., 2011; Zhang & Yu, 2008a, 2008b). However, mitochondrial transfer can be challenging due to the potential loss of mitochondrial function and viability following isolation from tissues or cells, and this approach may limit the transfer of exogenous mitochondria (Kubat et al., 2021; Ulger & Kubat, 2022). To overcome these limitations, it is necessary to develop technological devices that allow injecting substances into multiple cells in less time. Therefore, new technologies and methods that increase the efficiency of mitochondrial transplantation may have the potential to significantly advance the development of treatments for mitochondrial diseases.

Penetrating peptides

Recently, mitochondrial penetrating peptides (MPPs) have been employed as a means of targeting and delivering cargo into mitochondria. These peptides are composed of cationic and hydrophobic amino acids that can permeate the mitochondrial membrane to treat mitochondrial diseases. These peptides can both be used to deliver cargo into target mitochondria with cell-permeable properties or either used to facilitate cellular uptake of foreign mitochondria also known as mitochondrial transplantation. Despite its importance, monitoring mitochondrial morphology changes and dynamics over extended periods remains a challenge due to their continuous movement and changing shapes (Jendrach et al., 2008; Picard et al., 2013). To date, the longest reported time for live cell mitochondria imaging is 24 h (Jendrach et al., 2008; Picard et al., 2013). Hence, the development of a long-term tracking strategy would be of great value in studying mitochondrial morphology and exploring prospective drug delivery systems. Over the years, various mitochondrial targeting modules have been developed, with most being delocalized lipophilic cations that exhibit significant toxicity to mitochondria (Battogtokh et al., 2018; Lu et al., 2016). A recent advancement in this field is the emergence of MPPs, which are short and hydrophobic peptides with high mitochondrial uptake. The cationic and hydrophobic nature of MPPs generally enables permeation through the hydrophobic mitochondrial membrane and subsequent accumulation in the mitochondrial matrix

(Fonseca et al., 2011; Hidaka et al., 2017; Lei & Kelley, 2017). Due to the biocompatible properties and primary peptide structures of MPPs with short amino acid sequences (less than 10 amino acids), low cytotoxicity is expected, making them a potentially promising candidate for the development of new probes for mitochondrial investigation (He et al., 2018; Horton et al., 2008; Jean et al., 2016). A study published by Chang et al. (2013) was one of the earliest in the field, which used Pep-1, a cell-penetrating peptide consisting of three domains: a hydrophobic tryptophan-rich sequence, a hydrophilic lysine-rich domain (KKKRKV), and a spacer domain (SQP). The study found that Pep-1 can effectively transfer mitochondria into cells through electrostatic and hydrophobic interactions with the cell membrane. They isolated mitochondria from parent 143B cells and conjugated with Pep-1 to efficiently transfer mtDNA to mitochondrial disease myoclonic epilepsy with ragged red fibers (MERRF) cells. Pep-1 and mitochondria were conjugated by incubating for 30 min at 37°C to generate a combination using a weight ratio of Pep-1 to mitochondria. Mitochondria were transplanted into cells by the coincubation method. The transplanted mitochondria remained in the cells for at least 15 days, restored mitochondrial function, and decreased cell death. The recipient cells also expressed higher levels of mtDNA 11 days after transplantation (Chang et al., 2017). However, it was noted that free mitochondria were present (Chang et al., 2013) in the culture medium 48 h after transplantation, which suggests that larger organelles like mitochondria may not be suitable for Pep-1-mediated transplantation (Muñoz-Morris et al., 2007). In addition to Pep-1, a few other MPPs are used for targeting mitochondria possibly after mitochondrial transplantation for long-term tracking. For example, peptides like (FxrFxrFxr), which deliver therapeutic cargos into mitochondria with high efficiency or are used for long-term mitochondria tracking, have been found to have low cytotoxicity and high mitochondrial targeting ability. Furthermore, Zhao et al. (2019) reported that a novel peptide contai-ning six amino acid residues (ᴅ-argine-phenylalanine-ᴅ-argine-phenylalanine-ᴅ-argine-phenylalanine-NH2 (rFrFrF)), conjugated to a rhodamine derived fluorescent probe (Longmire et al., 2008) resulting successfully long-term in vivo mitochondrial tracking up to 3 days. In conclusion, ample evidence suggests that mitochondrial transplantation utilizing MPP conjugated mitochondria derived from healthy donor cells could potentially restore the impaired mitochondrial function of recipient cells. Moreover, both targeting mitochondria for delivering therapeutic cargo and imaging live cell with MPPs are crucial for understanding mitochondrial morphology and treating diseases rooted in mitochondrial dysfunction. Nevertheless, more studies are needed to use MPPs safely and efficiently in humans.

Mitoception

Mesenchymal stem cells (MSCs) are capable of transferring mitochondria to diverse target cells, such as cardiomyocytes, endothelial cells, pulmonary alveolar epithelial cells, renal tubular cells, and cancer cells, leading to modifications in their functional characteristics (Acquistapace et al., 2011; Islam et al., 2012; Liu et al., 2014; Pasquier et al., 2013; Plotnikov et al., 2008; Plotnikov et al., 2010). The exchange of mitochondria, the transfer of mitochondria between cells, has emerged as a widely used means of communication and biological property modification. This exchange can occur through the formation of TNTs that involve connexin 43 containing gap junctions (Islam et al., 2012) or by M-Sec/TNFaip2 and the exocyst complex (Hase et al., 2009). Additionally, arrestin domain-containing protein 1 mediated microvesicles were also demonstrated to mediate mitochondrial transfer (Phinney et al., 2015). Notably, the efficacy of mitochondrial transfer is correlated with the expression rate of Rho GTPase 1 MIRO1 (Ahmad et al., 2014), which helps explain differences in mitochondrial transfer efficiencies between induced pluripotent stem cell (iPSC)-MSCs and adult BM-MSCs (Zhang et al., 2016). Caicedo et al. developed a method called "MitoCeption" that quantifies the transfer of isolated mitochondria (from MSCs) to target cells (Andrés Caicedo et al., 2015). This method involves transferring mitochondria isolated from a donor cell to the recipient cell by centrifugation. Cells are coincubated with isolated mitochondria following isolation, and then centrifuged at a variable speed and certain temperature for the transplantation process. Then, quality control analyses (MitoTracker-Cell Tracker imaging, mitochondrial activity XF24 Flux analyzer, ATP measurement, fluorescence activated cell sorting (FACS) analysis, PCR analysis) are required for the efficiency of the transfer (Andrés Caicedo et al., 2015). Several studies demonstrated the ability of isolated mitochondria to be taken up by living cells both in vitro and in vivo (Andrés Caicedo et al., 2015; Kitani, Kami, Kawasaki, et al., 2014; Masuzawa et al., 2013). The MitoCeption method involves utilizing the capability to transfer mitochondria of human mesenchymal stem cells (hMSCs) to MDA-MB-231 cancer cells. The effective transfer of mitochondria was verified and quantified through FACS imaging and mtDNA analysis. Moreover, the researchers demonstrated that the MSCs mitochondria transferred through MitoCeption could enhance the endogenous mitochondrial pool and energetics of MDA-MB-231 cells (Andrés Caicedo et al., 2015). Similarly, Nzigou et al. (Nzigou Mombo et al., 2017) adapted this protocol to transfer isolated hMSCs mitochondria to glioblastoma stem cells. Mitochondrial activity is the key to the regulation of cell metabolism and dysregulation in this activity can lead to severe metabolic diseases. The mitochondria transfer between different types of living cells enables the opportunity of opening new research areas and MitoCeption can be used as a possible therapeutic approach for some diseases such as cancer, aging, type 2 diabetes, and more. This method could be employed to revitalize cells for regenerative medicine (autologous cell based therapies). However, further studies need to be done in vitro and in vivo models to successfully implement MitoCeption for the treatment of mitochondria related diseases.

Photothermal nanoblade and BLAST (Biophotonic laser assisted surgery tool)

A new method of mitochondrial transplantation, photothermal nanoblade, and BLAST (Biophotonic Laser Assisted Surgery Tool), offers some unique advantages for mitochondrial transfer through bypassing endocytosis and cell fusion resulting in increased effectiveness. Due to the fragile nature of mammalian cell membranes, delivering large cargo such as chromosomes, nuclei, DNA, mtDNA, mitochondria, RNA, etc. into the cell poses a challenge. Cellular delivery methods such as endocytosis can harm the cargo due to the low pH and lytic enzymes inside the endosome (Luo & Saltzman, 2000). On the other hand, viral and chemical transfer types pack desired cargo inside a virus or chemical complex which increases the uptake (Naldini et al., 1996). However, toxicity, high degree of specificity, and limited size to be transferred make these types of cargo delivery limited to a certain degree.

Besides the aforementioned delivery methods, physical delivery methods include electroporation (Chu et al., 1987), sonoporation (Mitragotri, 2005), optoporation (Tirlapur & König, 2002), etc. All these methods work by implementing pores on the cellular membrane and delivering the cargo by thermal diffusion or electric field. Because of the nature of the pores implemented by physical methods, transportation of large cargo remains to be inefficient due to low speed of diffusion and decreased cell viability (Stevenson et al., 2006). Another novel approach that combines metallic nanostructures with electron oscillations was shown to guide nanowire growth (Cao et al., 2007), micro and nanoscale fluids (Liu et al., 2006), cancer therapy (Jain et al., 2007) and efficient drug delivery (Skirtach et al., 2005). However, an important limitation comes from the very nature of the created cavitation bubbles which are used to transfer cargo through the membrane in which if the bubble is too large it can cause a rupture in the membrane and result in cell death (Lukianova-Hleb et al., 2010). The photothermal nanoblade method is a novel technique that has many advantages for mitochondrial transplantation. A titanium tip covered microcapillary pipette containing isolated mitochondria is attached to a micromanipulator arm and placed in light contact with a cell membrane in this method. A laser pulse of a certain wavelength warms the titanium, causing a localized explosive cavitation bubble to form and rupture the contacting cell membrane. Following a laser pulse and membrane opening, mitochondria are transferred into the cell by pressured delivery (Wu et al., 2016). The method is highly effective for the delivery of isolated mitochondria into the target cells. Some difficulties occur due to technical equipment, knowledge of the technician, and its application to very few cells. Because of these obstacles, only one cell can be transplanted at one time, and ~200 cells per hour (Wu et al., 2011). Wu et al. have taken the treatment to the next level with the technique named BLAST. This technique allowed the delivery of larger cargo into the cytosol of ~100,000 cells in ~1 min (Ulger & Kubat, 2022). A silicon chip with several microscopic holes is used by the BLAST. The cells are placed on the chip, and the titanium covering of the chip is heated using a laser. The liquid behind the perforations bubbles up and propels itself into the cell membrane, creating a little rupture. This occurs in a fraction of a second, allowing particles from the liquid underneath the perforations to enter the cells. The method works incredibly well for delivering isolated mitochondria to the target cells. Compared to microinjection approaches, this method lessens mechanical trauma and improves the effectiveness of transplantation. However, this method is rather limited in that it can only manipulate a small number of cells, and it is suitable for in vitro studies.

Magnetomitotransfer

The potential for new cell therapies would increase with the introduction of healthy or modified mitochondria into host cells. Anti TOM22 magnetic bead-labeled mitochondria with the use of a magnetic plate propose a novel method of mitochondrial translocation. The fact is that the usual method yields a crude cell extract with both intact and damaged mitochondria, as well as other organelle debris, which may be toxic to the host cell (Kitani, Kami, Matoba, & Gojo, 2014). The most common application to transfer isolated mitochondria to the host cells is coculturing cells with the isolated mitochondria in vitro (Andrés Caicedo et al., 2015). Interestingly, a novel method that utilizes anti TOM22 magnetic beads leads to mitochondria isolates from cells obtained from mouse tissues with higher purity and reproducibility in vitro (Macheiner et al., 2016). The process of magnetomitotransfer, whereby magnetic beads bind to TOM22, a protein of the mitochondrial outer membrane, is achieved through the use of magnetic forces. The usage of magnetic beads improved the transfer rate while preserving the quality of the mitochondria. Host cells could internalize mitochondria when the free mitochondria were marked with an anti TOM22 magnetic bead. Both the higher transfer ratio and density of transferred mitochondria within recipient cells were indicative of the enhanced mitochondrial transfer ability provided by magnetomitotransfer (Franko et al., 2013; Macheiner et al., 2016). Magnetomitotransfer is a more efficient way to deliver isolated mitochondria than passive coculturing in vitro. Importantly, magnetomitotransferred cells showed that transferred mitochondria looked to be functioning as they significantly increased respiration. A range of mitochondria associated disorders, such as different neurodegenerative diseases, may be treated with the innovative technology of magnetomitotransfer. On the other hand, a limitation of this method is the potential for unintended harmful consequences of magnetic beads. Therefore, this novel method is suitable for in vitro studies (Zhang & Miao, 2023).

MitoPunch

Several methods have been developed to restore mitochondrial formation by transferring isolated mitochondria into mtDNA deficient cells (Nzigou Mombo et al., 2017). The metabolism and other functions necessary for the survival and activity of mammalian cells are regulated by the genome coordination of the mtDNA and nuclear DNA (nDNA) (Singh et al., 2017). It is necessary but difficult to generate mammalian cells with specific mtDNA-nDNA combinations in conditions of disease or mitochondrial dysfunction (Sercel et al., 2021). Using the high throughput mitochondrial transfer method known as MitoPunch, isolated mitochondria from cells with certain mtDNA-nDNA combinations are transferred into cells with mtDNA deficiency (Patananan et al., 2020). MitoPunch is based on previous technologies and a device known as a photothermal nanoblade. This method utilizes force to guide an isolated mitochondrial suspension through a porous membrane covered with cells, as opposed to the photothermal nanoblade, which depends on complex lasers and optical systems to function. By rupturing cell membranes, this pressure gradient enables mitochondria to enter recipient cells directly. Dawson et al. demonstrated that transferring isolated mitochondria to mouse and human cells served as the first MitoPunch method (Dawson et al., 2020). In this study, they found that exogenous mtDNA is lost from metabolically compromised, mtDNA intact (+) cells while newly introduced mtDNA is persistently retained in mtDNA deficient (0) recipient cells following uridine free selection. Thereafter, they used a second selective pressure by introducing chloramphenicol resistant mitochondria into chloramphenicol sensitive, metabolically compromised p + murine cybrid cells (a cytoplasmic hybrid cell). Transferred mtDNA was retained by recipient cells with mismatched nDNA and mtDNA genomes, which increases cell respiration by replacing endogenous mutant mtDNA (Dawson et al., 2020). Patanan et al. reprogrammed nonimmortal stable isolated mitochondrial recipient (SIMR) fibroblasts into differentiation iPSCs that differentiated into MSCs, adipocytes, osteoblasts, and chondrocytes. They reported that SIMR fibroblasts were molecularly and phenotypically similar to nonmanipulated control fibroblasts after reprogramming and differentiation (Patananan et al., 2020). While coincubation of mitochondria with cells does not produce SIMR cells, the force-based method of mitochondrial transfer known as MitoPunch and MitoCeption does result in SIMR cells that indefinitely maintain exogenous mtDNA. Even though MitoPunch or MitoCeption produces a sizable number of immortalized SIMR clones with controlled oxidative phosphorylation, only MitoPunch can produce replication limited, nonimmortal human SIMR clones (Sercel et al., 2021). Future studies will continue to use methods like MitoPunch to examine cells with mutant mtDNAs in organs that consume considerable energy, including cardiomyocytes or neurons.

FluidFM

Single cell adhesion was measured to understand the mechanism of focal adhesions (Nagy et al., 2022). In this context, approaches applied to single cells include removing cellular segments or introducing other segments into the cell without impairing cell viability (Gäbelein et al., 2022). Fluidic force microscopy (FluidFM) has introduced a powerful multifunctional toolbox for studying mechanics at the single cell level (Meister et al., 2009). FluidFM is derived from atomic force microscopy (AFM), which can study living biological specimens in their native condition with nanoscale spatial resolution and millisecond temporal resolution under aqueous conditions (Li et al., 2019). FluidFM uses a cantilever that moves with changes in the depth of a particular surface either through direct contact, tapping, or oscillating just above the surface (Verbiest & Rost, 2016). FluidFM is used to apply fluid forces inside individual cells, allowing researchers to study organelle biology as well as subcellular sampling (Guillaume-Gentil et al., 2016). AFM studies were conducted on organelles (e.g., cell mitochondria) isolated from cell lysis, and thus the results only reflect the averaged behavior of organelles from cell populations (Tian et al., 2013; Wang et al., 2018). Smolyakov et al. optimized a novel multiparametric AFM mode to investigate the biophysical characteristics of the cardiomyocyte surface and they focused on the mechanical properties and topography of the heart lateral membrane. This innovation enables the identification of Z lines and M bands in subsarcolemmal structures such as mitochondria and sarcomeric apparatus (Smolyakov et al., 2017). FluidFM, a single cell method created by Gäbelein et al. permits the transplantation of mitochondria between living cells while preserving cellular and mitochondrial integrity (Gäbelein et al., 2022). The FluidFM method has important points in terms of mitochondria transplantation/transfer. For instance, they successfully transplanted mitochondria into the cytoplasm of recipient cells in 95% of them using FluidFM while maintaining cell viability. There were between 3 and 15 mitochondria transplanted into each cell, according to the quantification of the transplant. They monitored the progression of the transplanted mitochondria in 22 cells throughout time; 18 of the cells displayed complete mitochondrial fusion, while four displayed mitochondrial disintegration. In the majority of cases, fusion events were initially noticed within 30 min after transplantation. The majority of transplanted mitochondria are incorporated into the network of primary HEKa cells via mitochondrial fusion. They noticed that the majority of transplanted mitochondria had mtDNA (Gäbelein et al., 2022).

EV mediated mitochondrial transplantation/transfer

In the extracellular area, cells produce lipid-bound vesicles known as extracellular vesicles (EVs) (Yáñez-Mó et al., 2015). MVs, exosomes, and apoptotic bodies are the three main subtypes of EVs, and they are distinguished by their biogenesis, release pathways, size, content, and function (Zaborowski et al., 2015). EVs have evolved as crucial cell to cell communication vehicles in the last decade, transporting nucleic acids and proteins between cells. Although plasma membrane and endosomal proteins are present in EV cargo, EVs can include material from other cellular compartments, such as mitochondria (Amari & Germain, 2021). There are three forms of extracellular mitochondrial content: free mtDNA, functioning mitochondria, and mitochondrial content inside EVs (Keshari et al., 2012). Mitochondrial degradation and dysfunction are implicated in a variety of illnesses, including neurodegenerative disorders, cardiovascular disease, and metabolic disease, implying that EV mediated mitochondria delivery may have broad biological implications (Manickam, 2022). Two major studies investigated the innovative EV mediated mitochondrial transplanting approach. Luca et al. found that transferring mitochondria from EVs to mtDNA deficient L929 Rho0 cells restored mitochondrial function and enhanced Rho0 cell survival. Moreover, the integration of mitochondria from EVs into inflammatory mononuclear phagocytes returned cellular metabolism and mitochondrial dynamics to normal levels and also decreased the expression of proinflammatory markers in target cells. Exogenous neural stem cells (NSCs) actively transported mitochondria to mononuclear phagocytes after being implanted in an animal model of multiple sclerosis, which resulted in a notable improvement in clinical impairments. They are the first to show that NSCs transfer functioning mitochondria to target cells through EVs (Peruzzotti-Jametti et al., 2021). Ikeda et al. proposed that restoration of myocardial bioenergetics through EVs mediated transfer of autologous mitochondria enhances cardiac function. Mitochondria rich extracellular vesicles (MEVs) were extracted from pluripotent stem cell derived cardiomyocytes (iCMs). The shape and quantity of EVs were determined using nanoparticle tracking analysis. Flow cytometry was utilized to characterize MEVs. MEVs ($1 \times 10^8/30\ \mu L$ PBS) were injected in three locations around infarct regions. Healthy shaped mitochondria were found inside MEVs via electron microscopy. Confocal microscopy demonstrated that the transplanted iCMs received MEV derived mitochondria and merged them with their indigenous mitochondrial networks. Early on following treatment with MEVs, hypoxia-damaged iCMs showed better contractile profiles and a considerable restoration of intracellular ATP generation. At 24 h following treatment, recipient iCMs that had received MEVs with mitochondrial biogenesis related mRNAs initiated the process. Intramyocardial injection of MEVs exhibited considerably post MI cardiac function (Ikeda et al., 2021).

Protocol of coincubation method of Suvay et al. (Serpil Suvay, 2022)

Suvay et al. used the coincubation method to incubate isolated mitochondria with prostate cancer cells. 3-(4,5-dimethylthiazol-2-yl)-2,5-diphenyltetrazolium bromide (MTT) was applied to evaluate the effects of 3-BP, mitochondrial doses, and their combinations on cell viability in the study.

Materials: • Bromopyruvic acid (Sigma 16,490), • RPMI-1640 Cell medium (Sigma, Lot 116K2416), • Dulbecco's Modified Eagle's Medium (DMEM) (Capricorn DMEMHA), • Fetal Bovine Serum (FBS) (Capricorn FBS-11A), • Dulbecco's Phosphate Buffered Saline (PBS) (Capricorn PBS-1A), • Penicillin Streptomycin Antibiotic (PSA, HyClone), • Trypsin EDTA (Capricorn TRY-3B), • Dimethyl Sulfoxide (Sigma, D8418), • Problock Gold Mammalian Protein Inhibitor Cocktail (GOLDBIO • GB-331–1), • BCA Protein Measurement Kit (ABP BIO USA P011), • Cell-Quant MTT Cell Proliferation Kit (ABP BIO A015), • Mitochondria Isolation Kit (Biovision K288-50).

Methods: 1. Prepare 2×10^7 MSCs of human MSCs using the cell culture method for the assessment of mitochondrial dosage. Isolate mitochondria using the mitochondria isolation kit. 2. Measure the protein quantity of isolated mitochondria suspended in 500 μL storage buffer using the BCA technique and store the isolated mitochondria at +4°C. 3. Put the cells to be treated with mitochondria 24 h before isolation, with approximately 5×10^3–10^4 cells and eight duplicates in each well. 4. Incubate at 37°C in a CO_2 incubator (5% CO_2/95% air) for 24 h. 5. Discard the postincubation media (DMEM). 6. Incubate for 24 h for exposure in 3-BP groups (30 and 45 μM). 7. Discard the postincubation media (DMEM). 8. Coincubate cells and mitochondria (1 μg/mL, 5 μg/mL, and 10 μg/mL) at 37 °C with 5% CO_2 and 95% air for 24 h 9. Following incubation, remove the solutions from the wells and perform the Cell-Quant MTT Cell Proliferation Assay Kit protocol to determine MTT cell viability. 10. Read the absorbance using a microplate reader (Molecular Devices Filter Max F5) at 570 nm. 11. 30 μM doses of 3-BP and 5 μg/mL doses of mitochondria were used in proliferation experiments.

References

Acquistapace, A., Bru, T., Lesault, P.-F., Figeac, F., Coudert, A. E., le Coz, O., Christov, C., Baudin, X., Auber, F., Yiou, R., Dubois-Randé, J.-L., & Rodriguez, A.-M. (2011). Human mesenchymal stem cells reprogram adult cardiomyocytes toward a progenitor-like state through partial cell fusion and mitochondria transfer. *Stem Cells, 29*(5), 812–824. https://doi.org/10.1002/stem.632

Ahmad, T., Mukherjee, S., Pattnaik, B., Kumar, M., Singh, S., Rehman, R., Tiwari, B. K., Jha, K. A., Barhanpurkar, A. P., Wani, M. R., Roy, S. S., Mabalirajan, U., Ghosh, B., & Agrawal, A. (2014). Miro1 regulates intercellular mitochondrial transport and enhances mesenchymal stem cell rescue efficacy. *EMBO Journal, 33*(9), 994–1010. https://doi.org/10.1002/embj.201386030

Amari, L., & Germain, M. (2021). Mitochondrial extracellular vesicles – origins and roles. *Frontiers in Molecular Neuroscience, 14*. https://doi.org/10.3389/fnmol.2021.767219

Battogtokh, G., Choi, Y. S., Kang, D. S., Park, S. J., Shim, M. S., Huh, K. M., Cho, Y. Y., Lee, J. Y., Lee, H. S., & Kang, H. C. (2018). Mitochondria-targeting drug conjugates for cytotoxic, anti-oxidizing and sensing purposes: Current strategies and future perspectives. *Acta Pharmaceutica Sinica B, 8*(6), 862–880. https://doi.org/10.1016/j.apsb.2018.05.006

Caicedo, A., Aponte, P. M., Cabrera, F., Hidalgo, C., & Khoury, M. (2017). Artificial mitochondria transfer: Current challenges, advances, and future applications. *Stem Cells International, 2017*, Article 7610414. https://doi.org/10.1155/2017/7610414

Caicedo, A., Fritz, V., Brondello, J. M., Ayala, M., Dennemont, I., Abdellaoui, N., De Fraipont, F., Moisan, A., Prouteau, C. A., Boukhaddaoui, H., Jorgensen, C., & Vignais, M. L. (2015). MitoCeption as a new tool to assess the effects of mesenchymal stem/stromal cell mitochondria on cancer cell metabolism and function. *Scientific Reports, 5*. https://doi.org/10.1038/srep09073

Cao, L., Barsic, D. N., Guichard, A. R., & Brongersma, M. L. (2007). Plasmon-assisted local temperature control to pattern individual semiconductor nanowires and carbon nanotubes. *Nano Letters, 7*(11), 3523–3527. https://doi.org/10.1021/nl0722370

Chang, J. C., Hoel, F., Liu, K. H., Wei, Y. H., Cheng, F. C., Kuo, S. J., Tronstad, K. J., & Liu, C. S. (2017). Peptide-mediated delivery of donor mitochondria improves mitochondrial function and cell viability in human cybrid cells with the MELAS A3243G mutation. *Scientific Reports, 7*(1). https://doi.org/10.1038/s41598-017-10870-5

Chang, J. C., Liu, K. H., Li, Y. C., Kou, S. J., Wei, Y. H., Chuang, C. S., Hsieh, M., & Liu, C. S. (2013). Functional recovery of human cells harbouring the mitochondrial DNA mutation MERRF A8344G via peptide-mediated mitochondrial delivery. *Neurosignals, 21*(3–4), 160–173. https://doi.org/10.1159/000341981

Chu, G., Hayakawa, H., & Berg, P. (1987). Electroporation for the efficient transfection of mammalian cells with DNA. *Nucleic Acids Research, 15*(3), 1311–1326. https://doi.org/10.1093/nar/15.3.1311

Clark, M. A., & Shay, J. W. (1982). Mitochondrial transformation of mammalian cells. *Nature, 295*(5850), 605–607. https://doi.org/10.1038/295605a0

D'Amato, M., Morra, F., Di Meo, I., & Tiranti, V. (2023). Mitochondrial transplantation in mitochondrial medicine: Current challenges and future perspectives. *International Journal of Molecular Sciences, 24*(3). https://doi.org/10.3390/ijms24031969

Dawson, E. R., Patananan, A. N., Sercel, A. J., & Teitell, M. A. (2020). Stable retention of chloramphenicol-resistant mtDNA to rescue metabolically impaired cells. *Scientific Reports, 10*(1). https://doi.org/10.1038/s41598-020-71199-0

Ebert, K. M., Alcivar, A., Liem, H., Goggins, R., & Hecht, N. B. (1989). Mouse zygotes injected with mitochondria develop normally but the exogenous mitochondria are not detectable in the progeny. *Molecular Reproduction and Development, 1*(3), 156–163. https://doi.org/10.1002/mrd.1080010303

Elliott, R. L., Jiang, X. P., & Head, J. F. (2012). Mitochondria organelle transplantation: Introduction of normal epithelial mitochondria into human cancer cells inhibits proliferation and increases drug sensitivity. *Breast Cancer Research and Treatment, 136*(2), 347–354. https://doi.org/10.1007/s10549-012-2283-2

Fonseca, S. B., Pereira, M. P., Mourtada, R., Gronda, M., Horton, K. L., Hurren, R., Minden, M. D., Schimmer, A. D., & Kelley, S. O. (2011). Rerouting chlorambucil to mitochondria combats drug deactivation and resistance in cancer cells. *Chemistry & Biology, 18*(4), 445–453. https://doi.org/10.1016/j.chembiol.2011.02.010

Franko, A., Baris, O. R., Bergschneider, E., Von Toerne, C., Hauck, S. M., Aichler, M., Walch, A. K., Wurst, W., Wiesner, R. J., Johnston, I. C. D., & De Angelis, M. H. (2013). Efficient isolation of pure and functional mitochondria from mouse tissues using automated tissue disruption and enrichment with anti-TOM22 magnetic beads. *PLoS One, 8*(12). https://doi.org/10.1371/journal.pone.0082392

Gäbelein, C. G., Feng, Q., Sarajlic, E., Zambelli, T., Guillaume-Gentil, O., Kornmann, B., & Vorholt, J. A. (2022). Mitochondria transplantation between living cells. *PLoS Biology, 20*(3). https://doi.org/10.1371/journal.pbio.3001576

Guillaume-Gentil, O., Grindberg, R. V., Kooger, R., Dorwling-Carter, L., Martinez, V., Ossola, D., Pilhofer, M., Zambelli, T., & Vorholt, J. A. (2016). Tunable single-cell extraction for molecular analyses. *Cell, 166*(2), 506–516. https://doi.org/10.1016/j.cell.2016.06.025

Hase, K., Kimura, S., Takatsu, H., Ohmae, M., Kawano, S., Kitamura, H., Ito, M., Watarai, H., Hazelett, C. C., Yeaman, C., & Ohno, H. (2009). M-Sec promotes membrane nanotube formation by interacting with Ral and the exocyst complex. *Nature Cell Biology, 11*(12), 1427–1432. https://doi.org/10.1038/ncb1990

He, H., Wang, J., Wang, H., Zhou, N., Yang, D., Green, D. R., & Xu, B. (2018). Enzymatic cleavage of branched peptides for targeting mitochondria. *Journal of the American Chemical Society, 140*(4), 1215–1218. https://doi.org/10.1021/jacs.7b11582

Hidaka, T., Pandian, G. N., Taniguchi, J., Nobeyama, T., Hashiya, K., Bando, T., & Sugiyama, H. (2017). Creation of a synthetic ligand for mitochondrial DNA sequence recognition and promoter-specific transcription suppression. *Journal of the American Chemical Society, 139*(25), 8444–8447. https://doi.org/10.1021/jacs.7b05230

Horton, K. L., Stewart, K. M., Fonseca, S. B., Guo, Q., & Kelley, S. O. (2008). Mitochondria-penetrating peptides. *Chemistry & Biology, 15*(4), 375–382. https://doi.org/10.1016/j.chembiol.2008.03.015

Ikeda, G., Santoso, M. R., Tada, Y., Li, A. M., Vaskova, E., Jung, J. H., O'Brien, C., Egan, E., Ye, J., & Yang, P. C. (2021). Mitochondria-rich extracellular vesicles from autologous stem cell–derived cardiomyocytes restore energetics of ischemic myocardium. *Journal of the American College of Cardiology, 77*(8), 1073–1088. https://doi.org/10.1016/j.jacc.2020.12.060

Irwin, M. H., Johnson, L. W., & Pinkert, C. A. (1999). Isolation and microinjection of somatic cell-derived mitochondria and germline heteroplasmy in transmitochondrial mice. *Transgenic Research, 8*(2), 119–123. https://doi.org/10.1023/A:1008925419758

Islam, M. N., Das, S. R., Emin, M. T., Wei, M., Sun, L., Westphalen, K., Rowlands, D. J., Quadri, S. K., Bhattacharya, S., & Bhattacharya, J. (2012). Mitochondrial transfer from bone-marrow-derived stromal cells to pulmonary alveoli protects against acute lung injury. *Nature Medicine, 18*(5), 759–765. https://doi.org/10.1038/nm.2736

Jaffe, L. A., & Terasaki, M. (2004). Quantitative microinjection of oocytes, eggs, and embryos. *Methods in Cell Biology, 2004*(74), 219–242. https://doi.org/10.1016/s0091-679x(04)74010-8

Jain, P. K., ElSayed, I. H., & El-Sayed, M. A. (2007). Au nanoparticles target cancer. *Nano Today, 2*(1), 18–29. https://doi.org/10.1016/S1748-0132(07)70016-6

Jean, S. R., Ahmed, M., Lei, E. K., Wisnovsky, S. P., & Kelley, S. O. (2016). Peptide-mediated delivery of chemical probes and therapeutics to mitochondria. *Accounts of Chemical Research, 49*(9), 1893–1902. https://doi.org/10.1021/acs.accounts.6b00277

Jendrach, M., Mai, S., Pohl, S., Vöth, M., & Bereiter-Hahn, J. (2008). Short- and long-term alterations of mitochondrial morphology, dynamics and mtDNA after transient oxidative stress. *Mitochondrion, 8*(4), 293–304. https://doi.org/10.1016/j.mito.2008.06.001

Jinturkar, K. A., Rathi, M. N., & Misra, A. (2011). Gene delivery using physical methods. In *Challenges in delivery of therapeutic genomics and proteomics* (pp. 83–126). Elsevier Inc. https://doi.org/10.1016/B978-0-12-384964-9.00003-7

Katrangi, E., D'Souza, G., Boddapati, S. V., Kulawiec, M., Singh, K. K., Bigger, B., & Weissig, V. (2007). Xenogenic transfer of isolated murine mitochondria into human ρ0 cells can improve respiratory function. *Rejuvenation Research, 10*(4), 561–570. https://doi.org/10.1089/rej.2007.0575

Keshari, R. S., Jyoti, A., Kumar, S., Dubey, M., Verma, A., Srinag, B. S., Krishnamurthy, H., Barthwal, M. K., & Dikshit, M. (2012). Neutrophil extracellular traps contain mitochondrial as well as nuclear DNA and exhibit inflammatory potential. *Cytometry, Part A, 81*(3), 238–247. https://doi.org/10.1002/cyto.a.21178

Kesner, E. E., Saada-Reich, A., & Lorberboum-Galski, H. (2016). Characteristics of mitochondrial transformation into human cells. *Scientific Reports, 6*. https://doi.org/10.1038/srep26057

King, M. P., & Attardi, G. (1988). Injection of mitochondria into human cells leads to a rapid replacement of the endogenous mitochondrial DNA. *Cell, 52*(6), 811–819. https://doi.org/10.1016/0092-8674(88)90423-0

Kitani, T., Kami, D., Kawasaki, T., Nakata, M., Matoba, S., & Gojo, S. (2014). Direct human mitochondrial transfer: A novel concept based on the endosymbiotic theory. In *Transplantation proceedings* (Vol 46, pp. 1233–1236). USA: Elsevier. https://doi.org/10.1016/j.transproceed.2013.11.133. Issue 4.

Kitani, T., Kami, D., Matoba, S., & Gojo, S. (2014). Internalization of isolated functional mitochondria: Involvement of macropinocytosis. *Journal of Cellular and Molecular Medicine, 18*(8), 1694–1703. https://doi.org/10.1111/jcmm.12316

Kubat, G. B., Ulger, O., & Akin, S. (2021). Requirements for successful mitochondrial transplantation. *Journal of Biochemical and Molecular Toxicology, 35*(11). https://doi.org/10.1002/jbt.22898

Lei, E. K., & Kelley, S. O. (2017). Delivery and release of small-molecule probes in mitochondria using traceless linkers. *Journal of the American Chemical Society, 139*(28), 9455–9458. https://doi.org/10.1021/jacs.7b04415

Li, F., Ford, W. E., Duran, F. S., Castora, F. J., Jones, H. W., & Swanson, J. R. (2010). Mitochondrial changes in aged oocytes and improvement of fertility rate through autologous platelets mitochondrial microinjection. *Fertility and Sterility, 94*(4), S57. https://doi.org/10.1016/j.fertnstert.2010.07.221

Li, M., Xi, N., Wang, Y., & Liu, L. (2019). Advances in atomic force microscopy for single-cell analysis. *Nano Research, 12*(4), 703–718. https://doi.org/10.1007/s12274-018-2260-0

Liu, C., Xie, W., Gui, C., & Du, Y. (2013). Pronuclear microinjection and oviduct transfer procedures for transgenic mouse production. *Methods in Molecular Biology, 1027*, 217–232. https://doi.org/10.1007/978-1-60327-369-5_10

Liu, G. L., Kim, J., Lu, Y. U., & Lee, L. P. (2006). Optofluidic control using photothermal nanoparticles. *Nature Materials, 5*(1), 27–32. https://doi.org/10.1038/nmat1528

Liu, K., Ji, K., Guo, L., Wu, W., Lu, H., Shan, P., & Yan, C. (2014). Mesenchymal stem cells rescue injured endothelial cells in an in vitro ischemia-reperfusion model via tunneling nanotube like structure-mediated mitochondrial transfer. *Microvascular Research, 92*, 10–18. https://doi.org/10.1016/j.mvr.2014.01.008

Liu, Z., Sun, Y., Qi, Z., Cao, L., & Ding, S. (2022). Mitochondrial transfer/transplantation: An emerging therapeutic approach for multiple diseases. *Cell & Bioscience, 12*(1). https://doi.org/10.1186/s13578-022-00805-7

Longmire, M. R., Ogawa, M., Hama, Y., Kosaka, N., Regino, C. A. S., Choyke, P. L., & Kobayashi, H. (2008). Determination of optimal rhodamine fluorophore for in vivo optical imaging. *Bioconjugate Chemistry, 19*(8), 1735–1742. https://doi.org/10.1021/bc800140c

Lu, P., Bruno, B. J., Rabenau, M., & Lim, C. S. (2016). Delivery of drugs and macromolecules to the mitochondria for cancer therapy. *Journal of Controlled Release, 240*, 38–51. https://doi.org/10.1016/j.jconrel.2015.10.023

Lukianova-Hleb, E., Hu, Y., Latterini, L., Tarpani, L., Lee, S., Drezek, R. A., Hafner, J. H., & Lapotko, D. O. (2010). Plasmonic nanobubbles as transient vapor nanobubbles generated around plasmonic nanoparticles. *ACS Nano, 4*(4), 2109–2123. https://doi.org/10.1021/nn1000222

Luo, D., & Saltzman, W. M. (2000). Enhancement of transfection by physical concentration of DNA at the cell surface. *Nature Biotechnology, 18*(8), 893–895. https://doi.org/10.1038/78523

Macheiner, T., Fengler, V. H. I., Agreiter, M., Eisenberg, T., Madeo, F., Kolb, D., Huppertz, B., Ackbar, R., & Sargsyan, K. (2016). Magnetomitotransfer: An efficient way for direct mitochondria transfer into cultured human cells. *Scientific Reports, 6*. https://doi.org/10.1038/srep35571

Manickam, D. S. (2022). Delivery of mitochondria via extracellular vesicles – a new horizon in drug delivery. *Journal of Controlled Release, 343*, 400–407. https://doi.org/10.1016/j.jconrel.2022.01.045

Masuzawa, A., Black, K. M., Pacak, C. A., Ericsson, M., Barnett, R. J., Drumm, C., Seth, P., Bloch, D. B., Levitsky, S., Cowan, D. B., & McCully, J. D. (2013). Transplantation of autologously derived mitochondria protects the heart from ischemia-reperfusion injury. *American Journal of Physiology—Heart and Circulatory Physiology, 304*(7), H966–H982. https://doi.org/10.1152/ajpheart.00883.2012

McCully, J. D., del Nido, P. J., & Emani, S. M. (2022). Mitochondrial transplantation for organ rescue. *Mitochondrion, 64*, 27–33. https://doi.org/10.1016/j.mito.2022.02.007

Meister, A., Gabi, M., Behr, P., Studer, P., Vörös, J., Niedermann, P., Bitterli, J., Polesel-Maris, J., Liley, M., Heinzelmann, H., & Zambelli, T. (2009). FluidFM: Combining atomic force microscopy and nanofluidics in a universal liquid delivery system for single cell applications and beyond. *Nano Letters, 9*(6), 2501–2507. https://doi.org/10.1021/nl901384x

Mitragotri, S. (2005). Healing sound: The use of ultrasound in drug delivery and other therapeutic applications. *Nature Reviews Drug Discovery, 4*(3), 255–260. https://doi.org/10.1038/nrd1662

Mobarak, H., Heidarpour, M., Tsai, P. S. J., Rezabakhsh, A., Rahbarghazi, R., Nouri, M., & Mahdipour, M. (2019). Autologous mitochondrial microinjection; A strategy to improve the oocyte quality and subsequent reproductive outcome during aging. *Cell & Bioscience, 9*(1). https://doi.org/10.1186/s13578-019-0360-5

Mombo, B. N., Gerbal-Chaloin, S., Bokus, A., Daujat-Chavanieu, M., Jorgensen, C., Hugnot, J. P., & Vignais, M. L. (2017). MitoCeption: Transferring isolated human MSC mitochondria to glioblastoma stem cells. *Journal of Visualized Experiments, 2017*, 120. https://doi.org/10.3791/55245

Muñoz-Morris, M. A., Heitz, F., Divita, G., & Morris, M. C. (2007). The peptide carrier Pep-1 forms biologically efficient nanoparticle complexes. *Biochemical and Biophysical Research Communications, 355*(4), 877–882. https://doi.org/10.1016/j.bbrc.2007.02.046

Nagy, Á. G., Kanyó, N., Vörös, A., Székács, I., Bonyár, A., & Horvath, R. (2022). Population distributions of single-cell adhesion parameters during the cell cycle from high-throughput robotic fluidic force microscopy. *Scientific Reports, 12*(1). https://doi.org/10.1038/s41598-022-11770-z

Naldini, L., Blömer, U., Gallay, P., Ory, D., Mulligan, R., Gage, F. H., Verma, I. M., & Trono, D. (1996). In vivo gene delivery and stable transduction of nondividing cells by a lentiviral vector. *Science, 272*(5259), 263–267. https://doi.org/10.1126/science.272.5259.263

Pasquier, J., Guerrouahen, B. S., Al Thawadi, H., Ghiabi, P., Maleki, M., Abu-Kaoud, N., Jacob, A., Mirshahi, M., Galas, L., Rafii, S., Le Foll, F., & Rafii, A. (2013). Preferential transfer of mitochondria from endothelial to cancer cells through tunneling nanotubes modulates chemoresistance. *Journal of Translational Medicine, 11*(1). https://doi.org/10.1186/1479-5876-11-94

Patananan, A. N., Sercel, A. J., Wu, T. H., Ahsan, F. M., Torres, A., Kennedy, S. A. L., Vandiver, A., Collier, A. J., Mehrabi, A., Van Lew, J., Zakin, L., Rodriguez, N., Sixto, M., Tadros, W., Lazar, A., Sieling, P. A., Nguyen, T. L., Dawson, E. R., Braas, D., … Teitell, M. A. (2020). Pressure-driven mitochondrial transfer pipeline generates mammalian cells of desired genetic combinations and fates. *Cell Reports, 33*(13). https://doi.org/10.1016/j.celrep.2020.108562

Peruzzotti-Jametti, L., Bernstock, J. D., Willis, C. M., Manferrari, G., Rogall, R., Fernandez-Vizarra, E., Williamson, J. C., Braga, A., van den Bosch, A., Leonardi, T., Krzak, G., Kittel, Á., Benincá, C., Vicario, N., Tan, S., Bastos, C., Bicci, I., Iraci, N., Smith, J. A., … Pluchino, S. (2021). Neural stem cells traffic functional mitochondria via extracellular vesicles. *PLoS Biology, 19*(4). https://doi.org/10.1371/journal.pbio.3001166

Phinney, D. G., Di Giuseppe, M., Njah, J., Sala, E., Shiva, S., St Croix, C. M., Stolz, D. B., Watkins, S. C., Di, Y. P., Leikauf, G. D., Kolls, J., Riches, D. W. H., Deiuliis, G., Kaminski, N., Boregowda, S. V., McKenna, D. H., & Ortiz, L. A. (2015). Mesenchymal stem cells use extracellular vesicles to outsource mitophagy and shuttle micro-RNAs. *Nature Communications, 6*. https://doi.org/10.1038/ncomms9472

Picard, M., Shirihai, O. S., Gentil, B. J., & Burelle, Y. (2013). Mitochondrial morphology transitions and functions: Implications for retrograde signaling? *American Journal of Physiology—Regulatory, Integrative and Comparative Physiology, 304*(6), R393–R406. https://doi.org/10.1152/ajpregu.00584.2012

Pinkert, C. A., Irwin, M. H., Johnson, L. W., & Moffatt, R. J. (1997). Mitochondria transfer into mouse ova by microinjection. *Transgenic Research, 6*(6), 379–383. https://doi.org/10.1023/A:1018431316831

Plotnikov, E. Y., Khryapenkova, T. G., Galkina, S. I., Sukhikh, G. T., & Zorov, D. B. (2010). Cytoplasm and organelle transfer between mesenchymal multipotent stromal cells and renal tubular cells in co-culture. *Experimental Cell Research, 316*(15), 2447–2455. https://doi.org/10.1016/j.yexcr.2010.06.009

Plotnikov, E. Y., Khryapenkova, T. G., Vasileva, A. K., Marey, M. V., Galkina, S. I., Isaev, N. K., Sheval, E. V., Polyakov, V. Y., Sukhikh, G. T., & Zorov, D. B. (2008). Cell-to-cell cross-talk between mesenchymal stem cells and cardiomyocytes in co-culture. *Journal of Cellular and Molecular Medicine, 12*(5A), 1622–1631. https://doi.org/10.1111/j.1582-4934.2007.00205.x

Pour, P. A., Kenney, M. C., & Kheradvar, A. (2020). Bioenergetics consequences of mitochondrial transplantation in cardiomyocytes. *Journal of the American Heart Association, 9*(7). https://doi.org/10.1161/JAHA.119.014501

Sercel, A. J., Patananan, A. N., Man, T., Wu, T. H., Yu, A. K., Guyot, G. W., Rabizadeh, S., Niazi, K. R., Chiou, P. Y., & Teitell, M. A. (2021). Stable transplantation of human mitochondrial DNA by high-throughput, pressurized isolated mitochondrial delivery. *Elife, 10*, 1–45. https://doi.org/10.7554/ELIFE.63102

Singh, B., Modica-Napolitano, J. S., & Singh, K. K. (2017). Defining the momiome: Promiscuous information transfer by mobile mitochondria and the mitochondrial genome. *Seminars in Cancer Biology, 47*, 1–17. https://doi.org/10.1016/j.semcancer.2017.05.004

Sinha, P., Islam, M. N., Bhattacharya, S., & Bhattacharya, J. (2016). Intercellular mitochondrial transfer: Bioenergetic crosstalk between cells. *Current Opinion in Genetics & Development, 38*, 97–101. https://doi.org/10.1016/j.gde.2016.05.002

Skirtach, A. G., Dejugnat, C., Braun, D., Susha, A. S., Rogach, A. L., Parak, W. J., Möhwald, H., & Sukhorukovt, G. B. (2005). The role of metal nanoparticles in remote release of encapsulated materials. *Nano Letters, 5*(7), 1371–1377. https://doi.org/10.1021/nl050693n

Sluijter, J. P. G., Verhage, V., Deddens, J. C., Van Den Akker, F., & Doevendans, P. A. (2014). Microvesicles and exosomes for intracardiac communication. *Cardiovascular Research, 102*(2), 302–311. https://doi.org/10.1093/cvr/cvu022

Smolyakov, G., Cauquil, M., Severac, C., Lachaize, V., Guilbeau-Frugier, C., Sénard, J. M., Galés, C., & Dague, E. (2017). Biophysical properties of cardiomyocyte surface explored by multiparametric AFM. *Journal of Structural Biology, 198*(1), 28–37. https://doi.org/10.1016/j.jsb.2017.03.001

Stevenson, D., Agate, B., Tsampoula, X., Fischer, P., Brown, C. T. A., Sibbett, W., Riches, A., Gunn-Moore, F., & Dholakia, K. (2006). Femtosecond optical transfection of cells:viability and efficiency. *Optics Express, 14*(16), 7125–7133. https://doi.org/10.1364/OE.14.007125

Suvay, & Tekin, V. (2022). *The effects of hexokinase-2 inhibitor and mitochondrial transplantation in prostate cancer cells. Paper presented at the Acta Physiologica. Oral presentation retrieved from. Gizem Esra Koc, Mehmet Ozler.*

Tian, Y., Li, J., Cai, M., Zhao, W., Xu, H., Liu, Y., & Wang, H. (2013). High resolution imaging of mitochondrial membranes by in situ atomic force microscopy. *RSC Advances, 3*(3), 708–712. https://doi.org/10.1039/c2ra22166g

Tirlapur, U. K., & König, K. (2002). Targeted transfection by femtosecond laser. *Nature, 418*(6895), 290–291. https://doi.org/10.1038/418290a

Ulger, O., & Kubat, G. B. (2022). Therapeutic applications of mitochondrial transplantation. *Biochimie, 195*, 1–15. https://doi.org/10.1016/j.biochi.2022.01.002

Verbiest, G. J., & Rost, M. J. (2016). Resonance frequencies of AFM cantilevers in contact with a surface. *Ultramicroscopy, 171*, 70–76. https://doi.org/10.1016/j.ultramic.2016.07.018

Wang, X., Liu, H., Zhu, M., Cao, C., Xu, Z., Tsatskis, Y., Lau, K., Kuok, C., Filleter, T., McNeill, H., Simmons, C. A., Hopyan, S., & Sun, Y. (2018). Mechanical stability of the cell nucleus—Roles played by the cytoskeleton in nuclear deformation and strain recovery. *Journal of Cell Science, 131*(13). https://doi.org/10.1242/jcs.209627

Wu, T. H., Sagullo, E., Case, D., Zheng, X., Li, Y., Hong, J. S., Teslaa, T., Patananan, A. N., McCaffery, J. M., Niazi, K., Braas, D., Koehler, C. M., Graeber, T. G., Chiou, P. Y., & Teitell, M. A. (2016). Mitochondrial transfer by photothermal nanoblade restores metabolite profile in mammalian cells. *Cell Metabolism, 23*(5), 921–929. https://doi.org/10.1016/j.cmet.2016.04.007

Wu, T. H., Teslaa, T., Teitell, M. A., & Chiou, P. Y. (2011). Photothermal nanoblade for single cell surgery. In *Proceedings of the IEEE international conference on micro electro mechanical systems (MEMS)* (pp. 1055–1058). https://doi.org/10.1109/MEMSYS.2011.5734610

Yáñez-Mó, M., Siljander, P. R. M., Andreu, Z., Zavec, A. B., Borràs, F. E., Buzas, E. I., Buzas, K., Casal, E., Cappello, F., Carvalho, J., Colás, E., Cordeiro-Da Silva, A., Fais, S., Falcon-Perez, J. M., Ghobrial, I. M., Giebel, B., Gimona, M., Graner, M., Gursel, I., et al.De Wever, O. (2015). Biological properties of extracellular vesicles and their physiological functions. *Journal of Extracellular Vesicles, 4*(2015), 1–60. https://doi.org/10.3402/jev.v4.27066

Zaborowski, M. P., Balaj, L., Breakefield, X. O., & Lai, C. P. (2015). Extracellular vesicles: Composition, biological relevance, and methods of study. *BioScience, 65*(8), 783–797. https://doi.org/10.1093/biosci/biv084

Zhang, T.g., & Miao, C.y. (2023). Mitochondrial transplantation as a promising therapy for mitochondrial diseases. *Acta Pharmaceutica Sinica B, 13*(3), 1028–1035. https://doi.org/10.1016/j.apsb.2022.10.008

Zhang, Y., & Yu, L. C. (2008a). Microinjection as a tool of mechanical delivery. *Current Opinion in Biotechnology, 19*(5), 506–510. https://doi.org/10.1016/j.copbio.2008.07.005

Zhang, Y., & Yu, L. C. (2008b). Single-cell microinjection technology in cell biology. *BioEssays, 30*(6), 606–610. https://doi.org/10.1002/bies.20759

Zhang, Y., Yu, Z., Jiang, D., Liang, X., Liao, S., Zhang, Z., Yue, W., Li, X., Chiu, S. M., Chai, Y. H., Liang, Y., Chow, Y., Han, S., Xu, A., Tse, H. F., & Lian, Q. (2016). iPSC-MSCs with high intrinsic MIRO1 and sensitivity to TNF-α yield efficacious mitochondrial transfer to rescue anthracycline-induced cardiomyopathy. *Stem Cell Reports, 7*(4), 749–763. https://doi.org/10.1016/j.stemcr.2016.08.009

Zhao, T., Liu, X., Singh, S., Liu, X., Zhang, Y., Sawada, J., Komatsu, M., & Belfield, K. D. (2019). Mitochondria penetrating peptide-conjugated TAMRA for live-cell long-term tracking. *Bioconjugate Chemistry, 30*(9), 2312–2316. https://doi.org/10.1021/acs.bioconjchem.9b00465

CHAPTER

11

Mitochondrial replacement therapy

Duygu Dayanir

Department of Histology and Embryology, Faculty of Medicine, Gazi University, Ankara, Turkey

The mitochondrion is an essential part of the cell that is responsible for energy production through oxidative phosphorylation, takes part in cellular apoptosis mechanisms as well as balancing the amount of reactive oxygen species. Although each cell contains thousands of mitochondria of its own, each mitochondrion contains its own mitochondrial DNA (mtDNA). Mutations that occur on mtDNA or DNA in the nucleus with known efficacy on mitochondrial functions may cause clinical conditions that may result in serious or even death (Dougherty et al., 2018; Greenfield et al., 2017). Many individuals are homoplastic, with mitochondria and cells uniform. However, parents with a diagnosis of mitochondrial disease exhibit heteroplastic expression, that is, normal and mutated DNA structures coexist (Dougherty et al., 2018).

Mitochondrial diseases due to mtDNA mutations were first described in 1988 (Tachibana et al., 2018; Wallace et al., 1988; Zeviani et al., 1988). Since then, approximately 150 mutations have been discovered, including nearly 100 deletions and 50 point mutations. The prevalence of the aforementioned mutations among individuals with Alzheimer's, Parkinson's, Huntington's, obesity, and diabetes is another crucial characteristics (Tachibana et al., 2018). The fact that the number of individuals diagnosed with mitochondrial disease is increasing day by day and this diagnosis can be seen in all age groups from newborns to adults has led to an increasing interest in this field. Leber's hereditary optic neuropathy (LHON), which is caused by NADH-Q oxidoreductase gene mutation, is one of the important diseases in the mitochondrial diseases. In patients with this diagnosis, myoclonic epilepsy with ragged-red fibers ultimately leads to myoclonus formation. Patients may also have ataxia due to mutations in their tRNA genes (Tachibana et al., 2018; Wallace et al., 1988; Zeviani et al., 1988). Large deletions on mtDNA have been identified in some mitochondrial diseases. The best-known example in this group is Kearns-Sayre syndrome, which includes pigment retinopathy and cardiac symptoms. In individuals diagnosed with LHON, approximately >60% mutation is required for the phenotype to be evident; this rate is known to be >85% for symptoms such as MERFF. Currently, even at the last point reached in the treatment of mitochondrial diseases, there is no definitive and generally approved method. It can be said that

Mitochondrial Transplantation and Transfer
https://doi.org/10.1016/B978-0-443-18858-9.00011-1

individuals who are diagnosed are trying to be treated with the aim of slowing the progression of the disease only against the existing symptoms (Tachibana et al., 2018).

A wide range of patient phenotypes can be observed when the characteristics of individuals with pathological DNA structure are examined. The type of mutation, the degree of heteroplasmic mutations, and the energy requirement of the affected organ or system are determinative in this distribution. The brain, nerve–muscle tissues, kidney, heart, liver, eye, ear, or pancreas can be named among the structures with the highest energy needs (Tachibana et al., 2018). Although approximately one in every 22 individuals has an mtDNA mutation, symptomatic disease can be seen in only one of 10,000 adults (Tachibana et al., 2018; Tanaka et al., 2013).

Mitochondrial replacement therapy (MRT) is being identified as a method that aims to prevent mtDNA mutation transmission and thus to prevent mitochondrial disorder diseases. Embryos obtained by MRT method contain genetic material supported by one male and two females. For example, only nuclear DNA from the female with mutant mtDNA structure is transferred to the embryo, while the passage of cytoplasmic components is not allowed. While nuclear DNA is allowed to pass from the male individual to the same embryo, the healthy mtDNA structure of the embryo is obtained from the second (wild-type) female oocyte. In this way, it is aimed to prevent pathogenic expression due to mutant mitochondria structure (Dougherty et al., 2018; Noohi et al., 2022; Tachibana et al., 2018).

Two methods have been studied with the aim of diagnosing mitochondrial diseases which are prenatal diagnosis (PND) and pre-implantation genetic diagnosis (PGD) (Noohi et al., 2022). When both methods are considered together, it cannot be declared that transmission of mitochondrial mutations to the next generation can be prevented for a large number of women. Although Pnd is an invasive method, in the case of a mitochondrial defect, parents have to decide between terminating or continuing the pregnancy. With the PGD method, uncertainty is encountered about the mtDNA heteroplasmy to be transferred to the next generation and the possible degree of the disease (Noohi et al., 2022; Wai et al., 2008).

Barriers to mitochondrial replacement therapy

The only treatment option that can be offered to individuals affected by mitochondrial diseases is the symptomatic treatment. For this reason, the prevention of diseases stands out as an important option. However, there appear to be many barriers to prevent the transmission of mitochondrial diseases. In vitro fertilization practices combined with pre-implantation genetic testing (PGT) have been insufficient to eradicate mitochondrial diseases, although the transition has been reduced. The first reason why PGT application is suboptimal is that the amount of mutant mitochondria changes from cell to cell. This creates uncertainty about the gamete to be selected for use in the process of assisted reproductive techniques. As a second barrier, it is noteworthy that no threshold value has been defined to estimate the level of exposure. Most important, third barrier is that changes in heteroplasmy levels are observed during the embryogenesis process, which determines the final individual phenotype (Engelstad et al., 2016; Greenfield et al., 2017; Richardson et al., 2015).

Previous advancements

The structure of human mitochondria does not follow the classical Mendelian distribution, but is derived through a matrilineal pattern. Therefore, sperm do not contribute mitochondria on the zygote and embryonic mtDNA is formed entirely by strict maternal inheritance. However, there is no selection method between mutant and wild-type on a cell line. For this reason, it can be said that the mutagenic mitochondrial structure is very sensitive to the 'bottleneck effect' (Chinnery & Hudson, 2013; Stewart & Chinnery, 2015). This situation can be summarized as a woman with a pathogenic mutation causing a high prevalence of transmission for the next offspring (Amato et al., 2014). All these features have led to research on preventing mitochondrial diseases on the germline rather than embryonic alteration. Studies on the manipulation of mitochondria in individual cells seem to have started with cytoplasm transfer research in the 1990s. In these studies, it is known that a limited amount of ooplasm from young female donors was transferred to the oocyte of a female patient with recurrent implantation failure during the in vitro fertilization process. The main aim was not the treatment of mitochondrial diseases, but the potential improvement of low-quality oocytes for couples with infertility diagnosis treated by assisted reproductive techniques. Although the main goal of similar studies was to increase the number of young and healthy mitochondria of the oocyte and to prevent the loss of mitochondrial function that can be seen in age dependent oocytes, studies naturally created mtDNA mismatches (Cecchino et al., 2018; Dougherty et al., 2018). By the end of the 1990s, approximately 30 live births were reported using these methods in the United States and other developed countries. Although it is not possible to declare with certainty that this method increases the risk of miscarriage in older women, it is known that it contributes in this direction. With all its features, this method, which is being applied, played an inspiring role in the formation of the idea of MRT for upcoming studies (Cohen et al., 1998; Dougherty et al., 2018).

Other data indicate that attempts were made to advance the MRT method to isolate cellular components and to prevent mtDNA mismatches and heteroplasmy. In 1983, pronuclear transfer (PNT) was successfully completed in the rodent model (McGrath & Solter, 1983; Tanaka et al., 2013). In the PNT method, the oocyte belonging to the donor and the mother is fertilized with the sperm obtained from the father. In the following steps, it is ensured that the fertilized oocyte structure progresses to the early zygote stage. By removing the pronucleus in the zygote structure provided by the donor's oocyte, it is replaced with the pronucleus removed from the zygote obtained by the maternal and paternal gametes (Noohi et al., 2022). In the 1990s, a new method to prevent crossover contamination of mutagenic mitochondria was discovered: nuclear spindle transfer (Tachibana et al., 2009; Tanaka et al., 2013) or Maternal Spindle Transfer (Noohi et al., 2022). With this method, it is aimed to transfer from the 'meiotic spindle' of the carrier oocyte structure in the metaphase stage to the donor oocyte which is known to be wild-type. Since the process is performed with gametes that have not yet been fertilized, it can be considered that this technique is much more applicable when using Maternal spindle transfer, since the embryo can be considered as intact compared to PNT. Despite the fact that the Human Embryology and Fertilization Authority (HFEA) has prohibited the introduction of this technique as a therapeutic option for known carriers, polar body transfer in human cell lines has been reported in the literature with good results (Adashi & Cohen, 2017; Dougherty et al., 2018). Considering the steps of the MRT method, it is known that besides all the controversial issues presented, there are also accepted steps related to the application. However, active research on this topic has started with different research groups (Adashi & Cohen, 2017). According to the data shared in the literature, a total of 17 births have been reported with these methods (Noohi et al., 2022).

MRT is applied under five headings (Table 11.1) (Sendra et al., 2021).

Meiotic spindle transfer

By the increase in gonadotropins during the ovulation process, the secondary oocyte structure is maturated. Until fertilization takes place, the process has halted in Metaphase II. In metaphase II, the chromosomes are located in the meiotic spindle, although the structure of the nucleus cannot be completely observed. In the Mst process, the meiotic spindle structure is removed from the patient's oocyte. This meiotic spindle transfer to the oocyte belonging to the donor whose nucleus has been removed before is used for Mst. In the next step, fertilization and implantation of the reconstituted oocyte will be provided (Fig. 11.1) (Sendra et al., 2021).

There are positive results in human and animal models regarding the Mst method. Among these studies, the study performed by Tachibana et al. can be considered a milestone by its significance (Tachibana et al., 2009). Four live births in the rhesus macaques model with Mst were reported in this study. In the same study, Mst was also noted for its high efficacy and low mitochondrial carry-over from biological mother (Tachibana et al., 2009, 2013). There is very limited data on defining long-term results of mitochondrial carryover in the next generation with the method. For this reason, the need for further studies in this area continues (Sendra et al., 2021).

Satisfactory data from animal models have led to human studies, and the Mst method has been used in the progress. Studies on the five embryonic stem cells lines obtained following the Mst method also draw attention to their successful carry-over rates (Sendra et al., 2021; Tachibana et al., 2013). However, it was observed that the embryos obtained in the human model developed up to the blastocyst stage at a lower rate compared to animal studies (Kang et al., 2016; Paull et al., 2013; Yamada et al., 2016). The birth of a live male baby carrying a deleterious mutation in mtDNA has been reported following the application of the method. In light the data, the Mst method can be declared as a promising method in animal and human models, and has live birth results in human studies. Despite the fact that low rates of carry-over have been reported, it is possible that there will be an increase in carried mtDNA in the next generation. The low number of animal models related to the method constitutes a limiting feature to knowing the long-term effects of the method (Sendra et al., 2021).

Germinal vesicle transfer

The germinal vesicle can be defined as a large nuclear structure within the immature oocyte. With the onset of puberty, the immature oocyte waiting in the prophase 1 stage is stimulated. By germinal vesicle transfer (Gvt) method, the germinal vesicle structure in the patient's oocyte is transferred to the oocyte from the donor, whose germinal vesicle has

TABLE 11.1 Classification of MRT procedures and related properties.

Classification	**Nuclear transfer**				
Abbreviation	**Gvt**	**PNT**	**PB1t**[a]	**Pb2t**[b]	**Mst**[c]
Procedure name	Germinal Vesicle Transfer	Pronuclear transfer	Polar Body1 transfer[a]	Polar Body 2 transfer[b]	Meiotic spindle transfer[c]
Genetic material/ nuclear ploidy	GV/ tetraploid: 4n	PN/diploid: 2n	PB1/diploid: 2n	PB2/haploid: n	MII spindle-chromosomal complex/ diploid: 2n
Dilution/ replacement	Replacement	Replacement	Replacement	Replacement	Replacement
Type (stage) of recipient; patient or maternal	GV intact	2 PN	PB1	PB2	MII spindle
Type (stage) of donor; healthy donor	GV stage cytoplasm	PN stage cytoplasm	MII cytoplasm	Pn stage cytoplasm w/o female Pn	MII cytoplasm
Applied to domestic animals?/Live birth?	Yes/Yes	Yes/Yes	Yes/Yes	Yes/Yes	Yes/Yes
mtDNA heteroplasmy in offspring (mouse/ primate)	NA	Heteroplasmic (up to 69%; mouse)	Homoplasmy to donor mtDNA (undetectable mouse)	Nearly homoplasmic to donor mtDNA (1.7%; mouse)	Heteroplasmic ~ nearly
Applied to human? / live birth?	Yes/No	Yes/No	Yes/No	No/−	Yes/Yes
mtDNA heteroplasmy in embryos (human)	NA	Heteroplasmic (vary <2% ~ >5%)	NA	−	Nearly homoplasmic to donor mtDNA (vary <1 %–5%)

[a]*Also referred to as Polar Body Nuclear Transfer 1 (PBNT1).*
[b]*Also referred to as Polar Body Nuclear Transfer 2 (PBNT2).*
[c]*Also referred to as Spindle-Chromosomal complex Transfer (St).*
Gvt, germinal vesicle transfer.

1. Meiotic Spindle Transfer (Mst)
2. Germinal Vesicle Transfer (Gvt)
3. Pronuclear Transfer (PNT)
4. Polar Body Transfer
   a. First Polar Body Transfer (PB1t)
   b. Second Polar Body Transfer (PB2t)

Table 11.1 summarizes the mentioned methods within the scope of MRT, along with their specific features Tachibana, M., Kuno, T., Yaegashi, N. (2018). Mitochondrial replacement therapy and assisted reproductive technology: A paradigm shift toward treatment of genetic diseases in gametes or in early embryos. Reproductive medicine and biology, *17(4), 421–433.*

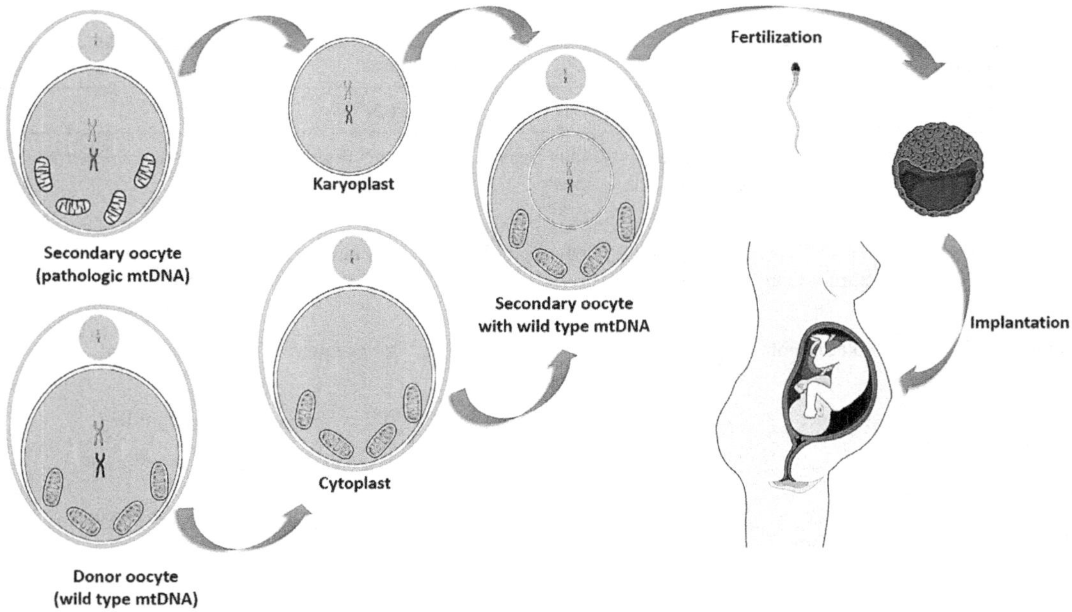

FIGURE 11.1 Meiotic spindle transfer (Mst). *Created with BioRender.com.*

been removed before. Since the oocyte obtained is immature, its maturation and fertilization are completed under in vitro conditions (Fig. 11.2) (Cheng et al., 2009; Sendra et al., 2021).

It is noteworthy that the data obtained by the method have a low chance of success. In the study shared by Neupane et al. (2014) it was shared that no embryo was observed as healthy until the blastocyst stage. Considering the low success rates obtained from animal studies related to the method, the application of the method in human without providing an optimized procedure in in vitro maturation seems disconcerting (Sendra et al., 2021).

Pronuclear transfer

After the oocyte in the metaphase II stage is fertilized by the spermatozoid, it is expected to complete the second meiosis process. The genetic structure of the resulting zygote is represented by two pronuclei from the mother and father. In the PNT method, the pronucleus of the patient's oocyte is isolated in the karyoplast and transferred to the donor oocyte whose pronucleus was previously removed (Fig. 11.3) (Sendra et al., 2021). Unlike Gvt and Mst methods, PNT is a process that requires the manipulation of the zygote. The first question mark about this feature is about the ethical approach. During the method process, two embryos are formed, but one of them is given a chance to survive. A change in carryover levels is reported among the groups after using this method (Cheng et al., 2009; McGrath & Solter, 1983; Neupane et al., 2014; Sato et al., 2005; Wang et al., 2014). This result can be interpreted as the effect of the operator and the micromanipulation procedure on the success of the PNT method (Dougherty et al., 2018).

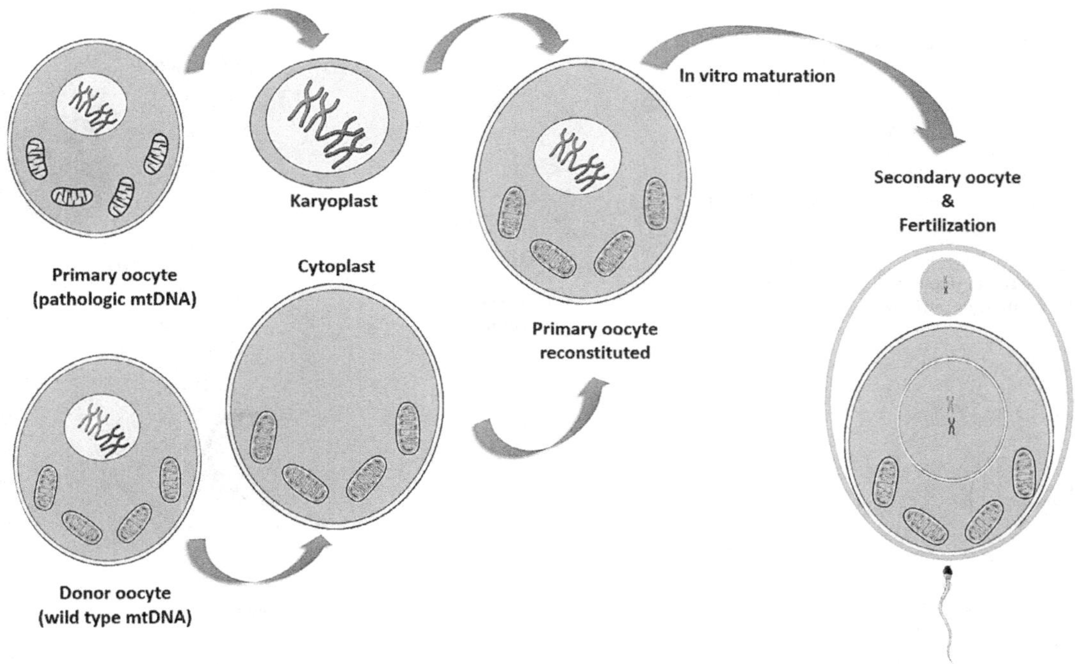

FIGURE 11.2 Germinal vesicle transfer (Gvt). *Created with BioRender.com.*

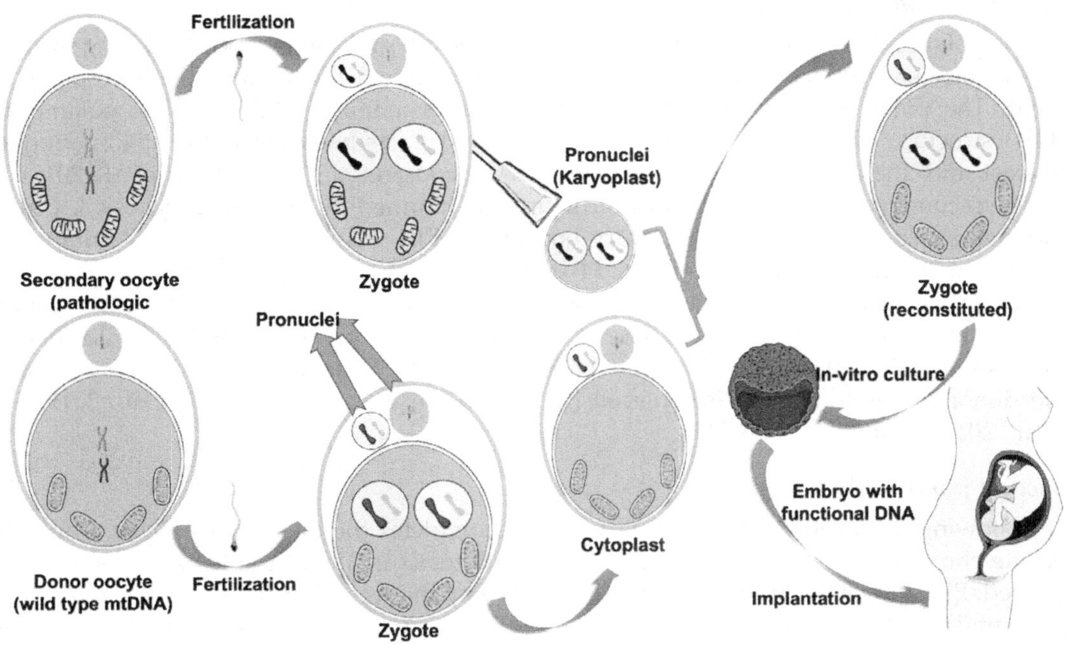

FIGURE 11.3 Pronuclear transfer (PNT). *Created with BioRender.com.*

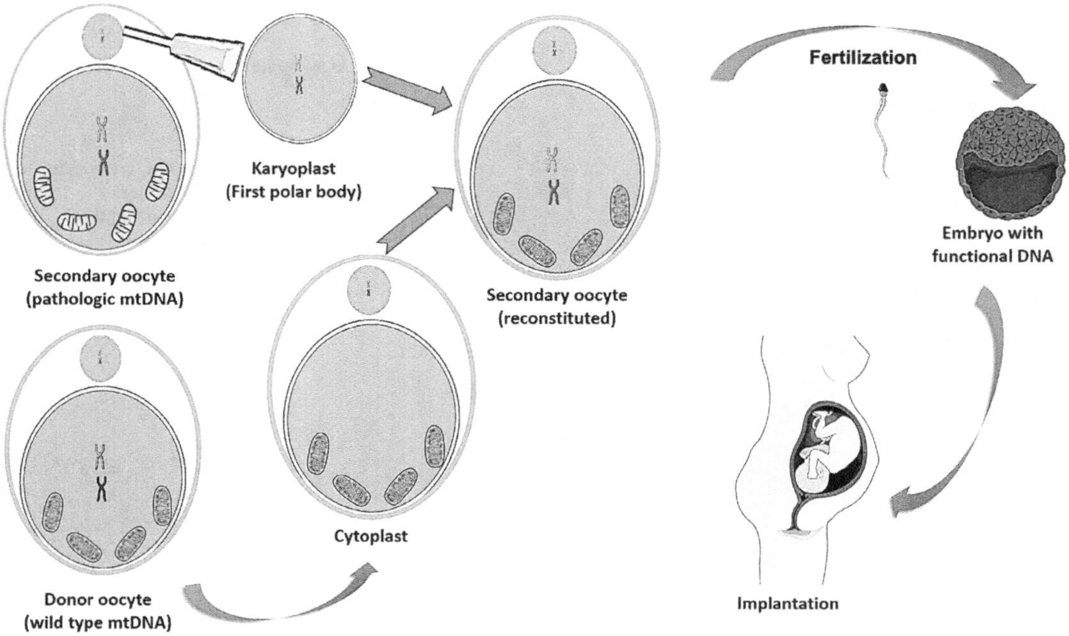

FIGURE 11.4 First polar body transfer (PB1T). *Created with BioRender.com.*

First polar body transfer

The polar body is formed as a result of asymmetric cell division that occurs during the process of meiosis and it is known that they are dispensable for subsequent embryo development. The polar body is composed of chromosomes mainly. While the first polar body (PB1) structure disappears within hours, the second polar body structure (PB2) continues to exist until the early blastocyst stage (Fig. 11.4) (Sendra et al., 2021; Zhang et al., 2017).

The secondary oocyte and the PB1 structure have formed by the end of meiosis I. Unlike the metaphase spindle of the secondary oocyte, the PB1 is surrounded by a plasma membrane with a low amount of cytoplasm. PB1t includes the transfer of the PB1 of the secondary oocyte in the metaphase II stage of the patient to the donor oocyte from which the meiosis spindle was previously removed. The continuation of the process includes fertilization of the oocyte and implantation. The difference between the PB1t method and the Mst is that the donor's meiotic spindle is replaced by the PB1 in the progress (Ma et al., 2017; Wu et al., 2017; Zhang et al., 2017).

Second polar body transfer

Following fertilization, the secondary oocyte completes the meiosis II process. Thus, the oocyte chromosome structure changes ((n, 2c) becomes (n, c)) and the female pronucleus is formed. PB2, which is the second structure formed by the second meiosis, has a genetic structure similar to the female pronucleus of the zygote (n, c). Like PB1, PB2 is located in the periphery and has less amount of cytoplasm. In the PB2t method, the PB2 in the zygote of the patient is transferred to the embryo formed by the donor whose pronucleus has been removed previously. It can be thought that the PB2t method is similar to the PNT method in terms of zygote manipulation in both methods (Fig. 11.5) (Sendra et al., 2021).

Wang et al. reported that the polar body carries the genetic information necessary for development in the mice model. When PB1t and PB2t methods were compared with Mst and PNT methods, it was observed that there was a significant increase in blastocyst development and birth rates, especially in the PB1t method. It is known that one of the positive aspects of polar body transfer is the low rate of carry-over. Moreover, studies on mice have shown that this scenario lasts for two generations (Wang et al., 2014).

Data from various studies in non human experimental animal models provided guidance for human studies. Although the first PNT application performed on humans in 2003 resulted in pregnancy, no live birth was observed (Greenfield et al., 2017; Zhang et al., 2016). In 2016, the first live birth which was planned by 'spindle transfer' following IVF was shared by Zhang. Despite the ethical criticism at which the study was pointed, the clinical success it produced gave rise to an opportunity to assess progress. Today, discussions about the application of the method and possible measures for the safest application continue to be discussed. The Institute of Medicine (IOM) Committee of Ethical and Social Policy Consideration of Novel Techniques for Prevention of Maternal Transmission of Mitochondrial DNA Diseases (United States) granted permission for the initiation of clinical trials. The center has set an important framework for safety precautions and efficacy questions by ensuring compliance for MRT applications only in the presence of certain conditions and principles. In addition, the necessary conditions for long-term follow-up were provided for the studies (Dougherty et al., 2018).

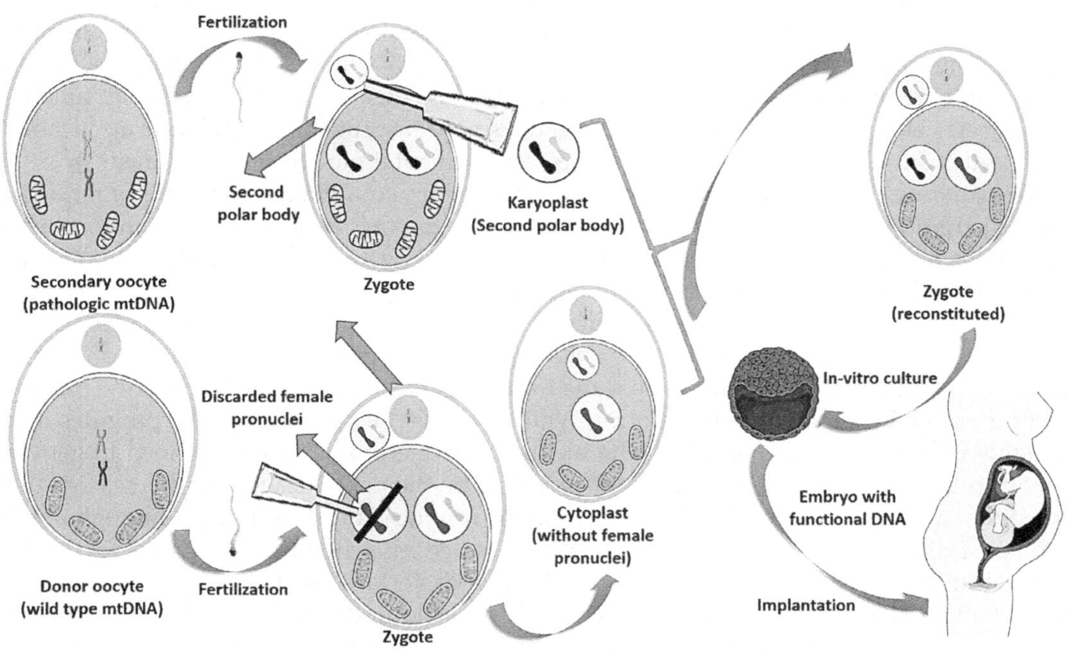

FIGURE 11.5 Second polar body transfer (PB2t). *Created with BioRender.com.*

Because there is no regulatory authority in the United States, professional organizations must collaborate with societies for FDA oversight to ensure research is in accordance with the IOM's recommendations. It is declared that a clear policy is created by the FDA to conduct clinical researches that can reduce the probability of disagreeable implementations of MRT (Claiborne et al., 2016; Falk et al., 2016). In light of the current shared data of Section 749 of the Consolidation Appropriation Act of 2016, MRT applications are declared inapplicable (Dougherty et al., 2018).

Consequently, it is known that MRT applications have been stopped in the United States. As a result of legal responsibilities, it was also restrictive for research on the subject. Especially, the long-term effects of the treatment on the new generation, the fertility level of the new generation—mitochondrial effects, and the possible consequences of nuclear—mitochondrial mismatch can be summarized as primarily threatening topics (Adashi & Cohen, 2017; Dougherty et al., 2018).

Related complications and risks of MRT

According to the data from animal experiments which are shared in literature, it shows that there are no side effects related to MRT application. Shared results on non human models similarly do not reveal increased offspring morbidity and mortality either. However, considering the stages of the method, nuclear and mtDNA incompatibility and the possible change in the fertility level in the next offspring can still be interpreted as disquieting. Henceforth, it was shared that in the follow-ups of the human stem cell line for approximately 1 year, the cells maintained both wild-type mitochondrial structure and normal mitochondrial functions and also they were able to differentiate into the human oocyte structure. Data from the studies planned between different species declared that MRT studies performed between species result in incompatibility, decreased fertility, and decreased mitochondrial functions (Dougherty et al., 2018). Although this result does not constitute data for clinical applications, it provides an important perspective for new experimental models to be planned.

Considering the unidirectional progression of heredity and the natural effect of mitochondrial distribution on genetics, it is obvious that the formation of original mutant mitochondria still poses a risk for the next generation. Although mutagenic mitochondria formation was observed following MRT application in animal models, this result was not phenotypically significant. As a result, it can be speculated that for the clinical applications in humans, only a male baby should be chosen for embryo transfer in order to prevent the transmission of heritable mutant mtDNA or to avoid the transfer of the abnormal mtDNA to the offspring during assisted reproductive techniques. While all the available information provides many insights for the application of MRT, the method will need to be further defined to get clear identification of the heritable effects of the method on the next lineage. For this reason, there are different concerns about MRT to be among the routine clinical methods. Although there is a distinct difference between MRT removal of mutated mitochondrial structures and editing of nuclear DNA with techniques such as CRISPR/Cas9, MRT is still a clinical intervention that changes the germline. For this reason, all possible undesirable results of the method should be clarified with further scientific studies (Dougherty et al., 2018).

Examples of regulatory framework for MRT

United Kingdom (UK) and Australia are known to be the first countries to provide legal regulations for clinical applications related to MRT treatments. In 2015, three independent reports were prepared by the Nuffield Council on Bioethics (2012), the Human Fertility and Embryology Authority (HFEA) (2012–13), and the Department of Health (2014). In this way, Human Fertilization and Embryology (Mitochondrial Donation) Regulations were published by the UK. In this way, important arrangements regarding MRT application have been achieved and it has been stated that the method can be performed in the prevention of serious mitochondrial diseases. In 2022, 27 MRT applications were completed for couples approved by the HFEA Statutory Approvals Committee (Castro, 2016; Fertilisation & Authority, 2022; Noohi et al., 2022).

Mitochondrial Donation Law Reform was submitted to parliament in 2020 with the publication of two reports for Australia (by the National Health and Medical Research Council). In 2021, the Australian Government shared that $10.3M is allocated for 10 years in (to) research and applications related to MRT. All necessary regulations and arrangements have now been provided for legal training and research related to MRT (Fertilisation & Authority, 2022; Noohi et al., 2022).

Different perspectives on MRT

MRT method has many different ethical concerns in terms of genetic and reproductive health. Mito-nuclear incompatibility, infertility, and possible effects on offspring may constitute common points of discussion that MRT is not appropriate in many situations (1–24–26). Apart from those gray areas, the need for serious time and financial resources for clinical research studies may also constitute a target for ethical criticism. Although there are different opinions, the limited number of approved patients suitable for MRT treatment can ethically criticize the planning of the mentioned time and financial resources for this finite group (Adashi & Cohen, 2017; Amato et al., 2014; Appleby, 2018; Baylis, 2013, 2017; Castro, 2016; Cecchino et al., 2018; Cheng et al., 2009; Chinnery & Hudson, 2013; Claiborne et al., 2016; Cohen et al., 1998; Dougherty et al., 2018; Engelstad et al., 2016; Falk et al., 2016; Fertilisation & Authority, 2022; Greenfield et al., 2017; Kang et al., 2016; Klitzman et al., 2015; Ma et al., 2017; McGrath & Solter, 1983; Neupane et al., 2014; Noohi et al., 2022; Paine & Jaiswal, 2016; Paull et al., 2013; Richardson et al., 2015; Sato et al., 2005; Sendra et al., 2021; Stewart & Chinnery, 2015; Tachibana et al., 2009, 2013, 2018; Tanaka et al., 2013; Wai et al., 2008; Wallace et al., 1988; Wang et al., 2014; Wu et al., 2017; Yamada et al., 2016; Zeviani et al., 1988; Zhang et al., 2016, 2017).

A different view questions the social consequences of germline changes. Although 'eugenics' and 'designer babies' titles can be considered more futuristic titles for today, the importance of legal regulations to prevent possible negative effects of the subject is already in question. The HFEA agency, which is in charge of overseeing MRT practices in the UK, controls both the government's authority over such facilities as well as the growth and development of these studies. In this respect, it is considered a great necessity to have similar pursuance and follow-up processes for ongoing research in each country to reach similar procedures (Dougherty et al., 2018).

The true genetic identity of the baby born as a result of MRT methods continues to get attention with increasing frequency. Two different views stand out in this regard: The first one is the opinion that the child born with the MRT method will be different from the child born with the diagnosis of mitochondrial disease. Thus, it is possible to take into consideration the worry that the child who will be born through the MRT procedure would have DNA from three parents. This concern can be encountered in many cases where assisted reproductive technologies are recommended (such as embryo selection using PGS-PGD methods, gamete donor) with the diagnosis of infertility. The units where these options are presented draw attention as an important point for transmission of accurate information to the couples as a matter of autonomous selection (Dougherty et al., 2018; Liao, 2017; Scully, 2017). The second, more concerning problem is the dominance of the three genetic parents. Considering the limited phenotypic effects of wild-type mitochondria in this regard, the genetic similarity of the unborn child should be expected to be similar to natural parents. Another issue is the consequences of known epigenetic effects on the embryo with multiple data. In this respect, the social, emotional, and personal characteristics of the new generations (offspring) will be expected to resemble their natural parents. The presence of oocyte donors is of great importance for the continuation of MRT therapy. This point has great importance for the feasibility of treatment. In a study, it was shown that 87% of individuals who volunteer as oocyte donors within the scope of assisted reproductive techniques can also volunteer for MRT applications (Dougherty et al., 2018). Beyond the required advanced technical implementation and monitoring processes, ethical restrictions on MRT practices are also of particular importance. It is known that there may be differences between countries on ethical issues. With the legal regulations introduced, MRT may be allowed in some countries in cases where alternative methods are insufficient or are not authorized (Adashi & Cohen, 2017; Dougherty et al., 2018). In this respect, the UK is the first country to allow MRT treatment in 2015, and Australia is the second country in this regard in 2021. The differences between regions with more strict reproductive laws and countries that do not have similar laws form the basis for addressing key ethical issues with MRT (Noohi et al., 2022).

Ethical issues related to MRT can be classified into two groups from different perspectives (Noohi et al., 2022).

(1) MRT is not a complete treatment or a cure for mitochondrial diseases; it aims to reduce the frequency
(2) The possibility of suggesting methods such as oocyte or embryo donation or adoption for parents who are candidates for MRT application (Noohi et al., 2022; Russell et al., 2020; Wolf et al., 2015).

MRT application is not a treatment method for mitochondrial diseases. It can be considered as a special assisted reproduction method to prevent the formation of mitochondrial diseases, which can have life-limiting effects and do not have a definitive treatment method. Considering that having a standard health level is a common right for every child, MRT can be considered as an opportunity for parents to offer this right to their children (Craven et al., 2018; Noohi et al., 2022).

A study planned to ask parents with a diagnosis of mitochondrial disease about their MRT preferences, the opinion that MRT would have an important place in the decision-making process of their children was observed more frequently (Castro, 2016). Mothers diagnosed with mitochondrial disease and experiencing painful processes as a result of changes in their own genetics stated that MRT application could be an important chance for their children to avoid these diseases under equal conditions. The statement belonging to one of the families who agreed to be included in the study is shared below with specific comments (Noohi et al., 2021, 2022):

> I'm gonna be honest with you. I have a daughter. I don't want her to live through what I lived through. God forbid if she doesn't make it to her child-bearing years. If she does, she will be informed that if she wants a child of her own, there are countries that could help her. We have money aside for that. I know that it's a very very sad thing to talk about, but I want her to have the opportunity to have and to carry a child of her own that will not land her in the hospital like I did. I think everybody is equally allowed to have that. [...] and what's great in her case is that she was born in 2011, she has a lot of time in front of her for this to become legal, or for research to advance. I didn't, but that's okay, we will deal with what it is now. But for her, there is a promising future, and a way to prevent it from continuing, right? ***Noohi et al. (2021, 2022)***

Although different approaches are observed in different geographical regions regarding the MRT application, the regulations made on this subject are of great importance. Adoption of the preventive approaches accepted by the UK regulations could be considered as a solution to the management of controversial topics on the subject (Noohi et al., 2022; Zhongming et al., 2020):

(a) Support from patient groups advocating technology
(b) Public participation on ethical aspects
(c) Safety and efficacy reviews conducted by independent expert panels
(d) Regulatory approval on a case-by-case basis

Conclusion

Mitochondrion, which is vital for the cellular cycle, is also known for its effect on oxidative stress mechanisms. Mitochondrial diseases, the frequency of which are increasing, are caused by mtDNA mutations and can affect all stages of life from birth to adult life. This diagnosis, which can be life-threatening, can progress with disabilities and constitutes a clinical problem of increasing importance. Despite all the information about mitochondrial diseases, the treatment process is still limited to approaches to alleviate symptoms. Early diagnosis methods applied on embryos PND and PGD. obtained within the scope of assisted reproductive techniques are insufficient to completely eradicate the disease due to their restrictive features.

It can be considered that the MRT method may be a novel choice for couples, who do not prefer gamete or embryo donation or adoption methods, to have a genetically related child away from mitochondrial diseases. The apparent desire of future parents to have children who are genetically similar to themselves and the fact that this issue concerns a wider area within the scope of assisted reproductive techniques brings about discussions on the subject. Therefore, it may be considered that MRT should not be a priority for the community when alternatives such as oocyte and/or embryo donation and/or adoption choices are available. However, as long as sufficient data on MRT applications cannot be collected, uncertainties regarding clinical applications in humans will continue to exist. Therefore, MRT should be carefully applied as a risk-reducing method in approved private couples. Considering the painful effects of mitochondrial diseases that can shorten life expectancy and cause disability, the necessity of the method emerges once again. With the increase in data on the MRT method, whose clinical applications are now starting, it is expected to create significant differences in the future of thousands of families with a diagnosis of mitochondrial disease (Noohi et al., 2022).

References

Adashi, E. Y., & Cohen, I. G. (2017). Mitochondrial replacement therapy: Born in the USA: The untold story of a conceptual breakthrough. *American Journal of Obstetrics and Gynecology, 217*(5), 561–563.

Amato, P., Tachibana, M., Sparman, M., & Mitalipov, S. (2014). Three-parent in vitro fertilization: Gene replacement for the prevention of inherited mitochondrial diseases. *Fertility and Sterility, 101*(1), 31–35.

Appleby, J. B. (Ed.). (2018). *Should mitochondrial donation be anonymous? The journal of medicine and philosophy: A forum for Bioethics and philosophy of medicine*. Oxford University Press US.

Baylis, F. (2017). Human nuclear genome transfer (So-Called mitochondrial replacement): Clearing the underbrush. *Bioethics, 31*(1), 7–19.

Baylis, F. (2013). The ethics of creating children with three genetic parents. *Reproductive BioMedicine Online, 26*(6), 531–534.

Castro, R. J. (2016). Mitochondrial replacement therapy: The UK and US regulatory landscapes. *Journal of Law and the Biosciences, 3*(3), 726–735.

Cecchino, G. N., Seli, E., da Motta, E. L. A., & García-Velasco, J. A. (2018). The role of mitochondrial activity in female fertility and assisted reproductive technologies: Overview and current insights. *Reproductive BioMedicine Online, 36*(6), 686–697.

Cheng, Y., Wang, K., Kellam, L. D., Lee, Y. S., Liang, C.-G., Han, Z., et al. (2009). Effects of ooplasm manipulation on DNA methylation and growth of progeny in mice. *Biology of Reproduction, 80*(3), 464–472.

Chinnery, P. F., & Hudson, G. (2013). Mitochondrial genetics. *British Medical Bulletin, 106*(1), 135–159.

Claiborne, A., English, R., & Kahn, J. (2016). *Mitochondrial replacement techniques: Ethical, social, and policy considerations.*

Cohen, J., Scott, R., Alikani, M., Schimmel, T., Munné, S., Levron, J., et al. (1998). Ooplasmic transfer in mature human oocytes. *Molecular Human Reproduction, 4*(3), 269–280.

Craven, L., Murphy, J., Turnbull, D. M., Taylor, R. W., Gorman, G. S., & McFarland, R. (2018). Scientific and ethical issues in mitochondrial donation. *The New Bioethics, 24*(1), 57–73.

Dougherty, M. P., Dolitsky, S., Chattopadhyay, R., & Sauer, M. V. (2018). Mitochondrial replacement therapy. *Current Opinion in Obstetrics and Gynecology, 30*(4), 217–222.

Engelstad, K., Sklerov, M., Kriger, J., Sanford, A., Grier, J., Ash, D., et al. (2016). Attitudes toward prevention of mtDNA-related diseases through oocyte mitochondrial replacement therapy. *Human Reproduction, 31*(5), 1058–1065.

Falk, M. J., Decherney, A., & Kahn, J. P. (2016). Mitochondrial replacement techniques—implications for the clinical community. *New England Journal of Medicine, 374*(12), 1103.

Fertilisation, H., & Authority, E. (2022). *Re: Mitochondrial donation_ 2022–23-excel-0111. Received by forough Noohi* (Vol. 23).

Greenfield, A., Braude, P., Flinter, F., Lovell-Badge, R., Ogilvie, C., & Perry, A. C. (2017). Assisted reproductive technologies to prevent human mitochondrial disease transmission. *Nature Biotechnology, 35*(11), 1059–1068.

Kang, E., Wu, J., Gutierrez, N. M., Koski, A., Tippner-Hedges, R., Agaronyan, K., et al. (2016). Mitochondrial replacement in human oocytes carrying pathogenic mitochondrial DNA mutations. *Nature, 540*(7632), 270–275.

Klitzman, R., Toynbee, M., & Sauer, M. V. (2015). Controversies concerning mitochondrial replacement therapy. *Fertility and Sterility, 103*(2), 344–346.

Liao, S. M. (2017). Do mitochondrial replacement techniques affect qualitative or numerical identity? *Bioethics, 31*(1), 20–26.

Ma, H., O'Neil, R. C., Gutierrez, N. M., Hariharan, M., Zhang, Z. Z., He, Y., et al. (2017). Functional human oocytes generated by transfer of polar body genomes. *Cell Stem Cell, 20*(1), 112–119.

McGrath, J., & Solter, D. (1983). Nuclear transplantation in the mouse embryo by microsurgery and cell fusion. *Science, 220*(4603), 1300–1302.

Neupane, J., Vandewoestyne, M., Ghimire, S., Lu, Y., Qian, C., Van Coster, R., et al. (2014). Assessment of nuclear transfer techniques to prevent the transmission of heritable mitochondrial disorders without compromising embryonic development competence in mice. *Mitochondrion, 18*, 27–33.

Noohi, F., Li, M., & Joly, Y. (2021). Clinical translation of mitochondrial replacement therapy in Canada: A qualitative study of stakeholders' attitudes. *Facets, 6*(1), 449–464.

Noohi, F., Ravitsky, V., Knoppers, B. M., & Joly, Y. (2022). Mitochondrial replacement therapy: In whose interests? *Journal of Law Medicine and Ethics, 50*(3), 597–602.

Paine, A., & Jaiswal, M. K. (2016). Promise and pitfalls of mitochondrial replacement for prevention and cure of heritable neurodegenerative diseases caused by deleterious mutations in mitochondrial DNA. *Frontiers in Cellular Neuroscience, 10*, 219.

Paull, D., Emmanuele, V., Weiss, K. A., Treff, N., Stewart, L., Hua, H., et al. (2013). Nuclear genome transfer in human oocytes eliminates mitochondrial DNA variants. *Nature, 493*(7434), 632–637.

Richardson, J., Irving, L., Hyslop, L. A., Choudhary, M., Murdoch, A., Turnbull, D. M., et al. (2015). Concise reviews: Assisted reproductive technologies to prevent transmission of mitochondrial DNA disease. *Stem Cells, 33*(3), 639–645.

Russell, O. M., Gorman, G. S., Lightowlers, R. N., & Turnbull, D. M. (2020). Mitochondrial diseases: Hope for the future. *Cell, 181*(1), 168–188.

Sato, A., Kono, T., Nakada, K., Ishikawa, K., Inoue, S.-I., Yonekawa, H., et al. (2005). Gene therapy for progeny of mito-mice carrying pathogenic mtDNA by nuclear transplantation. *Proceedings of the National Academy of Sciences, 102*(46), 16765–16770.

Scully, J. L. (2017). A mitochondrial story: Mitochondrial replacement, identity and narrative. *Bioethics, 31*(1), 37–45.

Sendra, L., García-Mares, A., Herrero, M. J., & Aliño, S. F. (2021). Mitochondrial DNA replacement techniques to prevent human mitochondrial diseases. *International Journal of Molecular Sciences, 22*(2), 551.

Stewart, J. B., & Chinnery, P. F. (2015). The dynamics of mitochondrial DNA heteroplasmy: Implications for human health and disease. *Nature Reviews Genetics, 16*(9), 530–542.

Tachibana, M., Amato, P., Sparman, M., Woodward, J., Sanchis, D. M., Ma, H., et al. (2013). Towards germline gene therapy of inherited mitochondrial diseases. *Nature, 493*(7434), 627–631.

Tachibana, M., Kuno, T., & Yaegashi, N. (2018). Mitochondrial replacement therapy and assisted reproductive technology: A paradigm shift toward treatment of genetic diseases in gametes or in early embryos. *Reproductive Medicine and Biology, 17*(4), 421–433.

Tachibana, M., Sparman, M., Sritanaudomchai, H., Ma, H., Clepper, L., Woodward, J., et al. (2009). Mitochondrial gene replacement in primate offspring and embryonic stem cells. *Nature, 461*(7262), 367–372.

Tanaka, A. J., Sauer, M. V., Egli, D., & Kort, D. H. (2013). Harnessing the stem cell potential: The path to prevent mitochondrial disease. *Nature Medicine, 19*(12), 1578–1579.

Wai, T., Teoli, D., & Shoubridge, E. A. (2008). The mitochondrial DNA genetic bottleneck results from replication of a subpopulation of genomes. *Nature Genetics, 40*(12), 1484–1488.

Wallace, D. C., Singh, G., Lott, M. T., Hodge, J. A., Schurr, T. G., Lezza, A. M., et al. (1988). Mitochondrial DNA mutation associated with Leber's hereditary optic neuropathy. *Science, 242*(4884), 1427–1430.

Wang, T., Sha, H., Ji, D., Zhang, H. L., Chen, D., Cao, Y., et al. (2014). Polar body genome transfer for preventing the transmission of inherited mitochondrial diseases. *Cell, 157*(7), 1591–1604.

Wolf, D. P., Mitalipov, N., & Mitalipov, S. (2015). Mitochondrial replacement therapy in reproductive medicine. *Trends in Molecular Medicine, 21*(2), 68–76.

Wu, K., Zhong, C., Chen, T., Zhang, X., Tao, W., Zhang, J., et al. (2017). Polar bodies are efficient donors for reconstruction of human embryos for potential mitochondrial replacement therapy. *Cell Research, 27*(8), 1069–1072.

Yamada, M., Emmanuele, V., Sanchez-Quintero, M. J., Sun, B., Lallos, G., Paull, D., et al. (2016). Genetic drift can compromise mitochondrial replacement by nuclear transfer in human oocytes. *Cell Stem Cell, 18*(6), 749–754.

Zeviani, M., Moraes, C. T., DiMauro, S., Nakase, H., Bonilla, E., Schon, E., et al. (1988). Deletions of mitochondrial DNA in kearns-sayre syndrome. *Neurology, 38*(9), 1339.

Zhang, J., Zhuang, G., Zeng, Y., Grifo, J., Acosta, C., Shu, Y., et al. (2016). Pregnancy derived from human zygote pronuclear transfer in a patient who had arrested embryos after IVF. *Reproductive BioMedicine Online, 33*(4), 529–533.

Zhang, S.-P., Lu, C.-F., Gong, F., Xie, P.-Y., Hu, L., Zhang, S.-J., et al. (2017). Polar body transfer restores the developmental potential of oocytes to blastocyst stage in a case of repeated embryo fragmentation. *Journal of Assisted Reproduction and Genetics, 34*, 563–571.

Zhongming, Z., Linong, L., Xiaona, Y., & Wei, L. (2020). *Heritable human genome editing*.

CHAPTER

12

Application route of mitochondrial transplantation

Meghnad G. Joshi[1,2], *Mrunal N. Damle*[1,2] *and Rakesh Kumar Sharma*[3]

[1]Stem Plus Biotech Pvt. Ltd., Research and Development, Sangli, Maharashtra, India; [2]Department of Stem Cell and Regenerative Medicine, D.Y. Patil Education Society (Deemed University), Kolhapur, Maharashtra, India; [3]Department of Obstetrics and Gynecology, D.Y. Patil Medical College, Kolhapur, Maharashtra, India

Introduction

Mitochondrion, a symbiotic organelle, has energy producing system composed of NADH (nicotinamide adenine dinucleotide (NAD) + hydrogen (H)) dehydrogenase, cytochrome, and cytochrome oxidase for most cells. During the process of oxidative phosphorylation carbohydrates, amino acids and lipids are imported into the cell and converted to pyruvate, fatty acids and amino acids, which are oxidized in the tricarboxylic acid cycle and ATP is produced on the mitochondrial matrix (Burke, 2017). Thus, mitochondria play important roles in energy production, but also in critical processes such as cell death, differentiation, signaling, metabolic homeostasis, and innate immunity Zhang, Liu et al. (2023). Now, mitochondria are no longer viewed only as the energy machinery of the cell but also as a vital source of dynamic microenvironmental signals (Bennett et al., 2022; Brand, 2020; Shen et al., 2022). Cells use ATP as an energy source to synthesize proteins, mechanical motion, sustain homeostasis, and perform other life supporting functions (Chakrabarty & Chandel, 2022).

Mitochondria contain genetic material (mitochondrial DNA, mtDNA), which only encodes for 13 electron transport chain components, 2 rRNAs, and 22 tRNAs (Gray, 2012). The nuclear genome is used to translate proteins essential for mitochondrial biogenesis. During the process of oxidative phosphorylation, reactive oxygen species (ROS) are generated as by products which are involved in diverse cellular processes. Interestingly, ROS are considered as

Mitochondrial Transplantation and Transfer
https://doi.org/10.1016/B978-0-443-18858-9.00012-3

secondary messengers in receptor mediated cell signaling during different physiological functions. Mitochondria also play a key role in Ca^{2+} homeostasis through regulating cytosolic concentration. Retrograde signaling from mitochondria to the nucleus, generally in the form of metabolites, translocated mitochondrial proteins, and mitochondrial ROS regulates a variety of cellular processes such as cell cycling, apoptosis, immunological response, and epigenetic control (Fig. 12.1) (Tan & Finkel, 2020; Weinberg et al., 2015).

The mitochondrial genome's high mutation rate is the most likely due to a less efficient damage repair system mixed with continuous and random replication of mtDNA molecules (Pham et al., 2022; D'Amato et al., 2023). Dysregulation of mitochondrial-nuclear communication has been demonstrated to play a direct role in aging and several pathologies (Liesa et al., 2009; Murphy & Hartley, 2018; Nunnari & Suomalainen, 2012). Mitochondria respond adaptively to their host cell's changing needs in order to preserve bioenergetic and oxidative equilibrium. Oxidative stress and protein denaturation are caused by a variety of physiological stresses. As a result, various stresses frequently induce cellular adaptation responses, such as the creation of stress proteins, the upregulation of oxidative stress protectors, and the accumulation of protective solutes. In many circumstances, mitochondria, particularly

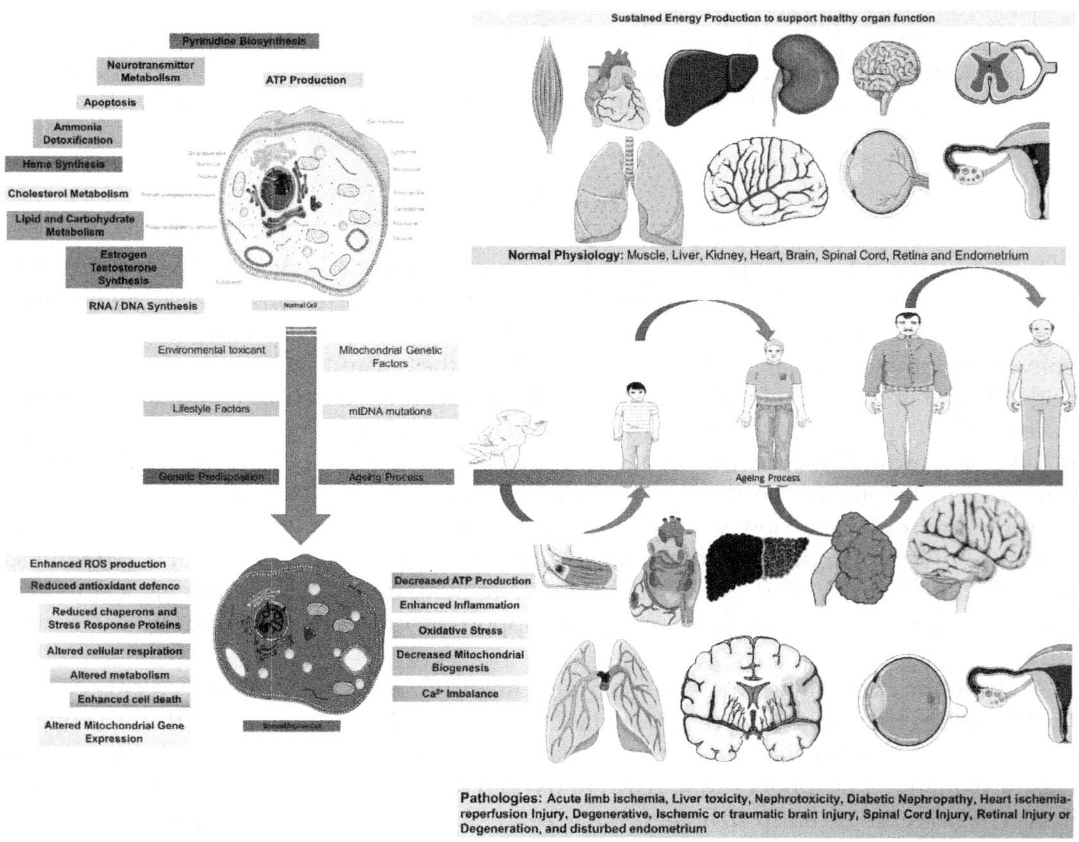

FIGURE 12.1 Mitochondria in homeostasis, and in dysfunctional mitochondria.

mitochondrial electron transport, are critical sites of damage during physiological stress. Hepatocytes, cardiomyocytes, neuronal, renal, and pancreatic cells are the examples of cells that require a lot of ATP molecules. They are particularly prone to destruction if their ability to synthesize ATP is compromised. Since mitochondria are involved in nearly every aspect of cellular function, it is not surprising that they have been associated with a variety of common disorders, including neurodegenerative diseases, cardiac diseases, and metabolic diseases (Fig. 12.1) (Blajszczak & Bonini, 2017; Chaudhari et al., 2023; Tan & Finkel, 2020; Wang et al., 2019; Wang & Gerdes, 2015; Zhang, Sun, et al., 2023).

In order to optimize cellular adaptability to energy and metabolic demands, mitochondrial biogenesis is regulated in a tissue and signal specific way (Hock & Kralli, 2009). Mitochondrial biogenesis can be activated by stimuli including cell division, development, exercise, thermogenesis, postnatal breathing, thyroid hormone and erythropoietin secretion, calorie restriction, oxidative stress, and inflammation (Piantadosi & Suliman, 2012). Cellular programs that protect the quality of the cell's mitochondria by eliminating of severely damaged organelles and replacing them with new ones are intended to mitigate the harmful effects of inflammation on mitochondria. This mitochondrial quality control process involves a complex bigenomic program that regulates mtDNA transcription, replication, and protein synthesis in the mitochondria. The precise coordination of nuclear DNA (nDNA) and mtDNA transcription enables mitochondrial adaptation to physiological and pathological changes, and also adequate performance (Piantadosi & Suliman, 2012). Major players in mitochondrial biogenesis regulation include peroxisome proliferator-activated receptor gamma coactivator-1 (PPARGC1A/PGC-1), nuclear respiratory factors (NRF1 and NFE2L2/NRF2), and mitochondrial transcription factor A, (TFAM), and posttranslational modifications of PGC-1α (Hock & Kralli, 2009). The biological program is intended to respond immediately to specific cell signals to attempt to adapt mitochondrial mass, number, size, distribution, and/or phenotype (Piantadosi & Suliman, 2012; Scarpulla, 2008). Mitochondrial biogenesis is a long term adaptive response that is not necessarily required to meet transiently increased energy demands. Temporary variations in energy demand can be met by increasing the expression of a selection of mitochondrial genes or essential regulators, as well as improving mitochondrial performance (Scarpulla, 2008).

Based on this evidence, the researchers proposed the innovative approach of directly transplanting viable mitochondria into diseased tissue in order to explore the therapeutic potential of transplanted mitochondria rather than pharmaceuticals or whole cells. In this chapter, we discuss the existing methods of mitochondrial transplantation (MT), the treatment modality of transferring entire healthy mitochondria to recipient tissue, and the therapeutic implications. In addition, we also discuss the clinical outcome of MT and ongoing clinical trials.

Application routes of mitochondrial transplantation

Mitochondrial disease triggers inefficient energy production and the loss of normal physiological ROS function. As a result, a new paradigm of mitochondrial therapy was devised, based on an organelle delivery approach. It was helpful to enhance energy production, to stop excessive ROS production, and to restore mitochondrial function by adding healthy mitochondria to cells that had previously damaged mitochondria. MT comprises the isolation of healthy mitochondria from autologous or allogenic source and injecting it into diseased tissue.

Mitochondrial transport is essential in many physiological and pathological situations. The most explored method supporting mitochondria internalization includes tunneling nanotubes/cytoplasmic bridges, gap junctions, cell fusion, vesicle endocytosis, or as free organelles (Caicedo et al., 2017; Dong et al., 2023; Herst et al., 2018; Islam et al., 2012; Wang & Gerdes, 2015). According to published reports, damaged cells can also accept exogenous functional mitochondria from a donor cell and integrate them into their endogenous mitochondrial networks to improve biological activities.

The fate of transplanted mitochondria can be efficiently monitored by membrane potential dependent florescent dyes such as Rhodamine 123, tetramethylrhodamine methyl ester, and tetramethylrhodamine ethyl ester dyes (Chazotte, 2011; Chen, 1988; Kubat et al., 2020). A commercially available fluorescent dye MitoTracker links to the thiol groups in the mitochondria. These dyes allow monitoring the dynamics of mitochondria after transplantation and integrity inside the cell over time (Kubat et al., 2020). In our studies, we used rhodamine 123 to label human placental derived mitochondria. The rhodamine 123 effectively showed transplanted human mitochondria cluster in endometrium till 72 h (Kshersagar et al., 2023).

The applications of autologous mitochondria in transplants have special therapeutic potential. Masuzawa et al. (2013) found that animals receiving MT had considerably lower levels of inflammatory markers (TNF, IL-6, IL-10, MCP-1, and hsCRP). There were similar findings which demonstrated that autologous MT did not trigger an immunological response in numerous animal models (Doulamis & McCully, 2021; McCalley, 2018; Shin et al., 2017; Yamada et al., 2020). Additionally, using single and repeated intraperitoneal injections in mice, different researchers studied the immunological response to isolated allogenic mitochondrial gastrocnemius and quadriceps femoris muscles. As a result, there was no direct or indirect, acute or chronic alloreactivity, immunological reactivity of certain classes of T lymphocytes against transplanted mitochondria, or damage-associated molecular pattern molecule (DAMP) reactions to single or serial injections of autogenetic or allogeneic mitochondria (Kshersagar et al., 2023; Ramirez-Barbieri et al., 2019). Ramirez-Barbieri et al. (2019) on the other hand, we have seen extensive skeletal muscle tissue destruction in some of our sickest patients, which makes it challenging to isolate live mitochondria. In these cases, serial injections may be helpful. In addition, the use of autologous mitochondria would not be feasible in patients with mitochondrial myopathies brought on by pathogenic mutations in mtDNA (Phadke, 2017; Wallace, 2016). In these circumstances, an allogeneic source would be a potential substitute for healthy mitochondria. Additionally, the repeated injection can enhance the use of MT in therapeutic treatments.

Our findings indicate that injecting 1×10^6 human placenta derived mitochondria into a damaged rat endometrium does not elicit an immunological response. The absence of any early signs of immunological reactions associated with MT was particularly noteworthy; this could be attributed to the relatively small number of mitochondria for transplants (Kshersagar et al., 2023). The use of allogenic mitochondria from sources like the placenta, which has the potential to produce enough mitochondria for patients, could be a future remedy.

Many promising breakthroughs in mitochondrial treatments have boosted work toward treating various diseases caused by tissue malfunction. A considerable amount of research is dedicated to advancing the understanding of organ regeneration following mitochondrial transplantation.

The routes of administration of mitochondria have been shown in different studies such as direct injection, systemic injection, and nasal delivery mainly. Reports of the application routes of MT used in animal models of disease are summarized in Tables 12.1–12.3.

Direct injection route for MT

Numerous animal studies demonstrated the positive effects of MT in disease animal models. The effective applications of MT to treat a variety of pathologies, including acute limb ischemia, liver toxicity, nephrotoxicity, heart ischemia/reperfusion (IR) injury, ischemic or traumatic brain injury, spinal cord injury (SCI), retinal injury or degeneration, diabetic nephropathy (DN), and disturbed endometrium, have demonstrated that it leads in the functional recovery of the tissues by enhancing mitochondrial function, reducing inflammatory levels, and lowering cell apoptosis.

In different models, the direct injection was employed to introduce isolated mitochondria into recipient cells using microinjection or tissues via needle. Direct injection into the tissue is desirable in the event of local damage since it allows for the quick delivery of many healthy mitochondria (Fig. 12.2).

The following section will specifically discuss the direct injection route for MT in animal models for various pathological conditions (Table 12.1).

Skeletal muscle

Skeletal muscle IR damage occurs when blood flow and oxygenation to the extremities are significantly reduced, resulting in muscle viability and limb function loss. Orfany et al. (2020) demonstrated use of allogenic muscle derived mitochondria for the treatment of acute limb ischemia in murine model. This study was carried out in C57BL/6J male mice (10–12 weeks) and ischemia was induced for 2 h with a 4.5 ounce rubber band (tourniquet) placed at the hip joint of the left hindlimb using a McGivney hemorrhoidal ligator. Perfusion in the hind limb was assessed by three dimensional (3D) ultrasound imaging using contrast enhanced ultrasound and 3D color Doppler ultrasound before and during ischemia, immediately after reperfusion, and at 15 min of reperfusion. After 15 min of reperfusion, isolated mitochondria in six injections were delivered to the hindlimb muscles individually by use of a tuberculin syringe (0.5 mL of buffer/per injection). Mitochondria were delivered separately at concentrations of 1×10^6, 1×10^7, 1×10^8, and 1×10^9 mitochondria per gram muscle wet weight.

Analysis of mitochondrial uptake revealed that injected mitochondria (^{18}F-rhodamine 6G-labeled) were distributed primarily within myofibers in soleus, vastus medialis, and gastrocnemius. A significant reduction in infarct size was observed between mice receiving 1×10^9 mitochondria which improved infarct size and ATP content. MT mice showed little muscle damage and no difference in cellular influx. Interestingly, MT mice showed reduction in

TABLE 12.1 Direct injection route of mitochondrial transplant for various pathological conditions.

No.	Source of mitochondria	Route of mitochondrial transplantation	Animal model	Mitochondrial Dose	Therapeutic Outcome	References
Skeletal muscle						
1	Muscle from donor C57BL/6J mice	Injected directly into all the muscles of the hindlimb	C57BL/6J mice model of acute limb ischemia	1×10^6 to 1×10^9 g/wet weight to each muscle injecting 50 mL per injection in the following volumes: 100 mL to gastrocnemius; 150 mL to quadriceps; 50 mL to soleus; and 100 mL to semimembranosus, semitendinosus, and biceps femoris	MT ameliorates skeletal muscle injury and enhances hindlimb function	Orfany et al. (2020)
Liver						
2	Fresh rat liver	Intrasplenic injection	Male Wistar rats partial liver ischemia	$7.7 \times 10^6 \pm 1.5 \times 10^6$/mL after 45 min of liver ischemia.	Attenuation of hepatic injury	Lin et al. (2013)
3	Rat mesenchymal stem cells	Subcapsular region of the spleen	Acetaminophen-induced liver toxicity, Sprague Dawley (SD) male rats	8.2×10^6 isolated mitochondria/100 μL respiration buffer/rat	Decreased plasma ALT levels, apoptotic cells, and total oxidant levels, increased GSH levels. Histological and functional improvement.	Ulger et al. (2021)
Kidney						
4	BM-MSCs	Directly injected under the renal capsule of the left kidney	Diabetic nephropathy induced with streptozotocin in male SD rats	From 1×10^6 BM-MSCs	Improved cellular morphology prevented apoptosis	Konari et al. (2019)
5	MSCs	Transplanted directly into the cortex of right and left kidneys	Doxorubicin (DOX) induced nephrotoxicity in male SD rats	8×10^6/mL isolated mitochondria in 500 μL of respiration buffer	Improved cellular OS regeneration of tubular cells, Reverted renal deficits, Bcl-2, caspase-3 levels	Kubat et al. (2020, 2021)

Heart						
6	Autologous pectoralis major muscle tissue	Prolene thread (3–0) used for the snare for 30 min of regional ischemia in heart	In situ blood-perfused regional ischemia in male New Zealand white rabbits	8×0.1 mL injections; $9.7 \times 10^6 \pm 1.7 \times 10^6$/mL	Reserving myocardial energetics, cell viability, and enhanced postinfarct cardiac function.	Masuzawa et al. (2013)
7	Autologous pectoralis major muscle tissue	Artery was temporarily occluded for 25 min and regional ischemia was induced	Yorkshire pigs	8×0.1 mL injections; $9.9 \times 10^7 \pm 1.4 \times 10^7$/mL; 1.3×10^7 mitochondria per injection site	Enhance myocardial cell viability, enhanced postinfarct cardiac function	Kaza et al. (2017)
8	Human umbilical cord-derived MSCs (hUC-MSCs) w	Icv into the right lateral ventricle (0.8 mm posterior, 1.5 mm lateral and 4.0 mm ventral to the bregma according to Paxinos atlas)	Adult Wistar rats	Mitochondria were isolated from 3×10^7 MSCs. Immediately after reperfusion (within 10 min), in 10 μL	Mitochondrial transplantation ameliorated the ischemia/reperfusion (IR) induced damages as reflected by declined blood creatine phosphokinase level, abolished apoptosis, decreased astroglyosis and microglia activation, reduced infarct size, and improved motor function	Pourmohammadi-Bejarpasi et al. (2020)
9	Ventricular cardiomyocytes	Myocardium of the left ventricle during perfusion at 4 different points.	DOX challenge, male C57BL/6 mice	Total of 50,000 isolated mitochondria suspended in 80 mL; 20 mL per injection site	Protects cardiac function in vivo and prevents myocardial apoptosis	Zhang, Sun et al. (2023), Zhang, Liu et al. (2023)
Brain						
10	Mitochondria derived from a hamster cell line	Ischemic rat brains and rat primary cortical neurons	Male SD rats	75 μg of mitochondria (1.2×10^6 mitochondria particles) directly into the ischemic striatum and 750 μg mitochondria (in 100 μL) was infused into the femoral artery	Attenuation of brain infarction area and neuronal death, restoration of motor performance	Huang et al., 2016

(Continued)

TABLE 12.1 Direct injection route of mitochondrial transplant for various pathological conditions.—cont'd

No.	Source of mitochondria	Route of mitochondrial transplantation	Animal model	Mitochondrial Dose	Therapeutic Outcome	References
11	PC12 cells	Medial forebrain bundle	Adult female SD rats	The 105 μg allogeneic mitochondria (Peptide-mediated allogeneic mitochondrial delivery)	Mitochondrial dynamics were restored, attenuation of oxidative damage, and alleviating 6-OHDA–induced degeneration of dopaminergic neurons.	Chang et al. (2016)
12	Autologous pectoralis major muscle tissue	Delivery of mitochondria through the lateral ventricles	Male SD rats	Unilateral injection with $5 \times 10^6/10\ \mu L$ mitochondrial particles for 10 min.	Neuroprotective effect after IR injury. Decreased brain infarct volume and reversed neurological deficits.	Zhang et al. (2019)
13	Mitochondria derived from allogeneic liver, and autogenic muscle.	Directly injecting mitochondria into cerebral cortex.	Traumatic brain injury, adult male C57BL mice	1.2 to 1.4×10^6	Mitochondria may be effective in reducing the secondary injury, may function independently or fuse with endogenous mitochondria in recipient cells to generate ATP, promoting cell survival and enhancing cellular resistance to stress.	Zhao, Z., et al. (2021)
14	The brain of young rats	Administered into the right cerebral ventricle	Male Wistar rats model of chronic mild stress	10 μL in saline suspension	Alleviated chronic stress –induced oxidative damage. Reduced apoptosis.	Javani et al. (2023)
Spinal cord injury						
15	PC-12 cells and the soleus muscle were dissected from naïve SD rats	The mediolateral gray matter at 4 circumferential sites of the injured cord separated by 2 mM in the rostral caudal direction	Female naïve SD rats	A total of 50, 100 or 150 μg in 3 μL volume	The mitochondrial transplantation did not yield long-term functional neuroprotection as assessed by overall tissue sparing or recovery of motor and sensory functions.	Gollihue et al. (2018)

16	Rat bone marrow mesenchymal stem cells	Spinal cord injury (SCI) injected into the epicenter of the injured spinal cord	SD male rats SCI model.	3×10^6 in 10 mL	Improved locomotor functional recovery improved the bioenergetics profile, decreased apoptosis, and promoted cell survival	Li et al. (2019)
17	Allogeneic mitochondria were freshly isolated from the bilateral soleus muscles of donor rats	Injured region of spinal cord of rats via the intraparenchymal route.	Adult SD rats, SCI	50 μg in 1.5 μL 1× PBS, each shot, two injections	SCI-induced somatosensory and functional impairments in rats. Repressed mitochondrial fragmentation, apoptosis, oxidative stress, and inflammation	Lin et al. (2022)
Retina						
18	Allogenic rat liver	Intravitreal injection	Females Lister Hooded rats, Optic nerve crush	0.25 mg/mL in saline.	Mitochondria incorporate into the retina, improve its oxidative metabolism and electrophysiological activity at 1 day after transplantation.	Nascimento-Dos-Santos et al. (2020)
Lung						
19	Freshly isolated airway epithelium rat	Intratracheally	SD rats airway hyperresponsiveness with AHR by cigarette smoke and LPS intratracheal instillation	3.75×10^8/mL	Rescued the chronic obstructive pathological alternations, including the AHR, the airway remodeling, the ciliary morphology, and the inflammatory cytokine expression.	Su et al. (2016)
Endometrium						
20	Human full-term placenta	Intrauterine infusion	Endometrial injury female, adult Wistar rats	1×10^6 suspended in 100 μL medium	Improve endometrial junction protein thus regeneration in the disturbed endometrium	Kshersagar et al. (2023)

TABLE 12.2 Systemic injection route of mitochondrial transplant for various pathological conditions.

No.	Source of mitochondria	Route of mitochondrial transplantation	Animal model	Mitochondrial dose	Therapeutic outcome	References
Skeletal muscle						
1	Liver mitochondria were isolated from PhAM B6; 129S-Gt(ROSA) $26Sor^{tm1.1(CAG-COX8A/Dendra2)Dcc}$/J or C57BL/6 mice	Injected in the tail vein	C57BL/6 mice $BaCl_2$to cause muscle injury	50 μg of MitoTracker-labeled mitochondria	Systemic mitochondria delivery can enhance the rate of muscle regeneration and restoration of muscle function following injury	Always et al. (2023)
Liver						
2	HepG2 cells	Intravenous tail injection	C57BL/6J mice administration of a high-lard-fat and high-cholesterol diet	0.5 mg/kg body weight	Mitotherapy decreased lipid content and restored cellular redox balance, consequently attenuated lipid deposit and prevented cell injury	Fu et al. (2017)
3	HepG2 cells	Intravenous tail injection	C57BL/6J mice acetaminophen-induced liver injury	10^{12} in saline	Attenuation of tissue injury and enhancement of hepatocyte metabolism	Shi et al. (2018)
4	CD-1 mice liver	Intravenous tail injection	SD rats	0.5 mL of purified at a steady rate of 160 μL/min.	Targetable mitochondrial complexes can be delivered to rat liver. 27% of the total injected mitochondria was detected in the liver, 2% was found in spleen, and <1% in lungs	Liu et al. (2020)
5	Donor SD rats	Portal vein administration	Adult male SD rats with acute liver IR injury	10 mg of isolated mitochondria	Transplanted mitochondria (pretreated with melatonin) effectively protected the liver against IR injury mainly through refreshment of the endogenous mitochondria which were at the gate of death in IR injury hepatocytes.	Ko et al. (2020)

6	Mice liver	Intravenous injection	Healthy Kunming mice carbon tetrachloride (CCl_4) liver injury	1 mg/mL equals to 1.6×10^8 mitochondria 0.2 mg/kg (mito-low) or 0.4 mg/kg (mito-high) in saline	Increasing bioenergy supply and diminishing free radicals. The protective response to cell survival. Prevents liver cell injury	Zhao, Z. et al. (2021)
Kidney						
7	Rat liver	Intraperitoneal injection in 1 mL of saline	Kidney IR (40 min) outbred white male rats	Dose of 10 mg mitochondrial protein per 100 g body weight.	Increased animal survival, decrease in markers of renal failure and inflammation	Plotnikov et al. (2020)
8	Autologous pectoralis major muscle cells.	Renal artery	Kidney IR (45 min) Wistar male rats	7.5×10^6 mitochondria/mL using a 32 gauge needle during 2 min.	Restores renal function, and enhances renal cells repair and proliferation capacity.	Jabbari et al. (2020)
9	Autologous sternocleidomastoid muscle	Renal artery	Female Yorkshire pig renal ischemia was achieved by balloon catheter occlusion of the renal arteries	A single bolus injection of 1×10^9 mitochondria in 6 mL vehicle or as serial injections with three injections of 1×10^9 in 6 mL vehicle each injection delivered every 20 min.	Renal protection from IR injury, significantly enhancing renal function and reducing renal damage.	Doulamis, Guariento, Duignan, Orfany, et al. (2020), Doulamis, Guariento, Duignan, Kido, et al. (2020)
Heart						
10	Human adult cardiac fibroblasts	Coronary artery injection	Female New Zealand white rabbits global ischemia was induced by cross-clamping the aorta	1×10^8 human mitochondria labeled with ^{18}F-rhodamine 6G (^{18}F-R6G) ± iron oxide nanoparticles	Reduction in infarct size and enhancement of cardiac function	Cowan et al. (2016)
11	C57BL/6J donors,	Coronary artery injection	Heterotopic heart transplantation	1×10^8 in respiration buffer delivered antegrade to the coronary arteries via injection to the coronary ostium. A second injection of either mitochondria (1×10^8) delivered antegrade to the coronary arteries 5 min after transplantation.	Beating score, calculated ejection fraction significantly enhanced and neutrophil infiltration significantly decreased.	Moskowitzova et al. (2019)

(Continued)

TABLE 12.2 Systemic injection route of mitochondrial transplant for various pathological conditions.—cont'd

No.	Source of mitochondria	Route of mitochondrial transplantation	Animal model	Mitochondrial dose	Therapeutic outcome	References
12	Heart pectoralis major muscle and human cardiac fibroblasts	Delivered retrograde to the coronary arteries through the aortic cannula	Zucker diabetic fatty rats and nondiabetic Zucker lean rats	1 mL of vehicle solution containing $2 \times 10^5 \pm 0.3 \times 10^5$ mitochondria/g wet weight was resuspended in 2 mL of 37°C Krebs–Ringer solution and administered antegradely to the coronary arteries through the aortic cannula over 10 s.	Cardioprotective, increased ATP content	Doulamis, Guariento, Duignan, Orfany, et al. (2020), Doulamis, Guariento, Duignan, Kido, et al. (2020)
13	Autologous heart pectoralis major muscle	Delivered as a 6-mL bolus, antegrade to the left main coronary	Yorkshire pigs	Single injections were delivered as a bolus antegrade into the left main coronary artery (1×10^9 in 6 mL). Serial injections (10 injections of 1×10^9 in 6 mL of respiration buffer in each injection) were delivered every 5 min	Myocardial protection, significantly decreasing infarct size and enhancing global and regional function.	Guariento et al. (2020)
14	Allogenic brain or muscle tissues	Left femoral artery and vein	Male SD rat cardiac arrest	0.01 mg/mL (final concentration) resuspended in 500 μL of fresh prewarmed medium	Improved survival and neurological recovery in post-CA rats	Hayashida et al. (2023)
15	Allogenic muscle tissues	Intravenously	Male Wistar rat sepsis-induced myocardial dysfunction	7.5×10^6 mitochondria/400 μL of respiration buffer	Improved mitochondrial function, biogenesis, and dynamic associated with SIRT-1/PGC-1α network and suppressed inflammatory response.	Mokhtari et al. (2023)

16	C57BL/6J and Cox4i1-GFP mice heart	Tail vein injection	Male C57BL/6J mice, Mice myocardial ischemia injury model	Polypeptide: CSTSMLKAC (PEP), bound to triphenylphosphonium cations (TPP+) bonded mitochondria (PEP–TPP –Mitochondria)7.5 -10×10^4/mL in 200 μL	PEP–TPP–mitochondrial compound effectively transported the mitochondria to the ischemic myocardium and promoting cardiomyocyte internalization. Effectively reduced immune cell infiltration, the inflammatory response, and ultimately cardiomyocyte apoptosis.	Sun et al. (2023)
Brain						
17	HepG2 cells	Intravenous injection	Experimental Parkinson's disease (PD), Male mice, C57BL/6J	0.5 mg/kg body weight.	Mitochondrial replacement prevented experimental PD progress through increasing the activity of electron transport chain, decreasing reactive oxygen species level, and preventing cell apoptosis and necrosis.	Shi et al. (2017)
18	HeLa DsRed2-mito cells.	Tail vein injection	Alzheimer's disease (AD) male C57BL/6 mice	200 mg/mouse, containing 50–55 mU (milliunits = nmol/min) CS and 46–51 mU COX equivalents, respectively, into the tail vein.	A significant decrease in neuronal loss and reduced gliosis in the hippocampus	Nitzan et al. (2019)
19	C57BL/6J mice hippocampus mitochondria (mice)	Intravenous injection	Male ICR mice, LPS-induced model of depression	1 mg/kg mitochondria	Ameliorated depression-like behaviors, restored the neurogenesis and BDNF expression, melioration of neuroinflammation, and oxidative stress	Wang et al. (2019)

(*Continued*)

TABLE 12.2 Systemic injection route of mitochondrial transplant for various pathological conditions.—cont'd

No.	Source of mitochondria	Route of mitochondrial transplantation	Animal model	Mitochondrial dose	Therapeutic outcome	References
20	Various tissues including snap-frozen placenta (E17 pregnant female mice), freshly isolated skeletal muscle, brain, and brown adipose tissues	Intravenous injection	Cerebral IR injury, Male C57BL6 mice	Mitochondrial fractions (100 μg protein/100 μL saline/mouse) immediately after reperfusion	Mitochondria can be transferred in the central nervous system. Reduce infarction after cerebral IR	Nakamura et al. (2020)
21	Male C57BL/6J mice hippocampus	Tail vein injection	Pilocarpine-induced status epilepticus in male C57BL/6J mice	1 mg/kg, 1 h after diazepam	Improve cognitive deficits and depressive and anxiety-like behaviors. Mitochondria blunted the production of reactive oxygen species, proliferation of microglia and astrocytes, and loss of neurons in the hippocampus	Jia et al. (2023)
Lung						
22	Rat bilateral soleus muscles	Intravenously transplanted via the internal jugular vein	Male SD rats	The isolated mitochondria (100 mg in 0.2 mL PBS) once weekly for 3 consecutive weeks	Restored the vascular smooth muscle cell phenotype and vasomotor function of pulmonary artery. Significantly reducing pulmonary artery pressure and improving right ventricular performance	Hsu et al. (2022)
23	Gastrocnemius muscle from C57BL/6J male mice	Pulmonary artery injection or nebulization	C57BL/6J male mice acute lung IR	Vascular delivery: 1×10^8 mitochondria suspended in 0.3 mL buffer nebulization 3×10^8 in 90 μL buffer	Improves lung mechanics and decreases lung tissue injury	Moskowitzova et al. (2020)

24	Mice mesenchymal stromal cells	Intravenously	C57BL/6 male mice experimental sepsis.	From 3×10^6 MSCs	Restored oxygen consumption rate and reduced total reactive oxygen species production. Reduced peritoneal bacterial load, improved lung mechanics and histology, and decreased the expression of interleukin-1β, keratinocyte chemoattractant, indoleamine 2,3-dioxygenase-2, and programmed cell death protein 1 in lung tissue	de Carvalho et al. (2021)
25	Allogenic from bilateral soleus muscles	The jugular veins	Adult male SD rats, acute lung injury	100 μg in 200 μL PBS	Mitochondria protects the integrity of endothelial lining of the alveolar–capillary barrier, thereby improving gas exchange	Pang et al. (2022)
Spinal cord injury						
26	Allogenic from bilateral soleus muscles	Intravenously transplanted via the jugular vein of the recipient rat	SCI was induced in adult SD male rats	Freshly isolated mitochondria (100 mg) were resuspended in 0.2 mL of PBS and	Significantly suppresses the ER stress and mitochondrial dysfunction in the ischemic cord, leading to attenuation in the inflammatory and neuroapoptotic reactions, and improvement in the motor function recovery	Fang et al. (2021)

TABLE 12.3 Nasally administered mitochondrial transplant for various pathological conditions.

No.	Source of mitochondria	Route of mitochondrial transplantation	Animal model	Mitochondrial dose	Therapeutic outcome	References
Brain						
1	Allogeneic rat livers	Intranasal (IN) drop infusion (ipsilateral to the lesioned side)	Adult female SD rat model of PD	200 μg mito or P-mito in 50 μL respiration buffer	Mitochondria penetrated the accessory olfactory bulb and doublecortin-positive neurons of the rostral migratory stream (RMS) on the ipsilateral sides of lesions	Chang et al. (2021)
2	Human MSC	IN administered (12 μL total volume per mouse)	Male C57BL/6J mice, Mice model for cognitive deficits	3 μL per nasal administration of mitochondria. For each mouse, up to 170 μg of mitochondria isolated from 4.5×10^6 human MSC in a total volume of 12 μL	Reversal of these neuropathologies is associated with resolution of cisplatin-induced deficits in myelination, synaptosomal mitochondrial integrity and neurogenesis.	Alexander et al. (2021)
3	Brain of NMRI mice	IN administered in 6 μL/mouse	NMRI male mice, spatial memory impairment in OBX mice.	2–20 μg of freshly isolated mitochondria in a volume of 3 μL into each nostril (total volume 6 μL/mouse).	Improved spatial memory in the olfactory bulbectomized mice with Alzheimer's type degeneration.	Bobkova et al. (2022)
4	Human MSC that had been lentivirally transduced with mitochondria-targeted *PDHA1-DsRed.*	IN administered (12 + 12 μL)	Male C57BL/6J mice Mice model for cognitive deficits (chemobrain) and peripheral neuropathy	Mitochondria (3 μg of mitochondrial protein per day obtained from an average 0.08 × 106 human MSC) or uncoated mitochondria (170 μg of protein on average 4.5×10^6 human MSC)	MT resulted in resolution of chemobrain and peripheral neuropathy, remyelination, synaptosomal mitochondrial integrity, and neurogenesis.	Alexander et al. (2021)
5	Mouse bone marrow mesenchymal stem cells	IN route every other day (3 days per week) for 1 week.	C57BL/6J mice, Photothrombotic mPFC ischemia	85 μg, 170 μg, and 340 μg of mitochondrial	Induction of ischemia decreased the levels of synaptic markers in mPFC while exogenous mitochondria (170 and 340 μg) significantly upregulated the expression of GAP-43 and PSD-95 after ischemic stroke.	Hosseini et al. (2022)

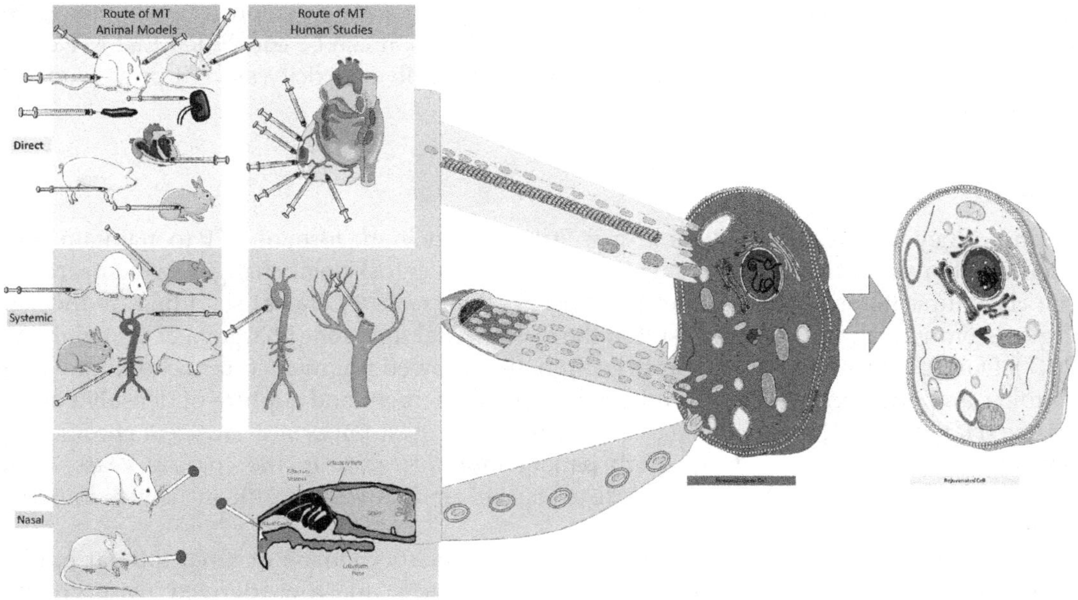

FIGURE 12.2 Route of mitochondrial transplantation (MT).

inflammation associated with increased interleukin 10. The study demonstrated that the mitochondria are rapidly taken in by the cells via actin dependent endocytosis.

Liver

Reperfusion injury is a clinically significant issue associated with liver transplantation and inflow controlled hepatectomy. Lin et al. (2013) demonstrated the protective effects of a splenic infusion of functional allogenic liver derived mitochondria on hepatic IR injury in a rat model.

In this study, the fresh liver mitochondria ($7.7 \times 10^6 \pm 1.5 \times 10^6$/mL) were labeled with MitoTracker and injected intrasplenic (100 μL) after 45 min of liver ischemia. The use of intrasplenic injection is justified because the mitochondria (4–6 nm) are smaller than the liver endothelium fenestrate (98 ± 13.0 nm), which would improve their dispersion within the liver parenchyma. The study offers two key conclusions. First, 240 min after infusion, the recipient animal's liver still had intact membrane potential from the donor mitochondria. Second, infusion of donor mitochondria reduced tissue damage, oxidative stress, and cell death in the livers after reperfusion injury.

Ulger and Kubat (2022) recently showed that MT is effective in the acetaminophen (APAP) induced liver poisoning model in male Sprague Dawley (SD) rats. Mitochondria were extracted from rat MSCs and tagged using MitoTracker Red. Following treatment of APAP, isolated mitochondria 8.2×10^6/100 mL respiration buffer/rat were injected into the subcapsular area of the spleen. Through an injection into the spleen, mitochondria were delivered to the liver via the portal circulation. It became apparent that MT enhanced liver function and tissue histological morphology. The MT group demonstrated higher glutathione (GSH) levels and the lowest overall oxidant levels. It can be concluded that MT could be a more successful therapy for treating drug induced liver toxicity since it prevented and repaired toxic liver damage.

Kidney

The kidney requires a considerable quantity of energy in the form of ATP to maintain body fluid equilibrium. Mitochondrial failure may result in cytoskeletal changes, which separate cell communications from endothelial connections, increase epithelial shedding, reduce energy generation, and ultimately result in the loss of renal function (Bhargava & Schnellmann, 2017). During ischemia, ultrastructural examination showed fragmented mitochondria within 15 min of reperfusion, resulting in mitochondria enlargement and rupture of densely packed cristae (Hall et al., 2013; Kubat et al., 2020). One of the most important causes of chronic kidney illness, nephrotoxicity, has a high patient mortality rate. In the coming years, more research into the role of MT in renal diseases will undoubtedly result in new therapeutic approaches.

Konari et al. (2019) explored underlying therapeutic mechanism of renal tubular epithelium repair of DN by mitochondria derived from bone marrow–derived mesenchymal stem cells (BM-MSCs). Male SD rats were used in the investigation, and diabetes was induced by streptozotocin (STZ). The mitochondrial-specific protein DsRed2 was used for labeling the mitochondria. Eleven weeks after STZ injection, 1×10^6 mitochondria in 100 mL (PBS) was injected under the renal capsule of the left kidney and 1×10^4 cells/g body weight of BM-MSCs was delivered intravenously. Study concluded that mitochondria incorporated into the damaged tubular basement membrane and restored renal tubule structures and induced cell proliferation. It also induced histological improvements of proximal tubules in STZ rats within 3 days which resulted in recovering the expression of megalin and SGLT2 structural markers.

Kubat et al. (2020) investigated mitochondrial regeneration potential in a doxorubicin (DOX) mediated kidney damage model in SD rats. Rat MSC derived mitochondria (8×10^6/mL) in 500 μL of respiration buffer were applied directly below the cortex capsule of each kidney. Study demonstrated that MT triggered antioxidative pathway, reduced protein accumulation of tubular cells and promoted regeneration of tubular cells after renal injury. These results show promise for the treatment of kidney injury by transplantation of MSC-derived mitochondria.

Heart

High levels of mitochondrial ATP synthesis are required for cardiomyocyte contractions in the circulatory system. Heart failure results from mitochondrial dysfunction. Previous investigations have revealed that at the commencement of ischemia, there is a quick decrease in high energy phosphate levels, as well as changes in mitochondrial structure, volume, oxygen consumption, and ATP synthesis. In recent years, mitochondrial transplantation emerged as a therapeutic option to restore adequate energy metabolism.

Masuzawa et al. (2013) presented data using the in situ blood perfused regional ischemic (RI) heart model in male New Zealand white rabbits. Following 29 min of RI, autologous pectoralis major muscle derived mitochondria ($9.7 \pm 1.7 \times 10^6$/mL) were injected directly into the RI zone. Each injection site in RI-mitochondria received 8×0.1 mL injections of sterile respiration buffer containing mitochondria and the hearts were allowed to recover for 4 week MT provided cardioprotection by enhancing ATP production and upregulation of cytokine (EGF, GRO, IL-6, and MCP-3). The MT did not generate proarrhythmic response or immune response.

Similarly, Kubat et al. (2017) provided evidence of MT induced RI recovery in female Yorkshire pigs. In the study, RI was induced for 24 min and confirmed visually by ST segment changes on the electrocardiogram. The pectoralis major muscle was used to isolate mitochondria. The injection was delivered subendocardially into predesignated sites 8×0.1 mL injections; $9.9 \times 10^7 \pm 1.4 \times 10^7$/mL; 1.3×10^7 mitochondria per injection site. MT provided cardioprotection through upregulation of energy synthesis, cell rescue, replacement of damaged mtDNA, upregulation of cytokines, immunomodulation, progenitor cell migration, prevention of apoptosis, enhanced cell salvage, and postischemic functional recovery.

Recent study of Zhang, Liu, et al. (2023) tested efficacy of MT in two models of heart failure. One was DOX induced heart failure model and a mutation driven dilated cardiomyopathic human-induced pluripotent stem cell-derived cardiomyocyte model (hiPSC-CM) in male C57BL/6 mice. Mitochondria were isolated from four different cell sources: human colon carcinoma cell line, mouse fibroblast, mouse skeletal muscle, and mouse neonatal cardiomyocytes. Mitochondria were labeled with MitoTracker Green and total of 50,000 isolated mitochondria suspended in 80 mL; 20 mL per injection was injected into the myocardium of the left ventricle during perfusion at four different points. Study demonstrated that MT confers protection in Duchenne muscular dystrophy and hiPSC-CM animal models. MT alleviated mitochondrial dysfunction and boosted myocardial function in dilated cardiomyopathic cardiomyocytes. Importantly, in hiPSC-CMs animals reestablished mitochondrial membrane potential faster compared with untreated.

Brain

The brain is a high metabolism organ that consumes 20% of the body's energy; mitochondria play an important homeostatic function in practically all the aspects of brain cell physiology, pathology, and calcium homeostasis maintenance. In addition to controlling neuronal activity and behavior, mitochondrial distribution and transport are crucial for the development of synaptogenesis and dendritic spines. Several studies have shown that mitochondrial dysfunction plays an important role in neurodegeneration. Indeed, nervous system cells require a significant amount of energy to carry out synaptic transmission, neurogenesis, and neuronal differentiation (Nascimento-Dos-Santos et al., 2021). Consequently, an MT approach has been tested in animal models to treat neurological disorders and explore mitochondrial availability in the brain.

In this context, a unique treatment strategy to save recipient cells and restore normal function may be to transplant healthy and viable mitochondria into the damaged brain regions. In this part, different in vivo experiments of MT that were directly routed in several different models of central nervous system (CNS) damage and brain disease are mustered.

Huang et al. (2016) investigated MT in adult male SD rats using a Middle Cerebral Artery Occlusion (MCAO) model for cerebral stroke. Mitochondria were extracted from BHK-21 cells (hamster kidney cell line), and their mass was calculated using the Bradford method. In order to quantify the number of mitochondria, the researchers labelled the mitochondria with MitoTracker prior to isolation. They then employed a flow cytometer to interpolate that 75 μg mass contains 1.2×10^6 mitochondria particles. Following 24 h of MCAO, rats with ischemic brains had intracerebral injections of 75 g of mitochondria in 10 μL of SEH buffer (0.25 M sucrose, 0.5 mM ethylene glycol tetraacetic acid, 3 mM N-(2-hydroxyethyl) piperazine-N¢-ethanesulfonic acid, pH 7.2) and femoral artery infusions of 750 g of mitochondria in 100 μL. This study is the first to report the application of intracerebral MT for the treatment of ischemic lesions after perfusion. It was demonstrated that systemic or cerebral treatment of xeno-hamster mitochondria reduced damage to the brain and led to significant functional recovery of their motor activities. MT effectively rescued forelimb and hindlimb motor functions at 7 days postoperatively and sustained this recovery for 1 month.

Pourmohammadi-Bejarpasi et al. (2020) showed the role of MT in IR damage by MCAO in adult Wistar rats. Mitochondria isolated from 3×10^7 human umbilical cord derived mesenchymal stem cells (hUC-MSCs) and followed by administration via intracerebroventricular injection into the right lateral ventricle (0.8 mm posterior, 1.5 mm lateral and 4.0 mm ventral to the bregma according to Paxinos atlas). They concluded that mitochondrial transplantation alleviated IR injury, prevented cell death, reduced astroglyosis and microglial activity, diminished infarct area, and enhanced motor function.

Chang et al. (2016) conducted their study using an adult female SD rat model of Parkinson's disease (PD). The study evaluated the performance of mitochondrial transfer in 6-hydroxydopamine (6-OHDA) induced PC12 cells and PD rat models with and without Pep-1 conjugation. Stable Green Fluorescence Protein (GFP) transfection was then carried out, followed by mitochondrial isolation. Pep-1 conjugated mitochondria (1.05 mg) from rat or human cells were injected locally into the medial forebrain bundle (4.4 mm behind the bregma, 1.2 mm to the side of the midline, and 7.8 mm below the dura) in a total volume of 5 mL PBS. The study demonstrated that allogeneic and xenogeneic sources significantly sustained mitochondrial function to resist the neurotoxin induced oxidative stress, apoptotic death, and a greater capability of neurite outgrowth. Furthermore, allogeneic and xenogeneic MT improved the locomotive activity in the PD rats after 3 months.

Zhang et al. (2019) reported application autologous MT for treating focal cerebral ischemia performed by MCAO in male SD rats. Autologous biopsies of muscle samples derived from the pectoralis major were used to isolate mitochondria. Freshly prepared mitochondria $5 \times 10^6/10$ μL were injected into the left lateral ventricle (0.8 mm posterior, 1.5 mm lateral, and 4.0 mm ventral to the bregma) after reperfusion. This study showed that mitochondrial transplantation was effective in alleviating the IR injury involved in enhanced ATP production, decreased ROS release, and reduced apoptosis. Additionally, study advocates the infusion of isolated mitochondria through left lateral ventricle (direct route). It was also pointed out the limitations intravenous route for treatment of brain injury due to aimless and uncontrolled delivery of mitochondria.

Recently, Zhao, J. et al. (2021) specified the use of MT in the controlled cortical impact model of traumatic brain injury (TBI) in adult male C57BL mice. The allogeneic liver and autogenic muscle were used to isolate the mitochondria, and $1.2-1.4 \times 10^6$ MitoTracker labeled mitochondria were injected into cerebral cortex. Experimental data showed that allogenic and autologous mitochondria triggered similar activation of microglia in cortex and easily internalized by neural cells indicated the low immunogenicity of mitochondria. Efficiency of mitochondrial isolation was high from liver. Data suggested that MT may be effective in reducing the secondary injury of TBI through multiple mechanisms. Further, MT improved the mice behavior, trauma associated anxiety, spatial memory, and cognitive functions.

Javani et al. (2023) presented recent reported MT for aged Wistar rats exposed to chronic stress model. Mitochondria were isolated from the brain of a young rat and labeled with MitoTracker Green FM and directly injecting mitochondria (10 mL in saline) into cerebral cortex. Treatment with young mitochondria alleviated chronic stress induced oxidative damage and reduced apoptosis.

These findings in animal studies suggest that mitochondria can be potentially transferred to cells. Moreover, MT significantly improved cellular and behavioral outcomes in all animal models. Additionally, reports suggested that autologous, allogenic, and xenogeneic MT did not trigger severe immune response. Studies confirmed that mitochondria enter neurons to support cell viability and recovery after CNS.

Spinal cord injury

SCI has devastating neurological repercussions, including partial or total paralysis. The pathophysiology of traumatic SCI begins with a mechanical injury and progresses through a number of widespread effects of the secondary injury cascades, including edema, ischemia, excitotoxicity, inflammation, oxidative damage, and activation of necrotic and apoptotic cell death signaling events (Anjum et al., 2020). In the event of an injury, necrotic tissue at the original damage site initiates secondary injury cascades, such as mitochondrial dysfunction, which results in decreased ATP and increased ROS at the injury site (Kwon et al., 2004).

Numerous clinical trials have been conducted aiming to alleviate SCI; however, none of these approaches have been successful to regenerate injured SCI. Therefore, it may be asserted that therapeutic modalities that concentrate on a certain biochemical pathway might not provide the best means of enhancing functional recovery from SCI. An MT has recently come to light as a crucial player not only in the energy supplementation but also in the processes of neural homeostasis and neuro regeneration (Cheng et al., 2012).

Gollihue et al. (2018) were the first to investigate mitochondrial transplantation into the damaged rat spinal cord and its implications on bioenergetics. In the study, mitochondria were isolated from transfected transgenically labeled turbo green fluorescent (tGFP) PC12 cells and soleus muscle of SD rats. Adult female SD rats received a T12 laminectomy to expose the L1/L2 spinal level and within 30 min after injury, each injection of 750 nL of mitochondria (for a total of 50, 100 or 150 mg) in a total of 3 mL volume was injected into the mediolateral gray matter at four circumferential sites of the injured cord. According to the study, adding healthy, tightly connected mitochondria to an injured spinal cord will boost energy generation, reduce oxidation, and mitigate the effects of glutamate excitotoxicity, resulting in the maintenance of tissue bioenergetics. Further, exogenous mitochondria were predicted to integrate into a number of cell types, with short-term benefits including tissue sparing and improvements in hindlimb functional recovery.

Li et al. (2019) demonstrated efficacy of cultured BMSC derived mitochondria in rat SCI model. The SCI model was performed by Allen's method (laminectomy was carried out at T10 level). Immediately, after SCI induction, mitochondria (3×10^6) were injected into the epicenter of the injured spinal cord using an electrode microneedle. The study demonstrated the transfer of mitochondria through tunneling nanotubes or direct contact, and in a paracrine manner. Study confirmed that transplanted mitochondria were internalized by neurons in vivo resulted in decreased the number of apoptotic cells in the early stage of SCI and reduced the area of lesion cavity, glial scar, and the number of glial fibrillary acidic protein (GFAP) positive cells in the late stage of SCI. The important finding of the study was that mitochondria treatment both improved blood brain barrier (BBB) score and footprint behavior at 3 and 6 weeks after transplantation.

Lin et al. (2022) demonstrated therapeutic effects of mitochondrial transplantation in the traumatic SCI rats. SCI was induced in adult SD rats by applying aneurysm clip for 20 s on the T10 spinal cord to elicit compressive trauma. Allogeneic mitochondria were freshly isolated from the bilateral soleus muscles of healthy donor rats and received two injections of mitochondria (50 μg in 1.5 μL 1× PBS) to the epicenter of the injured site at 2 mm rostral and caudal (0.6 mm in depth). Results showed that MT improved the recovery of somatosensory and locomotor functions after SCI and promoted the preservation of white matter. MT rats showed reduced mitochondrial fragmentation, cellular apoptosis, neuroinflammation, and oxidative stress. The authors were able to demonstrate that transplanted allogeneic mitochondria survived in the lesioned region for at least 28 days in a rat model of traumatic SCI.

Retina

The retinal pigment epithelium is a highly specialized pigmented monolayer of postmitotic cells found in the posterior portion of the eye between the neurosensory retina and the vascular choroid. It acts as a selective barrier while nourishing retinal vision cells. Due to their high energy dependence, these cells rely on a large number of healthy mitochondria and high oxygen consumption. Consequently, they are more susceptible to chronic oxidative stress and ROS. Numerous human ophthalmologic disorders, including glaucoma, Leber's hereditary optic neuropathy, autosomal dominant optic atrophy, and optic neuropathies, have been linked to the disturbance of mitochondrial processes. Elevated intraocular pressure and aging have well established linkages to axonal degeneration and consequent death of retinal ganglion cells; the population is responsible for visual input integration and transmission to the brain.

Nascimento-Dos-Santos et al. (2020) presented a new therapy approach to overcome glaucoma implications. The Lister Hooded rat experimental model of glaucoma was developed by optic nerve crush performed 1 mm from the ocular globe with tweezers pressed one time for 15 s. Functional mitochondria were isolated from the liver and injected intravitreally (0.25 mg/mL in saline). This investigation demonstrated first time that injected rat liver mitochondria into the rat vitreous body incorporate into retinal cells resulted in the improved tissue bioenergetics status and electroretinographic responses. MT resulted in rapid mitochondrial diffusion and a greater mitochondrial mass were seen in the mitochondria transplanted group, which was evidenced by an increase in the entire retina spare capacity. MT induced neuroprotection of retinal ganglion cells at 14 days and an increase in the number of optic nerve fibers outside the crush site at 28 days after transplantation. Current treatment modality has the potential to overcome the deficiencies of cell, gene, and other molecular therapies.

Lungs

The structure and functioning of the bronchial airways adjust in the course of a person's life in response to environmental factors and occupational demands. Major lung illnesses including asthma, bronchitis, and chronic obstructive pulmonary disease all present dysfunctional excessive airway constriction along with wall thickening from increased cell proliferation with varying degrees of fibrosis. Airway structural and functional changes in these conditions are a result of genetic, proteomic, and physiological processes within specific resident cell types (epithelial cells, airway smooth muscle, fibroblasts), in addition to the modulating effects of allergens, pollutants, cigarette smoke (CS), and inflammation. There has been increasing understanding that altered mitochondrial structure and function develop in response to disease processes and, in turn, adaptation of downstream cellular processes impact in lung disorders (Aggarwal et al., 2016; Lerner et al., 2016; Prakash et al., 2017).

Su et al. (2016) studied the role of MT in airway hyperresponsiveness (AHR) SD rat model induced by expos to CS and lipopolysaccharide (LPS) administration. The mitochondria were isolated from rat airway epithelium and labeled with DsRed and intratracheally administered 3.75×10^8/mL for an additional week after CS plus LPS treatment. The MT significantly attenuated AHR caused by 4 weeks treatment of CS plus LPS treatment through inhibition of ROS enhanced epithelial cholinergic hyperactivity. MT resulted in structural changes in airway epithelium through the elimination of excessive ROS or blockage of muscarinic M3 receptor but inhibited by M receptor enhancer.

Endometrium

The endometrium is a dynamic tissue, and estrogen and progesterone, ovarian hormones, regulate changes in the mucosal lining and thickness throughout each menstrual cycle. Three layers, referred to as the stratum basalis, stratum spongiosum, and stratum compactum, make up the endometrium. Each menstrual cycle regulates the specific and dynamic processes of the endometrium, including differentiation and regeneration. The stratum basalis, also referred to as the basal layer, does not change significantly over the course of a cycle, in contrast to the stratum spongiosum and stratum compactum layers, which undergo transforming into the functional endometrial layer (Gargett et al., 2012). Repeated implantation failure, spontaneous miscarriage, and female infertility are all known to be caused by endometrial abnormalities like adenomyosis, Asherman's syndrome, endometrial polyps, thin endometrium, luteal phase deficiency, and endometriosis (Bashiri et al., 2018; Hong, 2023).

Adequate thickness of the endometrium is one of the important factors associated with uterine receptivity. Mechanisms regulating this remarkable remodeling of endometrial thickness remain largely unclear, but endometrial resident stem cell populations play an important role in modulation of endometrial thickness. This is a possible direct involvement of endometrial stem cells in endometrial disease formation (Hong, 2023). Mitochondria are key regulators of stem cell self renewal, multilineage differentiation, and fate determination (Correia et al., 2022). Defects in mitochondrial function that affect the oxidative phosphorylation system make it difficult for cells to synthesize the intermediates needed for growth and increase the creation of oxygen free radicals that can damage both nDNA and mtDNA (Kim et al., 2018; Vizioli et al., 2020). Mitochondrial dysfunction and underlying energetic defects in endometrial cells result in cellular decline, altered cellular activity, and regenerative potential.

In our recently published study (Kshersagar et al., 2023) for the first time, we showed efficacy of human placenta derived mitochondria for endometrial regeneration. The adult female Wistar rats were used in the study and endometrium was disturbed by administration of 0.25 mL of 95% ethanol into uterine horns followed by intrauterine infusion of rhodamine labeled 1×10^6 mitochondria in 0.25 mL followed by 0.25 mL activated platelet rich plasma (PRP) (Figs. 12.3–12.5). We choose to isolate mitochondria from placenta considering the abundance of tissue availability without ethical constraint, nonsurgical source consideration toward the clinical translational prospects of the procedure. Importantly, activated PRP is able to hold transplanted mitochondria at the injury site and give stable growth factor support during initial period of engraftment. This result in an engraftment rate of 33.30% after 12 h (Fig. 12.5) which leads to the speculation that the mitochondria transplanted into the damaged endometrium provide the energy requirements of the cells, thus resulting

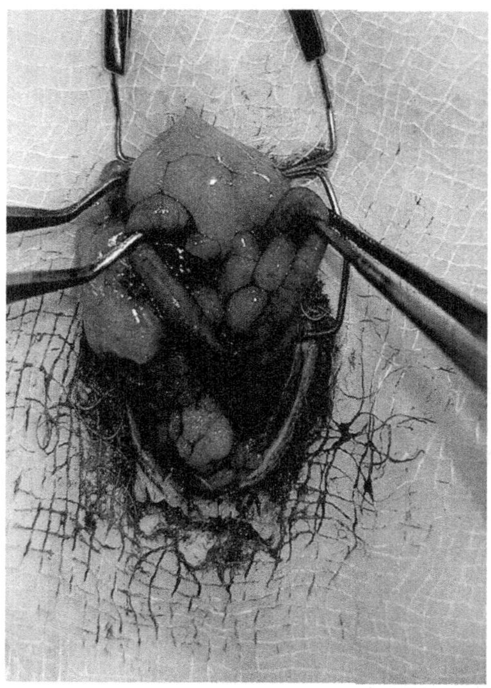

FIGURE 12.3 Rat model of disturbed endometrium.

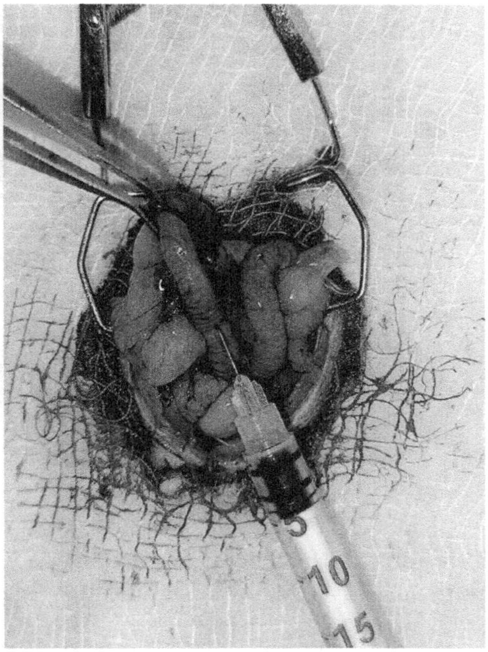

FIGURE 12.4 Intrauterine infusion of mitochondria.

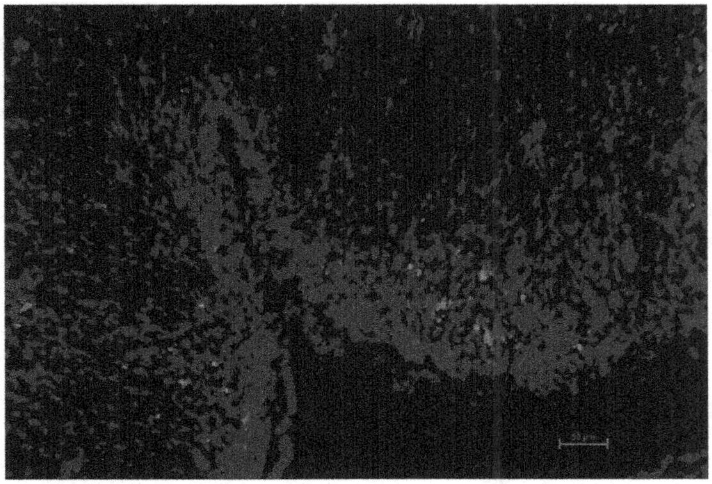

FIGURE 12.5 Engraftment of transplanted mitochondria after 12 h.

in regeneration and endometrial homeostasis (Fig. 12.6). The functionality of mitochondria accessed by the estimation of specific activities of NADH-cytochrome c reductase, NADPH-cytochrome c reductase, cytochrome *b*5, cytochrome P450, complex I, II, III, and IV enzymes demonstrated that MT group had significantly higher values compared with

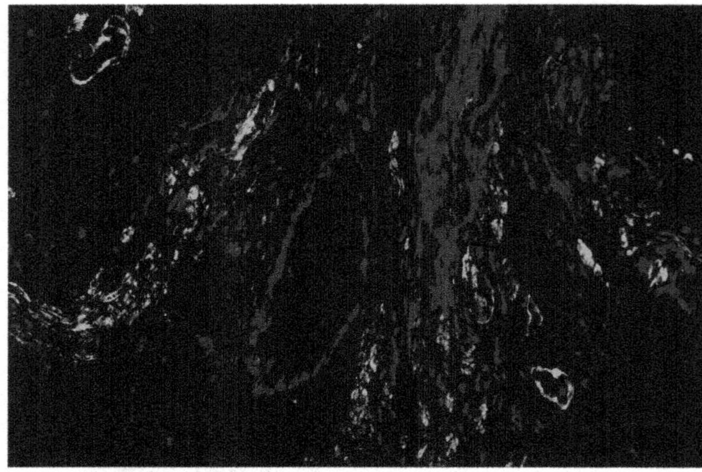

FIGURE 12.6 α SMA expression: 72 h after MT.

disturbed endometrium indicating the regeneration of damaged endometrial cells. Importantly, birth rate of live pups in MT group was significantly high compared to disturbed endometrium group.

Systemic administration route for MT

Systemic mitochondrial delivery is a promising therapeutic approach that involves the administration of exogenous mitochondria to target tissues or organs via systemic routes. Recent publications support the treatment strategy of mitochondrial transplantation via intravenous injection to diverse disorders since the translational application of mitochondrial transplantation is constrained owing to the difficulties of direct injection route. In recent studies, systemic mitochondrial delivery has shown encouraging results in improving tissue regeneration, protecting against organ injury, and enhancing overall physiological function (Fig. 12.2).

Systemic application routes of MT used in animal models of disease are summarized in Table 12.2.

Skeletal muscle

Alway et al. (2023) recently reported the effects of MT in restoring muscle function after neuromuscular injury. They used barium chloride ($BaCl_2$) induced injury into the gastrocnemius muscle of C57BL/6 mice. The injection of liver-derived mitochondria labeled with MitoTracker (50 μg/mL) into mice via the tail vein resulted in improved muscle regeneration and restoration of function compared to mice treated with PBS. The labeled mitochondria allowed for tracking and localization within live cells. This finding suggests that systemic delivery of exogenous mitochondria has the potential to enhance muscle healing and functional recovery following injury. Muscle mass and force had recovered to control levels by 14 days in MT treated mice, while PBS treated mice showed no improvement. These findings imply that systemic administration of mitochondria can enhance muscle repair and function resulting from injury.

Liver

The study was conducted by Fu et al. (2017) suggests that MT isolated from hepatoma cells were effective in treating fatty liver induced by high fat diets in C57BL/6J mice. Mitochondrial dysfunction is a common mechanism in nonalcoholic fatty liver disease. Mitochondria were intravenously injected at a dose of 0.5 mg/kg body weight. The outcomes demonstrated that MT prevented cellular damage in the animals, reducing lipid deposits, oxidative damage in the fatty liver, improving the production of energy, and restoring hepatocyte function. These results suggest that MT has promise as a treatment for fatty liver disease.

Shi et al. (2018) used human hepatoma cells to isolate mitochondria that were injected intravenously into the tails of C57BL/6J mice to treat mice model of acetaminophen induced liver damage. The results demonstrated that this treatment approach attenuated tissue injury and enhanced hepatocyte metabolism. MT improved hepatocyte energy supply, reduced oxidative stress, and ameliorated tissue injury in the model mice. These findings highlight the potential effectiveness of MT as a therapeutic strategy for treating acetaminophen induced liver damage. Mitochondrial damage is a common factor in liver injury, prompting the exploration of targeted delivery strategies.

In another study, Liu et al. (2020) aimed to deliver mitochondrial complexes specifically to CD-1 rat livers via intravenous injection. By combining mitochondria with a carrier protein (AsOR-PL), they successfully achieved targeted delivery, with a significant portion of injected mitochondria reaching the liver. The researchers administered 0.5 mL of purified mitochondria at a steady rate of 160 μL/min. The study revealed that the delivery method was successful, with 27% of the injected mitochondria detected in the liver. Additionally, 2% of the injected mitochondria were found in the spleen, while less than 1% were detected in the lungs. These results indicate the ability to deliver specific mitochondrial complexes to the liver, highlighting the potential for targeted MT for various medical applications.

Ko et al. (2020) after the successful results from previous study, conducted further experiment where they explored the protective effects of these mitochondria in the context of acute liver IR injury. Additionally, the objective of this investigation was to determine if 31P-magnetic resonance spectroscopy (31P-MRS) can reliably identify and assess the severity of acute liver IR injury. Mitochondria isolated from donor SD rats were administered through the portal vein to adult male SD rats with IR injury, following intraperitoneal melatonin administration to the donors. Remarkably, the transplantation of 10 mg of isolated mitochondria effectively shielded the liver against IR injury. This protection was primarily attributed to the rejuvenation of endogenous mitochondria, preventing hepatocytes deterioration in IR injury. The protective effects of melatonin pretreated mitochondria against liver IR injury were confirmed by microscopic investigation. The study found that melatonin pretreated mitochondria may successfully protect the liver from IR injury and that 31P-MRS is a viable method for assessing mitochondrial ATP consumption in living animals.

Zhao, Z., et al. (2021) continued with the exploration of MT therapy, a study investigating its potential for treating carbon tetrachloride (CCl_4) induced liver injury. When the cells grew up, CCl_4 (10 mM) were introduced inside cultured cell media and incubated for about 12 h. Further, the media was substituted by fresh DMEM, and various concentrations of exogenous healthy mitochondria were introduced inside the cell culture medium. According to mitochondrial number and protein level determination, mitochondria of 1 mg/mL was equivalent to 1.6×10^8 mitochondria per mL. Mice in mitochondrial therapy group were respectively intravenously (IV) injected with 0.2 mg/kg (Mito low) or 0.4 mg/kg (Mito high) mitochondrial solution in saline once every day for 7 days. The study showed that mitochondrial transplantation increased bioenergy supply, decreased free radicals, and increased cell survival. Additionally, it enhanced liver performance and reduced tissue fibrogenesis. The protective response and survival of liver cells were attributed, according to transcriptomic analysis, to the activation of the mitochondrial unfold protein response and the upregulation of antioxidant genes. This study highlights the potential of mitochondrial transplantation as a therapeutic intervention for liver injury.

The common trait of drug induced liver damage is mitochondrial dysfunction. A promising core strategy for treating drug mediated hepatotoxicity is MT, which involves replacing malfunctioning mitochondria with functional new mitochondria.

Kidney

The creation of novel treatment approaches is essential due to a lack of effective therapeutic options for acute kidney damage (AKI).

Plotnikov et al. (2020) showed the application of rat liver mitochondria in kidney IR injury model in white male rats. Additionally, the development of AKI and animal mortality was evaluated along with the effects of combined kidney IR injury and systemic delivery of mitochondria. In brief, IR injury was induced by nephrectomy of the right kidney, and circulation was restored in the left kidney by removing the clip after 40 min. The rat liver mitochondria were isolated by differential centrifugation at 4°C and administered dose was 10 mg of mitochondrial protein per 100 g body weight. Treatment with combination of mitochondria and mitochondria targeted antioxidants improved animal survival, indicating the involvement of mitochondrial ROS in systemic inflammatory response and death related to acute renal failure. This suggests that the intraperitoneal injection of rat liver mitochondria holds potential as a therapeutic intervention to mitigate the negative effects of kidney IR injury.

Jabbari et al. (2020) addressed mitochondrial dysfunction related renal tubular cell death in AKI (nephrectomy of the right kidney and, circulation was restored by removing the clip after 45 min in the left kidney). They performed autotransplantation of healthy muscle cell derived mitochondria in a kidney IR injury animal model through renal artery injection. The transplantation was performed via the renal artery using a 32 gauge needle, delivering a concentration of 7.5×10^6 mitochondria/mL over a period of 2 min. This method effectively minimized renal tubular cell death, recovered renal function, reduced kidney damage, enhanced the ability of the renal tubules to regenerate, and decreased IR induced apoptosis. These results demonstrate the therapeutic potential of autologous mitochondrial transplantation for the management of renal IR injury.

Doulamis, Guariento, Duignan, Kido, et al. (2020) conducted study in line with the previous study on mitochondrial transplantation therapy for AKI in female Yorkshire pigs. A small piece of muscle was surgically extracted from the sternocleidomastoid muscle using a 6 mm biopsy punch ($\sim$0.01 g) and used for mitochondrial isolation, which was performed 30 min before the injection of mitochondria. The transplantation was performed via the renal artery using either a single bolus injection of 1×10^9 mitochondria in 6 mL vehicle or serial injections, with three injections of 1×10^9 mitochondria in 6 mL vehicle, with each injection delivered every 20 min. Additionally, the cellular uptake of mitochondria was evaluated by the intravascular injection of 1×10^9 ^{18}F-rhodamine 6G (^{18}F-R6G) labelled mitochondria (isolated from human cardiac fibroblasts) into the left kidney. The findings demonstrated renal protection, improved renal function, and reduced renal damage, supporting the potential of mitochondrial transplantation as a novel therapeutic strategy for treating AKI.

These findings suggest a unique therapeutic approach for the treatment of AKI needs more research and clinical relevance with potential benefits including increased patient quality of life and decreased rates of morbidity and death.

Heart

Mitochondria occupy up to approximately 20%–35% of myocardial volume (Neubauer, 2007) and myocardial infarction caused by coronary artery obstruction frequently results in ischemic mitochondrial damage which causes imbalance in the energy supply and demand. The consequences will be irreversible cell death, harmful myocardial alterations, heart failure, electrical instability, and malignant arrhythmia. Effective methods for minimizing the ischemic insult after myocardial infarction including early vascular reperfusion through cardiac catheterization, coronary stenting, and coronary artery bypass surgery. However, the recovery of oxygen and substrates frequently results in cardiac IR damage, which is characterized by mitochondrial dysfunction, increased ROS, mitochondrial calcium dysregulation, and cell death (Chouchani et al., 2014; Murphy & Steenbergen, 2008).

Cowan et al. (2016) investigated the therapeutic potential of MT for global ischemia related cardiac injury by aortic cross clamping in female New Zealand White rabbits. This study was conducted using human adult cardiac fibroblasts for mitochondrial isolation labeled with ^{18}F–R6G iron oxide nanoparticles and 1×10^8 mitochondria were injected into the left anterior descending coronary artery. The results showed the increase in ventricular function and a reduction in infarct size, pointing to the possibility of mitochondrial transplantation as a therapeutic strategy to improve cardiac outcomes following global ischemia.

Moskowitzova et al. (2019) studied the role of mitochondria in enhancing graft function and reducing tissue injury during heart transplantation with prolonged cold ischemia time in male C57BL/6J mice. Mitochondria were isolated from gastrocnemius muscle and 1×10^8 mitochondria in 0.5 mL respiration buffer injected into the coronary arteries during harvesting and after transplantation. The mitochondria were implanted into the coronary arteries via injection at the coronary ostium, followed by a second injection 5 min later. MT resulted in improved beating score, ejection fraction, and shortening fraction, of the transplanted hearts, as well as a notable decrease in neutrophil infiltration. These findings highlight the potential of mitochondrial transplantation as a therapeutic strategy to improve cardiac function and mitigate inflammation in heart transplantation procedures.

In a related study, Doulamis, Guariento, Duignan, Orfany, et al. (2020) showed cardioprotective role of mitochondria in Zucker diabetic fatty rats (ZDFLeprfa/Crl; ZDF) and nondiabetic Zucker lean rats (Lean fa/+; ZL). The mitochondria were isolated from the pectoralis major muscle tissue, and 1 mL of a vehicle solution containing $2 \times 10^5 \pm 0.3 \times 10^5$ mitochondria/g of wet weight was resuspended in 2 mL of 37°C Krebs–Ringer solution before being injected antegrade into the coronary arteries through the aortic cannula over 10 s. For visualization of mitochondrial uptake in the heart, ^{18}F-R6G labeled mitochondria (1×10^9 in 1 mL respiration buffer) were delivered antegrade to the coronary arteries via injection to the coronary ostium using a tuberculin syringe with a 30 gauge needle. In a separate set of ZDF rats, 1×10^9 mitochondria were isolated from human cardiac fibroblasts and were delivered retrogradely to the coronary arteries through the aortic cannula. The outcomes demonstrated that this distribution strategy had cardioprotective benefits and increased the amount of ATP in the heart. These results suggest that this strategy may enhance heart function and offer damage prevention.

Guariento et al. (2020) reported a study on preischemic intracoronary autologous MT as a therapeutic strategy for prophylactic myocardial protection in a porcine model of regional IR injury. The left main coronary artery was cannulated in Yorkshire pigs and mitochondria (1×10^9) were delivered as a single bolus or serially (10 injections over 60 min). At 15 min after injection, the heart was subjected to temporary RI by snaring the left anterior descending artery. The treatment resulted in myocardial protection, substantial reduction in infarct size, and notable improvements in global and regional cardiac function.

In another study, by Hayashida et al. (2023) in male SD rats experiencing cardiac arrest were treated with allogenic mitochondria from brain or muscle tissues. Mitochondria were administered through the left femoral vein. Increased survival and neurological recovery were observed in the postcardiac arrest rats. This research reveals the possibility of a unique treatment strategy to increase postcardiac arrest rat neurological recovery and survival. The observed improvements in mitochondrial function and biogenesis suggest that MT approach holds promise for addressing various cardiac diseases, potentially offering a new avenue for therapeutic intervention.

Mokhtari et al. (2023) conducted a study with male Wistar rats with sepsis induced cardiac dysfunction after receiving allogenic muscle tissue mitochondria (7.5×10^6/400 mL of respiration buffer). The treatment effectively suppressed the inflammatory response, offering promising therapeutic potential for sepsis induced myocardial dysfunction. The observed improvements in mitochondrial function and biogenesis support the notion that this treatment modality could provide valuable insights into future therapeutic interventions for sepsis-related cardiac complications.

Recently, Sun et al. (2023) studied role of MT in mice myocardial ischemia established by ligation of the left anterior descending branch for 45 min ischemia and following 24 h reperfusion. PEP(Cy5)–TPP–mitochondria compound $7.5–10 \times 10^4$/mL in 200 mL was transplanted by tail vein before myocardial ischemia. Authors did not observe significant differences in the overall left ventricular ejection fraction (LVEF) and left ventricular fractional shortening (LVFS) parameters between the mice of IR injury that received intravenous mitochondrial transplantation and sham operated. MT group did not alter in the decrease of proapoptotic protein cleaved caspase 3 as well as the BAX/BCL2 ratio indicated increased apoptosis in response to IR injury. MT without modification elicited no therapeutic effects on myocardial IR injury. PEP–TPP–mitochondrial compound effectively protected against IR injury through decreased myocardial infarction, an increase in their overall LVEF and LVFS. PEP–TPP–mitochondrial compound effectively transported the mitochondria to the ischemic myocardium and promoting cardiomyocyte internalization. Additionally, authors observed that the transplanted mitochondria were also selectively transferred to cardiomyocytes via the endothelium. Transplantation with the PEP–TPP–mitochondrial compound effectively reduced the immune cell infiltration, the inflammatory response, and ultimately cardiomyocyte apoptosis.

Brain

Shi et al. (2017) studied an experimental PD model using male C57BL/6J mice. In the study, mitochondria were isolated from HepG2 cells transfected with GFP and administered via intravenous injection at a dosage of 0.5 mg/kg body weight. The findings of the study revealed that MT therapy demonstrated significant efficacy in halting the advancement of experimental PD. This was accomplished by enhancing electron transport chain activity, mitigating ROS levels, and averting cellular apoptosis and necrosis. These results underscore the potential of MT as a viable therapeutic approach in managing PD.

Nitzan et al. (2019) used Alzheimer's disease (AD) mice (amyloid-intracerebroventricularly injected) specifically male C57BL/6 mice and injected freshly isolated human mitochondria intravenously. These human mitochondria were separated from transfected HeLa DsRed2-mito cells (HeLa cells were previously transfected by researchers with the plasmid DsRed2-mito (encodes fusion of *Discosoma* sp), red fluorescent protein and the mitochondrial targeting sequence from subunit VIII of human cytochrome *c* oxidase, which targets the protein to the mitochondria). Mitochondria were administered at a dosage of 200 μg per mouse through the tail vein. This treatment resulted in a noteworthy reduction in neuronal loss and gliosis within the hippocampus, highlighting the potential of HeLa DsRed2-mito cell administration as a therapeutic approach for alleviating AD related pathology.

Wang et al. (2019) presented depression ICR mice model (LPS induced) to study antidepressant like properties of mitochondria. Normal mice hippocampus used for mitochondrial isolation and labeled with Mitotracker Red CMXRos and intravenously administrated at 1 mL/kg dose. MT alleviated depressive like behaviors induced by inflammation, as evidenced by decreased astrocyte and microglia activation, neuroinflammation markers, while increasing brain-derived neurotrophic factor (BDNF) expression, neurogenesis, and restoring ATP production and oxidative stress. These findings highlight the efficiency of mitochondrial transplantation as a new therapeutic strategy for major depressive disorder.

Nakamura et al. (2020) presented results of cerebral IR injury in male C57BL6 mice, in which mitochondria from a variety of tissues, including snap-frozen placenta from E17 pregnant female mice, freshly isolated skeletal muscle, brain, and brown adipose tissues, were successfully administered intravenously. Following reperfusion, intravenously administered mitochondrial fractions (100 mg protein/100 mL saline/mouse) with the MitoTracker Deep

Red label were delivered into the central nervous system. After a central nervous system injury, mitochondria may be actively released into extracellular space and may even be transported between neurons, astrocytes, microglia, and endothelial cells in the neurovascular unit. Mice treated with placental mitochondrial fractions soon after reperfusion experienced a considerable reduction in infarction following focal cerebral ischemia.

Jia et al. (2023) recently presented work on animal model for cognitive decline and emotional disorders in C57BL/6J mice. In a prior investigation, male C57BL/6J mice were administered with scopolamine methyl nitrate (2 mg/kg) to antagonize the peripheral effects of pilocarpine. After 30 min, pilocarpine hydrochloride (320 mg/kg) was injected to induce status epilepticus (SE). Mitochondria isolation was carried out from hippocampus tissue and administered once by the tail vein injection (1 mg/kg) 1 h after diazepam. Improvements in cognitive impairments and a decline in depressed and anxiety like behaviors were both brought on by MT. Notably, the delivery of mitochondria reduced the generation of ROS, the proliferation of microglia and astrocytes, and the degeneration of neurons in the hippocampus. These results imply that mitochondrial treatment may be effective in reducing neurological damage and related behavioral problems in the setting of SE.

Lungs

Hsu et al. (2022) presented promising results of intravenous transplantation of mitochondria derived from bilateral soleus muscles via the internal jugular vein in male SD rats. MT was administered once a week for three consecutive weeks for the dosage of 100 mg in 0.2 mL PBS; the treatment successfully restored the vascular smooth muscle cell phenotype and improved vasomotor function in the pulmonary artery. Notably, MT resulted in a considerable decrease in pulmonary artery pressure and an improvement in the function of the right ventricle, showing its potential as a treatment for pulmonary artery dysfunction and boosting cardiac function.

Moskowitzova et al. (2020) presented report on application of MT in acute lung IR injury model in C57BL/6J male mice. The gastrocnemius muscle was used to isolate mitochondria and was either injected into the pulmonary artery (1×10^8 suspended in 0.3 mL of buffer) or given by nebulization (3×10^8 in 90 mL of buffer). Remarkably, this mitochondrial intervention demonstrated significant improvements in lung mechanics and a reduction in lung tissue injury, underscoring its potential as a therapeutic strategy for alleviating the detrimental effects of acute lung IR injury.

de Carvalho et al. (2021) presented efficacy of MT in the experimental sepsis model in C57BL/6 male mice. Mitochondria were isolated from 3×10^6 mice mesenchymal stromal cells and intravenously administrated. The MT restored the oxygen consumption rate and reduced the production of total ROS. In addition, it reduced the number of bacteria in the peritoneal cavity, enhanced the mechanics and histology of the lungs, and reduced the expression of the protein like interleukin-1, indoleamine 2, 3-dioxygenase-2, keratinocyte chemoattractant, and programmed cell death protein 1 in the lung tissue. These results highlight the potential of mitochondria derived from mesenchymal stromal cell for attenuating sepsis-related complications, providing promising prospects for its clinical application.

Pang et al. (2022) presented a study on acute lung injury (LPS, 1 mg/kg) in SD rats. Male adult SD rats' bilateral soleus muscles were used to isolate mitochondria. 100 μg of mitochondria obtained from skeletal muscle were administered in 200 μL PBS intravenously (through the jugular veins). The results demonstrated that the mitochondrial intervention effectively preserved the integrity of the endothelial lining of the alveolar capillary barrier, leading to improved gas exchange. These findings underline the therapeutic potential of mitochondria in protecting lung function and alleviating the impact of acute lung injury.

Spinal cord

Fang et al. (2021) reported for the first time a promising therapeutic approach in SCI involves the intravenous transplantation of freshly isolated mitochondria (100 μg in 0.2 mL PBS) derived from bilateral soleus muscles. SCI was induced by cross clamping of thoracic aorta at T6 level for 25 min, followed by release of vascular clip to restore aortic blood flow. Remarkably, this mitochondrial intervention effectively suppressed endoplasmic reticulum stress and mitochondrial dysfunction in the ischemic spinal cord. As a result, it attenuated inflammatory and neuroapoptotic reactions, leading to improved motor function recovery. These findings show the therapeutic potential of freshly isolated mitochondria in alleviating the negative consequences of SCI and enhancing the restoration of motor function.

Intranasal administration route for MT

Intranasal (IN) administration is gaining popularity as a noninvasive medication delivery method at a rapid rate. The nasal mucosa is an ideal site for drug delivery simply because of a number of benefits, including its enormous surface area, quick onset, the possibility for central nervous system administration, and without hepatic route of metabolism (Costantino et al., 2007).

IN administration getting much attention as this route could bypass the BBB by migrating from the nasal mucosa through the cribriform plate along the olfactory neural pathway into the brain and cerebrospinal fluid. Importantly, this could minimize or eliminate the distribution of deliverables to peripheral organs and obviate the need for targeted implantation. This idea was tested by IN administration of mesenchymal stem cells to adult mice and administration of glioma cells to young rats (Danielyan et al., 2009).

Although the nasal mucosa serves as a permeation barrier for high molecular weight therapies such as peptides and proteins, the tight junctions that constitute this barrier to paracellular drug delivery can be opened reversibly and safely (Johnson & Quay, 2005). IN delivery is noninvasive, virtually painless, does not need to be sterilized, and may be simply and quickly delivered. Additionally, the nasal route may provide better biomolecule delivery (Danielyan et al., 2009). A variety of medications with low and large molecular weights (such as peptides and proteins) have IN formulations that have been approved for marketing, and other products are now being developed as a result of these variables (Fig. 12.2). Given these advantages, it seems sensible to take IN administration into account while creating novel treatments like MT.

IN administration routes of MT used in animal models of disease are summarized in Table 12.3.

Brain

Chang et al. (2021) proposed pursuing novel therapeutics for PD by IN delivery as a targeted approach. In study, allogeneic rat livers were used for isolation of mitochondria and mtDNA in hepatocytes and were labeled with 5-Bromodeoxyuridine 36 h prior to isolation. Administration of liver mitochondria was performed through an IN drop (on the ipsilateral side of the lesioned area) of 200 μg/50 μL MiR05 respiration buffer using a micropipette. Notably, mitochondrial function was restored, oxidative damage was attenuated in the lesioned substantia nigra, and plasma levels of inflammatory cytokines were suppressed in the MT group. The given mitochondria effectively entered the ipsilateral accessory olfactory bulb and doublecortin positive rostral migratory stream (RMS) neurons, indicating their possible participation in neurogenesis. These findings provide insights into the targeted delivery of allogeneic liver mitochondria for potential therapeutic applications in PD.

In a study by Alexander et al. (2021), human MSC derived mitochondria were injected into male C57BL/6J mice with a cisplatin induced cognitive impairments model at a dose of 2.3 mg/kg daily for 5 days (5 days of rest followed by an additional 5 days of cisplatin injection). Human MSC lentivirally transduced with mitochondria targeted PDHA1-DsRed to express PDHA1-DsRed mitochondria. Mitochondria were extracted from 4.5×10^6 human MSC (170 g of mitochondrial protein) and resuspended in PBS before being nasally delivered in 3 mL drops to each nostril (12 mL total volume per mouse). The minimal effective dose of mitochondria needed for reversal of cognitive deficits was 340 μg of human MSC derived mitochondria (obtained from on average 9×10^6 MSC). The study concluded that IN delivery of mitochondria resulted in regaining brain structure as well as function. It is significant that nasal delivery enables the mitochondria to penetrate the meninges and parenchyma of the brain, undergo fast cellular internalization, and modify the hippocampus transcriptome. These findings emphasize the potential of IN delivery of mitochondria and a therapeutic approach for the treatment of cognitive deficits and peripheral neuropathy.

Bobkova et al. (2022) studied spatial memory impairment by olfactory bulbectomized (OBX) in mice with AD type degeneration. Freshly isolated allogenic brain derived mitochondria were administered IN of a total volume of 6 μL per mouse three times, with 2, 10, 20 μg of mitochondrial protein distributed evenly into each nostril. Notably, this intervention resulted in a significant improvement in spatial memory in the OBX mice. Mitochondria penetrated the RMS and other parts of the brain, including the hippocampus, where they were colocalized. These findings underscore the potential therapeutic value of noninvasive IN delivery of mitochondria in mitigating cognitive deficits associated with AD.

Alexander et al. (2022) recently reported application of human MSCs derived mitochondria by IN delivery in cognitive deficits model in male C57BL/6J mice. Human MSC that had been lentivirally transduced with mitochondria targeted PDHA1-DsRed to express

PDHA1-DsRed mitochondria. To increase nasal mucosal permeability, hyaluronidase in PBS (50U in 3 μL) was administered once to each nostril (100 U total per mouse), 30 min prior to administration of mitochondria. The two types of mitochondria used in the therapy were dextran-TPP coated mitochondria (3 μg of mitochondrial protein per day obtained from an average 0.08×10^6 human MSC) or uncoated mitochondria (170 μg of mitochondrial protein per day obtained from on average 4.5×10^6 human MSC) were resuspended in PBS and nasally administered in 3 μL drops, twice alternately to each nostril (12 μL total volume per mouse) after 48 and 96 h. The study emphasized the therapeutic potential of isolated mitochondria in regaining brain structure as well as function. Interestingly, nasal delivery enabled the mitochondria were rapidly detectable in macrophages in the brain meninges but do not reach the brain parenchyma. The reversal of these neuropathologies was related to remyelination, synaptosomal mitochondrial integrity, and neurogenesis. These results highlight the possibility of IN administration, which at >55 times lower doses than uncoated mitochondria can correct cisplatin induced cognitive impairments and alleviate neuropathic pain.

Hosseini et al. (2022) conducted study involving photothrombotic medial prefrontal cortex (mPFC) ischemia in C57BL/6J mice. Mitochondria were isolated from BM-MSCs and administered via the IN route. The administration was performed every other day for 1 week, using three different doses of mitochondria (85 μg, 170 μg, and 340 μg). The results revealed that ischemic stroke led to decreased levels of synaptic markers in the mPFC. However, the introduction of exogenous mitochondria at doses of 170 and 340 μg significantly increased the expression of GAP-43 and PSD-95, indicating their potential in enhancing synaptic marker expression following ischemic stroke.

The results of these investigations highlight the importance of MT in controlling synaptic marker expression as well as stimulating optimal mitochondrial activity. These findings suggested that MT might be an effective technique for treating cognitive and memory deficits that occur following injury.

Clinical reality of mitochondrial transplantation

Numerous studies have revealed how important a role mitochondrial malfunction plays in a number of diseases and disorders and how healthy mitochondria help rescue these conditions. Importantly, MT from autologous, allogeneic, or xenogeneic sources administered directly, systemically, or IN did not elicit an immunological response or induce the formation of IgM, favoring the safety and application of MT in human subjects (McCully et al., 2017). MT approach has been subsequently tested on humans to treat a few pathological disorders, which are detailed in this section (Table 12.4).

The first attempt in this regard was performed in 2016 by the McCully research team at Harvard University's Boston Children's Hospital (Emani et al., 2017). Patients with myocardial IR are helped by a cutting edge method known as "mitochondrial autotransplantation." The investigation involved five pediatric patients with myocardial ischemia who required central extracorporeal membrane oxygenation (ECMO) support, which range in age from 2 days to 2 years.

TABLE 12.4 Human studies on mitochondrial transplant.

No.	Source of mitochondria	Route of mitochondrial transplantation	Human disease	Number of patients	Mitochondrial dose	Therapeutic outcome	References
Heart							
1	Rectus abdominis	In situ to ischemic area of heart	Central ECMO support for IR associated myocardial dysfunction after cardiac surgery	5 Pediatric patients	Ten 100 mL injections containing $1 \times 10^7 \pm 1 \times 10^4$	5 subjects, 4 demonstrated improvement in ventricular function and were successfully separated from ECMO support. All demonstrated improvement in ventricular function within several days after treatment.	Emani et al. (2017)
2	Rectus abdominis	Affected regions of myocardium by direct epicardial injection	Severe refractory cardiogenic shock after IR injury.	24 patients (MT, n = 10; control, n = 14).	10–20 separate injections of 50 μL to 100 μL of mitochondrial suspension using a 1-mL tuberculin syringe, with each injection containing approximately $1–5 \times 10^7$ mitochondria.	The safety of MT in pediatric patients following myocardial IR injury. The safety of MT in pediatric patients following myocardial IR injury. Enhanced myocardial cellular viability and function following ischemia.	Guariento et al. (2021)

In enrolled patients, the mediastinum was accessed and epicardial echocardiography was performed to identify regions of myocardial akinesis or hypokinesis. Autologous rectus abdominis muscle (6 × 6 mm) was used for mitochondria isolation and $1 \times 10^8 \pm 1 \times 10^5$ mitochondria were isolated under sterile conditions and suspended in 1 mL respiration buffer. Ten 100 mL injections containing $1 \times 10^7 \pm 1 \times 10^4$ mitochondria each were delivered by direct injection with a 1 mL tuberculin syringe to IR affected myocardium.

There was no detectable difference between preinjection and postinjection markers of systemic inflammatory response syndrome evidenced by stable respiratory and renal status.

Two of the patients did not survive, due to persistent pulmonary insufficiency and other complications. Three patients discharged from the hospital and echocardiogram after therapy showed global moderate, mild and borderline dysfunction respectively.

Subsequent to this, in 2020, the same team reported on their further experience injecting autologous mitochondria into the myocardium of pediatric patients at Boston Children's Hospital from May 2002 to December 2018, 10 pediatric patients (14 controls) with severe heart disease who required central ECMO support for IR injury associated myocardial dysfunction after cardiac surgery (Guariento et al., 2021). Autologous mitochondria were isolated from rectus abdominis muscle in 20–30 min under sterile circumstances in the cardiac intensive care unit or operating room. The mitochondria of a healthy rectus abdominis muscle sample measuring 6 mm by 6 mm were separated, generating 2×10^{10} live and respiration competent mitochondria from a tissue sample weighing 0.18 ± 0.04 g (wet weight). Approximately $1 \times 10^8 \pm 1 \times 10^9$ isolated mitochondria were suspended in 1 mL of respiration buffer. Using a tuberculin syringe, mitochondria were directly injected into the IR injury affected myocardium during the procedure. No individuals experienced any short term adverse reaction after mitochondrial injection, such as an arrhythmia, an intramyocardial hematoma, or scarring. Patients who received MT and controls did not have a difference in their blood parameters (white blood count, serum lactate, creatinine, and blood urea nitrogen). Importantly, patients who had MT had improved ventricular strain magnitude at the time of initial decannulation, a higher likelihood of successfully weaning from ECMO, and a reduced overall incidence of cardiovascular events.

One of the clinical applications of MT by Jacoby et al. (2022) worth mentioning in this section where a mitochondrial augmentation therapy (MAT) approach has been applied on six patients with single large scale mtDNA deletion syndromes. Autologous hematopoietic stem and progenitor cells (HSPCs) were augmented ex vivo with mitochondria obtained from maternal tissue.

It has been demonstrated that HSPCs can migrate to distant organs and prevent the progression of metabolic or mitochondrial diseases. In this case, $CD34^+$ cells were obtained from a patient, whereas peripheral blood mononuclear cells (PBMCs) were obtained from the maternal donor. Healthy mitochondria were implanted into the patients' $CD34^+$ cells using a coincubation technique and then reinfused intravenously. In four out of six patients, MAT resulted in decreased heteroplasmy of mtDNA deletion, increased mtDNA levels, and higher ATP content in PBMCs, as well as improved muscle strength and endurance in two others.

According to numerous preclinical investigations in animals, mitochondrial damage can be repaired by introducing healthy mitochondria into diseases cells. MT have been conducted for experimental human therapies targeted toward cerebral ischemia, ischemic heart damage, and polymyositis dermatomyositis. A list of four human trials mined from clinicaltrials.gov (Dated: 29-05–2023) is presented in Table 12.5.

The investigators at the University of Washington are recruiting subjects for a first in human brain trial to confirm the safety of autologous mitochondrial transplant to patients with brain ischemia. During treatment, investigators will obtain a very small biopsy from the muscle tissue adjacent to surgical access site. Muscle tissue will be processed at bedside to extract the autologous mitochondria. The mitochondria will be infused into the brain artery via microcatheter throughout the reperfusion. Outcome measures of the trial will be incidence

TABLE 12.5 Ongoing clinical trials of mitochondrial transplantation (https://clinicaltrials.gov/, Dated: 29-05–2023).

No.	Condition or disease	Intervention/Treatment	Phase (Participants)	Locations	ClinicalTrials.gov identifier:	Current status
1	Cerebral ischemia	Endovascular autologous mitochondrial transplantation	NA (20)	Harborview Medical Centre, University of Washington, USA	NCT04998357	Recruiting
2	Extracorporeal membrane oxygenation complication	Autologous mitochondria transplantation	NA (16)	Boston Children's Hospital Boston, Massachusetts, USA	NCT02851758	Recruiting
3	Myocardial infarction/ Ischemia/ Stunning	Autologous mitochondria and MSC-derived exosomes	1/2 (20)	Tehran University of Medical Sciences Tehran, Iran	NCT05669144	Recruiting
4	Polymyositis Dermatomyositis	Mitochondria isolated from allogeneic umbilical cord-derived mesenchymal stem cells.	1/2 (9)	Seoul National University Hospital, Seoul, Korea, Soonchunhyang University Seoul Hospital, Seoul, Korea, Hanyang University Seoul Hospital, Seoul, Korea,	NCT04976140	Enrolling by invitation

of severe adverse events during and postmitochondrial infusion and related to muscle biopsy.

An ongoing clinical trial in human hearts at Boston Children's Hospital will be demonstrating tolerability and safety of autologous transplantation of mitochondria via infusion or direct injection. Autologous mitochondria will be isolated from one to two, 6 mm biopsies from the exposed skeletal muscle of the chest wall. The tissue will be processed at bedside for autologous mitochondria isolation and will be injected via 5–10 injections of approximately 0.1 mL each to the damaged area via injection into the proximal aorta while cross clamped for clinically indicated surgery for global distribution of mitochondria via the coronary arteries. Following completion of surgical maneuvers, the mitochondria will be injected into the aorta and the cross clamp will be removed. Mitochondria will be infused in 5 mL of buffer via intracoronary infusion followed by a 5 mL flush with normal saline. If there will be no marked improvement in ventricular function following the infusion of autologous mitochondria, a second infusion will be administered. Primary outcome measures of the trial will be incidence of severe adverse events of MT and secondary outcome measures will be efficacy in improving in ventricular function and the ability to decannulate from ECMO support.

Tehran University of Medical Sciences is recruiting participants with myocardial infarction, ischemia, and stunning. The study will involve the administration of exosomes, mitochondria, exosomes and mitochondria together, and a placebo solution to five patients each. The treatments will be administered via intracoronary and intramyocardial injections. The dose of exosomes will be 100 μg in 1 mL, while the concentration of mitochondria will be 10 million in 1 mL. Exosomes and mitochondria will be administered concurrently, with a concentration of 100 μg of exosomes in 1 mL and 10 million mitochondria in 1 mL. Finally, the placebo solution will be 1 mL. Primary outcome measures of the trial will be incidence of allergic reactions including angioedema, hypotension, acute allergic reaction, and improvement in the left ventricle ejection fraction. Secondary outcome measures will be improvement cardiac functions, cardiac defect size analysis, on pump duration, ECMO duration, creatine kinase, and cardiac troponin levels.

Paean Biotechnology Inc. will be conducting clinical trial in collaboration with Seoul National University College of Medicine, Hanyang University College of Medicine, and Soonchunhyang University College of Medicine, Korea, in patients with refractory polymyositis or dermatomyositis. Mitochondria will be isolated from hUC-MSCs (referred as PN-101 mitochondria) and intravenously administered. Study did not reveal the dose of mitochondria and patient will be grouped according to dosage of PN-101 mitochondria. Low dose group (3 or 6 subjects) will be intravenously administered a single low dose, intermediate dose group (3 or 6 subjects) will be intravenously administered a single intermediate dose and high dose group (3 or 6 subjects) will be intravenously administered a single high dose. The primary goal of the trial will be to evaluate PN-101 mitochondria, confirm dosage, and restrict toxicity for 2 weeks administration. Additionally, International Myositis and Clinical Studies Group-Total Improvement Score (IMACS-TIS) will be assessed at Week 12 after the intravenous administration. Secondary outcome measures will be to access improvement in IMACS-TIS, 4, 8, and 12 weeks after the intravenous administration and Peak Pruritus Numeral Rating Scale changes on Day 1, and Weeks 4, 8, and 12 following intravenous administration.

The clinical outcomes suggest that MT has a promising future in the treatment of diseases involving mitochondria. The lack of uniform standards for all patients and mitochondrial numbers is the study's main constraint. The studies overlook the causal relationship between MT and subsequent clinical improvement, which turns out to be the most significant issue. For MT to be regarded as a therapeutic tool, a more uniform assessment of its potential in clinical research is necessary, and a large number of patients should be included in the studies.

Dosing considerations

It was observed a lot of variability in the therapeutic dose of MT per organ or tissue in preclinical investigations, and up to this point, an ideal dose of mitochondria for transplantation has not been optimized. Importantly, dose dependent treatment efficacy was not measured in many studies. According to Emani and McCully (Emani & McCully, 2018), mitochondrial dose of 2×10^5 mitochondria per gram of tissue wet weight was determined to be an optimal dose for myocardial IR injury studies and has been used for initial pilot trials in humans. There is still insufficient information about the dose of mitochondria required per organ or tissue for various disease pathologies.

In these preclinical experiments, injection dosages ranged from 1×10^6 to 1×10^9 for different disease models. Orfany et al. (2020) employed a dose of 1×10^6 to 1×10^9 g/wet weight to treat skeletal muscle damage. Lin et al. (2013) transplanted $7.7 \times 10^6 \pm 1.5 \times 10^6$/mL of mitochondria into the liver. In a lung ischemia model, Moskowitzova et al. (2020) transplanted 1×10^8/0.3 mL mitochondria. In order to treat the ischemic myocardium, Shin et al. (2019) transplanted 1×10^9/mL of mitochondria.

Various studies evaluated protein content to identify the appropriate dose of mitochondria. For various disease models in these preclinical trials, injection dosages ranged from 50 μg to 10 mg. 50 μg mitochondria were delivered systemically in a recent study by Alway et al. (2023). In rat brain ischemia injury, Huang et al. (2016) employed two separate dosages of 75 and 750 μg mitochondria. Sun et al. (2015) performed a similar procedure using 750 and 1500 μg of mitochondria on a model of acute respiratory distress syndrome. For the tendinopathy model, Lee et al. (2021) injected 10 and 50 μg of mitochondria into the Achilles tendon, and for the liver toxicity model, Zhao, Z., et al. (2021) injected 0.2 or 0.4 mg/kg of mitochondria once per day for 7 days. Although the therapeutic effectiveness of mitochondrial transplantation was shown in these investigations, the ideal dose was not evaluated.

Our study suggested that the number of mitochondria needed for endometrial regeneration is 1×10^6 for uterine cavities. Interestingly, we used cryopreserved mitochondria (10% DMSO in FBS) for the study. Viability of transplanted mitochondria was 97.7% and functionality of the mitochondria determined the specific activities of enzymes (NADH-cytochrome c reductase, NADPH-cytochrome c reductase, cytochrome *b*5, cytochrome P450, complex I, II, III, and IV). We found that the MT followed by PRP resulted in 33.33% engraftment. It is interesting to note that the viability, health, and function of isolated mitochondria are found to be considerably more critical than the administration dose (Kshersagar et al., 2023).

Ethical challenges

The legalization of MT as a therapeutic modality is significant both nationally and internationally. Technologies that can prevent a child from inheriting the condition have attracted widespread support because there are currently no cures and limited therapy options for mitochondrial disease (Caicedo et al., 2017). In March 2015, the UK government became the first to allow the clinical use of innovative in vitro fertilization techniques including mitochondrial donation. The National Guidelines for Accreditation, Supervision, and Regulation of Artificial Reproductive Technology (ART), published in 2005 by the Indian Council of Medical Research (ICMR) and the National Academy of Medical Sciences (NAMS) to regulate ART clinics, prohibit germline therapy, gene therapy for the enhancement of genetic characteristics (designer babies), and eugenic genetic engineering for selection against personality, character, formation of body organs, fertility, and intelligence (National Guidelines for Accreditation, 2005). Thus, regulations related to MT for inheriting mitochondrial disease are well placed in many countries and ethical review committees allow parents to decide if they wish to undergo the mitochondrial replacement technique. Although MT has been successfully practiced in clinic for limited number of patients, there are ongoing questions about the use of MT and benefits of this procedure to patients.

This section will explore ethical issues raised by MT in treating diseases such as concerns about safety of the procedure, the transfer of mitochondrial genetic material, source of mitochondria, and quality control of mitochondria.

The WHO Expert Advisory committee on developing global standards for governance and oversight of human genome editing noted that it was particularly difficult to articulate the ethical issues raised by cutting edge biological techniques like MT because these arguments are, at least in part, rooted in the sociocultural contexts from which they emerged (WHO Expert Advisory Committee on, 2021). Patients are often given the option to choose whether to undergo the MT procedure by ethical review boards in many different nations. The safety of MT and immune system behavior following autologous and allogeneic injections, single and serial injections, as well as a different number of isolated mitochondria (1×10^5, 1×10^6, or 1×10^7), was demonstrated in investigations conducted by McCully and his team (McCully et al., 2016). According to the results, the levels of immune system profiles such as IL-1, IL-4, IL-6, IL-12, IL-18, IP-10, and the macrophage inflammatory proteins MIP-1 alpha and MIP-1 beta remained stable. Injecting mitochondria one at a time or in series did not cause any DAMP to appear in the recipient tissues. The MT has been refined for cardiac applications; it should be transformed and optimized for other tissues as well. Although MT has been successfully applied to a wide range of animal research for diseases and human studies, there are concerns about the stress and invasiveness of the technique (Wang & Gerdes, 2015; Zhao, J., et al., 2021). Therapeutic application of mitochondria can be administered by direct application, systemic application, or IN route. However, the reports of in vivo studies raised the question of selecting the route of transplantation for uniform distribution of mitochondria at the site of injury (Liu et al., 2020).

In advance of evaluating MT in a therapeutic setting, this approach needs to undergo stringent ethical analysis. Both mitochondrial and nuclear genomes influence mitochondrial activity, and mtDNA is only inherited maternally (Tinker et al., 2021). The way that mtDNA and nDNA interact could have significant effects. An organism that is typically referred to as homoplasmy contains only one kind of mtDNA. Heteroplasmy is defined as the presence of different mtDNA types within a single organism (Parakatselaki & Ladoukakis, 2021), and transfer of genetic material of allogenic mitochondria can cause heteroplasmy (Clemente-Suárez et al., 2023). On the other hand, mitochondrial diseases (whether the recipient is heteroplasmic or homoplasmic) can be treated by transplantation or transfer of healthy mitochondria (D'Amato et al., 2023). The influence of the nuclear and mitochondrial genomes on mitochondrial function has been shown to have critical consequences (Rusecka et al., 2018). The genetic makeup of donor mitochondria and how recipient cell mitochondria function is an important issue. A study of McCully indicated that mtDNA concentrations increased in recipient supplied cardioprotection via damaged mtDNA substitute, mitochondrial protein production, and energy synthesis (Emani & McCully, 2018). In depth analysis of the heteroplasmy level in patient and tissues must be done to reveal the precise contribution of drift and selection in heteroplasmy function and dynamics.

Donor mitochondria's origin is one important concern. Because each donor source may have various ethical and biological implications, separate evaluations of autologous, allogenic, and xenogeneic sources are necessary. Through autologous mitochondria, a severely damaged organ in the same person could be treated with the mitochondria from a different tissue source with a minimal risk of mtDNA mutation. This donor source has high morbidity due to additional surgical procedures, but it does not create numerous ethical concerns. Mitochondrial donations from genetically close relatives would be used in allogeneic MT. The recipient and donor should ideally have the same haplotype (Burgstaller et al., 2015). One possibility that is ethically acceptability is the availability of other tissues sources like the placenta for the purpose of obtaining functional and more mitochondria. The last donor source approach, called xenogeneic, involves offering mitochondria from different animals to humans. Even though it may seem strange, investigations using xenogeneic MT did not result in any evident mismatch effects in animal to animal (McCully et al., 2023). Before establishing any potential clinical uses, various ethical issues relating to the effective in vitro or in vivo xenogeneic mitochondria between animals and humans would need to be addressed. Importantly, prior to clinical application of mitochondria, some quality control factors must be performed, including the number of mitochondria, mitochondrial viability, mitochondrial purity, and mitochondrial function.

Experts recommended long term follow up studies in order to monitor any possible long term effects of MT. The only way to solve the multiplicity of ethical and safety issues impeding the worldwide growth of this technology is to gather more evidence on these procedures and facilitate national and international conversations on these matters.

Conclusions

Therapeutics that enhance mitochondrial functionality in cells or tissues may be used to treat a variety of disorders when the mitochondria-related functions are lost. In this chapter, we detailed the methods by which MT can be used to evaluate its potential as a therapeutic agent for the treatment of mitochondrial dysfunction, considering direct, systemic, and nasal applications. Researchers are getting closer to healing mitochondrial diseases as our understanding of MT improves and technologies such as genome sequencing, large-scale screening, and bioconjugation increase. Many promising advancements in mitochondrial targeting treatments are accelerating studies toward defining the biology of this organelle and treating diseases caused by its malfunction.

In terms of the development of MT applications, preclinical and clinical methods have been explored in lights of the mitochondrial application approach used to date. The direct injection route of MT into target tissues for a large number of mitochondrial delivery has been supported by a significant amount of animal research and even human investigations. Although helpful in invasive surgical settings, it may not always be the ideal choice in clinical settings, especially in conditions affecting the brain such as ischemic strokes, PD, and AD. Numerous studies on animals and people have discussed the effectiveness of MT when administered systemically to treat a variety of diseases that involve mitochondrial damage. Interestingly, recently, the application of mitochondria through the IN route was investigated for brain disorders and ischemic brain disease.

It has been discovered that the requirement for mitochondria in the acute and chronic diseases differs, and the therapeutic efficacy of the treatment depends on this requirement. Intake efficiencies have also varied among different delivery routes. In light of these facts, the method of mitochondrial administration, the quantity of mitochondria, and the number of infusions must be tailored to the disease and its stage.

We would like to point out important conclusions which seem to influence the efficacy of MT:

1. The mitochondrial involvement in the disease and stage
2. The method of isolating mitochondria
3. The selection of the donor source of mitochondria
4. The functionality the donor source of mitochondria
5. The number of mitochondria to be transplanted
6. The number of infusions of mitochondria required and
7. The delivery route

We anticipate that mitochondrial transplantation will enhance the development of clinical translation to a range of medical conditions for patients and turn into a helpful treatment option. The effectiveness of these treatments will ultimately be determined by how well they work, how well they can be applied in therapeutic settings, and how well they can address ethical, legal, and biological issues. By overcoming these obstacles, we can unleash mitochondria's revolutionary potential, which was previously only recognized as the cell's source of energy.

Acknowledgments

Dr. Meghnad G. Joshi acknowledges the research funding support from the D.Y. Patil Education Society (Deemed to be University) (DYPES/DU/R&D/3104). The schematic art pieces used to produce the figures of this article were provided by Servier Medical Art. Servier Medical Art by Servier is licensed under a Creative Commons Attribution 3.0 Unported License.

References

Aggarwal, S., Mannam, P., & Zhang, J. (2016). Differential regulation of autophagy and mitophagy in pulmonary diseases. *American Journal of Physiology - Lung Cellular and Molecular Physiology, 311*, L433–L452. https://doi.org/10.1152/ajplung.00128.2016

Alexander, J. F., Mahalingam, R., Seua, A. V., Wu, S., Arroyo, L. D., Hörbelt, T., Schedlowski, M., Blanco, E., Kavelaars, A., & Heijnen, C. J. (April 2022). Targeting the meningeal compartment to resolve chemobrain and neuropathy via nasal delivery of functionalized mitochondria. *Advanced Healthcare Materials, 11*(8), 2102153.

Alexander, J. F., Seua, A. V., Arroyo, L. D., Ray, P. R., Wangzhou, A., Heiβ-Lückemann, L., Schedlowski, M., Price, T. J., Kavelaars, A., & Heijnen, C. J. (2021). Nasal administration of mitochondria reverses chemotherapy-induced cognitive deficits. *Theranostics, 11*, 3109–3130. https://doi.org/10.7150/THNO.53474

Alway, S. E., Paez, H. G., & Pitzer, C. R. (2023). The role of mitochondria in mediation of skeletal muscle repair. *Muscles, 2*(2), 119–163.

Alway, S. E., Paez, H. G., Pitzer, C. R., Ferrandi, P. J., Khan, M. M., Mohamed, J. S., Carson, J. A., & Deschenes, M. R. (2023). Mitochondria transplant therapy improves regeneration and restoration of injured skeletal muscle. *J Cachexia Sarcopenia Muscle, 14*, 493–507. https://doi.org/10.1002/jcsm.13153

Anjum, A., Yazid, M. D., Fauzi Daud, M., Idris, J., Ng, A. M. H., Selvi Naicker, A., Ismail, O. H. R., Athi Kumar, R. K., & Lokanathan, Y. (October 13, 2020). Spinal cord injury: Pathophysiology, multimolecular interactions, and underlying recovery mechanisms. *International Journal of Molecular Sciences, 21*(20), 7533. https://doi.org/10.3390/ijms21207533. PMID: 33066029; PMCID: PMC7589539.

Bashiri, A., Halper, K. I., & Orvieto, R. (2018). Recurrent Implantation Failure-update overview on etiology, diagnosis, treatment and future directions. *Reproductive Biology and Endocrinology, 16*, 1–19. https://doi.org/10.1186/s12958-018-0414-2

Bennett, C. F., Latorre-Muro, P., & Puigserver, P. (2022). Mechanisms of mitochondrial respiratory adaptation. *Nature Reviews Molecular Cell Biology, 23*(12), 817–835.

Bhargava, P., & Schnellmann, R. G. (October 2017). Mitochondrial energetics in the kidney. *Nature Reviews Nephrology, 13*(10), 629–646.

Blajszczak, C., & Bonini, M. G. (November 1, 2017). Mitochondria targeting by environmental stressors: Implications for redox cellular signaling. *Toxicology, 391*, 84–89. https://doi.org/10.1016/j.tox.2017.07.013. Epub 2017 Jul 24. PMID: 28750850; PMCID: PMC5939563.

Bobkova, N. V., Zhdanova, D. Y., Belosludtseva, N. V., Penkov, N. V., & Mironova, G. D. (2022). Intranasal administration of mitochondria improves spatial memory in olfactory bulbectomized mice. *Experimental Biology and Medicine, 247*, 416–425. https://doi.org/10.1177/15353702211056866

Brand, M. D. (2020). Riding the tiger - physiological and pathological effects of superoxide and hydrogen peroxide generated in themitochondrial matrix. *Critical Reviews in Biochemistry and Molecular Biology, 55*(6), 592–661.

Burgstaller, J. P., Johnston, I. G., & Poulton, J. (2015). Mitochondrial DNA disease and developmental implications for reproductive strategies. *Molecular Human Reproduction, 21*, 11–22. https://doi.org/10.1093/molehr/gau090

Burke, P. J. (2017). Mitochondria, bioenergetics and apoptosis in cancer. *Trends in Cancer, 3*, 857–870. https://doi.org/10.1016/j.trecan.2017.10.006

Caicedo, A., Aponte, P. M., Cabrera, F., Hidalgo, C., & Khoury, M. (October 2017). Artificial mitochondria transfer: Current challenges, advances, and future applications. *Stem Cells International, 2017*.

Chakrabarty, R. P., & Chandel, N. S. (2022). Beyond ATP, new roles of mitochondria. *The Biochemist, 44*(4), 2–8.

Chang, J. C., Chao, Y. C., Chang, H. S., Wu, Y. L., Chang, H. J., Lin, Y. S., Cheng, W. L., Lin, T. T., & Liu, C. S. (May 19, 2021). Intranasal delivery of mitochondria for treatment of Parkinson's Disease model rats lesioned with 6-hydroxydopamine. *Scientific Reports, 11*(1), 1–4.

Chang, J. C., Wu, S. L., Liu, K. H., Chen, Y. H., Chuang, C. S., Cheng, F. C., Su, H. L, Wei, Y. H, Kuo, S. J, & Liu, C. S. (2016). Allogeneic/xenogeneic transplantation of peptide-labeled mitochondria in Parkinson's disease: Restoration of mitochondria functions and attenuation of 6-hydroxydopamine-induced neurotoxicity. *Translational Research, 170*, 40–56.e3. https://doi.org/10.1016/j.trsl.2015.12.003

Chaudhari, L. R., Kawale, A. A., Desai, S. S., Kashte, S. B., & Joshi, M. G. (2023). Pathophysiology of spinal cord injury and tissue engineering approach for its neuronal regeneration: Current status and future prospects. *Advances in Experimental Medicine and Biology, 1409*, 51–81. https://doi.org/10.1007/5584_2022_731. PMID: 36038807.

Chazotte, B. (August 1, 2011). Labeling mitochondria with MitoTracker dyes. *Cold Spring Harbour Protocols, 2011*(8), pdb–rot5648.

Chen, L. B. (January 1, 1988). Fluorescent labeling of mitochondria. *Methods in Cell Biology, 29*, 103–123.

Cheng, G., Kong, R. H., Zhang, L. M., & Zhang, J. N. (October 2012). Mitochondria in traumatic brain injury and mitochondrial-targeted multipotential therapeutic strategies. *British Journal of Pharmacology, 167*(4), 699–719.

Chouchani, E. T., Pell, V. R., Gaude, E., Aksentijević, D., Sundier, S. Y., Robb, E. L., Logan, A., Nadtochiy, S. M., Ord, E. N. J., Smith, A. C., Eyassu, F., Shirley, R, Hu, C. H., Dare, A. J, James, A. M, Rogatti, S, Hartley, R. C., Eaton, S., Costa, A. S. H, … Murphy, M. P. (2014). Ischaemic accumulation of succinate controls reperfusion injury through mitochondrial ROS. *Nature, 515*, 431–435. https://doi.org/10.1038/nature13909

Clemente-Suárez, V. J., Martín-Rodríguez, A., Yáñez-Sepúlveda, R., & Tornero-Aguilera, J. F. (2023). Mitochondrial transfer as a novel therapeutic approach in disease diagnosis and treatment. *International Journal of Molecular Sciences, 24*(10). https://doi.org/10.3390/ijms24108848

Correia, B., Sousa, M. I., Branco, A. F., Rodrigues, A. S., & Ramalho-Santos, J. (2022). Leucine and arginine availability modulate mouse embryonic stem cell proliferation and metabolism. *International Journal of Molecular Sciences, 23*, 14286.

Costantino, H. R., Illum, L., Brandt, G., Johnson, P. H., & Quay, S. C. (2007). Intranasal delivery: Physicochemical and therapeutic aspects. *International Journal of Pharmacy, 337*, 1–24. https://doi.org/10.1016/j.ijpharm.2007.03.025

Cowan, D. B., Yao, R., Akurathi, V., Snay, E. R., Thedsanamoorthy, J. K., Zurakowski, D., Ericsson, M., Friehs, I., Wu, Y., Levitsky, S., Del Nido, P. J., Packard, A. B., & McCully, J. D. (2016). Intracoronary delivery of mitochondria to the ischemic heart for cardioprotection. *PLoS One, 11*(8), e0160889. https://doi.org/10.1371/journal.pone.0160889

D'Amato, M., Morra, F., Di Meo, I., & Tiranti, V. (2023). Mitochondrial transplantation in mitochondrial medicine: Current challenges and future perspectives. *International Journal of Molecular Sciences, 24*, 1969.

Danielyan, L., Schäfer, R., von Ameln-Mayerhofer, A., Buadze, M., Geisler, J., Klopfer, T., Burkhardt, U., Proksch, B., Verleysdonk, S., Ayturan, M., Buniatian, G. H., Gleiter, C. H., & Frey II, W. H. (2009). Intranasal delivery of cells to the brain. *European Journal of Cell Biology, 88*, 315–324. https://doi.org/10.1016/j.ejcb.2009.02.001

de Carvalho, L. R. P., Abreu, S. C., de Castro, L. L., Luísa Helen, A.da S., Silva, P. M., Vieira, J. B., Santos, R. T., Cabral, M. R., Khoury, M., Weiss, D. J., Lopes-Pacheco, M., Silva, P. L., Cruz, F. F., & Rocco, P. R. M. (September 2021). Mitochondria-rich fraction isolated from mesenchymal stromal cells reduces lung and distal organ injury in experimental sepsis. *Critical Care Medicine, 49*(9), e880–e890.

Dong, L. F., Rohlena, J., Zobalova, R., Nahacka, Z., Rodriguez, A. M., Berridge, M. V., & Neuzil, J. (March 6, 2023). Mitochondria on the move: Horizontal mitochondrial transfer in disease and health. *The Journal of Cell Biology, 222*(3), e20221104.

Doulamis, I. P., Guariento, A., Duignan, T., Kido, T., Orfany, A., Saeed, M. Y., Weixler, V., Blitzer, D., Shin, B., Snay, E. S., Inkster, J. A. H., Packard, A. B., Zurakowski, D., Rousselle, T., Bajwa, A., Parikh, S. M., Stillman, I., Nido, P. D., & McCully, J. D. (2020). Mitochondrial transplantation by intra-arterial injection for acute kidney injury. *American Journal of Physiology - Renal Physiology, 319*, F403–F413. https://doi.org/10.1152/AJPRENAL.00255.2020

Doulamis, I. P., Guariento, A., Duignan, T., Orfany, A., Kido, T., Zurakowski, D., Del Nido, P. J., & McCully, J. D. (2020). Mitochondrial transplantation for myocardial protection in diabetic hearts. *European Journal of Cardio-Thoracic Surgery, 57*, 836–845. https://doi.org/10.1093/ejcts/ezz326

Doulamis, I. P., & McCully, J. D. (June 3, 2021). Mitochondrial transplantation for ischemia reperfusion injury. In *Mitochondrial medicine: Volume 3: Manipulating mitochondria and disease-specific approaches* (pp. 15–37). New York, NY: Springer US.

Emani, S. M., & McCully, J. D. (2018). Mitochondrial transplantation: Applications for pediatric patients with congenital heart disease. *Translational Pediatrics, 7*(2), 169–175. https://doi.org/10.21037/tp.2018.02.02

Emani, S. M., Piekarski, B. L., Harrild, D., Del Nido, P. J., & McCully, J. D. (2017). Autologous mitochondrial transplantation for dysfunction after ischemia-reperfusion injury. *The Journal of Toracic and Cardiovascular Surgery, 154*(1), 286–289. https://doi.org/10.1016/j.jtcvs.2017.02.018

Fang, S. Y., Roan, J. N., Lee, J. S., Chiu, M. H., Lin, M. W., Liu, C. C., & Lam, C. F. (2021). Transplantation of viable mitochondria attenuates neurologic injury after spinal cord ischemia. *The Journal of Thoracic and Cardiovascular Surgery, 161*, e337–e347. https://doi.org/10.1016/j.jtcvs.2019.10.151

Fu, A., Shi, X., Zhang, H., & Fu, B. (2017). Mitotherapy for fatty liver by intravenous administration of exogenous mitochondria in male mice. *Frontiers in Pharmacology, 8*, 241.

Gargett, C. E., Nguyen, H. P. T., & Ye, L. (2012). Endometrial regeneration and endometrial stem/progenitor cells. *Reviews in Endocrine & Metabolic Disorders, 13*(4), 235–251. https://doi.org/10.1007/s11154-012-9221-9

Gollihue, J. L., Patel, S. P., Eldahan, K. C., Cox, D. H., Donahue, R. R., Taylor, B. K., Sullivan, P. G., & Rabchevsky, A. (2018). Effects of mitochondrial transplantation on bioenergetics, cellular incorporation, and functional recovery after spinal cord injury. *Journal of Neurotrauma, 35*, 1800–1818. https://doi.org/10.1089/neu.2017.5605

Gray, M. W. (2012). Mitochondrial evolution. *Cold Spring Harbor Perspectives in Biology, 4*(9), a011403.

Guariento, A., Blitzer, D., Doulamis, I., Shin, B., Moskowitzova, K., Orfany, A., Ramirez-Barbieri, G., Staffa, S. J., Zurakowski, D., Del Nido, P. J., & McCully, J. D. (August 1, 2020). Preischemic autologous mitochondrial transplantation by intracoronary injection for myocardial protection. *The Journal of Thoracic and Cardiovascular Surgery, 160*(2), e15–e29.

Guariento, A., Piekarski, B. L., Doulamis, I. P., Blitzer, D., Ferraro, A. M., Harrild, D. M., Zurakowski, D., Del Nido, P. J., McCully, J. D., & Emani, S. M. (September 1, 2021). Autologous mitochondrial transplantation for cardiogenic shock in pediatric patients following ischemia-reperfusion injury. *The Journal of Thoracic and Cardiovascular Surgery, 162*(3), 992–1001.

Hall, A. M., Rhodes, G. J., Sandoval, R. M., Corridon, P. R., & Molitoris, B. A. (2013). In vivo multiphoton imaging of mitochondrial structure and function during acute kidney injury. *Kidney International, 83*(1), 72–83.

Hayashida, K., Takegawa, R., Endo, Y., Yin, T., Choudhary, R. C., Aoki, T., Nishikimi, M., Murao, A., Nakamura, E., Shoaib, M., Kuschner, C., Miyara, S. J., Kim, J., Shinozaki, K., Wang, P., & Becker, L. B. (March 16, 2023). Exogenous mitochondrial transplantation improves survival and neurological outcomes after resuscitation from cardiac arrest. *BMC Medicine, 21*(1), 56.

Herst, P. M., Dawson, R. H., & Berridge, M. V. (2018). Intercellular communi- cation in tumor biology: A role for mitochondrial transfer. *Frontiers in Oncology, 8*, 344.

Hock, M. B., & Kralli, A. (2009). Transcriptional control of mitochondrial biogenesis and function. *Annual Review of Physiology, 71*(1), 177–203.

Hong, I. S. (2023). Endometrial stem/progenitor cells: Properties, origins, and functions. *Genes & Diseases, 10*(3), 931–947. https://doi.org/10.1016/j.gendis.2022.08.009

Hosseini, L., Karimipour, M., Seyedaghamiri, F., Abolhasanpour, N., Sadigh-Eteghad, S., Mahmoudi, J., & Farhoudi, M. (December 2022). Intranasal administration of mitochondria alleviated cognitive impairments and mitochondrial dysfunction in the photothrombotic model of mPFC stroke in mice. *Journal of Stroke and Cerebrovascular Diseases, 31*(12), 106801.

Hsu, C. H., Roan, J. N., Fang, S. Y., Chiu, M. H., Cheng, T. T., Huang, C. C., Lin, M. W., & Lam, C. F. (2022). Transplantation of viable mitochondria improves right ventricular performance and pulmonary artery remodeling in rats with pulmonary arterial hypertension. *The Journal of Thoracic and Cardiovascular Surgery, 163*(5), e361–e373.

Huang, P. J., Kuo, C. C., Lee, H. C., Shen, C. I., Cheng, F. C., Wu, S. F., Chang, J. C., Pan, H. C., Lin, S. Z., Liu, C. S., & Su, H. L. (2016). Transferring xenogenic mitochondria provides neural protection against ischemic stress in ischemic rat brains. *Cell Transplantation, 25*, 913–927. https://doi.org/10.3727/096368915X689785

Islam, M. N., Das, S. R., Emin, M. T., Wei, M., Sun, L., Westphalen, K., Rowlands, D. J., Quadri, S. K., Bhattacharya, S., & Bhattacharya, J. (2012). Mitochondrial transfer from bone- marrow-derived stromal cells to pulmonary alveoli protects against acute lung injury. *Nature Medicine, 18*, 759–765.

Jabbari, H., Roushandeh, A. M., Rostami, M. K., Razavi-Toosi, M. T., Shokrgozar, M. A., Jahanian-Najafabadi, A., Kuwahara, Y., & Habibi Roudkenar, M. (2020). Mitochondrial transplantation ameliorates ischemia/reperfusion-induced kidney injury in rat. *Biochimica et Biophysica Acta, Molecular Basis of Disease, 1866*, 165809. https://doi.org/10.1016/j.bbadis.2020.165809

Jacoby, E., Bar-Yosef, O., Gruber, N., Lahav, E., Varda-Bloom, N., Bolkier, Y., Bar, D., Blumkin, M. B., Barak, S., Eisenstein, E., & Ahonniska-Assa, J. (December 21, 2022). Mitochondrial augmentation of hematopoietic stem cells in children with single large-scale mitochondrial DNA deletion syndromes. *Science Translational Medicine, 14*(676), eabo3724.

Javani, G., Ghaffari-Nasab, A., Farajdokht, F., & Mohaddes, G. (2023). Chronic stress-induced apoptosis is mitigated by young mitochondria transplantation in the prefrontal cortex of aged rats. *Iran J Basic Med Sci, 26*, 725–730.

Jia, X., Wang, Q., Ji, J., Lu, W., Liu, Z., Tian, H., Guo, L., & Wang, Y. (2023). Mitochondrial transplantation ameliorates hippocampal damage following status epilepticus. *Anim Model Exp Med, 6*, 41–50. https://doi.org/10.1002/ame2.12310

Johnson, P. H., & Quay, S. C. (2005). Advances in nasal drug delivery through tight junction technology. *Expert Opinion on Drug Delivery, 2*(2), 281–298. https://doi.org/10.1517/17425247.2.2.281

Kaza, A. K., Wamala, I., Friehs, I., Kuebler, J. D., Rathod, R. H., Berra, I., Ericsson, M., Yao, R., Thedsanamoorthy, J. K., Zurakowski, D., Levitsky, S., Del Nido, P. J., Cowan, D. B., & McCully, J. D. (2017). Myocardial rescue with autologous mitochondrial transplantation in a porcine model of ischemia/reperfusion. *The Journal of Thoracic and Cardiovascular Surgery, 153*, 934–943. https://doi.org/10.1016/j.jtcvs.2016.10.077, 2016.

Kim, M. J., Hwang, J. W., Yun, C.-K., Lee, Y., & Choi, Y.-S. (2018). Delivery of exogenous mitochondria via centrifugation enhances cellular metabolic function. *Scientific Reports, 8*, 3330.

Ko, S. F., Chen, Y. L., Sung, P. H., Chiang, J. Y., Chu, Y. C., Huang, C. C., Huang, C. R., & Yip, H. K. (2020). Hepatic 31P-magnetic resonance spectroscopy identified the impact of melatonin-pretreated mitochondria in acute liver ischaemia-reperfusion injury. *Journal of Cellular and Molecular Medicine, 24*, 10088–10099. https://doi.org/10.1111/jcmm.15617

Konari, N., Nagaishi, K., Kikuchi, S., & Fujimiya, M. (2019). Mitochondria transfer from mesenchymal stem cells structurally and functionally repairs renal proximal tubular epithelial cells in diabetic nephropathy in vivo. *Scientific Reports, 9*, 5184.

Kshersagar, J., Pulgam, L., Damle, M. N., Tardalkar, K., Sharma, R., & Joshi, M. G. (2023). Transplantation of human placenta derived mitochondria promotes cell communication in endometrium in a murine model of disturbed endometrium. *Stem Cell Reviews and Reports, 19(5)*, 1384–1401. https://doi.org/10.1007/s12015-023-10516-2

Kubat, G. B., Ozler, M., Ulger, O., Ekinci, O., Atalay, O., Celik, E., Safali, M., & Budak, M. T. (2020). The effects of mesenchymal stem cell mitochondrial transplantation on doxorubicin-mediated nephrotoxicity in rats. *Journal of Biochemical and Molecular Toxicology, 35*, e22612.

Kubat, G. B., Ulger, O., & Akin, S. (2021). Requirements for successful mitochondrial transplantation. *Journal of Biochemical and Molecular Toxicology, 35*(11), e22898.

Kwon, B. K., Tetzlaff, W., Grauer, J. N., Beiner, J., & Vaccaro, A. R. (July 1, 2004). Pathophysiology and pharmacologic treatment of acute spinal cord injury. *The Spine Journal, 4*(4), 451–464.

Lee, J. M., Hwang, J. W., Kim, M. J., Jung, S. Y., Kim, K. S., Ahn, E. H., Min, K., & Choi, Y. S. (April 28, 2021). Mitochondrial transplantation modulates inflammation and apoptosis, alleviating tendinopathy both in vivo and in vitro. *Antioxidants, 10*(5), 696.

Lerner, C. A., Sundar, I. K., & Rahman, I. (2016). Mitochondrial redox system, dynamics, and dysfunction in lung inflammaging and COPD. *The International Journal of Biochemistry & Cell Biology, 81*, 294–306. https://doi.org/10.1016/j.biocel.2016.07.026

Li, H., Wang, C., He, T., Zhao, T., Chen, Y. Y., Shen, Y. L., Zhang, X., & Wang, L. L. (2019). Mitochondrial transfer from bone marrow mesenchymal stem cells to motor neurons in spinal cord injury rats via gap junction. *Theranostics, 9*, 2017–2035.

Liesa, M., Palacín, M., & Zorzano, A. (July 2009). Mitochondrial dynamics in mammalian health and disease. *Physiological Reviews, 89*(3), 799–845.

Lin, H. C., Liu, S. Y., Lai, H. S., & Lai, I. R. (2013). Isolated mitochondria infusion mitigates ischemia-reperfusion injury of the liver in rats. *Shock, 39*, 304–310.

Lin, M.-W., Fang, S.-Y., Hsu, J.-Y. C., Huang, C.-Y., Lee, P.-H., Huang, C.-C., Chen, H.-F., Lam, C.-F., & Lee, J.-S. (2022). Mitochondrial transplantation attenuates neural damage and improves locomotor function after traumatic spinal cord injury in rats. *Frontiers in Neuroscience, 16*, 800883.

Liu, X., Khouri-Farah, N., Wu, C. H., & Wu, G. Y. (2020). Targeted delivery of mitochondria to the liver in rats. *Journal of Gastroenterology and Hepatology, 35*, 2241–2247. https://doi.org/10.1111/jgh.15091

Masuzawa, A., Black, K. M., Pacak, C. A., Ericsson, M., Barnett, R. J., Drumm, C., Seth, P., Bloch, D. B., Levitsky, S., Cowan, D. B., & McCully, J. D. (2013). Transplantation of autologously derived mitochondria protects the heart from ischemia-reperfusion injury. *American Journal of Physiology - Heart and Circulatory Physiology, 304*, H966–H982, 2013.

McCalley, S. (2018). *Elucidating the mitochondrial architecture of branched-chain amino acid metabolism enzymes: Implications for treatment. Doctoral dissertation*. University of Pittsburgh.

McCully, J. D., Cowan, D. B., Emani, S. M., & Del Nido, P. J. (May 1, 2017). Mitochondrial transplantation: From animal models to clinical use in humans. *Mitochondrion, 34*, 127–134.

McCully, J. D., del Nido, P. J., & Emani, S. M. (2023). Mitochondrial transplantation: The advance to therapeutic application and molecular modulation. *Frontiers in Cardiovascular Medicine, 10*. https://doi.org/10.3389/fcvm.2023.1268814

McCully, J. D., Levitsky, S., Del Nido, P. J., & Cowan, D. B. (2016). Mitochondrial transplantation for therapeutic use. *Clinical and Translational Medicine, 5*(16). https://doi.org/10.1186/s40169-016-0095-4

Mokhtari, B., Hamidi, M., Badalzadeh, R., & Mahmoodpoor, A. (March 2023). Mitochondrial transplantation protects against sepsis-induced myocardial dysfunction by modulating mitochondrial biogenesis and fission/fusion and inflammatory response. *Molecular Biology Reports, 50*(3), 2147–2158.

Moskowitzova, K., Orfany, A., Liu, K., Ramirez-Barbieri, G., Thedsanamoorthy, J. K., Yao, R., Guariento, A., Doulamis, I. P., Blitzer, D., Shin, B., Snay, E. R., Inkster, J. A. H., Iken K, Packard, A. B., Cowan, D. B., Visner, G. A., Del Nido, P. J., & McCully, J. D. (2020). Mitochondrial transplantation enhances murine lung viability and recovery after ischemia-reperfusion injury. *American Journal of Physiology - Lung Cellular and Molecular Physiology, 318*, L78–L88. https://doi.org/10.1152/ajplung.00221.2019

Murphy, E., & Steenbergen, C. (2008). Mechanisms underlying acute protection from cardiac ischemia-reperfusion injury. *Physiological Reviews, 88*, 581–609. https://doi.org/10.1152/physrev.00024.2007

Moskowitzova, K., Shin, B., Liu, K., Ramirez-Barbieri, G., Guariento, A., Blitzer, D., Thedsanamoorthy, J. K., Yao, R., Snay, E. R., Inkster, J. A. H., Orfany, A., Zurakowski, D., Cowan, D. B., Packard, A. B., Visner, G. A., Del Nido, P. J., & McCully, J. D. (2019). Mitochondrial transplantation prolongs cold ischemia time in murine heart transplantation. *The Journal of Heart and Lung Transplantation, 38*, 92–99. https://doi.org/10.1016/j.healun.2018.09.025

Murphy, M. P., & Hartley, R. C. (December 2018). Mitochondria as a therapeutic target for common pathologies. *Nature Reviews Drug Discovery, 17*(12), 865–886.

Nakamura, Y., Lo, E. H., & Hayakawa, K. (2020). Placental mitochondria therapy for cerebral ischemia-reperfusion injury in mice. *Stroke*, 3142–3146. https://doi.org/10.1161/STROKEAHA.120.030152

Nascimento-Dos-Santos, G., de-Souza-Ferreira, E., Lani, R., Faria, C. C., Araújo, V. G., Teixeira-Pinheiro, L. C., Vasconcelos, T., Gonçalo, T., Santiago, M. F., Linden, R., Galina, A., & Petrs-Silva, H. (May 1, 2020). Neuroprotection from optic nerve injury and modulation of oxidative metabolism by transplantation of active mitochondria to the retina. *Biochimica et Biophysica Acta, Molecular Basis of Disease, 1866*(5), 165686.

Nascimento-Dos-santos, G., De-Souza-ferreira, E., Linden, R., Galina, A., & Petrs-Silva, H. (2021). Mitotherapy: Unraveling a promising treatment for disorders of the central nervous system and other systemic conditions. *Cells, 10*, 1–16. https://doi.org/10.3390/cells10071827

National Guidelines for Accreditation. (2005). *Supervision and regulation of ART clinics in India*. dated 09-08-2023 https://main.icmr.nic.in/sites/default/files/art/ART_Pdf.pdf.

Neubauer, S. (2007). The failing heart — an engine out of fuel. *New England Journal of Medicine, 356*, 1140–1151.

Nitzan, K., Benhamron, S., Valitsky, M., Kesner, E. E., Lichtenstein, M., Ben-Zvi, A., Ella, E., Segalstein, Y., Saada, A., Lorberboum-Galski, H., & Rosenmann, H. (2019). Mitochondrial transfer ameliorates cognitive deficits, neuronal loss, and gliosis in Alzheimer's disease mice. *J Alzheimers Dis, 72*, 587–604. https://doi.org/10.3233/JAD-190853

Nunnari, J., & Suomalainen, A. (March 16, 2012). Mitochondria: In sickness and in health. *Cell, 148*(6), 1145–1159.

Orfany, A., Arriola, C. G., Doulamis, I. P., Guariento, A., Ramirez-Barbieri, G., Moskowitzova, K., Shin, B., Blitzer, D., Rogers, C., Nido, PJ.d, & McCully, J. D. (2020). Mitochondrial transplantation ameliorates acute limb ischemia. *Journal of Vascular Surgery, 71*, 1014–1026.

Pang, Y. L., Fang, S. Y., Cheng, T. T., Huang, C. C., Lin, M. W., Lam, C. F., & Chen, K. B. (June 6, 2022). Viable allogeneic mitochondria transplantation improves gas exchange and alveolar-capillary permeability in rats with endotoxin-induced acute lung injuries. *International Journal of Medical Sciences, 19*(6), 1036–1046.

Parakatselaki, M. E., & Ladoukakis, E. D. (2021). mtDNA heteroplasmy: Origin, detection, significance, and evolutionary consequences. *Life, 11*. https://doi.org/10.3390/life11070633

Phadke, R. (July 4, 2017). Myopathology of adult and paediatric mitochondrial diseases. *Journal of Clinical Medicine, 6*(7), 64. https://doi.org/10.3390/jcm6070064. PMID: 28677615; PMCID: PMC5532572.

Pham, V. H., Nguyen, V. L., Jung, H.-E., Cho, Y.-S., & Shin, J.-G. (2022). The frequency of the known mitochondrial variants associated with drug-induced toxicity in a Korean population. *BMC Medical Genomics, 15*, 3.

Piantadosi, C. A., & Suliman, H. B. (April 1, 2012). Transcriptional control of mitochondrial biogenesis and its interface with inflammatory processes. *Biochimica et Biophysica Acta, General Subjects, 1820*(4), 532–541.

Plotnikov, E. Y., Jankauskas, S. S., Zinovkin, R. A., Zorova, L. D., Zorov, S. D., Pevzner, I. B., Silachev, D. N., & Zorov, D. B. (2020). A combination of kidney ischemia and injection of isolated mitochondria leads to activation of inflammation and increase in mortality rate in rats. *Bulletin of Experimental Biology and Medicine, 169*, 213–217. https://doi.org/10.1007/s10517-020-04853-2

Pourmohammadi-Bejarpasi, Z., Roushandeh, A. M., Saberi, A., Rostami, M. K., Toosi, S. M. R., Jahanian-Najafabadi, A., Tomita, K., Kuwahara, Y., Sato, T., & Roudkenar, M. H. (2020). Mesenchymal stem cells-derived mitochondria transplantation mitigates I/R-induced injury, abolishes I/R-induced apoptosis, and restores motor function in acute ischemia stroke rat model. *Brain Research Bulletin, 165*(2020), 70–80, 2020.

Prakash, Y. S., Pabelick, C. M., & Sieck, G. C. (2017). Mitochondrial dysfunction in airway disease. *Chest, 152*, 618–626. https://doi.org/10.1016/j.chest.2017.03.020

Ramirez-Barbieri, G., Moskowitzova, K., Shin, B., Blitzer, D., Orfany, A., Guariento, A., Iken, K., Friehs, I., Zurakowski, D., del Nido, P. J., & McCully, J. D. (2019). Alloreactivity and allorecognition of syngeneic and allogeneic mitochondria. *Mitochondrion, 46*, 103–115. https://doi.org/10.1016/j.mito.2018.03.002

Rusecka, J., Kaliszewska, M., Bartnik, E., & Tońska, K. (2018). Nuclear genes involved in mitochondrial diseases caused by instability of mitochondrial DNA. *Journal of Applied Genetics, 59*(1), 43–51. https://doi.org/10.1007/s13353-017-0424-3

Scarpulla, R. C. (2008). Transcriptional paradigms in Mammalian mitochondrial biogenesis and function. *Physiological Reviews, 88*, 611–638.

Shen, K., Pender, C. L., Bar-Ziv, R., Zhang, H., Wickham, K., Willey, E., Durieux, J., Ahmad, Q., & Dillin, A. (2022). Mitochondria as cellular and organismal signaling hubs. *Annual Review of Cell and Developmental Biology, 38*(1), 179–218.

Shi, X., Bai, H., Zhao, M., Li, X., Sun, X., Jiang, H., & Fu, A. (2018). Treatment of acetaminophen-induced liver injury with exogenous mitochondria in mice. *Translational Research, 196*, 31–41. https://doi.org/10.1016/j.trsl.2018.02.003

Shi, X., Zhao, M., Fu, C., & Fu, A. (2017). Intravenous administration of mitochondria for treating experimental Parkinson's disease. *Mitochondrion, 34*, 91–100. https://doi.org/10.1016/j.mito.2017.02.005

Shin, B., Cowan, D. B., Emani, S. M., Del Nido, P. J., & McCully, J. D. (2017). Mitochondrial transplantation in myocardial ischemia and reperfusion injury. *Mitochondrial Dynamics in Cardiovascular Medicine*, 595–619.

Sun, C. K., Lee, F. Y., Kao, Y. H., Chiang, H. J., Sung, P. H., Tsai, T. H., Lin, Y. C., Leu, S., Wu, Y. C., Lu, H. I., Chen, Y. L., Chung, S. Y., Su, H. L., & Yip, H. K. (March 2015). Systemic combined melatonin-mitochondria treatment improves acute respiratory distress syndrome in the rat. *Journal of Pineal Research, 58*(2), 137–150. https://doi.org/10.1111/jpi.12199. Epub 2014 Dec 27. PMID: 25491480.

Shin, B., Saeed, M. Y., Esch, J. E., Guariento, A., Blitzer, D., Moskowitzova, K., Ramirez-Barbieri, G., Orfany, A., Thedsanamoorthy, J. K., Cowan, D. B., Inkster, J. A., Snay, E. R., Staffa, S. J., Packard, A. B., Zurakowski, D., del Nido, P. J., & McCully, J. D. (2019). A Novel Biological Strategy for Myocardial Protection by Intracoronary Delivery of Mitochondria: Safety and Efficacy. *JACC Basic to Translational Science, 4*(8), 871–888. https://doi.org/10.1016/j.jacbts.2019.08.007

Su, Y., Zhu, L., Yu, X., Cai, L., Lu, Y., Zhang, J., Li, T., Li, J., Xia, J., Xu, F., & Hu, Q. (2016). Mitochondrial transplantation attenuates airway hyperresponsiveness by inhibition of cholinergic hyperactivity. *Theranostics, 6*, 1244–1260. https://doi.org/10.7150/thno.13804

Sun, X., Chen, H., Gao, R., Qu, Y., Huang, Y., Zhang, N., Hu, S., Fan, F., Zou, Y., Hu, K., Chen, Z., Ge, J., & Sun, A. (2023). Intravenous transplantation of an ischemic-specific peptide-TPP-mitochondrial compound alleviates myocardial ischemic reperfusion injury. *ACS Nano, 17*(2), 896–909.

Tan, J. X., & Finkel, T. (2020). Mitochondria as intracellular signaling platforms in health and disease. *The Journal of Cell Biology, 219*(5).

Tinker, R. J., Lim, A. Z., Stefanetti, R. J., & McFarland, R. (March 2021). Current and emerging clinical treatment in mitochondrial disease. *Molecular Diagnosis and Therapy, 25*, 181–206.

Ulger, O., Kubat, G. B., Cicek, Z., Celik, E., Atalay, O., Suvay, S., & Ozler, M. (2021). The effects of mitochondrial transplantation in acetaminophen-induced liver toxicity in rats. *Life Sciences, 279*, 119669.

Ulger, O., & Kubat, G. B. (April 2022). Therapeutic applications of mitochondrial transplantation. *Biochimie, 195*, 1–15. https://doi.org/10.1016/j.biochi.2022.01.002. Epub 2022 Jan 10. PMID: 35026361.

Vizioli, M. G., Liu, T., Miller, K. N., Robertson, N. A., Gilroy, K., Lagnado, A. B., Perez-Garcia, A., Kiourtis, C., Dasgupta, N., Lei, X., Kruger, P. J., Nixon, C., Clark, W., Jurk, D., Bird, T. G., Passos, J. F., Berger, S. L., Dou, Z., & Adams, P. D. (March 1, 2020). Mitochondria-to-nucleus retrograde signaling drives formation of cytoplasmic chromatin and inflammation in senescence. *Genes & Development, 34*(5–6), 428–445. https://doi.org/10.1101/gad.331272.119. Epub 2020 Jan 30. PMID: 32001510; PMCID: PMC7050483.

Wallace, D. C. (July 28, 2016). Genetics: Mitochondrial DNA in evolution and disease. *Nature, 535*(7613), 498–500. https://doi.org/10.1038/nature18902. Epub 2016 Jul 6. PMID: 27383787; PMCID: PMC6918947.

Wang, X., & Gerdes, H. H. (2015). Transfer of mitochondria via tunneling nanotubes rescues apoptotic PC12 cells. *Cell Death & Differentiation, 22*, 1181–1191. https://doi.org/10.1038/cdd.2014.211

Wang, Y., Ni, J., Gao, C., Xie, L., Zhai, L., Cui, G., & Yin, X. (2019). Mitochondrial transplantation attenuates lipopolysaccharide- induced depression-like behaviors. *Progress In Neuro-Psychopharmacology & Biological Psychiatry, 93*, 240–249. https://doi.org/10.1016/j.pnpbp.2019.04.010

Weinberg, S. E., Sena, L. A., & Chandel, N. S. (2015). Mitochondria in the regulation of innate and adaptive immunity. *Immunity, 42*(3), 406–417.

WHO Expert advisory committee on developing global standards for governance and oversight of human genome editing. Human genome editing: A framework for governance. (2021). Geneva: World Health Organization.

Yamada, Y., Ito, M., Arai, M., Hibino, M., Tsujioka, T., & Harashima, H. (September 2, 2020). Challenges in promoting mitochondrial transplantation therapy. *International Journal of Molecular Sciences, 21*(17), 6365.

Zhang, A., Liu, Y., Pan, J., Pontanari, F., Chang, A. C., Wang, H., Gao, S., Wang, C., & Chang, A. C. (2023). Delivery of mitochondria confers cardioprotection through mitochondria replenishment and metabolic compliance. *Molecular Therapy, 31*(5), 1468–1479. https://doi.org/10.1016/j.ymthe.2023.02.016

Zhang, Z., Ma, Z., Yan, C., Pu, K., Wu, M., Bai, J., Li, Y., & Wang, Q. (2019). Muscle-derived autologous mitochondrial transplantation: A novel strategy for treating cerebral ischemic injury. *Behavioural Brain Research, 356*, 322–331.

Zhang, X., Sun, J., Zhou, M., Li, C., Zhu, Z., & Gan, X. (2023). The role of mitochondria in the peri-implant microenvironment. *Experimental Physiology*, 1–14.

Zhao, J., Qu, D., Xi, Z., Huan, Y. U., Zhang, K., Yu, C., Yang, D., Kang, J., Lin, W., Wu, S., & Wang, Y. (2021). Mitochondria transplantation protects traumatic brain injury via promoting neuronal survival and astrocytic BDNF. *Translational Research, 235*, 102–114. https://doi.org/10.1016/j.trsl.2021.03.017

Zhao, Z., Hou, Y., Zhou, W., Keerthiga, R., & Fu, A. (2021). Mitochondrial transplantation therapy inhibit carbontetrachloride-induced liver injury through scavenging freeradicals and protecting hepatocytes. *Bioeng Transl Med, 6*, e10209.

CHAPTER

13

Limitations of mitochondrial transplantation

Leena Rajendra Chaudhari[1], *Mrunal N. Damle*[1,2], *Rakesh Kumar Sharma*[3] *and Meghnad G. Joshi*[1,2]

[1]Department of Stem Cells and Regenerative Medicine, D.Y. Patil Education Society (Deemed University), Kolhapur, Maharashtra, India; [2]Stem Plus Biotech Pvt. Ltd., Research and Development, Sangli, Maharashtra, India; [3]Department of Obstetrics and Gynecology, D.Y. Patil Medical College, Kolhapur, India

Introduction

Cellular fulfill energy metabolism depends heavily on mitochondria, which are vital organelles. They produce ATP via oxidative phosphorylation, which is required for cellular functions like protein synthesis, ion transport, and signal transmission. Numerous diseases, such as cancer, metabolic disorders, cardiovascular disease, and neurological disorders, have all been linked to dysfunctional mitochondria (Bhatti et al., 2017). In order to cure these disorders and restore mitochondrial function, MT has become a promising treatment strategy. In mitochondrial transplantation, healthy mitochondria are isolated from a donor cell or tissue and transferred to a recipient cell or tissue that has damaged mitochondria (Roushandeh et al., 2019). In the process of regaining mitochondrial function and enhancing cellular metabolism, the transplanted mitochondria then integrate into the recipient cell's mitochondrial network. There are various approaches for mitochondrial transplantation, such as onsite direct injection, systemic delivery, and intranasal delivery. Mitochondrial functionality is vital for transplantation studies due to its role in energy production, regulation of reactive oxygen species (ROS), apoptosis, and immune responses. Functional mitochondria ensure an adequate energy supply, minimize oxidative damage, promote cell survival, and reduce immunogenicity, enhancing the viability and success of transplanted organs. Monitoring mitochondrial function serves as a biomarker for organ

Mitochondrial Transplantation and Transfer
https://doi.org/10.1016/B978-0-443-18858-9.00013-5

quality assessment and guides therapeutic interventions in transplantation (Murphy et al., 2016; Weinrich et al., 2019).

According to a concept, an intercellular mitochondrial transfer occurs spontaneously in physiologic circumstances to support tissue development and homeostasis (Liu et al., 2021). Mitochondrial transfer restored damaged cells with nonfunctioning mitochondria by regaining their function, as seen by better oxidative phosphorylation and increased ATP generation (Tan et al., 2022). Although the precise method of isolated mitochondria uptake is uncertain, it was previously assumed that recipient cells might internalize mitochondria through micropinocytosis (Kitani et al., 2014). Extracellular substances are taken in vesicles called macropinosomes during the endocytic process of macropinocytosis. Daz-Carballo et al. showed that the integrity of the outer mitochondrial membrane and the fusion proteins, such as syncytin-1 and syncytin-2 on it, which may serve as ligands in the interaction between free mitochondria and the recipient cell, were shown to be essential for free mitochondria absorption (Díaz-Carballo et al., 2017).

In several mitochondrial diseases, replacing unhealthy mitochondria with functional ones may be able to reverse mitochondrial malfunction (Gollihue & Rabchevsky et al., 2017). Mitochondria play a crucial role in various diseases, and targeting mitochondrial dysfunction has shown potential as a therapeutic approach. Mitochondria can be used to treat various diseases such as neurodegenerative diseases (Chang et al., 2021), cardiovascular diseases (Sun et al., 2023), metabolic disorders (Bhatti et al., 2017), cancer (Chang et al., 2019), etc.

The use of MT to cure a variety of illnesses in the clinic is thought to be fascinating by the scientific community. Two of the many problems that have not been solved are safety and effectiveness. Isolating healthy, functional mitochondria is essential and crucial. It is notable that the therapeutic effects of mitochondria can only be exerted by functional mitochondria. The quality of healthy mitochondria must therefore be preserved. All recent investigations have demonstrated that the therapeutic benefit of a single mitochondrial treatment is transient, necessitating repeated delivery over an extended period of time (McCully et al., 2017). Although MT has therapeutic potential, there are a number of limitations and challenges that must be resolved before it can be widely used in clinical settings. These include issues with immunity, safety, ethics, and difficulty with the technology.

In this chapter, we will discuss potential technical challenges and their limitations. This could be the main hurdle of any therapy. Another potential limitation is safety concerns as well as adverse side effects. MT involve introducing foreign mitochondria into the recipient's cells, which can trigger an immune response. While the mitochondrial genome is relatively small and has limited immunogenic potential, there is still a risk of immune rejection, particularly in cases where the donor and recipient are not closely related (Ramirez-Barbieri et al., 2019).

Finally, there are ethical considerations to be addressed with the use of MT and transfer. For example, there may be concerns about the long term impact of introducing foreign mitochondria into an individual's cells, particularly with regard to potential off target effects or unintended consequences. Additionally, there may be concerns about the ethics of selecting or manipulating donor mitochondria based on their functional characteristics or genetic makeup. It is important not to forget about worldwide regulations to accept mitochondrial transplantation as a medical therapy to treat various diseases. All of these limitations are explored in Fig. 13.1 and Table 13.1 encompassing potential limitations, challenges, and mitigation strategies.

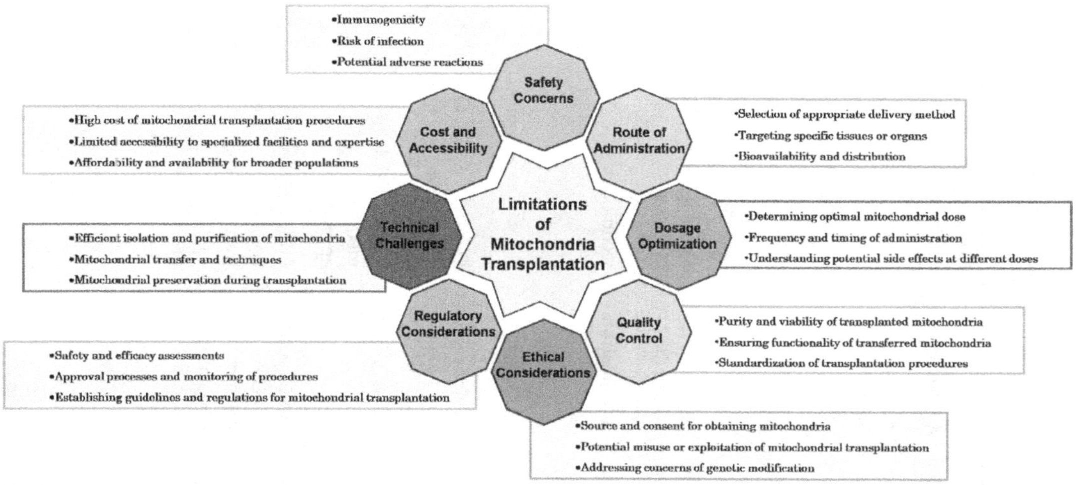

FIGURE 13.1 The figure illustrates the various limitations associated with mitochondrial transplantation as a cellular therapy for mitochondrial dysfunction. Mitochondrial transplantation involves the transfer of healthy mitochondria into cells with dysfunctional mitochondria to restore their normal function. However, several factors can hinder the effectiveness and widespread application of this approach.

Limitations

Technical challenge

MT is a technically challenging procedure that requires highly skilled personnel and specialized equipment. The isolation and transfer of mitochondria require the use of sophisticated techniques that may not be available in all clinical settings. Additionally, the procedure requires the availability of a healthy donor. MT faces several technical challenges that limit its application as a therapeutic approach. One of the primary limitations is the limited availability and quality of donor mitochondria in addition to their isolation and purification of functional mitochondria from donor cells. Another technical challenge is the delivery of transplanted mitochondria to the target cells or tissues. It is also a challenge to ensure that the transplanted mitochondria are integrated and functional within the host cells. The transplanted mitochondria must be able to produce ATP and carry out their normal functions within the host cells to provide any therapeutic benefit. Additionally, there may be issues with mitochondrial quality control, as the isolation and storage of mitochondria can result in damage to the organelles (McCully et al., 2017). All the potential technical challenges and limitations are discussed further in detail in this section.

Source of mitochondria

The source of donor mitochondria can be a limitation in mitochondrial transplantation. The mitochondria used for transplantation must be of high quality and function properly in order to successfully integrate into the host cells and carry out their normal functions. However, obtaining high quality donor mitochondria can be challenging. Currently, there

TABLE 13.1 Current mitochondrial transplantation status and their potential limitation, challenges, and their mitigating strategies in a few types of research are discussed.

Sr.	Target disease	Procedure	Mitochondrial dose	Limitation	Description	Implications	Challenges	Mitigation strategies	References
1	Kidney ischemia/reperfusion	Intraperitoneal injection	10 mg of MT per 100 g body wt.	Safety concerns	Potential adverse side effects, immune response, genetic changes	Requires careful monitoring and risk assessment	Ensuring safety	Strict selection criteria, and further preclinical studies needed	Plotnikov et al. (2020)
2	Myocardial ischemia −reperfusion injury	Intracoronary injection	Different doses used; 1×10^3, 1×10^5, 1×10^7, 1×10^9, and 1×10^{11}	Route of administration	Challenges in selecting the optimal delivery method. Invasive method and risk involved	Influences the efficiency and effectiveness of mitochondrial transfer	Identifying the most effective route of administration	Development of novel delivery techniques	Shin et al. (2019)
3	Skeletal muscle injury	Direct injection	Different doses from 1×10^6 to 1×10^9	Dosage optimization	Determining the appropriate number of transplanted mitochondria	Affects the therapeutic efficacy	Establishing optimal dosage guidelines	Dose −response studies, personalized medicine approaches	Orfany et al. (2020)
4	Damaged endometrium	Direct injection	1×10^6	Ethical considerations	Placenta collection needs informed consent, disease free donor selection needed	Necessitates ethical guidelines and responsible decision making	Addressing ethical dilemmas and societal concerns	Ethical review boards, transparent consent procedures	Kshersagar et al. (2023)

5	Liver	Intravenous	0.5 mL of purified mitochondria at a steady rate of 160 μL/min	Dose and route considerations	Method failure resulting ineffective or inadequate treatment	Compromised efficacy of the transplant procedure and hinder the potential benefits	Insufficient expertise or training in executing the method, limited availability and quality of donor mitochondria	Carefully evaluate and select appropriate dose and route	Liu et al. (2020)
6	Heart ischemia −reperfusion injury	Direct injection	10−20 separate injections of 1 -5×10^7 mitochondria per injection	Source, route of administration, and other technical challenges	Challenges in sample collection may occur. Limited efficiency, low no. of mitochondria, integration issues	Impact on the overall success and viability of transplantation	Enhancing mitochondrial transfer efficiency and	Novel approaches like nanocarriers, gene editing techniques	Guariento et al. (2020)
7	Parkinson disease	Direct injection at the medial forebrain bundle	105 mg allogeneic or xenogeneic mitochondria in 0.2 mL PBS	Route of administration and other technical challenges	Challenges in mitochondrial delivery,	Hinders accessibility and affordability of technique	Improving delivery system	New strategies are required for the delivery route.	Chang et al. (2016)

are several sources of donor mitochondria that have been explored for mitochondrial transplantation, including various organs, induced pluripotent stem cells (iPSCs), and other cell types. However, each of these sources has its own limitations and challenges.

There are mainly three sources from which mitochondria can be isolated such as autologous, allogenic, and xenogeneic source. For clinical studies, autologous MT was advised (Emani et al., 2017). The availability of recurrent dosages of mitochondria from the autologous tissue source may be severely restricted and the need for additional surgery to get autologous tissue could become a major challenge. This can limit the scalability of the procedure and its potential for widespread clinical use. For e.g., as a suitable option for mitochondrial transplantation, for tissue biopsies, nonischemic skeletal muscle cells have been suggested. The appropriate sources for obtaining mitochondria include the pectoralis major, rectus abdominis, gastrocnemius, and even neck strap muscles such as the sternohyoid, deltoid, and latissimus dorsi (Jabbari et al., 2020; Doulamis et al., 2020; Guariento et al., 2020; Hayashida et al., 2023; Hsu et al., 2022; Moskowitzova et al., 2020; Pang et al., 2022; Fang et al., 2021). Jabbari et al. harvested viable mitochondria from rat pectoralis major muscle cells to investigate the potential benefits of MT in a case of ischemia/reperfusion induced kidney injury (Jabbari et al., 2020). Similarly, clinical study by Mc Cully's research team at Boston Children's Hospital and Harvard University made their initial effort in 2016 to utilize the modern treatment for cardiac ischemia/reperfusion in five pediatric patients by autologous MT. They used autologous rectus abdominis muscle as the mitochondrial source (Emani et al., 2017). MT shows promising results in above mentioned studies. However, skeletal muscle cells could not be appropriate for therapeutic applications, though, and it might even be difficult, especially in the case of multiorgan disorders and/or for older people, which affects the quality of isolated mitochondria. Furthermore, it requires surgical intervention and repetitive collection is not ideal for patients due to cosmetic concerns (Emani et al., 2017; McCully et al., 2017). Further research is therefore required to identify a suitable source for the extraction of functional mitochondria to replace the skeletal muscle cells and to improve the scalability and safety of the procedure. Moreover, the preclinical studies utilize liver tissue as a source of mitochondria in various disease conditions like kidney ischemia/reperfusion, chemical induced liver injury, and $BaCl_2$ induced muscle injury (Plotnikov et al., 2020; Zhao, Hou, et al., 2021; Alway et al., 2023). But translation of these studies are not possible in regards to human patients. It is not ideal to take an autologous or allogenic liver biopsy to isolate mitochondria in clinical settings.

One possible source of mitochondria would be the iPSCs, which become transformed into a useful cell type just before the mitochondria isolation and transplantation. Another source for the mitochondrial isolation in a patient could be their younger siblings. However, they also need to be in good health. It would seem logical to create iPSCs from healthy siblings who share mitochondria because of maternal inheritance (Jabbari et al., 2020). Using iPSCs as a source of mitochondria in therapy has limitations. iPSCs are genetically modified using viral vectors, posing risks of viral contamination and unintended modifications (Herberts et al., 2011). Genetic, epigenetic, and transcriptional changes introduced during iPSC generation can interfere with mitochondrial DNA (mtDNA) (Merten et al., 2018). These modifications may impact iPSC function, differentiation, and mitochondrial integrity. To utilize iPSCs effectively, it is crucial to address these concerns and ensure safety while preserving mtDNA and mitochondrial function.

In several preclinical studies, mitochondria were isolated from cell lines such as HePG2 to treat chemical induced liver injury and fatty liver disease (Fu et al., 2017; Shi et al., 2018). It is a compatible source to treat liver diseases but it may limit to produce large number of cells in a short period of time along with further mitochondria isolation steps. Therefore, the use of cell lines in clinical settings may limit scalability. Additionally, it might be a time consuming procedure and need highly skilled personnel. These limitations can be overcome by replacing cell/tissue sources. The placenta has been identified as a promising source of healthy and functional mitochondria. In our recent study, we utilized the human placenta as a reliable source of mitochondria. The use of placental mitochondria was advantageous due to the abundant availability of placental tissue and the fewer ethical considerations. We isolated mitochondria from the placenta, considering the potential for future clinical applications. However, certain precautions must be taken into account, such as the risk of contamination and the possibility of using mitochondria from diseased or abnormal placentas. Therefore, before selecting the placenta as source tissue, it is important to test for any genetic abnormalities and infectious diseases such as HIV, hepatitis, etc. Despite these limitations, the placenta presents an attractive option as an ample and repeatable source of mitochondria (Kshersagar et al., 2023).

Quality control

Quality control is a crucial aspect of MT to ensure the safety and efficacy of the procedure. In advance of using mitochondria in a clinical setting, some quality control concerns must be judged, including those pertaining to the quantity, viability, purity, and function of the mitochondria.

The quality of isolated mitochondria can vary depending on the isolation method, the tissue source, and the handling and storage conditions. Therefore, a quick method for isolating mitochondria must be developed to enable clinical application, and Good Manufacturing Practice (GMP) and quality control and assurance standards must be created to guarantee consistency and quality of mitochondrial isolation to ensure the safety and efficacy of MT (Emani et al., 2018; Preble et al., 2014; McCully et al., 2017).

We can utilise a variety of quality control approaches to ensure the quality of isolated mitochondria. The methods are explained in the following sections.

Purity assessment

Cellular contaminants reduce the accuracy of the mitochondrial respiration measurement by decoupling the respiration chain or compromising the mitochondrial membrane integrity (Kiss et al., 2016). The purity of the isolated mitochondria can be assessed by measuring the levels of contaminating organelles and cellular debris, such as nuclei, lysosomes, and peroxisomes, using electron microscopy or flow cytometry. Differential centrifugation, one of several methods for the purification of mitochondria, is the strategy that is most frequently employed since it is rapid and affordable. The mitochondrial fraction produced by this technique, which was created in the 1950s (De Duve et al., 1995) and has since undergone numerous modifications, preserves the organelle's integrity and functionality. Previous studies suggest that the Percoll gradient method can greatly contribute to isolating purified mitochondrial fraction to minimize cellular debris contamination (Kiss et al., 2016). Other methods of purification that likewise yield extremely pure mitochondrial fractions, including the preparation of an isopycnic gradient followed by ultracentrifugation, are time and money consuming and demand access to an ultracentrifuge (Hartwig et al., 2015). Another fast and easy method to obtain pure functional mitochondria with high yield is based on magnetic sorting (Hornig-Do et al., 2009). Western blot can be used to test purity by finding the expression of mitochondrial marker COX IV (Lin et al., 2013).

Most of the studies did not discuss about the purity of mitochondrial fraction but it is necessary to discuss such points to ensure the safety and efficiency of mitochondrial transplantation in various disease conditions. If the mitochondrial preparation is impure and contains other cellular components, it may trigger stronger immune responses or unwanted reactions in the host. For e.g., in the case of an allogenic or xenogenic mitochondrial source, the mitochondrial fraction contaminated with nuclear DNA, may elicit an immune response or graft versus host reaction. Purifying the mitochondria minimizes the introduction of immunogenic or proinflammatory factors, reducing the risk of adverse immune reactions.

Therefore, the purity of mitochondrial preparations used in transplantation studies is crucial for optimizing functional outcomes, reducing unwanted immune responses, ensuring accurate interpretation of results, and ensuring safety in clinical applications.

Viability assessment

Various methods can be employed to assess the viability and functionality of isolated mitochondria. The JC-1 (5′,6,6′-tetrachloro-1,1′,3,3′- tetraethylbenzi-midazolylcarbocyanine iodide) assay, considered the gold standard for analyzing mitochondrial membrane potential ($\Delta\Psi m$), utilizes a fluorescent dye that forms aggregates in healthy mitochondria, emitting red fluorescence (Sivandzade et al., 2019). Conversely, damaged or depolarized mitochondria exhibit green fluorescence due to the monomeric form of JC-1. By quantifying the ratio of red to green fluorescence, the integrity of mitochondrial membrane potential can be determined (Marcondes et al., 2019). To assess mitochondrial mass and activity, MitoTracker Probes and rhodamine dyes are commonly used. MitoTracker dyes accumulate in active mitochondria and can be quantified by flow cytometry to provide information on mitochondrial health and function (Xiao et al., 2016). Similarly, rhodamine dyes offer insights into mitochondrial activity. Flow cytometry analysis enables the measurement of fluorescence intensity, aiding in the determination of mitochondrial health (Kshersagar et al., 2023; Léger et al., 2020; Lin et al., 2013).

The Clark-type oxygen electrode chamber provides another approach to evaluate mitochondrial viability. This method measures oxygen consumption, offering insights into mitochondrial activity and efficiency. By assessing the rate of oxygen consumption, mitochondrial permeability, and overall functionality can be inferred (Kiss et al., 2016). Morphological analysis through electron microscopy of crude and purified mitochondrial fractions provides valuable information on the integrity of isolated mitochondria. This technique allows for the visualization of mitochondrial structure and morphology, aiding in the determination of mitochondrial viability (Kiss et al., 2016).

While these methods are widely utilized, it is important to acknowledge their limitations. Factors such as dye concentration, mitochondrial size, and cellular environment can influence the accuracy of the JC-1 assay. MitoTracker probes and rhodamine dyes may exhibit nonspecific binding, potentially leading to false positive signals. The Clark-type oxygen electrode chamber requires a substantial number of mitochondria for accurate measurements, and isolation procedures can introduce artifacts. Electron microscopy, while providing detailed morphological analysis, requires specialized equipment and sample preparation, and does not offer real time information on mitochondrial function.

Researchers often employ multiple methods in combination, alongside appropriate controls, to overcome these limitations and obtain a comprehensive assessment of mitochondrial viability, membrane potential, mass, activity, and morphology.

Functional assessment

The functional capacity of the isolated mitochondria can be assessed by measuring their respiration rate, oxidative phosphorylation, or enzymatic activities. Assessing the functional capacity of isolated mitochondria is crucial to understand their overall performance. Various assays can be used to evaluate mitochondrial functionality, including activity assays for different mitochondrial complexes. These assays measure the enzymatic activities of specific complexes within the mitochondria, such as NADH-cytochrome C reductase, cytochrome P450, and complexes I, II, III, and IV. These spectrophotometer assays provide valuable information about the functional status of isolated mitochondria (Kshersagar et al., 2023).

In addition to complex specific assays, other techniques can be employed to assess mitochondrial functionality. One example is the mitochondrial integrity assay, and respiration assay, which evaluates the integrity of the mitochondrial membrane, respiration, and its ability to maintain proper function (Liu et al., 2020). Another method involves measuring the reduction of Janus green B (JG-B), a dye that is reduced by functional mitochondria. The reduction of JG-B can be quantified, providing an indication of mitochondrial functionality (Ahmad et al., 2018). In addition to the functionality test, other measurements included mitochondrial protein content, oxygen consumption, respiratory control index (RCI), and mtDNA and RNA (McCully et al., 2009). By examining mitochondrial protein content, oxygen consumption, RCI, and mtDNA and RNA, researchers can obtain a more comprehensive understanding of the effects of MT on mitochondrial function and cellular metabolism.

However, despite the availability of these assays, there is still a need for validated techniques that accurately assess mitochondrial functional capacity. Particularly in the context of mitochondrial transplantation, it is important to identify reliable assays that can determine the quality and functionality of isolated mitochondria. Researchers continue to explore and develop new methods to assess mitochondrial function accurately. By establishing consensus on optimal assays and validation protocols, it will be possible to reliably evaluate the functional capacity of isolated mitochondria and further our understanding of their role in various biological processes, including mitochondrial transplantation.

Sterility assessment

Ensuring sterility is an essential aspect of any biological material intended for transplantation or therapeutic use. However, as per our knowledge, there is limited information available regarding specific studies or protocols discussing sterility testing for isolated mitochondria in the context of mitochondrial transplantation therapy.

Sterility testing typically involves evaluating the absence of viable microorganisms, such as bacteria, fungi, or viruses, in a sample. These tests are crucial to ensure the safety and efficacy of biological materials used in transplantation or other therapeutic applications. The specific methods and requirements for sterility testing can vary depending on the nature of the material and the regulatory guidelines of the particular region or country.

In the case of mitochondrial transplantation, it is important to consider the potential sources of contamination, such as the isolation process, storage conditions, and any other steps involved in handling and preparing the mitochondria for transplantation. Developing appropriate sterility testing protocols specific to isolated mitochondria would be necessary to ensure their safety and prevent any potential adverse effects resulting from microbial contamination.

To address the issue of sterility, adhering to GMP and following regulatory guidelines is essential to ensure the safety and efficacy of biological materials used in transplantation therapies, for specific recommendations on sterility testing procedures and requirements. Additionally, collaborating with experts in the field of microbiology and transplantation can help in designing and implementing appropriate sterility testing protocols for isolated mitochondria prior to their therapeutic use.

Route of administration

The path of mitochondrial transplantation may present an additional challenge related to the ideal route of transplantation. MT can be administered using various routes, depending on the target tissue or organ. The most common routes of administration include direct injection, intravenous injection, intraperitoneal injection, intramyocardial injection, nasal route, and topical application. While there are several routes of administration for mitochondrial transplantation, each method has its limitations and challenges that must be addressed in order to ensure the safe and effective delivery of the mitochondria.

Direct injection

Direct injection is used in various models to transfer isolated mitochondria into recipient cells through microinjection or needle to defective tissue. In the event of local injury, direct injection into the tissue may be preferable since it allows for the rapid delivery of healthy mitochondria. Previous studies showed that researchers used direct injection to treat various disease conditions such as skeletal muscle ischemia–reperfusion damage, hepatic ischemia–reperfusion injury, diabetic nephropathy, regional heart ischemia, Parkinson's disease (PD), traumatic brain damage (TBD), spinal cord injury, lung injury, etc (Orfany et al., 2020; Lin et al., 2013; Konari et al., 2019; Masuzawa et al., 2013; Chang et al., 2016; Zhao, Qu, et al., 2021; Gollihue et al., 2017).

In a study by Lin et al. splenic injection was used to deliver the isolated mitochondria to the liver. While this method was shown to be effective in this study, it may not be the most practical or feasible delivery route in a clinical setting (Lin et al., 2013). Another study by Konari et al. used a rat model for the direct injection of mitochondria into a kidney capsule (Konari et al., 2019). However, the translation of this method to human applications may be challenging. The direct injection may involve intraperitoneal injection in which mitochondria are injected into the peritoneal cavity, which is the space surrounding the abdominal organs. The efficiency of this mode of administration in enhancing mitochondrial activity in muscle and liver tissues has been demonstrated in animal studies, although research on its efficacy for humans is currently ongoing. Plotnikov et al. found that intraperitoneal injection of isolated mitochondria can cause the appearance of mtDNA in the peripheral circulation (Plotnikov et al., 2020).

In heart related diseases, direct injection or intramyocardial injection involves injecting the mitochondria directly into the heart muscle. This route of administration has been used in animal studies and clinical trials to improve heart function in patients with heart failure and ischemic heart disease. Intramyocardial injection is a more invasive method as it involves direct injection of the mitochondria into the heart muscle. This method has been shown to be effective in improving heart function in animal studies and clinical trials, but it requires specialized equipment and expertise to perform safely and effectively (Masuzawa et al., 2013).

Further studies related to central nervous system (CNS) involve direct injection of mitochondria to defective parts of the brain or spinal cord to treat various conditions like stroke, ischemia, PD, TBD, etc. A previous study by Chang et al. injected mitochondria directly at the medial forebrain bundle of the PD rat model. Results show great improvement although it is difficult to perform multiple doses of treatment in clinical settings (Chang et al., 2016). Before MT reaches the peak of its potential in clinical settings, site specific delivery is a major obstacle that needs to be overcome. Direct mitochondrial injection into target tissues has been the emphasis of a significant number of both animal studies and human research. Although helpful in invasive surgical settings, this method may not be the best choice in most clinical circumstances when direct access is not possible, especially in CNS disorders like ischemic strokes, PD, and AD (Jain et al., 2023).

In our recent study, we investigated the direct injection of mitochondria into the damaged endometrium of rats through the uterine horns. This method involved transplanting mitochondria followed by platelet rich plasma to stabilize the mitochondria at the damaged site. The results showed an improvement in the endometrial lining, suggesting the potential for using the intrauterine injection of mitochondria in clinical studies related to infertility issues (Kshersagar et al., 2023). In a clinical setting, an intrauterine catheter can be used to directly inject mitochondria into the affected site in the uterus. Further clinical studies are warranted to explore the full potential and safety of this approach in treating infertility related issues.

The specific limitations and technical difficulties associated with direct injection in human patients are not addressed by many researchers. Further studies are needed to explore other routes of mitochondrial infusions, such as peripheral veins, which may be more reasonable and translationally significant.

Intravenous injection

Intravenous injection is a common route of administration for mitochondrial transplantation, especially for systemic mitochondrial disorders as well as multiple organ/tissue damage (Kubat et al., 2021). The mitochondria are typically isolated from a donor source, purified, and then injected directly into the bloodstream of the recipient. This method has been used in several animal studies and has shown promising results in improving mitochondrial function and ameliorating symptoms of mitochondrial diseases. Many preclinical studies have been done to understand the systemic delivery of mitochondria for various diseases/

conditions such as neuromuscular injury (Alway et al., 2023), fatty liver (Fu et al., 2017), liver injury (Shi et al., 2018), kidney ischemia–reperfusion (Plotnikov et al., 2020), acute kidney disease (Jabbari et al., 2020), heart ischemia (Cowan et al., 2016; Moskowitzova et al., 2020), PD (Shi et al., 2017), lung ischemia (Moskowitzova et al., 2020), spinal cord ischemia (Fang et al., 2021), etc.

In a recent study, Shin et al. used intracoronary injection to deliver isolated mitochondria to the myocardium in a pig model of cardiac ischemia–reperfusion. Their findings showed that the intracoronary injection of mitochondria was safe and had no impact on the patency of the coronary arteries. Although intracoronary injection is a promising approach for MT, it requires invasive procedures, such as cardiac catheterization. It involves threading a catheter through the blood vessels to reach the coronary arteries. This procedure carries inherent risks, including vessel damage, bleeding, infection, and adverse reactions to contrast agents used during imaging (Shin et al., 2019). Therefore, it is essential to carefully consider the potential limitations, risks, and benefits associated with intracoronary injection in each specific clinical or research scenario. Huang et al. conducted study in a cerebral stroke model, where they administered mitochondria through both by intravenously and direct tissue injection. On day 7 postperfusion, it was observed that delivering the mitochondria via systemic injection did not immediately restore the motor activity in the ischemic rats, in contrast to the direct injection (Huang et al., 2016). However, the study did not provide the reason behind it. Hippocampal mitochondria were isolated and administered through the tail vein. However, we still do not yet know the overall distribution and "pharmacokinetics" of exogenous mitochondria in the brain (Jia et al., 2023). However, the systemic administration of therapeutics intended to target the brain is hindered by the presence of the blood–brain barrier and blood–cerebrospinal fluid barriers. These physiological barriers pose challenges in delivering therapeutic substances from the bloodstream to the brain or cerebrospinal fluid, limiting the effectiveness of systemic treatments for brain related conditions (Ahlawat et al., 2020).

In certain cases, intravenous delivery of mitochondria can be advantageous for treating multiorgan injury or disease. This is because the systemic distribution of mitochondria through intravenous administration allows for their widespread delivery to multiple organs, potentially providing therapeutic benefits across different affected tissues.

Nasal delivery

Intranasal delivery of mitochondria refers to the administration of mitochondria through the nasal route. This concept is still a relatively new and developing area of research with potential applications in various fields, including medicine and biotechnology. However, due to some drawbacks, such as the difficulty to sustain mitochondrial activity after isolation, the lack of precision in targeted delivery, and the issue of poor delivery efficiency, noninvasive MT has rarely been addressed.

The idea behind the nasal delivery of mitochondria is to utilize the nasal cavity's unique properties for efficient and targeted delivery of mitochondria to specific tissues or organs. Simply said, the nasal mucosa is the optimal location for drug delivery because of numerous advantages, including its large surface area, rapid onset, and the potential for CNS administration. In one study, researchers assessed the feasibility of intranasal administration of mitochondria in treating neuropathology, and the results showed improvements by stimulating neurogenesis (Alexander et al., 2022). Another study conducted on NMRI male mice (Naval

Medical Research Institute, an outbred strain) with spatial memory impairment in an Alzheimer's like degeneration demonstrated that intranasal delivery of mitochondria led to improved memory function (Bobkava et al., 2022). Nasal delivery has been investigated as a potential method for delivering mitochondria in treating PD and other neurological disorders. While it may not be as successful as a direct injection in fully restoring mitochondrial function in dopaminergic neurons, it has shown promise in terms of improving behavioral symptoms and promoting neuronal survival in PD (Chang et al., 2021).

These studies mainly focused on issue related to the CNS, suggesting that delivering mitochondria through the nasal route may have specific advantages for neurodegenerative disorders and the growth of new brain cells in certain situation. However, its use may be more restricted to CNS related conditions. However, its effectiveness for delivering mitochondria to organs or tissues outside of the CNS is uncertain and requires further investigation. Another important point is that, while nasal delivery of mitochondria shows promise, it is still considered an experimental approach, and less research has been studied related to its stability, uptake, immune response, etc. and its clinical applications are yet to be established. As with any emerging field, further research is needed to validate its effectiveness and safety before it can be widely utilized for therapeutic purposes.

In addition to these above discussed limitations related to the route of administration, the optimal route of administration for MT may also depend on the specific disease or condition being treated, as well as the age, sex, and medical history of the patient. Further research is needed to determine the most effective and safe methods for delivering mitochondria to different tissues and organs. However, the effectiveness of each route of administration needs to be evaluated in preclinical and clinical studies to determine the optimal delivery method for each application.

Dosages

The optimal dosage for MT may vary depending on the disease or condition being treated, the route of administration, and the patient's individual characteristics. Additionally, because MT is still an emerging therapy, numerous preclinical investigations have shown significant variability in the therapeutic dosage of mitochondria per organ or tissue. However, to date, there is no optimized or ideal dose of mitochondria for transplantation. It is worth noting that many studies have not adequately measured the treatment efficacy in relation to the dosage administered, thus lacking a comprehensive understanding of the dose dependent response.

Direct injection

Since this technique is still in the early stages of development, there is no standardized dosage or protocol established for direct mitochondrial transplantation. The specific details regarding dosage, delivery method, and other factors would depend on the particular research study or experimental procedure being conducted.

The study (Orfany et al., 2020) conducted on MT in acute limb ischemia gave different dosages ranging from 1×10^6 to 1×10^9. The study revealed that the infusion of mitochondria resulted in a concentration dependent reduction in infarct size in all muscle types after 24 h of reperfusion. Interestingly, it was observed that different muscle types did not exhibit a preference for specific concentrations of mitochondria. Based on the findings, the recommended dosage of 1×10^9 mitochondria per gram wet weight demonstrated significant improvements in infarct size and ATP content, as well as widespread mitochondrial uptake in all muscle types. Initial investigations revealed that lower concentrations of mitochondria 1×10^5 per gram wet weight did not result in significant reductions in infarct size or enhancements in limb function, although specific outcomes were not presented. However, in the murine model utilized for the study, it was not feasible to administer higher amounts of mitochondria beyond 1×10^9 per gram wet weight due to limitations in the volume of injection allowed. The injection volume restriction limited the amount of mitochondria that could be delivered (Orfany et al., 2020).

Masuzawa et al. injected $9.7 \pm 1.7 \times 10^6$/mL mitochondria into a regional ischemic (RI) model in New Zealand white rabbits (Masuzawa et al., 2013). 0.1 mL of eight injections was given directly at the regional ischemia site. Similarly, Kaza AK et al. gives $9.9 \times 10^7 \pm 1.4 \times 10^7$/mL; 1.3×10^7 mitochondria per injection site of RI. This study uses the same procedure of MT therapy, in which 0.1 mL of eight injections was given intracardially at predefined sites (Kaza et al., 2017). Results of the above studies have found that this dose can provide cardioprotection, upregulation of cytokine, and apoptosis prevention without any significant immune response (Kaza et al., 2017; Masuzawa et al., 2013).

Based on preclinical research, and clinical trials focusing on cardiogenic shock in pediatric patients following ischemia–reperfusion injury, a specific method was employed to deliver mitochondria directly to the affected regions of the myocardium. This involved using a 1 mL tuberculin syringe to perform 10–20 separate injections. Each injection contained a mitochondrial suspension ranging from 50-μL to 100-μL in volume. It was estimated that each injection contained approximately $1-5 \times 10^7$ mitochondria. This technique allowed for precise targeting of the areas of the myocardium requiring treatment (Guariento et al., 2020). It is important to note that this study did not investigate the effectiveness of different dosages or compare the outcomes of different routes of administration. The dosages used in this clinical trial were determined based on preclinical studies, and their applicability to other clinical scenarios may vary.

Therefore, while the direct epicardial injection of mitochondria in this clinical trial offered a method for targeted delivery, further research is needed to explore the effectiveness of different dosages and routes of administration in various clinical applications of mitochondrial transplantation. Based on observations, we have found that an average dose of approximately 4.7×10^7 mitochondria for direct injection may be considered an optimal dose for various disease conditions. However, it is important to note that optimal dosages may vary depending on the specific disease, patient population, and other factors. Further research and clinical studies are necessary to validate and refine these dosage recommendations for different disease conditions.

Systemic route

Determining the ideal dosage for the systemic administration of mitochondria is a complex undertaking that depends on a number of variables. Subsequently, the dosage for systemic administration is calculated as the mitochondrial concentration or protein content of the mitochondria per unit of body weight or volume. However, preclinical and clinical studies aimed at assessing the safety, effectiveness, and tolerability of mitochondrial therapy would be required to identify the precise dosage.

Previous studies contributed to the understanding of MT in the treatment of various diseases (Fu et al., 2017; Zhao, Hou, et al., 2021; Jabbari et al., 2020; Plotnikov et al., 2020; Doulamis et al., 2020). For fatty liver disease, Fu et al. administered a dose of 0.5 mg/kg body weight of mitochondria. In the case of chemical induced liver injury, Shi et al. (2018) used a dosage of 1×10^{12} mitochondria in saline solution. This dosage was intended to address the specific injury induced by the chemicals. Another study by Zhao et al. (2021) investigated the use of mitochondria for liver injury, using two different dosages. They administered 1.6×10^{8} mitochondria for the low dose group and 0.2 mg/kg body weight, while the high dose group received 0.4 mg/kg body weight of mitochondria. These studies demonstrate the variation in dosage strategies employed for mitochondrial transplantation in liver injury models. Determining the optimal dosage for mitochondrial transplantation in liver injury requires further research and consideration of factors such as the specific liver disease or injury, delivery method, and desired therapeutic outcomes. These findings highlight the importance of tailoring the dosage of mitochondria to the specific condition being treated.

Similarly, in studies related to kidney ischemia–reperfusion injury, researchers have utilized different dosages of mitochondria. One study involved the injection of 7.5×10^{6} mitochondria per mL through systemic delivery (Jabbari et al., 2020). In contrast, another study administered 10 mg of mitochondria per 100 g of body weight through the renal artery (Plotnikov et al., 2020). Additionally, a different approach was taken in another study, where a dose of 1×10^{9} mitochondria in 6 mL was employed (Doulamis et al., 2020). These variations in dosages reflect the complexity of treating kidney injuries and the need to explore different administration methods and quantities. However, despite the valuable insights gained from these preclinical studies, a significant limitation arises when attempting to translate these findings into clinical settings.

As mentioned earlier, dosage for systemic delivery of mitochondria is often measured in terms of mitochondrial concentration or mitochondrial protein content per unit of body weight or volume. By quantifying the mitochondrial concentration or protein content or number of mitochondria, researchers and clinicians can calculate and adjust the dosage based on the specific needs of the patient, the target tissue, and the therapeutic objectives. Based on observations, it has been found that an average dose of approximately 3.2×10^{8} mitochondria for systemic delivery can be considered an optimal dose for various disease conditions. Additionally, previous studies have validated the use of mitochondria through systemic injection, ranging from the lowest dose of 0.01 mg to the higher dose of 100 mg of mitochondria (Hayashida et al., 2023; Plotnikov et al., 2020). In the process of determining the dosage for mitochondrial therapy, it is important to strike a balance between achieving sufficient mitochondrial uptake and minimizing potential side effects or complications. It is essential that we carefully analyze variables such as biodistribution, clearance rates, possible immunological reactions, and therapeutic response. Biodistribution makes sure that delivered mitochondria are delivered to the accurate target tissues, and clearance rates aid in determining how long they will survive. Potential immunological reactions and how they may affect stability and tolerability should be taken into consideration. Adjusting the dosage to maximize effectiveness is rendered possible by monitoring treatment response. The dosage of mitochondrial treatment can be optimized considering these aspects into account for maximal therapeutic benefit and least adverse effects.

Nasal delivery

The nasal delivery of mitochondria is an emerging field of research, and there may not be established optimal dosages for this specific route of administration. The nasal route has gained attention as a noninvasive and efficient method for delivering therapeutic agents to the brain and other target tissues. Till now, few studies show nasal delivery of mitochondria can be effective in CNS related conditions. The first study by Alexander et al. (2022) demonstrated that two doses of intranasal administration of 170 μg of mitochondria in 12 μL of buffer per mouse restored brain function in an animal model of chemotherapy induced cognitive defects. However, this study did not mention any limitations or side effects of the therapy. In another study by Chang et al. (2021), intranasal delivery of mitochondria was validated for PD. A higher dose of 200 μg of mitochondria in 50 μL of respiration buffer was administered via micropipette once a week for up to 3 months in a PD model. For Alzheimer's like degeneration, a study by Bobkova et al. (2022) used doses of 2–20 μg of mitochondria in 3 μL of buffer administered in each nostril. Administering a dose of 20 μg per animal, three times over a span of 3 weeks, resulted in improved spatial memory characteristics in mice with Alzheimer's type degeneration (OBX mice) compared to untreated control animals.

It is important to note that the concentration of mitochondrial doses used in these studies appears contradictory, indicating variations in dosage strategies among different research investigations. It is crucial to recognize that the number of mitochondria is not the sole determinant of their effectiveness, but rather their functionality plays a significant role. The functional capacity of mitochondria, including their ability to produce ATP and maintain cellular energy homeostasis, is of greater importance than their sheer quantity. Assessing and enhancing the functionality of mitochondria is crucial for evaluating their efficacy in various disease conditions. Therefore, it is important to prioritize this aspect in order to determine the effectiveness of mitochondria in different diseases.

The best mitochondrial therapy dosage and concentration for various diseases, as well as any potential drawbacks or side effects of these treatments, must be established through additional research. It is crucial for potential therapeutic uses of mitochondrial transplantation to establish the optimal dose for each delivery route for every disorder. Establishing the ideal dosage is crucial to ensuring efficacy and maximizing therapeutic results, even though each dose may be helpful in treating particular conditions. Further research is needed to establish precise dosage guidelines tailored to specific delivery routes and disease conditions.

Immunological reaction

One of the primary limitations of MT is the potential for developing immune responses in the recipient cells or tissues. Many different types of MT studies are based on levels found in *in vitro* studies due to environmental restrictions, which make it difficult to accurately replicate the environment of cells in vivo and the communication between different cells. Despite the fact that numerous studies indicate that mitochondria can be directed for translocation to the body's injured cells, absorption into other cells is still a possibility and may lead to unneeded cell activation. When mesenchymal stem cells (MSCs) deliver mitochondria to macrophages, microRNA containing exosomes are also produced to prevent macrophage activation by inhibiting toll like receptor (TLR) signaling, as demonstrated by Phinney et al. (2015). More testing is therefore required for patients receiving mitochondrial transfer, and it must be determined whether mitochondrial transfer encourages aberrant cell activation and proliferation. When cells are exposed to a harsh environment, more packed mitochondria can lead to harmful ROS amplification and higher cytochrome C production, which could result in insufficient cell clearance and cell apoptosis (Onodera et al., 2018). It is believed that when mitochondria are transplanted, they can travel to the affected tissues and replace dysfunctional mitochondria. However, when performing an intravenous mitochondrial transplantation procedure, there is a risk of damaging the mitochondrial membrane, and ROS released during this process may cause macrophages and immune cells to clear the mitochondria from the bloodstream rapidly and reduce their ability to reach and establish in the target tissues and make this procedure less effective.

On the other hand, studies have demonstrated that acquired mitochondria may help to raise the level of endogenous mtDNA (Caicedo et al., 2015). Extra ATP generation from surplus mitochondria could exceed the body's energy needs, which would be undesirable for the organism. Another potential safety concern is the immune response to transplanted mitochondria. Mitochondria contain unique antigens that can be recognized by the immune system, leading to the activation of T cells and the production of cytokines. This can reduce the efficiency of mitochondrial transplantation and cause tissue damage and inflammation. In inactivated immune cells, metabolic reprogramming directly enhances ROS production. For instance, the pentose phosphate pathway and active glycolysis lead to increased NADPH and ROS synthesis. A damaged mitochondrial respiratory chain is the root cause of mtROS production (Li et al., 2013). Immune responses can be reduced by matching up the recipients with the donor cells, taking immunosuppressive medications, or encapsulating the transplanted mitochondria in a barrier to shield them from immune detection.

Storage of mitochondria

Since there are very few studies about mitochondria storage, they must be administered as soon as they are isolated. Therefore, storage can be another limitation of mitochondrial transplantation. Mitochondria are delicate organelles that are sensitive to changes in temperature, pH, and osmolarity. Therefore, the storage conditions for transplanted mitochondria need to be carefully controlled to maintain their viability and functionality.

One challenge of mitochondrial storage is the preservation of mtDNA integrity. mtDNA is more susceptible to damage than nuclear DNA because of its close proximity to ROS produced by mitochondria during oxidative phosphorylation. Therefore, exposure to suboptimal storage conditions, such as temperature fluctuations, can cause mtDNA damage and impair mitochondrial function.

Another challenge is the limited shelf life of mitochondrial preparations. Mitochondria have a relatively short half life and can undergo rapid degradation after isolation. Therefore, mitochondrial preparations need to be used promptly after isolation or stored at low temperatures for short periods of time to maintain their functionality (Zhang & Miao, 2023). Therefore, it is essential to establish quality control measures for the storage and transport of isolated mitochondria to maintain their viability and function. According to Gnaiger et al. (2000) a buffer based on 4-(2-hydroxyethyl)-1-piperazineethanesulfonic acid (HEPES) and sucrose can sustain mitochondrial respiration capacity above 80% for a day during cold storage. It is important to know that the media containing Ca^{2+} makes the mitochondria lose their capacity to synthesize adenosine triphosphate (ATP) and respire on NAD-dependent substrate rapidly and permanently (Chernyak, 2020).

Several strategies have been proposed to overcome the limitations of mitochondrial storage. One approach is to use cryopreservation, which involves the freezing of mitochondrial preparations at very low temperatures to preserve their viability and functionality (Nukala et al., 2006). MT requires rapid operation due to the decreased activity of isolated mitochondria when stored on ice for over an hour (McCully et al., 2017). The preservation of mitochondrial function during refrigeration and freezing is crucial, and the choice of preservation buffer affects mitochondrial activity and structure. Studies have evaluated different solutions for cold storage, such as HEPES-sucrose-based buffer (Gnaiger et al., 2000), University of Wisconsin solution, and Euro Collins solution (Jassem et al., 2006), with varying success in maintaining respiratory capacity and preventing Cyt *c* release. Optimization of the preservation solution, including the addition of antioxidants and colloids, is necessary to protect mitochondrial structure and function during long term storage (Yamada et al., 2020).

Nukala et al. (2006) studied the storage of isolated mitochondria in dimethyl sulfoxide (DMSO) for up to 6 months. They discovered that freezing mitochondria in 10% DMSO preserved mitochondrial oxidative phosphorylation capacity but lowered mitochondrial activity. In our study, we employed cryopreservation as a method for storing mitochondria. Specifically, we used a cryopreservation medium that contained 10% dimethyl sulfoxide. Our findings indicate that this approach can successfully preserve functional mitochondria for a period exceeding 2 months without lossing mitochondrial activity (Kshersagar et al., 2023). Further research is required to ascertain the long term storage potential of mitochondria and investigate the optimal conditions for their preservation.

To increase the efficacy of mitochondrial transplantation, it is necessary to address the long term preservation of mitochondria, which is a key limitation. Therefore, resolving this issue would be a major advancement for mitochondrial therapy.

Risk

Method failure

Method failure in MT refers to situations in which the planned transport or incorporation of mitochondria into the target cells or tissues is unsuccessful or inefficient. Failure of the MT technique can be caused by a number of variables, including inadequate mitochondrial uptake, difficulties with delivery, poor mitochondrial quality and viability, etc. Different transplantation techniques can, in fact, each have their own benefits and drawbacks when it comes to procedure failures. Direct injection, as previously discussed, may face certain challenges. In the context of mitochondrial transplantation into cardiomyocytes, a study found that only a small percentage (approximately 3%–7%) of the transplanted mitochondria were observed to be integrated into the cardiomyocytes. The majority of the transplanted mitochondria remained extracellular, in close proximity to the cells, indicating limited integration with the cardiomyocytes.

These findings highlight the need for further optimization and improvement in transplantation techniques to enhance the integration and incorporation of transplanted mitochondria into the target cells. It suggests that the current method may not fully overcome the barriers that prevent efficient integration of the transplanted mitochondria with cardiomyocytes. In our investigation, we discovered that MT followed by activated platelet rich plasma (PRP) was able to stabilize mitochondria at the site of damage and facilitate 33.33% roust engraftment (Kshersagar et al., 2023).

The study of Orfany et al. used a mouse model of skeletal muscle damage and administered a greater dose of 1×10^9 mitochondria directly (Orfany et al., 2020). The overall efficacy and dosage of mitochondrial infusion may be affected due to the limited absolute volume of mitochondria that can be injected in mice, owing to their small muscle mass. This limitation emphasizes the advantage of performing research on larger animals, where a higher volume of mitochondria can be provided, potentially resolving this constraint (Orfany et al., 2020). In preclinical investigations, standardized animal models are frequently employed to examine the efficacy and safety of medicines. To explore the effects of therapies under controlled circumstances, researchers can tweak these models to imitate certain diseases or situations. However, it is important to recognize that replicating the exact same condition in human subjects for clinical studies is challenging.

In clinical studies, each human subject has unique medical conditions, disease severity, and individual variations. It is difficult to enroll subjects with the exact same pathological conditions found in preclinical models. For example, the size of an ischemic area in one patient may differ from another patient, and additional medical conditions may be present, making each case unique. Such variation in preclinical to clinical studies may result in method failure. Developing an efficient treatment requires considering the variability and complexity of human medical conditions. It is necessary to develop models and study designs that closely resemble the human situation. Clinical studies often involve diverse patient populations to capture the heterogeneity of the disease and its presentation in the human population. This may help in better MT therapy which can limit method failure.

Overcoming method failures in mitochondrial transplantation requires ongoing research and development to improve techniques for mitochondrial delivery, enhance mitochondrial survival and integration, optimize dosing strategies, and address potential immune reactions. Additionally, ensuring the quality and viability of transplanted mitochondria is vital for maximizing the success of MT as a therapeutic approach.

Adverse side effect

The possibility for MT to have adverse effects can be viewed as a limitation of the method and provide difficulties for its use and adoption globally, reducing its potential for success. Ongoing research attempts to improve compatibility, reduce immunological reactions, and improve procedures. Preclinical research and clinical trials are essential for assessing MT's safety and addressing issues that may arise (Guariento et al., 2021; McCully & Emani, 2022).

In a clinical study by Mc Cully and his team on the improvement of cardiac function in patients of ischemic heart diseases observed that one of their five pediatric patients who received MT as salvage therapy suffered renal damage and myocardial infarction due to prolonged ischemia (Emani & Mccully, 2018). In the study, it was noted that two patients did not survive the treatment. The causes of death were attributed to persistent pulmonary insufficiency and other complications. These unfortunate outcomes highlight the challenges and risks associated with the treatment and underscore the need for careful patient selection, monitoring, and management during MT. It emphasizes the importance of considering potential complications and ensuring patient safety in the clinical applications of mitochondrial therapy. Additionally, it is demonstrated that mitochondria may elicit an inflammatory response to the damage associated molecular patterns secreted by dying or necrotic cells (Grazioli & Pugin, 2018). There is a significant amount of scientific research documenting mitochondrial transfer/transplantation, as well as novel methods, to isolate and transfer mitochondrion in preclinical and clinical models but all the side effects are not fully described. The occurrence of adverse side effects can limit the overall success of MT, affecting patient outcomes and the feasibility of the procedure. Adverse effects may require additional medical intervention, prolong recovery times, or even result in treatment failure. It is essential to carefully assess and manage these potential risks to ensure the safety and well being of patients undergoing MT.

To mitigate the limitations posed by adverse side effects, ongoing research focuses on refining transplantation techniques, optimizing donorrecipient compatibility, and developing strategies to minimize immune responses and inflammation. Comprehensive preclinical studies and well designed clinical trials are essential for evaluating the safety profile of MT and identifying and addressing potential adverse effects.

Limited efficacy

Limited efficiency in preclinical and clinical MT refers to challenges and limitations encountered while translating beneficial findings from preclinical research to successful outcomes in actual patients. Despite encouraging results in preclinical models, there may be some difficulties in applying MT in clinical settings. The following are some causes of limited efficiency parameters.

Variability in preclinical models

Preclinical studies often use animal models to assess the feasibility and efficacy of MT. However, the effectiveness of the procedure in animals may not always accurately reflect its performance in humans. Preclinical animal models can never be entirely reliable because of the unknowns that are brought about by a variety of species. We argue that even if the next several decades were devoted to increasing the internal and external validity of animal models, their clinical use would ultimately only grow to a limited degree. This is because extrapolating results from animals to people would remain unreliable due to species differences. It is suggested that research should instead concentrate on human relevant research methodologies and technology in order to enhance clinical translation and eventually benefit patients. Variations in mitochondrial biology, immune responses, and disease pathophysiology between species can limit the predictive value of preclinical models, leading to difficulties in translating findings to human patients (Pound & Ritskes Hoitinga, 2018).

Lack of standardized protocols

MT is a complex procedure involving various steps, including mitochondrial isolation, characterization, and transplantation techniques. Recently, utilizing several filters, McCully and his team developed a quick isolation process that took less than 30 min. As a result, this is useful for clinical use. Meeting the requirements of GMP is crucial for the clinical use of MT (Preble et al., 2014). However, there is currently a lack of standardized protocols across research groups and clinical settings. The absence of standardized procedures hampers comparability between studies and poses challenges in reproducing consistent outcomes, limiting the overall efficiency of the technique.

Variability in transplantation outcomes

The outcomes of MT can vary among individuals, even when performed under similar conditions. Factors such as patient characteristics, underlying diseases, and overall health status can influence the success of transplantation. Additionally, individual differences in immune responses, mitochondrial heterogeneity, and the ability of transplanted mitochondria to integrate and function within host cells can contribute to variability in outcomes. While conducting studies on heart ischemia–reperfusion injury, animal models like rabbits were used (Masuzawa et al., 2013). However, it should be noted that these animal models may not fully replicate the pathophysiological processes observed in humans, particularly in the context of acute limb ischemia (Kin et al., 2020; Orfany et al., 2020). Although the findings from these studies offer valuable insights, further research using larger animal models, such as swine, is necessary to improve the translation of results to human patients (Kin et al., 2020; Orfany et al., 2020). Therefore, the applicability and potential limitations of mitochondrial transplantation in human patients require further investigation.

Challenges in scaling up

While preclinical studies may demonstrate promising results in small scale experiments, scaling up the transplantation process to treat larger tissue volumes or multiple organs poses additional challenges. Efficient delivery of transplanted mitochondria to target tissues, ensuring their proper distribution, and achieving consistent integration throughout the tissue are significant technical hurdles that need to be overcome for widespread clinical applications. When the results observed in preclinical models or small scale clinical studies do not translate effectively to larger scale clinical applications or produce consistent and desired outcomes, it can be seen as a limitation. The limited efficiency in preclinical and clinical studies may result from various factors, including the challenges and hurdles discussed earlier, such as the variability in preclinical models, lack of standardized protocols, variability in transplantation outcomes, and challenges in scaling up the procedure. These factors can hinder the translation of promising results from preclinical studies to successful outcomes in human patients, ultimately limiting the overall efficiency and effectiveness of MT.

Long term stability and persistence

The long term stability and persistence of transplanted mitochondria in host cells remain a concern. The efficiency of MT can be limited if the transplanted mitochondria are not stably maintained or fail to persist over extended periods. Factors such as immune responses, cellular turnover, and the presence of preexisting mitochondrial dysfunction can influence the long term viability and effectiveness of transplanted mitochondria. As discussed earlier, the study by Masuzawa et al., showed that only a small number of mitochondria have been up taken by cells while others still remain in the extracellular environment (Masuzawa et al., 2013). It implies that most cells were unable to uptake mitochondria; the cause of this is unknown. In a study conducted by Liu et al., it was observed that when mitochondria were delivered systemically, they did not completely reach the targeted organ. Instead, they were found predominantly in the liver, along with other organs. The liver accounted for a significant proportion, specifically 27%, of the injected mitochondria. The spleen, on the other hand, accounted for 2% of the detected mitochondria. In contrast, the lungs had a detection rate of less than 1% for the injected mitochondria (Liu et al., 2020). It still indicates some distribution of mitochondria to other organs.

Addressing the limitations in preclinical and clinical studies is essential to improve the efficiency of MT. By identifying and addressing the factors contributing to limited efficiency, researchers aim to enhance the success rate and clinical applicability of MT.

Safety concern

Safety concerns are one of the major limitations of MT that need to be addressed before this technique can be widely adopted in clinical practice. Mitochondria carry their own DNA, which is distinct from the nuclear DNA, and there is a risk of transmitting mtDNA mutations or other potentially harmful components along with the transplanted mitochondria. mtDNA is circular and encodes for 13 proteins involved in oxidative phosphorylation, as well as 22 transfer RNAs and 2 ribosomal RNAs (Taanman, 1999). One potential concern with MT is the transmission of mtDNA mutations from the donor to the recipient. There is a risk of transmitting mtDNA mutations or other potentially harmful components along with the transplanted mitochondria. Ramirez-Barbieri et al. (2019) noted that in severely ill patients, the presence of damaged skeletal muscle tissue poses challenges in isolating viable mitochondria. Furthermore, the use of autologous mitochondria is not feasible in individuals with mitochondrial myopathies caused by pathogenic mutations in mtDNA. mtDNA mutations are known to be associated with a wide range of diseases, including Leigh syndrome, Kearns

Sayre syndrome, and mitochondrial encephalomyopathy (Taylor & Turnbull, 2005). The transmission of these mutations through MT could potentially exacerbate these diseases or cause other adverse effects (Taylor & Turnbull, 2005). This risk can be minimized by selecting healthy donors with no known mtDNA mutations and conducting rigorous screening of the mitochondrial genome.

Additionally, there is a risk of immunological reactions that could lead to inflammation and tissue damage. Therefore, safety testing needs to be conducted thoroughly before MT can be used in humans. For instance, in a model of combined injury involving kidney ischemia/reperfusion and systemic delivery of mitochondria, the study investigated the development of acute kidney damage and animal mortality (Plotnikov et al., 2020). Intraperitoneal injection of isolated mitochondria led to the presence of mtDNA in peripheral blood, potentially triggering innate immune responses. The delivery of mitochondria, along with renal ischemia/reperfusion, resulted in proinflammatory changes, including increased levels of white blood cells and granulocytes. These inflammatory responses were associated with a high mortality rate in animals with combined injuries and multiple organ failure, potentially linked to the activation of TLR signaling pathways. The study also suggests that mitochondria along with antioxidants may provide beneficial effects in renal ischemia (Plotnikov et al., 2020). Understanding the underlying mechanisms of TLR activation and immune responses is crucial for improving the management of patients with combined injuries and multiple organ failure in intensive care units. As discussed earlier, direct injection of mitochondria may possess the risk of infection or inflammation at the injected site. For instance, intracoronary injection necessitates invasive techniques like cardiac catheterization, which involves threading a catheter through blood vessels to access the coronary arteries. This procedure carries inherent risks, including potential vessel damage, bleeding, infection, and adverse reactions to contrast agents employed during imaging procedures (Shin et al., 2019). In order to prioritize patient safety, researchers should carefully select clinically relevant mitochondrial treatments. This ensures that the chosen treatment approaches have been thoroughly evaluated for their safety and potential risks.

Despite these safety concerns, there have been some successful applications of MT in animal models and early clinical trials. However, further research is needed to optimize the technique and address safety concerns before MT can be used as a routine clinical treatment for mitochondrial disorders. While MT has shown promise in preclinical studies, its efficacy in human patients is still uncertain. The effects of the procedure may vary depending on the underlying disease, and long term follow up is needed to determine its effectiveness.

Worldwide regulations

MT is a relatively new technique and its regulatory status is still unclear in many countries. Regulatory issues may limit its availability in some countries or delay its approval for clinical use. While regulations can play a crucial role in ensuring the safety and ethical implementation of treatments like MT and other innovative medical procedures, they can also impose significant limitations on the advancement and widespread adoption. Regulatory bodies have stringent evaluation and approval processes for novel medical treatments such as mitochondrial transplantation. These processes involve preclinical and clinical trials to ensure patient safety.

Different countries have different sets of regulations on mitochondrial replacement therapy (MRT). The National Guidelines for Accreditation, Supervision, and Regulation of Assisted Reproductive Technology (ART) in India, jointly published by the Indian Council of Medical Research and the National Academy of Medical Sciences in 2005, establish regulations for ART clinics. These regulations explicitly prohibit germline therapy, gene therapy for enhancing genetic characteristics (known as "designer babies"), and eugenic genetic engineering aimed at selecting against certain traits. Consequently, regulations regarding MRT for addressing mitochondrial diseases are in place in many countries. Ethical review committees allow parents to make informed decisions regarding their choice to undergo MRT, ensuring adherence to ethical considerations and individual autonomy. The United Kingdom (UK) implemented regulations on MRT in 2015, allowing its use in certain cases only. In the UK, the Human Fertilization and Embryology Act was amended to permit mitochondrial donation techniques as part of IVF only. In contrast, the regulatory scenario in the United States of America is quite different, where the US Food and Drug Administration regulates the use of MRT and requires clinical trials under an investigational new drug application for approval (Castro, 2016). In 2004, Canada introduced the Assisted Human Reproduction Act (AHRA) which prohibited research on MRT. After passing all safety inspections, this Canadian regulation was later challenged to permit the usage of the MRT. However, due to the country's refusal to update the AHRA statute since 2004, MRT and germline changes are still prohibited (Cohen et al., 2015). The regulatory landscape in countries like Australia and Singapore is somewhat similar to the UK; they are currently working on consultation and decision making before legalizing mitochondria related therapies for the general public, stating the importance of public engagement in shaping regulations for mitochondrial donation. In 1999, the government of Israel passed a law that banned the use of human reproductive cloning and the use of genetically modified reproductive cells. While the law does not directly mention MRTs, it can be interpreted to include them as the law states that, "to prevent reproductive cloning in humans by establishing that certain kinds of genetic interventions shall not be performed on human beings in view of the moral, legal, social and scientific aspects of the prohibited forms of intervention and their implications on human dignity, and in order to assess public policy regarding those kinds of intervention in view of those aspects, considering freedom of scientific research for the advancement of medicine." However, changes in the current Israeli regulatory landscape would necessitate public opinion (Cohen et al., 2020).

MT related techniques face challenges such as ethical concerns (Bouzenita et al., 2017), cost and reimbursement issues, international regulatory variations, and safety requirements which have led to stringent regulations relating to their use (Wrigley et al., 2015). Financial coverage is influenced by regulatory actions, which affects patient affordability. Inconsistent policies across countries impede global adoption and collaboration. A balance between regulatory and innovative frameworks to address these concerns can increase the accessibility and acceptance of MT as a medical practice. MT is an emerging medical technique with the potential to treat certain mitochondrial diseases and other related conditions. While regulations can play a crucial role in ensuring the safety and ethical implementation of medical procedures, they can also impose limitations on the advancement and widespread adoption of innovative treatments like mitochondrial transplantation. Here are a few ways in which worldwide regulations can act as limitations: Approval and licensing processes, ethical considerations and guidelines, cost and reimbursement challenges, international variations in regulations, safety and monitoring requirements, etc. As scientific understanding and evidence surrounding MT continue to evolve, regulatory frameworks will need to adapt to foster its responsible development and implementation.

Ethical concern

Ethical considerations in mitochondrial transplantation revolve around various aspects of the procedure and its implications. These considerations include the following:

1. Informed Consent: Patients and/or their legal guardians must provide informed consent after receiving comprehensive information about the procedure, potential risks, benefits, and alternatives.
2. Genetic Modification: MT involving the transfer of mitochondria from a donor carries the potential for genetic modification. Ethical concerns arise regarding the transfer of genetic material to future generations and the implications of altering germline characteristics.
3. Safety and Efficacy: The safety and efficacy of mitochondrial transplantation need to be carefully evaluated through rigorous preclinical and clinical studies before widespread implementation. Adequate evidence of safety, long term effects, and potential risks should be established.
4. Allocation of Resources: Mitochondrial transplantation requires resources, including donor mitochondria and skilled medical personnel. Ethical considerations arise in ensuring equitable access to treatment, addressing potential inequalities, and prioritizing patients based on medical needs and available resources.
5. Donor Autonomy: Ethical considerations revolve around donor autonomy in making decisions about the use of these techniques to prevent the transmission of mitochondrial diseases to the recipient.
6. Regulatory Oversight: Ethical considerations also encompass the need for appropriate regulatory frameworks and oversight to ensure compliance with ethical principles, adherence to established guidelines, and protection of patient rights and welfare.
7. Long term Follow up: The MT has the possibility of transferring the DNA of exogenous mitochondria into recipient cells. Clinical trial data for MT are exceedingly limited, and no long-term follow-up has been recorded. From an ethical perspective, it is required to adopt mandatory long term follow up methods after MT.

It is essential to engage in robust ethical discussions, involve relevant stakeholders, and adhere to established ethical guidelines and regulatory frameworks to ensure that MT is conducted in an ethical manner, respects patient autonomy, and promotes the overall well being of individuals and society.

Conclusion

MT has significant potential as a treatment for disorders linked to mitochondrial dysfunction. However, it is critical to acknowledge the limitations and challenges that come with this ground breaking therapy. To safeguard patient well being, safety issues, such as possible negative effects and immunological reactions, must be properly addressed. The best dosage and administration route is still being investigated and improved. To ensure that transplanted mitochondria are viable and functional, quality control procedures must be implemented and established. Careful thought must be given to ethical issues like informed consent and long term follow ups. Worldwide regulations and variations in guidelines create hurdles for the widespread implementation of MT.

The widespread availability of this therapy is further constrained by technical difficulties, poor efficiency, and high costs. It requires ongoing research, protocol standardization, legislative harmonization, and technical developments to get over these constraints. To advance the subject, collaboration between researchers, doctors, and regulatory agencies is crucial. By overcoming these drawbacks, MT has the potential to revolutionize the treatment of diseases associated with mitochondrial dysfunction and other related conditions, leading to better health and patient outcomes.

It is important for studies involving mitochondrial transplantation or any therapeutic intervention to discuss both the positive and negative aspects of their findings. While it is common to emphasize the positive outcomes and benefits of treatment, it is equally crucial to address and discuss any method failures or limitations encountered during the study. By openly discussing method failures and limitations, researchers can contribute to the broader scientific understanding of the challenges and potential pitfalls associated with mitochondrial transplantation. This information can guide future research, help refine techniques and protocols, and provide a more comprehensive assessment of the therapy's efficacy and safety.

But not least, it is crucial to recognize the existing limitations, work toward their resolution, and refine the procedures associated with mitochondrial transplantation. With ongoing advancements and a comprehensive approach to addressing these challenges, MT can become a safe, effective, and accessible therapeutic option for patients affected by mitochondrial dysfunction.

References

Ahlawat, J., Guillama Barroso, G., Masoudi Asil, S., Alvarado, M., Armendariz, I., Bernal, J., Carabaza, X., Chavez, S., Cruz, P., Escalante, V., Estorga, S., Fernandez, D., Lozano, C., Marrufo, M., Ahmad, N., Negrete, S., Olvera, K., Parada, X., Portillo, B., … Narayan, M. (2020). Nanocarriers as potential drug delivery candidates for overcoming the blood–brain barrier: Challenges and possibilities. *ACS Omega, 5*(22), 12583–12595. https://doi.org/10.1021/acsomega.0c01592

Ahmad, F., Alamoudi, W., Haque, S., Salahuddin, M., & Alsamman, K. (2018). Simple, reliable, and time-efficient colorimetric method for the assessment of mitochondrial function and toxicity. *Bosnian Journal of Basic Medical Sciences, 18*(4), 367–374. https://doi.org/10.17305/bjbms.2018.3323

Alexander, J. F., Mahalingam, R., Seua, A. V., Wu, S., Arroyo, L. D., Hörbelt, T., Schedlowski, M., Blanco, E., Kavelaars, A., & Heijnen, C. J. (2022). Targeting the meningeal compartment to resolve chemobrain and neuropathy via nasal delivery of functionalized mitochondria. *Advanced Healthcare Materials, 11*(8). https://doi.org/10.1002/adhm.202102153

Bhatti, J., Bhatti, G., & Reddy, PH. 2017 Mitochondrial dysfunction and oxidative stress in metabolic disorders—A step towards mitochondria based therapeutic strategies. Biochimica et Biophysica (BBA)—Molecular Basis of Diseases (Vol. 1863). PMCID. https://doi.org/10.1016/j.bbadis.2016.11.010. (Epub).

Alway, S. E., Paez, H. G., Pitzer, C. R., Ferrandi, P. J., Khan, M. M., Mohamed, J. S., Carson, J. A., & Deschenes, M. R. (2023). Mitochondria transplant therapy improves regeneration and restoration of injured skeletal muscle. *Journal of Cachexia, Sarcopenia and Muscle, 14*(1), 493–507.

Bobkova, N. V., Zhdanova, D. Y., Belosludtseva, N. V., Penkov, N. V., & Mironova, G. D. (2022). Intranasal administration of mitochondria improves spatial memory in olfactory bulbectomized mice. *Experimental Biology and Medicine, 247*(5), 416–425. https://doi.org/10.1177/15353702211056866

Bouzenita, A. I., Mirghani, M. E. S., & Jaswir, I. (2017). The islamic ethics of mitochondria transplantation. *IIUM Engineering Journal, 18*(2), 42–46. https://doi.org/10.31436/iiumej.v18i2.731

Caicedo, A., Fritz, V., Brondello, J. M., Ayala, M., Dennemont, I., Abdellaoui, N., De Fraipont, F., Moisan, A., Prouteau, C. A., Boukhaddaoui, H., Jorgensen, C., & Vignais, M. L. (2015). MitoCeption as a new tool to assess the effects of mesenchymal stem/stromal cell mitochondria on cancer cell metabolism and function. *Scientific Reports, 5*. https://doi.org/10.1038/srep09073

Castro, R. J. (2016). Mitochondrial replacement therapy: The UK and US regulatory landscapes. *Journal of Law and the Biosciences, 3*(3), 726–735. https://doi.org/10.1093/jlb/lsw051

Chang, J. C., Chang, H. S., Wu, Y. C., Cheng, W. L., Lin, T. T., Chang, H. J., Kuo, S. J., Chen, S. T., & Liu, C. S. (2019). Mitochondrial transplantation regulates antitumour activity, chemoresistance and mitochondrial dynamics in breast cancer. *Journal of Experimental & Clinical Cancer Research, 38*, 1–6.

Chang, J. C., Chao, Y. C., Chang, H. S., Wu, Y. L., Chang, H. J., Lin, Y. S., Cheng, W. L., Lin, T. T., & Liu, C. S. (2021). Intranasal delivery of mitochondria for treatment of Parkinson's disease model rats lesioned with 6-hydroxydopamine. *Scientific Reports, 11*(1). https://doi.org/10.1038/s41598-021-90094-w

Chang, J. C., Wu, S. L., Liu, K. H., Chen, Y. H., Chuang, C. S., Cheng, F. C., Su, H. L., Wei, Y. H., Kuo, S. J., & Liu, C. S. (2016). Allogeneic/xenogeneic transplantation of peptide-labeled mitochondria in Parkinson's disease: Restoration of mitochondria functions and attenuation of 6-hydroxydopamine-induced neurotoxicity. *Translational Research, 170*, 40–56.e3. https://doi.org/10.1016/j.trsl.2015.12.003

Chernyak, B. V. (2020). Mitochondrial transplantation: A critical analysis. *Biochemistry (Moscow), 85*(5), 636–641. https://doi.org/10.1134/S0006297920050132

Cohen, I. G., Adashi, E. Y., Gerke, S., Palacios-González, C., & Ravitsky, V. (2020). The regulation of mitochondrial replacement techniques around the world. *Annual Review of Genomics and Human Genetics, 21*(1), 565–586. https://doi.org/10.1146/annurev-genom-111119-101815

Cohen, I. G., Savulescu, J., & Adashi, E. Y. (2015). Transatlantic lessons in regulation of mitochondrial replacement therapy. *Science, 348*(6231), 178–180. https://doi.org/10.1126/science.aaa8153

Cowan, D. B., Yao, R., Akurathi, V., Snay, E. R., Thedsanamoorthy, J. K., Zurakowski, D., Ericsson, M., Friehs, I., Wu, Y., Levitsky, S., del Nido, P. J., Packard, A. B., McCully, J. D., & Bishopric, N. H. (2016). Intracoronary delivery of mitochondria to the ischemic heart for cardioprotection. *PLoS One, 11*(8), e0160889. https://doi.org/10.1371/journal.pone.0160889

Díaz-Carballo, D., Klein, J., Acikelli, A. H., Wilk, C., Saka, S., Jastrow, H., Wennemuth, G., Dammann, P., Giger-Pabst, U., Khosrawipour, V., Rassow, J., Nienen, M., & Strumberg, D. (2017). Cytotoxic stress induces transfer of mitochondria-associated human endogenous retroviral RNA and proteins between cancer cells. *Oncotarget, 8*(56), 95945–95964. https://doi.org/10.18632/oncotarget.21606

de Duve, C., Pressman, B. C., Gianetto, R., Wattiaux, R., & Appelmans, F. (1955). Tissue fractionation studies. 6. Intracellular distribution patterns of enzymes in rat-liver tissue. *Biochemical Journal, 60*(4), 604–617. https://doi.org/10.1042/bj0600604

Doulamis, I. P., Guariento, A., Duignan, T., Kido, T., Orfany, A., Saeed, M. Y., Weixler, V. H., Blitzer, D., Shin, B., Snay, E. R., Inkster, J. A., Packard, A. B., Zurakowski, D., Rousselle, T., Bajwa, A., Parikh, S. M., Stillman, I. E., del Nido, P. J., & McCully, J. D. (2020). Mitochondrial transplantation by intra-arterial injection for acute kidney injury. *American Journal of Physiology—Renal Physiology, 319*(3), F403–F413. https://doi.org/10.1152/AJPRENAL.00255.2020

Emani, S. M., & McCully, J. D. (2018). Mitochondrial transplantation: Applications for pediatric patients with congenital heart disease. *Translational Pediatrics, 7*(2), 169–175. https://doi.org/10.21037/tp.2018.02.02

Emani, S. M., Piekarski, B. L., Harrild, D., del Nido, P. J., & McCully, J. D. (2017). Autologous mitochondrial transplantation for dysfunction after ischemia-reperfusion injury. *The Journal of Thoracic and Cardiovascular Surgery, 154*(1), 286–289. https://doi.org/10.1016/j.jtcvs.2017.02.018

Fang, S. Y., Roan, J. N., Lee, J. S., Chiu, M. H., Lin, M. W., Liu, C. C., & Lam, C. F. (2021). Transplantation of viable mitochondria attenuates neurologic injury after spinal cord ischemia. *The Journal of Thoracic and Cardiovascular Surgery, 161*(5), e337–e347. https://doi.org/10.1016/j.jtcvs.2019.10.151

Fu, Shi, X., Zhang, H., & Fu, B. 2017. Mitotherapy for fatty liver by intravenous administration of exogenous mitochondria in male mice. Frontiers in Pharmacology (Vol. 8).

Gnaiger, E., Kuznetsov, A. V., Schneeberger, S., Seiler, R.ü., Brandacher, G., Steurer, W., & Margreiter, R. (2000). *Mitochondria in the cold* (pp. 431–442). Springer Science and Business Media LLC. https://doi.org/10.1007/978-3-662-04162-8_45

Gollihue, J. L., Patel, S. P., Mashburn, C., Eldahan, K. C., Sullivan, P. G., & Rabchevsky, A. G. (2017). Optimization of mitochondrial isolation techniques for intraspinal transplantation procedures. *Journal of Neuroscience Methods, 287*, 1–12. https://doi.org/10.1016/j.jneumeth.2017.05.023

Gollihue, J. L., & Rabchevsky, A. G. (2017). Prospects for therapeutic mitochondrial transplantation. *Mitochondrion, 35*, 70–79. https://doi.org/10.1016/j.mito.2017.05.007

Grazioli, S., & Pugin, J. (2018). Mitochondrial damage-associated molecular patterns: From inflammatory signaling to human diseases. *Frontiers in Immunology, 9*(May). https://doi.org/10.3389/fimmu.2018.00832

Guariento, A., Piekarski, B., Doulamis, B., Ferraro, A., Harrild, Zurakowski, D., Nido, D., McCully, P., & Emani, J. (2020). Autologous MT for cardiogenic shock in pediatric patients following ischemia-reperfusion injury. *The Journal of Thoracic and Cardiovascular Surgery, 162*(3). https://doi.org/10.1016/j.jtcvs.2020.10.151. Epub.

Guariento, A., Piekarski, B. L., Doulamis, I. P., Blitzer, D., Ferraro, A. M., Harrild, D. M., Zurakowski, D., del Nido, P. J., McCully, J. D., & Emani, S. M. (2021). Autologous mitochondrial transplantation for cardiogenic shock in pediatric patients following ischemia-reperfusion injury. *The Journal of Thoracic and Cardiovascular Surgery, 162*(3), 992–1001. https://doi.org/10.1016/j.jtcvs.2020.10.151

Hartwig, S., Kotzka, J., & Lehr, S. (2015). Isolation and quality control of functional mitochondria. *Methods in Molecular Biology, 1264*, 9–23. https://doi.org/10.1007/978-1-4939-2257-4_2

Hayashida, K., Takegawa, R., Endo, Y., Yin, T., Choudhary, R. C., Aoki, T., Nishikimi, M., Murao, A., Nakamura, E., Shoaib, M., Kuschner, C., Miyara, S. J., Kim, J., Shinozaki, K., Wang, P., & Becker, L. B. (2023). Exogenous mitochondrial transplantation improves survival and neurological outcomes after resuscitation from cardiac arrest. *BMC Medicine, 21*(1). https://doi.org/10.1186/s12916-023-02759-0

Herberts, C. A., Kwa, M. S. G., & Hermsen, H. P. H. (2011). Risk factors in the development of stem cell therapy. *Journal of Translational Medicine, 9*. https://doi.org/10.1186/1479-5876-9-29

Hornig-Do, H. T., Günther, G., Bust, M., Lehnartz, P., Bosio, A., & Wiesner, R. J. (2009). Isolation of functional pure mitochondria by superparamagnetic microbeads. *Analytical Biochemistry, 389*(1), 1–5. https://doi.org/10.1016/j.ab.2009.02.040

Hsu, C. H., Roan, J. N., Fang, S. Y., Chiu, M. H., Cheng, T. T., Huang, C. C., Lin, M. W., & Lam, C. F. (2022). Transplantation of viable mitochondria improves right ventricular performance and pulmonary artery remodeling in rats with pulmonary arterial hypertension. *The Journal of Thoracic and Cardiovascular Surgery, 63*(5), 361–373.

Huang, P. J., Kuo, C. C., Lee, H. C., Shen, C. I., Cheng, F. C., Wu, S. F., Chang, J. C., Pan, H. C., Lin, S. Z., Liu, C. S., & Su, H. L. (2016). Transferring xenogenic mitochondria provides neural protection against ischemic stress in ischemic rat brains. *Cell Transplantation, 25*(5), 913–927. https://doi.org/10.3727/096368915X689785

Jabbari, H., Roushandeh, A. M., Rostami, M. K., Razavi-Toosi, M. T., Shokrgozar, M. A., Jahanian-Najafabadi, A., Kuwahara, Y., & Roudkenar, M. H. (2020). Mitochondrial transplantation ameliorates ischemia/reperfusion-induced kidney injury in rat. *Biochimica et Biophysica Acta—Molecular Basis of Disease, 1866*(8). https://doi.org/10.1016/j.bbadis.2020.165809

Jain, R., Begum, N., Tryphena, K. P., Singh, S. B., Srivastava, S., Rai, S. N., Vamanu, E., & Khatri, D. K. (2023). Inter and intracellular mitochondrial transfer: Future of mitochondrial transplant therapy in Parkinson's disease. *Biomedicine and Pharmacotherapy, 159*. https://doi.org/10.1016/j.biopha.2023.114268

Jassem, W., Armeni, T., Quiles, J. L., Bompadre, S., Principato, G., & Battino, M. (2006). Protection of mitochondria during cold storage of liver and following transplantation: Comparison of the two solutions, University of Wisconsin and Eurocollins. *Journal of Bioenergetics and Biomembranes, 38*(1), 49–55. https://doi.org/10.1007/s10863-006-9005-6

Jia, X., Wang, Q., Ji, J., Lu, W., Liu, Z., Tian, H., Guo, L., & Wang, Y. (2023). Mitochondrial transplantation ameliorates hippocampal damage following status epilepticus. *Animal Models and Experimental Medicine, 6*(1), 41–50. https://doi.org/10.1002/ame2.12310

Kaza, A. K., Wamala, I., Friehs, I., Kuebler, J. D., Rathod, R. H., Berra, I., Ericsson, M., Yao, R., Thedsanamoorthy, J. K., Zurakowski, D., Levitsky, S., del Nido, P. J., Cowan, D. B., & McCully, J. D. (2017). Myocardial rescue with autologous mitochondrial transplantation in a porcine model of ischemia/reperfusion. *Journal of Thoracic and Cardiovascular Surgery, 153*(4), 934–943. https://doi.org/10.1016/j.jtcvs.2016.10.077

Kim, W., Choi, D., Jang, Y., Nam, C. M., Hur, S. H., & Hong, M. K. (2020). Effect of intentional restriction of venous return on tissue oxygenation in a porcine model of acute limb ischemia. *PLoS One, 15*(12). https://doi.org/10.1371/journal.pone.0243033

Kiss, D. S., Toth, I., Jocsak, G., Sterczer, A., Bartha, T., Frenyo, L. V., & Zsarnovszky, A. (2016). Preparation of purified perikaryal and synaptosomal mitochondrial fractions from relatively small hypothalamic brain samples. *MethodsX, 3*, 417–429. https://doi.org/10.1016/j.mex.2016.05.004

Kitani, T., Kami, D., Matoba, S., & Gojo, S. (2014). Internalization of isolated functional mitochondria: Involvement of macropinocytosis. *Journal of Cellular and Molecular Medicine, 18*(8), 1694–1703. https://doi.org/10.1111/jcmm.12316

Konari, N., Nagaishi, K., Kikuchi, S., & Fujimiya, M. (2019). Mitochondria transfer from mesenchymal stem cells structurally and functionally repairs renal proximal tubular epithelial cells in diabetic nephropathy in vivo. *Scientific Reports, 9*(1). https://doi.org/10.1038/s41598-019-40163-y

Kshersagar, J., Pulgam, L., Damle, M. N., Tardalkar, K., Sharma, R., & Joshi, M. G. (2023). Transplantation of human placenta derived mitochondria promotes cell communication in endometrium in a murine model of disturbed endometrium. *Stem Cell Reviews and Reports, 19(5)*, 1384–1401. https://doi.org/10.1007/s12015-023-10516-2

Kubat, G. B., Ulger, O., & Akin, S. (2021). Requirements for successful mitochondrial transplantation. *Journal of Biochemical and Molecular Toxicology, 35*(11). https://doi.org/10.1002/jbt.22898

Léger, J. L., Jougleux, J. L., Savadogo, F., Pichaud, N., & Boudreau, L. H. (2020). Rapid isolation and purification of functional platelet mitochondria using a discontinuous Percoll gradient. *Platelets, 31*(2), 258–264. https://doi.org/10.1080/09537104.2019.1609666

Li, X., Fang, P., Mai, J., Choi, E. T., Wang, H., & Yang, X. F. (2013). Targeting mitochondrial reactive oxygen species as novel therapy for inflammatory diseases and cancers. *Journal of Hematology & Oncology, 6*, 1–9.

Lin, H. C., Liu, S. Y., Lai, H. S., & Lai, I. R. (2013). Isolated mitochondria infusion mitigates ischemia-reperfusion injury of the liver in rats. *Shock, 39*(3), 304–310. https://doi.org/10.1097/SHK.0b013e318283035f

Liu, D., Gao, Y., Liu, J., Huang, Y., Yin, J., Feng, Y., Shi, L., Meloni, B. P., Zhang, C., Zheng, M., & Gao, J. (2021). Intercellular mitochondrial transfer as a means of tissue revitalization. *Signal Transduction and Targeted Therapy, 6*(1). https://doi.org/10.1038/s41392-020-00440-z

Liu, X., Khouri-Farah, N., Wu, C. H., & Wu, G. Y. (2020). Targeted delivery of mitochondria to the liver in rats. *Journal of Gastroenterology and Hepatology, 35*(12), 2241–2247. https://doi.org/10.1111/jgh.15091

Marcondes, N. A., Terra, S. R., Lasta, C. S., Hlavac, N. R. C., Dalmolin, M. L., Lacerda, L.d. A., Faulhaber, G. A. M., & González, F. H. D. (2019). Comparison of JC-1 and MitoTracker probes for mitochondrial viability assessment in stored canine platelet concentrates: A flow cytometry study. *Cytometry, Part A, 95*(2), 214–218. https://doi.org/10.1002/cyto.a.23567

Masuzawa, A., Black, K. M., Pacak, C. A., Ericsson, M., Barnett, R. J., Drumm, C., Seth, P., Bloch, D. B., Levitsky, S., Cowan, D. B., & McCully, J. D. (2013). Transplantation of autologously derived mitochondria protects the heart from ischemia-reperfusion injury. *American Journal of Physiology-Heart and Circulatory Physiology, 304*(7), 966–982.

McCully, d, N., & Emani, S. M. (2022). Therapeutic mitochondrial transplantation. *Current Opinion in Physiology, 3*.

McCully, J. D., Cowan, D. B., Emani, S. M., & del Nido, P. J. (2017). Mitochondrial transplantation: From animal models to clinical use in humans. *Mitochondrion, 34*, 127–134. https://doi.org/10.1016/j.mito.2017.03.004

McCully, J. D., Cowan, D. B., Pacak, C. A., Toumpoulis, I. K., Dayalan, H., & Levitsky, S. (2009). Injection of isolated mitochondria during early reperfusion for cardioprotection. *American Journal of Physiology—Heart and Circulatory Physiology, 296*(1), H94–H105. https://doi.org/10.1152/ajpheart.00567.2008

Mertens, J., Reid, D., Lau, S., Kim, Y., & Gage, F. H. (2018). Aging in a dish: IPSC-Derived and directly induced neurons for studying brain aging and age-related neurodegenerative diseases. *Annual Review of Genetics, 52*, 271–293. https://doi.org/10.1146/annurev-genet-120417-031534

Moskowitzova, K., Orfany, A., Liu, K., Ramirez-Barbieri, G., Thedsanamoorthy, J. K., Yao, R., Guariento, A., Doulamis, I. P., Blitzer, D., Shin, B., Snay, E. R., Inkster, J. A. H., Iken, K., Packard, A. B., Cowan, D. B., Visner, G. A., del Nido, P. J., & McCully, J. D. (2020). Mitochondrial transplantation enhances murine lung viability and recovery after ischemia-reperfusion injury. *American Journal of Physiology—Lung Cellular and Molecular Physiology, 318*(1), L78–L88. https://doi.org/10.1152/ajplung.00221.2019

Murphy, E., Ardehali, H., Balaban, R. S., DiLisa, F., Dorn, G. W., Kitsis, R. N., Otsu, K., Ping, P., Rizzuto, R., Sack, M. N., Wallace, D., & Youle, R. J. (2016). Mitochondrial function, biology, and role in disease: A scientific statement from the American Heart Association. *Circulation Research, 118*(12), 1960–1991. https://doi.org/10.1161/RES.0000000000000104

Nukala, V. N., Singh, I. N., Davis, L. M., & Sullivan, P. G. (2006). Cryopreservation of brain mitochondria: A novel methodology for functional studies. *Journal of Neuroscience Methods, 152*(1–2), 48–54. https://doi.org/10.1016/j.jneumeth.2005.08.017

Onodera, Y., Nam, J. M., Horikawa, M., Shirato, H., & Sabe, H. (2018). Arf6-driven cell invasion is intrinsically linked to TRAK1-mediated mitochondrial anterograde trafficking to avoid oxidative catastrophe. *Nature Communications, 9*(1). https://doi.org/10.1038/s41467-018-05087-7

Orfany, A., Arriola, C. G., Doulamis, I. P., Guariento, A., Ramirez-Barbieri, G., Moskowitzova, K., Shin, B., Blitzer, D., Rogers, C., del Nido, P. J., & McCully, J. D. (2020). Mitochondrial transplantation ameliorates acute limb ischemia. *Journal of Vascular Surgery, 71*(3), 1014–1026. https://doi.org/10.1016/j.jvs.2019.03.079

Pang, Y. L., Fang, S. Y., Cheng, T. T., Huang, C. C., Lin, M. W., Lam, C. F., & Chen, K. B. (2022). Viable allogeneic mitochondria transplantation improves gas exchange and alveolar-capillary permeability in rats with endotoxin-induced acute lung injuries. *International Journal of Medical Sciences, 19*(6), 1036–1046. https://doi.org/10.7150/ijms.73151

Phinney, D. G., Di Giuseppe, M., Njah, J., Sala, E., Shiva, S., St Croix, C. M., Stolz, D. B., Watkins, S. C., Di, Y. P., Leikauf, G. D., Kolls, J., Riches, D. W. H., Deiuliis, G., Kaminski, N., Boregowda, S. V., McKenna, D. H., & Ortiz, L. A. (2015). Mesenchymal stem cells use extracellular vesicles to outsource mitophagy and shuttle microRNAs. *Nature Communications, 6*. https://doi.org/10.1038/ncomms9472

Plotnikov, E. Y., Jankauskas, S. S., Zinovkin, R. A., Zorova, L. D., Zorov, S. D., Pevzner, I. B., Silachev, D. N., & Zorov, D. B. (2020). A combination of kidney ischemia and injection of isolated mitochondria leads to activation of inflammation and increase in mortality rate in rats. *Bulletin of Experimental Biology and Medicine, 169*(2), 213–217. https://doi.org/10.1007/s10517-020-04853-2

Pound, P., & Ritskes-Hoitinga, M. (2018). Is it possible to overcome issues of external validity in preclinical animal research? Why most animal models are bound to fail. *Journal of Translational Medicine, 16*(1). https://doi.org/10.1186/s12967-018-1678-1

Preble, J., Kondo, H., Levitsky, S., & McCully, J. D. (2014). Quality control parameters for mitochondria transplant in cardiac tissue. *JSM Biochemistry and Molecular Biology, 2*(1).

Ramirez-Barbieri, G., Moskowitzova, K., Shin, B., Blitzer, D., Orfany, A., Guariento, A., Iken, K., Friehs, I., Zurakowski, D., del Nido, P. J., & McCully, J. D. (2019). Alloreactivity and allorecognition of syngeneic and allogeneic mitochondria. *Mitochondrion, 46*, 103–115. https://doi.org/10.1016/j.mito.2018.03.002

Roushandeh, A. M., Kuwahara, Y., & Roudkenar, M. H. (2019). Mitochondrial transplantation as a potential and novel master key for treatment of various incurable diseases. *Cytotechnology, 71*(2), 647–663. https://doi.org/10.1007/s10616-019-00302-9

Shi, X., Zhao, M., Fu, C., & Fu, A. (2017). Intravenous administration of mitochondria for treating experimental Parkinson's disease. *Mitochondrion, 34*, 91–100.

Shi, X., Bai, H., Zhao, M., Li, X., Sun, X., Jiang, H., & Fu, A. (2018). Treatment of acetaminophen-induced liver injury with exogenous mitochondria in mice. *Translational Research, 196*, 31–41. https://doi.org/10.1016/j.trsl.2018.02.003

Shin, B., Saeed, M. Y., Esch, J. J., Guariento, A., Blitzer, D., Moskowitzova, K., Ramirez-Barbieri, G., Orfany, A., Thedsanamoorthy, J. K., Cowan, D. B., Inkster, J. A., Snay, E. R., Staffa, S. J., Packard, A. B., Zurakowski, D., del Nido, P. J., & McCully, J. D. (2019). A novel biological strategy for myocardial protection by intracoronary delivery of mitochondria: Safety and efficacy. *Journal of the American College of Cardiology: Basic to Translational Science, 4*(8), 871–888. https://doi.org/10.1016/j.jacbts.2019.08.007

Sivandzade, F., Bhalerao, A., & Cucullo, L. (2019). Analysis of the mitochondrial membrane potential using the cationic JC-1 dye as a sensitive fluorescent probe. *Bio-Protocol, 9*(1). https://doi.org/10.21769/BioProtoc.3128

Sun, M., Jiang, W., Mu, N., Zhang, Z., Yu, L., & Ma, H. (2023). MTas a novel therapeutic strategy for cardiovascular diseases. *Journal of Translational Medicine, 21*(1).

Taanman, J. W. (1999). The mitochondrial genome: Structure, transcription, translation and replication. *Biochimica et Biophysica Acta (BBA)—Bioenergetics, 1410*(2), 103–123. https://doi.org/10.1016/S0005-2728(98)00161-3

Tan, Y. L., Eng, S. P., Hafez, P., Abdul Karim, N., Law, J. X., & Ng, M. H. (2022). Mesenchymal stromal cell mitochondrial transfer as a cell rescue strategy in regenerative medicine: A review of evidence in preclinical models. *Stem Cells Translational Medicine, 11*(8), 814–827. https://doi.org/10.1093/stcltm/szac044

Taylor, R. W., & Turnbull, D. M. (2005). Mitochondrial DNA mutations in human disease. *Nature Reviews Genetics, 6*(5), 389–402. https://doi.org/10.1038/nrg1606

Weinrich, T. W., Kam, J. H., Ferrara, B. T., Thompson, E. P., Mitrofanis, J., & Jeffery, G. (2019). A day in the life of mitochondria reveals shifting workloads. *Scientific Reports, 9*(1). https://doi.org/10.1038/s41598-019-48383-y

Wrigley, A., Wilkinson, S., & Appleby, J. B. (2015). Mitochondrial replacement: Ethics and identity. *Bioethics, 29*(9), 631–638. https://doi.org/10.1111/bioe.12187

Xiao, B., Deng, X., Zhou, W., & Tan, E. K. (2016). Flow cytometry-based assessment of mitophagy using mitotracker. *Frontiers in Cellular Neuroscience, 10*(2016). https://doi.org/10.3389/fncel.2016.00076

Yamada, Y., Ito, M., Arai, M., Hibino, M., Tsujioka, T., & Harashima, H. (2020). Challenges in promoting mitochondrial transplantation therapy. *International Journal of Molecular Sciences, 21*(17), 1–17. https://doi.org/10.3390/ijms21176365

Zhao, J., Qu, D., Xi, Z., Huan, Y., Zhang, K., Yu, C., Yang, D., Kang, J., Lin, W., Wu, S., & Wang, Y. (2021). Mitochondria transplantation protects traumatic brain injury via promoting neuronal survival and astrocytic BDNF. *Translational Research, 235*, 102–114. https://doi.org/10.1016/j.trsl.2021.03.017

Zhang, T. G., & Miao, C. Y. (2023). Mitochondrial transplantation as a promising therapy for mitochondrial diseases. *Acta Pharmaceutica Sinica B, 13*(3), 1028–1035.

Zhao, Z., Hou, Y., Zhou, W., Keerthiga, R., & Fu, A. (2021). Mitochondrial transplantation therapy inhibit carbon tetrachloride-induced liver injury through scavenging free radicals and protecting hepatocytes. *Bioengineering and Translational Medicine, 6*(2). https://doi.org/10.1002/btm2.10209

CHAPTER

14

The ethical issues of mitochondrial transplantation and transfer

John B. Appleby

Lancaster Medical School, Lancaster University, Lancaster, United Kingdom

Introduction

The use of mitochondrial transplantation and transfer techniques—otherwise known as mitochondrial replacement techniques (MRTs)—are one way to reduce the risk of bringing children into existence with mitochondrial DNA (mtDNA) diseases. In brief, MRTs work by combining the healthy DNA of intending genetic parents in a healthy (with disease-free mtDNA) enucleated donor egg (using maternal spindle transfer—MST) or embryo (using pronuclear transfer—PNT). Crucially, these techniques allow an intending mother carrying an mtDNA disease to create a child that they are genetically related to while greatly reducing the risk that their child will inherit this disease. MtDNA diseases are among the most common group of neuromuscular diseases and severe cases can result in extreme suffering and death for those who carry them (Nuffield Council on Bioethics, 2012; UK Department of Health, 2014). Therefore, the possibility of having children with a substantially reduced risk of inherited mtDNA disease has been seen as beneficial by many (Appleby, 2015), but ethical debates surrounding the use of MRTs nevertheless exist (Nuffield Council on Bioethics, 2012).

This chapter explores a range of ethical concerns surrounding the use and governance of MRTs. While this chapter is not an exhaustive account of the ethics of MRTs, it does provide an overview of many of the key domains of ethical debate in this topic. First, I discuss some of the key background details underpinning current ethical debates, including regulation, first in human uses, safety, the metaphysics of identity, and patient privacy. Second, I discuss the debate that surrounds the ethically contentious topic of what constitutes a 'parent' and 'parenthood', especially given the prevalent use of the term '3-parent IVF' to describe MRTs. Third, I discuss the key ethical and philosophical issues surrounding the idea of one's 'sense of self' and how this may be influenced by MRTs. Fourth, I explain the ethical debate surrounding anonymous and non-anonymous mitochondrial donation. Fifth, I

Mitochondrial Transplantation and Transfer
https://doi.org/10.1016/B978-0-443-18858-9.00014-7

investigate the ethical debate surrounding germline genome editing and concerns about so-called 'designer babies'. I also consider the ethical debate surrounding the use of sex selection to avoid intergenerational transmission of disease and the objection that this amounts to eugenics.

Background

The only point of using MRTs is if you are trying to create children that you share a genetic link with. In particular, MRTs allow for the preservation of genetic ties (of the nuclear variety) in situations where any resulting offspring would otherwise run the risk of maternally inheriting an mtDNA disease. Some have argued that having offspring that you are genetically related to is an important enough desire to warrant using MRTs (Schaefer & Labude, 2017). However, others have argued that the desire to be genetically related to one's children does not warrant the possible medical risks of using MRTs because donor eggs or embryos can instead be used to create offspring that avoid the inheritance of mitochondrial disease (Baylis, 2013). It has also been argued that there is no ethical reason not to allow MRTs to be used in cases where mitochondrial disease inheritance isn't a risk. For example, MRTs could be used for the sole purpose of creating genetic ties between offspring and two lesbian parents (e.g., one parent would be genetically related to the child via a nuclear DNA contribution and the other parent would be genetically related to the child via an mtDNA contribution) (Cavaliere & Palacios-González, 2018). Objections to the idea of solely using MRTs to create genetic links are, once again, primarily based on the idea that it would be unnecessarily medically risky to use MRTs simply to create a genetic tie (Baylis, 2018). However, it appears that it is philosophically unnecessary to advance an argument in favor of using MRTs to allow same-sex couples to be genetically related to their offspring. This is because there are no compelling arguments for denying same-sex couples genetically related children beyond the basic argument (which applies to all uses and candidates for MRTs) that the technology should be safe enough to use. So long as the technology is safe enough (however, what constitutes 'safe' is a debate on its own), there is nothing left to debate when it comes to whether or not same sex couples should be able to use it. If desiring genetic parenthood is a permissible reason for heterosexual couples to use MRTs, then it is evident that this permissibility should obviously extend to same-sex couples as well.

The first country to explicitly regulate MRTs (i.e., PNT and MST) was the United Kingdom when The Human Fertilisation and Embryology (Mitochondrial Donation) Regulations 2015 were enacted. These regulations allow for the licensed use of MRTs in cases where there is a particular and significant risk of offspring inheriting a serious mitochondrial disease. In the lead up to this law being passed, there was considerable debate surrounding whether or not MRTs were safe enough to legalize (Nuffield Council on Bioethics, 2012; UK Department of Health, 2014). While authorities did not go so far as to say the techniques were 'safe', they did at least say they were not 'unsafe' (Appleby, 2015), which led to considerable speculation about how risk should be quantified and described when introducing radically new assisted reproductive innovations like MRTs. More recently, Australia enacted The Mitochondrial Donation Law Reform (Maeve's Law) Act 2022 to regulate the use of MRTs. This law took a staged approach to ensuring the safety of MRTs by beginning with a first clinical trial stage which would then progress to further clinical use stages if the MRTs were deemed safe and effective (Parliament of Australia, 2022).

Ironically, in 2016, the first child created via MRTs involved an embryo being created using MST in New York City, USA, and that embryo then being sent to Mexico to be transferred into the intending gestational mother (Zhang et al., 2017). Dr. John Zhang oversaw this procedure and famously said that he traveled to Mexico to complete the procedure, because he said 'there are no rules' (Hamzelou, 2016) there, although researchers later argued that the legal picture in Mexico was far more complicated (Palacios-González & Medina-Arellano, 2017). Since then, other MRT-conceived children have been born in the Ukraine (Forster, 2017) and Greece (Gallagher, 2019), but MRTs in these cases were used for the purposes of boosting fertility rather than avoidance of inheritable disease. While boosting fertility would not be a permissible reason to use MRTs in the United Kingdom, according to existing law, it is unclear why this would be ethically or medically problematic if the technique being used is considered safe enough to use for other reasons, such as disease prevention.

The reason that it would not necessarily be ethically problematic to use MRTs for other reproductive purposes, once deemed safe, is because according to the non-identity problem, bringing a particular child into existence cannot harm that child (Parfit, 1986). This is because if a different reproductive choice had been made then it would have resulted in a different child coming into existence. Therefore, comparatively speaking, it does not make metaphysical sense to argue that a particular child would have been better or worse off as a result of being conceived at a different time or in a different way (including via MRTs), because it is essentially impossible to alter the outcome for a particular child (i.e., any change would almost certainly result in a different egg and sperm cell coming together). Wrigley, Wilkinson, and Appleby have argued that MST does not constitute a therapy because a future particular human's nuclear DNA has not been fixed (the eggs are unfertilised) and therefore no particular future entity is being treated with a therapy (Wrigley et al., 2015). In contrast, they argue that PNT is a therapy because it is carried out on an embryo with fixed nuclear DNA and therefore does amount to treating a possible future human entity with a particular identity that we might later associate with a person (Wrigley et al., 2015). Rulli (2016) and Lewens (2021) have separately argued against this view and have said that both MST and PNT affect numerical identity, due to concerns such as environmental factors or placing too much weight on 'origin essentialist' approaches to identity (which associate numerical identity with the fixed nuclear DNA of an embryo). Liao also argues that MST and PNT create a person with a different numerical identity, but takes a different approach by arguing this is because the process of using either technique disrupts the cellular or organismic continuity of the egg or embryo (Liao, 2017). Finally, Schaefer has argued that MRTs can be 'life saving' if the question is approached from a population health perspective (Schaefer, 2020).

If we move away from 'person affecting' accounts of identity and harm, we can instead think of 'non-person affecting' accounts of identity and harm and this can be useful for understanding how MRTs could be safe or cause suffering to anyone brought into existence. It seems that this non-person affecting harm is what most policy-makers and scientists are likely thinking of when determining if the use of an MRT will result in any future offspring being better or worse off (Appleby, 2015). If we take this approach, then it appears that using MRTs to help boost fertility would also not be problematic, as the technique should be deemed safe before use and the other alternative may be not having any children at all. Indeed, the same logic applies to the use of MRTs for the avoidance of mitochondrial diseases. So long as the techniques are considered safe and satisfy an agreed permissible welfare threshold of non-person affecting harm, then it will likely be challenging to argue against their use on the grounds of safety (e.g., for other uses beyond mtDNA disease prevention).

Another important ethical aspect of the translation of mitochondrial techniques from bench to bedside is that the first human uses of these techniques should be carried out in a way that protects the privacy of intending parents and their mitochondrial donor conceived offspring. In particular, it is important that the first children created this way are not harmed by privacy violations from the media, academia, or the public for that matter. This is something, for example, UK regulators have done by restricting the immediate release of information about new MRT pregnancies and their outcomes (Parliament of Australia, 2022).

Finally, a central ethical issue that underpins the transition of MRTs from bench-to-bedside is whether they should be treated as a new experimental technique that should be clinically introduced via a research trial, or if they are in fact an innovation that should be clinically introduced without a trial. Going back to the UK and Australian regulatory regimes, there is a difference in how each piece of law approaches this issue. While the United Kingdom tightly licenses the use MRT techniques, the law itself does not state that MRTs needed to be introduced as part of a research trial. However, Australian regulation (Maeve's Law) requires a staged introduction where MRTs would first be studied as part of a clinical trial before proceeding further. This is an important distinction, because the Australian approach exposes the techniques to a full battery of research ethics considerations and research oversight, whereas the UK approach generally restricts this oversight to whatever is required within the licensing restrictions that have been placed on a technique, but it is not framed as research. Given the extensive history of assisted reproductive technologies (ARTs) having been introduced into clinical practice without proper research oversight, as was the case with IVF and ICSI (Appleby, 2015; Bredenoord & Appleby, 2017), it appears the United Kingdom has come closer to perpetuating this norm, whereas Australia has broken with this tradition and taken a cautious approach of introducing the techniques through research. However, the reality is that when Dr. John Zhang created the first MRT baby in Mexico, it could be argued that he perpetuated the tradition of introducing new ARTs for the first time on humans without the level of research oversight that some would have wanted (Dondorp & de Wert, 2011).

Issues surrounding the concept of 'parent' and parentage

There has been considerable ethical debate about whether or not mitochondrial donors who contribute to '3-person IVF' should be considered 'parents', and if so, what their ethical responsibilities might be. In order to properly understand and respond to this question, we must first consider what is meant by 'parent'. If we are thinking about parents in the 'biological parent' sense, then this may refer to someone who makes a biological contribution to bringing a child into existence. Conventionally, this would typically mean the man contributing half of a compliment of genes via his sperm, while a woman contributes the other half via her egg (along with the mitochondria). Furthermore, the woman would make a biological contribution by gestating the child. However, in the realm of assisted reproduction, these biological contributions can be split up and reallocated to an array of participants. For instance, surrogacy allows the labor of gestation and delivery to be taken on by someone else. MRTs also mean that a new actor shares in the project of creating an egg or embryo by donating an egg or embryo, which can be enucleated to accommodate the preferred nuclear genetics from both a man and a woman. Therefore, while all of these different parental contributors are biological in the sense of providing biological materials or services, it is only the mitochondrial donors who are mtDNA genetic parents and only the contributors of the nuclear DNA from the sperm and intending mother's egg who are the nuclear genetic parents (Appleby, 2018).

The language of 'parent' may also be used to talk about someone who is fulfilling a parental role in someone's life. As far as the law in Australia and the United Kingdom is concerned, mitochondrial donors (essentially egg donors) have no legal parental responsibilities to any resulting child, in the same way that normal gamete donors have no such legal parental responsibilities or parental status. Therefore, it seems misleading to refer to mitochondrial donors as 'parents' in this sense. It is also possible that 'parent' may refer to a person that someone views as their parent, regardless of the nature of their biological connection (or lack thereof). It seems that it is entirely possible that mitochondrial donors could be perceived as parents, but we cannot be sure if this is likely until empirical research has been carried out on the first cohorts of mitochondrial donor conceived children and their attitudes toward their donors are properly understood (Appleby, 2018).

Therefore, the significance of being considered a parent in the context of mitochondrial donation seems to hinge on whether you have any legal or ethical responsibilities. While the legal answer is a simple 'no' from the point of view of UK and Australian legal contexts, it is less clear from an ethical standpoint. Mitochondrial donors do not simply donate mitochondria, they donate a full egg or embryo and everything in it, except for the nuclear DNA (which is removed during enucleation) (Sparrow et al., 2021). It is possible that someone conceived with the help of a mitochondrial donor may or may not view that person as a parent in one of the above senses. In one sense, they are biological and mitochondrial genetic parents to the child and this an objective fact. It is also possible that offspring will in fact subjectively interpret donors to be parents in some sense. This would not be surprising given that it has happened in the context of generic egg, sperm, and embryo donation (Appleby, 2018). However, as aforementioned, further evidence is needed to understand how MRT-conceived children will think of their donors and if they will in fact be perceived as parents. If offspring do come to see mitochondrial donors as parents in some way, it is unclear what their normative ethical expectations will be of them. It remains unclear to what the extent donors have ethical responsibilities toward the MRT-conceived offspring they helped to create.

Ethics and sense of self

Mitochondrial donation and MRTs also raise ethical questions about the impact they will have on the 'sense of self' of the offspring created with them (Appleby, 2018). Here I use the term 'sense of self' in the psychological sense, which is characterized by Brock as ' ... the properties or qualities that an individual considers important to who he is, to what kind of person he is, to what properties of himself he identifies with' (Brock, 2002). Central to this particular debate is the issue of whether or not an MRT-conceived person's sense of self is considerably impacted by the fact that they received their mitochondria—and the mtDNA in it—from a donor. If we consider the experiences of the next closest analogue—donor conceived children—we can quickly determine that being donor conceived is impactful to the sense of self of many offspring and this is well documented in the evidence (Appleby, 2018). It is possible that some might find significance in their donor's gesture to help their parents have a child without a life-threatening disease. Others may find meaning in relation to their sense of self that another person helped to bring them into existence. However, despite the fact that the existing evidence on donor-conceived persons could possibly (eventually) be reflected in how some MRT donor-conceived persons might also feel about how their sense of self has been impact by knowledge that they were conceived via a donor, there have been several objections about why this is unlikely to be the case due to the nature of mtDNA (Nuffield Council on Bioethics, 2012).

It has been argued that mtDNA is not the kind of DNA that would have an impact on a person's sense of self because it is too small in quantity and it is not of the right qualities to have an impact on a person's character or physical appearance. For the sake of clarity, in 2018, I coined the terms 'the quantity claim' and 'the quality claim' to help formalize these views about the nature of mtDNA (Appleby, 2018). To be clear, the quality claim states that: 'mitochondrial genes are different in quantity from the nuclear genes involved in gamete donation' (Appleby, 2018). In contrast, the quantity claim states: 'mitochondrial donation is different from gamete donation because mitochondrial genes are different in quantity from the nuclear genes involved in gamete donation' (Appleby, 2018). These distinctions are significant, because by distilling them in this way it becomes apparent that these claims only carry descriptive value, and are not evaluative about the normative significance of mtDNA in relation to a person's sense of self. In order to become evaluative, either claim would need to be combined with another linking premise. In other words, the quality and quantity claims do not on their own amount to an argument that dismisses the possibility that MRT-conceived people may feel that a donor's mtDNA has impacted their sense of self.

The difficulty is that simply explaining that mtDNA is of a quantity and quality that is different from nuclear DNA does not accurately portray the myriad of ways that people (including donor conceived persons) think about genes of any sort in relation to their sense of self. Perhaps the most obvious example is the way that individuals carrying a harmful homoplasmic mtDNA gene mutation may identify with a patient group or as having a sense of self that identifies with certain medical aspects about their health (especially in cases where their genetics have an impact on the state of their daily welfare). In such a case, it would be misleading to suggest that mtDNA was of a quality that did not or could not impact on a person's sense of self. Similarly, being someone's offspring and not inheriting a harmful mtDNA mutation because of mitochondrial donation could also be of significance to one's sense of self. It is also important to recognize that it often only takes as little as a small genetic mutation to cause life-changing effects to someone's health, and therefore a large number of genes do not need to be altered in order to have an impact on a person's sense of self (Leach

Scully, 2016). For example, something as simple as having BRCA1 and BRCA2 genes could be life changing and impact on one's sense of self.

However, due to the prevalence of the quantity and quality claims being used in discourse and publications, in order to discount the significance mtDNA on sense of self, the potential impacts of mitochondrial donation on the sense of self of MRT-conceived people have generally remained under explored in the ethics literature.

Donor anonymity and disclosure

Another area of considerable ethical debate in the context of MRTs and mitochondrial donation is the question of whether mitochondrial donors should be anonymous and whether parents should tell their offspring if they were MRT conceived. From a regulatory point of view, the UK's Human Fertilisation and Embryology (Mitochondrial Donation) Regulations 2015 do not allow mitochondrial donors to be identified, despite the Human Fertilisation and Embryology Act 1990 (as amended) allowing all children conceived by egg, sperm, and embryo donations made after April 1st, 2005, to access identifying information about their donors at the age of 18. In contrast, Maeve's Law in Australia allows for mitochondrial donor-conceived persons to access identifying information about their donors at the age of 18, in the same way that donor-conceived persons do (Parliament of Australia, 2022).

Some have mistakenly characterized this as a 'right to know' debate, in the sense that the UK law gives donor-conceived people a right to know who their donor is, but the mitochondrial donation regulations do not extend the same option to MRT-conceived persons (Appleby, 2016, 2018). However, this characterization of the UK law as providing a 'right to know' is misleading. In the United Kingdom, donor-conceived persons are not granted a right to know their donor in the straightforward way one might assume. They may have a right to request identifying information about their donor from the HFEA (Human Fertilisation and Embryology Authority), but they can only enact this right if their parents let them know that they were donor conceived, and there is nothing in UK law that requires this. In 2018, I argued that mitochondrial donation is not convincingly different from gamete donation in a way that would warrant treating mitochondrial donors differently and making them anonymous (Appleby, 2018). As I have already argued above, mitochondrial donation shares many of the same legal, identity affecting, and sense of self-affecting properties, as gamete donation. It appears that while UK regulators have not shared this view to-date, the Australian authorities have done so in their regulatory approach (Australian Government Department of Health and Aged Care, 2023; Parliament of Australia, 2022).

Of course, as I have mentioned, providing offspring access to information (identifying or non-identifying) about donors in a registry is of limited use to those offspring if they are not aware that they were donor conceived. Both the UK and Australian authorities recommend that any intending parents disclose to any of their resulting MRT-conceived children at an early age that they were conceived using a donor. This is mainly because it is viewed as being beneficial to offspring's development of their sense of self, from a developmental psychology perspective. While this approach and rational for disclosure corresponds with the Australian approach to regulating in favor of non-anonymous mitochondrial donation, it seems to clash with the UK's approach to regulation. What I mean is that Australian children who are disclosed to will be able to take this information and find out identifiable information about their donor and potentially meet them, if that is important for their sense of self. By contrast, a UK MRT-conceived person may be told they were conceived using a donor, but they will be denied the information needed to take that extra step of contacting their donor. If a UK MRT-conceived person feels that such donor information is important to access further knowledge that relates to their sense of self, then they may end up feeling ethically wronged (Appleby, 2018).

Germline genome modification, sex selection, and designer babies

Some might argue that because MRTs involve whole mitochondrial germline genome replacement, this amounts to creating 'designer babies' and that this is unnatural. In one sense, some might argue against MRT designer babies on the basis that they are unethical due to the fact that inheritable genomic changes are being put in place that will therefore impact future generations. Some might be concerned that there could be health risks associated with such major genomic interventions (National Academies of Sciences, Engineering, and Medicine, 2016). Others might be preoccupied with the idea that scientists are 'playing god' and crafting offspring in a way that is unnatural, or at least impossible outside of a clinic. Meanwhile, it is possible that some might be concerned that MRTs are the 'thin end of the wedge' and could lead to the progressive adoption of more experimental interventions and a laissez-faire societal attitude to other technologies in the pipeline that modify the genome, such as CRISPR-Cas9 (Nuffield Council on Bioethics, 2012).

However, it appears that concerns about 'designer babies' and corresponding slippery-slope type objections are rooted in speculation and depend on science progressing in an unregulated and irrational manner. First of all, there is no reason why the use of MRTs now, to modify whole mitochondrial genomes, will necessarily result in society permitting other genetic modification technologies in the future. There is no clear chain of causation that necessarily results in one leading to another. Second, the idea that humans making reproductive choices or medical interventions to ensure the welfare of future generation is in any way new, is wrong and misleading. Humanity has always intervened in different ways (ranging from choosing which partners to have offspring with, to picking the healthiest embryos for transfer in IVF). This undermines the argument that the arrival of 'designer babies' has upset the status quo and that it is therefore somehow morally problematic. Such arguments would need the addition of some convincing premises before they gained any argumentative traction.

In 2015, I argued that it would be ethically prudent to use sex selection to select for only male embryos during the use of the first MRTs with humans (Appleby, 2015). The reason for this is that mitochondrial diseases are maternally inherited. This way only male offspring are created and if there is a problem with their mtDNA health (e.g., unintended carry-over of mitochondrial with harmful mtDNA mutations), then these health complications would be limited to this generation only. This recommendation was cited and adopted by the

2016 US National Academies of Science, Engineering and Medicine report titled 'Mitochondrial Replacement techniques: ethical, social and policy considerations' (National Academies of Sciences, Engineering, and Medicine, 2016). However, this recommendation for sex selection was not adopted in the UK government or the Australian government regulations for MRTs, likely because the idea of sex selection is inherently controversial.

Of course, only a handful of MRT-conceived children will likely to be created each year in any given country (at least in relation to avoiding serious mitochondrial disease), and, therefore, the recommendation to select only male embryos does not fall foul of objections that such an intervention will cause harm by skewing the gender balance in society (Appleby, 2015; National Academies of Sciences, Engineering, and Medicine, 2016). This recommendation also does not carry a risk of conveying negative connotations toward the value of women in society (Appleby, 2015). Instead, the message is clear, focused, and simple: use sex selection to avoid the intergenerational risk of serious mitochondrial disease transmission (Appleby, 2015).

However, unfortunately, some such as Brandt have misconstrued this narrow application of sex selection to represent some sort of a modern eugenics agenda (Brandt, 2018). While my original argument made clear this was not the case, Brandt nevertheless insists that my recommendation for first-in-human use sex selection amounts to harmful eugenics because it is ultimately a case of choosing one type of person to bring into existence over another, and that this therefore suggests that one type of person is of greater value than another (Brandt, 2018). However, this type of argument is unsound and unhelpful. For example, if we extend Brandt's account of eugenics to everyday life, then his view appears to be that anyone making a decision that will result in creating one type of offspring over another (this could be as simple as picking a donor based on their looks or even setting preferences for physical appearance on a dating app) amounts to a harmful and dangerous form of eugenics. This not only seems absurd, but it also erodes the work done by the concept of eugenics to highlight genuine atrocities across history, such as those that occurred at the hands of the Nazis in WW2. Brandt's criticism misconstrues the original argument put forward, fails to appropriately deploy the concept of eugenics in a meaningful narrow sense, and makes an uncompelling case against the use of sex selection.

Conclusion

It is likely that the use of MRTs will continue and grow as society learns more about their application for both avoiding disease transmission and their use to improve fertility. As a result, it is likely that new ethical issues will continue to emerge over time. This chapter has aimed to address some of the central concerns that have dominated the ethical debate surrounding mitochondrial donation and these techniques. I began by discussing some of the key background details and debates that are necessary to understand the ethical landscape of this field. Next, I explained the differences between different types of 'parent' and, where relevant, the ethical and legal implications. I then discussed the ethics of MRTs and sense-of-self, before proceeding to discuss the debate surrounding mitochondrial donor anonymity and disclosure. Finally, I addressed some of the key objections to MRTs that revolved around concerns about germline genome modification, designer babies, sex selection and eugenics. All of these ethical debates are rich and deserve investigation on their own, however, I have attempted to give the reader a brief overview of what some of these central debates entail.

References

Appleby, J. B. (2016). Regulating the provision of donor information to donor-conceived children: Is there room for improvement? In S Golombok, R. Scott, J. B. Appleby, M. Richards, & S. Wilkinson (Eds.), *Regulating reproductive donation* (pp. 334–351). Cambridge University Press: Cambridge.

Appleby, J. B. (2018). Should mitochondrial donation be anonymous? *Journal of Medicine and Philosophy, 43*(2), 261–280.

Appleby, J. B. (2015). The ethical challenges of the clinical introduction of mitochondrial replacement techniques. *Medicine, Healthcare and Philosophy, 18*, 501–514.

Australian Government Department of Health and Aged Care. (2023). *Mitochondrial donation.* https://www.health.gov.au/our-work/mitochondrial-donation#:~:text=In%20March%202022%2C%20the%20Australian,effect%20on%201%20October%202022. (Accessed 10 June 2023).

Baylis, F. (2018). No to lesbian motherhood using human nuclear genome transfer. *Journal of Medical Ethics, 44*(12), 865–867.

Baylis, F. (2013). The ethics of creating children with three genetic parents. *Reproductive BioMedicine Online, 26*, 531–534.

Brandt, R. (2018). Mandatory sex selection and mitochondrial transfer. *Bioethics, 32*(7), 437–444.

Bredenoord, A. L., & Appleby, J. B. (2017). Mitochondrial replacement techniques: Remaining ethical challenges. *Cell Stem Cell, 21*(3), 301–304.

Brock, D. (2002). Human cloning and our sense of self. *Science, 296*, 314–316.

Cavaliere, G., & Palacios-González, C. (2018). Lesbian motherhood and mitochondrial replacement techniques: Reproductive freedom and genetic kinship. *Journal of Medical Ethics, 44*, 835–842.

Dondorp, W., & de Wert, G. (2011). Innovative reproductive technologies: Risks and responsibilities. *Human Reproduction, 26*, 1604–1608.

Forster, K. (2017). *Three-person IVF' allows infertile couple to have baby in Ukraine in world first.* https://www.independent.co.uk/life-style/health-and-families/health-news/three-person-baby-ivf-techinque-world-first-infertile-couple-ukraine-kiev-valery-zukin-give-birth-child-a7532636.html. (Accessed 21 February 2023).

Gallagher, J. (2019). *Three-person baby boy born in Greece.* http://www.bbc.co.uk/news/health-47889387. (Accessed 25 February 2023).

Hamzelou, J. (2016). *Exclusive: World's first baby born with new "3 parent" technique.* https://www.newscientist.com/article/2107219-exclusive-worlds-first-baby-born-with-new-3-parent-technique/. (Accessed 15 January 2023).

Leach Scully, J. (2016). A mitochondrial story: Mitochondrial replacement, identity and narrative. *Bioethics, 31*(1), 37–45.

Lewens, T. (2021). The fragility of origin essentialism: Where mitochondrial 'replacement' meets the non-identity problem. *Bioethics, 35*(7), 615–622.

Liao, S. M. (2017). Do mitochondrial replacement techniques affect qualitative or numerical identity? *Bioethics, 31*(1), 20–26.

National Academies of Sciences, Engineering, and Medicine. (2016). *Mitochondrial replacement techniques: Ethical, social, and policy considerations.* Washington, DC.: The National Academies Press. https://doi.org/10.17226/21871

Nuffield Council on Bioethics. (2012). *Novel techniques for the prevention of mitochondrial DNA disorders: An ethical review.* London: Nuffield Council on Bioethics.

Palacios-González, C., & Medina-Arellano, M.de J. (2017). Mitochondrial replacement techniques and Mexico's rule of law: On the legality of the first maternal spindle transfer case. *Journal of Law and the Biosciences*, 50–69. https://doi.org/10.1093/jlb/lsw065

Parfit, D. (1986). *Reasons and persons.* Oxford: Oxford University Press.

Parliament of Australia. (2022). *Mitochondrial donation law reform (Maeve's law) bill 2021: Explanatory memorandum.* https://parlinfo.aph.gov.au/parlInfo/search/display/display.w3p;query=Id%3A%22legislation%2Fems%2Fr6697_ems_34f56965-6288-4da6-9a02-096f5b58d3c1%22. (Accessed 10 February 2023).

Rulli, T. (2016). The mitochondrial replacement "therapy" myth. *Bioethics, 31*, 368–374.

Schaefer, G. O. (2020). Can reproductive genetic manipulation save lives? *Medicine, Healthcare and Philosophy, 23*, 381–386.

Schaefer, G. O., & Labude, M. K. (2017). Genetic affinity and the right to "three-parent IVF". *Journal of Assisted Reproduction and Genetics, 34*, 1577–1580.

Sparrow, R., Mills, C., & Carroll, J. (2021). Gendering the seed: Mitochondrial replacement techniques and the erasure of the maternal. *Bioethics, 35*(7), 608–614.

UK Department of Health. (2014). *Mitochondrial donation: A consultation on draft regulations to permit the use of new treatment techniques to prevent the transmission of a serious mitochondrial disease from mother to child*. https://www.gov.uk/government/uploads/system/uploads/attachment_data/file/285251/mitochondrial_donation_consultation_document_24_02_14_Accessible_V0.4.pdf. (Accessed 10 April 2023).

Wrigley, A., Wilkinson, S., & Appleby, J. B. (2015). Mitochondrial replacement: Ethics and identity. *Bioethics, 29*, 631–638.

Zhang, J., Liu, H., Luo, S., Lu, Z., Chavez-Badiola, A., Liu, Z., Yang, M., Merhi, Z., Silber, S. J., Munne, S., Konstantinidis, M., Wells, D., Tang, J. J., & Huang, T. (2017). Live birth derived from oocyte spindle transfer to prevent mitochondrial disease. *Reproductive BioMedicine Online, 34*(4), 361–368.

SECTION V

Cardiovascular diseases and mitochondrial transplantation

CHAPTER

15

Cardiovascular diseases and pathophysiology

Therese Jamison

Department of Nursing, Lawrence Technological University, Southfield, MI, United States

Concepts of cardiac output

To fully understand the pathophysiology of atrial fibrillation (AF), heart failure (HF), ischemic heart disease, and hypertension (HTN), one needs to review the concept of cardiac output. Cardiac output is a measure of the amount of blood pumped from the heart each minute (Banasik & Copstead, 2018). Cardiac output is determined by multiplying the heart rate by the stroke volume (HR x SV) as shown in Fig. 15.1.

Heart rate

Within the autonomic nervous system, the sympathetic and parasympathetic nervous system influences heart rate. Catecholamines, norepinephrine, and epinephrine are released from the sympathetic nervous system (SNS) which serves as neurotransmitters and hormones that regulate heart rate (Banasik & Copstead, 2018; Honan, 2019). The use of nicotine (tobacco products) and caffeine (coffee and tea) can increase sympathetic stimulation, increasing heart rate. As well, acute pain, fever, anxiety, and stress can increase heart rate. Under normal resting conditions, the heart rate is influenced by the parasympathetic

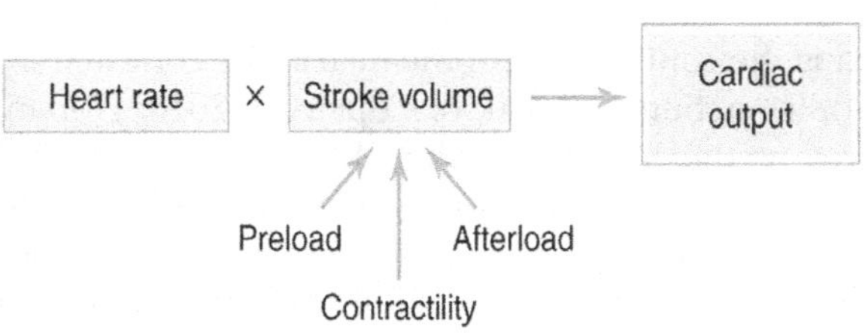

FIGURE 15.1 Determinates of cardiac output.

Mitochondrial Transplantation and Transfer
https://doi.org/10.1016/B978-0-443-18858-9.00015-9

tone, thus maintaining a normal resting heart rate of 70 (Banasik & Copstead, 2018; Honan, 2019). For the management of AF, heart rate control, along with rhythm control, is paramount with patients with ischemic heart disease, since angina could be exacerbated with fast heart rates. Anginal symptoms (such as chest pain, chest pressure, chest heaviness, and chest tightness) can occur since the coronary sinus is perfused during diastole, the relaxation cycle of the heart. The faster the heart rate, the shorter the time in diastole with less time for coronary artery perfusion to take place (Banasik & Copstead, 2018; Honan, 2019). Chronic tachycardia in conditions of AF or atrial flutter has been associated with tachycardia mediated cardiomyopathy which results in structural changes such as left ventricular dilation (Khan & Sanchez, 2017).

Stroke volume

Stroke volume (SV) has three components: Preload, afterload, and contractility. Preload is the volume of blood returning to the heart as affected by venous return and compliance. This concept is further described by the Frank–Starling law, which explains that the SV of the heart increases in response to an increase in the volume of blood in the ventricles. During diastolic filling, myocardial fibers stretch in response to circulating blood volume and this ultimately increases the force of systole or contraction. Thus, when the heart's volume increases, SV increases with a resultant increase in cardiac output (Banasik & Copstead, 2018; Honan, 2019). Preload can be referred to as the "filling pressures" of the heart as measured by central venous pressure (CVP) and pulmonary artery wedge pressure. Clinically, one can reduce preload by administering medications such as vasodilators and diuretics. Conversely, preload is increased with fluids or by increasing vascular tone with vasopressors that mimic norepinephrine and epinephrine.

The next component of SV is afterload. Simply defined, afterload is resistance. Left ventricular afterload is determined primarily by aortic blood pressure. Hence, afterload is the resistance that the left ventricle must overcome to eject its volume based on the pressure against the left ventricle from the aortic blood pressure (Banasik & Copstead, 2018; Honan, 2019). The three determinants that create this resistance are aortic impedance (for example, aortic stenosis), systemic vascular resistance (for example, HTN), and volume/viscosity (for example, dehydration). Thus, a patient with HTN has more resistance or a higher afterload to overcome to eject the blood from the left ventricle. The consequences of an increased afterload can manifest as left ventricular hypertrophy and diastolic HF. Conversely, patients with septic shock have decreased afterload due to low systemic vascular resistance in response to inflammatory mediators. Therefore, the primary management for septic shock is to maximize preload by giving fluids, while supporting systemic vascular resistance with vasopressors, and treating the underlying cause.

The final component of SV to be discussed is the contractility of the heart. The contractility of the heart is often referred to as its "inotropic" action (Banasik & Copstead, 2018; Honan, 2019). Clinically, contractility can be measured as the ejection fraction (EF) using a transthoracic echocardiogram, a transesophageal echocardiogram, or through direct visualization during cardiac catheterization. The EF is the percent of blood ejected from the ventricles with each beat. It is calculated by dividing SV by the end-diastolic volume (EDV). A normal

EF range is 45%–70% since not all blood is ejected from the ventricle at one time. Higher EFs are considered hyperdynamic and are often related to HTN. More detail of EF is described with regard to the pathophysiology of HF. SV is altered by changes in the venous return to the heart. An increase in the preload, second to the increase in the venous return, results in a stretching of the cardiac myocyte which increases the force generated during systole. This results in an increased ability of the heart to pump blood out of the ventricles and hence an increase in the SV.

Atrial fibrillation

Although there are other forms of arrhythmias, AF is the most common arrhythmia that healthcare providers manage. The European Society of Cardiology (Dai et al., 2021) estimated that in 2017, approximately 37 million individuals lived with AF and three million new AF cases occurred globally. The absolute number of AF cases has almost doubled from 1990 to 2017. The incidence of AF is predicted to increase second to age and comorbid conditions of diabetes, obesity, HTN, and congestive HF with risk factors of smoking, alcohol use, illicit drug use, and diets high in sodium (Chung et al., 2020; Dai et al., 2021). These comorbid conditions and risk factors cause structural heart changes to the chambers, valves, and lining of the heart, hence changing the heart's function, and ultimately changing the electrical excitability. Rather than memorizing the etiologies or risk factors, one should think of the "why" or "how" the heart changes based on changing the structure of the heart (Table 15.1). In addition, patients could present with a single risk factor or have multiple comorbidities, which could contribute to the development of new or persistent AF. One comorbid condition that requires specific attention is the incidence of AF in HF. Approximately one-third of all HF patients have AF. The prevalence of AF increases with HF severity, ranging from 5% in mild HF to 50% in severe HF. In HF with a preserved EF (diastolic failure), the prevalence of AF is 60% (Patel et al., 2020). Again, the concept of heart structure changing heart function, and electrical excitability assists the reader in identifying why AF is so prevalent in the HF population.

To understand why AF occurs with the resultant pathophysiology, a review of electrical cells is necessary. Automaticity is the property of cardiac electrical cells to generate a spontaneous action potential, also referred to as depolarization. Based on normal physiology, the sinoatrial (SA) node, located in the right atrium, is considered the pacemaker of the heart with cardiac myocytes firing at a rate of 60–100 beats per minute. Yet with AF, multiple wavelets of abnormal depolarizations take over the pacemaker function of the SA node. These depolarizations are rapid and irregular resulting in a loss of organized atrial mechanical contractions with a potential for a decrease in cardiac output and hemodynamic instability (Sauer & Zei, 2022). Active atrial contractions immediately before ventricular ejection acts to increase the efficiency of ventricular ejection second to acutely increasing preload. These active atrial contractions are known as "atrial kick." Atrial kick accounts for approximately 20% of cardiac output; hence, when patients develop AF with disorganized atrial contractions, they lose the atrial kick with resultant clinical manifestations of decreased cardiac output with hypotension, hemodynamic instability, dyspnea, presyncope, and rarely, syncope.

TABLE 15.1 Risk factors AF.

Factor (risk, precipitating)	Why/How
Age	Structural heart disease
Lifestyle Alcohol (holiday heart syndrome) Substance abuse Smoking Intense physical exercise Obesity	Holiday heart syndrome is related to binge drinking on weekends or holidays, typically in people without heart disease Substance abuse, such as cocaine and even over-the-counter cough and cold medications, can cause a sympathomimetic effect on the heart Intense physical exercise, as related to dehydration and electrolyte imbalance, can precipitate AF Obesity increases the workload on the heart, resulting in structural heart disease
Structural heart disease: o Sick sinus syndrome o Left atrial enlargement o Valvular heart disease/rheumatic heart disease o Cardiomyopathy o Atrial myxoma o Atrial septic defects o Pericardial disease o Hypertension	The concept here is that structural heart disease changes the heart's function and thus the heart's electrical activity, hence contributing to AF and other arrhythmias
Ischemic heart disease	The concept here is that ischemic tissue is irritable tissue, and irritable tissue is arrhythmic tissue
Noncardiac causes: o Acute infections (sepsis) o Electrolyte depletion o Diabetes mellitus o Pulmonary embolism o Obstructive Sleep Apnea (OSA) o Hyperthyroidism/thyrotoxicosis o Anemia o Postoperative o Idiopathic (lone AF)	Metabolic stress (sepsis, acute infections, thyroid) Activates the neurohumoral cascade Electrolyte depletion affects the Na-K-ATPase pump Magnesium is the cofactor for the Na-K-ATPase pump; hence, normal potassium can be obtained by normal magnesium
Cardiothoracic surgery	High catecholamine release is seen during the perioperative period and during critical illness. Volume overload during surgery increases preload
European ancestry and family history	Genetic variation
Family history	Genetic variation

Over time, the left atrium becomes stretched with left atrial enlargement and fibrosis (Walsh & Caple, 2018). Consequently, with the loss of organized atrial mechanical contractions, there is blood stasis, resultant thrombogenesis, embolization, and the potential for an ischemic stroke. The left atrial appendage is a small pouch extending off the side of the left atrium where thrombus formations occur. The left atrial appendage can be occluded or ligated with the use of interventional or surgical approaches to prevent the embolization of the thrombus. Furthermore, there is a risk of dementia in patients with AF related to vascular changes from embolic infarcts (Sauer & Zei, 2022).

Heart failure

Globally, more than 26 million people have HF (Givertz & Mehra, 2022). HF can be acute or chronic depending on the factors which cause HF. Acute HF is seen after acute coronary syndrome (ischemic heart disease), acute valvular dysfunction (mitral regurgitation), and hypertensive urgency/emergency (Givertz & Mehra, 2022). Chronic HF is a progressive disease involving a precipitating event leading to structural and functional cardiovascular remodeling with a chronic increase in heart pressure (increase in afterload) or volume overload (increase in preload). HF can be inherited, secondary to genetic cardiomyopathies or congenital disease (Givertz & Mehra, 2022). Despite advances in management strategies, the trajectory of chronic HF can be devastating with approximately 50% of patients dying within 5 years after the initial diagnosis (Lei & Mao, 2018; Riehle & Bauersachs, 2019).

Phenotypes of HF

The presence of HF is based on the measurement of left ventricular ejection fraction (LVEF). LVEF, or left ventricular systolic function, is the contractility of the heart. As previously noted, LVEF is the percent of blood ejected from the ventricles with each beat. It is calculated by dividing SV by the EDV. The different categories of HF are referred to as phenotypes. About 50% of patients with HF have a reduced EF (abbreviated as HFrEF) phenotype with an LVEF of less than 40%. In comparison, a preserved EF (abbreviated as HFpEF) is observed in the other one-half of patients with an LVEF greater than or equal to 50%. With HFpEF, the patient has normal systolic function, yet impaired relaxation during diastole. Patients with an LVEF in the range of 40%–49% are defined as HF with mid-range EF (abbreviated as HFmrEF) (Givertz & Mehra, 2022; Riehle & Bauersachs, 2019). The defining characteristic of HFrEF is a reduction in myocardial contractility or inotropy which impairs SV and reduces overall cardiac output (Pagel et al., 2021). HFpEF is becoming a more common phenotype of HF. With aging and the prevalence of HTN, diabetes, metabolic syndrome, and pulmonary diseases, structural changes in the heart, lungs, and vasculature lead to diastolic function (Matsushita et al., 2019; Normand et al., 2019; Pagel et al., 2021). Furthermore, the aging process is associated with systemic inflammatory states or geriatric syndromes that may lead to a decline in the physiologic reserve and function of multiple organ systems, even without specific diseases (Matsushita et al., 2019).

Clinically, left-sided versus right-sided HF can be seen. With normal physiology, blood is ejected out of the aortic valve. However, in left-sided HF, with systolic or diastolic dysfunction, the blood volume backs up into the lungs causing pulmonary vascular congestion with resultant acute or chronic dyspnea. The symptoms can occur at rest, with limited activity, or with exertion. In right-sided HF, the blood backs from the right atrium into the periphery with related symptoms of jugular vein distention, hepatomegaly, ascites, and sacral or lower extremity edema. Some individuals can have both left-sided and right-sided HF, also known as biventricular failure, exhibiting clinical manifestations of the lungs and the periphery (Givertz & Mehra, 2022).

Pathophysiology of HF

The pathophysiology associated with HFpEF is impaired LV systolic and diastolic relaxation and arterial stiffness as a consequence of altered actin–myosin cross-bridges which separate second to altered modulation of intracellular calcium hemostasis and decreased cyclic guanosine monophosphate (Pagel et al., 2021; Riehle & Bauersachs, 2019). Additionally, the patient with HFpEF can demonstrate elevated filling pressures with elevated CVP and PCWP, atrial HTN, pulmonary venous congestion, and ventricular dilation.

Cardiac output is impaired with increased preload (volume) and increased afterload (resistance). With regards to preload, the left atrial pressure and left ventricular end-diastolic pressures are increased which is seen in both HFrEF and HFpEF (Normand et al., 2019). Furthermore, acute decompensated HF may result in the presence of acute cardiogenic pulmonary edema with a rapid onset of fluid accumulation in the lungs. This rapid accumulation is related to *elevated pulmonary capillary hydrostatic pressure with* fluid buildup occurring in the space between the cells (interstitium) and within the alveoli impairing gas exchange (Collins & Martindale, 2018; Shiraishi et al., 2021).

Acute HF is often associated with HTN with blood pressure exceeding 140 mmHg. Patients with hypertensive acute HF may have HFrEF or HFpEF. Two pathways of hemodynamic deterioration are identified. Within the slow pathway, fluid buildup advances gradually via renal and dietary mechanisms. In contrast, the fast pathway is related to fluid redistribution and transient recruitment through direct stimulation of the SNS. The SNS causes vasoconstriction allowing for the mobilization of the venous reservoir from larger veins into the splanchnic vessels that supply the abdominal organs (Shiraishi et al., 2021). Hypertensive acute HF is related to vascular redistribution, and inadequate reduction of preload and afterload rather than inadequate diuresis (Collins & Martindale, 2018).

Ventricular remodeling in heart failure

The concept of ventricular remodeling is replete in the HF literature as a mechanism of its pathophysiology. Increases in myocardial wall stress, the inflammatory cascade, signal peptides, neuroendocrine signals, and oxidative stress are all mediators for ventricular remodeling. During remodeling, the myocytes, the cardiac muscle cells, are unable to adapt which triggers apoptosis, or programmed cell death. The result of remodeling is further impairment in ventricular function and myocardial wall stress (Givertz & Mehra, 2022). Ventricular remodeling, leading to hypertrophy, can occur as concentric and eccentric. Concentric left ventricular hypertrophy is noted when there is an increased myocardial mass out of proportion to chamber volume, which produces wall stress under conditions of pressure overload. With eccentric left ventricular hypertrophy, there is an increase in cavity size that occurs in volume overload conditions. Concentric ventricular left hypertrophy is seen with HFpEF and eccentric left ventricular hypertrophy is seen with HFrEF (Pagel et al., 2021). Ventricular remodeling, with an increase in muscle mass or an increase in cavity size, contributes to higher cardiovascular morbidity and mortality (Correale et al., 2018; Givertz & Mehra, 2022).

In HF, there is an increase in the incidence of arrhythmias which originate from the atrium and the ventricles second to the remodeling process with resultant changes in ion channel composition allowing abnormal electrical activity. Remodeling will perpetuate histological changes including fibrosis, and the formation of fibrous connective tissue, or scar tissue. Ventricular tachyarrhythmias (ventricular tachycardia and ventricular fibrillation) are responsible for sudden cardiac death in up to 50% of HF patients (Rahm et al., 2018). The prevalence of AF increases with HF severity, ranging from 5% in mild HF to 50% in severe HF (Patel et al., 2020).

Oxidative stress in heart failure

Oxidative stress is one mediator for ventricular remodeling which leads to endothelial dysfunction. Oxidative stress alters the balance between the formation of oxygen free radicals and the inactivation through the body's antioxidant defense system with the inactivation of nitric oxide. Nitric oxide is a gaseous, free radical molecule that can easily diffuse across membranes producing vasodilation and relaxation of the blood vessels. Nitric oxide production is enhanced by increases in vascular shear stress, with increases in acetylcholine neurotransmitters and increases in thrombin via the coagulation cascade (Paganini et al., 2017).

Under normal conditions, factors derived from the endothelium such as nitric oxide reduce coronary vascular resistance by relaxing the vascular smooth muscle. In HF, nitric oxide production is substantially impaired because of the downregulation of endothelial oxide synthase and inactivation by reactive oxygen species. This in turn produces vasoconstriction, vascular stiffness, fibrosis, and ventricular remodeling as seen in HF (Ghionzoli et al., 2022).

Endothelial dysfunction in heart failure

Endothelial dysfunction is seen in both HFrEF and HFpEF (Pagel et al., 2021; Zuchi et al., 2020). One of the most important biomarkers involved in oxidative stress is myeloperoxidase (MPO). MPO is an enzyme released during polymorphonuclear neutrophil degranulation. It is the catalyst for the formation of reactive oxidants, free radicals, and nitric oxide derived oxidants all of which produce myocardial tissue injury and the disruption of vascular tone in HF (Correale et al., 2018).

The neurohormonal effects on heart failure

The renin–angiotensin–aldosterone system (RAAS) is a hormone system that regulates blood pressure, blood volume, and sodium, and potassium balance. Renin, a potent vasoconstrictor, is released from the kidneys second to low blood volume/pressure or with low sodium levels. Renin stimulates angiotensin, another potent vasoconstrictor, to release aldosterone. Aldosterone, a mineralocorticoid hormone released from the adrenal cortex, functions to increase water and sodium reabsorption and potassium excretion. The RAAS under normal circumstances maintains hemostasis. In contrast, the overactivation of the RAAS leads to increased oxidative stress and cardiac remodeling with higher preload from volume retention and afterload from vasoconstriction (Lei & Mao, 2018). Aldosterone targets cardiac muscle cells called myocytes. It also targets fibroblasts, which are connective tissue cells, and endothelial cells potentiating sodium and water reabsorption in the nephrons. Resultantly, this can lead to cardiac enlargement, fibrosis, apoptosis (programmed cell death), and atherosclerosis in the vascular smooth muscle cells. Aldosterone stimulates leukocyte adhesion, thrombotic changes, and epithelial–mesenchymal transition in endothelial cells (Ghionzoli et al., 2022; Levi et al., 2018). These aforementioned pathophysiological effects increase the risk of mortality in patients with acute myocardial infarction (AMI) or HF. Another neurohormone response is generated by natriuretic peptides (NPs), a group of hormones that are important for fluid and electrolyte balance as well as hemostasis, as they counteract the overactivation of the RAAS. NPs are regulated by myocytes in response to volume overload and cardiac wall tension (Lei & Mao, 2018). NPs have diuretic, natriuretic, vasodilative, and antifibrotic actions. Atrial natriuretic peptide, produced mainly by atrial myocytes, is released following atrial stretch with acute changes in blood pressure and blood volume. B-type or brain natriuretic peptides (BNPs) are synthesized and released from an overload of volume in the left ventricle (Ghionzoli et al., 2022). Clinically, BNP is utilized as a diagnostic and prognostic biomarker for the presence of HF. High levels of BNP indicate higher preload states which can be managed with diuretics or vasodilators.

The stress response can have a consequential role in the progression of HF. Glucocorticoids, or cortisol, are steroid hormones released from the adrenal cortex. Glucocorticoids function to regulate metabolism and have antiinflammatory and immunosuppressive effects. During acute stress states, glucocorticoids can enhance cardiac muscle cell contractility and promote survival under stress. Sustained high levels of glucocorticoids, from chronic stress states, could be precarious to the cardiovascular system. Patients who have chronic elevated levels of glucocorticoids, as seen in Cushing's syndrome, can develop HTN and metabolic syndrome, which are risk factors for HF. Arginine vasopressin (AVP), also known as antidiuretic hormone, is a hormone released from the posterior pituitary in response to decreased blood pressure and increased plasma osmolality. AVP with its vasoconstrictive properties contributes to increased afterload, fibrosis, and enlargement of the heart muscle in the later stages of HF (Correale et al., 2018; Lei & Mao, 2018).

Heart failure and diabetes

HF is not typically seen in isolation. Other comorbid conditions, such as type 2 diabetes (T2DM), metabolic syndrome, obesity, and chronic kidney disease, are prevalent in the HF population. Patients with HF have a higher prevalence of T2DM (20%) than patients without HF (4%–6%). Patients with T2DM have a 75% higher risk of CV death or HF hospitalization compared with those without T2DM as related to HFrEF by increasing mortality associated with ischemia (Maack et al., 2018). In HF, neuroendocrine activation alters cellular metabolism, predisposing one to the development of diabetes through insulin resistance whereby glucose is not recognized by cells with resultant hyperglycemia. To describe the pathway of insulin resistance in HF, there is an impaired ability of cells to take up glucose from the bloodstream associated with increased breakdown of triglycerides, synthesis of lipids in the liver, and gluconeogenesis leading to myocardial lipid overload. Prolonged hyperglycemia induces macrovascular dysfunctions, such as atherosclerosis. Over time, atherosclerotic changes lead to myocardial ischemia/infarction and HFrEF. Microvascular dysfunction can also occur with damage to smaller vessels contributing to diabetic retinopathy, nephropathy, and neuropathy. In the absence of ischemia, HFpEF, through a combination of myocardial stiffness and fibrosis, will prevail (Maack et al., 2018; Swedberg, 2021).

Ischemic heart disease

Overview

Ischemic heart disease is the leading cause of morbidity and mortality worldwide (Frak et al., 2022; Paganini et al., 2017). It is the result of an imbalance in oxygen supply and demand where oxygen is supplied to the myocardium through the coronary blood vessels and then optimized by sufficient diastolic filling time when the coronary arteries are perfused. The blood's oxygen carrying capacity improves oxygen supply. Conversely, oxygen demands are related to changes in cardiac output: increases in heart rate, contractility, preload, and resistance all increase myocardial oxygen demand. Furthermore, ischemic heart disease has four main mechanisms related to its pathophysiology: atherosclerosis, coronary microvascular dysfunction, inflammation, and coronary artery vasospasm. In atherosclerosis, fatty streaks progress to plaque, which narrows the arteries. In coronary microvascular dysfunction, the symptoms of ischemic heart disease occur in the absence of significant blockages affecting the small vessels as seen within the arterioles and capillaries. Inflammation occurs with collagen vascular diseases. Coronary artery vasospasm is transient and is noted with the absence of coronary occlusion (Severino et al., 2020; Frak et al., 2022; Paganini et al., 2017).

Specific features of ischemic heart disease

The trajectory of ischemic heart disease ranges from reversible ischemia with a milder clinical presentation, known as angina, to irreversible ischemia with resultant cell death as seen with AMI (Paganini et al., 2017). Angina can furthermore be described as stable angina, unstable angina, and variant angina (Paganini et al., 2017). Clinically, one divides AMI based on the electrocardiogram into ST-elevation myocardial infarction (STEMI) and non-ST elevation myocardial infarction (NSTEMI). STEMI will present with total coronary occlusion with the degree of necrosis being significant as compared to an NSTEMI. NSTEMI has a lower incidence of total coronary occlusion. Despite the artery lumen not being totally obstructed, it has severe blockage with an intraluminal coronary thrombus.

Cardiovascular disease risk factors for ischemic heart disease are represented in two main categories: nonmodifiable and modifiable. Nonmodifiable risk factors are age, sex, ethnicity, and family history. Aging is associated with vascular fibrosis, collagen deposition, less elastin, thickening of the vascular intima, and storage of cholesterol and phospholipids. Aging promotes meditators of platelet aggregation including cyclooxygenase, thromboxane A2, von Willebrand factor, and factor VIII, which results in hypercoagulation (Severino et al., 2020).

Modifiable risk factors for atherosclerosis and the related progression to ischemic heart disease are abnormal lipids profiles, diet, the use of tobacco products, high blood pressure, diabetes, abdominal obesity, metabolic syndrome, psychosocial factors, alcohol, and limited physical activity (Kawamoto et al., 2019). Furthermore, conditional risk factors include elevated levels of homocysteine, lipoprotein(a), triglycerides, inflammatory markers (i.e., C-reactive protein), prothrombotic factors (i.e., fibrinogen), and small, dense LDL particles (Crawford, 2017).

The largest subset of patients with ischemic heart disease have atherosclerosis. Although coronary atherosclerosis is the main cause of the reduction of oxygen supply and increased demand other clinical conditions can be implicated, with anemia, there is a decrease in the oxygen carrying capacity of the blood which decreases the oxygen supply. Increases in the ventricular workload with a resultant increase in afterload are present with HTN and aortic stenosis (Paganini et al., 2017).

Furthermore, collagen vascular diseases, such as rheumatoid arthritis, systemic lupus erythematosus, and scleroderma, are common conditions leading to coronary artery disease. These diseases have altered cellular mechanisms that affect the vasculature's distribution, and size and accelerate atherosclerotic changes through chronic inflammation and necrosis of blood vessel walls. Specifically, in collagen vascular diseases, cell mediated inflammation, also known as cell mediated immunity, activates immune cells, called T lymphocytes (T cells) which then activate monocytes and macrophages leading to the release of interleukins and interferons. This inflammatory cascade thickens the intimal lining of the blood vessels with resultant stenosis, infarction, and hemorrhage (Frak et al., 2022).

Other forms of ischemic heart disease that reduce coronary blood flow are coronary artery dissection and aortic dissection, a tear in the inner layer of the heart. Less common are infiltrative diseases of the heart, such as amyloidosis, sarcoidosis, and tumor metastases. Embolic events from infectious endocarditis and AF will compromise blood flow (Crawford, 2017).

Coronary artery vasospasm was described by an American cardiologist, Dr. Prinzmetal in 1959. This syndrome of angina called Prinzmetal angina, referred to as vasospastic angina or variant angina, is related to transient coronary artery vasospasm with sudden constriction or narrowing of the coronary arteries. Here, there is an acute reduction in coronary blood supply provoking myocardial ischemia occurring in the absence of coronary obstruction, yet with the presence of smooth muscle endothelial dysfunction. With Prinzmetal angina, there is transient electrocardiographic ST-segment elevation occurring in an episodic nature, oftentimes at rest which is different from unstable angina occurring with physical exertion. The spasms can be precipitated by several factors, i.e., smoking, emotional stress, exposure to cold, and epicardial coronary hyperactivity. Coronary artery spasm frequently appears at sites of hemodynamic significant atherosclerotic stenosis (Severino et al., 2020).

Cardiac ischemia

Fatty streaks are the earliest visible lesions in the development of atherosclerosis and are found in areas exposed to increased shear stresses and turbulent flow (Kusumoto, 2019). This turbulent blood flow is seen in branch points, bifurcations, and the major curvatures of the coronary vessels. It is these areas that are more prone to atherosclerotic lesions (Frak et al., 2022).

The fatty streaks are raised and yellow. They arise from isolated macrophages which are drawn into sites of injury or inflammation in the arterial wall. Macrophages transform into foam cells. These foam cells migrate into areas of chronic intimal injury and can progress to smooth muscle cell proliferation, extracellular fat, and collagen deposition (Kusumoto, 2019). Over time, the inflammatory process, along with the accumulation of lipids and the formation of a fibrous cap, leads to the development of advanced atherosclerotic plaques which can be prone to erosion and subsequent rupture. When plaque is exposed to blood, two significant and interrelated processes occur: platelets are activated and aggregate, which in turn potentiates the coagulation cascade leading to the deposition of fibrin strands. A thrombus, consisting of platelets and fibrin strands, traps the red blood cells and reduces myocardial blood flow, leading to the clinical manifestations of myocardial ischemia (Antman & Loscalzo, 2022).

During cardiac ischemia, there is a decrease in the vascular tone, impairing vasodilation. The heart will try to compensate for the stenosis as a result of the fixed narrowing of the coronary blood vessel. This compensation is referred to as a coronary flow reserve that is reasonably adequate if the stenosis is not severe. However, if the fixed narrowing of the vessel is at a borderline level (50%–70% of the normal vessel diameter), the coronary flow reserve is barely able to maintain the blood supply to the heart at rest. If there is any increase in oxygen demand, as with exercise, the flow reserve is inadequate to provide sufficient oxygen to this portion of the heart and ischemia will develop. Furthermore, if the stenosis is 80%–90%, ischemia will develop at rest even with maximal dilatation of the distal vessels downstream from the obstruction (Antman & Loscalzo, 2022; Paganini et al., 2017).

Reversible to irreversible injury in ischemic heart disease

The myocardium is highly dependent on proper coronary perfusion. The sudden interruption of myocardial perfusion leads to anaerobic metabolism and the production of ATP in the absence of oxygen (Antman & Loscalzo, 2022). Both light microscopes and electron microscopes identify pathological features of the myocardium over time. Reversible injury to the myocardium is seen before $^{1}/_{2}$ to 1 h with myocardial fibers having a wavy appearance at the border along with distorted function with swelling of the mitochondria, distorted cristae, matrix densities, and relaxation of myofibrils (Paganini et al., 2017). Within 1 hour, there is an irreversible injury to the myocardium, with a beginning of coagulation necrosis, neutrophilic infiltration, edema, and hemorrhage. After 24 h, there is necrosis of the myocardium. From 1 to 14 days, the pathological features include resorption of sarcoplasm by macrophages and neutrophils. The macrophages remove debris with a disappearance of necrotic muscle cells. There is the beginning of fibrosis and the organization of granulation tissue with prominent fibrovascular reactions. Within 2 weeks, there is the development of fibrous scar tissue (Paganini et al., 2017).

Sex differences in ischemic heart disease

Men and women present differently with acute coronary syndrome. Plaque rupture with occlusive coronary thrombus is the most common cause of acute coronary syndrome. Plaque rupture is less common in women. Women presenting with acute coronary syndrome have a higher incidence of nonobstructive coronary artery disease with coronary microvascular dysfunction second to the smaller size of their coronary artery anatomy. In men, the plaque is more vulnerable to rupture with thrombosis and occlusion of the coronary vessel second to the fibrous cap thinning and becoming unstable (Kawamoto et al., 2019).

Hypertension

In this chapter, the pathophysiology of AF, HF, and ischemic heart disease has been reviewed. The last alteration in cardiovascular function to be addressed in this chapter is HTN. In adults, HTN is defined as a systolic blood pressure of 140 mm Hg and/or a diastolic blood pressure of 90 mm Hg (Grassi et al., 2019). In the nonadult population with younger age groups, HTN is defined as systolic and/or diastolic blood pressure consistently >95th percentile for age, sex, and height (Kotchen, 2022). In patients with HTN, peripheral resistance is increased with higher afterload and cardiac output is normal or decreased. In younger patients with HTN, cardiac output may be increased and peripheral resistance, or afterload, may be normal (Kotchen, 2022).

HTN is related to multiple factors: genetics and environmental toxins, such as air and noise pollution. HTN has also been linked to the social determinants of health, for example, access to healthcare, education, and employment opportunities. A sedentary lifestyle, use of nicotine products, excessive alcohol consumption, diets high in sodium or low in potassium, and certain forms of emotional or social stress may aggravate or precipitate HTN in genetically susceptible individuals (Mohrman & Heller, 2018).

HTN remains the leading preventable risk factor for coronary artery disease, renal insufficiency with hypertensive nephropathy; heart failure; peripheral vascular disease; hypertensive retinopathy; and the risk of cognitive decline as related to ischemic stroke, vascular dementia, and Alzheimer's disease (Grassi et al., 2019). Despite a stable trend, its prevalence remains high with increased mortality, especially in low economic status individuals and underserved populations (Felkle et al., 2022; Manosroi & Williams, 2019).

Primary versus secondary hypertension

Essential, or primary HTN, is the most common cause of HTN, implicated in 85%–95% of the cases. Essential HTN means that elevated blood pressure is not a result of any other medical condition or secondary cause (Manosroi & Williams, 2019; Kotchen, 2022). This form of HTN will develop gradually over time and is influenced by a combination of genetic, lifestyle, and environmental factors. Conversely, secondary HTN is a less common form of high blood pressure seen clinically in 5%–15% of the population. Unlike essential HTN, secondary HTN is caused by an underlying medical condition or specific factor that directly contributes to the increase in blood pressure. Secondary causes of HTN are seen with endocrine diseases such as primary aldosteronism, Cushing's syndrome, and hyperthyroidism. Secondary HTN is also noted with obstructive sleep apnea; renal parenchymal disease including glomerulonephritis and polycystic kidney disease; and renal artery stenosis. Pheochromocytoma, a benign adrenal medulla tumor, and coarctation of the aorta are also medical conditions associated with secondary HTN (Grassi et al., 2019; Kotchen, 2022).

Unique to women is polycystic ovarian disease which is characterized by dysfunctions in ovulation and hyperandrogenism. Patients with polycystic ovarian disease have an increased risk of multiple cardiovascular risk factors including HTN; T2DM; metabolic syndrome with insulin resistance; abnormal lipid profiles including elevated triglycerides, decreased high-density lipoproteins, and elevated low-density lipoprotein; elevated C-reactive protein and homocysteine levels all of which contributing to endothelial dysfunction (Kawamoto et al., 2019).

Substantial data reports that essential high blood pressure can be an interplay between genetics and environmental factors. Furthermore, genetic studies have identified numerous genes related to the RAAS: 18 genes are associated with salt-sensitive HTN, four genes with normal renin levels, eight genes with lower renin levels, and six with undefined renin levels (Manosroi & Williams, 2019).

Resistant hypertension

When the goal of blood pressure cannot be achieved, another classification exists called resistant HTN. Resistant HTN is the blood pressure of a hypertensive patient that remains elevated above goal despite lifestyle modification and adherence to pharmacological agents. Here, a minimum of three antihypertensive drug classes with different mechanisms of action are needed to achieve optimal control. Based on evidence-based practice guidelines, the pharmacological agents include diuretics, ACE inhibitors, angiotensin II receptor blockers, calcium channel blockers, beta-blockers, and direct-acting vasodilators (Grassi et al., 2019).

The pathophysiology of HTN

The pathophysiology of HTN involves interactions between the heart, blood vessels, and the nervous and endocrine systems. HTN leads to end organ damage, meaning damage to various organs and tissues in the body. The effects are secondary to structural changes of left ventricular hypertrophy leading to HF and increased atrial size contributing to cardiac arrhythmias, including AF. In essence, in hypertensive heart disease, each chamber within the heart is enlarged or hypertrophied which changes systolic and diastolic dysfunction and hence cardiac output. Furthermore, within the brain and nervous system conditions of stroke, transient ischemic attacks, and age related cognitive decline with the potential of dementia can occur. Both macrovascular diseases with atherosclerotic coronary artery disease and peripheral vascular disease, and microvascular diseases of hypertensive nephropathy, hypertensive retinopathy, and erectile dysfunction can occur (Kotchen, 2022; Nwabuo & Vasan, 2022). Specifically, the depletion of adenosine 5′ triphosphate and the impairment of sarcoplasmic–endoplasmic reticulum ATPase may arise, causing the reduction in the capacity of the pump to return calcium into the sarcoplasmic reticulum, consequently impeding both cardiac contraction and relaxation (Nwabuo & Vasan, 2022).

Additionally, there is neurohormonal activation through the sympathetic system and the RAAS. At the molecular level, there is mitochondrial dysfunction and endothelial dysfunction which contribute to cardiac fibrosis (Nwabuo & Vasan, 2022). Cardiac fibrosis results in changes in the dynamic equilibrium of the heart with abnormal pathways and the release of cytokines, growth factors, proteases, reactive oxygen species, and signaling proteins (Kotchen, 2022).

Conclusion

In this chapter, on the pathophysiology of cardiovascular disease, the reader has been exposed to the concepts of cardiac output as related to AF, HF, ischemic heart disease, and HTN. When cardiac output is altered related to high heart rates, diastolic filling time is impaired with decreased coronary artery perfusion in coronary artery disease. High levels of fluid volume with high preload are seen in HF leading to the clinical manifestation of pulmonary vascular congestion resulting in dyspnea. Impaired contractility with systolic dysfunction is seen in HF with reduced EF. Higher afterload, with increased resistance, prevents the ventricles from unloading their volume resulting in both left and right-sided HF. High afterload is also implicated in hypertensive heart disease and aortic stenosis.

References

Antman, E. M., & Loscalzo, J. (2022). *Harrison's principles of internal medicine*. McGraw Hill, 2022.

Banasik, J. L., & Copstead, L. E. (2018). *Pathophysiology*.

Chung, M. K., Refaat, M., Shen, W. K., Kutyifa, V., Cha, Y. M., Di Biase, L., Baranchuk, A., Lampert, R., Natale, A., Fisher, J., & Lakkireddy, D. R. (2020). Atrial fibrillation: JACC council perspectives. *Journal of the American College of Cardiology, 75*(14), 1689–1713. https://doi.org/10.1016/j.jacc.2020.02.025, 15583597.

Collins, S., & Martindale, J. (2018). Optimizing hypertensive acute heart failure management with afterload reduction. *Current Hypertension Reports, 20*(1). https://doi.org/10.1007/s11906-018-0809-7, 1522-6417.

Correale, M., Monaco, I., Brunetti, N. D., Di Biase, M., Metra, M., Nodari, S., Butler, J., & Gheorghiade, M. (2018). Redefining biomarkers in heart failure. *Heart Failure Reviews, 23*(2), 237–253. https://doi.org/10.1007/s10741-018-9683-2, 15737322.

Crawford, M. H. (2017). Chronic ischemic heart disease. In M. H. Crawford (Ed.), Current diagnosis and treatment: Cardiology (5th ed.). McGraw Hill.

Dai, H., Zhang, Q., Much, A. A., Maor, E., Segev, A., Beinart, R., Adawi, S., Lu, Y., Bragazzi, N. L., & Wu, J. (2021). Global, regional, and national prevalence, incidence, mortality, and risk factors for atrial fibrillation, 1990–2017: Results from the Global Burden of Disease Study 2017. *European Heart Journal—Quality of Care and Clinical Outcomes, 7*(6), 574–582. https://doi.org/10.1093/ehjqcco/qcaa061, 20581742.

Felkle, D., Jarczyński, M., Kaleta, K., Zieba, K., & Nazimek, K. (2022). The immunomodulatory effects of antihypertensive therapy: A review. *Biomedicine & Pharmacotherapy, 153*. https://doi.org/10.1016/j.biopha.2022.113287, 19506007 https://www.journals.elsevier.com/biomedicine-and-pharmacotherapy.

Frak, W., Wojtasinska, A., Lisinska, W., Mlynarska, E., Franczyk, B., & Rysk, J. (2022). Pathophysiology of cardiovascular diseases: New insights into molecular mechanisms of atherosclerosis, arterial hypertension, and coronary artery disease. *Biomedicines, 10*, 1938. https://doi.org/10.3390/biomedicines10081938.

Ghionzoli, N., Gentile, F., Del Franco, A. M., Castiglione, V., Aimo, A., Giannoni, A., Burchielli, S., Cameli, M., Emdin, M., & Vergaro, G. (2022). Current and emerging drug targets in heart failure treatment. *Heart Failure Reviews, 27*(4), 1119–1136. https://doi.org/10.1007/s10741-021-10137-2, 15737322.

Givertz, M. G., & Mehra, M. R. (2022). *Heart failure: Pathophysiology and diagnosis*. McGraw Hill, 2022.

Grassi, G., Calhoun, D. A., Mancia, G., & Carey, R. M. (2019). Resistant hypertension management: Comparison of the 2017 American and 2018 European high blood pressure guidelines. *Current Hypertension Reports, 21*(9). https://doi.org/10.1007/s11906-019-0974-3, 15343111 www.springer.com.

Honan, L. (2019). *Focus on adult health: Medical-surgical nursing*. Wolters Kluwer, 2019.

Kawamoto, K., Duvernoy, C. S., & Davis, M. B. (2019). *Sex differences in coronary artery disease. In* Role of Sex and Atherosclerosis. Elsevier. https://doi.org/10.1016/B978-0-12-813197-8.00012-9.

Khan, M. K., & Sanchez. (2017). *Congestive heart failure and management drugs. Cardiology: An integrated approach*. McGraw Hill.

Kotchen, T. A. (2022). Hypertension. In Harrison's Principles of Internal Medicine, *21e*. McGraw Hill.

Kusumoto, F. (2019). *Pathophysiology of disease: An introduction to clinical medicine*. McGraw Hill, 2019.

Lei, L., & Mao, Y. (2018). Hormone treatments in congestive heart failure. *Journal of International Medical Research, 46*(6), 2063–2081. https://doi.org/10.1177/0300060518761262, 14732300.

Levi, A., Leshem-Lev, D., Weissler-Snir, Hasin, T., Mats, I., Murninkas, D., Kornowski, R, Lev, E. L., & Ben-Gal, T. (2018). The effect of mineralocorticoid receptor antagonists on recruitment and function of endothelial progenitor cells in patients with congestive heart failure. *IMAJ, 20*, 233-238.

Maack, C., Lehrke, M., Backs, J., Heinzel, F. R., Hulot, J. S., Marx, N., Paulus, W. J., Rossignol, P., Taegtmeyer, H., Bauersachs, J., Bayes-Genis, A., Brutsaert, D., Bugger, H., Clarke, K., Cosentino, F., De Keulenaer, G., Cas, A. D., González, A., Huelsmann, M., ... Heymans, S. (2018). Heart failure and diabetes: Metabolic alterations and therapeutic interventions: A state-of-the-art review from the translational research committee of the heart failure association-European Society of Cardiology. *European Heart Journal, 39*(48), 4243–4254. https://doi.org/10.1093/eurheartj/ehy596, 15229645.

Manosroi, W., & Williams, G. H. (2019). Genetics of human primary hypertension: Focus on hormonal mechanisms. *Endocrine Reviews, 40*(3), 825–856. https://doi.org/10.1210/er.2018-00071, 0163-769X.

Matsushita, K., Harada, K., Miyazaki, T., Miyamoto, T., Kohsaka, S., Iida, K., Yamamoto, Y., Nagatomo, Y., Yoshino, H., Yamamoto, T., Nagao, K., & Takayama, M. (2019). Younger- vs older-old patients with heart failure with preserved ejection fraction. *Journal of the American Geriatrics Society, 67*(10), 2123–2128. https://doi.org/10.1111/jgs.16050, 0002-8614.

Mohrman, D. E., & Heller, L. (2018). *Cardiovascular function in pathological situations. Cardiovascular Physiology*. McGraw Hill, 2018.

Normand, C., Kaye, D. M., Povsic, T. J., & Dickstein, K. (2019). Beyond pharmacological treatment: An insight into therapies that target specific aspects of heart failure pathophysiology. *The Lancet, 393*(10175), 1045–1055. https://doi.org/10.1016/S0140-6736(18)32216-5, 1474547X.

Nwabuo, C. C., & Vasan, R. S. (2020). Pathophysiology of hypertensive heart disease: Beyond left ventricular hypertrophy. *Current Hypertension Reports, 22*, 11. https://doi.org/10.1007/s11906-020-1017-9.

Paganini, A., Blue, D., & Elmoselhi, A. (2017). Ischemic heart disease and management drugs. In A. Elmoselhi (Ed.), Cardiology: An integrated approach. McGraw Hill.

Pagel, P. S., Tawil, J. N., Boettcher, B. T., Izquierdo, D. A., Lazicki, T. J., Crystal, G. J., & Freed, J. K. (2021). Heart failure with preserved ejection fraction: A comprehensive review and update of diagnosis, pathophysiology, treatment, and perioperative implications. *Journal of Cardiothoracic and Vascular Anesthesia, 35*(6), 1839–1859. https://doi.org/10.1053/j.jvca.2020.07.016, 15328422.

Patel, R. B., Passman, R. S., & Shah, S. J. (2020). Embarking upon atrial fibrillation management in heart failure with preserved ejection fraction: Charting a course. *Journal of Cardiovascular Electrophysiology, 31*(9), 2284–2287. https://doi.org/10.1111/jce.14635, 15408167 http://onlinelibrary.wiley.com/journal/10.1111/(ISSN)1540-8167.

Rahm, A. K., Lugenbiel, P., Schweizer, P. A., Katus, H. A., & Thomas, D. (2018). Role of ion channels in heart failure and channelopathies. *Biophysical Reviews, 10*(4), 1097–1106. https://doi.org/10.1007/s12551-018-0442-3, 18672469.

Riehle, C., & Bauersachs, J. (2019). Key inflammatory mechanisms underlying heart failure. *Herz, 44*(2), 96–106. https://doi.org/10.1007/s00059-019-4785-8, 0340-9937.

Sauer, W. H., & Zei, P. C. (2022). *Harrison's principles of internal medicine*. McGraw Hill, 2022.

Severino, P., D'amato, A., Pucci, M., Infusino, F., Adamo, F., Birtolo, L. I., Netti, L., Montefusco, G., Chimenti, C., Lavalle, C., Maestrini, V., Mancone, M., Chilian, W. M., & Fedele, F. (2020). Ischemic heart disease pathophysiology paradigms overview: From plaque activation to microvascular dysfunction. *International Journal of Molecular Sciences, 21*(21), 1–30. https://doi.org/10.3390/ijms21218118, 14220067.

Shiraishi, Y., Kawana, M., Nakata, J., Sato, N., Fukuda, K., & Kohsaka, S. (2021). Time-sensitive approach in the management of acute heart failure. *ESC Heart Failure, 8*(1), 204–221. https://doi.org/10.1002/ehf2.13139, 2055-5822.

Swedberg, K. (2021). Heart failure subtypes Pathophysiology and definitions. *Diabetes Research and Clinical Practice, 175*, 1–5.

Walsh, K., & Caple, C. (2018). EKG series: EKG interpretation – Atrial fibrillation. *Cinahl Information Systems. EBSCO Publishing*.

Zuchi, C., Tritto, I., Carluccio, E., Mattei, C., Cattadori, G., & Ambrosio, G. (2020). Role of endothelial dysfunction in heart failure. *Heart Failure Reviews, 25*(1), 21–30. https://doi.org/10.1007/s10741-019-09881-3, 1382-4147.

CHAPTER

16

Cardiovascular diseases and mitochondrial transplantation: Basic and preclinical studies and transitions to clinical applications

Andrei B. Borisov[1], *Kevin McCusker*[2], *David McIntyre*[3] and *Jean D. Peduzzi-Nelson*[4]

[1]Department of Biomedical Engineering, Wayne State University, Detroit, MI, United States; [2]Lawrence Technological University, Southfield, MI, United States; [3]Department of Product Development, Terumo Cardiovascular, Ann Arbor, MI, United States; [4]Department of Neurosurgery, Department of Ophthalmology, Visual and Anatomical Sciences, School of Medicine, Wayne State University, Detroit, MI, United States

Introduction

In the cardiovascular system, mitochondria play the leading functional role in generating adenosine triphosphate (ATP) for the maintenance of cell homeostasis and survival and are responsible for different signal transduction pathways (Dia et al., 2020; Gustafsson & Gottlieb, 2008; Lenaz & Genova, 2010; Tilokani et al., 2018; Vásquez-Trincado et al., 2016; Zhang et al., 2020). Mitochondria are involved in the generation of reactive oxygen species (ROS), inducing oxidative stress and oxidative damage, controlling intracellular calcium balance, and regulating apoptosis by activating caspase cleavage pathways in different cell types, including cardiac and vascular cells in the cardiovascular system (Chen et al., 2016; Chen & Zweier, 2014; Del Re et al., 2019; Dey et al., 2018; Lee & Gustafsson, 2009; Reuter, 2022; Yu et al., 2012). Abnormal release of ROS can initiate damaging effects on cells in the heart and blood vessels (Dey et al., 2018).

Many pathological conditions in the heart and blood vessels affect mitochondria's metabolic activity and viability in these energy-demanding cells (Cherry, 2019; Zhou & Tian, 2018). Dysfunctional changes and injury in these organelles occur in hypoxic/ischemic and

https://doi.org/10.1016/B978-0-443-18858-9.00016-0

reperfused myocardial tissue, atherosclerosis, hypertension, progression of diabetes mellitus, and severe obesity (Ballinger, 2005; Hamilton, 2013; Kattoor et al., 2017; Madamanchi & Runge, 2007; Rocha et al., 2016). Damage to mitochondria is one of the main factors in the pathogenesis of congestive heart failure, cardiomyopathies, uncompensated myocardial hypertrophies, and other types of cardiovascular diseases (Boovarahan & Kurian, 2018; Hamilton, 2013; Ranjbarvaziri et al., 2021; Rosca & Hoppel, 2013; Siasos et al., 2018).

Another cause of mitochondrial dysfunction is the expression of mutated genes in mitochondrial and nuclear DNA that encodes mitochondrial proteins and ribosomal and transfer RNA molecules. Expression of these mutations through family generations leads to the impairment of normal cell and tissue functions resulting in the development of dilated and hypertrophic cardiomyopathies and other types of myocardial and vascular pathology (Alston et al., 2017; Davison & Rahman, 2017; Gorman et al., 2016; Marín-García & Goldenthal, 2002a; Pohjoismäki et al., 2010; Poznyak et al., 2020; Sazonova et al., 2018, 2021; Sobenin et al., 2013). Mutations in mitochondrial genes also aggravate other types of cardiovascular insufficiencies. The functions of mitochondria affected by mutations may include abnormal changes in energy metabolism and production of ATP, control of cell death, and calcium transport (Bray & Ballinger, 2017; Pohjoismäki et al., 2010; Sorrentino et al., 2018). These characteristics are essential for the support of normal functions of cells and organs in the cardiovascular system. Recently, mitochondria targeted gene delivery has been suggested as a strategy for prospective therapy of mitochondrial diseases caused by mutations (see review Jang & Lim, 2018). However, it is challenging to accomplish a significant efficiency of intramitochondrial gene transfer, given that individual cells can contain hundreds of mitochondria.

An alternative method to normalize the mitochondrial function is the intercellular transfer of intact mitochondria from normal cells into cells with dysfunctional mitochondria. Recent findings have shown that intercellular transfer and exchange of mitochondria occur in different tissues under physiological and pathological conditions (Gäbelein et al., 2022; Hu et al., 2022; Liu et al., 2021; Shanmughapriya and Langford, 2020). For this reason, the transfer of mitochondria from normal cells to injured cells containing defective organelles affected by mitochondrial dysfunction appears as an emerging novel therapeutic approach. Transplantation of mitochondria in this context would allow us to rescue the cardiovascular tissues affected by mitochondrial pathology by improving their respiration and survival, replenishing their numbers, and activating their metabolic activity would lead to improved functional outcomes and clinical prognosis for affected patients.

In this chapter, we review recent findings concerning mitochondrial transplantation within the cardiovascular system obtained in basic research and discuss translational applications of this technique in a clinical setting.

Pathology of the cardiovascular system associated with mitochondrial dysfunction

Pathologies not involving the primary structure of mitochondrial genome

The human myocardial tissue spends more energy per gram of mass than any other tissue in the body (Weiss et al., 2005). In energy demanding cardiac muscle cells, mitochondria are involved in fatty acid β-oxidation, respiratory electron chain, and oxidative phosphorylation and are responsible for generating large quantities of ATP-supporting myocardial contractions (Ventura-Clapier et al., 2011). These organelles are abundant in cardiac myocytes comprising 20%–40% of their volume (Marín-García & Goldenthal, 2002b). The numbers of mitochondria in cardiac muscle cells increase under conditions of a higher functional load, such as myocardial hypertrophy and physical exercise, and are controlled by endocrine factors, such as thyroid hormone (Marín-García & Goldenthal, 2002b). This process allows mitochondria to serve the increased requirements for cellular respiration and energy metabolism under elevated functional demands in cardiac muscle cells.

Mitochondria in cardiomyocytes generate more than 95% of the cardiac ATP, and cellular responses to myocardial ischemia are associated with a mitochondrial malfunction that persists and increases after reperfusion (Shin et al., 2017). Mitochondrial damage and necrosis accompanied by progressive energy deficiency is an essential factor in the pathogenesis of congestive heart failure, cardiomyopathies, diabetes mellitus, atherosclerosis, and other diseases of the cardiovascular system (Rossmann et al., 2021; Vásquez-Trincado et al., 2016; Wu et al., 2019). This damage considerably impairs mitochondrial homeostasis and diminishes mitochondrial fitness making these organelles less capable of performing their functions. Complete block or insufficiency of ATP production, disruption of signal transduction pathways, and the impaired mechanism of cell death control are only a few examples of cellular pathology associated with mitochondrial dysfunction. Activation of mitophagy of damaged mitochondria (mitochondrial autophagocytosis and subsequent digestion within lysosomes) is another critical factor in decreasing the numbers of these organelles during the progression of cardiovascular diseases (Luan et al., 2021; Morales et al., 2020; Vásquez-Trincado 2016).

Thus, a large group of pathological conditions associated with mitochondrial injury requiring rescue or replacement of these organelles is related to nongenetic causes. These conditions include the damage resulting from hypoxia, ischemia and reperfusion, the progression of heart failure, cardiomyopathies, and atherosclerosis. Dysfunctional changes in mitochondria lead to an imbalance between energy supply and demand in a failing heart. Reviews of mitochondrial diseases of different etiology in the cardiovascular system derived from several pathological processes have been presented in the recent literature (Davison & Rahman, 2017; Forte et al., 2021; Liu et al., 2022; Nan et al., 2017; Tian et al., 2019; Yu et al., 2012).

Pathologies associated with mutations in mitochondrial genome

During development, cell mitochondria originate from the maternal gametes (oocytes). Circular double-helix DNA molecules forming the mitochondrial chromosome can carry a variety of mutations. The intramitochondrial genome is small: it consists of only 37 genes encoding 13 proteins of the oxidative phosphorylation system, 2 ribosomal RNAs, and 22 transfer RNAs consisting in humans of 16,569 nucleotide base pairs (Bray & Ballinger, 2017; El-Hattab & Scaglia, 2016). The total number of genes serving the needs of mitochondria is considerably higher, and many of these genes are encoded by chromosomal DNA located within cell nuclei (Meyers et al., 2013). The nuclei contain up to 99% of the total mitochondrial genes (El-Hattab

& Scaglia, 2016). Proteins encoded by these nuclear genes are necessary for mitochondrial replication and DNA repair, the structure of mitochondrial ribosomes, and serve as structural/transport proteins necessary for the assembly of the respiratory chain complex contributing to ATP production. These proteins are translated from mRNA into protein molecules in the cell cytoplasm by cytoplasmic ribosomes and are translocated into mitochondria following their synthesis. It is important to note that the nuclear genome serving mitochondria's needs also accumulates and expresses mutations independently from mutations in the intramitochondrial genome. Some of these mutations, except neutral changes in DNA structure, can affect normal mitochondrial physiology.

Normal mitochondrial function is essential for maintaining healthy homeostasis in the cells of the cardiovascular system. Mutations in the genomes localized in both mitochondria and cell nuclei are responsible for the functional impairment of these organelles and cause mitochondrial diseases. Mitochondrial diseases resulting from these pathogenic mutations lead to the expression of dysfunctional versions of intramitochondrial and nuclear genes serving the needs of mitochondria (Bray & Ballinger, 2017; Lioncino et al., 2022). Numerous mitochondrial pathologies, including mitochondrial cardiomyopathies, result from mutations in mitochondrial and nuclear genes encoding mitochondrial proteins (Bray & Ballinger, 2017). Mitochondrial diseases are diagnosed in 1 in 5000 live births, and up to 40% of these patients with mitochondrial mutations show symptoms of cardiomyopathies (El-Hattab & Scaglia, 2016). Primary mitochondrial disorders represent a clinically heterogeneous group associated with impaired functions or low effectiveness of the mitochondrial respiratory chain in the heart and vascular cells (Finsterer & Kothari, 2014). The effects of different mutations expressed in the cardiovascular system affecting energy metabolism, calcium homeostasis, oxidative stress, inflammatory processes, cell survival, and abnormal control of cell death are discussed in a recent review (Bonora et al., 2019). It was demonstrated that mutations affecting aerobic energy metabolism, production of ATP molecules, and other mitochondrial functions significantly impair the normal functional activity of the cardiovascular system (Ballinger, 2005; El-Hattab & Scaglia, 2016; Poznyak et al., 2020).

Given that in cardiac and vascular muscle cells, the mitochondria are numerous, and somatic mutations in mitochondrial DNA are random, the intracellular populations of these organelles may contain a mix of normal and mutated genotypes. This phenomenon is called mitochondrial heteroplasmy (see Chapter 4 of this book). Multiple reports suggest heteroplasmy's role in the pathogenesis of cardiovascular diseases (Hefti et al., 2018; Mitrofanov et al., 2016; Sobenin et al., 2019; Stewart & Chinnery, 2021). The ratio of normal and mutated genotypes determines the expression of pathological phenotypes and the severity of clinical symptoms.

The mitochondrial mutations are observed in some cases of hypertrophic, dilated, and ischemic cardiomyopathies, congestive heart failure, dysfunctional changes in the cardiac conduction system, impaired apoptosis control, impaired calcium transport, and other pathological conditions. Mitochondrial cardiomyopathies caused by the expression of mutations in mitochondrial proteins encoded by mitochondrial and nuclear genes represent a heterogeneous group of diseases (reviewed by Meyers et al., 2013; El-Hattab & Scaglia, 2016; Bray & Ballinger, 2017; Yang et al., 2022). Depending on their clinical manifestations and symptoms, mitochondrial cardiomyopathies are classified into hypertrophic and dilated cardiomyopathies, arrhythmias, and left ventricular myocardial noncompaction, which can lead to congestive heart failure (El-Hattab, Scaglia, 2016). Cardiomyopathies are diagnosed in 20%–40% of children with mitochondrial mutations. The symptomatic conditions of these patients typically worsen under increased energy demand in the heart, imposing an uncompensated higher functional load on the activity of the electron transportation chain (El-Hattab, Scaglia, 2016; Finsterer & Kothari, 2014). A significant number of these patients are diagnosed with hypertrophic cardiomyopathy (Lioncino et al., 2022).

Mitochondrial mutations can activate cardiac cell loss following the onset of ischemia–reperfusion injury, progression of cardiomyopathies, and heart failure (Gustafsson & Gottlieb, 2008). This process is accompanied by the release of proteins from mitochondria, such as cytochrome *c*, Smac, Omi/Htr2A, and AIF, that play a significant functional role in activating the apoptotic cell death program (Gustafsson & Gottlieb, 2008). In addition, mitochondrial calcium overload resulting from mitochondrial mutations may trigger the opening of the mitochondrial permeability pores, causing uncoupling of oxidative phosphorylation, swelling of the mitochondria due to the influx of water, and rupture of the outer mitochondrial membrane (Gustafsson & Gottlieb, 2008). The pathological process involving mitochondrial mutations may involve several tissues and organs (Marin-Garcia et al., 2002), but it affects the heart muscle more profoundly as one of the most energy dependent tissue types with high ATP synthesis and consumption rates. The clinical manifestations of cardiomyopathies can be significantly different from case to case.

Mitochondrial mutations play a similar dysfunctional role in vascular smooth muscle cells and vascular endothelium. Impaired mitochondria are associated with lower energy production levels and abnormal signaling pathways controlling metabolism and maintenance of cell survival (Quintero et al., 2006). Thus, mutations in the mitochondrial genome are linked to different vascular pathologies, including atherosclerosis (Bray & Ballinger, 2017; Glanz et al., 2020; Kirichenko, 2020; Poznyak et al., 2020; Yu et al., 2012).

Intercellular mitochondrial transfer in the cardiovascular system

Studies using animal models and cell cultures

Mitochondrial transfer is an emerging novel therapy aiming to replace dysfunctionally impaired mitochondria under conditions of different types of nongenetic and genetic pathology. The new concept of replacing damaged mitochondria with healthy mitochondria by transplantation is emerging as a potential therapy for myocardial rescue and as a way to understand mitochondrial metabolism and functions (Shin et al., 2017). The cellular pathways leading to donor mitochondria uptake and functional integration by recipient cells in the cardiovascular system are discussed in recent publications (Ali Pour et al., 2021; Cowan et al., 2017; McCully et al., 2022).

Methods of mitochondrial transfer in vivo include direct injections of isolated mitochondria into targeted tissues and organs or intravascular infusions of mitochondrial suspensions (reviewed by Lim, 2020; McCully et al., 2022). Both methods include a series of injections of 40–100 μL of suspensions using a tuberculin syringe. The mitochondria delivered by these techniques typically remain within the boundaries of the targeted organs (Moskowitzova et al., 2019; Guariento, Blitzer, et al., 2020; Orfany et al., 2020). Mitochondria suspensions are injected directly into the lumens of blood vessels for intravascular delivery. Both methods resulted in comparable positive outcomes.

Ali Pour et al. (2021) present the methods of intercellular mitochondrial transfer in cell cultures. In the current literature, three main types of intercellular mitochondrial transfer in vitro have been described: (1) transfer through tunneling nanotubes, (2) extracellular vesicles (EVs), and (3) mitochondrial transfer through gap junction channels (GJCs) (see Chapter 00 of this book). It is important to note that these three ways of intercellular mitochondrial transportation are used by the cells in cell culture and in vivo under normal and pathological conditions. A fourth novel method of mitochondrial transfer, the "nanosyringe" technique, has recently been reported by Gäbelein et al. (2022). Using the combination of optical microscopy, atomic force microscopy, and nanofluidics, the authors found that the mitochondrial transplantation between cells was successful in 95% of cases, and mitochondrial transfer provided a satisfactory outcome. As a result of transplantation using these four techniques, the donor mitochondria are internalized and integrated into the recipient cells. The mitochondria delivered into the targeted organs in vivo through injections or intravascular transfer may be internalized into the host cells using GJCs, other openings in cell membranes, or endocytosis similar to transfer mechanisms in vitro. However, the cellular mechanisms controlling the transmembrane transfer of mitochondria and their incorporation in the recipient cells in vivo are incompletely understood.

The most frequent type of injury damaging the mitochondria in the cardiovascular system is associated with severe hypoxia, ischemia, and ischemia–reperfusion injury. Mitochondrial death and cell death typically follow tissue injury. Intercellular mitochondrial transplantation can replace damaged mitochondria in the tissue exposed to these three conditions. The methods of mitochondria transfer and their effectiveness for cell rescue following ischemia–perfusion injury are discussed in a review by Doulamis and McCully (2021). Zhou et al. (2021) have discussed the mitochondrial dynamics, quality control, and the mechanisms causing dysfunctional changes and injury in mitochondria under conditions of ischemia and reperfusion. Hayashida et al. (2021) recently reviewed 20 animal studies focused on transplanting autografts and allografts of mitochondria in different tissues. Fourteen of these studies examined the results of mitochondrial transplantation in the two most oxygen demanding organs—the heart and the brain. The difference between these experiments included the dosage of transplanted mitochondria, the timing following injury, and the mode of organelle delivery. In all reviewed papers, the authors found considerable beneficial effects of mitochondrial transplantation following ischemia–reperfusion injury, although pathological manifestations of this injury and different biomarkers varied in each of these studies.

In *in vivo* experiments, Cowan and coauthors (2017) found that autologous exogenous mitochondria can be delivered into the rabbit heart through intravascular coronary injections and that these organelles are internalized and incorporated into cardiac muscle cells. It was found that the mitochondria perfused through the coronary vasculature significantly decreased infarct size and considerably enhanced postischemic myocardial function. The authors concluded that the delivery of mitochondria through the coronary arteries resulted in widespread distribution throughout the heart and rapid integration into cardiac cells. Similar studies found the effectiveness and safety of intravascular mitochondrial transfer in cardiomyocytes in experiments using pigs. It was shown that intracoronary mitochondrial delivery after temporary, regional ischemia significantly improved myocardial function, perfusion, and decreased infarct size (Shin et al., 2017, 2019). The authors concluded that intracoronary delivery of mitochondria is safe and efficacious therapy for myocardial ischemia–reperfusion injury. These and other results demonstrated the feasibility of mitochondrial transfer delivered through coronary vessels and direct injections (Lim, 2020). The study of cold ischemia found that autologous mitochondrial transplantation in the heart transplants in mice replaced damaged mitochondria and enhanced myocardial function, tissue viability, and prolonged cold ischemia time (Moskowitzova et al., 2019). Under these conditions, necrosis and neutrophil infiltration were significantly decreased, reducing graft failure and improving transplantation outcomes after prolonged cold ischemia.

Another group of experiments on pigs (Blitzer et al., 2020) demonstrated that delayed mitochondrial transplantation by intracoronary injection appreciably decreased myocardial infarct size, increasing regional and global myocardial function. According to the authors, these results suggest that mitochondrial transplantation can be a viable treatment modality in ischemia–reperfusion injury, thus potentially aiming to reduce long-term morbidity and mortality in cardiac surgical patients.

Earlier, the same group reported that transplantation of autologously derived, respiration-competent mitochondria by direct injection into the heart after transient ischemia and reperfusion enhanced cell viability and contractile function (Masuzawa et al., 2013; McCully et al., 2009). It was found that both injected and intravascularly perfused mitochondria were initially observed in blood vessels and interstitial spaces before their incorporation into cardiomyocytes. A study on Yorkshire pigs (Guariento, Blitzer, et al., 2020) demonstrated that preischemic intracoronary autologous mitochondrial transplantation ensured myocardial protection from regional ischemia–reperfusion injury. A single injection of 1×10^9 mitochondria in the isotonic buffer or a series of 10 injections over 60 min were performed 15 min before the exposure to temporary, regional ischemia induced by snaring the left anterior descending artery for 30 min. After releasing the snare, the heart was perfused for 2 h. It was reported that both single and serial intramyocardial injections of mitochondria provided myocardial protection from ischemia–reperfusion injury. Moreover, it was shown that in perfused hearts obtained from pigs after circulatory death, mitochondrial transplantation significantly preserved function and oxygen consumption, providing a possible option for expanding the heart donor pool (Guariento, Doulamis, et al., 2020).

Another study on pigs (Kaza et al., 2017) found that injections of autologous mitochondria isolated from skeletal muscle (the pectoralis major) into the myocardium following ischemia–reperfusion significantly diminished tissue injury improving cell survival and the functional outcome. Working on rabbits, Masuzawa et al. (2013) found that the delivery of autologous mitochondria into the ischemia–reperfusion injury zone in the heart diminished the consequences of ischemia. It was reported that the hearts with transplanted mitochondria demonstrated enhanced oxygen consumption, higher energy phosphate synthesis, increased cell viability, and enhanced postinfarction cardiac function. The hearts injected with mitochondria returned to normal contraction 10 min after starting reperfusion, whereas control hearts possessed persistent hypokinesia in the injury zone even at 4 weeks of recovery. In addition, mitochondria-injected hearts showed the expression of proteomic pathways consistent with better preservation of myocardial energetics.

Interestingly, injection of mitochondria from different sources into the free wall of the right ventricle of the myocardium in piglets resulted in protection from activation of cardiomyocyte apoptosis occurring after ischemia–reperfusion injury, preserved contractility, and prevented the development of right ventricular failure (Weixler et al., 2021). It was also found that the internalization of mitochondria and increased ATP levels did not depend on the source of mitochondria. These findings confirmed the results of earlier studies showing that the transplantation of autologous, syngeneic, and allogenic mitochondria did not produce any detectable immune or autoimmune response (Ramirez-Barbieri et al., 2019). It was also reported that transplantation of viable mitochondria improved right ventricular performance and pulmonary artery remodeling in rats with pulmonary arterial hypertension (Hsu et al., 2022; Vasanthan et al., 2022). Thus, these results suggest that intracoronary delivery of mitochondria in animals of different taxonomic groups is a safe and efficacious therapy for myocardial ischemia–reperfusion injury.

The study of neonatal cardiomyocytes isolated from rats exposed to pregestational diabetes mellitus found that these cells could internalize transplanted donor mitochondria from normal healthy neonatal rats (Louwagie et al., 2021). The authors reported that mitochondrial transfer provided a respiratory boost to cardiomyocytes, including the control nondiabetic groups. It was also noted that normal mitochondria transferred through intracoronary delivery improved cellular respiration and reduced stress-induced apoptosis in male rats exposed to pregestational diabetes mellitus but had no effect on the rate of apoptosis in female animals. It is essential that pregestational diabetes mellitus can lead to the development of diabetic cardiomyopathy accompanied by progressive manifestations of cardiovascular pathology.

Diabetes mellitus is considered one of the pathological conditions increasing myocardial susceptibility to ischemia–reperfusion injury and causing mitochondrial and cardiac dysfunction. The experiments with isolated type 2 diabetic rat hearts demonstrated that transplantation of mitochondria isolated from healthy animals provided cardioprotection from 30 min of global warm ischemia and 120 min of reperfusion by significantly enhancing postischemic myocardial functional recovery and significantly decreasing myocellular injury in the diabetic hearts (Doulamis et al., 2020).

In experiments with cell cultures, the cell line H9c2 derived from embryonic rat hearts was transfected with autologous and nonautologous mitochondria isolated from the same cells and cell line L6 developed from rat skeletal muscle, respectively (Ali Pour et al., 2020). The authors observed a statistically significant improvement in basal respiration rate and ATP production 2 days posttransplantation, accompanied by an increase in maximal respiration and spare respiratory capacity (Ali Pour et al., 2020). Interestingly, that interspecies nonautologous transplantation of mitochondria also resulted in an improvement of respiratory capabilities similar to autologous mitochondria. However, the authors found that the improved bioenergetics returned to normal baseline levels in subsequent periods. This finding supports earlier results of Ramirez-Barbieri et al. (2019), suggesting the absence of immune response and the feasibility of transplantation of nonautologous mitochondria.

Another group of studies includes the transplantation of mitochondria isolated from stem cells into cardiac and vascular cells with dysfunctional mitochondria. Ikeda et al. (2021) examined the effects of mitochondria isolated from human cardiomyocytes derived from induced pluripotent stem cells on tissue survival and recovery in the murine heart following myocardial infarction. The mitochondria were packed into EVs were injected into the infarcted myocardium. It was found that this treatment significantly restored the intracellular ATP production and improved the contractile properties of cardiac cells as early as 3 h after treatment. In addition, intramyocardial injection of EVs containing mitochondria demonstrated significantly improved postinfraction cardiac function through restoration of bioenergetics and mitochondrial biogenesis. Thus, the results of this study demonstrated that EV-mediated mitochondrial transfer contributes to improved intracellular energetics, and intramyocardial injection of mitochondria enhanced postinfarction cardiac function in vivo. It was also shown that mitochondria from stem cells could be successfully internalized and integrated into cardiac myocytes and endothelial cells (Murray & Krasnodembskaya, 2019). These results suggest that donating normal healthy mitochondria from mesenchymal stem cells to replace damaged mitochondria in injured cells is a physiological process aimed at compensating for functional insufficiency of mitochondria.

One more recent study (Manshouri et al., 2022) examined the effect of transplantation of mitochondria isolated from human menstrual blood stem cells delivered transcardially into the rat hearts with myocardial infarction, improved cardiac function, and promoted the survival of cardiac myocytes and endothelial cells. The authors also found that the transplanted mitochondria and their progeny remained viable in rat hearts for at least 28 days following transplantation in the periinfarction area, as demonstrated by labeling with an antibody developed against human mitochondria.

The earlier studies focused on transferring mitochondria from stem cells into the cells of the cardiovascular system are presented in the publications cited above. A recent review by Chen et al. (2021) discusses earlier publications on this topic. Thus, the results of experiments on animal models show the feasibility of exogenous intracellular mitochondrial transfer in the cardiovascular system. This reflects the natural capacity of cells of different types to exchange mitochondria under normal and pathological conditions.

Studies on human tissue and prospects of clinical applications

A large number of experiments have been performed to examine the efficacy and safety of mitochondrial transplantation in animals of different species (reviews—Hosseinian et al., 2022; McCully et al., 2022). In these experiments, the internalization and successful functional integration of transplanted mitochondria by recipient cells were demonstrated, and the dosages of mitochondria for successful transplantation were selected. Based on the results of these findings, McCully and coauthors performed the first clinical trials of mitochondrial transfer in pediatric patients (Emani et al., 2017; Emani & McCully, 2018; Guariento et al., 2021). These experiments were aimed at the replacement or augmentation of native mitochondria damaged, by ischemia, with viable, respiration-competent mitochondria isolated from nonischemic tissue obtained from the patient's own body (Cowan et al., 2017; McCully et al., 2022).

The first clinical trial included five pediatric patients from 2 days to 2 years of age who were treated for the consequences of ischemia–reperfusion injury following removal from extracorporeal membranous oxygenation (ECMO). Mitochondrial transplantation was performed in the intensive care unit or the operating room. Autologous mitochondria were isolated 20–30 min before the transplantation from a 6 × 6 mm specimen of the rectus abdominis muscle of each patient. The isolated organelles were suspended in a respiration buffer and rapidly injected 10 times in different ischemic areas of the myocardium as 100 μL aliquots using a tuberculin syringe. The authors reported that this procedure considerably improved the functional activity of the heart as evaluated by echocardiography (Emani et al., 2017). It is important to note that following transplantation of mitochondria, four out of five patients were transitioned from the ECMO protocol, but only three individuals eventually survived. The autopsy of the diseased patients found no difference in the levels of preinjection and postinjection inflammatory markers. This indicates that there was no rejection of transplanted mitochondria. This study shows that mitochondrial transplantation in pediatric patients might be helpful at the time of their separation from ECMO. The subsequent publication discussed the methodological aspects and the protocol for optimizing this procedure (Emani & McCully, 2018).

The second pilot study concerned pediatric patients requiring postcardiotomy ECMO for severe refractory cardiogenic shock after ischemia–reperfusion injury. It involved 10 patients subjected to mitochondrial transfer and 14 control patients (Guariento et al., 2021). All patients were subjected to ECMO, revascularized, and recovering from ischemia–reperfusion injury. The experimental conditions were similar to the first experiments described above. It was found that the time required for functional recovery after revascularization was significantly shorter in patients treated with mitochondrial transfer: 2 days versus nine days in the control group. The mitochondrial transfer was associated with successful separation from ECMO and positively affected enhanced ventricular strain in patients requiring postcardiotomy ECMO for severe refractory cardiogenic shock after ischemia–reperfusion injury (Guariento, Blitzer, et al., 2020; Guariento, Doulamis et al., 2020). The authors reported that eight patients (80%) who received mitochondrial transplants were removed from ECMO within a week compared to four patients (29%) without mitochondrial transplants. These results show that mitochondrial transplantation in pediatric patients appears to be an effective therapeutic approach at the time of separating from ECMO.

These results suggest that mitochondrial transfer applications for therapeutic purposes are feasible and relatively safe. This work attracted considerable interest and was enthusiastically received by physicians and cardiovascular researchers (Abdullah & Kaushal, 2021; Barron & Coles, 2021; Buratto & Konstantinov, 2021; Cherry, 2023; Hammel, 2017; Kumar, 2017; Pitceathly et al., 2021). It is essential to note that mitochondrial transplantation may be applied to treat other forms of myocardial pathology, including ischemia–reperfusion injury if pilot studies on pediatric patients show positive results.

Conclusions

The topic of mitochondrial transfer is subject to legal and ethical regulations. The US National Academies of Science, Engineering, and Medicine recently provided a conditional endorsement for mitochondrial transfer (Craven et al., 2018; Newson et al., 2016). The United Kingdom and other countries accepted their rules regulating the procedure of mitochondrial transplantation (Cohen et al., 2020). Thus, the legal aspects of this procedure make it open for the development of translational and clinical applications.

Cells and tissues with damaged and dysfunctional mitochondria have a poor prognosis for survival. The intercellular mitochondrial transfer method is based on mitochondria's natural capacity to be exchanged between cells (Lee et al., 2022; Valenti et al., 2021). For this reason, an intracellular mitochondrial transfer can be considered a treatment method for cell injury and tissue revitalization (Liu et al., 2021).

It should be noted that most of the current studies concerning the transfer of mitochondria in the cardiovascular system are performed on rodent animal models (mice, rats, and rabbits) and pigs. To what extent these approaches and results could be directly extrapolated on humans is unclear. However, the successful and encouraging results of mitochondrial transfer in pediatric clinical trials (Emani et al., 2018; Emani & McCully, 2018; Guariento, Blitzer, et al., 2020; Guariento, Doulamis et al., 2020; Guariento et al., 2021) suggest the applicability of mitochondrial transfer in a human clinical setting. In addition, the positive outcomes of similar trials in other tissue and organs, including the nervous system, liver, and respiratory system, indicate that translational applications of this approach are expected to be developed in the near future. The discussion of these experiments is presented in the current literature (McCully et al., 2022; Garone and Viscomi, 2018; Hirano et al., 2018; Almannai et al., 2020; Moskowitzova et al., 2020; Orfany et al., 2020; Chen et al., 2021; Pitceathly et al., 2021; Xie et al., 2021; Fairley et al., 2022; Huang et al., 2022; Liu et al., 2022; McCully et al., 2022; Sendra et al., 2021; Zhou et al., 2022; Caicedo & Singh, 2023).

The applicability of mitochondrial transplantation as a technique for the treatment of mitochondrial deficiencies and dysfunction in various tissues such as the heart, liver, lungs, and brain has been confirmed in animal studies. Thus, the uptake and functional integration of the transplanted mitochondria into recipient cells appears to occur in all studied cell types, and efficacy and safety of this method have been demonstrated in cell culture, isolated perfused organs, in vivo extensive animal studies and first-human clinical study (McCully et al., 2022).

A few challenges, methodological problems, and considerations still need to be addressed to apply the technique of intercellular mitochondrial transplantation more effectively and reliably (Hosseinian et al., 2022; Yamada et al., 2020). For example, it remains unclear what percentage of transplanted mitochondria survive and becomes internalized by receiving cells and how rapidly these organelles become internalized. A considerable share of the transplanted mitochondria may die because of the action of calcium ions in the blood plasma. In these cases, the studies should be designed to increase the efficacy and rate of survival of transplanted mitochondria. Another critical question is to determine to what degree and for how long these mitochondria remain functionally active and are involved in oxidative phosphorylation or if they mainly act as a factor in improving the trophic environment in the cells. It would also be essential to examine whether isolated mitochondria could be stored or cryopreserved before their transplantation and how to improve the effectiveness of their cellular uptake and internalization into recipient cells. However, these questions do not undermine that the remarkable functional improvement of the myocardium and blood vessels occurs following the transplantation of mitochondria.

Together, these results suggest that mitochondrial transfer into the cardiac and vascular cells is a potential therapeutic strategy for treating different cardiovascular diseases.

References

Abdullah, M., & Kaushal, S. (2021). Commentary: Energizing with ATP may be critical in ischemic pediatric hearts. *The Journal of Thoracic and Cardiovascular Surgery, 162*, 1005–1006.

Ali Pour, P., Hosseinian, S., & Kheradvar, A. (2021). Mitochondrial transplantation in cardiomyocytes: Foundation, methods, and outcomes. *American Journal of Physiology: Cell Physiology, 321*, C489–C503.

Ali Pour, P., Kenney, M. C., & Kheradvar, A. (2020). Bioenergetics consequences of mitochondrial transplantation in cardiomyocytes. *Journal of American Heart Association, 9*, e014501.

Almannai, M., El-Hattab, A. W., Ali, M., Soler-Alfonso, C., & Scaglia, F. (2020). Clinical trials in mitochondrial disorders, an update. *Molecular Genetics and Metabolism, 131*, 1–13.

Alston, C. L., Rocha, M. C., Lax, N. Z., et al. (2017). The genetics and pathology of mitochondrial disease. *The Journal of Pathology, 241*, 236–250.

Ballinger, S. W. (2005). Mitochondrial dysfunction in cardiovascular disease. *Free Radical Biology and Medicine, 38*, 1278–1295.

Barron, D. J., & Coles, J. G. (2021). Commentary: Autologous mitochondrial transplantation: An exciting innovation that needs evidence of efficacy. *The Journal of Thoracic and Cardiovascular Surgery, 162*, 1004–1005.

Blitzer, D., Guariento, A., Doulamis, I. P., et al. (2020). Delayed transplantation of autologous mitochondria for cardioprotection in a porcine model. *The Annals of Thoracic Surgery, 109*, 711–719.

Bonora, M., Wieckowski, M. R., Sinclair, D. A., et al. (2019). Targeting mitochondria for cardiovascular disorders: Therapeutic potential and obstacles. *Nature Reviews Cardiology, 16*, 33–55.

Boovarahan, S. R., & Kurian, G. A. (2018). Mitochondrial dysfunction: A key player in the pathogenesis of cardiovascular diseases linked to air pollution. *Reviews on Environmental Health, 33*, 111–122.

Bray, A. W., & Ballinger, S. W. (2017). Mitochondrial DNA mutations and cardiovascular disease. *Current Opinion in Cardiology, 32*, 267–274.

Buratto, E., & Konstantinov, I. E. (2021). Commentary: Mitochondrial transplantation for ischemia-reperfusion injury-a new revolution or tilting at windmills? *The Journal of Thoracic and Cardiovascular Surgery, 162*, 1002–1003.

Caicedo, A., & Singh, K. K. (2023). Advancing mitochondria as a therapeutic agent. *Mitochondrion, 69*, 33–35.

Chen, J., Zhong, J., Wang, L. L., & Chen, Y. Y. (2021). Mitochondrial transfer in cardiovascular disease: From mechanisms to therapeutic implications. *Frontiers in Cardiovascular Medicine, 8*, 771298.

Chen, X., Zhang, Y., Xu, B., et al. (2016). The mitochondrial calcium uniporter is involved in mitochondrial calcium cycle dysfunction: Underlying mechanism of hypertension associated with mitochondrial tRNA(Ile) A4263G mutation. *The International Journal of Biochemistry & Cell Biology, 78*, 307–314.

Chen, Y. R., & Zweier, J. L. (2014). Cardiac mitochondria and reactive oxygen species generation. *Circulation Research, 114*, 524–537.

Cherry, A. D. (2019). Mitochondrial dysfunction in cardiac surgery. *Anesthesiology Clinics, 37*, 769–785.

Cherry, A. D. (2023). Mitochondrial transplantation in cardiac surgical patients: Optimism, caveats, and outstanding questions. *Current Opinion in Anaesthesiology, 36*, 5–10.

Cohen, I. G., Adashi, E. Y., Gerke, S., et al. (2020). The regulation of mitochondrial replacement techniques around the world. *Annual Review of Genomics and Human Genetics, 21*, 565–586.

Cowan, D. B., Yao, R., Thedsanamoorthy, J., et al. (2017). Transit and integration of extracellular mitochondria in human heart cells. *Scientific Reports, 7*, 17450.

Craven, L., Murphy, J., Turnbull, D. M., et al. (2018). Scientific and ethical issues in mitochondrial donation. *The New Bioethics, 24*, 57–73.

Davison, J. E., & Rahman, S. (2017). Recognition, investigation, and management of mitochondrial disease. *Archives of Disease in Childhood, 102*, 1082–1090.

Del Re, D. P., Amgalan, D., Linkermann, A., et al. (2019). Fundamental mechanisms of regulated cell death and implications for heart disease. *Physiological Reviews, 99*, 1765–1817.

Dey, S., Mazumder, D., Sidor, A., et al. (2018). Mitochondrial ROS drive sudden cardiac death and chronic proteome remodeling in heart failure. *Circulation Research, 123*, 356–371.

Dia, M., Gomez, L., Thibault, H., et al. (2020). Reduced reticulum-mitochondria Ca^{2+} transfer is an early and reversible trigger of mitochondrial dysfunctions in diabetic cardiomyopathy. *Basic Research in Cardiology, 115*, 74.

Doulamis, I. P., & McCully, J. D. (2021). Mitochondrial transplantation for ischemia-reperfusion injury. *Methods in Molecular Biology, 2277*, 15–37.

Doulamis, I. P., Guariento, A., Duignan, T., et al. (2020). Mitochondrial transplantation for myocardial protection in diabetic hearts. *European Journal of Cardio-Thoracic Surgery, 57*, 836–845.

El-Hattab, A. W., & Scaglia, F. (2016). Mitochondrial cardiomyopathies. *Front Cardiovasc Med, 3*, 25.

Emani, S. M., & McCully, J. D. (2018). Mitochondrial transplantation: Applications for pediatric patients with congenital heart disease. *Translational Pediatrics, 7*, 169–175.

Emani, S. M., Piekarski, B. L., Harrild, D., et al. (2017). Autologous mitochondrial transplantation for dysfunction after ischemia-reperfusion injury. *The Journal of Thoracic and Cardiovascular Surgery, 154*, 286–289.

Fairley, L. H., Grimm, A., & Eckert, A. (2022). Mitochondria transfer in brain injury and disease. *Cells, 11*, 3603.

Finsterer, J., & Kothari, S. (2014). Cardiac manifestations of primary mitochondrial disorders. *International Journal of Cardiology, 177*, 754–763.

Forte, M., Schirone, L., Ameri, P., et al. (2021). The role of mitochondrial dynamics in cardiovascular diseases. *British Journal of Pharmacology, 178*, 2060–2076.

Gäbelein, C. G., Feng, Q., Sarajlic, E., et al. (2022). Mitochondria transplantation between living cells. *PLoS Biology, 20*, e3001576.

Garone, C., & Viscomi, C. (2018). Towards a therapy for mitochondrial disease: An update. *Biochemical Society Transactions, 46*, 1247–1261.

Glanz, V. Y., Sobenin, I. A., Grechko, A. V., et al. (2020). The role of mitochondria in cardiovascular diseases related to atherosclerosis. *Frontiers in Bioscience, 12*, 102–112.

Gorman, G. S., Chinnery, P. F., DiMauro, S., et al. (2016). Mitochondrial diseases. *Nature Reviews Disease Primers, 20*, 16080.

Guariento, A., Doulamis, I.,P., Duignan, T., et al. (2020). Mitochondrial transplantation for myocardial protection in ex-situ–perfused hearts donated after circulatory death. *The Journal of Heart and Lung Transplantation, 39*, 1279–1288.

Guariento, A., Blitzer, D., Doulamis, I., et al. (2020). Preischemic autologous mitochondrial transplantation by intracoronary injection for myocardial protection. *The Journal of Thoracic and Cardiovascular Surgery, 160*, e15–e29.

Guariento, A., Piekarski, B. L., Doulamis, I. P., et al. (2021). Autologous mitochondrial transplantation for cardiogenic shock in pediatric patients following ischemia-reperfusion injury. *The Journal of Thoracic and Cardiovascular Surgery, 162*, 992–1001.

Gustafsson, A. B., & Gottlieb, R. A. (2008). Heart mitochondria: Gates of life and death. *Cardiovascular Research, 77*, 334–343.

Hamilton, D. J. (2013). Mechanisms of disease: Is mitochondrial function altered in heart failure? *Methodist Debakey Cardiovascular Journal, 9*, 44–48.

Hammel, J. M. (2017). Mitochondrial autotransplantation: A "shot" in the dark? *The Journal of Thoracic and Cardiovascular Surgery, 154*, 290.

Hayashida, K., Takegawa, R., Shoaib, M., et al. (2021). Mitochondrial transplantation therapy for ischemia-reperfusion injury: A systematic review of animal and human studies. *Journal of Translational Medicine, 19*, 214.

Hefti, E., & Blanco, J. G. (2018). Mitochondrial DNA heteroplasmy in cardiac tissue from individuals with and without coronary artery disease. *Mitochondrial DNA Part A, DNA Mapping, Sequencing, and Analysing, 29*, 587–593.

Hirano, M., Emmanuele, V., & Quinzii, C. M. (2018). Emerging therapies for mitochondrial diseases. *Essays in Biochemistry, 62*, 467–481.

Hosseinian, S., Ali Pour, P., & Kheradvar, A. (2022). Prospects of mitochondrial transplantation in clinical medicine: Aspirations and challenges. *Mitochondrion, 65*, 33–44.

Hsu, C. H., Roan, J. N., Fang, S. Y., et al. (2022). Transplantation of viable mitochondria improves right ventricular performance and pulmonary artery remodeling in rats with pulmonary arterial hypertension. *The Journal of Thoracic and Cardiovascular Surgery, 163*, e361–e373.

Hu, Q., Lu, J., Zhang, X., et al. (2022). Mitochondria transplantation/transfer between single cells. *Journal of Cerebral Blood Flow and Metabolism, 42*, 1748–1750.

Huang, T., Zhang, T., & Gao, J. (2022). Targeted mitochondrial delivery: A therapeutic new era for disease treatment. *Journal of Controlled Release, 343*, 89–106.

Ikeda, G., Santoso, M. R., Tada, Y., et al. (2021). Mitochondria-rich extracellular vesicles from autologous stem cell-derived cardiomyocytes restore energetics of ischemic myocardium. *Journal of the American College of Cardiology, 77*, 1073–1088.

Jang, Y. H., & Lim, K. I. (2018). Recent advances in mitochondria-targeted gene delivery. *Molecules, 23*, 2316.

Kattoor, A. J., Pothineni, N. V. K., Palagiri, D., & Mehta, J. L. (2017). Oxidative stress in atherosclerosis. *Current Atherosclerosis Reports, 19*, 42.

Kaza, A. K., Wamala, I., Friehs, I., et al. (2017). Myocardial rescue with autologous mitochondrial transplantation in a porcine model of ischemia/reperfusion. *The Journal of Thoracic and Cardiovascular Surgery, 153*, 934–943.

Kirichenko, T. V., Ryzhkova, A. I., Sinyov, V. V., et al. (2020). Impact of mitochondrial D.N.A. mutations on carotid intima-media thickness in the Novosibirsk region. *Life (Basel), 10*, 160.

Kumar, S. R. (2017). Mitochondrial transplantation: Another miracle of molecular medicine? *The Journal of Thoracic and Cardiovascular Surgery, 154*, 284–285.

Lee, S. E., Kang, Y. C., Kim, Y., et al. (2022). Preferred migration of mitochondria toward cells and tissues with mitochondrial damage. *International Journal of Molecular Sciences, 23*, 15734.

Lee, Y., & Gustafsson, A. B. (2009). Role of apoptosis in cardiovascular disease. *Apoptosis, 14*, 536–548.

Lenaz, G., & Genova, M. L. (2010). Structure and organization of mitochondrial respiratory complexes: A new understanding of an old subject. *Antioxidants and Redox Signaling, 12*, 961–1008.

Lim, G. B. (2020). Intracoronary mitochondrial transplantation. *Nature Reviews Cardiology, 17*, 131.

Lioncino, M., Monda, E., Caiazza, M., et al. (2022). Cardiovascular involvement in mtDNA disease: Diagnosis, management, and therapeutic options. *Heart Failure Clinics, 18*, 51–60.

Liu, D., Gao, Y., Liu, J., et al. (2021). Intercellular mitochondrial transfer as a means of tissue revitalization. *Signal Transduction and Targeted Therapy, 6*, 65.

Liu, Z., Sun, Y., Qi, Z., et al. (2022). Mitochondrial transfer/transplantation: An emerging therapeutic approach for multiple diseases. *Cell & Bioscience, 12*, 66.

Louwagie, E. J., Larsen, T. D., Wachal, A. L., et al. (2021). Mitochondrial transfer improves cardiomyocyte bioenergetics and viability in male rats exposed to pregestational diabetes. *International Journal of Molecular Sciences, 22*, 2382.

Luan, Y., Luan, Y., Feng, Q., et al. (2021). Emerging role of mitophagy in the heart: Therapeutic potentials to modulate mitophagy in cardiac diseases. *Oxidative Medicine and Cellular Longevity*, 3259963, 2021.

Madamanchi, N. R., & Runge, M. S. (2007). Mitochondrial dysfunction in atherosclerosis. *Circulation Research, 100*, 460–473.

Manshori, M., Kazemnejad, S., & Golshahi, H. (2022). Mitochondrial transfer from menstrual blood stromal/stem cells promotes survival of cardiomyocytes following myocardial infarction. *Avicenna Journal of Medical Biotechnology, 14*, 321–322.

Marín-García, J., & Goldenthal, M. J. (2002b). The mitochondrial organelle and the heart. *Revista Espanola de Cardiologia, 55*, 1293–1310.

Marín-García, J., & Goldenthal, M. J. (2002a). Understanding the impact of mitochondrial defects in cardiovascular disease: A review. *Journal of Cardiac Failure, 8*, 347–361.

Marín-García, J., Goldenthal, M. J., & Filiano, J. (2002). Cardiomyopathy associated with neurologic disorders and mitochondrial phenotype. *Journal of Child Neurology, 17*, 759–765.

Masuzawa, A., Black, K. M., Pacak, C. A., et al. (2013). Transplantation of autologously derived mitochondria protects the heart from ischemia-reperfusion injury. *American Journal of Physiology—Heart and Circulatory Physiology*, H966–H982.

McCully, J. D., Del Nido, P. J., & Emani, S. M. (2022). Mitochondrial transplantation for organ rescue. *Mitochondrion, 64*, 27–33.

McCully, J. D., Cowan, D. B., Pacak, C. A., et al. (2009). Injection of isolated mitochondria during early reperfusion for cardioprotection. *American Journal of Physiology—Heart and Circulatory Physiology, 296*, H94–H105.

Meyers, D. E., Basha, H. I., & Koenig, M. K. (2013). Mitochondrial cardiomyopathy: Pathophysiology, diagnosis, and management. *Texas Heart Institute Journal, 40*, 385–394.

Mitrofanov, K. Y., Zhelankin, A. V., Shiganova, G. M., et al. (2016). Analysis of mitochondrial DNA heteroplasmic mutations A1555G, C3256T, T3336C, C5178A;, G12315A, G13513A, G14459A, G14846A and G15059A in CHD patients with the history of myocardial infarction. *Experimental and Molecular Pathology, 100*, 87–91.

Morales, P. E., Arias-Durán, C., Ávalos-Guajardo, Y., et al. (2020). Emerging role of mitophagy in cardiovascular physiology and pathology. *Molecular Aspects of Medicine, 71*, 100822.

Moskowitzova, K., Orfany, A., Liu, K., et al. (2020). Mitochondrial transplantation enhances murine lung viability and recovery after ischemia-reperfusion injury. *American Journal of Physiology—Lung Cellular and Molecular Physiology, 318*, L78–L88.

Moskowitzova, K., Shin, B., Liu, K., et al. (2019). Mitochondrial transplantation prolongs cold ischemia time in murine heart transplantation. *The Journal of Heart and Lung Transplantation, 38*, 92–99.

Murray, L. M. A., & Krasnodembskaya, A. D. (2019). Concise review: Intercellular communication via organelle transfer in the biology and therapeutic applications of stem cells. *Stem Cells, 37*, 14–25.

Newson, A. J., Wilkinson, S., & Wrigley, A. (2016). Ethical and legal issues in mitochondrial transfer. *EMBO Molecular Medicine, 8*, 589–591.

Orfany, A., Arriola, C. G., Doulamis, I. P., et al. (2020). Mitochondrial transplantation ameliorates acute limb ischemia. *Journal of Vascular Surgery, 71*, 1014–1026.

Pitceathly, R. D. S., Keshavan, N., Rahman, J., & Rahman, S. (2021). Moving towards clinical trials for mitochondrial diseases. *Journal of Inherited Metabolic Disease, 44*, 22–41.

Pohjoismäki, J. L., Goffart, S., Taylor, R. W., et al. (2010). Developmental and pathological changes in the human cardiac muscle mitochondrial D.N.A. organization, replication and copy number. *PLoS One, 5*, e10426.

Poznyak, A. V., Ivanova, E. A., Sobenin, I. A., et al. (2020). The role of mitochondria in cardiovascular diseases. *Biology, 9*(6), 137. Jun 25.

Quintero, M., Colombo, S. L., Godfrey, A., & Moncada, S. (2006). Mitochondria as signaling organelles in the vascular endothelium. *Proceedings of the National Academy of Sciences of the United States of America, 103*, 5379–5384.

Ramirez-Barbieri, G., Moskowitzova, K., Shin, B., et al. (2019). Alloreactivity and allorecognition of syngeneic and allogeneic mitochondria. *Mitochondrion, 46*, 103–115.

Ranjbarvaziri, S., Kooiker, K. B., Ellenberger, M., et al. (2021). Altered cardiac energetics and mitochondrial dysfunction in hypertrophic cardiomyopathy. *Circulation, 144*, 1714–1731.

Reuter, S. (2022). Mitochondria—More than just batteries for cellular function. *Acta Physiology (Oxford), 235*, e13852.

Rocha, M., Diaz-Morales, N., Rovira-Llopis, S., et al. (2016). Mitochondrial dysfunction and endoplasmic reticulum stress in diabetes. *Current Pharmaceutical Design, 22*, 2640–2649.

Rosca, M. G., & Hoppel, C. L. (2013). Mitochondrial dysfunction in heart failure. *Heart Failure Reviews, 18*, 607–622.

Rossmann, M. P., Dubois, S. M., Agarwal, S., et al. (2021). Mitochondrial function in development and disease. *Dis. Model Mech., 14*, dmm048912.

Sazonova, M. A., Ryzhkova, A. I., Sinyov, V. V., et al. (2018). Mitochondrial genome mutations associated with myocardial infarction. *Disease Markers*, 9749457, 2018.

Sazonova, M. A., Ryzhkova, A. I., Sinyov, V. V., et al. (2021). Mutations of mtDNA in some vascular and metabolic diseases. *Current Pharmaceutical Design, 27*, 177–184.

Sendra, L., García-Mares, A., Herrero, M. J., & Aliño, S. F. (2021). Mitochondrial DNA replacement techniques to prevent human mitochondrial diseases. *International Journal of Molecular Sciences, 22*, 551.

Shanmughapriya, S., Langford, D., & Natarajaseenivasan, K. (2020). Inter and intracellular mitochondrial trafficking in health and disease. *Ageing Research Reviews, 62*, 101128.

Shin, B., Cowan, D. B., Emani, S. M., et al. (2017). Mitochondrial transplantation in myocardial ischemia and reperfusion injury. *Advances in Experimental Medicine and Biology, 982*, 595–619.

Shin, B., Saeed, M. Y., Esch, J. J., et al. (2019). A novel biological strategy for myocardial protection by intracoronary delivery of mitochondria: Safety and efficacy. *JACC: Basic to Translational Science, 4*, 871–888.

Siasos, G., Tsigkou, V., Kosmopoulos, M., et al. (2018). Mitochondria and cardiovascular diseases -from pathophysiology to treatment. *Annals of Translational Medicine, 6*, 256.

Sobenin, I. A., Sazonova, M. A., Postnov, A. Y., et al. (2013). Changes of mitochondria in atherosclerosis: Possible determinant in the pathogenesis of the disease. *Atherosclerosis, 227*, 283–288.

Sobenin, I. A., Zhelankin, A. V., Khasanova, Z. B., et al. (2019). Heteroplasmic variants of mitochondrial DNA in atherosclerotic lesions of human aortic intima. *Biomolecules, 9*, 455.

Sorrentino, V., Menzies, K. J., & Auwerx, J. (2018). Repairing mitochondrial dysfunction in disease. *Annual Review of Pharmacology and Toxicology, 58*, 353–389.

Stewart, J. B., & Chinnery, P. F. (2021). Extreme heterogeneity of human mitochondrial DNA. from organelles to populations. *Nature Reviews Genetics, 22*, 106–118.

Tian, R., Colucci, W.,S., Arany, Z., et al. (2019). Unlocking the secrets of mitochondria in the cardiovascular system: Path to a cure in heart failure—A report from the 2018 National Heart, Lung, and Blood Institute Workshop. *Circulation, 140*, 1205–1216.

Tilokani, L., Nagashima, S., Paupe, V., & Prudent, J. (2018). Mitochondrial dynamics: Overview of molecular mechanisms. *Essays in Biochemistry, 62*, 341–360.

Vásquez-Trincado, C., García-Carvajal, I., Pennanen, C., et al. (2016). Mitochondrial dynamics, mitophagy and cardiovascular disease. *Journal of Physiology, 594*, 509–525.

Valenti, D., Vacca, R. A., Moro, L., & Atlante, A. (2021). Mitochondria can cross cell boundaries: An overview of the biological relevance, pathophysiological implications and therapeutic perspectives of intercellular mitochondrial transfer. *International Journal of Molecular Sciences, 22*, 8312.

Vasanthan, V., Hassanabad, A. F., & Fedak, P. W. M. (2022). Commentary: Transplanting the powerhouse of the cell to enhance cardiopulmonary repair. *The Journal of Thoracic and Cardiovascular Surgery, 163*, e376–e377.

Ventura-Clapier, R., Garnier, A., Veksler, V., Joubert, F., et al. (2011). Bioenergetics of the failing heart. *Biochimica et Biophysica Acta, 1813*, 1360–1372.

Weiss, R. G., Gerstenblith, G., & Bottomley, P. A. (2005). ATP flux through creatine kinase in the normal, stressed, and failing human heart. *Proceedings of the National Academy of Sciences of the United States of America, 102*, 808–813.

Weixler, V., Lapusca, R., Grangl, G., et al. (2021). Autogenous mitochondria transplantation for treatment of right heart failure. *The Journal of Thoracic and Cardiovascular Surgery, 162*, e111–e121.

Wu, N. N., Zhang, Y., & Ren, J. (2019). Mitophagy, mitochondrial dynamics, and homeostasis in cardiovascular aging. *Oxidative Medicine and Cellular Longevity*, 9825061, 2019.

Xie, Q., Zeng, J., Zheng, Y., et al. (2021). Mitochondrial transplantation attenuates cerebral ischemia-reperfusion injury: Possible involvement of mitochondrial component separation. *Oxidative Medicine and Cellular Longevity*, 1006636, 2021.

Yamada, Y., Ito, M., Arai, M., et al. (2020). Challenges in promoting mitochondrial transplantation therapy. *International Journal of Molecular Sciences, 21*, 6365.

Yang, J., Chen, S., Duan, F., et al. (2022). Mitochondrial cardiomyopathy: Molecular epidemiology, diagnosis, models, and therapeutic management. *Cells, 11*, 3511.

Yu, E., Mercer, J., & Bennett, M. (2012). Mitochondria in vascular disease. *Cardiovascular Research, 95*, 173–182.

Zhang, Y., Murugesan, P., Huang, K., & Cai, H. (2020). NADPH oxidases and oxidase crosstalk in cardiovascular diseases: Novel therapeutic targets. *Nature Reviews Cardiology, 17*, 170–194.

Zhou, B., & Tian, R. (2018). Mitochondrial dysfunction in pathophysiology of heart failure. *Journal of Clinical Investigation, 128*, 3716–3726.

Zhou, M., Yu, Y., Luo, X., et al. (2021). Myocardial ischemia-reperfusion injury: Therapeutics from a mitochondria-centric perspective. *Cardiology, 146*, 781–792.

Zhou, M., Yu, Y., Luo, Y., et al. (2022). Mitochondrial transplantation: A unique treatment strategy. *Journal of Cardiovascular Pharmacology, 79*, 759–768.

Further reading

Headley, C., & Tsao, P. (2023). Building the case for mitochondrial transplantation as an anti-aging cardiovascular therapy. *Frontiers in Cardiovascular Medicine, 10*, 1141124.

Honda, H. M., Korge, P., & Weiss, J. N. (2005). Mitochondria and ischemia/reperfusion injury. *Annals of the New York Academy of Sciences, 1047*, 248–258.

McCully, J. D., Del Nido, P. J., & Emani, S. M. (2023). Mitochondrial transplantation: The advance to therapeutic application and molecular modulation. *Frontiers in Cardiovascular Medicine, 10*, 1268814.

Nadalutti, C. A., Ayala-Peña, S., & Santos, J. H. (2022). Mitochondrial DNA damage as driver of cellular outcomes. *American Journal of Physiology: Cell Physiology, 322*, C136–C150.

Nan, J., Zhu, W., Rahman, M. S., et al. (2017). Molecular regulation of mitochondrial dynamics in cardiac disease. *Biochimica et Biophysica Acta, Molecular Cell Research, 1864*, 1260–1273.

Sun, M., Jiang, W., Mu, N., Zhang, Z, Yu, L, & Ma, H. (2023). Mitochondrial transplantation as a novel therapeutic strategy for cardiovascular diseases. *Journal of Translational Medicine, 21*, 347.

SECTION VI

Neurological diseases and mitochondrial transplantation

CHAPTER

17

Neurological diseases and pathophysiology

Rıza Sonkaya[1] *and Elvin Malikov*[2]

[1]Department of Neurology, Gülhane School of Medicine, University of Health Sciences, Ankara, Turkey; [2]Department of Neurology, Ganja Military Hospital, Ganja, Azerbaijan

Introduction

Neurological disorders are diseases that affect the central and peripheral nervous systems. According to estimates, more than one billion people worldwide suffer from neurological diseases (Diego-Cordero et al., 2022), and one in every three people is affected by a neurological disease at some point in their lives (Olesen et al., 2012). However, neurological diseases are the most disabling diseases in the world and the second leading cause of death, accounting for 16.8% of all deaths (Feigin et al., 2020).

Many neurological disorders become more common as people age due to the fact that brain tissue is more susceptible to aging than other organ functions (Choi et al., 2018). Besides this, it is known that aging leads to progressive and detrimental changes in the brain (Liang et al., 2021). However, old age is accepted as the primary risk factor for many neurological disorders (Hou et al., 2019). Among these neurological disorders, neurodegenerative diseases, characterized by the progressive degeneration and death of central nervous system cells, have drawn more attention because of their irreversible nature, lack of effective treatments, and accompanied social and economic burdens. Currently, there are still no preventative or curative treatments for neurodegenerative diseases such as Alzheimer's disease (AD), Parkinson's disease (PD), and amyotrophic lateral sclerosis (ALS), despite intensive research and ongoing clinical trials (Katsuno et al., 2018).

Neurodegenerative diseases are a large group of debilitating nervous system disorders. These diseases manifest in diverse forms and arise from multiple underlying mechanisms; they are the result of distinct causes-induced neuronal population degeneration that occurs at different times (Cabral-Costa & Kowaltowski, 2020; Gao et al., 2017). However, many of them also share common features such as mitochondrial dysfunction, oxidative damage, and inflammation that are implicated in the gradual loss of neuronal function and cell death

Mitochondrial Transplantation and Transfer
https://doi.org/10.1016/B978-0-443-18858-9.00017-2

(Kodavati et al., 2020; Rey et al., 2022). It is postulated that various neurodegenerative diseases are predominantly caused by dysfunctional mitochondria, given the crucial role mitochondria play in the generation of reactive oxygen species (ROS) (Kumari et al., 2021). As a matter of fact, in a plethora of studies, it has been reported that impaired mitochondrial function is a central factor in the pathophysiology of neurodegenerative diseases (Briston & Hicks, 2018; Johri & Beal, 2012; Rey et al., 2022; Wu et al., 2019).

Mitochondria are the most distinct and important cell organelles in eukaryotes (Kumari et al., 2021). They make most of the adenosine triphosphate (ATP) through the electron transport chain (ETC), which is needed to set up the right electrochemical gradients and make sure that synaptic transmission works well through oxidative phosphorylation (Liang et al., 2021). Mitochondria, in addition to ATP, play an essential role in numerous codependent and interrelated functional processes in neurons, including biogenesis and cell death (Tapias, 2019). Notably, in neurons, mitochondria play additional significant roles in calcium homeostasis, control of membrane excitability and neurotransmission, as well as plasticity (Rose et al., 2017). Therefore, neurons are highly susceptible to mitochondrial defects. Besides this, neurons are dependent on mitochondrial oxidative phosphorylation to fulfill their energy demands (Zhao et al., 2019). Damage to mitochondria that lasts for a long time makes energy metabolism not work right. This results in less ATP production, more ROS, and less calcium buffering, all of which lead to neuronal loss that is a feature of both short-term and long-term degenerative neurological disorders (Johnson et al., 2021). Although, many neurological disorders are linked to problems with mitochondria, these problems include but are not limited to bioenergetics, dynamics, bioenergetics, and mitophagy that are not controlled properly (Bader & Winklhofer, 2020). Also, in many studies, it has been shown that these mitochondrial dysfunctions are generally observed prior to the onset of neurological symptoms (Collier et al., 2023).

In sum, various dysfunctions of the mitochondria, including membrane leakage and electrolyte imbalances, activation of the proapoptotic pathway, and mitophagy, have been identified in disorders of the central nervous system. These dysfunctions are among the mechanisms that contribute to the development of neurodegenerative diseases and ischemic stroke (Cabral-Costa & Kowaltowski, 2020). More recently, evidence has demonstrated that impaired mitochondrial dynamics have been associated with degenerative neurological diseases (Johnson et al., 2021; Rey et al., 2022; Wang et al., 2019). On the other hand, mitochondrial dysfunction is still a major hallmark of various neurodegenerative diseases such as AD, PD, and ALS, as well as ischemic stroke. However, the underlying mechanisms by which mitochondrial dysfunctions occur in such diseases remain elusive. In this chapter, it is aimed to provide a brief overview of recent advances and discuss the role of mitochondrial dysfunctions in the development of neurodegenerative diseases such as AD, PD, and ALS, as well as ischemic stroke, which have been extensively addressed in multiple studies.

Alzheimer's disease

AD is a progressive, chronic, irreversible neurodegenerative disease characterized by behavioral changes and loss of functional independence in addition to cognitive and functional deficits (Zvěřová, 2019). The disease was first described in 1906 by Alois Alzheimer, who found abnormal plaques and fibrillary pathology in the brain of his patient, Auguste Deter, who died due to an atypical mental illness (Peng et al., 2020).

AD is the most prevalent cause of dementia worldwide. It typically starts out as a progressive memory decline and progresses to other cognitive impairments like visuospatial abnormalities, executive issues, and language disturbances (Huang et al., 2020). Clinicopathological research provides evidence for the existence of a protracted preclinical phase of AD (Tarawneh & Holtzman, 2012). It is estimated that the initial phase of AD pathology commences around 10–15 years before the manifestation of clinical symptoms (Tarawneh & Holtzman, 2012).

AD is characterized by two pathological phenomena: aggregations of the amyloid beta peptide (Aβ) and phosphorylated tau protein tangles in the brain (Bell et al., 2021; Yiannopoulou & Papageorgiou, 2020). The exact mechanisms of disease pathology remain unclear despite comprehensive research to understand its pathophysiology (Misrani et al., 2021). There is still no clear answer to how AD works at the molecular level, but some ideas include beta amyloid and amyloid plaques, hyperphosphorylation of the tau protein and neurofibrillary tangles, oxidative stress, inflammation, and mitochondrial dysfunction (Peng et al., 2020; Reiss, 2022).

Many metabolic pathways have been found to be abnormal in AD, affecting both peripheral and nervous system cells (Bell et al., 2021). However, increasing evidence suggests that deficiencies in mitochondrial function, particularly how they control oxidative phosphorylation, may play a key role in the development and progression of AD (Moreira et al., 2010). When ETC does not work right, mitochondrial membrane potential and ATP production go down. This makes it harder for mitochondria to meet the active needs of the cell in AD (Hauptmann et al., 2009). Multiple studies and cell types have provided evidence that Complex IV's activity is diminished in AD; however, the functions of the other ETC proteins are also compromised (Bell et al., 2021). As the disease gets worse, AD mitochondria are more likely to make ROS, which raises the level of oxidative stress in cells and mitochondria and further damages ETC functions (Huang et al., 2020). Mitophagy is elevated in both AD brain and peripheral cells at the onset of the disease and deteriorates further as it advances (Tarawneh & Holtzman, 2012). The overproduction of ROS (Fig. 17.1) and a decrease in neuronal ATP levels, which suggest that mitochondria may not be able to sustain cellular energy, are indicative of mitochondrial dysfunction in AD (Misrani et al., 2021).

In many studies based on the observation of AD brains, it has been suggested that mitochondrial dysfunction is the initial trigger for AD (Monzio Compagnoni et al., 2020; Moreira et al., 2010; Swerdlow, 2018). In addition, animal studies conducted on AD brains also corroborate the view that the primary cause of AD is related to mitochondrial dysfunction (Fu et al., 2009; Hauptmann et al., 2009). In one study conducted by Haupmann et al. (2009), it was reported that mitochondrial dysfunction was an early event in mice bearing the human Swedish and London mutations, and these mitochondrial defects accumulate with age (Hauptmann et al., 2009). Additionally, it was shown in a different study conducted by Fu et al. (2009) that in YAC and C57B6/SJL transgenic mice, mitochondrial defects seem to be important characteristics throughout the maturation of AD-like disease (Fu et al., 2009). However, the precise role of mitochondrial dysfunction in the development of AD remains uncertain (Rey et al., 2022). It is possible that tau accumulation is the primary driver of

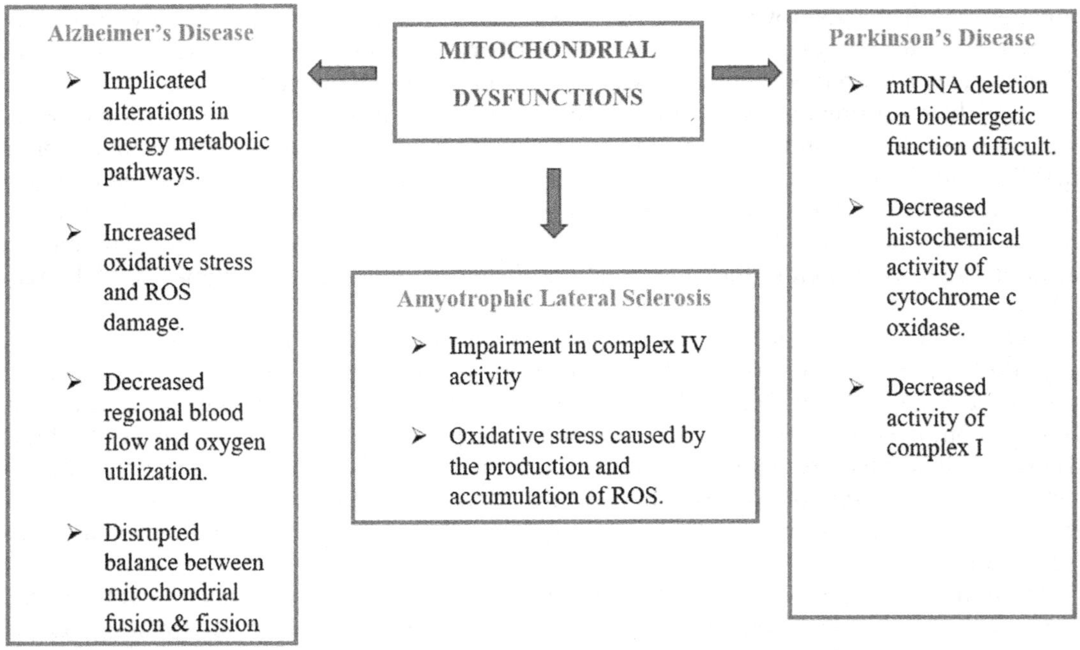

FIGURE 17.1 Mitochondrial dysfunctions related to neurodegenerative diseases.

amyloidogenic pathology, with mitochondrial dysfunction serving as a secondary consequence. Due to the fact that mitochondrial dysfunction is the primary factor in pathogenesis, further research is required to determine how mitochondria become dysfunctional.

In recent years, studies have been carried out on different strategies and different therapeutic agents in mitochondria-based research in AD. Baumanns et al. (2023) studied Aβ proteotoxicity on transgenic strain *Caenorhabditis elegans* which expresses human Aβ1-42 in muscle cells. They measured the Aβ-aggregation levels, ATP production, mitochondrial membrane potential, and oxygen consumption to analyze the effects of caprylic acid. They showed improvement in all these parameters. Moreover, they reported that CA reduced Aβ proteotoxicity by giving energy via FAO and CA might be used as a fuel for alternative energy sources (Bhatia et al., 2022).

An essential function in the elimination of harmful protein aggregates in AD is played by the demanding process known as microglial phagocytosis. Current research suggests that this crucial protective microglial function is disrupted when energy production shifts from mitochondrial respiration to glycolysis, which may provide treatment options for AD (Bhatia et al., 2022; Sonkaya et al., 2021). It has been reported that the translocator protein and hexokinase-2, a component of its mitochondrial complex, are essential for the respiratory-glycolytic metabolism and phagocytosis of microglia (Sonkaya et al., 2021).

It was considered that whether and how apolipoprotein E biology is influenced by mitochondrial biology. Gabrielli et al. (2023) reported the apolipoprotein E expression in human SH-SY5Y neuronal cells with various types of mitochondrial dysfunction, as well as mitochondrial dysfunction brought on by toxins. In a neural cell model, primary mitochondrial failure modifies transcription factors and stress signaling pathways to either directly or indirectly increase apolipoprotein E expression (Berg et al., 2014).

Increasing evidence points to several mitochondrial functions that are affected in AD (Kodavati et al., 2020; Misrani et al., 2021). Also, it is becoming apparent that mitochondria are a potential target for treatment in early-stage AD (Lee & Yankee, 2022). Further research is needed to elucidate the mechanisms underlying AD pathology and gain a robust understanding of mitochondrial dysfunction to facilitate the development of optimal therapeutic interventions.

Parkinson's disease

PD is the second most common chronic and neurodegenerative age-related disorder after AD. The disease is characterized by progressive loss of dopaminergic neurons, primarily in the substantia nigra pars compacta (Sonkaya et al., 2021). The presence of Lewy bodies and Lewy neuritis can also be shown to be the reason for PD (Berg et al., 2014).

PD was first described by James Parkinson in 1817. The disease is distinguished by motor symptoms such as bradykinesia, rigidity, resting tremor, postural instability, and gait problems, as well as nonmotor symptoms such as sleep problems, anxiety, and cognitive impairments (Sonkaya et al., 2021). Despite impacting a global population of over six million individuals, the etiology of PD is still unknown (Lee & Yankee, 2022; Sancho-Bielsa, 2022). However, findings obtained from postmortem studies have shown that mitochondrial dysfunction is suspected to be a trigger in the pathology of PD (Norat et al., 2020). Furthermore, research conducted on in vitro cell cultures derived from human patients diagnosed with PD, as well as autopsy tissues and other tissue samples, has provided further support for the hypothesis that mitochondrial dysfunction may initiate PD (Henchcliffe & Beal, 2008).

Mitochondrial dysfunction in PD is mainly characterized by the generation of ROS, a decrease in mitochondrial complex I enzyme activity (Fig. 17.1), cytochrome *c* release, ATP depletion, and caspase-3 activation (Moon & Paek, 2015). There is substantial evidence to suggest that mitochondrial dysfunction is a fundamental factor in the development of PD (Lee & Yankee, 2022; Liang et al., 2021). The first link between mitochondrial dysfunction and PD was found when MPTP, which stands for 1-methyl-4-phenyl-1,2,3,6-tetrahydropyridine, was shown to make drug users with Parkinson's-like symptoms (Chen et al., 2019). The study was expanded over the years, and animal models were quickly developed. Because of this, many studies show that mitochondrial dysfunction plays a big part in the development of PD. This was first seen as a common feature or even a cause of PD after MPTP-induced Parkinsonism was found (Buneeva et al., 2020).

The notion that respiratory chain complexes exhibited impaired functionality in brain sections obtained from deceased patients with Parkinson's disease (Sancho-Bielsa, 2022) provided further support for the notion that mitochondrial dysfunction contributes to the onset of the disease, shortly after the "frozen addicts" were described. ROS production is associated with disturbances in respiratory chain complexes, suggesting that oxidative stress may serve as a pathogenic mechanism in Parkinson's disease where mitochondrial dysfunction is involved. Further evidence has emerged, however, linking oxidative stress to dopamine metabolism which emphasizes the significance of ROS (Henchcliffe & Beal, 2008).

As it is well known, the primary hallmark of PD is the loss of dopaminergic neurons in the substantia nigra. Mitochondrial dysfunction has also been linked to substantia nigra neuronal loss (Compagnoni et al., 2020). In one study, it was revealed that the mitochondrial complex I of the ETC was not functioning properly in the substantia nigra of Parkinson's disease patients (Schapira, 2008). In another study, reduced activities of complexes I, II, and IV of the ETC were determined in postmortem substantia nigra homogenates of PD patients (Grünewald et al., 2019). Likewise, in a postmortem study, varying degrees of complex I and complex II deficiency have been found in individual substantia nigra neurons from patients with PD (Grünewald et al., 2016). Nevertheless, ETC dysfunction negatively impacts neuronal viability in PD by not only decreasing ATP production but also inducing apoptosis or impaired mitophagy (Lee & Yankee, 2022). Moreover, no discernible impairment in mitochondrial activity has been observed in any other cerebral region affected by Parkinson's disease, such as the substantia innominata, caudate putamen, globus pallidus, tegmentum, cortex, or cerebellum (Schapira, 2008). In Parkinson's disease, the susceptibility of dopaminergic neurons is now known to be significantly influenced by mitochondrial dysfunction. A number of critical genes, such as PARK7 (which encodes DJ-1), α-synuclein, parkin, PINK1, and LRRK2, are afflicted with pathogenic mutations that induce PD and disrupt the dynamics and functionality of mitochondria (Wang et al., 2019).

Lai et al. (2023) investigated that the protective impact of hispidin in a model of PD induced by 1-methyl-4-phenylpyridinium (MPP+). MPP+ resulted in ROS production that diminished cell viability and dopamine content. The inhibition of activity in the mitochondrial respiratory chain complex I, the collapse of potential in the mitochondrial transmembrane, and the liberation of mitochondrial cytochrome *c* were signs of mitochondrial dysfunction in cells exposed to MPP+ (Shi et al., 2010).

Transient receptor potential channels have recently been demonstrated to be crucial in the regulation of mitochondrial and neuronal functioning in PD. ROS production, cell viability, and mitochondrial membrane potential were all decreased in MPP+-treated SH-SY5Y cells. In comparison to MPP+-treated SH-SY5Y cells, cotreatment with 2-aminoethoxydiphenyl borate (transient receptor potential channel blocker) increased cell survival, decreased intracellular and mitochondrial ROS, and improved mitochondrial membrane potential (Feldman, 2022).

For people with PD, exercise has been suggested as a beneficial nonpharmacological therapy strategy. Exercise elevates irisin, a newly discovered myokine that is essential to energy metabolism. Exogenous irisin therapy may enhance motor function and slow the death of dopaminergic neurons in PD models. A decrease in oxidative stress, an increase in mitochondrial complex I activity and content, an increase in mitochondrial biogenesis, and corrected mitochondrial shape are all signs that irisin can reduce cell death in PD models (Abe et al., 2017).

Although comprehensive research to understand main cause of the PD, the mechanisms underlying pathophysiology of disease are still unclear. However, direct experimental validation is necessary to ascertain whether these scenarios are applicable to mitochondrial dysfunction in Parkinson's disease. The role and mechanisms of mitochondrial defects in PD need to be better understood in therapeutic interventions.

Amyotrophic lateral sclerosis

ALS primarily impacts the upper and lower motor neurons located in the cortex, brainstem, and spinal cord (Candelise, 2022). It is a progressive neurodegenerative disorder that impacts motor neurons. Progressive muscular atrophy, palsy of the medulla oblongata, fibrillation of the muscle bundles, progressive limb weakness, and pyramidal signs are frequently observed in patients with ALS (Chen, 2022; Zhao, 2022). There are two main types of ALS, namely familial and sporadic, with approximately 10%–20% of the cases being familial, whereas the majority of them are sporadic (Shi et al., 2010). To date, more than 50 genes have been reported to be associated with the development of ALS (Chen, 2022). Among them, superoxide dismutase 1 (SOD1), TAR DNA binding protein (TARDBP; TDP-43), fused in sarcoma (FUS), and C9orf72 are the most common genetic variants related to the onset of ALS (Feldman, 2022). Like other neurodegenerative diseases, there is no cure, and therapies are currently available to halt the disease progression. Two medications, edaravone and riluzole, have been granted marketing approval (Abe et al., 2017); nevertheless, their effects are limited to postponing functional decline and extending survival time by a few months (Zhao, 2022).

ALS was first described in 1869 by the French neurologist Jean-Martin Charcot. Since the description of the clinical and pathologic features of ALS, studies about the underlying pathophysiology of the disease have expanded over the years. However, despite intensive research to understand ALS, the mechanisms underlying the pathophysiology of the disease are elusive. Moreover, current management of ALS remains suboptimal from diagnosis to prognosis (Anderson, 2023).

Increasing evidence suggests that mitochondrial dysfunction is a crucial component of ALS, despite the fact that the precise pathophysiology of the disease remains unknown (Chen, 2022). Indeed, evidence from in vitro and in vivo disease models and patient studies strongly suggests that many of the ALS genes identified have a role in mitochondrial-associated functions and thus mitochondrial dysfunction as a core component of disease (Feldman, 2022). Furthermore, analysis of human postmortem or biopsy specimens has revealed the presence of aberrant mitochondria within the cell bodies of skeletal muscle, proximal axons, motor neurons, and intramuscular nerves (Sasaki et al., 2005).

Defective oxidative phosphorylation, the generation of ROS, a diminished calcium buffering capacity, and impaired mitochondrial dynamics are all manifestations of mitochondrial dysfunction associated with ALS (Smith et al., 2019). In fact, abnormalities in oxidative metabolism have been linked to alterations in ETC activity and impaired ATP production in cell culture models and transgenic animal studies (Muyderman & Chen, 2014). In one study, Onesto et al. (2016) initially found that mitochondrial functions in TDP-43 and C9orf72 fibroblasts were affected under oxidative conditions and impaired mitochondrial activity in ALS neurons (Onesto et al., 2016). Besides, in another study, dysregulation in ETC complexes was demonstrated in both G93A SOD1 transgenic mice and patients with ALS (Browne et al., 2006). Also, Delic et al. (2018) demonstrated impairment in complex IV activity and an aberrant profile of mitochondrial distribution throughout motor neurons, which is compatible with widespread mitochondrial integrity loss in the disease (Delic et al., 2018). Indeed, both familial and sporadic patients with ALS display widespread mitochondrial alterations in fibroblasts (Walczak et al., 2019).

However, decreased activities of oxidative phosphorylation complexes I + III, II + III, IV, and citrate synthase were noticed in mitochondria from the spinal cords of ALS patients (Wu et al., 2015). Also, mitochondrial dysfunction, evidenced by reduced respiration and ATP synthesis, precedes rather than follows behavioral deficits, indicating an important role of mitochondrial dysfunction in disease progression (Mattiazzi et al., 2002). However, it has been reported that many ALS-associated mutations in SOD1 result in the loss of antioxidant activity and the overproduction of ROS (Anderson, 2023; Feldman, 2022), and not surprisingly, a plethora of studies have demonstrated increased oxidative stress or oxidative damage in the spinal cords of ALS patients (Delic et al., 2018; Kihira et al., 2013; Mattiazzi et al., 2002). Elevated Ca^{2+} levels in mitochondria were also reported in ALS patients and the ALS SOD1 transgenic mouse model as an early event preceding cytosolic Ca^{2+} increase and mutant SOD1 aggregation, further supporting the critical role of mitochondrial dysfunction in ALS pathogenesis (Wu et al., 2015). In addition, it is not known whether mitochondrial dysfunction is a primary or secondary event in these processes.

According to Magri et al. (2023), the onset of the most prevalent ALS motor symptoms was correlated with a decline in mitochondrial respiration, namely oxygen fluxes associated to ATP synthesis and maximal capacity. They unequivocally demonstrated that translocator protein levels are markedly elevated in ALS animals and translocator protein is a crucial regulator of mitochondrial homeostasis in ALS (Chen et al., 2020).

Upper motor neurons of ALS patients have been observed to have a damaged inner mitochondrial membrane. Cardiolipin, a phospholipid present in IMM, is stabilized by SBT-272, a well-tolerated brain-penetrant small molecule, restoring mitochondrial structure and respiratory activity. SBT-272 greatly enhanced mitochondrial structural integrity and restored mitochondrial motility and function, according to research by Gautam et al. (2023) (Huang et al., 2016).

Despite extensive research, the mechanisms underlying etiology and pathophysiology of disease remain elusive. Because the etiology and pathophysiology of ALS is not fully understood, effective treatments cannot be developed.

Ischemic stroke

Stroke is one of the main causes of mortality and disability worldwide. The disease is widely known as a medical emergency in which the diminution of blood flow to the brain ends in the death of brain cells (Andrabi et al., 2020). According to the World Health Organisation, 16.9 million people suffer a stroke each year (Jia et al., 2021). There are two distinct types of stroke: ischemic and hemorrhagic. More than 80% of ischemic strokes are attributed to cerebral artery occlusion (He et al., 2020).

Ischemic stroke is characterized by localized inflammation, cellular death, and oxidative stress; it is defined as a sustained lack of blood flow to a specific region of the brain (Monsour et al., 2023). The FDA has only approved tissue plasminogen activator (tPA) as the most effective pharmacological therapy for acute ischemic stroke, and thrombolytic therapy is still the most effective treatment for acute ischemic stroke to date (Jia et al., 2021). However, the use of tPA remains limited due to the narrow therapeutic window (Liu et al., 2018). Therefore, it is highly important to develop a novel therapeutic strategy for the disease.

It is well documented that mitochondria play a central role in the pathophysiological processes of acute ischemic stroke (An et al., 2021). In addition to inhibiting ATP synthesis, these organelles also participate in cell death pathways including necrosis and apoptosis (Andrabi et al., 2020), which make them the primary contributors to ischemic cell death. Ischemia cascades, including membrane ion pump failure, efflux of cellular potassium, influx of sodium, chloride, and water, and membrane depolarization, are frequently initiated by ATP depletion (Liu et al., 2018).

Mitochondria play a central role in ischemic neuronal death. Ischemia-induced neuronal death is caused by a cascade of events that is initiated by mitochondrial respiratory function and membrane potential impairment (An et al., 2021). Depolarized mitochondria initiate excessive ROS production, decreased ATP generation, PTEN-induced putative kinase 1 accumulation, as well as an unfolded protein response (Liu et al., 2018). Cytochrome *c* is liberated, effector caspases are triggered, and apoptotic death is ultimately carried out (Chen et al., 2020), as the membrane permeability transition pore is unobstructed by the elevated levels of ROS and calcium that have become overloaded. PTEN-induced putative kinase 1 selectively recruits Parkin and phosphorylates both Parkin and ubiquitin to trigger mitophagy (Huang et al., 2016). Interruption of the signaling pathways is apparently able to disrupt, or in some instances, greatly delay, the spiral of increasingly vicious reactions that culminate in cell death (Liu et al., 2018).

Increasing evidence suggests that maintaining mitochondrial function is crucial for neuron survival and neurological improvement (An et al., 2021). Therefore, it is thought that mitochondria-based treatments may represent an effective therapeutic approach for ischemic stroke.

A multitude of recent studies have confirmed that the transfer of mitochondria from one cell to another is a protective mechanism that prevents stressed-induced mitochondrial dysfunctions in damaged cells (He et al., 2020; Jia et al., 2021; Chen et al., 2020). However, the transfer of healthy mitochondria into the injured cells was also regarded as a new therapeutic strategy for ischemic stroke (Chen et al., 2020). In one study conducted by Liu et al. (2018), mitochondrial transplantation was demonstrated to recover ATP levels and decrease energy deficits following ischemic stroke (Liu et al., 2018). Likewise, in another study by Huang et al. (2016), it was shown that transplantation of exogenous mitochondria into brain-ischemic rats can restore motor performance, attenuate brain infarct area, and decrease neuronal cell death (Huang et al., 2016). Similarly, many studies have hypothesized that mitochondrial transfer triggers the release of ROS in injured cells (Collier et al., 2023; Chen et al., 2020).

Stem cell therapy has also emerged as a mitochondrial targeting approach besides antioxidant and pharmacological strategies (Babenko et al., 2015; Jia et al., 2021). Therefore, in recent years, many studies have been carried out on the application of stem cell repair in cerebral ischemia (Jia et al., 2021; Li et al., 2021; Li, 2020). One metaanalysis study focused on preclinical research on mesenchymal stem cells for ischemic stroke showed that mesenchymal stem cells significantly improved all functional outcomes regardless of dose, intravenously administered (Li, 2020). Likewise, in another study conducted on rats with middle carotid artery occlusion, it was demonstrated that administration of human mesenchymal stem cells decreased infarction area and improved neurological indexes, which is caused by mitochondrial transfer from human mesenchymal stem cells to the injured neurons (Babenko et al., 2015). In addition, in an animal study conducted by Nakamura et al. (2020), it was found that mitochondrial transplantation from the placenta can also reduce infarction after cerebral ischemia and reperfusion in mice (Nakamura et al., 2020). Besides, current findings highlight the importance of mitochondrial transfer in stem cell repair, and the transfer of healthy mitochondria from astrocytes to ischemic neurons provides a theoretical basis for the application of stem cell therapy in ischemic stroke (Babenko et al., 2015; Hayakawa et al., 2018; Jia et al., 2021). Although the use of stem cells to treat mitochondria-related disorders has garnered much interest in ischemic stroke therapy, the mechanism of mitochondrial transfer is still unclear.

The development and discovery of new pharmacological therapies is a major challenge for ischemic stroke research. Therefore, it is thought that mitochondria-based treatments may represent an effective therapeutic approach for ischemic stroke.

Conclusion

Mitochondrial dysfunction is a major feature of various neurological disorders, especially neurodegenerative diseases such as AD, PD, and ALS, as well as ischemic stroke. However, the underlying mechanisms by which mitochondrial dysfunction occurs in such diseases are still unclear. In addition, this organelle is involved in multiple processes fundamental to homeostasis maintenance, such as energy production, ROS, etc. Besides, nowadays, there are limited pharmacological interventions to delay, prevent, or revert neurological disorders related to mitochondrial dysfunction. Moreover, effective treatment via at multiple dimensions, as for example, mitochondrial transplantation and stem cell therapy, might be effective for these disorders. Therefore, there is a requirement to develop innovative strategies that include both morphological studies and functional assessments of mitochondrial functionality for a more detailed study of this organelle.

References

Abe, K., Aoki, M., Tsuji, S., Itoyama, Y., Sobue, G., Togo, M., Hamada, C., Tanaka, M., Akimoto, M., Nakamura, K., Takahashi, F., Kondo, K., Yoshino, H., Abe, K., Aoki, M., Tsuji, S., Itoyama, Y., Sobue, G., Togo, M., ... Yoshino, H. (2017). Safety and efficacy of edaravone in well defined patients with amyotrophic lateral sclerosis: A randomised, double-blind, placebo-controlled trial. *The Lancet Neurology, 16*(7), 505–512. https://doi.org/10.1016/s1474-4422(17)30115-1

An, H., Zhou, B., & Ji, X. (2021). Mitochondrial quality control in acute ischemic stroke. *Journal of Cerebral Blood Flow and Metabolism, 41*(12), 3157–3170. https://doi.org/10.1177/0271678X211046992

Anderson, G. (2023). Amyotrophic lateral sclerosis pathoetiology and pathophysiology: Roles of astrocytes, gut microbiome, and muscle interactions via the mitochondrial melatonergic pathway, with disruption by glyphosate-based herbicides. *International Journal of Molecular Sciences, 24*(1). https://doi.org/10.3390/ijms24010587

Andrabi, S. S., Parvez, S., & Tabassum, H. (2020). Ischemic stroke and mitochondria: Mechanisms and targets. *Protoplasma, 257*(2), 335–343. https://doi.org/10.1007/s00709-019-01439-2

Babenko, V. A., Silachev, D. N., Zorova, L. D., Pevzner, I. B., Khutornenko, A. A., Plotnikov, E. Y., Sukhikh, G. T., & Zorov, D. B. (2015). Improving the post-stroke therapeutic potency of mesenchymal multipotent stromal cells by cocultivation with cortical neurons: The role of crosstalk between cells. *Stem Cells Translational Medicine, 4*(9), 1011–1020. https://doi.org/10.5966/sctm.2015-0010

Bader, V., & Winklhofer, K. F. (2020). Mitochondria at the interface between neurodegeneration and neuroinflammation. *Seminars in Cell & Developmental Biology, 99*, 163–171. https://doi.org/10.1016/j.semcdb.2019.05.028

Baumanns, S., Muehlemeyer, F., Miesbauer, L. C., Baake, J., Roloff, E. M., Beis, D. M., & Wenzel, U. (2023). Caprylic acid attenuates amyloid-β proteotoxicity by supplying energy via β-oxidation in an Alzheimer's disease model of the nematode *Caenorhabditis elegans*. *Nutritional Neuroscience*, 1–10.

Bell, S. M., Barnes, K., De Marco, M., Shaw, P. J., Ferraiuolo, L., Blackburn, D. J., Venneri, A., & Mortiboys, H. (2021). Mitochondrial dysfunction in Alzheimer's disease: A biomarker of the future? *Biomedicines, 9*(1). https://doi.org/10.3390/biomedicines9010063

Berg, D., Postuma, R. B., Bloem, B., Chan, P., Dubois, B., Gasser, T., Goetz, C. G., Halliday, G. M., Hardy, J., Lang, A. E., Litvan, I., Marek, K., Obeso, J., Oertel, W., Olanow, C. W., Poewe, W., Stern, M., & Deuschl, G. (2014). Time to redefine PD? Introductory statement of the MDS task force on the definition of Parkinson's disease. *Movement Disorders, 29*(4), 454–462. https://doi.org/10.1002/mds.25844

Bhatia, S., Rawal, R., Sharma, P., Singh, T., Singh, M., & Singh, V. (2022). Mitochondrial dysfunction in Alzheimer's disease: Opportunities for drug development. *Current Neuropharmacology, 20*(4), 675–692. https://doi.org/10.2174/1570159x19666210517114016

Briston, T., & Hicks, A. R. (2018). Mitochondrial dysfunction and neurodegenerative proteinopathies: Mechanisms and prospects for therapeutic intervention. *Biochemical Society Transactions, 46*(4), 829–842. https://doi.org/10.1042/BST20180025

Browne, S. E., Yang, L., DiMauro, J. P., Fuller, S. W., Licata, S. C., & Beal, M. F. (2006). Bioenergetic abnormalities in discrete cerebral motor pathways presage spinal cord pathology in the G93A SOD1 mouse model of ALS. *Neurobiology of Disease, 22*(3), 599–610. https://doi.org/10.1016/j.nbd.2006.01.001

Buneeva, O., Fedchenko, V., Kopylov, A., & Medvedev, A. (2020). Mitochondrial dysfunction in Parkinson's disease: Focus on mitochondrial DNA. *Biomedicines, 8*(12). https://doi.org/10.3390/biomedicines8120591

Cabral-Costa, J., & Kowaltowski, A. (2020). Neurological disorders and mitochondria. *Molecular Aspects of Medicine, 71*, 100826 (Original work published 2020).

Candelise, N. (2022). Mechanistic insights of mitochondrial dysfunction in amyotrophic lateral sclerosis: An update on a lasting relationship. *Metabolites, 12*(3), 233.

Chen, C., Turnbull, D. M., & Reeve, A. K. (2019). Mitochondrial dysfunction in Parkinson's disease—Cause or consequence? *Biology, 8*(2). https://doi.org/10.3390/biology8020038

Chen, W., Huang, J., Hu, Y., Khoshnam, S. E., & Sarkaki, A. (2020). Mitochondrial transfer as a therapeutic strategy against ischemic stroke. *Translational Stroke Research, 11*(6), 1214–1228. https://doi.org/10.1007/s12975-020-00828-7

Chen, W. (2022). The pathogenesis of amyotrophic lateral sclerosis: Mitochondrial dysfunction, protein misfolding and epigenetics. *Brain Research, 1786*, 147904.

Choi, J. G., Kim, S. Y., Jeong, M., & Oh, M. S. (2018). Pharmacotherapeutic potential of ginger and its compounds in age-related neurological disorders. *Pharmacology & Therapeutics, 182*, 56–69. https://doi.org/10.1016/j.pharmthera.2017.08.010

Collier, J. J., Oláhová, M., McWilliams, T. G., & Taylor, R. W. (2023). Mitochondrial signalling and homeostasis: From cell biology to neurological disease. *Trends in Neurosciences, 46*(2), 137–152. https://doi.org/10.1016/j.tins.2022.12.001

Compagnoni, M., Giacomo, Di F., Alessio, C., Stefania, C., Giacomo, P., Bresolin, N., & Masliah, E. (2020). The role of mitochondria in neurodegenerative diseases: The lesson from Alzheimer's disease and Parkinson's disease. *Molecular Neurobiology, 57*(7), 2959–2980. https://doi.org/10.1007/s12035-020-01926-1

Delic, V., Kurien, C., Cruz, J., Zivkovic, S., Barretta, J., Thomson, A., Hennessey, D., Joseph, J., Ehrhart, J., Willing, A. E., Bradshaw, P., & Garbuzova-Davis, S. (2018). Discrete mitochondrial aberrations in the spinal cord of sporadic ALS patients. *Journal of Neuroscience Research, 96*(8), 1353–1366. https://doi.org/10.1002/jnr.24249

Diego-Cordero, de, Martos-Lorite, I., & Vega-Escaño, J. (2022). Spiritual dimension in neurological and neurodegenerative diseases: A systematic mapping review. *Journal of Religion and Health*, 1–19.

Feigin, V. L., Vos, T., Nichols, E., Owolabi, M. O., Carroll, W. M., Dichgans, M., Deuschl, G., Parmar, P., Brainin, M., & Murray, C. (2020). The global burden of neurological disorders: Translating evidence into policy. *The Lancet Neurology, 19*(3), 255–265. https://doi.org/10.1016/S1474-4422(19)30411-9

Feldman, E. L. (2022). Amyotrophic lateral sclerosis. *The Lancet* (Original work published 2022).

Fu, Y. J., Xiong, S., Lovell, M. A., & Lynn, B. C. (2009). Quantitative proteomic analysis of mitochondria in aging PS-1 transgenic mice. *Cellular and Molecular Neurobiology, 29*(5), 649–664. https://doi.org/10.1007/s10571-009-9359-5

Gabrielli, A. P., Weidling, I., Ranjan, A., Wang, X., Novikova, L., Chowdhury, S. R., Menta, B., Berkowicz, A., Wilkins, H. M., Peterson, K. R., & Swerdlow, R. H. (2023). Mitochondria profoundly influence Apolipoprotein E Biology. *Journal of Alzheimer's Disease (Preprint)*, 1–14.

Gao, J., Wang, L., Liu, J., Xie, F., Su, B., & Wang, X. (2017). Abnormalities of mitochondrial dynamics in neurodegenerative diseases. *Antioxidants, 6*(2). https://doi.org/10.3390/antiox6020025

Gautam, M., Genç, B., Helmold, B., Ahrens, A., Kuka, J., Makrecka-Kuka, M., Günay, A., & Koçak, N. (2023). SBT-272 improves TDP-43 pathology in ALS upper motor neurons by modulating mitochondrial integrity, motility, and function. *Neurobiology of Disease*, 106022.

Grünewald, A., Kumar, K. R., & Sue, C. M. (2019). New insights into the complex role of mitochondria in Parkinson's disease. *Progress in Neurobiology, 177*, 73–93. https://doi.org/10.1016/j.pneurobio.2018.09.003

Grünewald, A., Rygiel, K. A., Hepplewhite, P. D., Morris, C. M., Picard, M., & Turnbull, D. M. (2016). Mitochondrial DNA depletion in respiratory chain–deficient Parkinson disease neurons. *Annals of Neurology, 79*(3), 366–378. https://doi.org/10.1002/ana.24571

Hauptmann, S., Scherping, I., Dröse, S., Brandt, U., Schulz, K. L., Jendrach, M., Leuner, K., Eckert, A., & Müller, W. E. (2009). Mitochondrial dysfunction: An early event in Alzheimer pathology accumulates with age in AD transgenic mice. *Neurobiology of Aging, 30*(10), 1574–1586. https://doi.org/10.1016/j.neurobiolaging.2007.12.005

Hayakawa, K., Chan, S. J., Mandeville, E. T., Park, J. H., Bruzzese, M., Montaner, J., Arai, K., Rosell, A., & Lo, E. H. (2018). Protective effects of endothelial progenitor cell-derived extracellular mitochondria in brain endothelium. *Stem Cells, 36*(9), 1404–1410. https://doi.org/10.1002/stem.2856

He, Z., Ning, N., Zhou, Q., Khoshnam, S. E., & Farzaneh, M. (2020). Mitochondria as a therapeutic target for ischemic stroke. *Free Radical Biology and Medicine, 146*, 45–58. https://doi.org/10.1016/j.freeradbiomed.2019.11.005

Henchcliffe, C., & Beal, F. M. (2008). Mitochondrial biology and oxidative stress in Parkinson disease pathogenesis. *Nature Clinical Practice Neurology, 4*(11), 600–609. https://doi.org/10.1038/ncpneuro0924

Hou, Y., Dan, X., Babbar, M., Wei, Y., Hasselbalch, S. G., Croteau, D. L., & Bohr, V. A. (2019). Ageing as a risk factor for neurodegenerative disease. *Nature Reviews Neurology, 15*(10), 565–581. https://doi.org/10.1038/s41582-019-0244-7

Huang, L. K., Chao, S. P., & Hu, C. J. (2020). Clinical trials of new drugs for Alzheimer disease. *Journal of Biomedical Science, 27*(1). https://doi.org/10.1186/s12929-019-0609-7

Huang, P. J., Kuo, C. C., Lee, H. C., Shen, C. I., Cheng, F. C., Wu, S. F., Chang, J. C., Pan, H. C., Lin, S. Z., Liu, C. S., & Su, H. L. (2016). Transferring xenogenic mitochondria provides neural protection against ischemic stress in ischemic rat brains. *Cell Transplantation, 25*(5), 913–927. https://doi.org/10.3727/096368915X689785

Jia, J., Jin, H., Nan, D., Yu, W., & Huang, Y. (2021). New insights into targeting mitochondria in ischemic injury. *Apoptosis, 26*(3–4), 163–183. https://doi.org/10.1007/s10495-021-01661-5

Johnson, J., Mercado-Ayon, E., Mercado-Ayon, Y., Dong, Y. N., Halawani, S., Ngaba, L., & Lynch, D. R. (2021). Mitochondrial dysfunction in the development and progression of neurodegenerative diseases. *Archives of Biochemistry and Biophysics, 702*. https://doi.org/10.1016/j.abb.2020.108698

Johri, A., & Beal, M. F. (2012). Mitochondrial dysfunction in neurodegenerative diseases. *Journal of Pharmacology and Experimental Therapeutics, 342*(3), 619–630. https://doi.org/10.1124/jpet.112.192138

Katsuno, M., Sahashi, K., Iguchi, Y., & Hashizume, A. (2018). Preclinical progression of neurodegenerative diseases. *Nagoya Journal of Medical Science, 80*(3), 289–298. https://doi.org/10.18999/nagjms.80.3.289

Kihira, T., Okamoto, K., Yoshida, S., Kondo, T., Iwai, K., Wada, S., Kajimoto, Y., Kondo, T., Kokubo, Y., & Kuzuhara, S. (2013). Environmental characteristics and oxidative stress of inhabitants and patients with amyotrophic lateral sclerosis in a high-incidence area on the Kii Peninsula, Japan. *Internal Medicine, 52*(13), 1479–1486. https://doi.org/10.2169/internalmedicine.52.9521

Kodavati, M., Wang, H., & Hegde, M. L. (2020). Altered mitochondrial dynamics in motor neuron disease: An emerging perspective. *Cells, 9*(4). https://doi.org/10.3390/cells9041065

Kumari, R., Shekhar, N., Tyagi, S., & Thakur, A. K. (2021). Mitochondrial dysfunctions and neurodegenerative diseases: A mini-review. *Journal of Analytical & Pharmaceutical Research, 10*(4), 147–149. https://doi.org/10.15406/japlr.2021.10.00378

Lai, M. C., Liu, W. Y., Liou, S. S., & Liu, I. M. (2023). Hispidin in the medicinal fungus protects dopaminergic neurons from JNK activation-regulated mitochondrial-dependent apoptosis in an MPP+-induced in vitro model of Parkinson's disease. *Nutrients, 15*(3), 549.

Lee, T. K., & Yankee, E. L. (2022). A review on Parkinson's disease treatment. *Neuroimmunology and Neuroinflammation, 8*. https://doi.org/10.20517/2347-8659.2020.58

Li, J., Zhang, Q., Wang, W., Lin, F., Wang, S., & Zhao, J. (2021). Mesenchymal stem cell therapy for ischemic stroke: A look into treatment mechanism and therapeutic potential. *Journal of Neurology, 268*(11), 4095–4107. https://doi.org/10.1007/s00415-020-10138-5

Li, Z. (2020). Stem cell-based therapies for ischemic stroke: A systematic review and meta-analysis of clinical trials. *Stem Cell Research & Therapy, 11*, 1–13 (Original work published 2020).

Liang, Z., Currais, A., Soriano-Castell, D., Schubert, D., & Maher, P. (2021). Natural products targeting mitochondria: Emerging therapeutics for age-associated neurological disorders. *Pharmacology & Therapeutics, 221*. https://doi.org/10.1016/j.pharmthera.2020.107749

Liu, F., Lu, J., Manaenko, A., Tang, J., & Hu, Q. (2018). Mitochondria in ischemic stroke: New insight and implications. *Aging and Disease, 9*(5). https://doi.org/10.14336/ad.2017.1126

Mattiazzi, M., D'Aurelio, M., Gajewski, C. D., Martushova, K., Kiaei, M., Beal, M. F., & Manfredi, G. (2002). Mutated human SOD1 causes dysfunction of oxidative phosphorylation in mitochondria of transgenic mice. *Journal of Biological Chemistry, 277*(33), 29626–29633. https://doi.org/10.1074/jbc.m203065200

Misrani, A., Tabassum, S., & Yang, L. (2021). Mitochondrial dysfunction and oxidative stress in Alzheimer's disease. *Frontiers in Aging Neuroscience, 13*. https://doi.org/10.3389/fnagi.2021.617588

Monsour, M., Gordon, J., Lockard, G., Alayli, A., & Borlongan, C. V. (2023). Mitochondria in cell-based therapy for stroke. *Antioxidants, 12*(1). https://doi.org/10.3390/antiox12010178

Moon, H. E., & Paek, S. H. (2015). Mitochondrial dysfunction in Parkinson's disease. *Experimental Neurobiology, 24* (Original work published 2015).

Moreira, P. I., Carvalho, C., Zhu, X., Smith, M. A., & Perry, G. (2010). Mitochondrial dysfunction is a trigger of Alzheimer's disease pathophysiology. *Biochimica et Biophysica Acta – Molecular Basis of Disease, 1802*(1), 2–10. https://doi.org/10.1016/j.bbadis.2009.10.006

Muyderman, H., & Chen, T. (2014). Mitochondrial dysfunction in amyotrophic lateral sclerosis – A valid pharmacological target? *British Journal of Pharmacology, 171*(8), 2191–2205. https://doi.org/10.1111/bph.12476

Nakamura, Y., Lo, E. H., & Hayakawa, K. (2020). Placental mitochondria therapy for cerebral ischemia-reperfusion injury in mice. *Stroke, 51*(10), 3142–3146. https://doi.org/10.1161/STROKEAHA.120.030152

Norat, P., Soldozy, S., Sokolowski, J. D., Gorick, C. M., Kumar, J. S., Chae, Y., Yağmurlu, K., Prada, F., Walker, M., Levitt, M. R., Price, R. J., Tvrdik, P., & Kalani, M. Y. S. (2020). Mitochondrial dysfunction in neurological disorders: Exploring mitochondrial transplantation. *Npj Regenerative Medicine, 5*(1). https://doi.org/10.1038/s41536-020-00107-x

Olesen, J., Gustavsson, A., Svensson, M., Wittchen, H.-U., & Jönsson, B. (2012). The economic cost of brain disorders in Europe. *European Journal of Neurology, 19*(1), 155–162. https://doi.org/10.1111/j.1468-1331.2011.03590.x

Onesto, E., Colombrita, C., Gumina, V., Borghi, M. O., Dusi, S., Doretti, A., Fagiolari, G., Invernizzi, F., Moggio, M., Tiranti, V., Silani, V., & Ratti, A. (2016). Gene-specific mitochondria dysfunctions in human TARDBP and C9ORF72 fibroblasts. *Acta Neuropathologica Communications, 4*(1). https://doi.org/10.1186/s40478-016-0316-5

Peng, Y., Gao, P., Shi, Le, Chen, L., Liu, J., & Long, J. (2020). Central and peripheral metabolic defects contribute to the pathogenesis of Alzheimer's disease: Targeting mitochondria for diagnosis and prevention. *Antioxidants and Redox Signaling, 32*(16), 1188–1236. https://doi.org/10.1089/ars.2019.7763

Reiss, A. B. (2022). The role of mitochondrial dysfunction in Alzheimer's disease: A potential pathway to treatment. *Experimental Gerontology, 164*, 111828.

Rey, F., Ottolenghi, S., Zuccotti, G. V., Samaja, M., & Carelli, S. (2022). Mitochondrial dysfunctions in neurodegenerative diseases: Role in disease pathogenesis, strategies for analysis and therapeutic prospects. *Neural Regeneration Research, 17*(4). https://doi.org/10.4103/1673-5374.322430

Rose, J., Brian, C., Woods, J., Pappa, A., Panayiotidis, M. I., Powers, R., & Franco, R. (2017). Mitochondrial dysfunction in glial cells: Implications for neuronal homeostasis and survival. *Toxicology, 391*, 109–115. https://doi.org/10.1016/j.tox.2017.06.011

Sancho-Bielsa, F. J. (2022). Parkinson's disease: Present and future of cell therapy. *Neurology Perspectives, 2*, S58–S68. https://doi.org/10.1016/j.neurop.2021.07.006

Sasaki, S., Warita, H., Murakami, T., Shibata, N., Komori, T., Abe, K., Kobayashi, M., & Iwata, M. (2005). Ultrastructural study of aggregates in the spinal cord of transgenic mice with a G93A mutant SOD1 gene. *Acta Neuropathologica, 109*(3), 247–255. https://doi.org/10.1007/s00401-004-0939-7

Schapira, A. H. V. (2008). Mitochondrial dysfunction in neurodegenerative diseases. *Neurochemical Research, 33*(12), 2502–2509. https://doi.org/10.1007/s11064-008-9855-x

Shi, P., Gal, J., Kwinter, D. M., Liu, X., & Zhu, H. (2010). Mitochondrial dysfunction in amyotrophic lateral sclerosis. *Biochimica et Biophysica Acta – Molecular Basis of Disease, 1802*(1), 45–51. https://doi.org/10.1016/j.bbadis.2009.08.012

Smith, E. F., Shaw, P. J., & De Vos, K. J. (2019). The role of mitochondria in amyotrophic lateral sclerosis. *Neuroscience Letters, 710.* https://doi.org/10.1016/j.neulet.2017.06.052

Sonkaya, Z. Z., Ceylan, M., & Sonkaya, A. R. (2021). Speech characteristics of Parkinson disease. *Medical Science and Discovery, 8*(12), 735–739. https://doi.org/10.36472/msd.v8i12.645

Swerdlow, R. H. (2018). Mitochondria and mitochondrial cascades in Alzheimer's disease. *Journal of Alzheimer's Disease, 62*(3), 1403–1416. https://doi.org/10.3233/JAD-170585

Tapias, V. (2019). *Mitochondrial dysfunction and neurodegeneration.* Frontiers Media SA (Original work published 2019).

Tarawneh, R., & Holtzman, D. M. (2012). The clinical problem of symptomatic Alzheimer disease and mild cognitive impairment. *Cold Spring Harbor Perspectives in Medicine, 2*(5), a006148. https://doi.org/10.1101/cshperspect.a006148

Walczak, J., Debska-Vielhaber, G., Vielhaber, S., Szymański, J., Charzyńska, A., Duszyński, J., & Szczepanowska, J. (2019). Distinction of sporadic and familial forms of ALS based on mitochondrial characteristics. *Federation of American Societies for Experimental Biology Journal, 33*(3), 4388–4403. https://doi.org/10.1096/fj.201801843r

Wang, Y., Xu, E., Musich, P. R., & Lin, F. (2019). Mitochondrial dysfunction in neurodegenerative diseases and the potential countermeasure. *CNS Neuroscience and Therapeutics, 25*(7), 816–824. https://doi.org/10.1111/cns.13116

Wu, L., Jiang, L., Ding, C., Sheng, Q., Ding, X., & Yao, J. (2015). Effect of optical pump on the dielectric properties of $LiTaO_3$ in Terahertz range. *Journal of Infrared, Millimeter and Terahertz Waves, 36*(1), 1–6. https://doi.org/10.1007/s10762-014-0118-4

Wu, Y., Chen, M., & Jiang, J. (2019). Mitochondrial dysfunction in neurodegenerative diseases and drug targets via apoptotic signaling. *Mitochondrion, 49,* 35–45. https://doi.org/10.1016/j.mito.2019.07.003

Yiannopoulou, K. G., & Papageorgiou, S. G. (2020). Current and future treatments in Alzheimer disease: An update. *Journal of Central Nervous System Disease, 12.* https://doi.org/10.1177/1179573520907397

Zhao, J. (2022). The impact of mitochondrial dysfunction in amyotrophic lateral sclerosis. *Cells, 11,* 2049.

Zhao, X. Y., Lu, M. H., Yuan, D. J., Xu, D. E., Yao, P. P., Ji, W. L., Chen, H., Liu, W. L., Yan, C. X., Xia, Y. Y., Li, S., Tao, J., & Ma, Q. H. (2019). Mitochondrial dysfunction in neural injury. *Frontiers in Neuroscience, 13.* https://doi.org/10.3389/fnins.2019.00030

Zvěřová, M. (2019). Clinical aspects of Alzheimer's disease. *Clinical Biochemistry, 72,* 3–6. https://doi.org/10.1016/j.clinbiochem.2019.04.015

CHAPTER

18

Preclinical studies

Pasquale Picone and Domenico Nuzzo

Istituto per la Ricerca e l'Innovazione Biomedica, Consiglio Nazionale delle Ricerche, Palermo, Italy

Introduction

The central nervous system (CNS) is rich in mitochondria and exhibits high levels of cellular respiration (Onyango et al., 2017). Mitochondrial activity is essential to support higher brain functions, such as long-term potentiation and synaptic plasticity. Furthermore, mitochondrial dynamics is very important in synaptic transmission (synthesis and storage of neurotransmitters) and is involved in the intracellular homeostasis of $Ca2^{+}$ and in neurogenesis during brain development (Cheng et al., 2010; Kann et al., 2007). In support of this notion, a mitochondrial dysfunction is an event implicated in brain diseases such as Parkinson's disease (PD), Alzheimer's disease, multiple sclerosis, amyotrophic lateral sclerosis , as well as following stroke and traumatic brain or spinal cord injury (SCI) (Di Carlo et al., 2012; P. Picone et al., 2014). Many studies have revealed anomalies of the mitochondrial structure, alteration of mitochondrial fission and fusion processes, deficiency of the enzymes of the Krebs cycles, alteration of the mitochondrial transport chain, elevated reactive oxygen species (ROS) production, alteration of calcium homeostasis, excessive mitochondrial fragmentation, and alteration of mitochondrial membrane potential in different brain pathologies (Bertholet et al., 2016; Johnson et al., 2021). A new biomedical approach that refers to a medical intervention aimed at mitochondria is called "mitochondrial medicine". Although preclinical studies have proposed various therapeutic agents against mitochondrial dysfunction, to date, no clinical protocols are available.

Furthermore, in order to improve mitochondrial brain targeting, new biotechnological strategies have been developed, such as nanoparticles, liposomes, dendrimers, and nanogels (Alexander et al., 2019; Nuzzo & Picone, 2021; Picone & Nuzzo, 2023).

Mitochondrial transplantation (MT) is a new potential treatment involving mitochondrial transfer aimed at replacing damaged mitochondria by arresting mitochondrial dysfunction in injury and disease states. Evidence has shown that mitochondrial transfer is a physiological brain process; in fact, it has been shown that when neurons are damaged by ischemic stroke, astrocytes generously deliver their healthy mitochondria via vesicles containing

Mitochondrial Transplantation and Transfer
https://doi.org/10.1016/B978-0-443-18858-9.00018-4

mitochondria to neurons to induce neuroprotection. Vesicles are internalized by a mechanism dependent on the membrane receptor CD38 (Hayakawa et al., 2016). Transferring healthy mitochondria into cells containing compromised mitochondria improved energy production, reduced ROS, and restored diverse mitochondrial functions. The results on MT for brain diseases application have shown promise using in vitro and in vivo models (Espino De La Fuente-Muñoz & Arias, 2021; Norat et al., 2020; Pasquale Picone & Nuzzo, 2022).

Administration routes for brain mitochondrial transplantation

Different routes of administration have been used to transfer mitochondria to the CNS. Direct injection into the brain allows localization of mitochondria to the target site; however, this method is invasive and multiple administrations are complex. In contrast, systemic intravenous injection is a simple, less invasive and safe method, which offers the benefits of to inject larger volumes and to be able to carry out multiple doses (Espino De La Fuente-Muñoz & Arias, 2021; Norat et al., 2020; Pasquale Picone & Nuzzo, 2022). Due to their small size ($\sim 1\ \mu m$) mitochondria can be administered intravenously, and it has also been shown that they are not captured by red blood cells and do not alter oxygen transport (Shi et al., 2017). The main limitations for intravenous administration are the presence of the blood brain barrier (BBB), which limits the entry of the drug to the CNS, and the diffusion of the drug throughout the body. Recently, several studies have indicated intranasal administration as a possible alternative route for brain MT (Chang et al., 2021). Fig. 18.1 shows the different administration routes used for brain MT.

Mitochondrial transplantation in the nervous system: preclinical studies

Recently, several preclinical studies on different models of brain pathologies have demonstrated that MT has great therapeutic potential. In one study, xenogenic mitochondria isolated from hamster cells were injected by local intracerebral injection or by systemic intraarterial injection, into an ischemic rat model. The results revealed that the transplanted mitochondria significantly restored the motor function of the ischemic brain. Moreover, apoptotic cells and brain infarct area were both reduced by mitochondria administration. Based on these observations, MT has neuroprotective effects after ischemia (Huang et al., 2016).

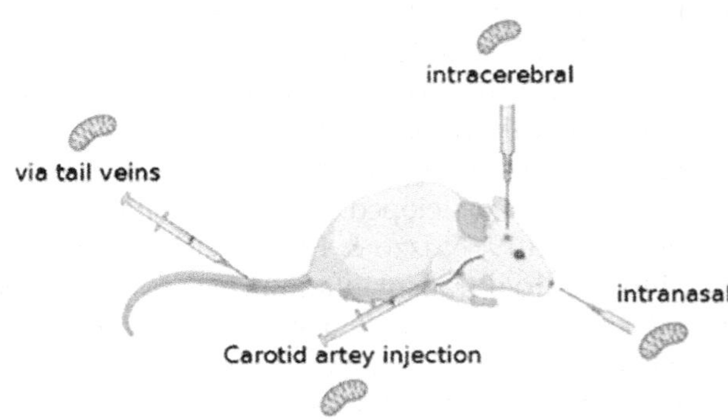

FIGURE 18.1 Brain mitochondrial injection routes. Different administration routes of mitochondria brain delivery: In situ (intracerebral), intravenous (via tail veins and carotid artery), and intranasal.

Moreover, exogenous MT reduced infarct size in cerebral ischemia–reperfusion injury (Xie et al., 2021). The authors hypothesized that when exogenous mitochondria entered host cells, separation of mitochondrial components occurred in the presence of oxygen. Some components fuse with host mitochondria, while others enter lysosomes and undergo lysosomal degradation. This process increased cell viability in an oxygen dependent manner, reduced ROS levels and apoptosis, and ameliorated neurobehavioral deficits, which in turn attenuated ischemia reperfusion injury (Xie et al., 2021).

The mitochondria extracts from the human hepatoma cell were able to penetrate the neuroblastoma SH-SY5Y cells (Shi et al., 2017). Cell viability increased in a time and concentration dependent manner, the ATP content increased, while the ROS level did not undergo significant changes. Furthermore, in the cellular model of PD, data showed that mitochondria transfer induced an increase in cell survival rate in a concentration-dependent manner of administered mitochondria (Shi et al., 2017). When mitochondria were injected intravenously into mice, not only did they survive and no obvious abnormalities appeared, but the mice improved their resistance by increasing energy production in the forced swim test. Furthermore, In vivo biodistribution of mitochondria showed that after 2h of administration, by systemic intravenous (via tail veins), they appear distributed to various tissues including the brain, liver, kidney, muscle, and heart. Finally, mitochondrial administration to the mouse model of respiratory chain inhibitor (1-methyl-4-phenyl-1,2,3,6-tetrahydropyridine) induced 1-methyl-4-phenyl-1, 2, 3, 6-tetrahydropyridine (MPTP) prevented PD progression (Shi et al., 2017).

Gollihue and collaborators performed in vitro and in vivo experiments they labeled isolated mitochondria with turbo-green fluorescent protein (tGFP). Using confocal imaging, they revealed the incorporation of exogenous mitochondria into PC-12 cells after 1 hour of coincubation. Furthermore, the authors demonstrated that tGFP mitochondria injected at spinal L1/L2 were located within both microglia/macrophages and endothelial cells even 48 h after injection (Gollihue et al., 2017). Using a rat model of L1/L2 contusion SCI, the transplantation of PC12 derived tGFP mitochondria allowed visualization of their incorporation into specific cells up to 7 days after injection. In particular, tGFP mitochondria colocalized with multiple resident cell types but were absent in neurons. Despite their contribution to the maintenance of bioenergetic normality, MT did not produce long term functional neuroprotection or recovery of motor and sensory functions (Gollihue et al., 2018). The authors speculate that the failure in long term functional neuroprotection is possibly due to a relatively low incorporation rate of mitochondria ($\leq 6\%$) and the absence of incorporation of mitochondria into specific target cells, such as neurons (Gollihue et al., 2017, 2018). Lin and coworkers demonstrated that transplanted mitochondria in traumatic SCI rats were present in the injured spinal cord up to 28 days after treatment. In rats receiving MT locomotor and sensory functions were recovered. In addition, the expression of dynamin-related protein 1, the severity of demyelination, cell apoptosis, and inflammatory responses were reduced (Lin et al., 2022). Although the authors provide evidence for the possible translational application of TM in traumatic SCI, they indicated, like Gollihue (Gollihue et al., 2017, 2018), that time course studies are needed to better define long-term therapeutic effects. In particular, the expression of proinflammatory cytokines and proteins associated with oxidative stress at different times should be analyzed (Lin et al., 2022).

Chien and collaborators tested the effect of mitochondrial transfer on neuronal regeneration. Specifically, the authors administered mitochondria to damaged hippocampal neurons and found a restoration of membrane potential and an increase in neurite regrowth. These findings support the role of mitochondria in the regeneration of damaged hippocampal neurons and highlight the therapeutic prospect of MT for the damaged CNS (Chien et al., 2018). Moreover, MT of mitochondria isolated from primary rat astrocytes enhances cell viability of H_2O_2-damaged neurons. Interestingly, mitochondria increased cell numbers in damaged neurons but not in normal neurons, although MT efficiency did not differ between them. Moreover, this effect was demonstrated in rat middle cerebral artery occlusion models (Lee et al., 2022).

Kuo and collaborators showed that transferred mitochondria into injured sciatic nerves after sciatic nerve crushing in a rat model allowed the integrity of the cytoskeleton to be maintained and improved nerve conduction properties and animal behavior (Kuo et al., 2017). Autologously derived mitochondria (isolated from autologous pectoralis major) were infused directly into the lateral ventricle, in a stroke model of middle cerebral artery occlusion, and rats were analyzed after 4 weeks. MT reduced cellular oxidative stress and apoptosis, attenuated reactive astrogliosis, promoted neurogenesis, reduced brain infarction volume, and reversed neurological deficits (Zhang et al., 2019). Injection of exogenous mitochondria into mice undergoing traumatic brain injury (TBI) significantly reduced apoptosis, promoted angiogenesis, relieved cerebral edema and loss of the BBB, and reduced excessive inhibition of long term depression in the hippocampus 7 days after TBI (Zhang et al., 2020).

Mitochondria isolated from young mice and injected intravenously into aged mice resulted in increased ATP production, reduced ROS, and improved cognitive and motor functions (Zhao et al., 2020). Using a similar approach, Adlimoghaddam and coworkers investigated whether the administration of healthy mitochondria, obtained from the livers of juvenile mice to aged mice, improves mitochondrial function in aged brain tissue. Specifically, the authors demonstrated that mitochondria intravenous injection improves mitochondrial function through upregulation of the mitochondrial complex II protein subunit Succinate dehydrogenase [ubiquinone] iron-sulfur (SDHB) in the hippocampus of aged mice (Adlimoghaddam et al., 2022).

Mice were subjected to sepsis-associated brain dysfunction followed by intracerebroventricular injection of exogenous functional mitochondria. The authors found that MT induced microglial M2 (antiinflammatory) rather than M1 (proinflammatory phenotype) response 24 h after sepsis and improved behavioral deficits. These findings support the potential use of exogenous MT therapy against the sepsis-associated brain dysfunction (Yan et al., 2020). Microglial cells represent approximately 15% of all brain cells and, like macrophage cells, represent the main form of active immune defense in the CNS. The finding that MT induced M2 rather than M1 microglial response indicates not only that MT has an inhibitory effect on inflammatory activity, but also that mitochondria do not activate the immune defenses of the brain. Platelets are abundant in the blood and can be easily isolated from peripheral blood. Therefore, such cells can be considered an easily accessible source of mitochondria. Platelet derived mitochondria transplantation (Mito-Plt) improved Diabetes-Associated Cognitive Impairment (DACI). Results showed that Mito-Plt was internalized in hippocampal neurons 24 h after mitochondrial injection. After 1 month, DACI was relieved in mice and increased mitochondrial number, mitochondrial function, decreased oxidative stress, neuronal apoptosis, and accumulation of Aβ and Tau in the hippocampus were observed (Ma et al., 2020). The data showed that MT of mitochondria derived from allogeneic liver or autogenous muscle significantly reduced neuronal death, relieved anxiety, and improved spatial memory following head trauma. In vitro experiments showed that exogenous mitochondria could be easily internalized by neurons, astrocytes, and microglia, with the exception of oligodendrocytes (J. Zhao et al., 2021). Mice undergoing photothrombosis induced stroke in the medial prefrontal cortex received mitochondria intranasally on alternate days (3 days per week) for 1 week. Behavioral findings revealed that mitochondrial therapy has alleviated ischemia-induced memory impairment. Exogenous mitochondria transplantation reduced ROS, restored ATP generation, improved mitochondrial membrane potential, and significantly upregulated the expression of GAP-43 and PSD-95 after ischemic stroke (Hosseini et al., 2022). Some authors have demonstrated that functional mitochondria isolated from the liver have emerged as a new approach to counteract the effects of optic nerve damage. In particular, mitochondria transplanted intravitreally incorporated into the retina, improving its oxidative metabolism and electrophysiological activity. Mitochondrial therapy increases the ganglion cell layer survival at 14 days and generated a higher number of axons extending beyond the injury site at 28 days, effects that depend on the structural integrity of the organelles (Nascimento-dos-Santos et al., 2020) (Table 18.1).

TABLE 18.1 List of some preclinical studies regarding brain MT.

Mitochondria source	Mitochondria delivery	Mitochondrial transplantation routes	Results	References
Hamster cells	Ischemic model rat	Intracerebral or intraarterial injection	Restoration of motor function, reduction of apoptotic cells, and the area of cerebral infarction	Huang (2016)
Human hepatoma cell	PD mouse model	Intravenous	Increasing the activity of the electron transport chain, decrease ROS levels and cell death	Shi (2017)
Exogenous mitochondria	Rat model of L1/L2 contusion SCI	Injected into the medial lateral gray matter	Maintained the acute bioenergetics of the damaged spinal cord	Gollihue (2018)
Pectoralis major	Stroke model of middle cerebral artery occlusion rat	Lateral ventricles	Reduced cellular oxidative stress, apoptosis, reactive astrogliosis, brain infarction volume, reversed neurological deficits, and promoted neurogenesis	Zhang (2019)
Young mice	Older mice	Intravenously injected	Increased ATP, decreased ROS levels, and improved cognitive and motor functions.	Zhao (2020)
Pectoralis major	Sepsis-associated brain dysfunction	Intracerebroventricular	Induction microglial M2 response and improved behavioral deficits	Yan (2020)

(*Continued*)

TABLE 18.1 List of some preclinical studies regarding brain MT.—cont'd

Mitochondria source	Mitochondria delivery	Mitochondrial transplantation routes	Results	References
Platelets	Diabetes-associated cognitive impairment	Intracerebroventricular	Decreased oxidative stress, neuronal apoptosis, and accumulation of Aβ and Tau in the hippocampus	Ma (2020)
Peptide-mediated allogeneic mitochondrial delivery	PD rat model	Medical forebrain bundle	Improve locomotor activity	Chang (2016)
Allogeneic mitochondria conjugated with Pep-1	PD rat model	Intranasal	Improvement of rotational and locomotor behaviors in PD rats	Chang (2021)

In the current situation, the mitochondrial isolation method, mitochondrial source, administration route, and doses are essential factors that could influence the brain MT performance efficiency. For these reasons, the optimal standard protocols for MT will need to be developed.

Brain mitochondrial transfer strategies

MT has been called a "magic" cure, because although injected mitochondria are able to penetrate damaged cells and recover ATP and mitochondrial function, it has several weaknesses (Bertero et al., 2018). Mitochondria must resist the transition from an intracellular environment to an inhospitable extracellular environment and cross cellular and bodily barriers (Bertero et al., 2018; Roushandeh et al., 2019). The data report that only a small percentage ($\leq 6\%$) of the administered mitochondria reach the brain cells and the transfer is often not specific to the target cells (Gollihue et al., 2017, 2018). It is important to stabilize mitochondrial suspension as high concentrations of injected mitochondria can cause aggregation of mitochondria (Gollihue et al., 2018). This phenomenon could be responsible for the low uptake efficiency or even physical damage neuronal cells and could cause an intensified inflammatory response. To improve MT, the development of mitochondrial delivery systems is of prime importance. Biotechnological approaches can offer the necessary tools to overcome the limitations and problems related to MT. Conjugation with the carrier peptide Pep-1 (a cell-penetrating peptide) was recently used to improve mitochondrial transfer. Peptide-mediated allogeneic mitochondrial delivery into medical forebrain bundle of 6-OHDA-unilaterally infused PD rats has enhancing mitochondrial function in the soma of the substantia nigra neuron through mitochondrial transport dynamics in the nigrostriatal circuit. Furthermore, allogeneic and xenogeneic transplantation of peptide-labeled mitochondria after 3 months improved the locomotive activity in the PD rats (Chang et al., 2016). Intranasal injection of allogeneic mitochondria conjugated with Pep-1 (P-Mito) or unconjugated (Mito) was performed once a week for 3 months for the PD treatment. The treatment effect was stronger in the P-Mito group than in the Mito group, but the difference was insignificant (Chang et al., 2021).

Recently, to improve neuronal mitochondrial release, synaptosomes (structures obtained from synaptic terminations by physical processes) have been proposed as vesicles for mitochondrial transfer. Synaptosomes were able to restore mitochondrial function in neuronal cells containing damaged mitochondria (P. Picone et al., 2021). Peruzzotti-Jamett and collaborators recently showed that mitochondria carried by extracellular vesicles, restoring mitochondrial activity in mice with experimental autoimmune encephalomyelitis, a model of multiple sclerosis (Peruzzotti-Jametti et al., 2021).

To improve the MT strategy, Patel and colleagues developed an erodible thermogelling hydrogel that protects isolated mitochondria from the extracellular environment as they are delivered to the injured spinal cord. The authors estimated that 70% of mitochondria are released from the hydrogel within 20 min at 37°C, and the respiration capacity of mitochondria released from the hydrogel was greater than those without gel. Therefore, the hydrogel allows protection of the mitochondria from the external environment, their local diffusion, and their incorporation into the recipient cells (Patel et al., 2022) (Fig. 18.2).

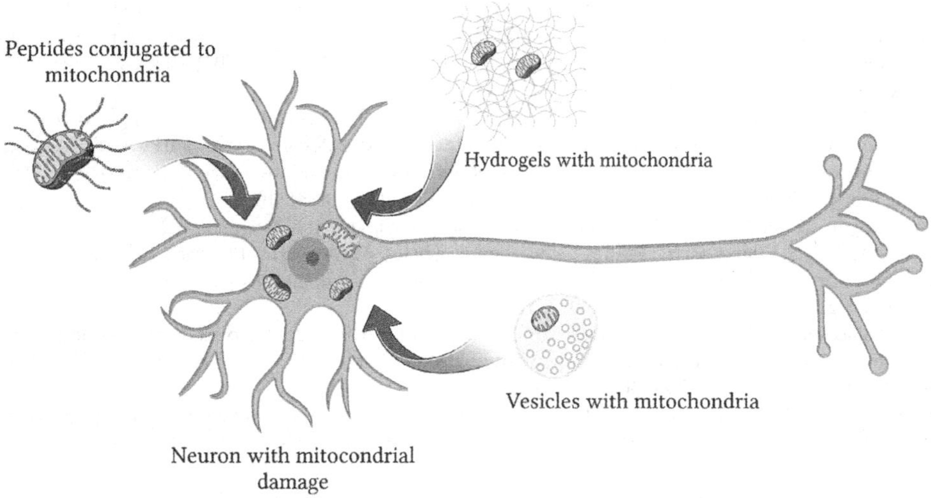

FIGURE 18.2 Strategies to improve neuronal mitochondrial transplantation. Use of vesicles, cell-penetrating peptides, and hydrogels as strategies to protect and enhance neuronal mitochondrial transplantation.

Conclusions

Interesting results on brain MT have been highlighted through preclinical studies. However, evidence on the efficacy and safety of MT, used to combat brain disorders, is still in its early stages and further studies are needed to understand its neuroprotective mechanisms. Furthermore, few studies have focused on the ability to isolated mitochondria to cross the BBB and more research is needed to confirm this permeability. To improve MT efficacy, delivery approaches to the brain and cellular uptake in specific brain cells should be investigated in future studies. Therefore, a fundamental next step in the treatment of many mitochondrial neurodegenerative diseases will be the research of new mitochondria pharmaceutical formulations and delivery system strategies to improve brain MT therapy efficiently.

References

Adlimoghaddam, A., Benson, T., & Albensi, B. C. (2022). Mitochondrial transfusion improves mitochondrial function through up-regulation of mitochondrial complex II protein subunit SDHB in the Hippocampus of aged mice. *Molecular Neurobiology, 59*(10), 6009–6017. https://doi.org/10.1007/s12035-022-02937-w, 15591182.

Alexander, A., Agrawal, M., Uddin, A., Siddique, S., Shehata, A. M., Shaker, M. A., Rahman, S. A. U., Abdul, M. I. M., & Shaker, M. A. (2019). Recent expansions of novel strategies towards the drug targeting into the brain. *International Journal of Nanomedicine, 14*, 5895–5909. https://doi.org/10.2147/IJN.S210876, 11782013.

Bertero, E., Maack, C., & O'Rourke, B. (2018). Mitochondrial transplantation in humans: "magical" cure or cause for concern? *Journal of Clinical Investigation, 128*(12), 5191–5194. https://doi.org/10.1172/JCI124944, 15588238.

Bertholet, A. M., Delerue, T., Millet, A. M., Moulis, M. F., David, C., Daloyau, M., Arnauné-Pelloquin, L., Davezac, N., Mils, V., Miquel, M. C., Rojo, M., & Belenguer, P. (2016). Mitochondrial fusion/fission dynamics in neurodegeneration and neuronal plasticity. *Neurobiology of Disease, 90*, 3–19. https://doi.org/10.1016/j.nbd.2015.10.011, 1095953X.

Chang, J. C., Chao, Y. C., Chang, H. S., Wu, Y. L., Chang, H. J., Lin, Y. S., Cheng, W. L., Lin, T. T., & Liu, C. S. (2021). Intranasal delivery of mitochondria for treatment of Parkinson's Disease model rats lesioned with 6-hydroxydopamine. *Scientific Reports, 11*(1). https://doi.org/10.1038/s41598-021-90094-w, 20452322 http://www.nature.com/srep/index.html.

Chang, J. C., Wu, S. L., Liu, K. H., Chen, Y. H., Chuang, C. S., Cheng, F. C., Su, H. L., Wei, Y. H., Kuo, S. J., & Liu, C. S. (2016). Allogeneic/xenogeneic transplantation of peptide-labeled mitochondria in Parkinson's disease: Restoration of mitochondria functions and attenuation of 6-hydroxydopamine-induced neurotoxicity. *Translational Research, 170*, 40–56.e3. https://doi.org/10.1016/j.trsl.2015.12.003, 18781810.

Cheng, A., Hou, Y., & Mattson, M. P. (2010). Mitochondria, neuroplasticity. *Neuro* (Vol. 2).

Chien, L., Liang, M. Z., Chang, C. Y., Wang, C., & Chen, L. (2018). Mitochondrial therapy promotes regeneration of injured hippocampal neurons. *Biochimica et Biophysica Acta—Molecular Basis of Disease, 1864*(9), 3001–3012. https://doi.org/10.1016/j.bbadis.2018.06.012, 1879260X.

Di Carlo, M., Giacomazza, D., Picone, P., Nuzzo, D., & San Biagio, P. L. (2012). Are oxidative stress and mitochondrial dysfunction the key players in the neurodegenerative diseases? *Free Radical Research, 46*(11), 1327–1338. https://doi.org/10.3109/10715762.2012.714466, 10292470.

Espino De La Fuente-Muñoz, C., & Arias, C. (2021). The therapeutic potential of mitochondrial transplantation for the treatment of neurodegenerative disorders. *Reviews in the Neurosciences, 32*(2), 203–217. https://doi.org/10.1515/revneuro-2020-0068, 03341763.

Gollihue, J. L., Patel, S. P., Eldahan, K. C., Cox, D. H., Donahue, R. R., Taylor, B. K., Sullivan, P. G., & Rabchevsky, A. G. (2018). Effects of mitochondrial transplantation on bioenergetics, cellular incorporation, and functional recovery after spinal cord injury. *Journal of Neurotrauma, 35*(15), 1800–1818. https://doi.org/10.1089/neu.2017.5605, 15579042.

Gollihue, J. L., Patel, S. P., Mashburn, C., Eldahan, K. C., Sullivan, P. G., & Rabchevsky, A. G. (2017). Optimization of mitochondrial isolation techniques for intraspinal transplantation procedures. *Journal of Neuroscience Methods, 287*, 1–12. https://doi.org/10.1016/j.jneumeth.2017.05.023, 1872678X.

Hayakawa, K., Esposito, E., Wang, X., Terasaki, Y., Liu, Y., Xing, C., Ji, X., & Lo, E. H. (2016). Transfer of mitochondria from astrocytes to neurons after stroke. *Nature, 535*(7613), 551–555. https://doi.org/10.1038/nature18928, 14764687.

Hosseini, L., Karimipour, M., Seyedaghamiri, F., Abolhasanpour, N., Sadigh-Eteghad, S., Mahmoudi, J., & Farhoudi, M. (2022). Intranasal administration of mitochondria alleviated cognitive impairments and mitochondrial dysfunction in the photothrombotic model of mPFC stroke in mice. *Journal of Stroke and Cerebrovascular Diseases, 31*(12), 106801. https://doi.org/10.1016/j.jstrokecerebrovasdis.2022.106801, 10523057.

Huang, P. J., Kuo, C. C., Lee, H. C., Shen, C. I., Cheng, F. C., Wu, S. F., Chang, J. C., Pan, H. C., Lin, S. Z., Liu, C. S., & Su, H. L. (2016). Transferring xenogenic mitochondria provides neural protection against ischemic stress in ischemic rat brains. *Cell Transplantation, 25*(5), 913–927. https://doi.org/10.3727/096368915X689785, 15553892.

Johnson, J., Mercado-Ayon, E., Mercado-Ayon, Y., Dong, Y. N., Halawani, S., Ngaba, L., & Lynch, D. R. (2021). Mitochondrial dysfunction in the development and progression of neurodegenerative diseases. *Archives of Biochemistry and Biophysics, 702*, 108698. https://doi.org/10.1016/j.abb.2020.108698

Kann, o, Kovács, R., & Mitochondria. (2007). *American Journal of Physiology: Cell Physiology, 292*.

Kuo, C. C., Su, H. L., Chang, T. L., Chiang, C. Y., Sheu, M. L., Cheng, F. C., Chen, C. J., Sheehan, J., & Pan, H. C. (2017). Prevention of axonal degeneration by perineurium injection of mitochondria in a sciatic nerve crush injury model. *Neurosurgery, 80*(3), 475–488. https://doi.org/10.1093/neuros/nyw090, 15244040.

Lee, E.-H., Kim, M., Ko, S. H., Lee, M., & Park, C.-H. (2022). The primary astrocytic mitochondrial transplantation ameliorates ischemic stroke. *The FASEB Journal, 36*(S1). https://doi.org/10.1096/fasebj.2022.36.s1.0r802, 0892-6638.

Lin, M. W., Fang, S. Y., Hsu, J. Y. C., Huang, C. Y., Lee, P. H., Huang, C. C., Chen, H. F., Lam, C. F., & Lee, J. S. (2022). Mitochondrial transplantation attenuates neural damage and improves locomotor function after traumatic spinal cord injury in rats. *Frontiers in Neuroscience, 16*. https://doi.org/10.3389/fnins.2022.800883, 1662453X https://www.frontiersin.org/journals/neuroscience#.

Ma, H., Jiang, T., Tang, W., Ma, Z., Pu, K., Xu, F., Chang, H., Zhao, G., Gao, W., Li, Y., & Wang, Q. (2020). Transplantation of platelet-derived mitochondria alleviates cognitive impairment and mitochondrial dysfunction in db/db mice. *Clinical Science, 134*(16), 2161–2175. https://doi.org/10.1042/CS20200530, 14708736.

Nascimento-dos-Santos, G., de-Souza-Ferreira, E., Lani, R., Faria, C. C., Araújo, V. G., Teixeira-Pinheiro, L. C., Vasconcelos, T., Gonçalo, T., Santiago, M. F., Linden, R., Galina, A., & Petrs-Silva, H. (2020). Neuroprotection from optic nerve injury and modulation of oxidative metabolism by transplantation of active mitochondria to the retina. *Biochimica et Biophysica Acta—Molecular Basis of Disease, 1866*(5). https://doi.org/10.1016/j.bbadis.2020.165686, 1879260X http://www.elsevier.com/locate/bbadis.

Norat, P., Soldozy, S., Sokolowski, J. D., Gorick, C. M., Kumar, J. S., Chae, Y., Yağmurlu, K., Prada, F., Walker, M., Levitt, M. R., Price, R. J., Tvrdik, P., & Kalani, M. Y. S. (2020). Mitochondrial dysfunction in neurological disorders: Exploring mitochondrial transplantation. *Npj Regenerative Medicine, 5*(1). https://doi.org/10.1038/s41536-020-00107-x, 20573995 https://www.nature.com/npjregenmed/.

Nuzzo, D., & Picone, P. (2021). Multiple Sclerosis: Focus on extracellular and artificial vesicles, nanoparticles as potential therapeutic approaches. *International Journal of Molecular Sciences, 22*(16), 8866. https://doi.org/10.3390/ijms22168866, 1422-0067.

Onyango, I. G., Khan, S. M., & Bennett, J. P. (2017). Mitochondria in the pathophysiology of Alzheimer's and Parkinson's diseases. *Frontiers in Bioscience—Landmark, 22*(5), 854–872. https://doi.org/10.2741/4521, 27686698.

Patel, S. P., Michael, F. M., Arif Khan, M., Duggan, B., Wyse, S., Darby, D. R., Chaudhuri, K., Pham, J. T., Gollihue, J., DeRouchey, J. E., Sullivan, P. G., Dziubla, T. D., & Rabchevsky, A. G. (2022). Erodible thermogelling hydrogels for localized mitochondrial transplantation to the spinal cord. *Mitochondrion, 64*, 145–155. https://doi.org/10.1016/j.mito.2022.04.002, 18728278.

Peruzzotti-Jametti, L., Bernstock, J. D., Willis, C., Manferrari, G., Rogall, R., Fernandez-Vizarra, E., Williamson, J. C., Braga, A., van den Bosch, A., & Leonardi, T. (2021). Neural stem cells traffic functional mitochondria via extracellular vesicles. *PLoS Biology*.

Picone, P., & Nuzzo, D. (2022). Promising treatment for multiple Sclerosis: Mitochondrial transplantation. *International Journal of Molecular Sciences, 23*(4), 2245. https://doi.org/10.3390/ijms23042245, 1422-0067.

Picone, P., & Nuzzo, D. (2023). Biofabrication of nanovesicles for brain diseases. *Neural Regeneration Research, 18*(3), 525–526. https://doi.org/10.4103/1673-5374.346473, 18767958.

Picone, P., Nuzzo, D., Caruana, L., Scafidi, V., & Di Carlo, M. D. (2014). Mitochondrial dysfunction: Different routes to Alzheimer's disease therapy. *Oxidative Medicine and Cellular Longevity, 2014*. https://doi.org/10.1155/2014/780179, 19420994 http://www.hindawi.com/journals/oximed/.

Picone, P., Porcelli, G., Bavisotto, C. C., Nuzzo, D., Galizzi, G., Biagio, Bulone, D., & Carlo. (2021). Synaptosomes: New vesicles 255 for neuronal mitochondrial transplantation. *Journal of Nanobiotechnology, 6*.

Roushandeh, A. M., Kuwahara, Y., & Roudkenar, M. H. (2019). Mitochondrial transplantation as a potential and novel master key for treatment of various incurable diseases. *Cytotechnology, 71*(2), 647–663. https://doi.org/10.1007/s10616-019-00302-9, 15730778.

Shi, X., Zhao, M., Fu, C., & Fu, A. (2017). Intravenous administration of mitochondria for treating experimental Parkinson's disease. *Mitochondrion, 34*, 91–100. https://doi.org/10.1016/j.mito.2017.02.005, 18728278.

Xie, Q., Zeng, J., Zheng, Y., Li, T., Ren, J., Chen, K., Zhang, Q., Xie, R., Xu, F., & Zhu, J. (2021). Mitochondrial transplantation attenuates cerebral ischemia-reperfusion injury: Possible involvement of mitochondrial component separation. *Oxidative Medicine and Cellular Longevity, 2021*. https://doi.org/10.1155/2021/1006636, 19420994 http://www.hindawi.com/journals/oximed/.

Yan, C., Ma, Z., Ma, H., Li, Q., Zhai, Q., Jiang, T., Zhang, Z., & Wang, Q. (2020). Mitochondrial transplantation attenuates brain dysfunction in sepsis by driving microglial M2 polarization. *Molecular Neurobiology, 57*(9), 3875–3890. https://doi.org/10.1007/s12035-020-01994-3, 15591182.

Zhang, B., Gao, Y., Li, Q., Sun, D., Dong, X., Li, X., Xin, W., & Zhang, J. (2020). Effects of brain-derived mitochondria on the function of neuron and vascular endothelial cell after traumatic brain injury. *World Neurosurgery, 138*, e1–e9. https://doi.org/10.1016/j.wneu.2019.11.172, 18788769.

Zhang, Z., Ma, Z., Yan, C., Pu, K., Wu, M., Bai, J., Li, Y., & Wang, Q. (2019). Muscle-derived autologous mitochondrial transplantation: A novel strategy for treating cerebral ischemic injury. *Behavioural Brain Research, 356*, 322–331. https://doi.org/10.1016/j.bbr.2018.09.005, 18727549.

Zhao, J., Qu, D., Xi, Z., Huan, Y., Zhang, K., Yu, C., Yang, D., Kang, J., Lin, W., Wu, S., & Wang, Y. (2021). Mitochondria transplantation protects traumatic brain injury via promoting neuronal survival and astrocytic BDNF. *Translational Research, 235*, 102–114. https://doi.org/10.1016/j.trsl.2021.03.017, 18781810.

Zhao, Z., Yu, Z., Hou, Y., Zhang, L., & Fu, A. (2020). Improvement of cognitive and motor performance with mitotherapy in aged mice. *International Journal of Biological Sciences, 16*(5), 849–858. https://doi.org/10.7150/ijbs.40886, 14492288.

SECTION VII

Other diseases and mitochondrial transplantation

CHAPTER

19

Hereditary mitochondrial diseases and mitochondrial transplantation

Aycan Celik[1] *and Sevim Karakas Celik*[2]

[1]Department of Molecular Biology and Genetics, Faculty of Sciences, Zonguldak Bulent Ecevit University, Zonguldak, Turkey; [2]Department of Medical Genetics, Faculty of Medicine, Zonguldak Bulent Ecevit University, Zonguldak, Turkey

Mitochondria are cellular organelles that contain their circular mitochondrial DNA (mtDNA) that are critical for many cellular processes, such as maintaining calcium homeostasis and a stable body temperature, regulation of metabolism, and apoptosis. Changes in mitochondria-related genes impair the oxidative phosphorylation processes of the mitochondrial respiratory chain, reducing ATP production. As a result, tissues with high energy needs, such as the brain, optic nerve, retina, muscle, heart, and endocrine glands, are frequently affected than the others. Mitochondrial disease can present with a diverse range of clinical features that can often resemble other neurological or systemic diseases. However, clinicians are more likely to encounter a nonspecific collection of clinical features. Although many symptoms associated with mitochondrial disease, such as deafness, myopathy, diabetes and gastrointestinal symptoms, are also common individually in the general population, the combination of these symptoms in the same person should raise suspicion of mitochondrial disease. To identify other system involvement, a detailed system based examination coupled with extensive investigations is necessary. This may include short stature, sensorineural deafness, diabetes mellitus and/or other endocrine disorders, cardiac involvement, renal tubulopathy, optic atrophy, pigmentary retinopathy, and others. In the early stages of the disease, systemic symptoms may not be observed or clinically indistinguishable from other disorders. Disease burden and progression are determined by disease rating scales (Hameed & Tadi, 2022; Howell et al., 1991). Mitochondrial diseases in humans are caused by mutations, deletions, or duplications in mtDNA or mitochondria related genes encoded by nuclear DNA (nDNA). Therefore, these diseases can be inherited through both Mendelian inheritance and maternal inheritance.

Since histone proteins do not protect mtDNA as in nDNA, DNA repair mechanisms are limited compared to nDNA, and mutagenic reactive oxygen species (ROS) produced by

Mitochondrial Transplantation and Transfer
https://doi.org/10.1016/B978-0-443-18858-9.00019-6

cellular respiration accumulate in mitochondria; the mtDNA mutation rate is higher than nDNA (Klug et al., 2014). The inheritance, diagnosis, and management of diseases caused by mtDNA mutations can be more complex due to mitochondrial heteroplasmy. In addition, the heteroplasmy level determines the degree of cellular dysfunction resulting from mitochondrial damage.

Mitochondrial diseases may show various clinical manifestations. These diseases affect organs highly dependent on aerobic metabolism and are often associated with high morbidity and mortality (McFarland et al., 2010). Some clinical symptoms associated with mitochondrial disorders can be grouped into specific syndromes such as Leigh syndrome, depletion syndromes, Kearns–Sayre syndrome, Leber's hereditary optic neuropathy, and Pearson syndrome.

Hereditary mitochondrial diseases

Leigh Syndrome

Leigh syndrome is a complex and rare neurodegenerative disease described by Denis Leigh in 1951 (Leigh, 1951). Hommes and colleagues first associated Leigh syndrome with defects in mitochondrial energy metabolism in 1968 (Hommes et al., 1968). Leigh syndrome, called subacute necrotizing encephalomyelopathy, is an early onset mitochondrial disorder with phenotypic and genetic heterogeneity. Leigh syndrome usually affects the brain, such as the brainstem, cerebellum, basal ganglia, oculomotor, and cranial nerves (Chang et al., 2020). The most common clinical features of Leigh syndrome are ataxia, seizures, hypotonia, developmental delay, dystonia, motor weakness, optic atrophy, nystagmus, strabismus, ophthalmoplegia, ptosis, and retinopathy. However, dilated cardiomyopathy, hypertrophic cardiomyopathy, arrhythmia, diabetes mellitus, gastritis, hepatomegaly, renal dysfunction, and hematological disorders are also seen less frequently (Bakare et al., 2021).

Leigh syndrome is the result of pathogenic mutations that lead to defects in complexes involved in the electron transport chain, mitochondrial translation, thiamine transport, deficiency in pyruvate dehydrogenase enzyme, and depletion of mtDNA. These mutations can occur in nuclear genes (such as *BCS1L, COX10, GFM1, NDUFA1, NDUFA2, NDUFA4, NDUFA9, NDUFV1*) and mitochondrial genes (such as *MT-ATP6, MT-ND1, MT-ND2, MT-ND3, MT-TK, MT-TV, MT-TW, MT-TL1*). Mutations in nuclear genes can lead to deficiencies in various subunits of the respiratory chain complexes, causing decreased activity of the chain and decreased ATP production.

- *BCS1L* encodes a protein required for the assembly and stability of complex III of the respiratory chain.
- *COX10* encodes a protein involved in the final step of heme A biosynthesis, which is required for the function of complex IV (cytochrome *c* oxidase) of the respiratory chain.
- *GFM1* is involved in the translation of mtDNA into protein. Mutations in *GFM1* can cause decreased levels of mitochondrial translation and, consequently, decreased levels of proteins required for the respiratory chain.

- *NDUFA1, NDUFA2, NDUFA4,* and *NDUFA9,* as well as the genes *MT-ND1, MT-ND2,* and *MT-ND3,* all encode subunits of complex I of the respiratory chain.
- *NDUFV1* encodes a subunit of complex I that is required for the assembly and stability of the complex.

Besides these, all mitochondrial gene mutations can lead to decreased ATP production and increased ROS production.

- *MT-ATP6* encodes a subunit of complex V (ATP synthase) of the respiratory chain. Mutations in *MT-ATP6* can cause deficiencies in complex V, leading to decreased ATP production.
- *MT-TK, MT-TV, MT-TW,* and *MT-TL1* genes all encode transfer RNAs (tRNAs) that are required for the translation of mtDNA into protein. Mutations in any of these tRNA genes can cause decreased levels of mitochondrial translation and, consequently, decreased levels of proteins that are required for the respiratory chain.

Overall, mutations in the nuclear and mitochondrial genes associated with Leigh syndrome can cause deficiencies in the mitochondrial respiratory chain, leading to decreased ATP production and increased ROS production. These effects can ultimately result in damage to brain and nervous system tissues, leading to the symptoms of Leigh syndrome (Baracca et al., 2007; Elpeleg et al., 2005; Kucharczyk et al., 2009; Lenaz et al., 2004; Leng et al., 2015; Levy et al., 2014; Rahman et al., 1996; Saneto & Singh, 2010; Santorelli et al., 1993). It is known that Leigh syndrome can inherit through maternal inheritance, autosomal recessive inheritance, very rare X-linked or autosomal dominant inheritance, and via sporadic de novo mutation (Lake et al., 2016; Leigh et al., 2015; Saneto & Ruhoy, 2014). Mitochondrial inheritance occurs in about 20% of people with Leigh syndrome (MedlinePlus [Internet], 2020).

There is no specific treatment for Leigh syndrome and related mitochondrial disorders. Currently available options are either symptomatic treatments or regulating energy metabolism by lowering lactate levels and optimizing ATP production (Saneto & Ruhoy, 2014). Here are some examples of current treatment options for Leigh syndrome:

1. Thiamine (vitamin B1) supplementation: Thiamine is a cofactor in the pyruvate dehydrogenase complex, which is necessary for energy production in the mitochondria. Some patients with Leigh syndrome have been shown to benefit from high dose thiamine supplementation, which can improve the activity of the pyruvate dehydrogenase complex and increase energy production. However, thiamine supplementation is not effective for all patients with Leigh syndrome.
2. Ketogenic diet: A ketogenic diet is a high fat, low carbohydrate diet that can increase the production of ketones, which can serve as an alternative energy source for the brain. Some studies have shown that a ketogenic diet can improve the symptoms and quality of life of patients with Leigh syndrome. However, the diet can be difficult to follow and may not be effective for all patients.
3. Coenzyme Q10 (CoQ10) supplementation: CoQ10 is a key component of the mitochondrial respiratory chain and is involved in energy production. Some patients with Leigh syndrome have been shown to benefit from CoQ10 supplementation, which can improve energy production and reduce the production of ROS. However, the

effectiveness of CoQ10 supplementation can vary depending on the specific genetic mutation causing the disease.

4. EPI-743: EPI-743 is a small molecule that can improve the activity of the NADPH oxidase enzymes, which can reduce oxidative stress and inflammation in the mitochondria. Some studies have shown that EPI-743 can improve the symptoms and quality of life of patients with Leigh syndrome, although the treatment is still experimental (Frazier et al., 2019; Rahman, 2013; Yubero et al., 2020).

Depletion syndromes

mtDNA depletion syndromes are a group of mitochondrial disorders that are caused by mutations affecting the replication and maintenance of mtDNA. These disorders are characterized by a significant reduction in the amount of mtDNA present in affected tissues, leading to impaired mitochondrial function and energy production. The amount of mtDNA in a child changes significantly after birth. However, the effect of this change on human health has yet to be fully explained. It is known that the decrease in the amount of mtDNA causes some diseases in humans. These diseases are grouped as mtDNA depletion syndromes. Although mtDNA depletion syndromes are more common in childhood, they can also occur in adults. Symptoms observed include muscle weakness, ataxia, seizures, liver impairment, polyneuropathy, lactic acidosis, hypotonia, and dystonia.

mtDNA depletion syndrome is genetically heterogeneous. Defects in proteins required for mtDNA replication (DNA polymerase gamma encoded by *POLG1*; Twinkle-helicase encoded by *C10orf2*) or genes affecting and encoding the mitochondrial deoxyribonucleotide pool (*TK2, DGUOK, RRM2B, TYMP*) cause these syndromes (Suomalainen & Isohanni, 2010). This results in depleted mitochondrial genome delivery to daughter cells.

Twenty three types of mtDNA depletion syndromes have been described. mtDNA depletion syndromes can be divided into nine clinical groups: encephalomyopathic, myopathic, MNGIE (mitochondrial neurogastrointestinal encephalomyopathy), hepatocerebral, Alpers, neuro-ophthalmic, cardiomyopathic, hepatic, and cardioencephalomyopathic. Each of these clinical groups can be caused by defects in different genes.

For example, mutations in the nuclear gene *RRM2B* have been found to cause mtDNA depletion syndrome type 8 (OMIM 612075), which is characterized by decreased mtDNA content and impaired mitochondrial function. *RRM2B* encodes a subunit of ribonucleotide reductase, which is required for the synthesis of deoxyribonucleotides that are used to build new DNA during replication.

Another gene associated with mtDNA depletion syndromes is *POLG*, which encodes the mtDNA polymerase gamma. Mutations in this gene have been associated with several different types of mtDNA depletion syndromes, including Alpers syndrome (OMIM 203700), which is characterized by liver disease, seizures, and developmental regression, and progressive external ophthalmoplegia (PEO) (OMIM 157640), which is characterized by weak eye muscles.

mtDNA depletion syndrome 1 (MNGIE type) (OMIM 603041) is caused by a homozygous or compound heterozygous mutation in the *TYMP* gene on chromosome 22q13. The *TYMP* gene encodes the enzyme thymidine phosphorylase (TP), which is responsible for the breakdown of thymidine, a building block of DNA. Mutations in the *TYMP* gene can lead to a decrease in TP activity, causing the accumulation of thymidine in the mitochondria. The excess thymidine is then converted into deoxythymidine triphosphate, which can incorporate into the mtDNA, leading to depletion of mtDNA. In contrast, mtDNA depletion syndrome 20 (MNGIE type) (OMIM 619780) is caused by a compound heterozygous mutation in the *LIG3* gene on chromosome 17q12. The *LIG3* gene encodes the DNA ligase III protein, which plays a role in the process of joining DNA strands together. Compound heterozygous mutations in the *LIG3* gene can lead to defects in mtDNA replication and maintenance, resulting in a depletion of mtDNA in affected tissues.

In addition, disorders in the same clinical group may show different symptoms. For example, the symptoms of mtDNA depletion syndrome 9 (OMIM 245400), the encephalomyopathic type, are accompanied by methylmalonic aciduria. In contrast, the symptoms of the same type of mtDNA depletion syndrome 8a (OMIM 612075) are accompanied by renal tubulopathy (Online Mendelian Inheritance in Man, 2000).

Although no treatment can cure all the symptoms of mtDNA depletion syndromes, treatments are applied to increase the quality and duration of life and to prevent the progression of symptoms. For example, MPV17 related mtDNA defect presents in most affected individuals as an early onset encephalohepatopathic (hepatocerebral) disease typically associated with mtDNA depletion, particularly in the liver. As a result of liver transplantation in patients with liver disorders, liver functions improved, while neurological symptoms developed (Suomalainen & Isohanni, 2010). As the plasma thymidine level in MNGIE is proportional to the severity of the phenotype, patients were infused with platelets from a healthy donor to lower the circulating nucleotide level. In these patients, circulating thymidine level decreased, and TP activity partially recovered. However, the short half lives of platelets limited this treatment (Lara et al., 2006). Since 2006, some MNGIE patients have been tried to be treated with allogeneic stem cell transplantation. This application resulted in an improvement in TP activity and thymidine levels in plasma (Halter et al., 2011; Hirano et al., 2006; Rahman et al., 2007). In addition, a study conducted in 2012 showed that this treatment was ineffective in neurological findings. At the same time, it has been reported that transplantation is risky because it may cause earlier mortality in patients with poor prognosis, and allogeneic stem cell transplantation treatment has been recommended in relatively healthy patients (Filosto et al., 2012).

Kearns–Sayre syndrome

Kearns–Sayre syndrome is a rare disease characterized by the following three clinical signs: PEO that impairs eye movement and causes drooping eyelids, pigmentary retinopathy, and onset before 20 years of age (Kearns, 1958). Additionally, individuals with this syndrome have cardiac conduction defects, cerebellar ataxia, hearing loss, neuropathy, dementia, and endocrine abnormalities (Krishna, 2017).

Large scale mtDNA deletions have been detected in the majority of patients with Kearns–Sayre syndrome (Zeviani et al., 1988). The mutations occur in the noncoding regions of mtDNA, particularly in the D-loop region, which is responsible for regulating mtDNA replication and transcription. The most common deletion is a 4977 bp deletion, found in over a third of cases (Schon et al., 1989), which results in a loss of several essential genes required for oxidative phosphorylation, a process that generates energy in the mitochondria. This deletion can occur at different positions in the mtDNA, but most commonly occurs between nucleotides 8471 and 13,708. The loss of these genes leads to a deficiency in mitochondrial function, particularly in tissues with high energy demands such as the muscles, heart, and brain. As a result, patients with Kearns–Sayre syndrome exhibit a range of symptoms including PEO, a condition where the muscles controlling eye movements become weak, and ptosis, a drooping of the eyelids. Other symptoms may include weakness, fatigue, hearing loss, heart block, and ataxia.

The pathogenesis of Kearns–Sayre syndrome is thought to involve several pathways, including oxidative stress, mitochondrial calcium handling, and altered mitochondrial dynamics. The loss of oxidative phosphorylation leads to an accumulation of ROS which can damage mitochondrial components and cause oxidative stress. Mitochondria play a critical role in regulating calcium homeostasis in the cell. The loss of mitochondrial function in Kearns–Sayre syndrome can result in an accumulation of calcium ions, which can further damage mitochondrial components. High levels of calcium can also lead to mitochondrial permeability transition, which can cause the release of proapoptotic factors, leading to cell death. Finally, the altered mitochondrial dynamics can lead to an accumulation of damaged mitochondria, which can exacerbate mitochondrial dysfunction (Schon et al., 1989).

There is no specific treatment for Kearns–Sayre syndrome. Treatment is generally symptomatic and supportive. Surgical treatment can be applied by monitoring ophthalmological symptoms. Pacemaker implantation is recommended in patients with cardiac conduction blocks. Hormone replacement therapy (folic acid supplementation) is highly recommended in endocrinopathies (Cisneros & Henderson, 2021). Kearns–Sayre syndrome is a heterogeneous syndrome affecting the musculoskeletal, cardiovascular, and endocrine systems. It remains a complex disease to diagnose due to the diversity of clinical symptoms and its rarity.

Pearson syndrome

Pearson syndrome is a mitochondrial disease that occurs in infancy. It was first described in 1979 (Pearson et al., 1979). This syndrome causes problems in the development of hematopoietic cells in the bone marrow. mtDNA deletions are generally more common in blood leukocytes than in other tissues, and in Pearson syndrome, these deletions can be identified from leukocyte DNA. Most patients present with anemia, some with neutropenia and thrombopenia. Although the initial symptom is usually anemia, vomiting, diarrhea, and failure to thrive are also seen in newborns. It also occurs with multiorgan involvement, including exocrine pancreatic, liver, and renal tubular defects (Rötig et al., 1990). Children with this disease die early in life from severe lactic acidosis or liver failure. Most survivors develop symptoms of Kearns–Sayre syndrome (Rötig et al., 1995).

Single large scale mtDNA deletions of varying sizes and locations can cause Pearson syndrome. For example, almost half of the patients have a length of 4977 bp deletion called the "common deletion." These single large scale mtDNA deletions have also been observed in patients with Kearns–Sayre syndrome or chronic PEO (Broomfield et al., 2015; Rötig et al., 1995). Pearson syndrome, known to occur mostly with de novo mutations, is rarely inherited from the mother. However, Pearson syndrome has been observed in children with deleted mtDNA inheritance in children of mothers with chronic PEO in a few cases (Bernes et al., 1993; Sara Shanske et al., 2002).

As with most mitochondrial diseases, Pearson syndrome has no specific treatment. Supplements with limited effectiveness, such as vitamins, CoQ10, and L-carnitine, are used to prevent symptoms and treat organ-specific complications. Sodium bicarbonate supplementation can treat metabolic acidosis and balance electrolytes (Parikh et al., 2017; Yoshimi et al., 2022).

Chronic progressive external ophthalmoplegia

Chronic progressive external ophthalmoplegia (CPEO) is the most common myopathy among mitochondrial myopathies in adults. Losing extraocular muscle functions responsible for eye and eyelid movement characterizes it. Ptosis and strabismus are observed following slowly progressive ophthalmoparesis. Ptosis is usually the first finding. In addition to ophthalmoplegia, pigmentary retinopathy and skeletal muscle weakness may also be observed (Emery, 1997; Lee & Brazis, 2002; Michaelides & Moore, 2004). Patients may experience muscle weakness. Rarely, central nervous system diseases such as stroke, seizures, and dementia may be observed (Caballero et al., 2007). CPEO can be distinguished from Kearns–Sayre syndrome by showing ocular dysfunction, pigmentary retinopathy, and complete heart block. In addition, the onset of symptoms in Kearns–Sayre syndrome is usually before the age of 20, and the clinical course is more severe (Berenberg et al., 1977).

Mitochondrial and nuclear genome mutations, deletions, or duplications cause CPEO. These lead to the disruption of mitochondrial oxidative phosphorylation. In most cases of CPEO, sporadically "common deletion" of 4977 bp length is observed, as in Kearns–Sayre syndrome and Pearson syndrome. More than half of sporadically or maternal inherited mtDNA point mutations are found in genes encoding mt-tRNAs (Schon et al., 2012; Yarham et al., 2010). The A3243G mutation is the most common point mutation in patients with mitochondrial inherited CPEO (Moraes et al., 1993). Nuclear defects cause disruptions in mtDNA replication, leading to this disease. Mutations in the *POLG 1, POLG 2, SPG 7, DGUOK,* and *RRM2B* genes that cause CPEO are inherited as autosomal recessive and autosomal dominant (McClelland et al., 2016). These mutations lead to multiple mtDNA deletions. Therefore, CPEO may occur sporadically or be inherited by both mitochondrial inheritance, autosomal recessive and autosomal dominant Mendelian inheritance (Schon et al., 2012).

There is no proven treatment for CPEO. To improve the patient's quality of life, the symptoms are tried to be improved. Ocular defects can be treated with surgical approaches. As with other mitochondrial diseases, supplements are used to increase mitochondrial activity, but the effectiveness of these supplements has yet to be proven.

Leber's hereditary optic neuropathy

Leber hereditary optic neuropathy (LHON, OMIM 535000) was first described by Theodore Leber (1840–1917) in 1871 (Leber, 1871). In 1988, Wallace et al. determined that the disease was maternally inherited (Wallace et al., 1988). The first symptoms observed are blurred vision or clouding in one eye. Usually, visual impairment develops in the other eye within a few weeks or months (Riordan-Eva & Harding, 1995). It usually occurs in the teens or twenties. As time passes, there is a severe loss of visual acuity and color vision, and vision worsens in both eyes. Vision loss results from the death of optic nerve cells that transmit visual information from the eyes to the brain (MedlinePlus [Internet], 2020a). Typically, vision loss is the only symptom of LHON. However, additional symptoms such as tremors, movement disorder, juvenile onset encephalopathy, and cardiac conduction defects were observed in some families, and the disease was defined as "LHON+" in these individuals (Man et al., 2002; MedlinePlus [Internet], 2020a). Some affected individuals, particularly females with the G11778A mutation, show clinical features similar to multiple sclerosis, including muscle weakness, poor coordination, numbness, and even unmatched oligoclonal bands in the cerebrospinal fluid (Jansen et al., 1996; Kellar-Wood et al., 1994; Mojon et al., 1999; Olsen et al., 1995).

The three most common point mutations were determined to be G3460A, G11778A, and T14484C, respectively, in *MT-ND1, MT-ND4,* and *MT-ND6,* which contain genes encoding mitochondrial respiratory complex I (Carelli et al., 2004; Fraser et al., 2010). The mutations G3460A, G11778A, and T14484C in the genes *MT-ND1, MT-ND4,* and *MT-ND6,* respectively, all affect the functioning of complex I in the mitochondrial respiratory chain, which is involved in the production of energy within the mitochondria. These mutations alter the amino acid sequence of the proteins encoded by these genes, which in turn impairs the ability of complex I to function properly. Specifically, these mutations lead to the substitution of a highly conserved amino acid in each of the proteins: *MT-ND1* changes from alanine to threonine, *MT-ND4* changes from arginine to histidine, and *MT-ND6* changes from leucine to proline. These changes can have functional consequences for the mitochondrial respiratory chain, specifically complex I. Complex I is responsible for the transfer of electrons from NADH to ubiquinone, a process that generates a proton gradient that is used to generate ATP. The mutations in *MT-ND1, MT-ND4,* and *MT-ND6* can disrupt this process by interfering with the proper functioning of complex I. This results in a decrease in energy production and an accumulation of ROS, which can damage the mtDNA and cause cell death, particularly in the cells of the optic nerve (Carelli et al., 2004; Wallace, 2010; Yu-Wai-Man et al., 2011).

These three mutations account for approximately 95% of LHON cases. Although these mutations are rarely de novo, they are usually inherited from the mother via mitochondria (Biousse et al., 1997). The LHON + syndrome, which is a more severe form of LHON, on the other hand, has been associated with different mtDNA mutations (A11696G and/or T14596A (De Vries et al., 1996), T4160C (Howell et al., 1991), and G14459A (Jun et al., 1994)) apart from the three primary mutations.

Alcohol consumption can increase the production of ROS, which can damage the mitochondria and exacerbate the effects of LHON mutations. Similarly, smoking has been shown to increase the production of ROS and impair mitochondrial function, which can also exacerbate the effects of LHON mutations. So carriers with family members of LHON syndrome are advised to reduce their alcohol intake and stop smoking. Although there is no effective treatment in people with symptoms, it has been shown that supportive supplements such as vitamin B12, vitamin C, and idebenone can accelerate healing in the eyes (Mashima et al., 2000). Vitamin B12, vitamin C, and idebenone have been suggested as potential treatments for LHON syndrome due to their ability to support mitochondrial function and reduce oxidative stress, which can be a contributing factor to vision loss in LHON.

Vitamin B12 is involved in the synthesis of DNA and the production of energy within cells, including mitochondria. It has been demonstrated to enhance mitochondrial functionality and decrease oxidative stress in cells, which may contribute in protecting against the effects of LHON mutations. Additionally, vitamin B12 has been shown to have neuroprotective effects, which may help to preserve vision in individuals with LHON (Kumar et al., 2015).

Vitamin C is an antioxidant that can help to reduce oxidative stress and protect against cellular damage. It is reported that vitamin C can improve mitochondrial function and decrease oxidative stress in cells, potentially protecting against the effects of LHON mutations. Additionally, vitamin C can help to reduce inflammation, which can contribute to vision loss in LHON (Carr & Frei, 1999).

Idebenone is a synthetic form of CoQ10, which is an important component of the mitochondrial respiratory chain. Idebenone has been shown to improve mitochondrial function and reduce oxidative stress in cells, which may help to protect against the effects of LHON mutations. Additionally, idebenone has been shown to have neuroprotective effects and can improve blood flow to the eyes, which may help to preserve vision in individuals with LHON (Carr & Frei, 1999).

Overall, these supplements can support mitochondrial function and reduce oxidative stress, which may help to protect against the effects of LHON mutations and preserve vision in individuals with LHON. However, it is important to note that these supplements are not a cure for LHON and should be used in conjunction with other treatments under the guidance of a healthcare professional.

Myoclonic epilepsy with ragged red fibers

Myoclonic epilepsy with ragged red fibers (MERRF) was described by Tsairis et al., in 1973 and defined as a specific clinical syndrome in 1980 by Fukuhara et al. (Fukuhara et al., 1980; Tsaris, 1973). It is a rare mitochondrial disease characterized by progressive myoclonus (muscle twitches) and seizures (Fukuhara et al., 1980). In addition to these, cerebellar ataxia, myopathy, cardiac arrhythmia, sensorineural hearing loss, optic atrophy, and dementia are other symptoms associated with MERRF (Lorenzoni et al., 2014). Affected individuals may also develop hearing loss or optic atrophy. In most cases, symptoms appear during childhood or adolescence (MedlinePlus [Internet], 2020b). MERRF is distinguished from other mitochondrial diseases by the predominant symptom of progressive myoclonic epilepsy (Hameed & Tadi, 2022). However, MERRF symptoms are highly variable, even among patients in the same family. One of the reasons for this variability is the degree of mitochondrial heteroplasmy (the ratio of normal and mutant mtDNA in each tissue) (Lorenzoni et al., 2014).

MERRF is an inherited mitochondrial encephalomyopathy transmitted by nonMendelian maternal inheritance and occurs due to genetic mutations in mtDNA. The A8344G mutation in the *tRNA (Lys)* gene of mtDNA is the most common mutation and is found in more than three quarters of patients. This mutation affects the processing of tRNA, leading to a decrease in mitochondrial protein synthesis and ATP production (Schaefer et al., 2008). Other mutations associated with MERRF syndrome include the G8363A mutation in the *tRNA (Lys)* gene and the T8356C mutation in the *tRNA (Ser)* gene. Some of the other mtDNA mutations identified in MERRF are T8356C, G8363A, A3243G, G3255A, and T3291C (Hameed & Tadi, 2022).

MERRF, like other mitochondrial disorders, cannot be cured. CoQ10, vitamin B complex supplementation, and L-carnitine are used to reduce progression. It is also essential to avoid drugs and toxins that have adverse mitochondrial effects, such as aminoglycoside antibiotics, smoking, and alcohol (Lorenzoni et al., 2014). Antiepileptic drugs are given for myoclonic seizures. Valproate is an antiepileptic drug that is used to treat seizures and myoclonic epilepsy associated with MERRF syndrome. Valproate is thought to work by increasing the levels of GABA (gamma aminobutyric acid), which is an inhibitory neurotransmitter in the brain that helps reduce neuronal excitability and prevent seizures. Additionally, valproate has been shown to have an effect on mitochondrial function, which is relevant to MERRF syndrome. Valproate has been shown to increase the expression of mitochondrial proteins and enhance mitochondrial function. This could help compensate for the mitochondrial dysfunction caused by the mtDNA mutations in MERRF syndrome (Mantoan & Walker, 2011; Sitarz et al., 2014). Sodium valproate is generally the recommended drug for myoclonic epilepsies. However, it should be used cautiously in mitochondrial disorders as it inhibits carnitine uptake and may accelerate liver failure (Finsterer & Zarrouk-Mahjoub, 2017).

Neuropathy, ataxia, and retinitis pigmentosa

Neuropathy, ataxia, and retinitis pigmentosa (NARP) is a syndrome that affects the nervous system. Symptoms of NARP usually begin in childhood or early adulthood and get progressively worse. The most common symptoms of this syndrome are numbness or pain in the arms and legs, muscle weakness, and difficulty with balance and coordination. Most people have vision loss due to retinitis pigmentosa, which causes the retina's light sensitive cells to deteriorate. Affected children may have learning difficulties, and the elderly may have dementia. In addition, seizures, hearing loss, and cardiac conduction defects are other symptoms of NARP (MedlinePlus [Internet], 2020c).

MT-ATP6 mutations are frequently observed in NARP syndrome. In terms of the m.8993T > G mutation in the *MT-ATP6* gene, a combination of NARP has been described in a heteroplasmic family (Holt et al., 1990). This mutation has also been commonly observed in patients with Leigh syndrome, a similar but more severe disorder to NARP. The mutation load determines which of these two diseases will develop (Tatuch et al., 1992). A heteroplasmy level greater than 90% of mutant mtDNA tends to result in Leigh syndrome. Below that percentage, the phenotype often manifests as a NARP syndrome (Carelli et al., 2002; Mäkelä-Bengs et al., 1995; Rojo et al., 2006). MT-ATP mutations are generally associated with Leigh syndrome, but m.8993T > G and m.8993T > C are associated with the classic NARP phenotype. Two other mutations, m.8618insT, and m.9185T > C, have been associated with a NARP-like phenotype (Childs et al., 2007; Lopez-Gallardo et al., 2008), but the clinical spectrum appears to be highly variable.

The molecular mechanism of the m.8993T > G mutation is not fully understood, but it is thought to disrupt the function of ATP synthase by affecting its assembly and stability. The mutation leads to a decrease in ATP synthase activity, which results in a decrease in ATP production and an increase in ROS generation in affected cells.

At the cellular level, the m.8993T > G mutation has been shown to cause a number of functional changes in affected cells, including altered mitochondrial morphology, impaired mitochondrial calcium uptake, and increased susceptibility to oxidative stress. These changes can lead to mitochondrial dysfunction and cell death in affected tissues. In addition, it has been reported that different mutations in the *MT-ATP6* gene are associated with the NARP phenotype. The m.8839G > C mutation (Blanco-Grau et al., 2013), m.8989G > C mutation (Duno et al., 2013), and m.9127-9128delAT microdeletion of 2 bp, which was identified for the first time (Mordel et al., 2017), have been associated with NARP syndrome. Apart from these, it has been reported that the combination of retinitis pigmentosa and ataxia in an adolescent is associated with the m.1606G > A variant in the mitochondrial *tRNA (Val) gene* (Sacconi et al., 2002). Furthermore, it has been shown that the m.5789T > C variant in the *tRNA (Cys)* gene of mtDNA is also related to late onset NARP syndrome (Hippen et al., 2022).

There is no specific treatment for NARP. Treatment is mainly supportive, including treatments for acute acidosis, dystonia, and cardiomyopathy. Regular follow up (every 6–12 months) and psychological support are recommended. Sodium valproate, barbiturates, dichloroacetate, and anesthetic drugs that may worsen the NARP syndrome should be avoided (Couser & Gucsavas-Calikoglu, 2017, pp. 167–190).

Sodium valproate is a commonly used medication for the treatment of epilepsy and bipolar disorder. It is known to inhibit mitochondrial function by reducing mitochondrial respiration and ATP production, which may exacerbate the energy deficit in NARP syndrome. Some studies have suggested that sodium valproate can also interfere with mtDNA replication and repair, leading to an increase in mtDNA mutations and further mitochondrial dysfunction.

Barbiturates are a class of drugs that are used as sedatives and anesthetics. They have been shown to impair mitochondrial respiration and decrease ATP production, which may worsen the energy deficit in NARP syndrome. Barbiturates have also been shown to cause mtDNA damage and reduce the expression of mitochondrial genes, which can further impair mitochondrial function.

Dichloroacetate is a medication that is used to treat metabolic disorders by stimulating mitochondrial function. However, in NARP syndrome, dichloroacetate has been reported to exacerbate symptoms, possibly due to its inhibitory effect on the pyruvate dehydrogenase complex, which is required for the conversion of pyruvate to acetyl CoA, a critical step in mitochondrial ATP production.

In summary, sodium valproate, barbiturates, dichloroacetate, and anesthetic drugs can worsen NARP syndrome by impairing mitochondrial function, reducing ATP production, and causing mtDNA damage. These effects can exacerbate the energy deficit in affected cells and lead to worsening of symptoms. Therefore, caution should be provided when using these drugs in patients with NARP syndrome. It is recommended that patients with mitochondrial disorders should be closely monitored when using any drugs that affect mitochondrial function (Anderson et al., 2002; Wong, 2007).

Mitochondrial encephalopathy, lactic acidosis, and stroke-like episodes

Mitochondrial encephalomyopathy, lactic acidosis, and stroke-like episodes (MELAS) is a genetically heterogeneous mitochondrial disorder. MELAS syndrome was first described by

Pavlakis in 1984 (Pavlakis et al., 1984). It is known that symptoms usually begin in childhood (Michio Hirano & Pavlakis, 1994; Sproule & Kaufmann, 2008; Yatsuga et al., 2012). Early symptoms include muscle weakness and pain, recurrent headaches, loss of appetite, vomiting, and seizures. Most individuals affected by this syndrome experience stroke-like episodes before the age of 40 (MedlinePlus [Internet], 2020d). Repeated stroke-like episodes can progressively damage the brain, leading to vision loss, problems with movement, and dementia. 40%–90% of MELAS syndrome cases have dementia, and 71%–96% have epilepsy (Michio Hirano & Pavlakis, 1994; Sproule & Kaufmann, 2008; Yatsuga et al., 2012). Other neurological symptoms are recurrent headaches, ataxia, developmental delay, learning disorders, and memory loss. Additionally, lactic acidosis is observed in most patients. This condition can cause abdominal pain, vomiting, fatigue, and muscle weakness (MedlinePlus [Internet], 2020d).

MELAS syndrome can result from mutations in one of several genes, including *MT-ND1, MT-ND5, MT-TH, MT-TL1,* and *MT-TV* (MedlinePlus [Internet], 2020d). *MT-ND1* and *MT-ND5* are both subunits of complex I of the mitochondrial respiratory chain. Mutations in these genes can lead to a defect in complex I activity, resulting in reduced ATP production and increased production of ROS. This can cause mitochondrial dysfunction and damage to the brain, leading to the symptoms of MELAS syndrome. *MT-TH* encodes for the tRNA for histidine, which is involved in protein synthesis within the mitochondria. *MT-TL1* encodes for the tRNA for leucine, which is also involved in protein synthesis within the mitochondria. *MT-TV* encodes for the tRNA for threonine, which is involved in protein synthesis within the mitochondria. Mutations in these genes can impair mitochondrial protein synthesis and lead to a deficiency in oxidative phosphorylation, which can cause energy failure and the symptoms of MELAS syndrome (El-Hattab et al., 2015; Goto et al., 1990).

A3243G mutation of mtDNA, which is known to be associated with other mitochondrial syndromes such as MERRF syndrome and Leigh syndrome, has also been associated with MELAS syndrome (Goto et al., 1990). In addition, several mtDNA mutations such as T3271C, A3252G, T3291C, G3959A, A10134C, T10191C, G10197A (Vodopivec et al., 2016), G13513A (Shanske et al., 2008), and T10158C (Wang et al., 2020) and mutations in nuclear genes such as *POLG1* (Deschauer et al., 2007) or pyruvate dehydrogenase complex deficiency can also cause MELAS syndrome (Wilichowski et al., 1998). How these changes lead to the specific signs and symptoms of MELAS syndrome has yet to be determined.

There is no cure for this syndrome. The treatment of symptoms includes adequate fluid intake, nutrition, the use of antipsychotics or tranquilizers, and rehabilitation. It has been suggested that prolonged seizures in patients with MELAS syndrome seriously damage the brain, cause dementia, and damage the blood–brain barrier (Fryer et al., 2016; Michio Hirano & Pavlakis, 1994; Sproule & Kaufmann, 2008; Yatsuga et al., 2012). Therefore, it is thought that using antiepileptic drugs will reduce the occurrence of stroke-like episodes (Hue et al., 2015). Antiepileptic drugs such as valproic acid, levetiracetam, and carbamazepine can also help control seizures and improve neurological symptoms.

Steroid response is observed during the treatment of MELAS syndrome, and discontinuation of steroids causes the worsening of the condition (Finsterer, 2019b). However, the mechanism of action of steroids in this treatment is not fully known. It can be mentioned the following as possible mechanism of action. Steroids are known to have potent antiinflammatory effects and can reduce inflammation and tissue damage caused by oxidative stress in

MELAS syndrome. This may help improve the symptoms and slow the progression of the disease. Also, there is some evidence to suggest that steroids can promote mitochondrial biogenesis, which is the process of generating new mitochondria in the cell. This may help compensate for the impaired mitochondrial function in MELAS syndrome and improve energy production. In addition, steroids may have neuroprotective effects and help preserve brain function in MELAS syndrome. This may be important in cases where the disease is associated with neurological symptoms such as seizures and stroke-like episodes (Finsterer & Frank, 2015; Rossi et al., 2002; Weber et al., 2002).

Patients with MELAS syndrome often have feeding difficulties, malabsorption, and nutrient deficiencies. Nutritional support, including enteral or parenteral feeding, vitamin and mineral supplementation, and dietary modification, can help improve nutritional status and overall health. Also supplements such as CoQ10, vitamin C, folic acid, vitamin E, vitamins B1, B2, B3, B12, creatine, biotin, and levocarnitine are used to regulate energy metabolism in the diet. L-arginine is a precursor of nitric oxide (NO), which plays a crucial role in vasodilation and blood flow regulation. Patients with MELAS syndrome often have impaired NO production, leading to decreased blood flow and tissue hypoxia. L-arginine supplementation can improve blood flow and oxygen delivery to the tissues. Patients with MELAS syndrome often experience muscle weakness and impaired motor function. In addition to these supplements, an appropriate physical therapy and rehabilitation can help improve muscle strength, coordination, and physical function of patients with MELAS syndrome (Finsterer, 2019a; Venturelli et al., 2019).

Mitochondrial transplantation in hereditary mitochondrial diseases

In recent years, mitochondrial transplantation has become a distinctive approach to treating mitochondrial diseases. Mitochondrial transplantation aims to transfer functional exogenous mitochondria into mitochondria defective cells. The mitochondrial transplantation was first performed by Clark and Shay to demonstrate the mitochondrial mediated transfer of antibiotic resistance via endocytosis. This study showed that purified mitochondria from chloramphenicol resistant and efrapeptin resistant cells were taken up by endocytosis and transferred antibiotic resistance to chloramphenicol sensitive and efrapeptin sensitive cells (Clark & Shay, 1982). This study of the mitochondria mediated transfer of antibiotic resistance brought to mind mitochondrial transplantation, a new approach to studying mitochondrial genetics in mammalian cells. However, in this study, it has been shown that in vitro uptake of purified mitochondria occurs with low efficacy and can be sustained under selective pressure (Clark & Shay, 1982). Subsequent studies have shown that transferring exogenous normal mitochondria to mitochondria defective human cells reconstitutes mtDNA content by changing the endogenous mtDNA in the cells. In the study published by King and Atardi in 1988, two different types of human cells were partially defected in terms of mtDNA by ethidium bromide application. Then, chloramphenicol resistant mtDNA was transferred to these cells by microinjection. As a result of DNA polymorphism analyses performed on total DNA extracted from cells grown in selective media, it was observed that exogenous mtDNA replaced endogenous mtDNA in most of the transformants (King & Attardi, 1988). In King and Atardi's study, published in 1989, it was observed that mitochondria defective (ρ^0)

human cells become dependent on uridine or pyruvate, unlike their parental cells. Mitochondria transfer was achieved by the fusion of ρ^0 cells with cytoplasts. Transformants were seeded in a medium lacking either pyruvate or uridine. Thus, only ρ^0 cells that received functional mitochondria survived. ρ^0 cells fused with cytoplasts are called cytoplasmic hybrid or cybrid cells (King & Attardi, 1989). In a 2007 study, mitochondria to be transplanted were isolated from mouse liver and mouse cell line. It has been shown that exogenous mitochondria stained with mitochondria specific fluorescent dye (MitoFluor green) can enter human and mouse cells within 1 h at 37°C. Thus, it was proved for the first time that the xenogenic transfer of mitochondria is possible. Mitochondria isolated from mouse liver tissue were transferred to ρ^0 human cells. These cells were incubated with murine mitochondria in calcium free media. Cells that receive mitochondria (ρ^*) and ρ^0 cells were tested for oxygen consumption, and it was found that oxygen consumption increased significantly in ρ^* cells. However, ρ^* cells did not show respiration levels as high as normal cells. This study demonstrated that xenogeneic transfer of mouse mitochondria into human cells lacking functional mitochondria is possible while at the same time that these human cells regain significant respiratory function. However, Katrangi et al. suggested another approach in the same study that could allow the transfer of exogenous DNA to mammalian mitochondria and the reversal of mitochondrial dysfunction (Katrangi et al., 2007). Given these successful studies, mitochondrial transplantation has been a promising approach that can improve defective mitochondria and treat mitochondrial disorders.

A study published in 2013 investigated mtDNA transfer between cells using the cell-penetrating peptide Pep-1 in a cybrid cell model of MERRF syndrome (Chang et al., 2013). Pep-1 (KETWWETWWTEWSQPKKKRKV—cysteamine) is a cell-penetrating protein. It can transport various peptides and proteins into the cell through hydrophobic and electrostatic interactions with the cell membrane (Morris et al., 2001). Pep-1 conjugated wild type mitochondria were isolated and coincubated with MERRF cybrid and ρ^0 cells. After 48 h, Pep-1 conjugated mitochondria were observed to have entered the MERRF cybrid and ρ^0 cells. Improvement was observed in the mitochondrial functions, and exogenous mitochondria remained in these cells for at least 15 days. In addition, it has been reported that mitochondria related cell death is prevented in cells. Peptide mediated mitochondrial delivery (PMD) has been observed to increase mitochondrial biogenesis in cells. Inhibition of mitochondrial fission was observed in rescued MERRF cybrid cells, but mitochondrial fusion was not improved. However, the promotion of mitochondrial fusion was observed in ρ^0 cells. It has been suggested that the reason for this result may be due to the insufficient functioning of mitochondria in MERRF cybrid cells. In general, combining damaged mitochondria in normal cells with neighboring exogenous mitochondria restores the functionality of endogenous mitochondria. Considering the results of this research, the efficacy of mitochondrial therapy with PMD is thought to depend on the properties of damaged endogenous mitochondria (Chang et al., 2013). These findings indicate that the translocation of foreign mitochondria inhibits endogenous mitofusin-2 and endoplasmic reticulum connection and reduces the excessive uptake of mitochondrial calcium. Thus, mitochondrial permeability transition pore opening and cell death are prevented (Jou et al., 2010).

In 2017, Chuang et al. investigated whether Wharton's jelly mesenchymal stem cells (WJMSCs) via mitochondria transfer had therapeutic potential for treating MERRF syndrome. In this study, MERRF cybrid cells with high mt.8344A $>$ G mutation were used.

They showed that mitochondria were transferred from WJSMC cells to MERRF cybrid cells by coculture experiments. After coculture with WJMSCs, the mt.8344A > G mutation load was partially reduced in MERRF cybrid cells. Also, ROS level and oxidative damage were reduced, mitochondrial bioenergetics was improved, and resistance to apoptotic stress developed in these cells. In addition, it was reported in this study that WJSMCs did not transfer mitochondria to healthy cells with normal mitochondria. This study demonstrated that WJMSCs exert bioenergetic therapeutic effects via mitochondrial transfer (Chuang et al., 2017). A similar study was carried out for MELAS syndrome in 2019. In this study, it was investigated whether mitochondrial damage in the cells of a MELAS patient could be treated by mitochondria transfer via WJMSCs. Coculture experiments show that WJMSCs transfer healthy mitochondria to rotenone stressed MELAS fibroblasts. The source of mitochondria transferred by sequencing was confirmed to be WJMSCs. After mitochondrial transfer, mt.3243A > G mutation load decreased, and mitochondrial functions improved in MELAS fibroblasts (Lin et al., 2019).

In a study published in 2018, Pearson syndrome patient derived hematopoietic stem cells were enriched with wild type healthy mitochondria by mitochondrial transplantation called mitochondrial augmentation therapy. Then ex vivo enriched autologous hematopoietic stem cells were transferred to three patients with Pearson syndrome. Patients were followed up for 1 year. Adverse events noted after treatment were related to leukapheresis, including anemia, hypocalcemia, and alkalosis. Mitochondrial enrichment began to be observed 3–4 months after treatment in two patients and continued to increase throughout the follow up period. In addition, improvement in lymphocyte ATP content, O2 consumption, and TMRE: MTG (tetramethylrhodamine ethyl ester: MitoTracker green) ratio was observed 5 months after treatment. The patient's quality of life has been reported to have dramatically improved (Elad Jacoby et al., 2018).

Another mitochondrial augmentation therapy study published in 2020 tried to improve the physiological and metabolic status of a patient with Kearns–Sayre syndrome. In this study, normal mitochondria from the patient's mother were transferred to the patient's $CD34^+$ hematopoietic stem cells and transplanted into the patient. Improvement in seizures was observed 4 months after treatment. After 7 months of follow up, the patient improved appetite, sitting, walking, as well as self-expression and the ATP content of peripheral blood lymphocytes increased (Yosef et al., 2020).

In a 2021 study, cord blood derived $CD34^+$ cells from a patient with Pearson syndrome were transplanted into humanized NOD-SCID-Gamma-SGM3 (NSGS) mice to evaluate the potential effects of mitochondrial augmentation therapy on human hematopoietic stem cells. Although complete human hematopoiesis is not expected, mitochondrial enriched $CD34^+$ cells improved long term engraftment, and a higher proportion of human $CD3^+$ T cells was observed in NSGS mice. In addition, it has been observed that in vivo transfer of exogenous mitochondria to host hematopoietic cells continues for up to 4, 5 months after transplantation. These findings indicate that mitochondrial augmentation therapy is feasible and improves the mitochondrial content and oxygen consumption of hematopoietic stem cells (Jacoby et al., 2021).

Mitochondrial transplantation aims to change defective mitochondria with healthy mitochondria in receiving cells and also improve cellular energy production. In vitro studies have confirmed that exogen mitochondria can be transferred to cells by mitochondrial transplantation. In addition, many studies have shown evidence of improvement in cellular bioenergetics. Although clinical studies are still limited when these data are evaluated, it is predicted that mitochondrial transplantation therapies will be a potential therapeutic method for mtDNA disorders. However, most of the available information on the applications of mitochondrial transplantation in the treatment of hereditary mitochondrial diseases is based on in vitro experiments. Nevertheless, as a result of the studies examined, it is clear that mitochondrial transplantation is a promising therapeutic approach in the clinical treatment of hereditary mitochondrial diseases.

References

Anderson, C. M., Norquist, B. A., Vesce, S., Nicholls, D. G., Soine, W. H., Duan, S., & Swanson, R. A. (2002). Barbiturates induce mitochondrial depolarization and potentiate excitotoxic neuronal death. *Journal of Neuroscience, 22*(21), 9203–9209. https://doi.org/10.1523/jneurosci.22-21-09203.2002

Bakare, A. B., Lesnefsky, E. J., & Iyer, S. (2021). Leigh syndrome: A tale of two genomes. *Frontiers in Physiology, 12*. https://doi.org/10.3389/fphys.2021.693734

Baracca, A., Sgarbi, G., Mattiazzi, M., Casalena, G., Pagnotta, E., Valentino, M. L., Moggio, M., Lenaz, G., Carelli, V., & Solaini, G. (2007). Biochemical phenotypes associated with the mitochondrial ATP6 gene mutations at nt8993. *Biochimica et Biophysica Acta – Bioenergetics, 1767*(7), 913–919. https://doi.org/10.1016/j.bbabio.2007.05.005

Berenberg, R. A., Pellock, J. M., DiMauro, S., Schotlan, D. L., Bonilla, E., Eastwood, A., Hays, A., Vicale, C. T., Behrens, M., Chutorian, A., & Rowland, L. P. (1977). Lumping or splitting? "Ophthalmoplegia-plus" or Kearns-Sayre syndrome? *Annals of Neurology, 1*(1), 37–54. https://doi.org/10.1002/ana.410010104

Bernes, S. M., Bacino, C., Prezant, T. R., Pearson, M. A., Wood, T. S., Fournier, P., & Fischel-Ghodsian, N. (1993). Identical mitochondrial DNA deletion in mother with progressive external ophthalmoplegia and son with Pearson marrow-pancreas syndrome. *The Journal of Pediatrics, 123*(4), 598–602. https://doi.org/10.1016/S0022-3476(05)80962-X

Biousse, V., Brown, M. D., Newman, N. J., Allen, J. C., Rosenfeld, J., Meola, G., & Wallace, D. C. (1997). De novo 14484 mitochondrial DNA mutation in monozygotic twins discordant for Leber's hereditary optic neuropathy. *Neurology, 49*(4), 1136–1138. https://doi.org/10.1212/wnl.49.4.1136

Blanco-Grau, A., Bonaventura-Ibars, I., Coll-Cantí, J., Melià, M. J., Martinez, R., Martínez-Gallo, M., Andreu, A. L., Pinós, T., & García-Arumí, E. (2013). Identification and biochemical characterization of the novel mutation m.8839G>C in the mitochondrial ATP6 gene associated with NARP syndrome. *Genes, Brain and Behavior, 12*(8), 812–820. https://doi.org/10.1111/gbb.12089

Broomfield, A., Sweeney, M. G., Woodward, C. E., Fratter, C., Morris, A. M., Leonard, J. V., Abulhoul, L., Grunewald, S., Clayton, P. T., Hanna, M. G., Poulton, J., & Rahman, S. (2015). Paediatric single mitochondrial DNA deletion disorders: An overlapping spectrum of disease. *Journal of Inherited Metabolic Disease, 38*(3), 445–457. https://doi.org/10.1007/s10545-014-9778-4

Caballero, P. E. J., Candela, M. S., Álvarez, C. I. C., & Tejerina, A. A. (2007). Chronic progressive external ophthalmoplegia: A report of 6 cases and a review of the literature. *The Neurologist, 13*(1), 33–36. https://doi.org/10.1097/01.nrl.0000252953.49721.f5

Carelli, V., Baracca, A., Barogi, S., Pallotti, F., Valentino, M. L., Montagna, P., Zeviani, M., Pini, A., Lenaz, G., Baruzzi, A., & Solaini, G. (2002). Biochemical-clinical correlation in patients with different loads of the mitochondrial DNA T8993G mutation. *Archives of Neurology, 59*(2), 264–270. https://doi.org/10.1001/archneur.59.2.264

Carelli, V., Ross-Cisneros, F. N., & Sadun, A. A. (2004). Mitochondrial dysfunction as a cause of optic neuropathies. *Progress in Retinal and Eye Research, 23*(1), 53–89. https://doi.org/10.1016/j.preteyeres.2003.10.003

Carr, A. C., & Frei, B. (1999). Toward a new recommended dietary allowance for vitamin C based on antioxidant and health effects in humans. *American Journal of Clinical Nutrition, 69*(6), 1086–1107. https://doi.org/10.1093/ajcn/69.6.1086

Chang, J. C., Liu, K. H., Li, Y. C., Kou, S. J., Wei, Y. H., Chuang, C. S., Hsieh, M., & Liu, C. S. (2013). Functional recovery of human cells harbouring the mitochondrial DNA mutation MERRF A8344G via peptide-mediated mitochondrial delivery. *Neurosignals, 21*(3–4), 160–173. https://doi.org/10.1159/000341981

Chang, X., Wu, Y., Zhou, J., Meng, H., Zhang, W., & Guo, J. (2020). A meta-analysis and systematic review of Leigh syndrome: Clinical manifestations, respiratory chain enzyme complex deficiency, and gene mutations. *Medicine, 99*(5), e18634. https://doi.org/10.1097/MD.0000000000018634

Childs, A. M., Hutchin, T., Pysden, K., Highet, L., Bamford, J., Livingston, J., & Crow, Y. J. (2007). Variable phenotype including Leigh syndrome with a 9185T>C mutation in the MTATP6 gene. *Neuropediatrics, 38*(6), 313–316. https://doi.org/10.1055/s-2008-1065355

Chuang, Y.-C., Liou, C.-W., Chen, S.-D., Wang, P.-W., Chuang, J.-H., Tiao, M.-M., Hsu, T.-Y., Lin, H.-Y., & Lin, T.-K. (2017). Mitochondrial transfer from Wharton's jelly mesenchymal stem cell to MERRF cybrid reduces oxidative stress and improves mitochondrial bioenergetics. *Oxidative Medicine and Cellular Longevity, 2017*, 1–22. https://doi.org/10.1155/2017/5691215

Cisneros, A., & Henderson, J. O. (2021). Kearns Sayre syndrome: A mitochondrial disorder. *Journal of Student Research, 10*(1). https://doi.org/10.47611/jsr.v10i1.1099

Clark, M. A., & Shay, J. W. (1982). Mitochondrial transformation of mammalian cells. *Nature, 295*(5850), 605–607. https://doi.org/10.1038/295605a0

Couser, N., & Gucsavas-Calikoglu, M. (2017). *Mitochondrial disorders*. Elsevier BV. https://doi.org/10.1016/b978-0-12-802896-4.00008-0

De Vries, D. D., Went, L. N., Bruyn, G. W., Scholte, H. R., Hofstra, R. M. W., Bolhuis, P. A., & Van Oost, B. A. (1996). Genetic and biochemical impairment of mitochondrial complex I activity in a family with Leber hereditary optic neuropathy and hereditary spastic dystonia. *The American Journal of Human Genetics, 58*(4), 703–711.

Deschauer, M., Tennant, S., Rokicka, A., He, L., Kraya, T., Turnbull, D. M., Zierz, S., & Taylor, R. W. (2007). MELAS associated with mutations in the POLG1 gene. *Neurology, 68*(20), 1741–1742. https://doi.org/10.1212/01.wnl.0000261929.92478.3e

Duno, M., Wibrand, F., Baggesen, K., Rosenberg, T., Kjaer, N., & Frederiksen, A. L. (2013). A novel mitochondrial mutation m.8989G>C associated with neuropathy, ataxia, retinitis pigmentosa – The NARP syndrome. *Gene, 515*(2), 372–375. https://doi.org/10.1016/j.gene.2012.12.066

El-Hattab, A. W., Adesina, A. M., Jones, J., & Scaglia, F. (2015). MELAS syndrome: Clinical manifestations, pathogenesis, and treatment options. *Molecular Genetics and Metabolism, 116*(1–2), 4–12. https://doi.org/10.1016/j.ymgme.2015.06.004

Elpeleg, O., Miller, C., Hershkovitz, E., Bitner-Glindzicz, M., Bondi-Rubinstein, G., Rahman, S., Pagnamenta, A., Eshhar, S., & Saada, A. (2005). Deficiency of the ADP-forming succinyl-CoA synthase activity is associated with encephalomyopathy and mitochondrial DNA depletion. *The American Journal of Human Genetics, 76*(6), 1081–1086. https://doi.org/10.1086/430843

Emery, A. E. (1997). *Diagnostic criteria for neuromuscular disorders*. Royal Society of Medicine Press.

Filosto, M., Scarpelli, M., Tonin, P., Lucchini, G., Pavan, F., Santus, F., Parini, R., Donati, M. A., Cotelli, M. S., Vielmi, V., Todeschini, A., Canonico, F., Tomelleri, G., Padovani, A., & Rovelli, A. (2012). Course and management of allogeneic stem cell transplantation in patients with mitochondrial neurogastrointestinal encephalomyopathy. *Journal of Neurology, 259*(12), 2699–2706. https://doi.org/10.1007/s00415-012-6572-9

Finsterer, J. (2019a). Commentary: Neuromuscular and muscle metabolic functions in melas before and after resistance training: A case study. *Frontiers in Physiology, 10*.

Finsterer, J. (2019b). Mitochondrial metabolic stroke: Phenotype and genetics of stroke-like episodes. *Journal of the Neurological Sciences, 400*, 135–141. https://doi.org/10.1016/j.jns.2019.03.021

Finsterer, J., & Frank, M. (2015). Glucocorticoids for mitochondrial disorders. *Singapore Medical Journal, 56*(2), 122–123. https://doi.org/10.11622/smedj.2015026

Finsterer, J., & Zarrouk-Mahjoub, S. (2017). Management of epilepsy in MERRF syndrome. *Seizure, 50*, 166–170. https://doi.org/10.1016/j.seizure.2017.06.010

Fraser, J. A., Biousse, V., & Newman, N. J. (2010). The neuro-ophthalmology of mitochondrial disease. *Survey of Ophthalmology, 55*(4), 299–334. https://doi.org/10.1016/j.survophthal.2009.10.002

Frazier, A. E., Thorburn, D. R., & Compton, A. G. (2019). Mitochondrial energy generation disorders: Genes, mechanisms, and clues to pathology. *Journal of Biological Chemistry, 294*(14), 5386–5395. https://doi.org/10.1074/jbc.R117.809194

Fryer, R. H., Bain, J. M., & De Vivo, D. C. (2016). Mitochondrial encephalomyopathy lactic acidosis and stroke-like episodes (MELAS): A case report and critical reappraisal of treatment options. *Pediatric Neurology, 56*, 59–61. https://doi.org/10.1016/j.pediatrneurol.2015.12.010

Fukuhara, N., Tokiguchi, S., Shirakawa, K., & Tsubaki, T. (1980). Myoclonus epilepsy associated with ragged-red fibres (mitochondrial abnormalities): Disease entity or a syndrome? Light- and electron-microscopic studies of two cases and review of literature. *Journal of the Neurological Sciences, 47*(1), 117–133. https://doi.org/10.1016/0022-510X(80)90031-3

Goto, Y. I., Nonaka, I., & Horai, S. (1990). A mutation in the tRNALeu(UUR) gene associated with the MELAS subgroup of mitochondrial encephalomyopathies. *Nature, 348*(6302), 651–653. https://doi.org/10.1038/348651a0

Halter, J., Schüpbach, W. M. M., Casali, C., Elhasid, R., Fay, K., Hammans, S., Illa, I., Kappeler, L., Krähenbühl, S., Lehmann, T., Mandel, H., Marti, R., Mattle, H., Orchard, K., Savage, D., Sue, C. M., Valcarcel, D., Gratwohl, A., & Hirano, M. (2011). Allogeneic hematopoietic SCT as treatment option for patients with mitochondrial neurogastrointestinal encephalomyopathy (MNGIE): A consensus conference proposal for a standardized approach. *Bone Marrow Transplantation, 46*(3), 330–337. https://doi.org/10.1038/bmt.2010.100

Hameed, & Tadi, P. (2022). *Myoclonic epilepsy and ragged red fibers. StatPearls [Internet]*. StatPearls Publishing.

Hippen, M., Zsurka, G., Peeva, V., Machts, J., Schwiecker, K., Debska-Vielhaber, G., Wiesner, R. J., Vielhaber, S., & Kunz, W. S. (2022). Novel pathogenic sequence variation m.5789T>C causes NARP syndrome and promotes formation of deletions of the mitochondrial genome. *Neurology: Genetics, 8*(2). https://doi.org/10.1212/NXG.0000000000000660. http://ng.neurology.org/

Hirano, M., & Pavlakis, S. G. (1994). Topical review: Mitochondrial myopathy, encephalopathy, lactic acidosis, and strokelike episodes (MELAS): Current concepts. *Journal of Child Neurology, 9*(1), 4–13. https://doi.org/10.1177/088307389400900102

Hirano, M., Marti, R., Casali, C., Tadesse, S., Uldrick, T., Fine, B., Escolar, D. M., Valentino, M. L., Nishino, I., Hesdorffer, C., Schwartz, J., Hawks, R. G., Martone, D. L., Cairo, M. S., DiMauro, S., Stanzani, M., Garvin, J. H., & Savage, D. G. (2006). Allogeneic stem cell transplantation corrects biochemical derangements in MNGIE. *Neurology, 67*(8), 1458–1460. https://doi.org/10.1212/01.wnl.0000240853.97716.24

Holt, I. J., Harding, A. E., Petty, R. K. H., & Morgan-Hughes, J. A. (1990). A new mitochondrial disease associated with mitochondrial DNA heteroplasmy. *The American Journal of Human Genetics, 46*(3), 428–433.

Hommes, F. A., Polman, H. A., & Reerink, J. D. (1968). Leigh's encephalomyelopathy: An inborn error of gluconeogenesis. *Archives of Disease in Childhood, 43*(230), 423–426. https://doi.org/10.1136/adc.43.230.423

Howell, N., Kubacka, I., Xu, M., & McCullough, D. A. (1991). Leber hereditary optic neuropathy: Involvement of the mitochondrial ND1 gene and evidence for an intragenic suppressor mutation. *The American Journal of Human Genetics, 48*(5), 935–942.

Hue, C. D., Cho, F. S., Cao, S., Dale Bass, C. R., Meaney, D. F., & Morrison, B. (2015). Dexamethasone potentiates in vitro blood-brain barrier recovery after primary blast injury by glucocorticoid receptor-mediated upregulation of ZO-1 tight junction protein. *Journal of Cerebral Blood Flow and Metabolism, 35*(7), 1191–1198. https://doi.org/10.1038/jcbfm.2015.38

Jacoby, E., Ben Yakir-Blumkin, M., Blumenfeld-Kan, S., Brody, Y., Meir, A., Melamed-Book, N., Napso, T., Pozner, G., Saadi, E., Shabtay-Orbach, A., Yivgi-Ohana, N., Sher, N., & Toren, A. (2021). Mitochondrial augmentation of CD34+ cells from healthy donors and patients with mitochondrial DNA disorders confers functional benefit. *npj Regenerative Medicine, 6*(1). https://doi.org/10.1038/s41536-021-00167-7. https://www.nature.com/npjregenmed/

Jacoby, E., Blumkin, M., Anikster, Y., Varda-Bloom, N., Pansheen, J., Bar Yoseph, O., Gruber, N., Lahav, E., Besser, M. J., Schachter, J., Sher, N., Yivgi Ohana, N., & Toren, A. (2018). First-in-human mitochondrial augmentation of hematopoietic stem cells in pearson syndrome. *Blood, 132*(Suppl. 1), 1024. https://doi.org/10.1182/blood-2018-99-113773

Jansen, P. H. P., Van Der Knaap, M. S., & De Coo, I. F. M. (1996). Leber's hereditary optic neuropathy with the 11778 mtDNA mutation and white matter disease resembling multiple sclerosis: Clinical, MRI and MRS findings. *Journal of the Neurological Sciences, 135*(2), 176–180. https://doi.org/10.1016/0022-510X(95)00287-C

Jou, M. J., Peng, T. I., Hsu, L. F., Jou, S. B., Reiter, R. J., Yang, C. M., Chiao, C. C., Lin, Y. F., & Chen, C. C. (2010). Visualization of melatonin's multiple mitochondrial levels of protection against mitochondrial Ca2+-mediated permeability transition and beyond in rat brain astrocytes. *Journal of Pineal Research, 48*(1), 20–38. https://doi.org/10.1111/j.1600-079X.2009.00721.x

Jun, A. S., Brown, M. D., & Wallace, D. C. (1994). A mitochondrial DNA mutation at nucleotide pair 14459 of the NADH dehydrogenase subunit 6 gene associated with maternally inherited Leber hereditary optic neuropathy and dystonia. *Proceedings of the National Academy of Sciences, 91*(13), 6206–6210. https://doi.org/10.1073/pnas.91.13.6206

Katrangi, E., D'Souza, G., Boddapati, S. V., Kulawiec, M., Singh, K. K., Bigger, B., & Weissig, V. (2007). Xenogenic transfer of isolated murine mitochondria into human ρ^0 cells can improve respiratory function. *Rejuvenation Research, 10*(4), 561–570. https://doi.org/10.1089/rej.2007.0575

Kearns, T. P. (1958). Retinitis pigmentosa, external ophthalmoplegia, and complete heart block. *AMA Archives of Ophthalmology, 60*(2), 280. https://doi.org/10.1001/archopht.1958.00940080296016

Kellar-Wood, H., Robertson, N., Govan, G. G., Compston, D. A. S., & Harding, A. E. (1994). Leber's hereditary optic neuropathy mitochondrial DNA mutations in multiple sclerosis. *Annals of Neurology, 36*(1), 109–112. https://doi.org/10.1002/ana.410360121

King, M. P., & Attardi, G. (1989). Human cells lacking mtDNA: Repopulation with exogenous mitochondria by complementation. *Science, 246*(4929), 500–503. https://doi.org/10.1126/science.2814477

King, M. P., & Attardi, G. (1988). Injection of mitochondria into human cells leads to a rapid replacement of the endogenous mitochondrial DNA. *Cell, 52*(6), 811–819. https://doi.org/10.1016/0092-8674(88)90423-0

Klug, W. S., Cummings, M. R., Spencer, C. A., & Palladino, M. A. (2014). *Concepts of genetics* (11th ed.). Pearson Education, Inc.

Krishna, M. R. (2017). Kearns sayre syndrome: Looking beyond A-V conduction. *Indian Pacing and Electrophysiology Journal, 17*(3), 78–80. https://doi.org/10.1016/j.ipej.2017.03.003

Kucharczyk, R., Zick, M., Bietenhader, M., Rak, M., Couplan, E., Blondel, M., Caubet, S. D., & di Rago, J. P. (2009). Mitochondrial ATP synthase disorders: Molecular mechanisms and the quest for curative therapeutic approaches. *Biochimica et Biophysica Acta – Molecular Cell Research, 1793*(1), 186–199. https://doi.org/10.1016/j.bbamcr.2008.06.012

Kumar, V., Abbas, A. K., & Aster, J. C. (2015). *Vitamin B12 and mitochondrial function. Robbins and Cotran pathologic basis of disease*. Elsevier.

Lake, N. J., Compton, A. G., Rahman, S., & Thorburn, D. R. (2016). Leigh syndrome: One disorder, more than 75 monogenic causes. *Annals of Neurology, 79*(2), 190–203. https://doi.org/10.1002/ana.24551. http://onlinelibrary.wiley.com/journal/10.1002/(ISSN)1531-8249

Lara, M. C., Weiss, B., Illa, I., Madoz, P., Massuet, L., Andreu, A. L., Valentino, M. L., Anikster, Y., Hirano, M., & Martí, R. (2006). Infusion of platelets transiently reduces nucleoside overload in MNGIE. *Neurology, 67*(8), 1461–1463. https://doi.org/10.1212/01.wnl.0000239824.95411.52

Leber, T. (1871). Ueber hereditäre und congenital-angelegte Sehnervenleiden. *Albrecht von Græfe's Archiv für Ophthalmologie, 17*(2), 249–291. https://doi.org/10.1007/BF01694557

Lee, A. G., & Brazis, P. W. (2002). Chronic progressive external ophthalmoplegia. *Current Neurology and Neuroscience Reports, 2*(5), 413–417. https://doi.org/10.1007/s11910-002-0067-5

Leigh, D. (1951). Subacute necrotizing encephalomyelopathy in an infant. *Journal of Neurology, Neurosurgery, and Psychiatry, 14*(3), 216–221. https://doi.org/10.1136/jnnp.14.3.216

Leigh, P. N., Al-Sarraj, S., & DiMauro, S. (2015). Subacute necrotising encephalomyelopathy (Leigh's disease; Leigh syndrome). *Journal of Neurology, Neurosurgery and Psychiatry, 86*(4), 363–365. https://doi.org/10.1136/jnnp-2012-304601

Lenaz, G., Baracca, A., Carelli, V., D'Aurelio, M., Sgarbi, G., & Solaini, G. (2004). Bioenergetics of mitochondrial diseases associated with mtDNA mutations. *Biochimica et Biophysica Acta – Bioenergetics, 1658*(1–2), 89–94. https://doi.org/10.1016/j.bbabio.2004.03.013

Leng, Y., Liu, Y., Fang, X., Li, Y., Yu, L., Yuan, Y., & Wang, Z. (2015). The mitochondrial DNA 10197 G > A mutation causes MELAS/Leigh overlap syndrome presenting with acute auditory agnosia. *Mitochondrial DNA, 26*(2), 208–212. https://doi.org/10.3109/19401736.2014.905860

Levy, R. J., Ríos, P. G., Akman, H. O., Sciacco, M., Vivo, D. C. D., & Dimauro, S. (2014). Long survival in patients with leigh syndrome and the m.10191T>C mutation in MT-ND3: A case report and review of the literature. *Journal of Child Neurology, 29*(10), NP105–NP110. https://doi.org/10.1177/0883073813506783

Lin, T.-K., Chen, S.-D., Chuang, Y.-C., Lan, M.-Y., Chuang, J.-H., Wang, P.-W., Hsu, T.-Y., Wang, F.-S., Tsai, M.-H., Huang, S.-T., Wang, X.-W., Tsai, P.-C., Lin, H.-Y., & Liou, C.-W. (2019). Mitochondrial transfer of Wharton's jelly mesenchymal stem cells eliminates mutation burden and rescues mitochondrial bioenergetics in rotenone-stressed MELAS fibroblasts. *Oxidative Medicine and Cellular Longevity, 2019*, 1–17. https://doi.org/10.1155/2019/9537504

Lopez-Gallardo, E., Solano, A., Herrero-Martin, M. D., Martinez-Romero, I., Castano-Perez, M. D., Andreu, A. L., Herrera, A., Lopez-Perez, M. J., Ruiz-Pesini, E., & Montoya, J. (2008). NARP syndrome in a patient harbouring an insertion in the MT-ATP6 gene that results in a truncated protein. *Journal of Medical Genetics, 46*(1), 64–67. https://doi.org/10.1136/jmg.2008.060616

Lorenzoni, P. J., Scola, R. H., Kay, C. S. K., Silvado, C. E. S., & Werneck, L. C. (2014). When should MERRF (myoclonus epilepsy associated with ragged-red fibers) be the diagnosis? *Arquivos de Neuro-Psiquiatria, 72*(10), 803–811. https://doi.org/10.1590/0004-282X20140124

Mäkelä-Bengs, P., Suomalainen, A., Majander, A., Rapola, J., Kalimo, H., Nuutila, A., & Pihko, H. (1995). Correlation between the clinical symptoms and the proportion of mitochondrial DNA carrying the 8993 point mutation in the NARP syndrome. *Pediatric Research, 37*(5), 634–639. https://doi.org/10.1203/00006450-199505000-00014

Man, P. Y. W., Turnbull, D. M., & Chinnery, P. F. (2002). Leber hereditary optic neuropathy. *Journal of Medical Genetics, 39*(3), 162–169. https://doi.org/10.1136/jmg.39.3.162

Mantoan, L., & Walker, M. (2011). Treatment options in juvenile myoclonic epilepsy. *Current Treatment Options in Neurology, 13*(4), 355–370. https://doi.org/10.1007/s11940-011-0131-z

Mashima, Y., Kigasawa, K., Wakakura, M., & Oguchi, Y. (2000). Do idebenone and vitamin therapy shorten the time to achieve visual recovery in Leber hereditary optic neuropathy? *Journal of Neuro-Ophthalmology, 20*(3), 166–170. https://doi.org/10.1097/00041327-200020030-00006

McClelland, C., Manousakis, G., & Lee, M. S. (2016). Progressive external ophthalmoplegia. *Current Neurology and Neuroscience Reports, 16*(6). https://doi.org/10.1007/s11910-016-0652-7. http://www.springer.com

McFarland, R., Taylor, R. W., & Turnbull, D. M. (2010). A neurological perspective on mitochondrial disease. *The Lancet Neurology, 9*(8), 829–840. https://doi.org/10.1016/S1474-4422(10)70116-2

MedlinePlus [Internet]. (2020a). Bethesda (MD): National Library of Medicine (US). https://medlineplus.gov/genetics/condition/leber-hereditary-optic-neuropathy/. (Accessed 13 December 2022).

MedlinePlus [Internet]. (2020b). Bethesda (MD): National Library of Medicine (US). https://medlineplus.gov/genetics/condition/leigh-syndrome/. (Accessed 20 November 2022).

MedlinePlus [Internet]. (2020c). Bethesda (MD): National Library of Medicine (US). https://medlineplus.gov/genetics/condition/mitochondrial-encephalomyopathy-lactic-acidosis-and-stroke-like-episodes/. (Accessed 20 December 2022).

MedlinePlus [Internet]. (2020d). Bethesda (MD): National Library of Medicine (US). https://medlineplus.gov/genetics/condition/neuropathy-ataxia-and-retinitis-pigmentosa/. (Accessed 15 December 2022).

MedlinePlus [Internet]. (2020e). Bethesda (MD): National Library of Medicine (US). https://medlineplus.gov/genetics/condition/myoclonic-epilepsy-with-ragged-red-fibers/. (Accessed 10 December 2022).

Michaelides, M., & Moore, A. T. (2004). The genetics of strabismus. *Journal of Medical Genetics, 41*(9), 641–646. https://doi.org/10.1136/jmg.2004.021667

Mojon, D. S., Fujihara, K., Hirano, M., Miller, C., Lincoff, N. S., Jacobs, L. D., & Greenberg, S. J. (1999). Leber's hereditary optic neuropathy mitochondrial DNA mutations in familial multiple sclerosis. *Graefe's Archive for Clinical and Experimental Ophthalmology, 237*(4), 348–350. https://doi.org/10.1007/s004170050243

Moraes, C. T., Ciacci, F., Silvestri, G., Shanske, S., Sciacco, M., Hirano, M., Schon, E. A., Bonilla, E., & DiMauro, S. (1993). Atypical clinical presentations associated with the MELAS mutation at position 3243 of human mitochondrial DNA. *Neuromuscular Disorders, 3*(1), 43–50. https://doi.org/10.1016/0960-8966(93)90040-Q

Mordel, P., Schaeffer, S., Dupas, Q., Laville, M. A., Gérard, M., Chapon, F., & Allouche, S. (2017). A 2 bp deletion in the mitochondrial ATP 6 gene responsible for the NARP (neuropathy, ataxia, and retinitis pigmentosa) syndrome. *Biochemical and Biophysical Research Communications, 494*(1–2), 133–137. https://doi.org/10.1016/j.bbrc.2017.10.066

Morris, M. C., Depollier, J., Mery, J., Heitz, F., & Divita, G. (2001). A peptide carrier for the delivery of biologically active proteins into mammalian cells. *Nature Biotechnology, 19*(12), 1173–1176. https://doi.org/10.1038/nbt1201-1173

Olsen, N. K., Hansen, A. W., Nørby, S., Edal, A. L., Jørgensen, J. R., & Rosenberg, T. (1995). Leber's hereditary optic neuropathy associated with a disorder indistinguishable from multiple sclerosis in a male harbouring the mitochondrial DNA 11778 mutation. *Acta Neurologica Scandinavica, 91*(5), 326–329. https://doi.org/10.1111/j.1600-0404.1995.tb07016.x

Online Mendelian Inheritance in Man, OMIM®. (2000). *McKusick-Nathans Institute of Genetic Medicine.* Baltimore, MD: Johns Hopkins University. https://www.omim.org/search?index=entry&start=1&limit=10&sort=score+desc%2C+prefix_sort+desc&search=%22mtdna+depletion+syndrome%22. (Accessed 27 November 2022).

Parikh, S., Goldstein, A., Karaa, A., Koenig, M. K., Anselm, I., Brunel-Guitton, C., Christodoulou, J., Cohen, B. H., Dimmock, D., Enns, G. M., Falk, M. J., Feigenbaum, A., Frye, R. E., Ganesh, J., Griesemer, D., Haas, R., Horvath, R., Korson, M., Kruer, M. C., … Chinnery, P. F. (2017). Patient care standards for primary mitochondrial disease: A consensus statement from the mitochondrial medicine society. *Genetics in Medicine, 19*(12), 1–18. https://doi.org/10.1038/gim.2017.107

Pavlakis, S. G., Phillips, P. C., DiMauro, S., De Vivo, D. C., & Rowland, L. P. (1984). Mitochondrial myopathy, encephalopathy, lactic acidosis, and strokelike episodes: A distinctive clinical syndrome. *Annals of Neurology, 16*(4), 481–488. https://doi.org/10.1002/ana.410160409

Pearson, H. A., Lobel, J. S., Kocoshis, S. A., Naiman, J. L., Windmiller, J., Lammi, A. T., Hoffman, R., & Marsh, J. C. (1979). A new syndrome of refractory sideroblastic anemia with vacuolization of marrow precursors and exocrine pancreatic dysfunction. *The Journal of Pediatrics, 95*(6), 976–984. https://doi.org/10.1016/S0022-3476(79)80286-3

Rötig, A., Cormier, V., Blanche, S., Bonnefont, J. P., Ledeist, F., Romero, N., Schmitz, J., Rustin, P., Fischer, A., & Saudubray, J. M. (1990). Pearson's marrow-pancreas syndrome. A multisystem mitochondrial disorder in infancy. *Journal of Clinical Investigation, 86*(5), 1601–1608. https://doi.org/10.1172/JCI114881

Rötig, A., Bourgeron, T., Chretien, D., Rustin, P., & Munnich, A. (1995). Spectrum of mitochondrial DNA rearrangements in the pearson marrow-pancreas syndrome. *Human Molecular Genetics, 4*(8), 1327–1330. https://doi.org/10.1093/hmg/4.8.1327

Rahman, S., Blok, R. B., Dahl, H. H. M., Danks, D. M., Kirby, D. M., Chow, C. W., Christodoulou, J., & Thorburn, D. R. (1996). Leigh syndrome: Clinical features and biochemical and DNA abnormalities. *Annals of Neurology, 39*(3), 343–351. https://doi.org/10.1002/ana.410390311

Rahman, S., Hargreaves, I. P., Hirano, M., Marti, R. A., DiMauro, S., Chinnery, P. F., & Vissing, J. (2007). Allogeneic stem cell transplantation corrects biochemical derangements in MNGIE. *Neurology, 68*(21), 1872–1873. https://doi.org/10.1212/01.wnl.0000265356.98295.be

Rahman, S. (2013). Mitochondrial disease and its treatment. *Journal of Inherited Metabolic Disease, 36*(4), 669–684.

Riordan-Eva, P., & Harding, A. E. (1995). Leber's hereditary optic neuropathy: The clinical relevance of different mitochondrial DNA mutations. *Journal of Medical Genetics, 32*(2), 81–87. https://doi.org/10.1136/jmg.32.2.81

Rojo, A., Campos, Y., Sánchez, J. M., Bonaventura, I., Aguilar, M., García, A., González, L., Rey, M. J., Arenas, J., Olivé, M., & Ferrer, I. (2006). NARP-MILS syndrome caused by 8993 T > G mitochondrial DNA mutation: A clinical, genetic and neuropathological study. *Acta Neuropathologica, 111*(6), 610–616. https://doi.org/10.1007/s00401-006-0040-5

Rossi, F. H., Okun, M., Yachnis, A., Quisling, R., & Triggs, W. J. (2002). Corticosteroid treatment of mitochondrial encephalomyopathies. *The Neurologist, 8*(5), 313–315. https://doi.org/10.1097/00127893-200209000-00004

Sacconi, S., Salviati, L., Gooch, C., Bonilla, E., Shanske, S., & DiMauro, S. (2002). Complex neurologic syndrome associated with the G1606A mutation of mitochondrial DNA. *Archives of Neurology, 59*(6), 1013–1015. https://doi.org/10.1001/archneur.59.6.1013

Saneto, R. P., & Singh, K. K. (2010). Illness-induced exacerbation of Leigh syndrome in a patient with the MTATP6 mutation, m. 9185 T>C. *Mitochondrion, 10*(5), 567–572. https://doi.org/10.1016/j.mito.2010.05.006

Saneto, R., & Ruhoy, I. (2014). The genetics of Leigh syndrome and its implications for clinical practice and risk management. *The Application of Clinical Genetics, 7*, 221. https://doi.org/10.2147/TACG.S46176

Santorelli, F. M., Shanske, S., Macaya, A., DeVivo, D. C., & DiMauro, S. (1993). The mutation at nt 8993 of mitochondrial DNA is a common cause of Leigh's syndrome. *Annals of Neurology, 34*(6), 827–834. https://doi.org/10.1002/ana.410340612

Schaefer, A. M., McFarland, R., Blakely, E. L., He, L., Whittaker, R. G., Taylor, R. W., Chinnery, P. F., & Turnbull, D. M. (2008). Prevalence of mitochondrial DNA disease in adults. *Annals of Neurology, 63*(1), 35–39. https://doi.org/10.1002/ana.21217

Schon, E. A., Dimauro, S., & Hirano, M. (2012). Human mitochondrial DNA: Roles of inherited and somatic mutations. *Nature Reviews Genetics, 13*(12), 878–890. https://doi.org/10.1038/nrg3275

Schon, E. A., Rizzuto, R., Moraes, C. T., Nakase, H., Zeviani, M., & DiMauro, S. (1989). A direct repeat is a hotspot for large-scale deletion of human mitochondrial DNA. *Science, 244*(4902), 346–349. https://doi.org/10.1126/science.2711184

Shanske, S., Tang, Y., Hirano, M., Nishigaki, Y., Tanji, K., Bonilla, E., Sue, C., Krishna, S., Carlo, J. R., Willner, J., Schon, E. A., & DiMauro, S. (2002). Identical mitochondrial DNA deletion in a woman with ocular myopathy and in her son with pearson syndrome. *The American Journal of Human Genetics, 71*(3), 679–683. https://doi.org/10.1086/342482

Shanske, S., Coku, J., Lu, J., Ganesh, J., Krishna, S., Tanji, K., Bonilla, E., Naini, A. B., Hirano, M., & DiMauro, S. (2008). The G13513A mutation in the ND5 gene of mitochondrial DNA as a common cause of MELAS or Leigh syndrome: Evidence from 12 cases. *Archives of Neurology, 65*(3), 368–372. https://doi.org/10.1001/archneurol.2007.67

Sitarz, K. S., Elliott, H. R., Karaman, B. S., Relton, C., Chinnery, P. F., & Horvath, R. (2014). Valproic acid triggers increased mitochondrial biogenesis in POLG-deficient fibroblasts. *Molecular Genetics and Metabolism, 112*(1), 57–63. https://doi.org/10.1016/j.ymgme.2014.03.006

Sproule, D. M., & Kaufmann, P. (2008). Mitochondrial encephalopathy, lactic acidosis, and strokelike episodes: Basic concepts, clinical phenotype, and therapeutic management of MELAS syndrome. *Annals of the New York Academy of Sciences, 1142*, 133–158. https://doi.org/10.1196/annals.1444.011

Suomalainen, A., & Isohanni, P. (2010). Mitochondrial DNA depletion syndromes - many genes, common mechanisms. *Neuromuscular Disorders, 20*(7), 429–437. https://doi.org/10.1016/j.nmd.2010.03.017

Tatuch, Y., Christodoulou, J., Feigenbaum, A., Clarke, J. T. R., Wherret, J., Smith, C., Rudd, N., Petrova-Benedict, R., & Robinson, B. H. (1992). Heteroplasmic mtDNA mutation (T→G) at 8993 can cause Leigh disease when the percentage of abnormal mtDNA is high. *The American Journal of Human Genetics, 50*(4), 852–858.

Tsaris, P. (1973). Familial myoclonic epilepsy syndrome associated with skeletal muscle mitochondrial abnormalities. *Neurology, 23*.

Venturelli, M., Villa, F., Ruzzante, F., Tarperi, C., Rudi, D., Milanese, C., Cavedon, V., Fonte, C., Picelli, A., Smania, N., Calabria, E., Skafidas, S., Layec, G., & Schena, F. (2019). Neuromuscular and muscle metabolic functions in melas before and after resistance training: A case study. *Frontiers in Physiology, 10*. https://doi.org/10.3389/fphys.2019.00503. http://www.frontiersin.org/Physiology/archive/

Vodopivec, I., Cho, T. A., Rizzo, J. F., Frosch, M. P., & Sims, K. B. (2016). Mitochondrial encephalopathy and optic neuropathy due to m.10158 MT-ND3 complex I mutation presenting in an adult patient. *The Neurologist, 21*(4), 61–65. https://doi.org/10.1097/nrl.0000000000000084

Wallace, D. C. (2010). Mitochondrial DNA mutations in disease and aging. *Environmental and Molecular Mutagenesis, 51*(5), 440–450. https://doi.org/10.1002/em.20586

Wallace, D. C., Singh, G., Lott, M. T., Hodge, J. A., Schurr, T. G., Lezza, A. M. S., Elsas, L. J., & Nikoskelainen, E. K. (1988). Mitochondrial DNA mutation associated with Leber's hereditary optic neuropathy. *Science, 242*(4884), 1427–1430. https://doi.org/10.1126/science.3201231

Wang, S., Song, T., & Wang, S. (2020). Mitochondrial DNA 10158t> c mutation in a patient with mitochondrial encephalomyopathy with lactic acidosis, and stroke-like episodes syndrome: A case-report and literature review (care-complaint). *Medicine, 24*.

Weber, K., Brück, P., Mikes, Z., Küpper, J. H., Klingenspor, M., & Wiesner, R. J. (2002). Glucocorticoid hormone stimulates mitochondrial biogenesis specifically in skeletal muscle. *Endocrinology, 143*(1), 177–184. https://doi.org/10.1210/endo.143.1.8600

Wilichowski, E., Christoph Korenke, G., Ruitenbeek, W., De Meirleir, L., Hagendorff, A., Janssen, A. J. M., Lissens, W., & Hanefeld, F. (1998). Pyruvate dehydrogenase complex deficiency and altered respiratory chain function in a patient with Kearns-Sayre/MELAS overlap syndrome and A3243G mtDNA mutation. *Journal of the Neurological Sciences, 157*(2), 206–213. https://doi.org/10.1016/S0022-510X(98)00068-9

Wong, L. J. C. (2007). Diagnostic challenges of mitochondrial DNA disorders. *Mitochondrion, 7*(1–2), 45–52. https://doi.org/10.1016/j.mito.2006.11.025

Yarham, J. W., Elson, J. L., Blakely, E. L., Mcfarland, R., & Taylor, R. W. (2010). Mitochondrial tRNA mutations and disease. *Wiley Interdisciplinary Reviews: RNA, 1*(2), 304–324. https://doi.org/10.1002/wrna.27

Yatsuga, S., Povalko, N., Nishioka, J., Katayama, K., Kakimoto, N., Matsuishi, T., Kakuma, T., & Koga, Y. (2012). MELAS: A nationwide prospective cohort study of 96 patients in Japan. *Biochimica et Biophysica Acta – General Subjects, 1820*(5), 619–624. https://doi.org/10.1016/j.bbagen.2011.03.015

Yosef, O. B., Jacoby, E., Gruber, N., Varda-Bloom, N., Azaria, E., Eisenstein, E., & Toren. (2020). *Promising results for Kearns-Sayre syndrome of first in man treatment by mitochondrial augmentation therapy*.

Yoshimi, A., Ishikawa, K., Niemeyer, C., & Grünert, S. C. (2022). Pearson syndrome: A multisystem mitochondrial disease with bone marrow failure. *Orphanet Journal of Rare Diseases, 17*(1). https://doi.org/10.1186/s13023-022-02538-9. http://www.ojrd.com/

Yu-Wai-Man, P., Griffiths, P. G., & Chinnery, P. F. (2011). Mitochondrial optic neuropathies — Disease mechanisms and therapeutic strategies. *Progress in Retinal and Eye Research, 30*(2), 81–114. https://doi.org/10.1016/j.preteyeres.2010.11.002

Yubero, D., Montero, R., & Jou, C. (2020). Therapeutic approaches for mitochondrial diseases: A comprehensive review. *Frontiers in Bioscience, 25*, 1259–1297.

Zeviani, M., Moraes, C. T., DiMauro, S., Nakase, H., Bonilla, E., Schon, E. A., & Rowland, L. P. (1988). Deletions of mitochondrial DNA in Kearns-Sayre syndrome. *Neurology, 38*(8), 1339. https://doi.org/10.1212/WNL.38.8.1339

CHAPTER

20

Mitochondrial transplantation in kidney injury

Patricia Crespo Braga[1,2], *Anabela Rodrigues*[3], *Pedro Fontes Oliveira*[4] and *Marco G. Alves*[5]

[1]Institute of Biomedical Sciences Abel Salazar (ICBAS), University of Porto, Porto, Portugal; [2]ITR-Laboratory for Integrative and Translational Research in Population Health, Porto, Portugal; [3]Hospital Center of Vila Nova de Gaia/Espinho, Vila Nova de Gaia, Portugal; [4]LAQV-REQUIMTE and Department of Chemistry, University of Aveiro, Aveiro, Portugal; [5]Institute of Biomedicine - iBiMED and Department of Medical Sciences, University of Aveiro, Aveiro, Portugal

Introduction

Kidney disease is a public health problem that has grown. In 2020, Kidney Disease: Improving Global Outcomes, an independent organization with the mission to improve care and outcomes of patients with kidney disease worldwide, organized a consensus conference to harmonize existing acute kidney disease (AKD) and chronic kidney disease (CKD) definitions.

CKD is defined as a progressive and irreversible loss of kidney function. The principal hallmarks are an estimated glomerular filtration rate (eGFR) < 60 mL/min per 1.73 m^2, the presence of markers of kidney damage (proteinuria, active urine sediments, structural abnormalities, histological damages), for more than 3 months, and/or a history of kidney transplantation (Lameire et al., 2021). AKD and CKD can be distinguished by the duration of the renal outcomes. The term acute defines a condition of a sudden episode (6 h–7 days) of injury that is short and can be reversible, while chronic has a long term and with persistent episodes associated. On the other hand, acute kidney injury (AKI) is a subset of AKD associated with high morbidity and mortality, including an increase in adverse renal outcomes. It is also suggested that AKI is not self-limited, as it serves as an incoming for subsequent AKI episodes and potentially to CKD (Heung et al., 2016). Furthermore, an episode of AKI in patients with CKD can cause further renal deterioration to end-stage kidney disease

Mitochondrial Transplantation and Transfer
https://doi.org/10.1016/B978-0-443-18858-9.00020-2

(ESKD). Multiple factors can contribute to AKI-related renal disease progression, including age, diabetes, hypertension and cardiac problems, severe episodes of AKI, and even delayed renal function recovery from AKI (Yan et al., 2021). Several studies already demonstrated that mitochondrial morphological alterations are present during AKI to CKD, including swelling, fragmentation, and loss of cristae. Moreover, impaired mitochondrial functioning results in decreased electron transport chain (ETC) activity, a decrease in membrane potential, and an increase in reactive oxygen species (ROS) generation, particularly in podocytes (Aparicio-Trejo et al., 2019; Funk & Schnellmann, 2012; Li, Li, et al., 2020; Liang et al., 2022). Notably, some of these studies suggest that independently if the renal outcomes are from sudden or persistent episodes, mitochondria functioning is highly affected by kidney diseases. In fact, as the kidney is an extremely metabolically active organ, it needs a high amount of energy to accomplish its functions, such as filtration and reabsorption (Bhargava & Schnellmann, 2017). As there is no effective treatment for kidney disease, and as mitochondria are signalized as having a key role in disease progression, the interest in therapeutic mitochondria targeted therapy has increased, with a high interest in mitochondrial transplantation. In this chapter, we will highlight the importance of the mitochondria in the kidney, unveil, and summarize the principal techniques found in mitochondrial transplantation, particularly those applied or with possible relevance for application in kidney disease.

An overview of acute kidney injury

William MacNider used the term AKI for the first time in 1918 to describe a case of acute mercury poisoning. Recently, the term acute renal failure has been replaced by AKI (Bellomo et al., 2004). AKI is characterized by a sudden loss of renal function, uremia, and dysregulation of extracellular volume and electrolytes, resulting from multiple insults, including sepsis, cardiovascular surgery, and ischemia–reperfusion injury (IRI). AKI has additional causes, such as direct kidney injury and urinary tract obstruction, in addition to decreased blood flow (Lameire et al., 2006). Patients with AKI typically have a heterogeneous etiology, involving extrarenal pathology (such as prerenal azotemia and acute postrenal obstructive nephropathy) and particular kidney diseases (like acute interstitial nephritis, acute glomerular, and vasculitic renal disorders) (Chawla & Kimmel, 2012). As previously mentioned, the diagnosis of AKI is predicated on a rapid decrease in glomerular filtration rate (GFR), which is also accompanied by a decrease in urine production or an increase in serum creatinine levels (Khwaja, 2012). After this first step of diagnosis, AKI can be classified into three main categories: prerenal, intrinsic, and postrenal AKI. Prerenal and postrenal AKI result from a decline in GFR caused by extrarenal disease, whereas "intrinsic" AKI represents true kidney disease (Agarwal et al., 2016). AKI is considered an independent risk factor for CKD and ESKD, where severe AKI may be followed by an irreversible loss in kidney function (Bucaloiu et al., 2012; Coca et al., 2012). But currently, it is known to be strongly linked to interrelated syndromes, where CKD predisposes or sensitizes people to AKI (James et al., 2015), and AKI is one of the most notable variables to speed up the progression of CKD (Palomba et al., 2017). Among these causes of AKI, IRI is the most striking, where renal tissue is compromised.

In addition to mitochondrial damage, energy depletion, and tubular apoptosis, IRI induced AKI (IRI-AKI) also causes a disturbance in the cellular redox balance and intensifies ROS formation in the kidneys during reperfusion, which can lead to renal failure or dysfunction (Bhargava & Schnellmann, 2017; Li, Liao, et al., 2020; Yu et al., 2020). AKI has been described as a state of tubular ATP depletion (Thadhani et al., 1996). In addition, the opening of mitochondrial permeability transition pores due to mitochondrial swelling and dysfunction is a key event that contributes to AKI progression since the release of proapoptotic mediators, such as cytochrome C, can occur and induce renal cell apoptosis (Wakabayashi & Karbowski, 2001). Mitochondria are the main source of intracellular ROS, which are mainly produced at complexes I and III of the ETC when electrons derived from either NADH or $FADH_2$ react with O_2 (Chouchani et al., 2016); thus, the mitochondrial dysfunction caused by AKI potentiates the production of ROS. Thereby, mitochondrial (dys)function has been strongly associated with the pathogenesis of AKI, causing renal tubular dysfunction and cell death (Venkatachalam & Weinberg, 2012); however, targeting mitochondria therapies to improve mitochondria fitness is still a field barely explored and debatable.

Mitochondria function and its importance in the kidney

In mammalian cells, mitochondria are the intracellular organelles in the control of biological oxidation. The primary purpose of mitochondria is to produce ATP through electron transport via oxidative phosphorylation (OXPHOS). This is done in conjunction with metabolite oxidation through the tricarboxylic acid cycle and fatty acid catabolism by β-oxidation to provide energy to kidney basal cell function (Duann & Lin, 2017; Nourbakhsh & Singh, 2014). Indeed, one of the human organs that uses up the most energy is the kidney, which also consumes the most oxygen and has the second-highest mitochondrial content after the heart (Emma et al., 2016).

Mitochondria are crucial for metabolic control, production of ROS, preservation of intracellular calcium homeostasis, thermogenesis, and control of intrinsic apoptotic and proliferative pathways (Brooks & Dong, 2007; Che et al., 2014).

Proximal tubular cells generate ATP mainly through mitochondrial OXPHOS, while podocytes, endothelial, and mesangial cells exhibit a higher glycolytic capacity to generate energy (Bhargava & Schnellmann, 2017; Wirthensohn & Guder, 1986). A growing body of research indicates that a variety of acute and chronic injuries may result in oxidative stress derived from the ETC, ultrastructural defects, abnormal activation of the mitochondrial pathway of apoptosis, unstable mitochondrial DNA (mtDNA), and inadequate clearance of damaged mitochondria (Duann & Lin, 2017). Mitochondrial homeostasis is the balance between the control of mitochondrial biogenesis, mitochondrial dynamics, including mitochondrial fragmentation and mitophagy. An imbalance may induce ROS production and apoptosis. These alterations will, in turn, lead to the development and progression of diseases related to mitochondrial dysfunction. In the kidney, it leads to an increased risk of tubular interstitial disease, cystic kidney disease, podocytopathy, and nephropathy syndromes (Duann & Lin, 2017).

Thus, it is important to understand mitochondrial pathophysiology that is addressed in AKI and CKD. A deeper understanding of the cellular molecular processes that control mitochondrial activities in kidney disorders ought to contribute to the improvement of treatment approaches.

Mitochondrial quality control and biogenesis in AKI

Mitochondrial turnover is a mechanism adopted to assure mitochondrial quality. In this process, dysfunctional mitochondria are selectively eliminated and replaced through an increase in the number of preexisting mitochondria.

In order to increase the production of metabolic enzymes and increase the mitochondrial metabolic capacity, the nuclear and mitochondrial genomes must coordinate. The mtDNA is a double stranded circular molecule that is roughly 16 kb in size and encodes 37 genes, comprising 13 structural subunits of the mitochondrial respiratory chain (ETC complexes I, III, IV, and V) (Brown et al., 2010). Furthermore, mitochondrial biogenesis can be influenced by multiple external factors, such as exercise, caloric restriction, and oxidative stress (Brown et al., 2010). The peroxisome proliferator-activated receptor gamma coactivator-1 alpha (PGC-1α) protein family is responsible for activating mtDNA transcription. PGC-1α, the key regulator of mitochondrial biogenesis in this family, is foremost expressed in proximal tubules. During tissue damage and healing, the induction of mitochondrial biogenesis is linked to the activation of transcription and several signaling pathways, which includes AMP-activated protein kinase/PGC-1α (AMPK/PGC-1α) pathways used by C1q/tumor necrosis factor related protein-3 (CTRP3) to promote biogenesis, and the Gby-Akt-eNOS-sGC (soluble guanylatecyclase) pathways stimulated by the β2 adrenergic receptor agonists (Puigserver et al., 1998).

Unsurprisingly, the dysregulation of mitochondrial biogenesis has been associated to AKI (Fig. 20.1). Nuclear respiratory factor 1 (NRF1) and nuclear respiratory factor 2 (NRF2) are transcriptional coactivators of PGC-1α that promote mitochondrial biogenesis, and are responsible to control of OXPHOS, fatty acid oxidation (FAO), and the synthesis of nicotinamide adenine dinucleotide (NAD^+), which is a key metabolic coenzyme/cosubstrate involved in cellular energy, that confers renal protection (Gureev et al., 2019; Lynch et al., 2018; Tran et al., 2016). PGC-1α coordinately upregulates the enzymes that synthesize NAD de novo from amino acids, however in AKI there is an attenuation of this de novo pathway (Fig. 20.1). These results are supported by data obtained in PGC-1α knockout mice following IRI. Tran and collaborators demonstrated that PGC-1α knockout mice experienced substantial fat storage, a local deficit of the NAD precursor niacinamide, and an inability to restore normal renal function (Tran et al., 2016). Furthermore, in sepsis associated AKI, there is a suppression of PGC-1α which negatively affects the recruitment of NRF1 and NRF2, leading to an impairment of OXPHOS (Parikh et al., 2015; Tran et al., 2011) (Fig. 20.1).

According to a different study, PGC-1α protein levels rise and PGC-1α mRNA expression decreases on the third and tenth days following unilateral ureteral obstruction, and these changes are strongly linked to fibrosis (Hickey et al., 2011). However, extra advantages are provided by PGC-1α expression. For example, conditional ablation of the renal proximal tubular cells of mammalian target of rapamycin complex 1 causes an increase in the progression of fibrosis following acute ischemic kidney damage, resulting in unfavorable outcomes related to the kidney (Grahammer et al., 2014).

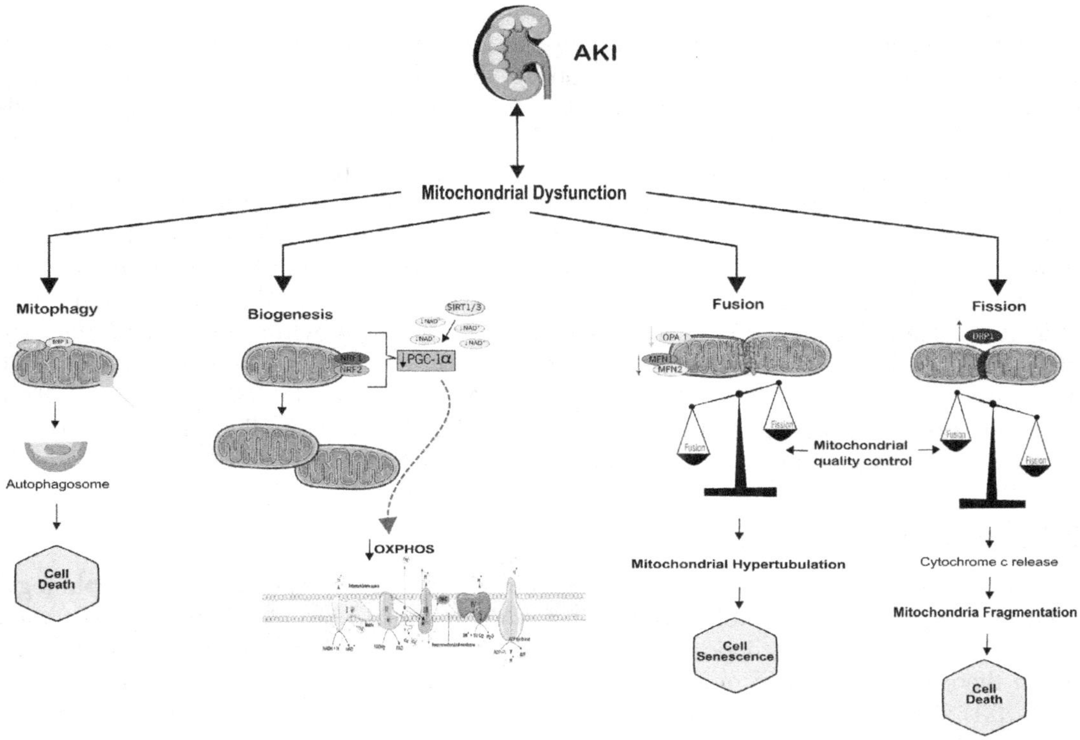

FIGURE 20.1 Schematic illustration of the pathological processes of mitochondrial dysfunction including biogenesis, fusion, fission, and mitophagy in AKI. Balance of fission and fusion processes is necessary to maintain the normal mitochondria quality control. In fusion, there is a dysregulation in the GTPases, where OPA1, Mfn1, and Mfn2 have a decreased activity. Excessive fusion is associated with mitochondrial hypertubulation, which leads to cell senescence. Biogenesis is also affected in AKI. There is a downregulation of PGC-1α which is related with the downregulation of the SIRT1 and 3, that will result in a downregulation of NRF1 and NRF2, culminating in a reduced activity of OXPHOS. Fission event is also altered in AKI, where it can be observed an upregulation of Drp1, that results in an increase of cytochrome C release, and further in mitochondrial fragmentation. Mitophagy protects kidney from IRI by removing damaged mitochondria. The two main mechanisms of mitophagy are the PTEN-induced putative kinase 1 (PINK1) and E3 ubiquitin ligase Parkin-dependent pathway and Bcl2/adenovirus E1B 19-kDa protein interacting protein 3 (BNIP3). An imbalance in mitochondrial control can also occur by excessive of fission over fusion, leading the mitochondria to death. Abbreviations: *AKI*, acute kidney injury; *IRI*, ischemia–reperfusion injury; *OPA1*, optic atrophy 1; *Mfn1*, mitofusin 1; *Mfn2*, mitofusin 2; *NRF1*, nuclear respiratory factor 1; *NRF2*, nuclear respiratory 2; *SIRT1/3*, sirtuin 1 and 3; *NAD+*, nicotinamide adenine dinucleotide; *PGC-1α*, PPArgama-coactivator-1 alpha; and *Drp1*, dynamin-related protein 1.

The family of NAD^+- dependent deacetylases known as sirtuins (SIRT1-7) can regulate mitochondrial biogenesis and play a key role in kidney disease (Fig. 20.1).

It has been shown that SIRT1, a NAD-dependent deacetylase, can control PGC-1α expression and activity. When a SIRT1 activator (SRT1720) was administered to patients with IRI-AKI, the kidneys' expression of PGC-1α was restored. This improved mitochondrial biogenesis, as observed by an increase in the mass and ATP levels of the mitochondria, and ultimately improved renal function (Funk & Schnellmann, 2013). Besides, another activator of mitochondrial biogenesis can be addressed to recover renal function after an AKI episode (Jesinkey et al., 2014). Another example is formoterol, which is a potent agonist of β2-adrenoreceptor, that can induce mitochondria biogenesis by increasing mtDNA copy numbers, oxygen consumption rate, and PGC-1α expression in renal proximal tubular cells (Wills et al., 2012). SIRT3 is present in the mitochondrial matrix, and is responsible to maintain OXPHOS efficiency (Haigis et al., 2012), mtDNA integrity (Osman et al., 2011), as well as mitochondrial biogenesis and fission (Morigi et al., 2015). In sepsis and cisplatin-induced AKI, SIRT3 overexpression has been shown to have a protective effect (Zhao et al., 2018; Zhao et al., 2016). Moreover, mitochondrial FAO in proximal tubules represents a major source of ATP production. In cisplatin induced AKI, a downregulation of PGC-1α was able to decrease the free FAO genes (like carnitine O-palmitoyltransferase and medium chain-specific acyl-CoA dehydrogenase), resulting in a decreased mitochondrial FAO. Thus, this impairment was reported as linked to ATP depletion in AKI (Jang et al., 2020). There are not many studies, but the ones found in the literature highlight a very important role for mitochondrial quality control and biogenesis in AKI.

Mitochondria dynamics in AKI

Mitochondria, as dynamic organelles, possess the ability to regulate their morphology, orchestrating the formation of a tubular network through coordinated fusion and fission events. An excess of mitochondria fission leads into mitochondrial fragmentation, whereas when there is an enhanced mitochondrial fusion results in mitochondrial hypertubulation (Fig. 20.1). A precise balance between these two events is crucial to maintain mitochondrial fitness (Galvan et al., 2017). The coordination between these two opposite processes is primarily mediated by GTPases belonging to the dynamin family. Mitochondrial fission preserves the genetic, biochemical, and functional integrity of mitochondria. Additionally, it makes it easier for them to communicate to each other and contributes to cell homeostasis. It is a process that allows the division of one mitochondrion in two daughter mitochondria, through dynamin-related protein 1 (Drp1). During the mitochondrial fission, Drp1 is recruited and triggers the recruitment of dynamin 2 to perform the membrane scission. On the other hand, the number and distribution of the mitochondria are regulated by mitochondrial fission (Fig. 20.1) (Roushandeh et al., 2019). This process is also responsible to remove the damaged mitochondria through mitophagy. Mechanistically, mitochondrial fusion is caused by mitofusin 1 (Mfn1) and mitofusin 2 (Mfn2) that contribute to the outer mitochondrial membrane, while optic atrophy 1 (Opa1) contributes to the inner membrane fusion.

It is significant to highlight that changes in the expression of these proteins and the disruption of the balance between fission and fusion can impact the integrity and functionality of mitochondria, which have been highly associated with AKI. Several studies have documented a correlation between AKI and an early-stage rise in mitochondrial fission following renal damage (Nowak et al., 2011; Zhan et al., 2013). In fact, an impaired regulation of Drp1 results in increased mitochondrial fission and subsequent release of cytochrome *c*, contributing to end organ damage (Galvan et al., 2017). Mitochondrial fragmentation results from AKI induced reduction in Mfn and Opa1, fusion regulators, whereas the protein Drp1, which controls mitochondrial fission, is quickly activated. To revert this event, Yan and collaborators demonstrated in a murine model that modulation of Mfn2 through microRNAs successfully prevented the disturbance of mitochondrial dynamics and the death of tubular cells in ischemic AKI (Yan et al., 2020). In agreement, Perry and collaborators, in an IR induced mouse model, demonstrated that Drp1 causes mitochondrial dysfunction and increases maladaptive repair pathways, resulting in fibrosis (Perry et al., 2018). Additionally in a rodent model with IRI in which was modulated the expression of Drp1 through an inhibitor, mdivi-1, there was an attenuation in mitochondrial fragmentation, cytochrome *c* release, caspase activation, and apoptosis (Brooks et al., 2009).

Subsequently, mdivi-1 treatment showed a renoprotective effect on mitochondrial function in sepsis-induced AKI. This was evident through an elevation in ATP production, a decrease in mitochondrial fragmentation, and a decrease of the inflammasome mediated pyroptosis of renal tubular epithelial cells caused by NOD-like receptor pyrin domain-3 (Liu et al., 2020). The NOD-like receptor is closely related to renal inflammation and fibrosis. These studies were further consolidated with research demonstrating that mitochondrial dynamics are critical to mediate AKI and the progression of fibrosis (Ludwig-Portugall et al., 2016).

Mitophagy is a sort of selective autophagy that targets and destroys damaged or redundant mitochondria. This process prevents cell death from mitochondrial oxidative stress and the release of proapoptotic proteins. There are two main mechanisms of mitophagy identified (Fig. 20.1). One is the PTEN-induced putative kinase 1 and E3 ubiquitin ligase Parkin-dependent pathway and the other is Bcl2/adenovirus E1B 19-kDa protein interacting protein 3 and the Nip3-like protein X (NIX) pathway (Parikh et al., 2015). Through the removal of damaged mitochondria, mitophagy protects the kidney from IRI. Moreover, some studies have shown that protecting mice against IRI-AKI involves the activation of mitophagy through AMPK (Livingston et al., 2019; Tang et al., 2018; Tran et al., 2011).

Furthermore, renal function, tissue damage, and apoptosis were more severe in proximal tubules with a deletion of autophagy gene-related 7 (Atg7), a mitophagy-related protein, in comparison to the wild type mice (Jiang et al., 2012). Furthermore, renal tubular cells have an increase in autophagosomes in septicemia patients' kidney tissues, which is in contrast with hydropic mitochondria with cristae destruction. This pattern shows that damaged mitochondria are removed via mitophagy (Takasu et al., 2013). Further research revealed that p53 binds to Parkin protein, leading to an impairment of its translocation in mouse heart, thereby inhibiting mitophagy (Hoshino et al., 2013). Notably, the activation of p53 has been documented in AKI, with suggestions that the p53 mediated inhibition of mitophagy might play a role in the pathophysiology of AKI (Ishimoto & Inagi, 2016). Furthermore, Gibbs and collaborators, discovered that when serotonin [5-hydroxytryptamine (5-HT)]1F receptor, which is a physiological regulator of mitochondrial biogenesis, is deleted, the renal damage is exacerbated leading to worse recovery during AKI (Gibbs et al., 2018). Regarding Mfn2, Gall and collaborators, demonstrated that a knockdown of this gene in cultured primary proximal tubular epithelial cells resulted in increased apoptosis under metabolic stress (Gall et al., 2012). However, in contrast to this, the same authors later demonstrated that a conditional knockdown of Mfn2 in mice proximal tubule could accelerate renal recovery in IR induced AKI, by increasing epithelial cell proliferation, rather than apoptosis (Gall et al., 2015). All of these reports demonstrated, as reviewed by Braga and collaborators (Braga et al., 2022), that mitochondrial homeostasis is critical to the maintenance of normal kidney function. Thus, mitochondria dynamics represent possible mediators in AKI and can be modulated to counteract or develop effective therapies to this disease.

Mitochondria as a therapeutic target in kidney transplantation

AKI has been associated to the development of CKD and ESKD. For ESKD, kidney transplantation is the recommended treatment option. Kidney transplantation increases patient survival compared to maintenance dialysis and confers a better quality of life (Lo et al., 2020). However, due to a shortage of transplantable kidneys, a higher number of patients continue to be treated by dialysis while they are waiting to a compatible kidney.

The rate of acute transplant rejection and short term graft failure has significantly decreased thanks to advancements in tissue type matching and immunosuppressive regimes; however, maximizing long term graft performance is still difficult, particularly with kidneys from deceased donors (Lo et al., 2020). The outcomes from living donors are better than those from deceased donors. Nowadays, the universal method for preserving kidneys from donors to recipients is cold storage (CS) (Debout et al., 2015; Hosgood et al., 2011; Salahudeen & May 2008). If the donor is deceased, the kidney undergoes long hours of CS, while kidney from living donors is exposed to only a few hours. Regardless of the technique chosen to preserve the kidney (hypothermic machine perfusion or static CS), the kidney decreases its metabolic rate (Kay et al., 2009; O'callaghan et al., 2013).

Ischemia—a condition in which tissues lack blood—is associated with a decrease of oxygen and nutrients supplies that mark the beginning of the early events that cause harm in the deceased donor kidney (Bonventre, 1988). Warm ischemia and reperfusion, also referred to as ischemia–reperfusion damage, are introduced during transplantation via vascular anastomosis. Short and long term graft outcomes are impacted by IRI events during kidney transplantation (Stewart et al., 2017). Through the activation of enzymes such as xanthine oxidase, NADPH oxidase, uncoupled nitric oxide synthase (NOS), and mitochondrial respiratory complexes, IRI itself encourages and enhances ROS generation. Furthermore, the cessation of supplied oxygen during ischemia triggers renal tissue hypoxia (Haase, 2006). Renal function recovery requires enough reperfusion in the transplant to prevent hypoxic tissue harm. Even when molecular oxygen is restored by reperfusion, the vascular injury from IRI causes a drop in renal blood flow of up to 50% following reperfusion (Hellberg et al., 1991; Mason et al., 1984; Schrier et al., 2004; Thiel et al., 1982). This will, in turn, limit the ability to exchange solutes (Rosen et al., 1992) and produce a high rate of aerobic glycolysis (Cohen, 1979). In fact, it has been suggested that one typical route to renal failure following transplantation is chronic hypoxia in the tubulointerstitium (Nangaku, 2006). Few studies highlight the importance of mitochondrial dynamics during renal CS combined with transplantation (CS/Tx). Salahudeen and collaborators, using primary human proximal tubular cells, tried to elucidate the impact in the mitochondria pathophysiology in transplanted kidneys, when exposed to CS and after a rewarming process that is associated with the transplantation event (Salahudeen et al., 2001). The authors observed that, after 24 h of CS exposure, the cells presented mitochondrial swelling and a loss of inner mitochondrial membrane. Given the continued necrosis, the rewarming was linked to significant levels of apoptosis. Additionally, this study found that the antioxidants in the storage solution shielded the cells from the effects of CS and rewarning, which can cause necrosis and apoptosis (Salahudeen et al., 2001). Later on, Yin and collaborators developed a study in Wistar rats, where the main goal was to determinate which renal cells lose their viability (observed using Trypan blue dye to identify the nuclei of nonviable cells) after warm and cold ischemia. Warm ischemia led to injury of proximal tubular cells, while cold ischemia caused renal failure at podocytes and peritubular endothelial cells level. Podocytes were blebbed, swelled, and their pedicels separated from the basement membrane in cold ischemia-induced damage, while proximal tubule disruption was less severe in comparison to the warm ischemia induced. Furthermore, in warm ischemic, electron microscopy demonstrated shedding of microvilli and marked swelling of mitochondria (Yin et al., 2002). Later on, Lewis rat kidneys were exposed to CS alone or CS combined with transplantation and proteins related to mitochondrial function and mitochondrial dynamics were evaluated (Parajuli et al., 2017). The results demonstrated that the treatment only with CS reduced respiratory complex I and II activities along with reduced expression of Drp1 protein. It also induced the loss of the OPA1 and an alteration of mitochondrial protease regulating OPA1 processing. Furthermore, in CS/Tx group, it was observed a reduction of complex I, II, and III activities as well as a loss of the short forms of OPA1, Mfn1 and Mfn2 proteins. It was also observed that after CS/Tx the expression of Drp1 protein was greatly reduced, along with the mitochondrial fission factor. Remarkably, Drp1 protein levels were unchanged in rats receiving autotransplants (transplants without CS exposure) (Parajuli et al., 2017).

Despite a lack of consistent studies, it is already noticed that mitophagy after CS/Tx is also affected. Indeed, a study performed in rat tubular epithelial cells demonstrated that autophagy is highly affected during CS rewarming. The inhibition of autophagy, through Atg5 siRNA, resulted in increased cleaved caspase-3 and increased apoptosis, demonstrating that this induction of apoptosis may be deleterious as a therapy to decrease delayed graft function (Jain et al., 2015).

Notably, lower ATP levels and ROS accumulation during renal CS/Tx are caused by mitochondrial respiratory dysfunction, which suggests the possibility that mitochondrial function is where "damage" signals from CS arise (Mitchell et al., 2011; Parajuli et al., 2017; Salahudeen, 1999; Salahudeen et al., 2003). Moreover, this mitochondrial respiratory dysfunction impairs the ion exchange/transport channel resulting in calcium accumulation (Devarajan, 2005; Kosieradzki & Rowiński, 2008), which in turn will activate several enzymes that are responsible for the occurrence of apoptosis (Orrenius et al., 2003). To overcome the issues of CS and rewarming, some solutions have been pointed out, also with repercussions on the mitochondrial level.

Schleef and colleagues conducted a comprehensive study involving mice subjected to various protocols. They employed a bilateral renal IRI model to evaluate acute injuries within the initial 24 h, and a less severe unilateral renal IRI model to assess chronic lesions at 1 month. Their objective was to investigate whether the application of mild therapeutic hypothermia (mTH) at 34°C during the ischemic period could offer protective effects on the structure and function of the kidneys against injuries (Schleef et al., 2022). An in vivo bilateral renal IR model led to an increase in plasma urea and acute tubular necrosis at 24 h. Moreover, 1 month following unilateral IR, there was a notable reduction in kidney atrophy and fibrosis. In terms of mitochondria fitness evaluation, mTH protected from mitochondrial dysfunction, through preservation of calcium retention capacity and OXPHOS. mTH effect was maintained in the long term (1 month), by decreasing fibrosis and renal atrophy, and abrogating chronic renal inflammation (IL-6, IL1-β and IL-10) (Schleef et al., 2022).

Since mitochondrial health is highly regulated to maintain kidneys physiology and function, exploring interventions to safeguard renal mitochondria during CS emerges as a valuable avenue for investigation, holding promise as potential therapeutic options. One example is the use of the pharmacological activation (NS11021) of the mitochondrial large conductance Ca^{2+} activated K^{+} (mitoBK) during CS in rats (Shrum et al., 2020). This activator immobilized the impaired renal mitochondrial respiration at the complex's levels, by providing protection against respiratory dysfunction at Complex II and partially protects Complexes III and IV. It also offers protection against protein tyrosine nitration as well as apoptosis (Shrum et al., 2020). All of these findings show that one of the main initiators of CS/Tx-induced kidney impairment is mitochondrial damage, leading to a significant decrease of mitochondrial fitness, that will in turn, result in the aggressiveness of the renal outcomes in kidney diseases.

Mitochondrial "medicine" is a new therapeutic approach focused on clinical use of mitochondrion (Picard et al., 2016). In fact, previous reports, already demonstrated that mitochondrial transplantation can improve the outcomes of multiple diseases, such as Parkinson's disease (Chang et al., 2016), liver IR injury (Hayakawa et al., 2016), brain stroke (Hayakawa et al., 2016), and heart diseases (Emani et al., 2017). Unfortunately, in kidney diseases, the best therapy is limited to preventive strategies, and lately dialysis and fluid management, which is not an effective solution. Therefore, the establishment of new therapeutic approaches is required to enhance the quality of patients' life with renal failure to solve the issues related with hemodialysis and organ transplantation (Chapman et al., 2016; Ibrahim et al., 2016).

Mitochondrial transplantation: A novel technique in evolution

Mitochondrial dysfunction is known to be related in multiple pathophysiology diseases, such as cancer, neurodegenerative diseases, cardiovascular, and renal failure. Thus, mitochondrial medicine reveals to be a novel and innovating therapeutic modality, which the treatment resembles only at the mitochondrial level.

Given that mitochondria are actually transferrable from one cell to another, the idea of mitochondrial transplantation is based on this fact. When mitochondria are harmed, in order to sustain the wounded cells, this can happen (Roushandeh et al., 2019). There are already reports documenting successful cases of mitochondrial transplantation. For instance, in response to cerebral ischemia, astrocytes have been found to secrete mitochondria via membrane vesicles (Falchi et al., 2013), which were subsequently successfully transported into neurons, improving cell survival signaling (Hayakawa et al., 2016).

Different contact modes, such as junction, cell fusion, and tunneling nanotube (TNT) development, allow for the migration of mitochondria between cells (Lu et al., 2017; Paliwal et al., 2018). In a work led by Islam and collaborators, mitochondrial transfer from mesenchymal stem cells (MSCs) to alveolar type II epithelial cells (ATII) restored the bioenergetic profile of ATII and improved survival in a mouse model of lipopolysaccharide-induced pneumonia (Islam et al., 2012). Concerning TNT formation, the transfer of functional mitochondria from healthy pheochromocytoma cells (PCs) to stressed PC cells via F-actin based TNTs reversed stressed cells in early stages of apoptosis in culture (Wang & Gerdes, 2015). In the context of acute respiratory distress syndrome, the transfer of mitochondria from MSCs to alveolar macrophages through TNTs is recognized as a crucial pathway contributing to the in vivo antimicrobial effect of MSCs. Jackson and collaborators, in 2016, taken advantage of MSCs being self renewing multipotent adult stem cells, which can be isolated from different tissues and organs (Alhadlaq & Mao, 2004; Hoogduijn et al., 2007) where, in vivo conditions, they can be differentiated into multiple cell types (Pittenger et al., 1999).

Furthermore, acute respiratory distress syndrome preclinical models have demonstrated the protective benefits of MSCs. Jackson and collaborators demonstrated that coculture of MSCs with monocyte-derived macrophages enhanced the phagocytic capacity in MSC treatment in mouse with pneumonia *Escherichia coli* associated (Jackson et al., 2016). Additionally, they showed that MSCs underwent a substantial mitochondrial translocation to macrophages, at least in part, through structures like TNT. Hence, with this study, they demonstrated that mitochondrial donation via direct cell contact from MSCs to innate immune cells can be considered as a novel mechanism adopted on MSCs in acute respiratory distress syndrome (Jackson et al., 2016). In vivo models of asthma and COPD have also demonstrated the protective effects of mitochondrial translocation from MCS to bronchial epithelial cells via TNT (Ahmad et al., 2014; Li et al., 2014) as well as in vitro model of reperfusion injury in human umbilical vein endothelial cell culture (Liu, Ji, et al., 2014).

Theoretically, the process of mitochondrial transference seems simple; however it is questionable if the delivering of mitochondria into the cells is active or passive. To overcome some issues related to the mitochondrial internalization, some strategies are being implemented to improve the mitochondrial delivery. Mitochondria transfer can be based on the transfer RNA (tRNA) targeting therapy (Kolesnikova et al., 2004; Mahata et al., 2006; Wang et al., 2012), cell–cell fusion (Ma et al., 2005), microinjection (Pinkert et al., 1997; Yi et al., 2007), or even through peptide mediated mitochondrial delivery (PMD) (Liu, Chang, et al., 2014) (Fig. 20.2). Enhancing mitochondrial uptake and improving transplanting efficiency are the main goals of PMD. Moreover, the PMD and microinjection share some similarities. Because this technique does not involve artificial gene recombination, which is required in the tRNA mitochondrial targeting method, it is simple to apply in clinical practice (Liu, Chang, et al., 2014). Given that human MSCs lack the capacity to transfer mitochondria or fuse with cells that have mitochondrial disease and pathogenic mtDNA mutations, the PMD model seems to be more effective and productive than the cell–cell fusion intervention. Indeed, Cho and collaborators demonstrated that human MSCs can transfer mitochondria to the cells with compromised mitochondrial function. In a coculture of MSCs and 143B derived cells, they induced irreversible damage to the mtDNA and it was noticed that the function was restored through a direct transfer of intact mitochondrial particles rather than mtDNA alone (Cho et al., 2012).

PMD was firstly implemented in cell lines exhibiting myoclonic epilepsy with ragged red fibers (MERF) disease, carrying mtDNA 8344np A to G mutation, in the attempt to rescue mitochondrial disease. In this study, isolated healthy mitochondria from donor were delivered into MERF fibroblast though Pep-1 conjugation (Pep-1-Mito). After 3 days of treatment with Pep-1-Mito, it was observed a restoration of OXPHOS subunits (complex I, II, and IV), mitochondrial membrane potential, ATP synthesis, and a reduction of ROS production (Chang et al., 2013). Therefore, a promising treatment to prevent the pathogenesis of mitochondrial disorders is the transplantation of mitochondria using PMD (Fig. 20.2).

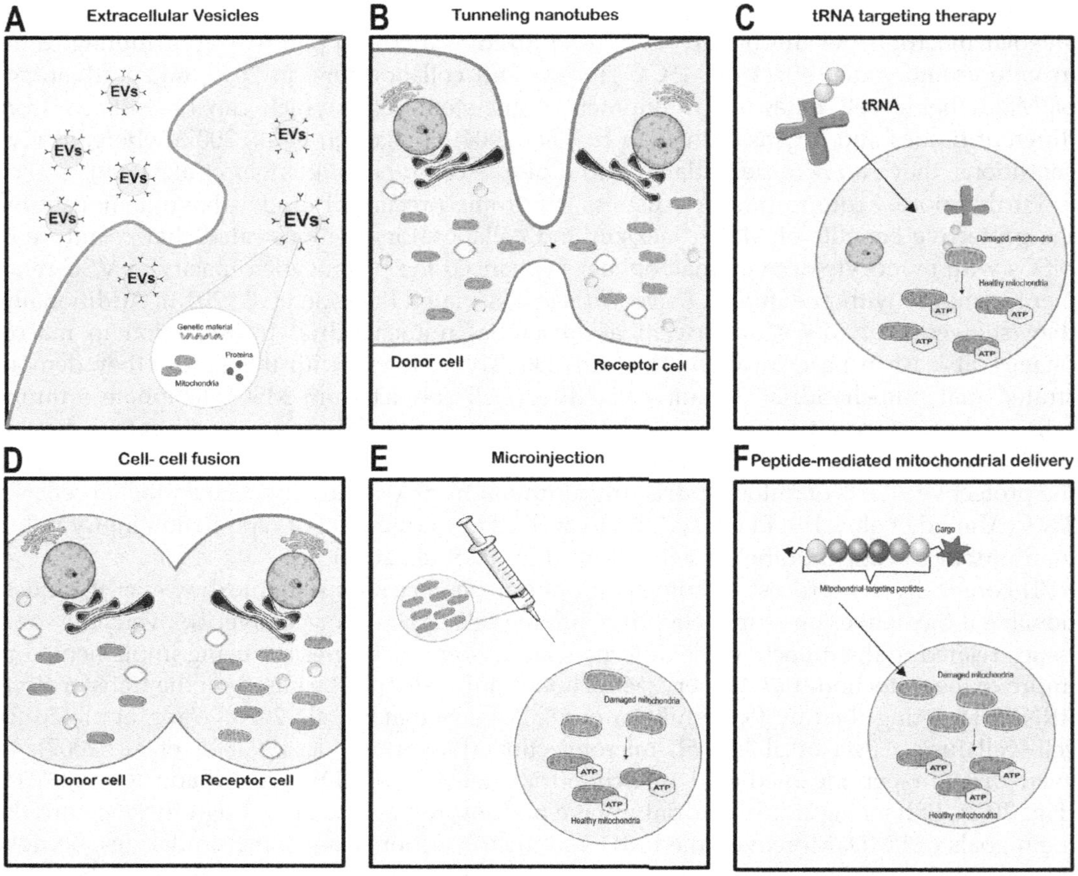

FIGURE 20.2 Schematic illustration of the potential mechanisms by which mitochondria can be transplanted, including extracellular vesicles (Panel A), tunneling nanotubes (Panel B), tRNA targeting therapy (Panel C), cell–cell fusion (Panel D), microinjection (Panel E), and peptide-mediated mitochondrial delivery (Panel F). The main objective of all the mechanisms is the turnout of damaged mitochondria into healthy mitochondria to reestablish mitochondrial quality control.

Furthermore, Wu and associates created a biocompatible polymer to improve in vivo applications for cellular transplantation, particularly mitochondrial transplantation into cardiovascular and cancer cells. Herein, the authors demonstrated that the polymer conjugate composed of dextran with triphenylphosphonium onto isolated mitochondria protects the organelles and facilitates cellular internalization compared with uncoated mitochondria, with an increase delivery rate of isolated mitochondria. Additionally, they noticed that in heart and breast cancer cell lines, the coating to preserve the mitotic respiratory function caused a notable change in the bioenergetic phenotype from glycolysis to OXPHOS (Wu et al., 2018).

The kidney, as with cancer and neurologic cells, requires a high amount of energy, and thus a correct mitochondrial functioning. In addition, considering renal diseases, it is important to address to the mitochondria a higher importance, since it has a role in the maintenance of cellular function and homeostasis. The reabsorption and control of water, glucose, and electrolytes like sodium ions are carried out by proximal tubules by active transport, which consumes a significant amount of ATP (Wang & Kestenbaum, 2018). Additionally, cytoskeletal changes brought on by mitochondria degeneration result in brush border destruction, a loss of cell to cell connections, and increased epithelial shedding (Ralto & Parikh, 2016; Tang & Dong, 2016; Zhan et al., 2013). Renal tubule healing requires morphologically improved renal tubular epithelial cells that are complemented by functional endogenous mitochondria (Konari et al., 2019), and so mitochondrial transplantation can be valuable solution to improve mitochondria fitness, which in turn will be reflected in kidney normal function recovery.

Mitochondrial transplantation in the kidney

Renal dysfunction or failure can result from a disruption or cessation in oxygen delivery because mitochondrial function and viability depend on a steady supply of oxygen delivered by the blood. The resulting tissue ischemia and disruption of oxygen delivery impacts mitochondrial energy production (Bhargava & Schnellmann, 2017; Bonventre & Weinberg, 2003; Ishimoto & Inagi, 2016). Mitochondrial dysfunction in the kidney has a key role in the onset of diseases including diabetic nephropathy (DN) and acute and chronic renal failure (Galvan et al., 2017; Tábara et al., 2014).

There are already some studies demonstrating that mitochondrial transplantation has protective outcomes in the heart and brain ischemia, resulting in an enhanced of cardiac function (Huang et al., 2013) and restoration of motor function in an ischemic rat brain (Huang et al., 2016). Furthermore, a growing number of studies indicate that mitochondrial dysfunction plays a significant role in AKI (Forbes, 2016; Ishimoto & Inagi, 2016; Ralto & Parikh, 2016). Several preclinical studies have used mitochondrial transplantation, including in a case of IRI in pig hearts (Guariento et al., 2020), spinal cord injury in rat (Gollihue et al., 2017), and IRI of brain rat (Zhang et al., 2019). Therefore, the rescue of cells and the return of normal function will be facilitated by supplementing or replacing the damaged mitochondria with transplanted viable mitochondria that have been separated from a healthy tissue and then delivered to the injured organ, tissue, or cells.

In stem cell based interventions, Togel and associates looked into the safety and long-term management of autologous and allogenic MSCs following AKI during a 3-month period, which is comparable to 3 to 4 years of follow-up. They showed that both autologous and allogeneic MSCs are effective in the treatment of AKI, particularly in retarding interstitial renal fibrosis and loss of renal function. Furthermore, the protection from chronic renal damage provided by MSCs was shown only in rats with severe AKI, suggesting that the degree of the initial injury may have an impact on this protection (Tögel et al., 2009).

Renal tubular epithelial cells can be damaged by renal ischemia, which can result in the loss of renal function (Emma et al., 2016). Nevertheless, following kidney injury, these tubular epithelial cells have some capacity to repair; however, the source of the cells responsible for this recovery remains unclear. Some studies suggested that there is a group of renal resident scattered tubular cells that undergo proliferation after injury (Ebrahimi et al., 2013; Emma et al., 2016). Extracellular vesicles (EVs) that are released from donor cells to target cells carry proteins and genetic material, making them a key intercellular communication mechanism. It is known that some EV can have and deliver mitochondrial materials that can increase the bioenergetic function in recipient cells (Fig. 20.3) (Boudreau et al., 2014; Guescini et al., 2010; Phinney et al., 2015). Zou and collaborators explored if the EV release could have a positive impact on scattered tubular cells in the repairing of injured tubular epithelial cells. They tested this hypothesis, using a proximal tubular epithelial cell line (PK1 cells), that was cocultured with stem-like cells (STC) isolated from fresh pig kidneys in three different conditions: STC-like cells, STC-like cell derived EV or EV free conditioned medium during 3 days. STC-like cells were intraarterially injected into mice 2 weeks after the induction of unilateral renal artery stenosis. They reported that STC-like cells EV carried and transferred mitochondrial material to injured tubular epithelial cells improvement in renal function and alleviated fibrosis in ischemic kidney injury. Furthermore, oxidative stress and mitochondrial function in tubular epithelial cells have been repaired in unilateral renal artery stenosis model in vivo and antimycin-induced injured proximal tubular epithelial cells in vitro (Zou et al., 2018).

Moreover, the administration of microvesicles derived from MSCs, reported by Bruno and collaborators, enhanced the recovery from glycerol-induced AKI in mice through stimulation of tubular cell proliferation (Bruno et al., 2009). In support to this, Souidi et al. provided evidence in favor of this by showing that the microvesicular structures of MSCs enabled mitochondrial transplantation to contribute to the regeneration of ischemic kidney tissue (Souidi et al., 2013).

Concerning mitochondria transplantation, Yuan and collaborators have also taken advantage from the MSCs and for the first time reported the mechanisms that are involved in the kidney protection in DN through MSCs mitochondrial transference. In streptozotocin (STZ) induced DN mice, it was demonstrated that PGC-1α mediated mitochondria synthesis and PGC-1α/TFEB mediated autophagy, which was able to reduce inflammation and ameliorate kidney damage by simply mitochondria transference from MSCs to macrophages (Yuan et al., 2021). Regarding autologous mitochondria transfer through intraarterial injection, there are few studies on experimental models of kidney focal ischemia. This method involves the injection of viable and respiration-competent autologous mitochondria extracted from the patient's healthy skeletal muscle to restore the functionality of native mitochondria that have been damaged by IRI (Fig. 20.3) (Roushandeh et al., 2019). Doulamis and collaborators investigated the effectiveness and safety of intraarterial autologous mitochondrial transplantation for renal protection in bilateral renal IRI (Doulamis et al., 2020). Following 60 min of ischemia, female Yorkshire pigs received 1×10^9 mitochondria in a single injection into each renal artery. Within the tubular epithelium of the cortex and medulla, the injected mitochondria were rapidly absorbed by the kidney. After 24 h of reperfusion, urine output and eGFR were significantly increased along with a decrease in serum creatine and blood urea nitrogen in pigs that received mitochondria. Furthermore, histopathological analysis revealed that mitochondrial transplanted kidney presented only patchy mild acute tubular injury, while the kidney treated only with vehicle had extensive coagulative necrosis of the proximal tubules (Doulamis et al., 2020). Additionally, still in the kidney treated only with the vehicle, the renal cortex IL-6 expression was elevated in comparison to those who received mitochondrial transplantation. The authors thus suggested that their results obtained from a clinically relevant model allow a translation to humans, enhancing the knowledge of clinicians for the treatment of AKI (Doulamis et al., 2020). After this study, Jabbari and collaborators followed the same approach in mice, by examined the effect of auto transplantation of healthy mitochondria from muscle cells to injured kidney cells through injection into the renal artery.

They proved that renal tubular cell death could be avoided by mitochondrial transplantation (7.5×10^6 mitochondria/mL), which also increased renal tubules' capacity to regenerate, reduced IRI-induced apoptosis, and restored renal function (Jabbari et al., 2020). Kubat and collaborators explored the beneficial effects of isolated mitochondria transplantation of MSCs in vivo, to mitigate renal damage. The authors used rats as animal model and the kidney damage/renal injury was induced with the administration of a single dose of doxorubicin (6 mg/kg). The mitochondrial transplantation was performed in two separated injections (8×10^6 mitochondrial in 500 μL) on the lateral side of the renal cortex. After 10 days, the animals were sacrificed. The findings showed that proteinuria levels started to decrease 6 days after isolated mitochondria from MSCs had been inserted into the renal cortex. Furthermore, after transplanting mitochondria, tubular cells showed histological evidence of tubular regeneration and it was observed a rising antioxidant stress enzyme levels (SOD and GPx) against oxidative stress (Kubat, Ozler, et al., 2021). Later, using a similar methodology, the same authors investigated the outcomes of the direct injection of mitochondria isolated from MSCs into the renal cortex of rats experiencing renal ischemia (Kubat, Kartal, et al., 2021). In this study, it was administered a single dose of 2×10^7 mitochondria/mL and the rats were monitored for 48 h. This study demonstrated that mitochondrial transplantation, reduced apoptotic markers, including TUNEL and Caspase-3 in injured renal. Moreover, it promoted proliferation and regeneration of tubular cells. In terms of mitochondrial dynamics, the expression of Drp1 protein was decreased in tubular cells after injury (Kubat, Kartal, et al., 2021) (Fig. 20.3).

Konari et al. conducted a study where for the first time it was unveiled the potential therapeutic effects of mitochondrial transplantation from bone marrow derived MSCs in DN. They observed that mitochondria (1×10^4 mitochondria per gram of body weight), within 24 h, were incorporated into proximal tubular cells and affected SOD2, Bcl-2, Bax, and PGC-1α expression in STZ-induced diabetic mice and rats, which suppressed ROS production, and inhibition of apoptosis. Furthermore, it was noted that within 3 days, STZ rats' proximal tubules showed an induced histological improvement that led to the recovery of megalin and SGLT2 expression (Fig. 20.3). Thus, this study demonstrates that mitochondria transplantation plays a key role in the therapeutics of bone marrow–derived MSCs toward DN treatment (Konari et al., 2019).

Overall, all the studies so far undoubtedly highlight that mitochondrial transplantation can have a role in kidney disease. Still, further studies are needed to determine the conditions and the mechanisms by which it can be useful for each particular case of the plethora of kidney diseases.

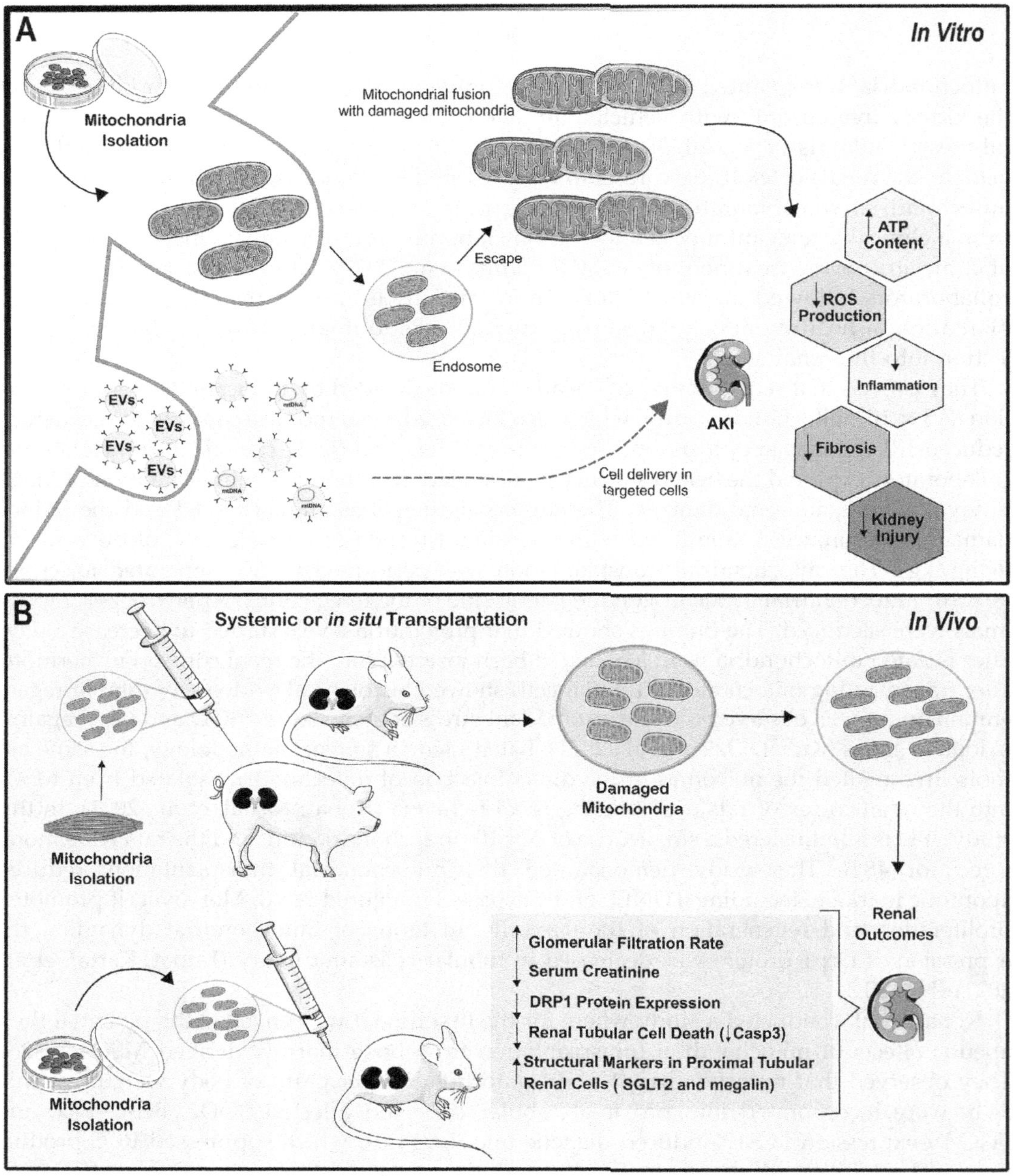

FIGURE 20.3 Mitochondrial transplantation in different kidney cell models in vitro (Panel A) and in vivo (Panel B). In vitro, the main mitochondria derived from MSCs, which enter in the target cell through endocytosis, and then there is a fusion between the healthy and the damaged mitochondria. Furthermore, EVs are released from donor cells to target cells that carry proteins and genetic and mitochondrial material that will contribute to an increase of renal function. With this approach, a decrease was observed in kidney injury, associated with an increase of ATP content, ROS production, and with a decrease in fibrosis and inflammation. Mitochondrial transplantation in vivo, in kidney disease, is only reported in mice, rat, and swine model. The isolation of the mitochondria can be from MSCs or even from skeletal muscle. The mitochondria are transplanted through microinjection technique, where their delivery can be systemic or in situ. After the injection of healthy mitochondria, the damaged ones are replaced, leading to a higher mitochondrial quality. The principal renal outcomes observed with this novel technique are an increase in glomerular filtration along with a decrease in serum creatine. A decrease was also observed in the expression of the Drp1 protein, and the renal tubular apoptotic markers, such as Casp3. Furthermore, an increase is observed in structural markers in proximal tubular renal cells including SGLT-2 and megalin. Abbreviations: *AKI*, acute kidney injury; *casp 3*, caspase 3; *EVs*, extracellular vesicles; *MSCs*, mesenchymal stem cells; *SGLT2*, sodium glucose transporter 2.

Conclusion

Mitochondrial dysfunction can be traced in multiple diseases including diabetes, ischemia, liver injury, aging, cancer, skeletal muscles disorders, neurodegenerative diseases, infertility, and even renal failure (Ma et al., 2017; McCully et al., 2016). Since tubular epithelial cells are among the most metabolically active epithelia in the human body, it is not surprising that the first pathological anomalies in kidney illnesses may manifest in tubular epithelial mitochondria following acute renal damage (Bhargava & Schnellmann, 2017). Thus, mitochondrial dysfunction is related to typical pathology in AKI (Thadhani et al., 1996). The proteins related to mitochondrial homeostasis, including biogenesis, fragmentation and mitophagy, which are processes that play important roles in AKI progression, highlight the importance to understanding the pathophysiology of AKI, and the mechanisms that link AKI to the progression of CKD (Wakabayashi & Karbowski, 2001) and mitochondria (dys)function. More specifically, as there is no effective treatment in kidney disease, despite dialysis and kidney transplantation, there is an urgent need to find new approaches that can rely on the mechanisms that link all the described processes in this chapter. Taking in consideration the role of mitochondria physiology and function in kidney, mitochondrial transplantation appears to be a safe approach, without major risks associated, with substantial improvements in mitochondrial functioning, and thus, it can be suggested as an alternative for the treatment of kidney diseases. Nevertheless, more studies are needed to understand how different mitochondrial transplantation procedures can be useful or implemented to treat distinct kidney diseases since this technique has limitations that need a special attention to be overcome. We are only now starting to unveil and understand how these mechanisms occur and can be useful and thus, exciting new works are expected in the years to come.

Acknowledgments

This research was funded by "Fundação para a Ciência e a Tecnologia" — FCT to LAQV-REQUIMTE (UIDB/50006/2020), UMIB (UIDB/00215/2020, and UIDP/00215/2020), ITR - Laboratory for Integrative and Translational Research in Population Health (LA/P/0064/2020), IBIMED (UIDP/04501/2020 and UIDB/04501/2020), Patrícia C. Braga PhD fellowship (UI/BD/150750/2020) and Pedro F Oliveira (CEECINST/00026/2018). The work was also co-funded by FEDER through the COMPETE/QREN, FSE/POPH and POCI - COMPETE 2020 (POCI-01-0145-FEDER-007491) funds.

References

Agarwal, A., Dong, Z., Harris, R., Murray, P., Parikh, S. M., Rosner, M. H., Kellum, J. A., & Ronco, C. (2016). Cellular and molecular mechanisms of AKI. *Journal of the American Society of Nephrology, 27*(5), 1288–1299, 1046-6673.

Ahmad, T., Mukherjee, S., Pattnaik, B., Kumar, M., Singh, S., Kumar, M., Rehman, R., Tiwari, B. K., Jha, K. A., & Barhanpurkar, A. P. (2014). Miro1 regulates intercellular mitochondrial transport and enhances mesenchymal stem cell rescue efficacy. *The EMBO Journal, 33*(9), 994–1010, 0261-4189.

Alhadlaq, A., & Mao, J. J. (2004). Mesenchymal stem cells: Isolation and therapeutics. *Stem Cells and Development, 13*(4), 436–448, 1547-3287.

Aparicio-Trejo, O. E., Reyes-Fermín, L. M., Briones-Herrera, A., Tapia, E., León-Contreras, J. C., Hernández-Pando, R., Sánchez-Lozada, L. G., & Pedraza-Chaverri, J. (2019). Protective effects of N-acetyl-cysteine in mitochondria bioenergetics, oxidative stress, dynamics and S-glutathionylation alterations in acute kidney damage induced by folic acid. *Free Radical Biology and Medicine, 130*, 379–396, 0891-5849.

Bellomo, R., Ronco, C., Kellum, J. A., Mehta, R. L., & Palevsky, P. (2004). Acute renal failure—definition, outcome measures, animal models, fluid therapy and information technology needs: The second international consensus conference of the acute dialysis quality initiative (ADQI) group. *Critical Care, 8*(4), 1–9, 1364-8535.

Bhargava, P., & Schnellmann, R. G. (2017). Mitochondrial energetics in the kidney. *Nature Reviews Nephrology, 13*(10), 629–646, 1759-507X.

Bonventre, J. V. (1988). Mediators of ischemic renal injury. *Annual Review of Medicine, 39*(1), 531–544, 0066-4219.

Bonventre, J. V., & Weinberg, J. M. (2003). Recent advances in the pathophysiology of ischemic acute renal failure. *Journal of the American Society of Nephrology, 14*(8), 2199–2210, 1046-6673.

Boudreau, L. H., Duchez, A.-C., Cloutier, N., Soulet, D., Martin, N., Bollinger, J., Paré, A., Rousseau, M., Naika, G. S., & Lévesque, T. (2014). Platelets release mitochondria serving as substrate for bactericidal group IIA-secreted phospholipase A2 to promote inflammation. *Blood, The Journal of the American Society of Hematology, 124*(14), 2173–2183, 0006-4971.

Braga, P. C., Alves, M. G., Rodrigues, A. S., & Oliveira, P. F. (2022). Mitochondrial pathophysiology on chronic kidney disease. *International Journal of Molecular Sciences, 23*(3), 1776, 1422-0067.

Brooks, C., & Dong, Z. (2007). Regulation of mitochondrial morphological dynamics during apoptosis by Bcl-2 family proteins: A key in Bak? *Cell Cycle, 6*(24), 3043–3047, 1538-4101.

Brooks, C., Wei, Q., Cho, S.-G., & Dong, Z. (2009). Regulation of mitochondrial dynamics in acute kidney injury in cell culture and rodent models. *The Journal of Clinical Investigation, 119*(5), 1275–1285, 0021-9738.

Brown, G. C., Murphy, M. P., Jornayvaz, F. R., & Shulman, G. I. (2010). Regulation of mitochondrial biogenesis. *Essays in Biochemistry, 47*, 69–84, 0071-1365.

Bruno, S., Grange, C., Deregibus, M. C., Calogero, R. A., Saviozzi, S., Collino, F., Morando, L., Busca, A., Falda, M., Bussolati, B., Tetta, C., & Camussi, G. (2009). Mesenchymal stem cell-derived microvesicles protect against acute tubular injury. *Journal of the American Society of Nephrology, 20*(5), 1053–1067.

Bucaloiu, I. D., Kirchner, H. L., Norfolk, E. R., Hartle II, J. E., & Perkins, R. M. (2012). Increased risk of death and de novo chronic kidney disease following reversible acute kidney injury. *Kidney International, 81*(5), 477–485, 0085-2538.

Chang, J.-C., Liu, K.-H., Chuang, C.-S., Su, H.-L., Wei, Y.-H., Kuo, S.-J., & Liu, C.-S. (2013). Treatment of human cells derived from MERRF syndrome by peptide-mediated mitochondrial delivery. *Cytotherapy, 15*(12), 1580–1596, 1465-3249.

Chang, J.-C., Wu, S.-L., Liu, K.-H., Chen, Y.-H., Chuang, C.-S., Cheng, F.-C., Su, H.-L., Wei, Y.-H., Kuo, S.-J., & Liu, C.-S. (2016). Allogeneic/xenogeneic transplantation of peptide-labeled mitochondria in Parkinson's disease: Restoration of mitochondria functions and attenuation of 6-hydroxydopamine–induced neurotoxicity. *Translational Research, 170*, 40–56, 1931-5244.

Chapman, J., Craig, J., Strippoli, G., Wong, G., Wyburn, K., Lim, W. H., Lo, P., Palmer, S., & Sharma, A. (2016). *Preconditioning therapy in ABO-incompatible living kidney transplantation: A systematic review and meta-analysis*. Lippincott Williams and Wilkins.

Chawla, L. S., & Kimmel, P. L. (2012). Acute kidney injury and chronic kidney disease: An integrated clinical syndrome. *Kidney International, 82*(5), 516–524, 0085-2538.

Che, R., Yuan, Y., Huang, S., & Zhang, A. (2014). Mitochondrial dysfunction in the pathophysiology of renal diseases. *American Journal of Physiology—Renal Physiology, 306*(4), F367–F378, 1931-857X.

Cho, Y. M., Kim, J. H., Kim, M., Park, S. J., Koh, S. H., Ahn, H. S., Kang, G. H., Lee, J.-B., Park, K. S., & Lee, H. K. (2012). Mesenchymal stem cells transfer mitochondria to the cells with virtually no mitochondrial function but not with pathogenic mtDNA mutations. *PLoS One, 7*(3), e32778, 1932-6203.

Chouchani, E. T., Pell, V. R., James, A. M., Work, L. M., Saeb-Parsy, K., Frezza, C., Krieg, T., & Murphy, M. P. (2016). A unifying mechanism for mitochondrial superoxide production during ischemia-reperfusion injury. *Cell Metabolism, 23*(2), 254–263, 1550-4131.

Coca, S. G., Singanamala, S., & Parikh, C. R. (2012). Chronic kidney disease after acute kidney injury: A systematic review and meta-analysis. *Kidney International, 81*(5), 442–448, 0085-2538.

Cohen, J. J. (1979). Is the function of the renal papilla coupled exclusively to an anaerobic pattern of metabolism? *American Journal of Physiology—Renal Physiology, 236*(5), F423–F433, 1931-857X.

Debout, A., Foucher, Y., Trébern-Launay, K., Legendre, C., Kreis, H., Mourad, G., Garrigue, V., Morelon, E., Buron, F., & Rostaing, L. (2015). Each additional hour of cold ischemia time significantly increases the risk of graft failure and mortality following renal transplantation. *Kidney International, 87*(2), 343–349, 0085-2538.

Devarajan, P. (2005). Cellular and molecular derangements in acute tubular necrosis. *Current Opinion in Pediatrics, 17*(2), 193–199, 1040-8703.

Doulamis, I. P., Guariento, A., Duignan, T., Kido, T., Orfany, A., Saeed, M. Y., Weixler, V. H., Blitzer, D., Shin, B., & Snay, E. R. (2020). Mitochondrial transplantation by intra-arterial injection for acute kidney injury. *American Journal of Physiology—Renal Physiology, 319*(3), F403–F413, 1931-857X.

Duann, P., & Lin, P.-H. (2017). Mitochondria damage and kidney disease. *Mitochondrial Dynamics in Cardiovascular Medicine*, 529–551.

Ebrahimi, B., Eirin, A., Li, Z., Zhu, X.-Y., Zhang, X., Lerman, A., Textor, S. C., & Lerman, L. O. (2013). Mesenchymal stem cells improve medullary inflammation and fibrosis after revascularization of swine atherosclerotic renal artery stenosis. *PLoS One, 8*(7), e67474, 1932-6203.

Emani, S. M., Piekarski, B. L., Harrild, D., Pedro, J., & McCully, J. D. (2017). Autologous mitochondrial transplantation for dysfunction after ischemia-reperfusion injury. *The Journal of Thoracic and Cardiovascular Surgery, 154*(1), 286–289, 0022-5223.

Emma, F., Montini, G., Parikh, S. M., & Salviati, L. (2016). Mitochondrial dysfunction in inherited renal disease and acute kidney injury. *Nature Reviews Nephrology, 12*(5), 267–280, 1759-507X.

Falchi, A. M., Sogos, V., Saba, F., Piras, M., Congiu, T., & Piludu, M. (2013). Astrocytes shed large membrane vesicles that contain mitochondria, lipid droplets and ATP. *Histochemistry and Cell Biology, 139*(2), 221–231, 1432-119X.

Forbes, J. M. (2016). Mitochondria–power players in kidney function? *Trends in Endocrinology and Metabolism, 27*(7), 441–442, 1043-2760.

Funk, J. A., & Schnellmann, R. G. (2013). Accelerated recovery of renal mitochondrial and tubule homeostasis with SIRT1/PGC-1α activation following ischemia–reperfusion injury. *Toxicology and Applied Pharmacology, 273*(2), 345–354, 0041-008X.

Funk, J. A., & Schnellmann, R. G. (2012). Persistent disruption of mitochondrial homeostasis after acute kidney injury. *American Journal of Physiology—Renal Physiology, 302*(7), F853–F864, 1931-857X.

Gall, J. M., Wang, Z., Bonegio, R. G., Havasi, A., Liesa, M., Vemula, P., & Borkan, S. C. (2015). Conditional knockout of proximal tubule mitofusin 2 accelerates recovery and improves survival after renal ischemia. *Journal of the American Society of Nephrology, 26*(5), 1092–1102, 1046-6673.

Gall, J. M., Wang, Z., Liesa, M., Molina, A., Havasi, A., Schwartz, J. H., Shirihai, O., Borkan, S. C., & Bonegio, R. G. B. (2012). Role of mitofusin 2 in the renal stress response. *PLoS One, 7*(1), e31074, 1932-6203.

Galvan, D. L., Green, N. H., & Danesh, F. R. (2017). The hallmarks of mitochondrial dysfunction in chronic kidney disease. *Kidney International, 92*(5), 1051–1057, 0085-2538.

Gibbs, W. S., Collier, J. B., Morris, M., Beeson, C. C., Megyesi, J., & Schnellmann, R. G. (2018). 5-HT1F receptor regulates mitochondrial homeostasis and its loss potentiates acute kidney injury and impairs renal recovery. *American Journal of Physiology—Renal Physiology, 315*(4), F1119–F1128, 1931-857X.

Gollihue, J. L., Patel, S. P., Mashburn, C., Eldahan, K. C., Sullivan, P. G., & Rabchevsky, A. G. (2017). Optimization of mitochondrial isolation techniques for intraspinal transplantation procedures. *Journal of Neuroscience Methods, 287*, 1–12, 0165-0270.

Grahammer, F., Haenisch, N., Steinhardt, F., Sandner, L., Roerden, M., Arnold, F., Cordts, T., Wanner, N., Reichardt, W., & Kerjaschki, D. (2014). mTORC1 maintains renal tubular homeostasis and is essential in response to ischemic stress. *Proceedings of the National Academy of Sciences, 111*(27), E2817–E2826, 0027-8424.

Guariento, A., Blitzer, D., Doulamis, I., Shin, B., Moskowitzova, K., Orfany, A., Ramirez-Barbieri, G., Staffa, S. J., Zurakowski, D., & Pedro, J. (2020). Preischemic autologous mitochondrial transplantation by intracoronary injection for myocardial protection. *The Journal of Thoracic and Cardiovascular Surgery, 160*(2), e15–e29, 0022-5223.

Guescini, M., Genedani, S., Stocchi, V., & Agnati, L. F. (2010). Astrocytes and Glioblastoma cells release exosomes carrying mtDNA. *Journal of Neural Transmission, 117*(1), 1–4, 1435-1463.

Gureev, A. P., Shaforostova, E. A., & Popov, V. N. (2019). Regulation of mitochondrial biogenesis as a way for active longevity: Interaction between the Nrf2 and PGC-1α signaling pathways. *Frontiers in Genetics, 10*, 435, 1664-8021.

Haase, V. H. (2006). Hypoxia-inducible factors in the kidney. *American Journal of Physiology—Renal Physiology, 291*(2), F271–F281, 1931-857X.

Haigis, M. C., Deng, C.-X., Finley, L. W. S., Kim, H.-S., & Gius, D. (2012). SIRT3 is a mitochondrial tumor suppressor: A scientific tale that connects aberrant cellular ROS, the Warburg effect, and carcinogenesis. *Cancer Research, 72*(10), 2468–2472, 0008-5472.

Hayakawa, K., Esposito, E., Wang, X., Terasaki, Y., Liu, Y., Xing, C., Ji, X., & Lo, E. H. (2016). Transfer of mitochondria from astrocytes to neurons after stroke. *Nature, 535*(7613), 551–555, 1476-4687.

Hellberg, P. O. A., Källskog, Ö., & Wolgast, M. (1991). Red cell trapping and postischemic renal blood flow. Differences between the cortex, outer and inner medulla. *Kidney International, 40*(4), 625–631, 0085-2538.

Heung, M., Steffick, D. E., Zivin, K., Gillespie, B. W., Banerjee, T., Hsu, C.-yuan, Powe, N. R., Pavkov, M. E., Williams, D. E., & Saran, R. (2016). Acute kidney injury recovery pattern and subsequent risk of CKD: An analysis of veterans health administration data. *American Journal of Kidney Diseases, 67*(5), 742–752, 0272-6386.

Hickey, F. B., Corcoran, J. B., Docherty, N. G., Griffin, B., Bhreathnach, U., Furlong, F., Martin, F., Godson, C., & Murphy, M. (2011). IHG-1 promotes mitochondrial biogenesis by stabilizing PGC-1α. *Journal of the American Society of Nephrology, 22*(8), 1475–1485, 1046-6673.

Hoogduijn, M. J., Crop, M. J., Peeters, A. M. A., Van Osch, G. J. V. M., Balk, A. H. M. M., Ijzermans, J. N. M., Weimar, W., & Baan, C. C. (2007). Human heart, spleen, and perirenal fat-derived mesenchymal stem cells have immunomodulatory capacities. *Stem Cells and Development, 16*(4), 597–604, 1547-3287.

Hosgood, S. A., Bagul, A., & Nicholson, M. L. (2011). Minimising cold ischaemic injury in an experimental model of kidney transplantation. *European Journal of Clinical Investigation, 41*(3), 233–240, 0014-2972.

Hoshino, A., Mita, Y., Okawa, Y., Ariyoshi, M., Iwai-Kanai, E., Ueyama, T., Ikeda, K., Ogata, T., & Matoba, S. (2013). Cytosolic p53 inhibits Parkin-mediated mitophagy and promotes mitochondrial dysfunction in the mouse heart. *Nature Communications, 4*(1), 1–12, 2041-1723.

Huang, P.-J., Kuo, C.-C., Lee, H.-C., Shen, C.-I., Cheng, F.-C., Wu, S.-F., Chang, J.-C., Pan, H.-C., Lin, S.-Z., & Liu, C.-S. (2016). Transferring xenogenic mitochondria provides neural protection against ischemic stress in ischemic rat brains. *Cell Transplantation, 25*(5), 913–927, 0963-6897.

Huang, X., Sun, L., Ji, S., Zhao, T., Zhang, W., Xu, J., Zhang, J., Wang, Y., Wang, X., & Franzini-Armstrong, C. (2013). Kissing and nanotunneling mediate intermitochondrial communication in the heart. *Proceedings of the National Academy of Sciences, 110*(8), 2846–2851, 0027-8424.

Ibrahim, A., Ahmed, M. M., Kedir, S., & Bekele, D. (2016). Clinical profile and outcome of patients with acute kidney injury requiring dialysis—an experience from a haemodialysis unit in a developing country. *BMC Nephrology, 17*(1), 1–5, 1471-2369.

Ishimoto, Y., & Inagi, R. (2016). Mitochondria: A therapeutic target in acute kidney injury. *Nephrology Dialysis Transplantation, 31*(7), 1062–1069, 1460-2385.

Islam, M. N., Das, S. R., Emin, M. T., Wei, M., Sun, L., Westphalen, K., Rowlands, D. J., Quadri, S. K., Bhattacharya, S., & Bhattacharya, J. (2012). Mitochondrial transfer from bone-marrow–derived stromal cells to pulmonary alveoli protects against acute lung injury. *Nature Medicine, 18*(5), 759–765, 1546-170X.

Jabbari, H., Roushandeh, A. M., Rostami, M. K., Razavi-Toosi, M. T., Shokrgozar, M. A., Jahanian-Najafabadi, A., Kuwahara, Y., & Roudkenar, M. H. (2020). Mitochondrial transplantation ameliorates ischemia/reperfusion-induced kidney injury in rat. *Biochimica et Biophysica Acta—Molecular Basis of Disease, 1866*(8), 165809, 0925-4439.

Jackson, M. V., Morrison, T. J., Doherty, D. F., McAuley, D. F., Matthay, M. A., Kissenpfennig, A., O'Kane, C. M., & Krasnodembskaya, A. D. (2016). Mitochondrial transfer via tunneling nanotubes is an important mechanism by which mesenchymal stem cells enhance macrophage phagocytosis in the in vitro and in vivo models of ARDS. *Stem Cells, 34*(8), 2210–2223, 1066-5099.

Jain, S., Keys, D., Nydam, T., Plenter, R. J., Edelstein, C. L., & Jani, A. (2015). Inhibition of autophagy increases apoptosis during re-warming after cold storage in renal tubular epithelial cells. *Transplant International, 28*(2), 214–223, 0934-0874.

James, M. T., Grams, M. E., Woodward, M., Elley, C. R., Green, J. A., Wheeler, D. C., De Jong, P., Gansevoort, R. T., Levey, A. S., & Warnock, D. G. (2015). A meta-analysis of the association of estimated GFR, albuminuria, diabetes mellitus, and hypertension with acute kidney injury. *American Journal of Kidney Diseases, 66*(4), 602–612, 0272-6386.

Jang, H.-S., Noh, M. R., Kim, J., & Padanilam, B. J. (2020). Defective mitochondrial fatty acid oxidation and lipotoxicity in kidney diseases. *Frontiers of Medicine, 7*, 65, 2296-858X.

Jesinkey, S. R., Funk, J. A., Stallons, L. J., Wills, L. P., Megyesi, J. K., Beeson, C. C., & Schnellmann, R. G. (2014). Formoterol restores mitochondrial and renal function after ischemia-reperfusion injury. *Journal of the American Society of Nephrology, 25*(6), 1157–1162, 1046-6673.

Jiang, M., Wei, Q., Dong, G., Komatsu, M., Su, Y., & Dong, Z. (2012). Autophagy in proximal tubules protects against acute kidney injury. *Kidney International, 82*(12), 1271–1283, 0085-2538.

Kay, M. D., Hosgood, S. A., Bagul, A., & Nicholson, M. L. (2009). Comparison of preservation solutions in an experimental model of organ cooling in kidney transplantation. *Journal of British Surgery, 96*(10), 1215–1221, 0007-1323.

Khwaja, A. (2012). KDIGO clinical practice guidelines for acute kidney injury. *Nephron Clinical Practice, 120*(4), c179–c184, 1660-2110.

Kolesnikova, O. A., Entelis, N. S., Jacquin-Becker, C., Goltzene, F., Chrzanowska-Lightowlers, Z. M., Lightowlers, R. N., Martin, R. P., & Tarassov, I. (2004). Nuclear DNA-encoded tRNAs targeted into mitochondria can rescue a mitochondrial DNA mutation associated with the MERRF syndrome in cultured human cells. *Human Molecular Genetics, 13*(20), 2519–2534, 1460-2083.

Konari, N., Nagaishi, K., Kikuchi, S., & Fujimiya, M. (2019). Mitochondria transfer from mesenchymal stem cells structurally and functionally repairs renal proximal tubular epithelial cells in diabetic nephropathy in vivo. *Scientific Reports, 9*(1), 1–14, 2045-2322.

Kosieradzki, M., & Rowiński, W. (2008). Ischemia/reperfusion injury in kidney transplantation: Mechanisms and prevention. *Transplantation Proceedings, 40*(10), 3279–3288, 0041-1345.

Kubat, G. B., Kartal, Y., Atalay, O., Ulger, O., Ekinci, O., Celik, E., Safali, M., Urkan, M., Karahan, S., & Ozler, M. (2021). Investigation of the effect of isolated mitochondria transplantation on renal ischemia-reperfusion injury in rats. *Toxicology and Applied Pharmacology, 433*, 115780, 0041-008X.

Kubat, G. B., Ozler, M., Ulger, O., Ekinci, O., Atalay, O., Celik, E., Safali, M., & Budak, M. T. (2021). The effects of mesenchymal stem cell mitochondrial transplantation on doxorubicin-mediated nephrotoxicity in rats. *Journal of Biochemical and Molecular Toxicology, 35*(1), e22612, 1095-6670.

Lameire, N. H., Levin, A., Kellum, J. A., Cheung, M., Jadoul, M., Winkelmayer, W. C., Stevens, P. E., Caskey, F. J., Farmer, C. K. T., & Fuentes, A. F. (2021). Harmonizing acute and chronic kidney disease definition and classification: Report of a kidney disease: Improving Global outcomes (KDIGO) consensus conference. *Kidney International, 100*(3), 516–526, 0085-2538.

Lameire, N., Van Biesen, W., & Vanholder, R. (2006). The changing epidemiology of acute renal failure. *Nature Clinical Practice Nephrology, 2*(7), 364–377, 1745-8331.

Li, G., Flodby, P., Luo, J., Kage, H., Sipos, A., Gao, D., Ji, Y., Beard, L. M. L., Marconett, C. N., & DeMaio, L. (2014). Knockout mice reveal key roles for claudin 18 in alveolar barrier properties and fluid homeostasis. *American Journal of Respiratory Cell and Molecular Biology, 51*(2), 210–222, 1044-1549.

Li, M., Li, C.-ming, Ye, Z.-chun, Huang, J., Li, Y., Lai, W., Peng, H., & Lou, T.-qi (2020). Sirt3 modulates fatty acid oxidation and attenuates cisplatin-induced AKI in mice. *Journal of Cellular and Molecular Medicine, 24*(9), 5109–5121, 1582-1838.

Li, X., Liao, J., Su, X., Li, W., Bi, Z., Wang, J., Su, Q., Huang, H., Wei, Y., & Gao, Y. (2020). Human urine-derived stem cells protect against renal ischemia/reperfusion injury in a rat model via exosomal miR-146a-5p which targets IRAK1. *Theranostics, 10*(21), 9561.

Liang, N.-N., Zhao, Y., Guo, Y.-Y., Zhang, Z.-H., Gao, L., Yu, D.-X., Xu, D.-X., & Xu, S. (2022). Mitochondria-derived reactive oxygen species are involved in renal cell ferroptosis during lipopolysaccharide-induced acute kidney injury. *International Immunopharmacology, 107*, 108687, 1567-5769.

Liu, C.-S., Chang, J.-C., Kuo, S.-J., Liu, K.-H., Lin, T.-T., Cheng, W.-L., & Chuang, S.-F. (2014). Delivering healthy mitochondria for the therapy of mitochondrial diseases and beyond. *The International Journal of Biochemistry & Cell Biology, 53*, 141–146, 1357-2725.

Liu, K., Ji, K., Guo, L., Wu, W., Lu, H., Shan, P., & Yan, C. (2014). Mesenchymal stem cells rescue injured endothelial cells in an in vitro ischemia–reperfusion model via tunneling nanotube like structure-mediated mitochondrial transfer. *Microvascular Research, 92*, 10–18, 0026-2862.

Liu, R., Wang, S.-cong, Li, M., Ma, X.-hui, Jia, X.-nan, Bu, Y., Sun, L., & Yu, K.-jiang (2020). An inhibitor of DRP1 (Mdivi-1) alleviates LPS-induced septic AKI by inhibiting NLRP3 inflammasome activation. *BioMed Research International, 2020*, 2314-6133.

Livingston, M. J., Wang, J., Zhou, J., Wu, G., Ganley, I. G., Hill, J. A., Yin, X.-M., & Dong, Z. (2019). Clearance of damaged mitochondria via mitophagy is important to the protective effect of ischemic preconditioning in kidneys. *Autophagy, 15*(12), 2142–2162, 1554-8627.

Lo, S. B., Blaszak, R. T., & Parajuli, N. (2020). Targeting mitochondria during cold storage to maintain proteasome function and improve renal outcome after transplantation. *International Journal of Molecular Sciences, 21*(10), 3506, 1422-0067.

Lu, J., Zheng, X., Li, F., Yu, Y., Chen, Z., Liu, Z., Wang, Z., Xu, H., & Yang, W. (2017). Tunneling nanotubes promote intercellular mitochondria transfer followed by increased invasiveness in bladder cancer cells. *Oncotarget, 8*(9), 15539.

Ludwig-Portugall, I., Bartok, E., Dhana, E., Evers, B. D. G., Primiano, M. J., Hall, J. P., Franklin, B. S., Knolle, P. A., Hornung, V., & Hartmann, G. (2016). An NLRP3-specific inflammasome inhibitor attenuates crystal-induced kidney fibrosis in mice. *Kidney International, 90*(3), 525–539, 0085-2538.

Lynch, M. R., Tran, M. T., & Parikh, S. M. (2018). PGC1α in the kidney. *American Journal of Physiology—Renal Physiology, 314*(1), F1–F8, 1931-857X.

Ma, S., Feng, J., Zhang, R., Chen, J., Han, D., Li, X., Yang, B., Li, X., Fan, M., & Li, C. (2017). SIRT1 activation by resveratrol alleviates cardiac dysfunction via mitochondrial regulation in diabetic cardiomyopathy mice. *Oxidative Medicine and Cellular Longevity, 2017*, 1942-0900.

Ma, Y.-S., Chen, Y.-C., Lu, C.-Y., Liu, C.-Y., & Wei, Y.-H. (2005). Upregulation of matrix metalloproteinase 1 and disruption of mitochondrial network in skin fibroblasts of patients with MERRF syndrome. *Annals of the New York Academy of Sciences, 1042*(1), 55–63, 0077-8923.

Mahata, B., Mukherjee, S., Mishra, S., Bandyopadhyay, A., & Adhya, S. (2006). Functional delivery of a cytosolic tRNA into mutant mitochondria of human cells. *Science, 314*(5798), 471–474, 0036-8075.

Mason, J., Torhorst, J., & Welsch, J. (1984). Role of the medullary perfusion defect in the pathogenesis of ischemic renal failure. *Kidney International, 26*(3), 283–293, 0085-2538.

McCully, J. D., Levitsky, S., Del Nido, P. J., & Cowan, D. B. (2016). Mitochondrial transplantation for therapeutic use. *Clinical and Translational Medicine, 5*(1), 1–13, 2001-1326.

Mitchell, T., Rotaru, D., Saba, H., Smith, R. A. J., Murphy, M. P., & MacMillan-Crow, L. A. (2011). The mitochondria-targeted antioxidant mitoquinone protects against cold storage injury of renal tubular cells and rat kidneys. *Journal of Pharmacology and Experimental Therapeutics, 336*(3), 682–692, 0022-3565.

Morigi, M., Perico, L., Rota, C., Longaretti, L., Conti, S., Rottoli, D., Novelli, R., Remuzzi, G., & Benigni, A. (2015). Sirtuin 3–dependent mitochondrial dynamic improvements protect against acute kidney injury. *The Journal of Clinical Investigation, 125*(2), 715–726, 0021-9738.

Nangaku, M. (2006). Chronic hypoxia and tubulointerstitial injury: A final common pathway to end-stage renal failure. *Journal of the American Society of Nephrology, 17*(1), 17–25, 1046-6673.

Nourbakhsh, N., & Singh, P. (2014). Role of renal oxygenation and mitochondrial function in the pathophysiology of acute kidney injury. *Nephron Clinical Practice, 127*(1–4), 149–152, 1660-2110.

Nowak, G., Bakajsova, D., & Samarel, A. M. (2011). Protein kinase C-ε activation induces mitochondrial dysfunction and fragmentation in renal proximal tubules. *American Journal of Physiology—Renal Physiology, 301*(1), F197–F208, 1931-857X.

O'callaghan, J. M., Morgan, R. D., Knight, S. R., & Morris, P. J. (2013). Systematic review and meta-analysis of hypothermic machine perfusion versus static cold storage of kidney allografts on transplant outcomes. *Journal of British Surgery, 100*(8), 991–1001, 0007-1323.

Orrenius, S., Zhivotovsky, B., & Nicotera, P. (2003). Regulation of cell death: The calcium–apoptosis link. *Nature Reviews Molecular Cell Biology, 4*(7), 552–565, 1471-0080.

Osman, M. M., Lulic, D., Glover, L., Stahl, C. E., Lau, T., van Loveren, H., & Borlongan, C. V. (2011). Cyclosporine-A as a neuroprotective agent against stroke: Its translation from laboratory research to clinical application. *Neuropeptides, 45*(6), 359–368, 0143-4179.

Paliwal, S., Chaudhuri, R., Agrawal, A., & Mohanty, S. (2018). Regenerative abilities of mesenchymal stem cells through mitochondrial transfer. *Journal of Biomedical Science, 25*(1), 1–12, 1423-0127.

Palomba, H., Castro, I., Yu, L., & Burdmann, E. A. (2017). The duration of acute kidney injury after cardiac surgery increases the risk of long-term chronic kidney disease. *Journal of Nephrology, 30*(4), 567–572, 1724-6059.

Parajuli, N., Shrum, S., Tobacyk, J., Harb, A., Arthur, J. M., & MacMillan-Crow, L. A. (2017). Renal cold storage followed by transplantation impairs expression of key mitochondrial fission and fusion proteins. *PLoS One, 12*(10), e0185542, 1932-6203.

Parikh, S. M., Yang, Y., He, L., Tang, C., Zhan, M., & Dong, Z. (2015). Mitochondrial function and disturbances in the septic kidney. *Seminars in Nephrology, 35*(1), 108–119, 0270-9295.

Perry, H. M., Huang, L., Wilson, R. J., Bajwa, A., Sesaki, H., Yan, Z., Rosin, D. L., Kashatus, D. F., & Okusa, M. D. (2018). Dynamin-related protein 1 deficiency promotes recovery from AKI. *Journal of the American Society of Nephrology, 29*(1), 194–206, 1046-6673.

Phinney, D. G., Di Giuseppe, M., Njah, J., Sala, E., Shiva, S., St Croix, C. M., Stolz, D. B., Watkins, S. C., Di, Y. P., & Leikauf, G. D. (2015). Mesenchymal stem cells use extracellular vesicles to outsource mitophagy and shuttle microRNAs. *Nature Communications, 6*(1), 1–15, 2041-1723.

Picard, M., Wallace, D. C., & Burelle, Y. (2016). The rise of mitochondria in medicine. *Mitochondrion, 30*, 105–116, 1567-7249.

Pinkert, C. A., Irwin, M. H., Johnson, L. W., & Moffatt, R. J. (1997). Mitochondria transfer into mouse ova by microinjection. *Transgenic Research, 6*(6), 379–383, 1573-9368.

Pittenger, M. F., Mackay, A. M., Beck, S. C., Jaiswal, R. K., Douglas, R., Mosca, J. D., Moorman, M. A., Simonetti, D. W., Craig, S., & Marshak, D. R. (1999). Multilineage potential of adult human mesenchymal stem cells. *Science, 284*(5411), 143–147, 1095-9203.

Puigserver, P., Wu, Z., Park, C. W., Graves, R., Wright, M., & Spiegelman, B. M. (1998). A cold-inducible coactivator of nuclear receptors linked to adaptive thermogenesis. *Cell, 92*(6), 829–839, 0092-8674.

Ralto, K. M., & Parikh, S. M. (2016). Mitochondria in acute kidney injury. *Seminars in Nephrology, 36*(1), 8–16, 0270-9295.

Rosen, S., Epstein, F. H., & Brezis, M. (1992). Determinants of intrarenal oxygenation: Factors in acute renal failure. *Renal Failure, 14*(3), 321–325, 0886-022X.

Roushandeh, A. M., Kuwahara, Y., & Habibi Roudkenar, M. (2019). Mitochondrial transplantation as a potential and novel master key for treatment of various incurable diseases. *Cytotechnology, 71*(2), 647–663.

Salahudeen, A. K. (1999). Free radicals in kidney disease and transplantation. *Saudi Journal of Kidney Diseases and Transplantation, 10*(2), 137, 1319-2442.

Salahudeen, A. K., Huang, H., Joshi, M., Moore, N. A., & Jenkins, J. K. (2003). Involvement of the mitochondrial pathway in cold storage and rewarming-associated apoptosis of human renal proximal tubular cells. *American Journal of Transplantation, 3*(3), 273–280, 1600-6135.

Salahudeen, A. K., Joshi, M., & Jenkins, J. K. (2001). Apoptosis versus necrosis during cold storage and rewarming of human renal proximal tubular cells. *Transplantation, 72*(5), 798–804, 0041-1337.

Salahudeen, A. K., & May, W. (2008). Reduction in cold ischemia time of renal allografts in the United States over the last decade. *Transplantation Proceedings, 40*(5), 1285–1289, 0041-1345.

Schleef, M., Gonnot, F., Pillot, B., Leon, C., Chanon, S., Vieille-Marchiset, A., Rabeyrin, M., Bidaux, G., Guebre-Egziabher, F., & Juillard, L. (2022). Mild therapeutic hypothermia protects from acute and chronic renal ischemia-reperfusion injury in mice by mitigated mitochondrial dysfunction and modulation of local and systemic inflammation. *International Journal of Molecular Sciences, 23*(16), 9229, 1422-0067.

Schrier, R. W., Wang, W., Poole, B., & Mitra, A. (2004). Acute renal failure: Definitions, diagnosis, pathogenesis, and therapy. *The Journal of Clinical Investigation, 114*(1), 5–14, 0021-9738.

Shrum, S., Tobacyk, J., Lo, S., Parajuli, N., & MacMillan-Crow, L. A. (2020). The BK activator NS11021 partially protects rat kidneys from cold storage and transplantation-induced mitochondrial and renal injury. *Archives of Biochemistry and Biophysics, 688*, 108410, 0003-9861.

Souidi, N., Stolk, M., & Seifert, M. (2013). Ischemia–reperfusion injury: Beneficial effects of mesenchymal stromal cells. *Current Opinion in Organ Transplantation, 18*(1), 34–43, 1087-2418.

Stewart, D. E., Garcia, V. C., Rosendale, J. D., Klassen, D. K., & Carrico, B. J. (2017). Diagnosing the decades-long rise in the deceased donor kidney discard rate in the United States. *Transplantation, 101*(3), 575–587, 0041-1337.

Tábara, L. C., Poveda, J., Martin-Cleary, C., Selgas, R., Ortiz, A., & Sanchez-Niño, M. D. (2014). Mitochondria-targeted therapies for acute kidney injury. *Expert Reviews in Molecular Medicine, 16*, 1462-3994.

Tögel, F., Cohen, A., Zhang, P., Yang, Y., Hu, Z., & Westenfelder, C. (2009). Autologous and allogeneic marrow stromal cells are safe and effective for the treatment of acute kidney injury. *Stem Cells and Development, 18*(3), 475–486, 1547-3287.

Takasu, O., Gaut, J. P., Watanabe, E., To, K., Fagley, R. E., Sato, B., Jarman, S., Efimov, I. R., Janks, D. L., & Srivastava, A. (2013). Mechanisms of cardiac and renal dysfunction in patients dying of sepsis. *American Journal of Respiratory and Critical Care Medicine, 187*(5), 509–517, 1535-4970.

Tang, C., & Dong, Z. (2016). Mitochondria in kidney injury: When the power plant fails. *Journal of the American Society of Nephrology, 27*(7), 1869–1872, 1046-6673.

Tang, C., Han, H., Yan, M., Zhu, S., Liu, J., Liu, Z., He, L., Tan, J., Liu, Y. U., & Liu, H. (2018). PINK1-PRKN/PARK2 pathway of mitophagy is activated to protect against renal ischemia-reperfusion injury. *Autophagy, 14*(5), 880–897, 1554-8627.

Thadhani, R., Pascual, M., & Bonventre, J. V. (1996). Acute renal failure-Reply. *New England Journal of Medicine, 335*(17), 1321–1322, 0028-4793.

Thiel, G., Rougemont, D de, Kriz, W., Mason, J., Torhorst, J., & Wolgast, M. (1982). The role of reduced medullary perfusion in the genesis of acute ischemic renal failure. *Nephron, 31*, 321.

Tran, M. T., Zsengeller, Z. K., Berg, A. H., Khankin, E. V., Bhasin, M. K., Kim, W., Clish, C. B., Stillman, I. E., Karumanchi, S. A., & Rhee, E. P. (2016). PGC1α drives NAD biosynthesis linking oxidative metabolism to renal protection. *Nature, 531*(7595), 528–532, 1476-4687.

Tran, M., Tam, D., Bardia, A., Bhasin, M., Rowe, G. C., Kher, A., Zsengeller, Z. K., Akhavan-Sharif, M. R., Khankin, E. V., & Saintgeniez, M. (2011). PGC-1α promotes recovery after acute kidney injury during systemic inflammation in mice. *The Journal of Clinical Investigation, 121*(10), 0021-9738.

Venkatachalam, M. A., & Weinberg, J. M. (2012). The tubule pathology of septic acute kidney injury: A neglected area of research comes of age. *Kidney International, 81*(4), 338–340, 0085-2538.

Wakabayashi, T., & Karbowski, M. (2001). Structural changes of mitochondria related to apoptosis. *Neurosignals, 10*(1–2), 26–56, 1424-862X.

Wang, G., Shimada, E., Zhang, J., Hong, J. S., Smith, G. M., Teitell, M. A., & Koehler, C. M. (2012). Correcting human mitochondrial mutations with targeted RNA import. *Proceedings of the National Academy of Sciences, 109*(13), 4840–4845, 0027-8424.

Wang, K., & Kestenbaum, B. (2018). Proximal tubular secretory clearance: A neglected partner of kidney function. *Clinical Journal of the American Society of Nephrology, 13*(8), 1291–1296, 1555-9041.

Wang, X., & Gerdes, H.-H. (2015). Transfer of mitochondria via tunneling nanotubes rescues apoptotic PC12 cells. *Cell Death and Differentiation, 22*(7), 1181–1191, 1476-5403.

Wills, L. P., Trager, R. E., Beeson, G. C., Lindsey, C. C., Peterson, Y. K., Beeson, C. C., & Schnellmann, R. G. (2012). The β2-adrenoceptor agonist formoterol stimulates mitochondrial biogenesis. *Journal of Pharmacology and Experimental Therapeutics, 342*(1), 106–118, 0022-3565.

Wirthensohn, G., & Guder, W. G. (1986). Renal substrate metabolism. *Physiological Reviews, 66*(2), 469–497, 0031-9333.

Wu, S., Zhang, A., Li, S., Chatterjee, S., Qi, R., Segura-Ibarra, V., Ferrari, M., Gupte, A., Blanco, E., & Hamilton, D. J. (2018). Polymer functionalization of isolated mitochondria for cellular transplantation and metabolic phenotype alteration. *Advanced Science, 5*(3), 1700530, 2198-3844.

Yan, M.-T., Chao, C.-T., & Lin, S.-H. (2021). Chronic kidney disease: Strategies to retard progression. *International Journal of Molecular Sciences, 22*(18), 10084, 1422-0067.

Yan, Y., Ma, Z., Zhu, J., Zeng, M., Liu, H., & Dong, Z. (2020). miR-214 represses mitofusin-2 to promote renal tubular apoptosis in ischemic acute kidney injury. *American Journal of Physiology—Renal Physiology, 318*(4), F878–F887, 1931-857X.

Yi, Y.-C., Chen, M.-J., Ho, J. Y.-P., Guu, H.-F., & Ho, E. S.-C. (2007). Mitochondria transfer can enhance the murine embryo development. *Journal of Assisted Reproduction and Genetics, 24*(10), 445–449, 1573-7330.

Yin, M., Currin, R. T., Peng, X.-X., Mekeel, H. E., Schoonhoven, R., & Lemasters, J. J. (2002). Different patterns of renal cell killing after warm and cold ischemia. *Renal Failure, 24*(2), 147–163, 0886-022X.

Yu, H., Jin, F., Liu, Di, Shu, G., Wang, X., Qi, J., Sun, M., Yang, P., Jiang, S., & Ying, X. (2020). ROS-responsive nano-drug delivery system combining mitochondria-targeting ceria nanoparticles with atorvastatin for acute kidney injury. *Theranostics, 10*(5), 2342.

Yuan, Y., Yuan, L., Li, L., Liu, F., Liu, J., Chen, Y., Cheng, J., & Lu, Y. (2021). Mitochondrial transfer from mesenchymal stem cells to macrophages restricts inflammation and alleviates kidney injury in diabetic nephropathy mice via PGC-1α activation. *Stem Cells, 39*(7), 913–928, 1066-5099.

Zhan, M., Brooks, C., Liu, F., Sun, L., & Dong, Z. (2013). Mitochondrial dynamics: Regulatory mechanisms and emerging role in renal pathophysiology. *Kidney International, 83*(4), 568–581, 0085-2538.

Zhang, Z., Ma, Z., Yan, C., Pu, K., Wu, M., Bai, J., Li, Y., & Wang, Q. (2019). Muscle-derived autologous mitochondrial transplantation: A novel strategy for treating cerebral ischemic injury. *Behavioural Brain Research, 356*, 322–331, 0166-4328.

Zhao, W.-Y., Zhang, L., Sui, M.-X., Zhu, Y.-H., & Zeng, L. (2016). Protective effects of sirtuin 3 in a murine model of sepsis-induced acute kidney injury. *Scientific Reports, 6*(1), 1–11, 2045-2322.

Zhao, W., Zhang, L., Chen, R., Lu, H., Sui, M., Zhu, Y., & Zeng, L. (2018). SIRT3 protects against acute kidney injury via AMPK/mTOR-regulated autophagy. *Frontiers in Physiology, 9*, 1526, 1664-042X.

Zou, X., Kwon, S. H., Jiang, K., Ferguson, C. M., Puranik, A. S., Zhu, X., & Lerman, L. O. (2018). Renal scattered tubular-like cells confer protective effects in the stenotic murine kidney mediated by release of extracellular vesicles. *Scientific Reports, 8*(1), 1–12, 2045-2322.

CHAPTER

21

Cancer and mitochondrial transplantation, and novel insights into its application in the cancer-like phenotype of pulmonary arterial hypertension

Olakunle Sanni, Carmen Payne and Gerald Maarman

Centre for Cardio-Metabolic Research in Africa (CARMA), Division of Medical Physiology, Department of Biomedical Sciences, Faculty of Medicine & Health Sciences, Stellenbosch University, Stellenbosch, South Africa

Mitochondrial transplantation

Naturally, cells can exchange their mitochondria with other cells and take up exogenous mitochondria from other tissues or organs. The transfer of mitochondria from one cell to the other is well documented and includes the following mechanisms: tunneling nanotubes (Qin et al., 2021; Rustom et al., 2004), gap junctions (Shanmughapriya et al., 2020), microvesicles, and mitochondrial extrusions (Maeda & Fadeel, 2014; Zappulli et al., 2016). Cell-to-cell mitochondrial transfer can integrate healthy and functional mitochondria into a dysfunctional cell as a mechanism of expelling dysfunctional mitochondria through mitophagy (Phinney et al., 2015). The above phenomenon gave rise to the concept of mitochondrial artificial transfer, called mitochondrial transplantation.

Mitochondrial transplantation has recently been seen as a potential therapeutic approach in the treatment of diseases that are associated with mitochondrial dysfunction. Due to the specificity of the function of mitochondria in different tissues, mitochondria dysfunction contributes to numerous diseases. For instance, overexpression of Drp1 (the gene that controls apoptosis and necroptosis) in the brain is known to produce excessive mitochondrial fission in

Mitochondrial Transplantation and Transfer
https://doi.org/10.1016/B978-0-443-18858-9.00021-4

Alzheimer's disease and other neurodegenerative illnesses, resulting in mitochondrial dysfunction and neuronal damage (Joshi et al., 2018). Likewise, the impaired function of the electron transport system coupled with an increase in reactive oxygen species (ROS) has been attributed to heart failure (Brown et al., 2017; Kang & Hamasaki, 2005). Mitochondrial transplantation from a healthy cell to a diseased cell is possible as a therapeutic mechanism for increasing cell wellness or avoiding death (Hayakawa et al., 2016; Ibanez & Villena-Gutierrez, 2021; Islam et al., 2012; Levoux et al., 2021; Torralba et al., 2016). For example, a recent study demonstrated that cardiomyocytes outsource mitophagy through a network of cardiac-resident macrophages in order to maintain heart homeostasis (Phinney et al., 2015). Furthermore, mitochondrial transplantation has received great attention over the years in both translational and clinical research as a novel strategy for the treatment of diseases (Maarman, 2022) related to mitochondrial dysfunction such as Parkinson's disease and other neurodegenerative diseases (Chang et al., 2021; De La Fuente-Munoz & Arias, 2020), heart failure (Weixler et al., 2021), ischemic heart disease (Kaza et al., 2017; McCully et al., 2016), and cancer (Elliott et al., 2012).

Recent research has demonstrated the novel approach of mitochondrial transplantation for cancer treatment through the distinct regulation of mitochondrial function such as ROS production, imbalance fission and fusion dynamics; rescue dysregulated aerobic respiration and metabolism; and apoptotic signaling in cancer cells. Therefore, restoring mitochondrial quality in pathological conditions (via mitochondrial transplantation) is regarded as a novel strategy for the treatment of a variety of diseases.

Cancer

Cancer has gained significant attention as an important research niche, and this is important as millions of people die each year due to this dreadful disease, globally. The prominent feature of cancer cells is the continuous uncontrolled proliferation of cells within the body (Hatzikirou et al., 2012; Warth et al., 2014), and at final stages, it affects organ function as metastasis occurs via diffuse lymphatic systems and the circulatory system.

Most cancer cells are classified into three broad categories and these include carcinomas, sarcomas, and leukemia or lymphomas (Roy et al., 2021). Carcinomas are malignant tumors associated with epithelia cells and represent 90% of cancer in humans (Ling et al., 2021; Ridky et al., 2010). Sarcomas are cases of tumor in human (Kato et al., 2023) and associated with connective tissue like muscle, bone, cartilage, and fibrous tissue (Yiallouros et al., 2022). Leukemia or lymphomas account for approximately 8% of all human cancers (Rodriguez-Abreu et al., 2007), and are associated with the tumors from blood-forming cells and immune system cells, respectively.

The development of cancers is a result of DNA changes that disrupt gene structure or function (DeBerardinis et al., 2008) and a key hallmark of cancer is the increase of mitosis (Chan et al., 2012). This is often used when determining whether a tumor is benign or malignant as the existence of cancer increases mitosis (Serrano-Del Valle et al., 2021). This is followed by its ability to evade a potent growth inhibitory signaling of the cell (48). The main component of growth inhibitory signaling is called tumor suppressor genes (Gamudi & Blundell, 2010). Its inactivation causes cells to lose their regulatory control of cell proliferation (Chan, 2020).

Mitochondria and cancer

Aside from its ATP production action, mitochondria are also involved in the regulation of cellular redox status, calcium homeostasis, ROS production, cell signaling, and apoptosis (Brown et al., 2017). The diverse functions of mitochondria make them robust cellular stress sensors, allowing for cellular adaptability to physiological or pathological changes. Mitochondria are dynamic organelles that can modulate themselves and their functions to maintain a delicate balance between morphological fusion and fission and relating mitochondrial actions/responses to e.g., physiological stress (Mishra & Chan, 2016).

In cancer, mitochondrial functions are different based on the genetic and environmental stressors, or the origin of the tumor cells. Mitochondrial functions, such as biogenesis and turnover, apoptosis, metabolites and ROS production, fission and fusion, signaling, metabolism bioenergetics, and maintaining the quality of mitochondrial DNA, contribute immensely to tumorigenesis (De La Fuente-Munoz & Arias, 2020; Roy et al., 2017; Torralba et al., 2016). In fact, cancer can cause metabolic reprogramming in resident cells (DeBerardinis et al., 2008).

Mitochondrial biogenesis in the context of cancer

Mitochondrial biogenesis is the formation of new healthy mitochondria to meet cell requirements. During this process, the cell replicates the mitochondrial genome organizes the expression of the mitochondrial proteins and encodes the nuclear genome that participates in the functionality of the mitochondria (Boland et al., 2013). Mitochondrial biogenesis is, to a great extent, controlled by transcriptional programs that are further coordinated by the nuclear transcription factors the localized gene in the nucleus that express proteins in the mitochondria (McGuirk et al., 2013). In turn, nuclear transcription factors such as the mitochondrial peroxisome proliferator-activated receptor gamma coactivator 1-alpha (PGC-1α) are centrally regulated by transcription programs (McGuirk et al., 2013).

Under normal physiological conditions, mitochondrial biogenesis is dependent on the synthesis of various mitochondrial proteins, which are tightly controlled at the transcription and translational levels of both nuclear and mitochondrial genes (Senft & Ronai, 2016). The expression of these proteins is key to mitochondrial biogenesis. These proteins can act directly or indirectly with other transcription factors to influence the expression of many genes that encode mitochondrial proteins (Vazquez et al., 2013). Mitochondrial biogenesis is also a regulator of the pathogenesis of cancer. Mitochondrial proteins are critical for the metabolic reprogramming that occurs in cancer cells. To this end, metabolic reprogramming refers to the ability of cancer cells to alter their metabolism in order to support the increased energy request due to continuous growth, rapid proliferation, and other characteristics typical of neoplastic cells (Cazzaniga & Bonanni, 2015).

As earlier stated, PGC-1α is critical to mitochondrial biogenesis and the connection with tumorigenesis has been established. For instance, high expression of PGC-1α confers significant proliferation and survival capacity on cancer cells in melanoma (Vazquez et al., 2013). Also, in breast cancer, overexpression of PGC-1α causes the activation of estrogen-related receptor α that regulates glutamine metabolism enzyme expression, increasing glutamine absorption and hence Krebs cycle flux thus providing growth and proliferation advantages (McGuirk et al., 2013).

In addition, cellular-myelocytomatosis (c-Myc), a transcription factor, has been identified as a key activator of mitochondrial biogenesis and more than 400 mitochondrial genes are known to be c-Myc targets. Under normal physiological conditions, mitochondrial biogenesis is coupled with cell-cycle progression by c-Myc; therefore, the increase in c-Myc translates to an increased mitochondrial biomass (Li et al., 2005). This facilitates the rapid growth of oncogenic cells through an increase in respiratory capacity and biosynthetic pathway. Also, it could lead to upregulation of the increase of mitochondrial biomass to favor rapid proliferation through the stimulation of cell cycle progression and glycolytic pathway (Li et al., 2005).

Moreover, the mammalian target of rapamycin (mTOR) signaling pathway is another effector of mitochondria biogenesis. mTOR is a role player in the PI3K/AKT pathway, playing an important role in stimulating anabolic pathways to promote the synthesis of proteins, lipids, and nucleic acids required for cell proliferation and energy homeostasis (Maracci et al., 2022). mTOR activity is tightly regulated in normal cells such that it is only activated by physiological growth signals. This control is necessary for cellular, tissue-specific, and organismal homeostasis. However, one of the first events in the change of normal cells to cancer cells is the disconnection of growth-promoting pathways from their reliance on physiological signals (Maracci et al., 2022).

Mitophagy in the context of cancer

Even under normal conditions, mitochondrial defects may occur, which results in a build-up of defective mitochondria that can impair the functions of the electron transport system and promote oxidative stress. Mitophagy is necessary to keep a healthy mitochondrial pool, a process described as the selective degradation of mitochondria that are defective or damaged (Ding & Yin, 2012). There are two primary mechanisms that regulate mitochondrial mitophagy: (A) the Pink1/Parkin pathway, which activates Pink1 kinase after mitochondrial depolarization results in Pink1 phosphorylating mitofusin, which then serves as a Parkin receptor and (B) the activation of BNIP3 and NIX, both of which are hypoxia inducible but also controlled by FoxOs. In normal cells, these pathways are activated when the mitochondrial membrane depolarizes, which is an indication of mitochondrial dysfunction caused by a variety of factors such as a lack of antioxidants, hypoxia, and poor mitochondrial electron transfer. Reduced mitophagy could result in a threshold of mitochondrial dysfunction to persist, thus generating a considerable increase in tumor-promoting ROS or triggering other tumorigenic mitochondrial signals. However, in cancer, tumors use mitophagy as an adaptative mechanism against stress and for survival. In fact, to maintain uncontrolled growth rates, cancer cells use mitophagy as a means to achieve this (Song et al., 2022). In line with this, researchers have shown that inhibiting mitophagy, specifically, the inhibition of PINK1/Parkin-dependent mitophagy, sensitizes multidrug-resistant cancer cells to treatment (Yao et al., 2019).

Fission and fusion dynamics in the context of cancer

Mitochondrial fission and fusion are highly dynamic processes that help regulate the mitochondrial differentiation of stem cells, proliferation, migration, and apoptosis

(Boulton & Caino, 2022). The mitochondrial morphology is determined by the balance of fission and fusion. An important stage in mitochondrial membrane fission is the recruitment of dynamic-related protein-1 (Drp1) to outer mitochondria membrane receptors where it interacts and causes membrane constriction powered by GTPase Drp1 overexpression improved endometrial cancer epithelial–mesenchymal transition and invasion by altering glucose metabolism (Guo et al., 2020). Mitochondrial fusion is a selective process that is mediated by polarized mitochondria and promoted by oxidative carbon sources that stimulate the respiratory chain proteins expression increased cristae density and formation of respiratory chain complexes (Boland et al., 2013). Fusion process plays a vital role in the activation of oxidative phosphorylation through the control of mitochondrial DNA (Eisner et al., 2014) and is mediated by three GTAPase such as Mitofusin 1, Mitofusin 2 (MFN2) and optic atrophy-1 (Mishra & Chan, 2016). Senft and Ronai (Senft & Ronai, 2016) showed that the imbalance of fission and fusion supports tumorigenesis. They demonstrated that elevated fission activity coupled with decreased fusion resulted in a fragmented mitochondrial network that causes the remodeling of the mitochondrial network, ROS generation through decreasing membrane potential, and increased metabolites.

Apoptosis in the context of cancer

Apoptosis is a normal process and homeostatic mechanism that occurs during development and aging to maintain cell populations in tissues. Apoptosis can also occur as a defense mechanism, for example, in immunological responses or when cells are harmed by disease or toxic chemicals. Apoptosis occurs when Bax and Bak, proapoptotic Bcl-2 family, are recruited to the outer mitochondrial membrane to mediate mitochondrial outer membrane permeabilization oligomerization (Aranovich et al., 2012) as a result of upstream effectors such as DNA damage, ROS, stress, cell death signals, etc. Following mitochondrial outer membrane permeabilization, the outer membrane becomes porous and allows high-molecular-weight proteins cytochrome-C to escape into the cytosol to activate caspases (programmed cell death executors) (Tait et al., 2014).

In normal physiology, antiapoptotic members such as Bcl-2 and Bcl-XL located in the outer mitochondrial membrane bind and inhibit Bax/Bak and inhibit apoptosis. Tumor cells, however, evade apoptosis by downregulating proapoptotic Bcl-2 genes and/or upregulating antiapoptotic Bcl-2 genes, thus preventing mitochondrial outer membrane permeabilization and the activation of caspases (Lopez & Tait, 2015). Also, overexpression of Bcl-XL has been found in many cancers where it contributes to tumor initiation, progression, and resistance to therapy (Mayer & Oberbauer, 2003). Therefore, drugs that activate the intrinsic and extrinsic apoptotic pathways have been highlighted as therapeutic targets in cancer (Burz et al., 2009). These drugs achieve this through death receptors, inhibitors for BCL-2, XIAP inhibition, and alkyl-phospholipid analogs which can act as apoptotic signals (Pfeffer & Singh, 2018). However, this is not as straightforward forward and many challenges exist, therefore, current literature suggests that proapoptotic drugs should be used in combination with agents targeting oncogenic survival pathways, immunotherapy, and classical chemotherapy and radiotherapy (Carneiro & El-Deiry, 2020).

Metabolites and ROS production in the context of cancer

Most of the cellular glucose is metabolized via aerobic glycolysis and the pentose phosphate pathway in normal cells. This metabolism is attributed to the level of ROS which is related to the activity of the mitochondrial electron transfer system and is closely controlled by the endogenous antioxidant system. Cancer cells exhibit higher basal levels of ROS compared with normal cells, and at low to moderate levels, ROS act as signal transducers to activate cell proliferation, migration, invasion, and angiogenesis (Nakamura & Takada, 2021). Increased levels of ROS cause oxidation of macromolecules, such as lipids, proteins, and DNA, and result in genomic instability to promote mutation. Many mutations in mitochondrial genes encoded by nuclear DNA and the mitochondrial genome result in mitochondrial metabolic reprogramming. Metabolic reprogramming in cancer occurs through the overexpression of ROS-producing nicotinamide adenine dinucleotide phosphate (NADPH) oxidase and the ROS-eliminating antioxidant enzymes, both of which enzyme systems share NADPH as a reducing power source (Nakamura & Takada, 2021).

The potential benefits of mitochondrial transplantation in cancer

An imbalance in mitochondrial functions and dynamics that could be metamorphous to increased aerobic glycolysis, elevated levels of ROS, decreased apoptosis, and dysregulated fission and fusion dynamics can be classified as mitochondrial dysfunctions of cancer cells. Therefore, targeting mitochondrial dynamics has recently been proposed as a potential anticancer treatment. For instance, suppression of immersed mitochondrial fission has been shown to suppress breast cancer metastasis (Zhao et al., 2013) and cancer stemness (Peiris-Pagès et al., 2018).

Decreased ROS production

ROS stimulates the proliferation and signaling of tumor cells through the upregulation of the antioxidant pathways that are important to prevent ROS-mediated cytotoxicity of tumor cells thus promoting tumor survival. Jui-Chih Chang and coworkers transported the whole mitochondrial into MCF-7 breast cancer cells through coincubation and observed reduced mitochondrial ROS production as a result of increased antioxidant enzyme levels (Chang et al., 2019a, 2019b). Also, Skildum and his team observed that mitochondrial amplification in mammary epithelia MCF-12A cells through transplantation enhances the capacity of the electron transfer system to reduce ROS production (Skildum et al., 2011).

Rescuing of mitochondrial fission and fusion dynamics

The density of mitochondrial cristae consistently upregulates the mitochondrial fusion such that is independent of mitochondrial respiratory capacity (Chang et al., 2020). Previous studies have demonstrated the regulatory role of mitochondrial fusion in preventing cancer cell growth (Braganza et al., 2019). One of the mechanisms proposed is that the increased mitochondrial fusion ameliorates mitochondrial oxidative damage, by complementing damaged content and assisting in the maintenance of mitochondrial homeostasis

(Senft & Ronai, 2016). Mitochondrial transplantation promotes mitochondrial fusion, which is a critical regulator of cancer therapy, as demonstrated in experimental models (Chang et al., 2020). Also, mitochondrial transplantation decreases Drp-1 and consequently enhances the tubular mitochondrial network and increases the expression of Parkin and mitochondrial fusion proteins in breast cancer (Chang et al., 2020). In another study, protein levels of MFN2 and Opa-1 in the cells significantly increased after the mitochondrial transplantation in breast cancer cells (Chang et al., 2019a, 2019b). These findings support the notion that mitochondrial transplantation can improve mitochondrial fission and fusion in cancer.

Rescue metabolic reprogramming

Metabolic reprogramming of cells is the hallmark of oncogenesis (Liang et al., 2016). Endocytosis-mediated mitochondrial transplantation of normal human astrocytes to human glioma cells showed that most genes associated with glycolysis were upregulated, while genes associated with the tricarboxylic acid cycle were downregulated in human glioma cells (Sun et al., 2019). Consequently, mitochondrial transplantation salvages aerobic respiration and attenuates the use of glycolysis to support oncogenesis. In an interesting study, mitochondria were transplanted from healthy mice to those with metastatic melanoma (Yu et al., 2021). The transplanted mitochondria inhibited energy production through glycolysis in the metastatic melanoma hosts, suggesting that it had beneficial effects. Data are scant, but future studies should investigate the benefits of mitochondrial transplantation on metabolic reprogramming or energy substrate utilization pathways in a cancer context. It would be helpful to know if mitochondrial transplantation can rescue metabolic reprogramming in cancer.

Rescued apoptotic signaling

The ability of cancer cells to evade apoptosis arises when the balance of apoptotic protein is shifted in favor of its survival. Mitochondrial transplantation rescues apoptotic signaling through dysregulation of oxidative stress and mitochondrial membrane potential (Lee et al., 2021). Lee and coworker transplanted mitochondria into damaged tenocytes that mimic mitochondrial dysfunction in cancer and found out that apoptosis signaling pathway proteins (Bid, Bax, and Bcl-2) were restored to the predamage levels 2 weeks after transplantation (Lee et al., 2021). Also, it was demonstrated that mitochondria that were transplanted to human breast cancer cell lines inhibited cancer cell proliferation and increased cancer cells' sensitivity to anticancer agents (Elliott et al., 2012). To better investigate if mitochondrial transplantation can improve the apoptosis signaling of cancer cells, studies could assess the impact of mitochondrial transplantation on apoptosis gene and protein expression. A study by Chan et al. has demonstrated exactly this by showing that mitochondrial transplantation initiates apoptosis and suppresses cancer cell growth via an increase in apoptosis-inducing factor. In line with this, it was demonstrated that mitochondria from different genders that are transplanted on hepatocellular carcinoma inhibit proliferation (Zhou et al., 2022).

Cancer-like phenotype of pulmonary artery hypertension

Pulmonary arterial hypertension (PAH) is characterized by excessive remodeling and destruction of pulmonary arterioles (Teichert-Kuliszewska et al., 2006). For many years it was thought to have developed due to an imbalance between vasoconstriction and vasodilation pathways (Maarman et al., 2017). However, treatment with vasodilators does not cure or completely counteract PAH, suggesting that the pathophysiology is more complex. We now know that it includes an imbalance between vasoconstriction and vasodilation pathways as well as an elevated proliferation of the pulmonary artery smooth muscle cell (PASMC), PASMC resistance to apoptosis and the excessive proliferation of pulmonary artery endothelial cells (PAECs), giving PAH its cancer-like character (Humbert et al., 2019). PASMC, PAECs, fibroblasts, and at the peak of the disease, myofibroblasts display a proliferative and antiapoptotic phenotype which results in the destruction of the lumen. The destruction is characterized by the thickening of the intima and media and plexogenic arteriopathy (Fig. 21.1) (Hunt et al., 2012).

Thus, PAH comprises abnormal cell proliferation and vasoconstriction of pulmonary arterioles; however, there is also the involvement of multiple factors and proteins that play a role in PAH pathophysiology (Maarman, 2018). Various factors initiate pulmonary vascular changes such as concentric hypertrophy of the vascular walls and intimal thickening, thereby obliterating pulmonary arterioles and elevating the mean pulmonary pressure (Maarman, 2018). The pathology includes lesions in distal arterioles and disorganized growth of endothelial cells (Hunt et al., 2012). The formation of plexiform lesions results in reduced distal perfusion, thereby causing increased right ventricular afterload and right ventricular failure. Several mechanisms play a role in pulmonary arteriole remodeling, including PAEC dysfunction, PASMC hypertrophy, proliferation, migration, and apoptosis, and there is a plethora of contributory factors that underpin the pathophysiology such as inflammation, genetic factors, growth factors, metabolic remodeling, hypoxia, and oxidative stress (Fig. 21.2) (Rajagopal & Yu, 2022).

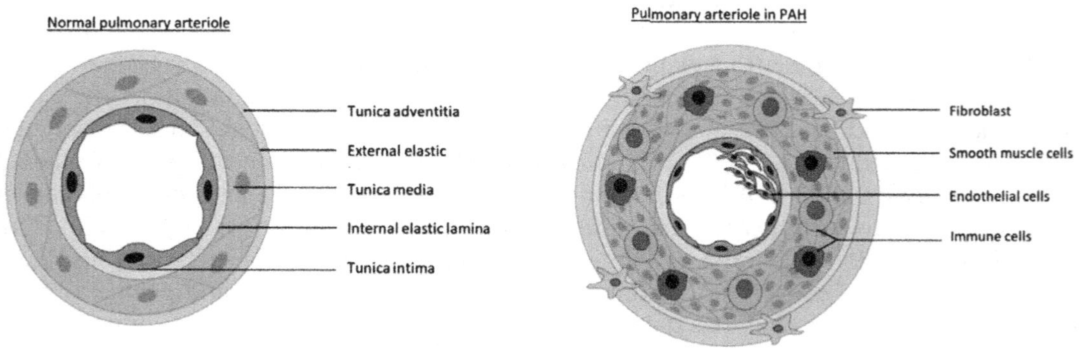

FIGURE 21.1 **Remodeling of the pulmonary vasculature in PAH.** The left figure represents a normal pulmonary arteriole, and the right figure below represents the remodeling of the pulmonary arteriole that occurs during PAH. In PAH, the intimal lining and media become thicker with increased proliferation of PAECs and PASMCs, respectively, resulting in the narrowing of the pulmonary arteriole. *Figure adapted from Rajagopal, S., & Yu, Y.R.A. (2022). The pathobiology of pulmonary arterial hypertension.* Cardiology Clinics, *40(1), 1–12. https://doi.org/10.1016/j.ccl.2021.08.001.*

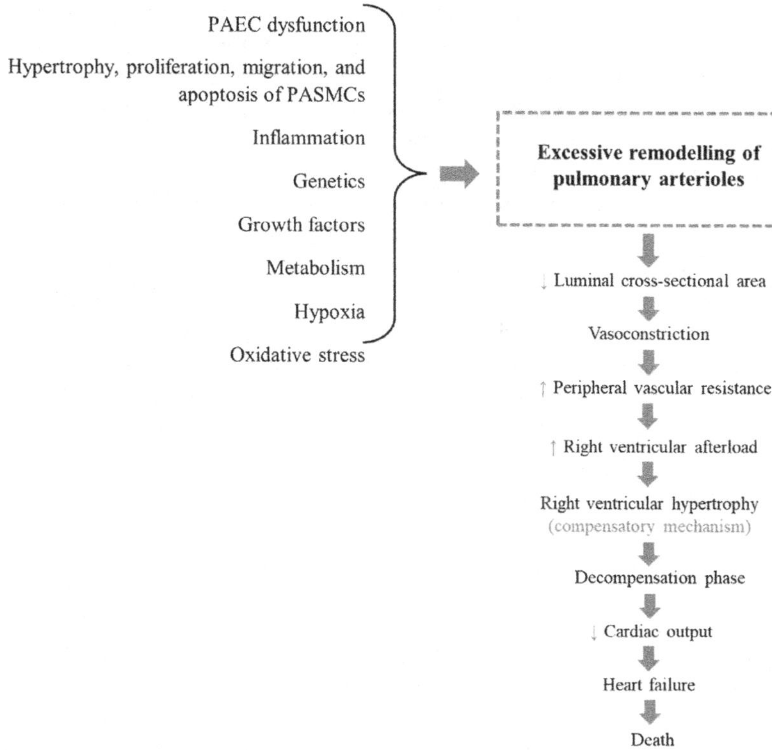

FIGURE 21.2 **Factors that contribute toward pulmonary arteriole remodeling in PAH.** Contributory factors toward pulmonary arteriole remodeling in PAH include PAEC dysfunction and apoptosis, PASMC hypertrophy, proliferation, migration, and apoptosis. Other factors include inflammation, genetic factors, growth factors, metabolic remodeling, hypoxia, and oxidative stress.

Triggers for PAH include inflammatory or oxidative injury of PAECs, which leads to PAEC dysfunction and apoptosis (Teichert-Kuliszewska et al., 2006). Apoptosis of PAEC is central to the pathophysiology of PAH as it leads to loss of microvasculature (Kuebler et al., 2017, p. 315). This occurs either directly, as a result of degeneration of fragile distal lung arterioles, or indirectly, by resulting in the emergence of dysregulated, a hyperproliferative state (Kuebler et al., 2017, p. 315). PAEC apoptosis has been shown in PAH models and the inhibition thereof shown to prevent PAH (Jurasz et al., 2010; Taraseviciene-Stewart et al., 2001). It is not simply PAEC apoptosis but also increased PAEC proliferation that has been shown to play a role in PAH. PAECs isolated and cultured from PAH human patients had increased proliferation in comparison to PAEC of controls with no PAH (Kim et al., 2013). In rodent models, PAH was improved when microRNAs (Kim et al., 2013) and matrix stiffening (Bertero et al., 2016) were altered, two modulating factors of PAEC proliferation. Literature shows that the trigger of PAEC apoptosis occurs in the initial stage of the disease, followed by the hyperproliferative and pathogenic PAEC which drive the advanced stage of PAH and apoptosis resistance (Michelakis, 2006; Sakao et al., 2005).

The proliferation and apoptosis of these cells are involved in the development of PAH (Voelkel et al., 2012). PASMCs can take on a proliferative phenotype, when under a disease or injury environment, resulting in vessel remodeling. This differs from the healthy state of quiescent, nonmigratory, and contractile PASMCs in healthy individuals. The proliferation of PASMCs can result in pathological muscularization of distal arterioles and recapitulates numerous aspects of arterial wall development, namely PASMC dedifferentiation, distal migration, proliferation, and redifferentiation (Sheikh et al., 2014). Thus, hypertrophy, migration, proliferation, and apoptosis resistance of PASMCs are central to PAH pathophysiology.

These features are what provide a cancer-like characteristic to PAH (Boucherat et al., 2017). Factors like inflammation cause metabolic reprogramming toward a glycolytic metabolism, the Warburg effect. This is mainly driven by the overactivation of hypoxia-inducible factors (mediated via upregulation of glucose transporters, pyruvate dehydrogenase kinase, and lactate dehydrogenase). Moreover, this is associated with mitochondrial membrane hyperpolarization and opening of the mitochondria transition pore that reduces sensitivity to cell death inducers. Elevated glycolysis in PAH relies on other metabolic fuels that stimulate glutamine metabolism and resistance to apoptosis.

Mitochondrial aspects of PAH

There is accumulating evidence that metabolic changes contribute to the pathogenesis of PAH (Chan & Rubin, 2017). Healthy endothelial cells produce most of their energy from aerobic glycolysis, whereas PAECs in PAH patients display a shift toward lactate synthesis (Caruso et al., 2017). By studying the dysregulated metabolic mechanisms in PAH, improved diagnostic and therapeutic therapies can be developed to potentially prevent or reverse disease progression (Chan & Rubin, 2017). It is the metabolic changes in the affected vascular and cardiac tissue that are of interest for research as it is observed not only at the end stages of the disease but also during its development (Chan & Rubin, 2017; Saygin et al., 2017; Xu et al., 2007). Abnormalities have been found in the endothelial cells of PAH patients which show changes in glucose uptake and utilization as well as a reduction in mitochondrial oxidative phosphorylation (Caruso et al., 2017). PASMC mitochondria in PAH models and human tissues have also shown hyperpolarization, upregulated glycolysis, suppressed glucose oxidation, and therefore suppressed respiration (Dromparis & Michelakis, 2013).

The exact molecular interplay between pathophysiological pathways of PAH is not understood (Hemnes & Humbert, 2017). The underlying mechanisms of PAH include the involvement of abnormal gene expression, elevated inflammatory markers, increased growth factor circulation, and increased ROS and oxidative stress, and although mitochondrial signaling integrates many triggers of these underlying mechanisms, the exact role of mitochondria in PAH remains unknown. Do mitochondria play a causal role? Or is it a compensatory mechanism?

Mitochondria are oxygen sensors and can therefore change the production of mitochondria-derived ROS in turn triggering PASMC contraction (Sutendra & Michelakis, 2014). Mitochondria from PASMCs in patients with PAH are more hyperpolarized than mitochondria in patients without PAH. This is important, as hyperpolarization is linked to an increase in the threshold to open the mitochondrial transition pore, which essentially then contributes to resistance of apoptosis, a key feature in PAH (Dromparis & Michelakis, 2013; Sutendra & Michelakis, 2014). Although literature states these changes in mitochondria, the cause and the trigger of these changes remain unknown. *Why is it that some patients with PAH have these changes and others do not?*

The systemic circulation can differ in response to that of pulmonary circulation (Sutendra & Michelakis, 2014). A clear example to indicate this is under the condition of hypoxia whereby pulmonary arteries constrict and the systemic arteries dilate (Sutendra & Michelakis, 2014). Upon further investigation, it was shown that these differences came down to that of mitochondria in PASMCs in the pulmonary circulation (Sutendra & Michelakis, 2014; Weir et al., 2005). PASMCs mitochondria had reduced function and expression of the electron transfer system, specifically complex I, II, and III in comparison to mitochondria in systemic vascular smooth muscle cells (Sutendra & Michelakis, 2014). These findings show that mitochondria do play a role in PASMCs, which are largely affected in PAH; however, this does not elucidate the role of mitochondria in PAH. *Does this contribute toward the development of PAH or does this occur because PAH is already present?* If one takes the view that mitochondria play a causal role in PAH, the question remains, *What causes these changes? What is the trigger factor that causes the reduction of complex I, II, and III in PASMC mitochondria?*

Another underlying mechanism in PAH is metabolic remodeling, with metabolic changes seen in the RV and pulmonary arteries (Paulin & Michelakis, 2014). Suppression of mitochondrial function has detrimental downstream effects which results in characteristic features of PAH. These implications include suppressed apoptosis, disturbed downstream signaling affecting proproliferative factors, and increased availability of nonoxidized sugars (Paulin & Michelakis, 2014). Additional literature shows that PAECs in healthy individuals produce most of their energy from aerobic glycolysis, whereas PAECs in PAH patients display a shift toward lactate synthesis. This phenomenon, known as the Warburg effect, is observed in cancer cells as well as in highly proliferative cells (Caruso et al., 2017; Dromparis & Michelakis, 2013). Abnormalities in PASMCs have also been proposed to be due to mitochondrial remodeling shown by upregulated glycolysis and suppressed glucose oxidation and respiration (Dromparis & Michelakis, 2013). Although changes have been found in the PAECs and PASMCs of PAH patients, yet again there is the unanswered question of what causes these abnormalities.

The mitochondria and mitochondrial respiratory pathways are key metabolic pathways in both a healthy and diseased state (Spinelli & Haigis, 2018). Metabolic changes in the pulmonary vessels, as well as the right ventricle, are considered crucial components of PAH (Chan & Rubin, 2017). Many studies have shown dysregulated metabolic pathways play a role in PAH; however, to improve diagnostic and therapeutic strategies, the importance lies in what triggers this dysregulation of mitochondrial pathways and this remains unknown.

The novel role of mitochondrial pathways and bioenergetics in the cancer-like phenotype of PAH

PAH is a complex disease with a complex pathophysiology. Due to this complexity, PAH is not well understood. There is a plethora of underlying mechanisms, including genetics, vascular factors, angiogenic factors, inflammatory, and endothelial factors; it is not simply the mitochondria that play a role. Studies in the mitochondrial field have investigated a range of mitochondrial factors such as mitochondrial genes, DNA, haplotypes, and proteins. Although it has been shown that mitochondria play a role in this disease, there is a gap in the literature as to what this role is exactly.

Do disrupted mitochondrial pathways play a causal role? There could be impairments in these enzymes that perhaps predispose to dysregulated mitochondrial pathways which may lead to PAH. Further and detailed investigations are required to understand the role of mitochondrial pathways and bioenergetics in the cancer-like phenotype of PAH.

Mitochondria transplantation in PAH: What is missing

In studies from a pure cardiovascular context (and not PAH), mitochondrial transplantation allows for the transferred benefits of metabolic reprogramming from human mesenchymal stem cells to adult cardiomyocytes (Coudert et al., 2011). Similar transferred benefits were seen with mitochondrial transplantation from mesenchymal cells to vascular endothelial cells, with benefits transferred being summarized as increased mitochondria biogenesis, decreased cell apoptosis, increased proliferation, and finally promoted cell survival (Liu et al., 2014). A recent manuscript from our group reviewed the usefulness of mitochondrial transplantation in a PAH context (Maarman, 2022). To reiterate, experimental PAH models have demonstrated through several studies that mitochondrial transplantation provides health benefits in PAH to the effect of reduced pulmonary vasoconstriction, reduced pulmonary vascular remodeling, improved right ventricular function, and reduced proliferation of pulmonary artery smooth muscle cells.

Although mitochondrial transplantation requires further investigation in a PAH context, there are different ways to transplant mitochondria (Sun et al., 2023). Naked mitochondrial transplantation is a technique in which mitochondria isolated from healthy tissue or cells are injected directly into the myocardium or into coronary arteries or veins with the aid of microinjection techniques to improve cardiac function (Sun et al., 2023). (1) The same may be useful for PAH. (2) Tunnel nanotubes comprise cell-based mitochondrial transplantation where one can inject mesenchymal stem cells and progenitor cells into the myocardium or perhaps the lungs for PAH. This technique is suggested to yield high-quality mitochondria (Sun et al., 2023). (3) Partial (temporary) or complete (permanent) cell fusion is the gradual sharing of organelle and cytoplasmic components by merging plasma membranes (Sun et al., 2023). Here, there is coculture of e.g., cardiomyocytes and human pluripotent adipose stem cells and this mediates the exchange of materials and mitochondria and facilitates cardiomyocyte reprogramming; this may therefore also be a useful application of mitochondrial transplantation to lung tissue in a PAH context. (4) The use of extracellular vesicle–mediated mitochondrial transfer has become apparent in recent research, which uses extracellular vesicles for mitochondrial transfer. A previous study showed that extracellular vesicles can secrete mitochondria from human-induced pluripotent stem cell–derived cardiomyocytes to damaged cardiomyocytes and attenuate myocardial infarction and ischemia–reperfusion injury (Sun et al., 2023). It is possible that this approach could provide benefits in PAH models too.

Needless to say, there are several key challenges with this relatively new therapeutic approach. Several mechanisms or potential long-term risks are still unknown, but they are surmountable and should not deter researchers from further investigating its effectiveness and trying to overcome these challenges in creative ways. However, as reviewed in the aforementioned sections, the benefits of mitochondrial transplantation in PAH could include, decreased ROS production, rescuing of mitochondrial fission and fusion dynamics, improved metabolic programming, and reduced apoptotic signaling. More research is warranted.

Conclusion

Mitochondrial transplantation is an emerging field that has a wide array of applications in diseases that are associated with mitochondrial dysfunction. It has a promising benefit in cancer treatment and as we highlight in this paper, also against the cancer-like phenotype of PAH via the rescuing of mitochondrial biogenesis, metabolic reprogramming, improved apoptotic signaling, and decreased ROS production. However, further research is warranted to fully understand what the risks and benefits are of mitochondrial transplantation against cancers, or the cancer-like phenotype of PAH.

Acknowledgments

The authors would like to express their gratitude to the Faculty of Medicine and Health Sciences, Stellenbosch University, the National Research Foundation of South Africa, the South African Medical Research Council, the South African Rooibos Council, and the Harry Crossley Foundation.

Conflict of interest

The authors have no conflict of interest to declare.

Author contributions

All authors meet the criteria to qualify as authors, have contributed sufficiently to the manuscript, and have seen/approved the final version of the manuscript.

References

Aranovich, A., Liu, Q., Collins, T., Geng, F., Dixit, S., Leber, B., & Andrews, D. W. (2012). Differences in the mechanisms of proapoptotic BH3 proteins binding to Bcl-XL and Bcl-2 quantified in live MCF-7 cells. *Molecular Cell, 45*(6), 754–763. https://doi.org/10.1016/j.molcel.2012.01.030

Bertero, T., Oldham, W. M., Cottrill, K. A., Pisano, S., Vanderpool, R. R., Yu, Q., Zhao, J., Tai, Y., Tang, Y., Zhang, Y. Y., Rehman, S., Sugahara, M., Qi, Z., Gorcsan, J., Vargas, S. O., Saggar, R., Saggar, R., Wallace, W. D., Ross, D. J., ... Chan, S. Y. (2016). Vascular stiffness mechanoactivates YAP/TAZ-dependent glutaminolysis to drive pulmonary hypertension. *Journal of Clinical Investigation, 126*(9), 3313–3335. https://doi.org/10.1172/JCI86387

Boland, M. L., Chourasia, A. H., & Macleod, K. F. (2013). Mitochondrial dysfunction in cancer. *Frontiers in Oncology, 3*. https://doi.org/10.3389/fonc.2013.00292

Boucherat, O., Vitry, G., Trinh, I., Paulin, R., Provencher, S., & Bonnet, S. (2017). The cancer theory of pulmonary arterial hypertension. *Pulmonary Circulation, 7*(2), 285–299. https://doi.org/10.1177/2045893217701438

Boulton, D. P., & Caino, M. C. (2022). *Mitochondrial fission and fusion in tumor progression to metastasis.*

Braganza, A., Quesnelle, K., Bickta, J., Reyes, C., Wang, Y., Jessup, M., St Croix, C., Arlotti, J., Singh, S. V., & Shiva, S. (2019). Myoglobin induces mitochondrial fusion, thereby inhibiting breast cancer cell proliferation. *Journal of Biological Chemistry, 294*(18), 7269–7282. https://doi.org/10.1074/jbc.RA118.006673

Brown, D. A., Perry, J. B., Allen, M. E., Sabbah, H. N., Stauffer, B. L., Shaikh, S. R., Cleland, J. G. F., Colucci, W. S., Butler, J., Voors, A. A., Anker, S. D., Pitt, B., Pieske, B., Filippatos, G., Greene, S. J., & Gheorghiade, M. (2017). Mitochondrial function as a therapeutic target in heart failure. *Nature Reviews Cardiology, 14*(4), 238–250. https://doi.org/10.1038/nrcardio.2016.203

Burz, C., Berindan-Neagoe, I., Balacescu, O., & Irimie, A. (2009). Apoptosis in cancer: Key molecular signaling pathways and therapy targets. *Acta Oncologica, 48*(6), 811–821. https://doi.org/10.1080/02841860902974175

Carneiro, B. A., & El-Deiry, W. S. (2020). Targeting apoptosis in cancer therapy. *Nature Reviews Clinical Oncology, 17*(7), 395–417. https://doi.org/10.1038/s41571-020-0341-y

Caruso, P., Dunmore, B. J., Schlosser, K., Schoors, S., Dos Santos, C., Perez-Iratxeta, C., Lavoie, J. R., Zhang, H., Long, L., Flockton, A. R., Frid, M. G., Upton, P. D., D'Alessandro, A., Hadinnapola, C., Kiskin, F. N., Taha, M., Hurst, L. A., Ormiston, M. L., Hata, A., … Morrell, N. W. (2017). Identification of microRNA-124 as a major regulator of enhanced endothelial cell glycolysis in pulmonary arterial hypertension via PTBP1 (polypyrimidine tract binding protein) and pyruvate kinase M2. *Circulation, 136*(25), 2451–2467. https://doi.org/10.1161/CIRCULATIONAHA.117.028034

Cazzaniga, M., & Bonanni, B. (2015). Relationship between metabolic reprogramming and mitochondrial activity in cancer cells. Understanding the anticancer effect of metformin and its clinical implications. *Anticancer Research, 35*(11), 5789–5796. http://ar.iiarjournals.org/content/35/11/5789.full.pdf.

Chan, S. Y., & Rubin, L. J. (2017). Metabolic dysfunction in pulmonary hypertension: From basic science to clinical practice. *European Respiratory Review, 26*(146). https://doi.org/10.1183/16000617.0094-2017

Chan, K., Koh, C. G., & Li, H. Y. (2012). Mitosis-targeted anti-cancer therapies: Where they stand. *Cell Death and Diseases, 3.*

Chan, D. C. (2020). Mitochondrial dynamics and its involvement in disease. *Annual Review of Pathology: Mechanisms of Disease, 15*, 235–259. https://doi.org/10.1146/annurev-pathmechdis-012419-032711

Chang, J.-C., Chang, H.-S., Wu, Y.-C., Cheng, W.-L., Lin, T.-T., Chang, H.-J., Kuo, S.-J., Chen, S.-T., & Liu, C.-S. (2019a). Mitochondrial transplantation regulates antitumour activity, chemoresistance and mitochondrial dynamics in breast cancer. *Journal of Experimental and Clinical Cancer Research, 38*, 30.

Chang, J.-C., Chang, H.-S., Wu, Y.-C., Cheng, W.-L., Lin, T.-T., Chang, H.-J., Kuo, S.-J., Chen, S.-T., & Liu, C.-S. (2019b). Mitochondrial transplantation regulates antitumour activity, chemoresistance and mitochondrial dynamics in breast cancer. *Journal of Experimental and Clinical Cancer Research, 38*, 1–16 (Original work published 2019).

Chang, J. C., Chang, H. S., Wu, Y. C., Cheng, W. L., Lin, T. T., Chang, H. J., Chen, S. T., & Liu, C. S. (2020). Antitumor actions of intratumoral delivery of membrane-fused mitochondria in a mouse model of triple-negative breast cancers. *OncoTargets and Therapy, 13*, 5241–5255. https://doi.org/10.2147/OTT.S238143

Chang, J.-C., Chao, Y.-C., Chang, H.-S., Wu, Y.-L., Chang, H.-J., Lin, Y.-S., Cheng, W.-L., Lin, T.-T., & Liu, C.-S. (2021). Intranasal delivery of mitochondria for treatment of Parkinson's disease model rats lesioned with 6-hydroxydopamine. *Scientific Reports, 11*, 1–14.

Coudert, A. E., Coz, Christov, C., Baudin, X., Auber, F., Yiou, R., Dubois-Randé, J.-L., & Rodriguez, A. (2011). Human mesenchymal stem cells reprogram adult cardiomyocytes towards a progenitor-like state through partial cell fusion and mitochondria transfer. *Stem Cells, 29* (Original work published 2011).

De La Fuente-Munoz, C. E., & Arias, C. (2020). The therapeutic potential of mitochondrial transplantation for the treatment of neurodegenerative disorders. *Reviews in Neurosciences, 32*, 203–217.

DeBerardinis, R. J., Lum, J. J., Hatzivassiliou, G., & Thompson, C. B. (2008). The biology of cancer: Metabolic reprogramming fuels cell growth and proliferation. *Cell Metabolism, 7*(1), 11–20. https://doi.org/10.1016/j.cmet.2007.10.002

Ding, W. X., & Yin, X. M. (2012). Mitophagy: Mechanisms, pathophysiological roles, and analysis. *Biological Chemistry, 393*(7), 547–564. https://doi.org/10.1515/hsz-2012-0119

Dromparis, P., & Michelakis, E. D. (2013). Mitochondria in vascular health and disease. *Annual Review of Physiology, 75*, 95–126. https://doi.org/10.1146/annurev-physiol-030212-183804

Eisner, V., Lenaers, G., & Hajnóczky, G. (2014). Mitochondrial fusion is frequent in skeletal muscle and supports excitation–contraction coupling. *Journal of Cell Biology, 205*(2), 179–195. https://doi.org/10.1083/jcb.201312066

Elliott, R. L., Jiang, X. P., & Head, J. F. (2012). Mitochondria organelle transplantation: Introduction of normal epithelial mitochondria into human cancer cells inhibits proliferation and increases drug sensitivity. *Breast Cancer Research and Treatment, 136*(2), 347–354. https://doi.org/10.1007/s10549-012-2283-2

Gamudi, D., & Blundell, R. (2010). Tumor suppressor genes. *Research Journal of Medical Sciences, 4*(4), 280–284. https://doi.org/10.3923/rjmsci.2010.280.284

Guo, J., Ye, F., Jiang, X., Guo, H., Xie, W., Zhang, Y., & Sheng, X. (2020). Drp1 mediates high glucose-induced mitochondrial dysfunction and epithelial-mesenchymal transition in endometrial cancer cells. *Experimental Cell Research, 389*.

Hatzikirou, H., Basanta, D., Simon, M., Schaller, K., & Deutsch, A. (2012). "Go or grow": The key to the emergence of invasion in tumour progression? *Mathematical Medicine and Biology, 29*(1), 49–65. https://doi.org/10.1093/imammb/dqq011

Hayakawa, K., Esposito, E., Wang, X., Terasaki, Y., Liu, Y., Xing, C., Ji, X., & Lo, E. H. (2016). Transfer of mitochondria from astrocytes to neurons after stroke. *Nature, 535*(7613), 551–555. https://doi.org/10.1038/nature18928

Hemnes, A. R., & Humbert, M. (2017). Pathobiology of pulmonary arterial hypertension: Understanding the roads less travelled. *European Respiratory Review, 26*(146). https://doi.org/10.1183/16000617.0093-2017

Humbert, M., Guignabert, C., Bonnet, S., Dorfmüller, P., Klinger, J. R., Nicolls, M. R., Olschewski, A. J., Pullamsetti, S. S., Schermuly, R. T., Stenmark, K. R., & Rabinovitch, M. (2019). Pathology and pathobiology of pulmonary hypertension: State of the art and research perspectives. *European Respiratory Journal, 53*(1). https://doi.org/10.1183/13993003.01887-2018

Hunt, J., Gandjeva, A., Groshong, S., McLaughlin, V., Jessup, M., Grizzle, W., Aldred, M., Cool, C., Tuder, R., EjajoR, S., et al. (2012). *Modern age pathology of pulmonary arterial hypertension* (Original work published 2012).

Ibanez, B., & Villena-Gutierrez, R. (2021). *Cardiac Mitochondrial Transplantation: The Force Awakens, 77* (pp. 1089–1092). Washington DC: American College of Cardiology Foundation.

Islam, M. N., Das, S. R., Emin, M. T., Wei, M., Sun, L., Westphalen, K., Rowlands, D. J., Quadri, S. K., Bhattacharya, S., & Bhattacharya, J. (2012). Mitochondrial transfer from bone-marrow-derived stromal cells to pulmonary alveoli protects against acute lung injury. *Nature Medicine, 18*(5), 759–765. https://doi.org/10.1038/nm.2736

Joshi, A. U., Saw, N. L., Shamloo, M., & Mochly-Rosen, D. (2018). Drp1/Fis1 interaction mediates mitochondrial dysfunction, bioenergetic failure and cognitive decline in Alzheimer's disease. *Oncotarget, 9*(5), 6128–6143. https://doi.org/10.18632/oncotarget.23640

Jurasz, P., Courtman, D., Babaie, S., & Stewart, D. J. (2010). Role of apoptosis in pulmonary hypertension: From experimental models to clinical trials. *Pharmacology & Therapeutics, 126*(1), 1–8. https://doi.org/10.1016/j.pharmthera.2009.12.006

Kang, D., & Hamasaki, N. (2005). Alterations of mitochondrial DNA in common diseases and disease states: Aging, neurodegeneration, heart failure, diabetes and cancer. *Current Medicinal Chemistry, 12*(4), 429–441. https://doi.org/10.2174/0929867053363081

Kato, R., Kibudde, S., Mwondha, Y., & Farah, S. (2023). A rare case of soft tissue sarcoma in the supraclavicular region: A case report. *Clinical Case Reports, 11*.

Kaza, Wamala, I., Friehs, I., Kuebler, J. D., Rathod, R. H., Berra, I., Ericsson, M., Yao, R., Thedsanamoorthy, J. K., & Zurakowski, D. (2017). Myocardial rescue with autologous mitochondrial transplantation in a porcine model of ischemia/reperfusion. *The Journal of Thoracic and Cardiovascular Surgery, 153*, 934–943.

Kim, J., Kang, Y., Kojima, Y., Lighthouse, J. K., Hu, X., Aldred, M. A., McLean, D. L., Park, H., Comhair, S. A., Greif, D. M., Erzurum, S. C., & Chun, H. J. (2013). An endothelial apelin-FGF link mediated by miR-424 and miR-503 is disrupted in pulmonary arterial hypertension. *Nature Medicine, 19*(1), 74–82. https://doi.org/10.1038/nm.3040

Kuebler, WM, Nicolls, MR, Olschewski, A, Abe, K, Rabinovitch, M, Stewart, D, Chan, SA, Morrell, N , Archer, SI, (2017). 315. (Original work published 2017).

Lee, J. M., Hwang, J. W., Jung, S. Y., Ahn, E. H., Min, K., & Choi, S. Y. (2021). Mitochondrial transplantation modulates inflammation and apoptosis, alleviating tendinopathy both in vivo and in vitro. *Antioxidants, 10*.

Levoux, J., Prola, A., Lafuste, P., Gervais, M., Chevallier, N., Koumaiha, Z., Kefi, K., Braud, L., Schmitt, A., & Yacia, A. (2021). Platelets facilitate the wound-healing capability of mesenchymal stem cells by mitochondrial transfer and metabolic reprogramming. *33*, 283–299.

Li, F., Wang, Y., Zeller, K.I., Potter, J.J., Wonsey, D.R., O'Donnell, K.A., Kim, J.W., Yustein, J.T., Lee, L.A., & Dang, C.V. (2005). Myc stimulates nuclearly encoded mitochondrial genes and mitochondrial biogenesis. Molecular Cell Biology, 25(14), 6225-6234. https://doi.org/10.1128/MCB.25.14.6225-6234.2005. PMID: 15988031; PMCID: PMC1168798.

Liang, J., Cao, r, Zhang, y, Xia, y, Zheng, Y., Li, X., Wang, l, Yang, W., & Lu, Z. (2016). PKM2 dephosphorylation by Cdc25A promotes the Warburg effect and tumorigenesis. *Nature Communications, 7* (Original work published 2016).

Ling, Z., Cheng, B., & Tao, X. (2021). Epithelial-to-mesenchymal transition in oral squamous cell carcinoma: Challenges and opportunities. *International Journal of Cancer, 148*(7), 1548–1561. https://doi.org/10.1002/ijc.33352

Liu, K., Ji, K., Guo, L., Wu, W., Lu, H., Shan, P., & Yan, C. (2014). Mesenchymal stem cells rescue injured endothelial cells in an in vitro ischemia-reperfusion model via tunneling nanotube like structure-mediated mitochondrial transfer. *Microvascular Research, 92*, 10–18. https://doi.org/10.1016/j.mvr.2014.01.008

Lopez, J., & Tait, S. W. G. (2015). Mitochondrial apoptosis: Killing cancer using the enemy within. *British Journal of Cancer, 112*(6), 957–962. https://doi.org/10.1038/bjc.2015.85

Maarman, G. J., Schultz, R., Sliwa, K., Schermuly, R. T., & Lecour, S. (2017). Novel putative pharmacological therapies to protect the right ventricle in pulmonary hypertension: A review of current literature. *British Journal of Pharmacology, 174*, 497–511. https://doi.org/10.1111/bph.13721

Maarman, G. J. (2018). Pulmonary arterial hypertension and the potential roles of metallothioneins: A focused review. *Life Sciences, 214*, 77–83. https://doi.org/10.1016/j.lfs.2018.10.039

Maarman, G. J. (2022). Reviewing the suitability of mitochondrial transplantation as therapeutic approach for pulmonary hypertension in the era of personalized medicine. *American Journal of Physiology. Lung Cellular and Molecular Physiology, 322*.

Maeda, A., & Fadeel, B. (2014). Mitochondria released by cells undergoing TNF-α-induced necroptosis act as danger signals. *Cell Death and Diseases, 5* (Original work published 2014).

Maracci, C., Motta, S., Romagnoli, A., Costantino, M., Perego, P., & Di Marino, D. (2022). The mTOR/4E-BP1/eIF4E signalling pathway as a source of cancer drug targets. *Current Medicinal Chemistry, 29*(20), 3501–3529. https://doi.org/10.2174/0929867329666220224112042

Mayer, B., & Oberbauer, R. (2003). Mitochondrial regulation of apoptosis. *News in Physiological Sciences, 18*(3), 89–94. http://nips.physiology.org/.

McCully, J. D., Levitsky, S., Nido, P. J., & Cowan, D. B. (2016). Mitochondrial transplantation for therapeutic use. *Clinical and Translational Medicine, 5*(1). https://doi.org/10.1186/s40169-016-0095-4

McGuirk, S., Gravel, S.-P., Deblois, G., Papadopoli, D. J., Faubert, B., Wegner, A., Hiller, K., Avizonis, D., Akavia, U. D., Jones, R. G., Giguère, V., & St-Pierre, J. (2013). PGC-1α supports glutamine metabolism in breast cancer. *Cancer and Metabolism, 1*(1). https://doi.org/10.1186/2049-3002-1-22

Michelakis, E. D. (2006). Spatio-temporal diversity of apoptosis within the vascular wall in pulmonary arterial hypertension: Heterogeneous BMP signaling may have therapeutic implications. *Circulation Research, 98*(2), 172–175. https://doi.org/10.1161/01.RES.0000204572.65400.a5

Mishra, P., & Chan, D. C. (2016). Metabolic regulation of mitochondrial dynamics. *Journal of Cell Biology, 212*(4), 379–387. https://doi.org/10.1083/jcb.201511036

Nakamura, H., & Takada, K. (2021). Reactive oxygen species in cancer: Current findings and future directions. *Cancer Science, 112*(10), 3945–3952. https://doi.org/10.1111/cas.15068

Paulin, R., & Michelakis, E. D. (2014). The metabolic theory of pulmonary arterial hypertension. *Circulation Research, 115*(1), 148–164. https://doi.org/10.1161/CIRCRESAHA.115.301130

Peiris-Pagès, M., Bonuccelli, G., Sotgia, F., & Lisanti, M. P. (2018). Mitochondrial fission as a driver of stemness in tumor cells: mDIVI1 inhibits mitochondrial function, cell migration and cancer stem cell (CSC) signalling. *Oncotarget, 9*(17), 13254–13275. https://doi.org/10.18632/oncotarget.24285

Pfeffer, C. M., & Singh, A. T. K. (2018). Apoptosis: A target for anticancer therapy. *International Journal of Molecular Sciences, 19*(2). https://doi.org/10.3390/ijms19020448

Phinney, D. G., Di Giuseppe, M., Njah, J., Sala, E., Shiva, S., St Croix, C. M., Stolz, D. B., Watkins, S. C., Di, Y. P., Leikauf, G. D., Kolls, J., Riches, D. W. H., Deiuliis, G., Kaminski, N., Boregowda, S. V., McKenna, D. H., &

Ortiz, L. A. (2015). Mesenchymal stem cells use extracellular vesicles to outsource mitophagy and shuttle microRNAs. *Nature Communications, 6*. https://doi.org/10.1038/ncomms9472

Qin, Y., Jiang, X., Yang, Q., Zhao, J., Zhou, Q., & Zhou, Y. (2021). The functions, methods, and mobility of mitochondrial transfer between cells. *Frontiers in Oncology, 11*. https://doi.org/10.3389/fonc.2021.672781

Rajagopal, S., & Yu, Y. R. A. (2022). The pathobiology of pulmonary arterial hypertension. *Cardiology Clinics, 40*(1), 1–12. https://doi.org/10.1016/j.ccl.2021.08.001

Ridky, T. W., Chow, J. M., Wong, D. J., & Khavari, P. A. (2010). Invasive three-dimensional organotypic neoplasia from multiple normal human epithelia. *Nature Medicine, 16*(12), 1450–1456. https://doi.org/10.1038/nm.2265

Rodriguez-Abreu, D., Bordoni, A., & Zucca, E. (2007). Epidemiology of hematological malignancies. *Annals of Oncology, 18*(1), i3–i8. https://doi.org/10.1093/annonc/mdl443

Roy, N. K., Bordoloi, D., Monisha, J., Anip, A., Padmavathi, G., & Kunnumakkara, A. B. (2017). Cancer—An overview and molecular alterations in cancer. *Fusion Genes and Cancer, 2017*, 1–15.

Rustom, A., Saffrich, R., Markovic, I., Walther, P., & Gerdes, H. H. (2004). Nanotubular highways for intercellular organelle transport. *Science, 303*(5660), 1007–1010. https://doi.org/10.1126/science.1093133

Sakao, S., Taraseviciene-Stewart, L., Lee, J. D., Wood, K., Cool, C. D., & Voelkel, N. F. (2005). Initial apoptosis is followed by increased proliferation of apoptosis-resistant endothelial cells. *Federation of American Societies for Experimental Biology Journal, 19*(9), 1178–1180. https://doi.org/10.1096/fj.04-3261fje

Saygin, D., Highland, K. B., Farha, S., Park, M., Sharp, J., Roach, E. C., Tang, W. H. W., Thomas, J. D., Erzurum, S. C., Neumann, D. R., & DiFilippo, F. P. (2017). Metabolic and functional evaluation of the heart and lungs in pulmonary hypertension by gated 2-[18F]-fluoro-2-deoxy-D-glucose positron emission tomography. *Pulmonary Circulation, 7*(2), 428–438. https://doi.org/10.1177/2045893217701917

Senft, D., & Ronai, Z. A. (2016). Regulators of mitochondrial dynamics in cancer. *Current Opinion in Cell Biology, 39*, 43–52. https://doi.org/10.1016/j.ceb.2016.02.001

Serrano-Del Valle, A., Reina-Ortiz, C., Benedi, A., Anel, A., Naval, J., & Marzo, I. (2021). Future prospects for mitosis-targeted antitumor therapies. *Biochemical Pharmacology, 190* (Original work published 2021).

Shanmughapriya, S., Langford, D., & Natarajaseenivasan, K. (2020). Inter and Intracellular mitochondrial trafficking in health and disease. *Ageing Research Reviews, 62*. https://doi.org/10.1016/j.arr.2020.101128

Sheikh, A. Q., Lighthouse, J. K., & Greif, D. M. (2014). Recapitulation of developing artery muscularization in pulmonary hypertension. *Cell Reports, 6*(5), 809–817. https://doi.org/10.1016/j.celrep.2014.01.042

Skildum, A., Dornfeld, K., & Wallace, K. (2011). Mitochondrial amplification selectively increases doxorubicin sensitivity in breast cancer cells with acquired antiestrogen resistance. *Breast Cancer Research and Treatment, 129*(3), 785–797. https://doi.org/10.1007/s10549-010-1268-2

Song, C., Pan, S., Zhang, J., Li, N., & Geng, Q. (2022). Mitophagy: A novel perspective for insighting into cancer and cancer treatment. *Cell Proliferation, 55*(12), e13327. https://doi.org/10.1111/cpr.13327

Spinelli, J. B., & Haigis, M. C. (2018). The multifaceted contributions of mitochondria to cellular metabolism. *Nature Cell Biology, 20*(7), 745–754. https://doi.org/10.1038/s41556-018-0124-1

Sun, C., Liu, X., Wang, B., Wang, Z., Liu, Y., Di, C., Si, J., Li, H., Wu, Q., & Xu, D. (2019). Endocytosis-mediated mitochondrial transplantation: Transferring normal human astrocytic mitochondria into glioma cells rescues aerobic respiration and enhances radiosensitivity. *Theranostics, 9*.

Sun, M., Jiang, W., Mu, N., Zhang, Z., Yu, L., & Ma, H. (2023). Mitochondrial transplantation as a novel therapeutic strategy for cardiovascular diseases. *Journal of Translational Medicine, 21*.

Sutendra, G., & Michelakis, E. D. (2014). The metabolic basis of pulmonary arterial hypertension. *Cell Metabolism, 19*(4), 558–573. https://doi.org/10.1016/j.cmet.2014.01.004

Tait, S. W. G., Ichim, G., & Green, D. R. (2014). Die another way - non-apoptotic mechanisms of cell death. *Journal of Cell Science, 127*(10), 2135–2144. https://doi.org/10.1242/jcs.093575

Taraseviciene-Stewart, L., Kasahara, Y., Alger, L., Hirth, P., Mc Mahon, G., Waltenberger, J., Voelkel, N. F., & Tuder, R. M. (2001). Inhibition of the VEGF receptor 2 combined with chronic hypoxia causes cell death-dependent pulmonary endothelial cell proliferation and severe pulmonary hypertension. *Federation of American Societies for Experimental Biology Journal, 15*(2), 427–438. https://doi.org/10.1096/fj.00-0343com

Teichert-Kuliszewska, K., Kutryk, M. J. B., Kuliszewski, M. A., Karoubi, G., Courtman, D. W., Zucco, L., Granton, J., & Stewart, D. J. (2006). Bone morphogenetic protein receptor-2 signaling promotes pulmonary arterial endothelial cell survival: Implications for loss-of-function mutations in the pathogenesis of pulmonary hypertension. *Circulation Research, 98*(2), 209–217. https://doi.org/10.1161/01.RES.0000200180.01710.e6

Torralba, D., Baixauli, F., & Sánchez-Madrid, F. (2016). Mitochondria know no boundaries: Mechanisms and functions of intercellular mitochondrial transfer. *Frontiers in Cell and Developmental Biology, 4*. https://doi.org/10.3389/fcell.2016.00107

Vazquez, F., Lim, J. H., Chim, H., Bhalla, K., Girnun, G., Pierce, K., Clish, C. B., Granter, S. R., Widlund, H. R., Spiegelman, B. M., & Puigserver, P. (2013). PGC1α expression defines a subset of human melanoma tumors with increased mitochondrial capacity and resistance to oxidative stress. *Cancer Cell, 23*(3), 287–301. https://doi.org/10.1016/j.ccr.2012.11.020

Voelkel, N. F., Gomez-Arroyo, J., Abbate, A., Bogaard, H. J., & Nicolls, M. R. (2012). Pathobiology of pulmonary arterial hypertension and right ventricular failure. *European Respiratory Journal, 40*(6), 1555–1565. https://doi.org/10.1183/09031936.00046612

Warth, A., Cortis, J., Soltermann, A., Meister, M., Budczies, J., Stenzinger, A., Goeppert, B., Thomas, M., Herth, F. J. F., Schirmacher, P., Schnabel, P. A., Hoffmann, H., Dienemann, H., Muley, T., & Weichert, W. (2014). Tumour cell proliferation (Ki-67) in non-small cell lung cancer: A critical reappraisal of its prognostic role. *British Journal of Cancer, 111*(6), 1222–1229. https://doi.org/10.1038/bjc.2014.402

Weir, E. K., López-Barneo, José, B., Keith, J., & Archer, S. L. (2005). Acute oxygen-sensing mechanisms. *New England Journal of Medicine, 353*(19), 2042–2055. https://doi.org/10.1056/nejmra050002

Weixler, V., Lapusca, R., Grangl, G., Guariento, A., Saeed, M. Y., Cowan, D. B., Del Nido, P. J., McCully, J. D., & Friehs, I. (2021). Autogenous mitochondria transplantation for treatment of right heart failure. *Journal of Thoracic and Cardiovascular Surgery, 162*.

Xu, W., Koeck, T., Lara, A. R., Neumann, D., DiFilippo, F. P., Koo, M., Janocha, A. J., Masri, F. A., Arroliga, A. C., Jennings, C., Dweik, R. A., Tuder, R. M., Stuehr, D. J., & Erzurum, S. C. (2007). Alterations of cellular bioenergetics in pulmonary artery endothelial cells. *Proceedings of the National Academy of Sciences of the United States of America, 104*(4), 1342–1347. https://doi.org/10.1073/pnas.0605080104

Yao, N., Wang, C., Hu, N., et al. (2019). Inhibition of PINK1/Parkin-dependent mitophagy sensitizes multidrug-resistant cancer cells to B5G1, a new betulinic acid analog. *Cell Death & Disease, 10*, 232. https://doi.org/10.1038/s41419-019-1470-z

Yiallouros, M., Koscielniak, P., & Sparber-Sauer, M. (2022). *Translation E: Soft tissue sarcoma and rare soft tissue tumours—Brief information* (Original work published 2022).

Yu, Z., Zhou, Y., Zhou, W., Zhao, Z., Liu, Z., & Fu, A. (May 11, 2021). The effect of mitochondrial transplantation therapy from different gender on inhibiting cell proliferation of malignant melanoma. *International Journal of Biological Sciences, 17*(8), 2021–2033. https://doi.org/10.7150/ijbs.59581. PMID: 34131403; PMCID: PMC8193273.

Zappulli, V., Pagh Friis, K., Fitzpatrick, Z., Maguire, C. A., & Breakefield, X. O. (2016). Extracellular vesicles and intercellular communication within the nervous system. *Journal of Clinical Investigation, 126*(4), 1198–1207. https://doi.org/10.1172/JCI81134

Zhao, J., Zhang, J., Yu, M., Xie, Y., Huang, Y., Wolff, D. W., Abel, P. W., & Tu, Y. (2013). Mitochondrial dynamics regulates migration and invasion of breast cancer cells. *Oncogene, 32*(40), 4814–4824. https://doi.org/10.1038/onc.2012.494

Zhou, W., Zhao, Z., Yu, Z., Hou, Y., Keerthiga, R., & Fu, A. (July 2022). Mitochondrial transplantation therapy inhibits the proliferation of malignant hepatocellular carcinoma and its mechanism. *Mitochondrion, 65*, 11–22. https://doi.org/10.1016/j.mito.2022.04.004. Epub 2022 Apr 30. PMID: 35504558.

CHAPTER

22

Diabetes and mitochondrial transplantation

Eli John Louwagie[1] *and Michelle Leigh Baack*[1,2]

[1]**Sanford School of Medicine, University of South Dakota, Sioux Falls, SD, United States;**
[2]**Environmental Influences on Health and Disease Group, Sanford Research, Sioux Falls, SD, United States**

Introduction

Diabetes is a complex, metabolic disease with various pathogeneses. Classically, diabetes has two common types: type 1 diabetes mellitus (T1DM), also referred to as "insulin-dependent diabetes," results from autoimmune destruction of beta-islet cells in the pancreas (Atkinson et al., 1994; Committee, 2021a). It typically presents in childhood or adolescence, and its standard-of-care treatment is life-long insulin replacement therapy (Committee, 2021a, 2021b). Type 2 diabetes mellitus (T2DM) is caused by insulin insensitivity or tissue unresponsiveness related to lifestyle-related hyperinsulinemia on a background of genetic susceptibility (Committee, 2021a). T2DM treatment varies by severity and individual variables, but common treatments include metformin, insulin, and other medications that increase glucose clearance, decrease glucose absorption, or increase insulin secretion and sensitivity (Buse et al., 2019; Davies et al., 2018). Diabetic pregnancy has additional considerations because it affects both maternal and fetal health. When T1DM or T2DM are present before and throughout pregnancy, the condition is termed pregestational diabetes mellitus (PGDM) and exposes the developing offspring to excess circulating fuels during embryogenesis and throughout pregnancy. More commonly, gestational diabetes mellitus (GDM) results when insulin resistance, a normal physiologic state associated with late-gestation pregnancy, exceeds the ability of the maternal pancreas to produce insulin. GDM arises later in pregnancy and is influenced by a complex interplay of patient lifestyle, genetics, and placental hormones (Committee, 2021c). Regardless of type, diabetic pregnancy exposes the developing offspring to a triad of maternal hyperglycemia, hyperlipidemia, and fetal hyperinsulinemia which can impair cellular metabolism, alter gene and protein expression, and perturb normal structure and function of the developing organs (Louwagie, 2023).

Mitochondrial Transplantation and Transfer
https://doi.org/10.1016/B978-0-443-18858-9.00022-6

Important in diabetes, mitochondria are the organelles that increase or decrease energy output in response to metabolic cues (i.e., oxygen versus ischemia, fuel excess versus starvation, types of available fuels) (Anderson et al., 2019; Kodde et al., 2007; Murphy et al., 2016). Nearly all tissues of the human body rely on highly functioning mitochondria for cellular respiration and ATP generation, hence mitochondria's nickname "the powerhouse of the cell." Under exposure to hyperglycemia, hyperlipidemia, or the combination (termed "glucolipotoxicity" (Cerf, 2018)), mitochondria undergo electron transport chain (ETC) uncoupling, leading to overproduction of reactive oxygen species (ROS) (Bugger et al., 2014; Duncan, 2011a; Eshaq et al., 2014; Nishikawa et al., 2000; Rovira-Llopis et al., 2018; Ruegsegger et al., 2018). Tissues with the greatest reliance on mitochondria, such as cardiac, nervous, and renal systems, are the most severely affected. In conditions of excess ROS production such as diabetes, hydroxyl radicals damage mitochondrial membranes, ETC complexes, and mitochondrial DNA, leading to inefficient respiration and worsening fatty acid oxidation (Aon et al., 2014; Murphy, 2009); this creates a vicious cycle leading to mitochondria-mediated apoptosis, necrosis, and ultimately tissue death (Kim-Campbell et al., 2019; Vakifahmetoglu-Norberg et al., 2017). Hence, mitochondria are central players in the pathology of diabetes.

Insulin resistance, a central component of T2DM and GDM, is associated with pathologic cellular signaling which has far-reaching and diverse effects on cell functions (Meyts et al., 2016; Shaham et al., 2008), particularly through the central insulin/insulin receptor substrate/AKT pathway which exerts powerful control over mitochondrial function (Ruegsegger et al., 2018). Insulin deficiency stimulates peroxisome proliferator–activated receptor gamma coactivator 1-alpha (PGC-1α), one of the prominent transcriptional regulators of mitochondrial biogenesis and bioenergetics (Dorn et al., 2015; Duncan, 2011b; Fontecha-Barriuso et al., 2020). With insulin deficiency, upregulation of PGC-1α leads to mitochondrial biogenesis and increased expression of enzymes involved in fatty oxidation, possibly in a cellular attempt to "scavenge" what fuels are present and ensure adequate ATP generation. However, exaggerated mitochondrial biogenesis can exacerbate ROS accumulation that further damages mitochondria (Seifert et al., 2010; Zabielski et al., 2014, 2016). In contrast, others have shown that insulin resistance due to chronic hyperglycemia and hyperlipidemia is associated with downregulation of the PGC-1α pathways (Mootha et al., 2003; Patti et al., 2003; Zhu et al., 2009), leading to decreased mitochondrial biogenesis and diminished fatty acid oxidation despite higher circulating lipid levels. Unfortunately, both insulin deficiency and insulin resistance increase mitochondrial uncoupling, albeit by different mechanisms (Boudina et al., 2007; Karakelides et al., 2007; Montaigne et al., 2014; Schrauwen et al., 2004; Zabielski et al., 2016). Although by different means, insulin deficiency and insulin resistance ultimately converge into one predominant outcome: mitochondrial dysfunction which drives dangerously high ROS levels, leading to cell damage and death to cause a variety of diabetes-associated morbidities.

As mitochondrial dysfunction is fundamental to the pathology of diabetes, several strategies have been developed to therapeutically target mitochondria. One of these novel approaches aims to transfer or transplant healthy mitochondria into diseased tissue, in theory augmenting or replacing dysfunctional with healthy mitochondria and facilitating repair through renewed mitochondrial populations. This chapter summarizes information regarding the use of mitochondrial transfer/transplantation in diabetic disease, paying special attention to studies involving cardiac, nervous, renal, hepatic, and pancreatic tissues. The studies discussed in depth in this chapter are summarized in Table 22.1.

TABLE 22.1 Studies evaluating the application of mitochondrial transplantation/transfer to diabetes mellitus.

	Study	Mitochondrial source/Donor	Model; mitochondrial recipient/Host	Outcome summary/Therapeutic relevance
Heart	Durr (2022)	Mito::mKate2 mice (Barrasso et al., 2018) ventricular tissue	HL-1 CM; FVB/NJ $Lepr^{db/db}$ mice	In vitro, HL-1 CM superficially interacted with isolated mitochondria within 8 h of coincubation, increasing up through 12 h, and retained this pattern for at least 24 h; successful transfer/internalization not observed. In vivo, isolated mitochondria spread across ~50% volume surrounding injection site but over <1% total ventricular volume. CM internalized mitochondria at injection sites.
	Doulamis et al. (2020)	Zucker rat pectoralis major from *control or diabetic donors*; human cardiac fibroblasts	IRI per Langendorff; Zucker diabetic fatty rats (Crl:$Lepr^{fa}$ strain)	Isolated mitochondria from diabetic or control donors perfused to IRI-damaged diabetic hearts ↑ myocardial function, ↑ ATP content, and ↓ infarct size. Mitochondria from control donors provided a greater ATP boost than diabetic donors. Both donor sources similarly reduced infarct size.
	Rosengart et al. (2022)	Zucker rat pectoralis major from *control or diabetic donors*	In vivo IRI (surgically induced); Zucker diabetic fatty rats (Crl:$Lepr^{fa}$ strain)	In vivo transplantation of control and diabetic mitochondria ↑ systolic function and ↑ ATP content at 1 day and 4 weeks; ATP boost was greater from control than diabetic donors. Control and diabetic mitochondria similarly ↓ infarct size and ↓ fibrosis at 4 weeks. Control and diabetic mitochondria distinct transcriptomic and proteomic regulation changes, suggesting variable benefits from control versus diabetic donors.
	Louwagie et al. (2021)	Sprague–Dawley rat newborn ventricular tissue	In utero exposure to maternal diabetes (STZ-induced), high-fat diet, or the combination; newborn Sprague–Dawley rat primary CM	Rates of mitochondrial transfer to diabetes-exposed CM were higher in the first 4 h of coincubation compared to controls. Combination-exposed CM exhibited transfer-stimulated mitochondrial biogenesis. Internalized mitochondria underwent high levels of mitophagy by host lysosomes in all groups. Transfer ↓ mitophagy of host mitochondria, membrane potential, and rates of loss under stress. Transfer ↑↑ respiration in diabetes-exposed CM, particularly when also exposed to maternal high-fat diet; mtDNA appeared essential to this benefit. Mitochondrial transfer exhibited an antiapoptotic effect in male CM.
Brain	Ma et al. (2020)	Platelets from male Sprague–Dawley rats	Leptin receptor knock-out ($Lepr^{db/db}$) diabetic mice; hippocampal neurons	Transfer of isolated mitochondria was higher in diabetic versus control hippocampal neurons. One month later, ↑ mitochondrial volume, ↓ numbers of damaged mitochondria, ↑ ATP levels, ↓ free radical damage, ↓ apoptosis, ↑ numbers of hippocampal neurons, and ↑ *cognitive performance* (spatial learning and long-term memory).

(*Continued*)

TABLE 22.1 Studies evaluating the application of mitochondrial transplantation/transfer to diabetes mellitus.—cont'd

	Study	Mitochondrial source/Donor	Model; mitochondrial recipient/Host	Outcome summary/Therapeutic relevance
Kidney	Konari et al. (2019)	Renal PTEC; rat bone marrow-derived MSC	STZ-induced DN; Sprague−Dawley rats and C57BL/6 mice	In vitro, mitochondrial transfer from MSC ↓ apoptosis, ↑ expression of mitochondrial SOD2 and Bcl-2, ↓ ROS production, ↓ nuclear translocation of PGC-1α under high glucose. In vivo, mitochondrial transfer restored cellular morphology and structure of basement membrane and brush border in diabetic kidneys.
	Yuan et al. (2021)	MSC isolated from mouse bone marrow	STZ-induced diabetes; C57BL/6 mice; primary and RAW264.7 macrophages	Macrophages can transfer mitochondria from MSC or internalize isolated mitochondria; both routes ↑ macrophage polarization to M2-type (antiinflammatory). Transfer ↓ ROS, ↑ mitochondrial membrane potentials, ↑ ATP production, and ↓ serum levels of IL-1β and TNF-α. Infusion of M2 macrophages (who received mitochondria in vitro) ↓ systemic inflammation and alleviated DN. PGC-1α was required for therapeutic benefits.
Liver	Bi et al. (2021)	Mouse bone marrow−derived MSC	NAFLD from high-fat diet-induced diabetes; C57BL/6 mice; hepatocytes	In vivo, mitochondrial transfer ↑ mitochondrial numbers, improved mitochondrial structures, ↑ respiration, ↓ ROS, ↓ lipid peroxidation, and ultimately reversed hepatic steatosis and hypertransaminasemia, in essence healing NAFLD. In vitro controls demonstrated that transfer of whole intact mitochondria (rather than paracrine factors or a component of the vehicle solution) was needed for ↓ ROS and ↑ ATP production.
Pancreas	Rackham et al. (2020)	C57BL/6 mice; human adipose-derived MSC	Nondiabetic human and C57BL/6 mice	In coculture, mitochondrial transfer from MSC to human islet was highly selective for β-cells, ↑ respiration, and ↑ insulin sensitivity. Human islets had a higher propensity for accepting transfer than mouse islets. Hypoxia exposure ↑ mitochondrial transfer.

Bcl-2, B-cell lymphoma 2; *CM*, cardiomyocyte; *DN*, diabetic nephropathy; *IL-1β*, interleukin-1beta; *IRI*, ischemia−reperfusion injury; *MSC*, mesenchymal stem cell; *mtDNA*, mitochondrial DNA; *NAFLD*, nonalcoholic fatty liver disease; *PGC-1α*, peroxisome proliferator−activated receptor-gamma coactivator-1alpha; *PTEC*, proximal tubular epithelial cells; *ROS*, reactive oxygen species; *SOD2*, superoxide dismutase 2, mitochondrial; *STZ*, streptozotocin; *TNF-α*, tumor necrosis factor-alpha.

Mitochondrial transplantation in the diabetic heart

Cardiovascular disease (CVD) is the most common cause of death in patients with T2DM (Joseph et al., 2022). CVD encompasses heart failure, myocardial infarction, cerebrovascular disease, and a host of other conditions, and mitochondrial dysfunction and insulin resistance are central pathophysiological players (Bugger et al., 2014; Duncan, 2011a). The diabetic myocardium develops a characteristic pattern of myocardial fibrosis, structural remodeling, and diastolic followed by systolic dysfunction, termed diabetic cardiomyopathy (Jubaidi et al., 2020). Importantly, this process occurs in the absence of coronary artery disease or hypertension, suggesting direct effects on the myocardium. Interestingly, diabetic pregnancy incites similar pathogenic mechanisms in the developing fetal heart leading to clinically overlapping findings at birth (Kozák-Bárány et al., 2004; Ren et al., 2011; Zablah et al., 2017). Several investigations have demonstrated a key role for mitochondria in developmentally programmed heart disease found in offspring exposed to diabetic pregnancy, both in the newborn and adult time points (Depla et al., 2021; Louwagie, 2023; Stuart et al., 2013). As the contractile cells of the myocardium, cardiomyocytes (CMs) have a very high mitochondria make up of approximately 35% of CM volume (Page et al., 1973). It is not surprising then that mitochondria are essential players in CM development and contractile function (Dorn et al., 2015; Murphy et al., 2016), and even slight inefficiencies can have detrimental effects on cardiac structure and function (Ashrafian et al., 2007; Bugger et al., 2014; Fukushima et al., 2016).

As with diabetic cardiomyopathy, mitochondrial dysfunction is a major component in the pathology of myocardial infarction. Indeed, many studies show that mitochondria are mediators of ischemia-reperfusion injury (IRI) (Marin et al., 2021). ROS produced by mitochondria damage crucial proteins, lipids, and nucleic acids in CM, leading to cell death through mitochondria-mediated mechanisms including necrosis, apoptosis, and ferroptosis (Galluzzi et al., 2018; Gan, 2021). In theory, by limiting the spread of mitochondria-mediated damage in the setting of IRI, one may limit CM death and myocardial remodeling to reduce morbidity and mortality associated with IRI. Recent studies have investigated mitochondrial transplantation for diabetic cardiomyopathy (Durr, 2022), mitochondria-mediated IRI (Doulamis et al., 2020; Rosengart et al., 2022), and developmentally programmed heart disease (Louwagie et al., 2021).

Mitochondrial transplantation for diabetic cardiomyopathy

A 2022 exploratory study by Durr et al. investigated mitochondrial transfer as a preemptive treatment for mitochondrial dysfunction in the diabetic myocardium (Durr, 2022). Fluorescent mitochondria isolated from healthy mito::mKate2 mice (Barrasso et al., 2018) were transplanted into cultured HL-1 CM and hearts of wild-type and leptin receptor knock-out ($Lepr^{db/db}$) mice, a strain shown to develop severe hyperglycemia, hyperinsulinemia, and insulin resistance (Chua et al., 2002; Wang et al., 2005). Authors aimed to assess mitochondrial uptake, distribution, and localization of mitochondrial transplantation when injected directly into the left ventricular myocardium via ultrasound-guided echocardiography. Authors did not evaluate functional outcomes, such as by echocardiogram, electrocardiogram, or bioenergetics assays, but their results led to support for feasibility studies aimed at future therapeutic intervention.

In their study, Durr et al. isolated mitochondria from healthy, 25-week-old ventricular tissue from mKate2 mice by differential centrifugation, resuspended them in respiration buffer (Masuzawa et al., 2013), and coincubated the isolated mitochondria with mouse CM. After a nonstained control, CM were stained for host mitochondria and host cell membranes and imaged after 24 and 48 h of coincubation. Using videography, authors found that mouse CM superficially interacted with mKate2 mitochondria after 8 h, increasing up through 12 h coincubation, and retained this pattern for at least 24 h. Further, mKate2 mitochondria, or at least their protein components, were not internalized by 48 h but rather were still present on the CM membranes; authors found more success in vivo.

These in vivo experiments used three mice: two diabetic mice received myocardial injections of either mKate2 mitochondria or buffer placebo (no mitochondria) and one control mKate2 mouse received no injection. Immediately following injection by ultrasound-guided echocardiography, CT imaging confirmed presence of mKate2 mitochondria within the heart (Durr, 2022). 3D reconstruction demonstrated that injected mitochondria may reach nearly 50% volumetric coverage at or near injection sites, but this drops to less than 1% over the left ventricle's total volume. 24 h after mitochondria or placebo (vehicle) injection, hearts were histologically examined for cytoskeleton, nuclei, and host mitochondria. Authors found mKate2 mitochondria colocalized with host mitochondria, indicating internalization and possible host:transplant interactions.

These experiments suggest that mitochondria directly injected into the diabetic murine heart are internalized into the myocardium where they, or at least their protein components, colocalize with host mitochondria (Durr, 2022). However, this study was not without limitations. For one, this exploratory study used a small number of animals, and authors did not delineate differences in uptake between control and diabetic mouse hearts; more experiments are needed to confirm results and determine functional outcomes, such as systolic and diastolic function and rates of progression to symptomatic diabetic cardiomyopathy. Further, risks of these procedures must also be addressed, such as the potential for cardiac arrhythmias and higher mortality briefly mentioned by authors. If substantiated, such risks could discount or limit the therapeutic use of mitochondrial transplantation, at least in the treatment of diabetic cardiomyopathy.

Mitochondrial transplantation reduces ischemia–reperfusion injury

Of direct clinical relevance, recent studies by Doulamis et al. and Rosengart et al. have investigated the therapeutic potential of mitochondrial transplantation in reducing IRI in diabetic patients (Doulamis et al., 2020; Rosengart et al., 2022). It is known that patients with diabetes are not only more likely to suffer an acute myocardial infarction (AMI) (Peng et al., 2011), but they are also at greater risk of severe myocardial damage and mortality from IRI (Lejay et al., 2016). The standard of care following AMI is percutaneous coronary intervention or fibrinolysis to restore blood flow to the infarcted tissue and prevent further cardiac death (Jacobs et al., 2021). The McCully research group at Boston Children's Hospital in Massachusetts, USA, have demonstrated that mitochondrial transplantation alongside reperfusion provides additional benefit, i.e., it reduces the spread of necrosis and improves cardiac function in animal models (Emani et al., 2017; Masuzawa et al., 2013; McCully et al., 2009; Moskowitzova et al., 2020; Pacak et al., 2015). They found that mitochondrial transplants following severe IRI damage even helped transition pediatric patients off extracorporeal membranous oxygenation (Guariento et al., 2021). Driven by these findings, authors applied the principles of mitochondrial transplantation to a murine model of IRI in T2DM.

Doulamis et al. used Zucker diabetic obese rats (ZDF Crl:Lepr^{fa} strain (Shiota et al., 2012)) and Zucker lean rats to determine (1) whether mitochondrial transplantation could reduce IRI in animals with T2DM and (2) if mitochondria from healthy donors were more effective than those from diabetic donors (Doulamis et al., 2020). Using ex vivo

Langendorff perfusion, hearts from diabetic Zucker rats underwent warm global ischemia for 30 min and then were reperfused to replicate IRI. With reperfusion, hearts were given placebo (no mitochondria) or mitochondria isolated from skeletal muscle of diabetic or control Zucker rats via differential filtration (Preble et al., 2014). Cardiac function was assessed before and through induced ischemia and for 2 hours of reperfusion with or without mitochondrial transplant. Outcomes included cell death, inflammation, fibrosis, infarct size, and ATP and protein content. Cardiac function between diabetic and control Zucker rats was not different before IRI, and authors found that mitochondrial transplantation improved systolic function and diastolic function (Doulamis et al., 2020). Further, treatment significantly increased cardiac ATP content and decreased infarct size without increasing cardiac fibrosis. Remarkably, transplantation decreased the average infarct size from approximately 50% to 10% of the cardiac muscle mass.

Interestingly, mitochondria from both diabetic and control Zucker rats significantly reduced the detrimental effects of IRI, at least at the 120-minute timepoint evaluated. However, hearts receiving mitochondria from diabetic versus control rats did vary significantly in one parameter: ATP content. Diabetic hearts receiving mitochondria from control rats received the greatest boost (Doulamis et al., 2020). This was unsurprising considering the impaired mitochondrial ATP production in rodents and humans with diabetes (Minet et al., 2010; Raza et al., 2015), and it highlights the significant impact of diabetes on mitochondrial function. Although promising, further experiments must evaluate in vivo efficacy and at later timepoints. For example, various surgical approaches have been well described and may be used to induce IRI in living animals to examine both short- and long-term effects (Lindsey et al., 2018), and additional studies are needed to evaluate the mechanisms of benefit (Bertero et al., 2018, 2020a, 2020b; McCully et al., 2020).

Ongoing study results by the McCully group were recently presented at the American Association for Thoracic Surgery 2022 Annual Meeting (Rosengart et al., 2022). Specifically, the long(er)-term effects of mitochondrial transplantation for the treatment of IRI were investigated in *living* Zucker diabetic rats. Here, investigators induced 30 min of regional ischemia by snaring the left anterior descending artery, and then rats received either placebo injection or mitochondria isolated from diabetic or control Zucker rat skeletal muscle. Cardiac function was assessed by echocardiography at baseline, on postoperative day one (POD1), and then weekly from one to four weeks. Hearts were harvested on either POD1 or four weeks later for ATP content, RNA sequencing, and proteomic analyses. Similar to their earlier experiments via Langendorff apparatus (Doulamis et al., 2020), mitochondrial transplants improved ATP levels and systolic function, and mitochondrial source dictated tissue ATP content but not systolic function. Overall, transplant recipients of both sources demonstrated significantly improved recovery, evident from POD1 onwards, and at four weeks they had significantly smaller infarct sizes and lower levels of fibrosis than rats receiving placebo.

Although Rosengart et al. found similarly improved functional and structural improvements regardless of mitochondrial source, i.e., diabetic versus control donors, they did observe interesting differences in RNA and protein expression (Rosengart et al., 2022). On POD1, compared to recipients of "diabetic" mitochondria, recipients of control mitochondria had a more robust transcriptomic upregulation of pathways related to cardiac metabolism and downregulation of pathways related to immune response, proteolysis, DNA damage, and mitochondrial apoptosis. At four weeks, three pathways remained significantly upregulated, those related to muscle contraction, skeletal muscle development, and calcium transport. On both POD1 and four weeks later, recipients of control versus diabetic mitochondria also had distinct changes in proteomics that reflected these transcriptomic differences. While these latest results have yet to undergo full peer-review and should be taken with caution, they show promise in unraveling the mechanistic benefits of mitochondrial transplantation in the diabetic heart.

Mitochondrial transplantation in developmentally programmed heart disease

The consequences of diabetes extend beyond those directly afflicted. The offspring of pregnant women with either PGDM (i.e., T1DM or T2DM present before conception and throughout gestation) or GDM display higher rates of heart disease at birth and later in life (Agarwal et al., 2018; Louwagie, 2023), and these consequences may even extend multiple generations (Ferey et al., 2019; Saben et al., 2016). A series of studies from the Baack lab demonstrated mitochondria-mediated cardiac dysfunction in newborn and adult offspring exposed to diabetes during fetal life (Larsen et al., 2019; Louwagie et al., 2020; Mdaki et al., 2016; Preston et al., 2020; Upadhyaya et al., 2017). Late-gestation diabetes impaired mitochondria number, dynamics, membrane potential, and respiration in newborn offspring CM and a maternal high-fat diet exacerbated cardiac dysfunction and mortality through PGC-1α activation, higher mitochondria number, and oxidative stress (Larsen et al., 2019; Louwagie et al., 2020; Mdaki et al., 2016). Like in humans, cardiac disease in surviving newborns improved by weaning, but with aging diabetes-exposed offspring are unable to increase respiratory capacity to the same extent as controls, leading to exaggerated mitochondrial biogenesis between 6 and 12 months of age, oxidative stress, poor respiratory spare capacity, and faster cell death under metabolic stress (Louwagie et al., 2020). Recently, they extended their experimental model to include offspring exposed to PGDM and used mitochondrial transfer to (1) delineate the role of mitochondria in CM bioenergetics and cell fate and (2) determine if transplantation could improve mitochondrial function in exposed CM (Louwagie et al., 2021).

Louwagie et al. used a Sprague–Dawley rat model with streptozotocin-induced maternal PGDM, with or without addition of maternal high-fat diet, to study the effects of maternal hyperglycemia and hyperlipidemia and fetal hyperinsulinemia (classical triad of exposure in human diabetic pregnancies) on the newborn heart and cardiac mitochondria (Louwagie et al., 2021). Mitochondria were isolated from healthy newborn rat ventricles via differential centrifugation and transferred by coincubation to CM from control or exposed newborn offspring; outcome measures included mitochondrial uptake, mitophagy, and changes in CM bioenergetics, stress responses, and cell death. Unsurprisingly, exposure to PGDM, especially in combination with maternal high-fat diet, led to cardiac systolic dysfunction and decreased cardiac output in male offspring.

Authors found that diabetes-exposed CM exhibited faster rates of mitochondrial uptake in the first four hours of coincubation, and there was an increase in host mitochondrial count in combination-exposed CM, at least in part due to transfer-stimulated mitochondrial biogenesis

(Louwagie et al., 2021). Mitophagy is a vital process used to cull dysfunctional mitochondria to maintain respiration and prevent cell death (Kubli et al., 2012; Nah et al., 2017). In all groups, rates of mitophagy of donor mitochondria increased with internalization, and there was a concurrent decrease in mitophagy of host mitochondria, potentially reflecting turnover. CM bioenergetics were determined by extracellular flux analyses, and diabetes-exposed CM treated with mitochondrial transfer received respiratory boosts; this effect was most pronounced in CM exposed to both diabetes and maternal high-fat diet. Mitochondrial transfer did not affect glycolysis. To evaluate responses of CM to stress, i.e., rates of mitochondrial membrane potential loss and cell death, CMs with or without mitochondrial transplants were subjected to chemically induced oxidative stress (Louwagie et al., 2020). In both control and diabetes-exposed CM, mitochondrial transplants reduced baseline levels of membrane potential as well as the rate of loss. Most importantly, diabetes-exposed CM exhibited more apoptosis than controls (approximately 17 versus 41% cells), and mitochondrial transfer significantly reduced apoptosis in males (Louwagie et al., 2021). In a series of controlled experiments designed to rupture cell membranes and denature protein complexes of donor mitochondria, with or without further DNase treatment to degrade donor mitochondrial (mt)DNA, authors found that intact mtDNA was an important component of respiratory benefit.

This study by the Baack lab demonstrates that mitochondrial transfer may play a role in alleviating mitochondria-mediated developmental programming (Louwagie et al., 2021). However, additional in vivo, long-term studies are needed to test this hypothesis and determine if mitochondrial transplantation can reduce the long-term developmentally programmed effects of PGDM and GDM (Ferey et al., 2019; Louwagie, 2023; Louwagie et al., 2020; Saben et al., 2016). The decrease in CM death seen in their model may reflect the decreased cardiac necrosis in IRI-subjected hearts treated with mitochondrial transplantation (Cowan et al., 2016; Doulamis et al., 2020; Guariento et al., 2020; Kaza et al., 2017; Masuzawa et al., 2013; McCully et al., 2009), but further investigation is needed to confirm this relationship.

Mitochondrial transplantation in the diabetic brain

Diabetes is associated with neural cell damage and death in both the central and peripheral nervous systems, classically in a sequence of hyperglycemia generating advanced glycation end products which in turn generate ROS and neuron apoptosis (Agarwal et al., 2018). T2DM is a well-established risk factor for cognitive decline and dementia (Biessels et al., 2006; Livingston et al., 2017), and the hyperglycemia and hyperinsulinemia seen with T2DM have been linked to pathology that overlaps Alzheimer's disease (Baker et al., 2011; Matsuzaki et al., 2010; Silva et al., 2011; Wijesekara et al., 2017). The hyperlipidemia and impaired insulin signaling associated with T2DM impair mitochondria dynamics and mitophagy, essential functions in mitochondrial health, and cellular homeostasis in the nervous system (Luo et al., 2022; Sherkhane et al., 2021). Several techniques directly or indirectly targeting mitochondria are currently under study in the prevention and treatment of diabetes-associated cognitive decline (Mostafa Tork et al., 2019; Sekhar, 2021; Wang et al., 2017), and considering the pathologic centrality of mitochondrial dysfunction it comes as no surprise that mitochondrial transplantation is among them (Ma et al., 2020).

In Ma et al., authors hypothesized that transplantation of healthy mitochondria into the hippocampus of diabetic rats would improve mitochondrial function and reduce neuron apoptosis (Ma et al., 2020). Authors used adult $Lepr^{db/db}$ mice, a well-characterized model of hyperglycemia, obesity, and hippocampal atrophy leading to cognitive decline (Suriano et al., 2021; Zheng et al., 2017). Mitochondria were isolated from adult male Sprague–Dawley rat platelets by gradient centrifugation. Authors transplanted mitochondria via intracerebroventricular injection to control or diabetic mice, euthanized animals after one - day to evaluate mitochondrial internalization or aged recipients for one month to conduct behavioral testing before evaluation of neural and mitochondrial structure (via electron microscopy), apoptosis, oxidative stress, and ATP levels. Their results were promising. Both control and diabetic mice demonstrated hippocampal internalization of donor mitochondria, and uptake was higher in the diabetic mice, i.e., 9.4% versus 5.3% of structural proteins colocalized with donor mitochondria (Ma et al., 2020). Further, mitochondrial transplantation led to quantifiable histologic and cytologic changes. At one month, electron microscopy studies showed the hippocampal neurons of diabetic mice had fewer mitochondria filling less neuronal area and a significantly higher ratio of damaged mitochondria; mitochondrial transplantation improved both of these parameters nearly to control levels. The lower count and area of mitochondria in diabetic mice was reflected in expression of mitochondrial complexes I, III, and IV as well as lower ATP levels; these parameters were also corrected by mitochondrial transplantation. Unsurprisingly, diabetic hippocampal neurons had lower levels of ATP and higher levels of ROS with widespread free radical damage (per H_2O_2, malondialdehyde, nitrotyrosine, 8-hydroxydeoxyguanosine, and total antioxidant capacity) (Ma et al., 2020). These too were improved by mitochondrial transplantation. Similar to changes seen in CM, there were significantly higher levels of apoptosis and cleaved caspase 3 proteins in diabetic mice along with fewer total neurons suggesting more neuronal death in diabetic mice. Mitochondrial transplants decreased levels of apoptosis and increased neuron numbers by one month.

Two well-characterized histologic changes of Alzheimer's disease are the accumulation of extracellular amyloid-beta and intracellular hyperphosphorylated Tao protein (Elahi et al., 2017). In Ma et al.'s study, both proteins were significantly higher in hippocampal tissue of diabetic mice, and transplantation brought levels of each down, nearly to control levels (Ma et al., 2020). These changes were reflected in improved neurocognitive behavior testing. Compared to controls, diabetic mice exhibited markedly poorer spatial learning and long-term memory per Morris water maze testing; these were significantly improved by mitochondrial transplantation. Although focused on aging-related cognitive decline rather than diabetes, these results were supported by Zhang et al. who found enhanced hippocampal neuron signaling, increased neurogenesis, and alleviated cognitive impairment in aged mice following mitochondrial transplantation (Zhang et al., 2022). Collectively, results of Ma et al.'s study not only support the central roles of mitochondrial dysfunction in T2DM-associated cognitive decline but also demonstrate potential for improving dementia and cognitive decline due to Alzheimer's disease.

Mitochondrial transplantation in the diabetic kidney

Diabetic nephropathy (DN) or diabetic kidney disease is the leading cause of end-stage renal disease and affects approximately one-third to one-half of patients with diabetes (Hoogeveen, 2022; Reutens et al., 2011; Selby et al., 2020). Consequences of DN are severe, both reducing quality of life (Kim et al., 2022; Zimbudzi et al., 2020) and substantially shortening lives (Wen et al., 2017). Pathogenesis of DN is complex and not yet fully understood, but it is well-known that hyperglycemia and overproduction of ROS lead to vascular damage, podocytopathy, fibrosis, sclerosis, and tubular injury (Mottl & Tuttle, 2022; Sulaiman, 2019). This is clinically evident in reduced glomerular filtration rates, rising creatinine and blood urea nitrogen level, and electrolyte disturbances, all of which are components of diagnosis (Mottl et al., 2022). The ROS seen with DN also contribute to mitochondrial dysfunction, particularly in the renal tubular epithelial cells, leading to reduced ATP production and further decline in renal function (Bhargava et al., 2017; Zhang et al., 2021; Zhu et al., 2022). It has been previously shown that treatment with mesenchymal stem cells (MSCs) may improve DN (Zhu et al., 2022), possibly through mitochondrial transfer (Yasuda et al., 2011).

Konari et al. investigated the in vitro efficacy of mitochondrial transfer in alleviating DN and in vivo capacity for transplantation (Konari et al., 2019). Authors sought to determine whether mitochondrial transfer from healthy rat bone marrow–derived MSC to mouse renal proximal tubular epithelial cells (PTECs) could repair damage due to DN. Diabetes in experimental C57BL/6 mice was induced by streptozotocin (Damasceno et al., 2014). Study outcomes included PTEC apoptosis, ROS production, expression of PTEC transport proteins, and PTEC and tubular epithelial structure.

In vitro, MSC readily transferred mitochondria to diabetic and control PTEC with uptake noted as early as 12 h postcoculture and increasing for at least 24 h. Interestingly, compared to controls, a greater ratio of diabetic PTEC accepted mitochondria from MSC. Similarly, isolated mitochondria were internalized by diabetic and control PTEC and were internalized by a greater proportion of diabetic than control PTEC. Diabetic PTEC displayed slower colony growth, abnormal cytology, and deformed, shrunken nuclei compared to control PTEC, but these abnormalities were all reduced by coculture with MSC or coincubation with isolated mitochondria (Konari et al., 2019). When comparing numbers of apoptotic cells in culture via terminal deoxynucleotidyl transferase dUTP nick end labeling (TUNEL), the percentage of apoptotic diabetic PTEC was significantly higher than control PTEC (>40% versus <2% PTEC) but decreased to <10% when treated with MSC or isolated mitochondria. Further, no diabetic PTECs that had experienced mitochondrial transfer were TUNEL-positive. Additionally, mitochondrial transfer increased expression of superoxide dismutase 2 and B-cell-lymphoma protein 2 (Redza-Dutordoir et al., 2016), reduced total ROS levels, and improved expression of protein and glucose transporters under high glucose conditions. Collectively, these in vitro studies support an antiapoptotic effect of mitochondrial transfer in the diabetic kidney (Konari et al., 2019).

In preliminary experiments designed to determine the capacity for mitochondrial transplantation in vivo, Konari et al. injected isolated mitochondria under the renal capsule of diabetic or control rats and histologically characterized distribution patterns (Konari et al., 2019). Compared to controls, diabetic kidneys displayed widespread degeneration, disorganized brush border, nuclear shrinkage, and eosinophilic cytoplasmic structures within PTEC, consistent with changes seen in human patients with DN (Pourghasem et al., 2015). Injected mitochondria accumulated specifically in these regions rather than becoming widely internalized by all PTEC. In contrast, control kidneys did not accumulate injected mitochondria anywhere outside of the injection sites. Authors did not evaluate functional or therapeutic outcomes. Considering that (1) in vitro, diabetic PTEC accepted transfer of mitochondria from MSC and internalized isolated mitochondria at higher rates than controls, and (2) in vivo, transplanted mitochondria primarily accumulated in regions of degeneration or destruction, a major question emerges: Is there a "signal" molecule being released or a change in membrane receptor expression that triggers mitochondrial transfer/ transplant? This and many additional questions remain, but demonstrating the ability for mitochondrial transplantation in the diabetic kidney was necessary before evaluating potential therapeutic benefits.

Taking a different approach at alleviating DN, Yuan et al. sought to determine if mitochondrial transfer can increase numbers of M2-type macrophages, thereby employing their anti-inflammatory abilities to repair DN-damaged kidney (Mills, 2012; Yuan et al., 2021). Briefly, macrophages can be activated into a proinflammatory M1-type or an antiinflammatory M2-type in a process termed "polarization" (Santa et al., 2019). M2 macrophages are reliant on mitochondrial oxidative phosphorylation for energy production and their antiinflammatory roles include reducing ROS levels as well as promoting angiogenesis and healing (Mills, 2012; Santa et al., 2019). Yuan et al. previously found that mice with DN have increased rates of M1 polarization (Yuan et al., 2019), which sets up a positive cycle of inflammation. Thus, Yuan et al. hypothesized that mitochondria could be transferred from MSC to macrophages to improve mitochondrial function, promote polarization to the M2-type, and alleviate DN (Yuan et al., 2021).

Authors investigated their hypothesis using a C57BL/6 mouse model of streptozotocin-induced diabetes and MSC isolated from bone marrow (Yuan et al., 2021). In vitro, high-glucose conditions, as would be the case in diabetic circulation, efficiently stimulated mitochondrial transfer from MSC to isolated macrophages. Similarly, isolated mitochondria (from MSC) were internalized by an immortalized macrophage line, RAW264.7 cells, indicating that macrophages can either transfer (internalize) mitochondria from MSC or the surrounding environment. In vivo, MSCs with fluorometrically labeled mitochondria were injected into tail veins. 24 h later, these mitochondria were detected in peritoneal macrophages to confirm transfer. Mitochondrial transfer to macrophages significantly increased the proportion of M2 macrophages, which was reflected in decreased ROS production, higher membrane potential and ATP production, and decreased serum levels of the inflammatory markers interleukin-1beta (IL-1β) and tumor necrosis factor-alpha (TNF-α). These cellular changes were reaffirmed in RAW264.7 transfer recipients, which also had enhanced mitochondrial respiration (Yuan et al., 2021), demonstrating that mitochondrial transfer effectively triggers M2 polarization of macrophages.

To evaluate potential therapeutic effects, transfer-recipient M2 macrophages were infused into mice with DN (Yuan et al., 2021). This "treatment" not only reduced systemic inflammation (per reduced IL-1β and TNF-α), but it also decreased severity of DN as evidenced by lower creatinine, blood urea nitrogen, and urine albumin:creatinine ratio, restored mesangial and renal tubular structure, and attenuated renal fibrosis. Going one mechanistic step further, authors used a series of knockdown experiments to show that therapeutic benefits of mitochondrial transfer in DN required renal activation of PGC-1α, a well-known mediator of mitochondrial biogenesis (Fontecha-Barriuso et al., 2020).

Mitochondrial transplantation in the diabetic liver

The liver plays many essential synthetic and homeostatic roles, and hepatic mitochondrial dysfunction caused by T2DM leads to impaired fatty acid oxidation, lipid accumulation, and ROS production causing wide-spread cellular damage and apoptosis (Begriche et al., 2006; Nassir et al., 2014). This lipid accumulation is termed nonalcoholic fatty liver disease (NAFLD) and encompasses a spectrum of pathology ranging from simple steatosis to nonalcoholic steatohepatitis (NASH) to cirrhosis. Corresponding with the rise of T2DM (Khan et al., 2020), NAFLD now affects approximately 25% of our world's adults (Diehl et al., 2017). Diabetes is a major risk factor for developing NAFLD; over 70% of patients with diabetes have NAFLD, and over 40% of patients with NAFLD have diagnosed diabetes (Targher et al., 2021; Younossi et al., 2016). As in other diabetes-associated conditions resulting from mitochondrial dysfunction, mitochondria serve as a promising therapeutic target in the prevention and treatment of NAFLD (Li et al., 2020; Xu et al., 2021). A recent study by Bi et al. investigated this application of mitochondrial transplantation (Bi et al., 2021).

Bi et al. proposed that NAFLD could be alleviated by mitochondrial transfer from bone marrow–derived MSC (Bi et al., 2021). Authors used a T2DM mouse model which selected insulin-resistant mice fed high-fat (60% fat content) diet, and they conducted both in vivo and in vitro experiments to interrogate this hypothesis. In vivo, diet-induced T2DM led to severe insulin resistance, weight gain, hypertransaminasemia, hyperlipidemia, severe steatosis, reduced numbers of mitochondria with aberrant structure, poor respiratory function, impaired ATP production, and increased ROS levels with lipid peroxidation, as expected in a model of NAFLD (Pafili et al., 2021). Transplantation of mitochondria using their MSC vector did not completely restore the nondiabetic phenotype, but it did improve every one of these parameters, many to nondiabetic levels. Remarkably, transplantation completely reversed hepatic steatosis and hypertransaminasemia, reduced mitochondrial ROS production, and improved mitochondrial bioenergetics and calcium activity (Bi et al., 2021). In a series of in vitro controls, they used a mitochondrial filter to show that transfer of mitochondria, rather than paracrine factors or a component of the vehicle, is essential to reduce ROS levels and increase ATP production.

The findings of Bi et al. are supported by the work of Fu et al. (2017) and Hsu et al. (2020). These supporting studies did not focus specifically on diabetes (Fu et al. used a model of high-fat diet-induced NAFLD and Hsu et al. used a model of methionine-choline-deficient diet-induced NASH), but both groups similarly found that mitochondrial intravenous infusion reduces steatosis and oxidative injury in the liver while restoring function to hepatocytes

(Fu et al., 2017; Hsu et al., 2020). Taken together, the extensive experiments of Bi et al. demonstrate that mitochondrial transplantation in mice may reverse the hallmark pathology of diabetes-induced NAFLD (Bi et al., 2021).

Mitochondrial transplantation in pancreatic beta-islet cells

As discussed earlier in this chapter, the pathophysiology of T1DM is due to autoimmune destruction of pancreatic beta-islet cells leading to inability to produce insulin (Committee, 2021a). This creates a state of insulinopenia requiring lifelong insulin-based therapy to prevent hyperglycemia, diabetic ketoacidosis, and mortality (Committee, 2021b). Although it requires lifelong systemic immunosuppression due to its autoimmune nature, one potential "cure" for T1DM is pancreatic or islet transplant which allows for restored synthesis of endogenous insulin (Dean et al., 2017; Holt et al., 2021). Pancreatic transplantation is often done simultaneously with kidney transplant as this improves chances of success (Gruessner et al., 2012). That said, these treatments for T1DM have room for improvement, particularly islet transplantation (Brooks et al., 2013; Speight et al., 2010). Several studies have demonstrated that preculturing pancreatic islet with MSC results in improved islet function and transplantation outcomes (Arzouni et al., 2019; Jung et al., 2011; Lu et al., 2010; Rackham et al., 2013, 2014, 2018), and one recent study specifically investigated the role of mitochondrial transfer in these processes (Rackham et al., 2020).

One of the challenges of islet transplantation is that the needed culture period before surgery leads to a large degree of islet cell loss and a reduction in secretory function in the cells that do survive (Kin et al., 2008; Shapiro et al., 2017). In their 2020 study, Rackham et al. hypothesized that the in vitro benefits islets receive from coculture with MSC are due to mitochondrial transfer (Rackham et al., 2020). Indeed, this is reasonable considering that mitochondrial respiration is a strong predictor of islet transplantation outcomes (Papas et al., 2007; Sweet et al., 2008), and many studies have demonstrated mitochondrial transfer from MSC (Li et al., 2019a). In Rackham et al.'s study, islets were isolated from nondiabetic human donors, and both human and mouse MSC served as mitochondrial donors. Study outcomes included islet bioenergetics, insulin secretion, extent and rates of mitochondrial transfer from MSC to islets, differences in transfer by donor/recipient species, and effect of hypoxia on transfer.

In all reported outcomes, mitochondrial transfer conveyed benefits to recipient islets (Rackham et al., 2020). Bioenergetically, mitochondrial transfer provided significant respiratory boosts when cells were stimulated by glucose or subjected to chemical respiratory uncoupling. Functionally, insulin secretion in response to glucose was enhanced by mitochondrial transfer. Islet beta-cells demonstrated extensive transfer of mitochondria from MSC, and transfer was strongly selective toward human beta-cells relative to alpha-, delta-, or endothelial-cells. Mitochondrial transfer occurred through extracellular vesicles and MSC-derived tunneling nanotubes, and most of the mitochondrial transfer from MSC to beta-cells occurred in the first day; it increased up to two days of coculture when it plateaued at approximately 81% of beta-cells containing MSC mitochondria. Interestingly, authors included experiments to determine whether the donor/recipient species or levels of hypoxia affected transfer capacity of MSC. Human-to-human and mouse-to-human transfers were more extensive and more selective for beta-cells than mouse-to-mouse and human-to-mouse mitochondrial transfer, indicating species-specific beta-cell differences with human beta-cells having higher capacity or disposition for accepting mitochondria. Although done with mouse rather than human islets, preexposure to hypoxia enhanced mitochondrial transfer rates several fold (Rackham et al., 2020). Similar to other organs where mitochondrial transplantation was more selective for regions of damage or degeneration (Konari et al., 2019), islets exposed to hypoxia demonstrated enhanced transfer rates. These experiments by Rackham et al. support their hypothesis that mitochondrial transfer is responsible, at least in part, for the improved islet beta-cell survival, insulin secretion, and transplantation outcomes following pretransplant coincubation with MSC (Rackham et al., 2020).

Remaining questions and future applications for mitochondrial transplantation in the treatment of diabetes

These preclinical studies show great promise for mitochondrial transplantation in alleviating the systemic consequences of diabetes. Looking across tissues, several themes emerge and highlight remaining questions. In diabetes-related studies spanning the heart (Ma et al., 2020), kidney (Konari et al., 2019), and pancreas (Rackham et al., 2020), damaged or dysfunctional cells displayed a higher propensity to internalize donor mitochondria. Therefore, these damaged or diseased cells may be releasing or expressing some form of signaling molecule or protein that attracts mitochondria, a "mitoattractant" per se. In support of such a signal, mitochondrial transfer/transplantation has been enhanced in other conditions including hypoxia (Desir et al., 2016), inflammation (Luz-Crawford et al., 2019), oxygen–glucose deprivation (Li et al., 2019b), severe cellular stress (Wang et al., 2015), and even damaged mitochondria that were released from cells (Maeda et al., 2014; Mahrouf-Yorgov et al., 2017). Identifying the exact triggers and regulatory factors will not only guide further applications but also allow us to optimize transplant efficacy.

The studies discussed in this chapter support future preclinical and clinical study of mitochondrial transplantation in the context of diabetes, but there is no guarantee the therapeutic benefits seen in animal studies will translate to humans. For example, the study by Bi et al. demonstrated near complete resolution of the hallmark pathologic components of NAFLD (Bi et al., 2021), but this is (as of now) not possible in advanced cases like cirrhosis. On the other hand, due to unknown species-specific and or evolutionary differences, human cells may make even better recipients than animal models, such as in Rackham et al. (2020). To date, there exists one human trial utilizing mitochondrial transplantation, which fortunately benefitted enrolled patients (Guariento et al., 2021). However, there remain many important, unanswered questions as to the mechanisms of transplantation (Bertero et al., 2018) and the short- and long-term effects on both cellular and systemic levels. Additional information is needed to understand the effects of heteroplasmy and risks of "rejection" over aging. Additional questions remain regarding technical translation and applicability of this relatively new technique (Table 22.2), but there now exists an exciting foundation to guide the experiments needed to answer these questions.

TABLE 22.2 Remaining questions in the preclinical and clinical application of mitochondrial transplantation/transfer.

Remaining questions that may guide future study
Which is the superior "delivery" method of transplantation: Direct internalization of isolated mitochondria or transfer from a vector such as MSC?
Although there exist data demonstrating no immunologic responses against injected syngeneic and allogeneic mitochondria (Ramirez-Barbieri et al., 2019), in the context of T1DM and its underlying autoimmune pathophysiology, would these patients face greater risk of reacting to mitochondria therapy?
Considering the contrasting pathogeneses of T1DM and T2DM, when and to which organ systems would treatments be catered to each disease?
The inclusion of pregnancy in any trial adds several ethical hurdles; assuming in vivo experiments show benefit of mitochondrial therapy in the treatment of developmentally programmed heart disease, when would be the optimal (gestational) age for preventing heart disease in offspring? Would the fetus be treated directly, or indirectly through treatment of the mother?
Mitochondrial dysfunction with diabetes is systemic; in the case of mitochondrial transplants to diabetic patients, what characteristics (e.g., age, sex, health status) would be sought in the optimal (nondiabetic) donor, and from what tissue source?
Considering the mounting studies demonstrating mitochondrial transplantation/transfer in tissues across the body, how could transfer/transplants be catered to specific cell types or tissues? If mitochondria were perfused rather than directly injected into tissue, would the potential for transfer to unintended tissues/cell types limit this application route?
If transplanted mitochondria accumulate primarily in damaged or diseased tissue, e.g., DN (Konari et al., 2019), would transplantation be more effective in earlier stages of disease? In other words, would transplantation be better at preventing *progression* of disease rather than in treating *already-damaged* tissue?

References

Agarwal, P., et al. (2018). Maternal obesity, diabetes during pregnancy and epigenetic mechanisms that influence the developmental origins of cardiometabolic disease in the offspring. *Critical Reviews in Clinical Laboratory Sciences, 55*(2), 71–101.

Anderson, A. J., et al. (2019). Mitochondria—Hubs for regulating cellular biochemistry: Emerging concepts and networks. *Open Biology, 9*(8), 190126.

Aon, M. A., Bhatt, N., & Cortassa, S. C. (2014). Mitochondrial and cellular mechanisms for managing lipid excess. *Frontiers in Physiology, 5*, 282.

Arzouni, A. A., et al. (2019). Characterization of the effects of mesenchymal stromal cells on mouse and human islet function. *Stem Cells Translational Medicine, 8*(9), 935–944.

Ashrafian, H., Frenneaux, M. P., & Opie, L. H. (2007). Metabolic mechanisms in heart failure. *Circulation, 116*(4), 434–448.

Atkinson, M. A., & Maclaren, N. K. (1994). The pathogenesis of insulin-dependent diabetes mellitus. *New England Journal of Medicine, 331*(21), 1428–1436.

Baker, L. D., et al. (2011). Insulin resistance and Alzheimer-like reductions in regional cerebral glucose metabolism for cognitively normal adults with prediabetes or early type 2 diabetes. *Archives of Neurology, 68*(1), 51–57.

Barrasso, A. P., Tong, X., & Poché, R. A. (2018). The mito::mKate2 mouse: A far-red fluorescent reporter mouse line for tracking mitochondrial dynamics in vivo. *Genesis, 56*(2).

Begriche, K., et al. (2006). Mitochondrial dysfunction in NASH: Causes, consequences and possible means to prevent it. *Mitochondrion, 6*(1), 1–28.

Bertero, E., O'Rourke, B., & Maack, C. (2020b). Mitochondria do not survive calcium overload during transplantation. *Circulation Research, 126*(6), 784–786.

Bertero, E., O'Rourke, B., Maack, C., & Response by Bertero et al to Letter Regarding Article. (2020a). Mitochondria do not survive calcium overload. *Circulation Research, 126*(8), e58–e59.

Bertero, E., Maack, C., & O'Rourke, B. (2018). Mitochondrial transplantation in humans: "magical" cure or cause for concern? *Journal of Clinical Investigation, 128*(12), 5191–5194.

Bhargava, P., & Schnellmann, R. G. (2017). Mitochondrial energetics in the kidney. *Nature Reviews Nephrology, 13*(10), 629–646.

Bi, Y., et al. (2021). Bone marrow derived-mesenchymal stem cell improves diabetes-associated fatty liver via mitochondria transformation in mice. *Stem Cell Research and Therapy, 12*(1), 602.

Biessels, G. J., et al. (2006). Risk of dementia in diabetes mellitus: A systematic review. *The Lancet Neurology, 5*(1), 64–74.

Boudina, S., et al. (2007). Mitochondrial energetics in the heart in obesity-related diabetes: Direct evidence for increased uncoupled respiration and activation of uncoupling proteins. *Diabetes, 56*(10), 2457–2466.

Brooks, A. M., et al. (2013). Attainment of metabolic goals in the integrated UK islet transplant program with locally isolated and transported preparations. *American Journal of Transplantation, 13*(12), 3236–3243.

Bugger, H., & Abel, E. D. (2014). Molecular mechanisms of diabetic cardiomyopathy. *Diabetologia, 57*(4), 660–671.

Buse, J. B., et al. (2019). Update to: Management of hyperglycemia in type 2 diabetes, 2018. A consensus report by the American Diabetes Association (ADA) and the European Association for the Study of Diabetes (EASD). *Diabetes Care, 43*(2), 487–493, 2020.

Cerf, M. E. (2018). Cardiac glucolipotoxicity and cardiovascular outcomes. *Medicina (Kaunas), 54*(5).

Chua, S., Jr., et al. (2002). Differential beta cell responses to hyperglycaemia and insulin resistance in two novel congenic strains of diabetes (FVB-Lepr (db)) and obese (DBA-Lep (ob)) mice. *Diabetologia, 45*(7), 976–990.

Committee, A. D. A. P. P. (2021a). Classification and diagnosis of diabetes: Standards of medical care in diabetes—2022. *Diabetes Care, 45*(Suppl_1), S17–S38. *2.*

Committee, A. D. A. P. P. (2021c). Management of diabetes in pregnancy: Standards of medical care in diabetes—2022. *Diabetes Care, 45*(Suppl_1), S232–S243. *15.*

Committee, A. D. A. P. P. (2021b). Pharmacologic approaches to glycemic treatment: Standards of medical care in diabetes—2022. *Diabetes Care, 45*(Suppl_1), S125–S143, 9.

Cowan, D. B., et al. (2016). Intracoronary delivery of mitochondria to the ischemic heart for cardioprotection. *PLoS One, 11*(8), e0160889.

Damasceno, D. C., et al. (2014). Streptozotocin-induced diabetes models: Pathophysiological mechanisms and fetal outcomes. *BioMed Research International*, 819065, 2014.

Davies, M. J., et al. (2018). Management of hyperglycemia in type 2 diabetes, 2018. A consensus report by the American Diabetes Association (ADA) and the European Association for the Study of Diabetes (EASD). *Diabetes Care, 41*(12), 2669–2701.

Dean, P. G., et al. (2017). Pancreas transplantation. *BMJ, 357*, j1321.

Depla, A. L., et al. (2021). Effect of maternal diabetes on fetal heart function on echocardiography: Systematic review and meta-analysis. *Ultrasound in Obstetrics and Gynecology, 57*(4), 539–550.

Desir, S., et al. (2016). Tunneling nanotube formation is stimulated by hypoxia in ovarian cancer cells. *Oncotarget, 7*(28), 43150–43161.

Diehl, A. M., & Day, C. (2017). Cause, pathogenesis, and treatment of nonalcoholic steatohepatitis. *New England Journal of Medicine, 377*(21), 2063–2072.

Dorn, G. W., 2nd, Vega, R. B., & Kelly, D. P. (2015). Mitochondrial biogenesis and dynamics in the developing and diseased heart. *Genes and Development, 29*(19), 1981–1991.

Doulamis, I. P., et al. (2020). Mitochondrial transplantation for myocardial protection in diabetic hearts. *European Journal of Cardio-Thoracic Surgery, 57*(5), 836–845.

Duncan, J. G. (2011a). Mitochondrial dysfunction in diabetic cardiomyopathy. *Biochimica et Biophysica Acta, 1813*(7), 1351–1359.

Duncan, J. G. (2011b). Peroxisome proliferator activated receptor-alpha (PPARα) and PPAR gamma coactivator-1alpha (PGC-1α) regulation of cardiac metabolism in diabetes. *Pediatric Cardiology, 32*(3), 323–328.

Durr, A. (2022). Amelioration of mitochondrial bioenergetic dysfunction in diabetes mellitus: Delving into specialized and non-specific therapeutics for the ailing heart. In *Exercise physiology*. West Virginia University: Online.

Elahi, F. M., & Miller, B. L. (2017). A clinicopathological approach to the diagnosis of dementia. *Nature Reviews Neurology, 13*(8), 457–476.

Emani, S. M., et al. (2017). Autologous mitochondrial transplantation for dysfunction after ischemia-reperfusion injury. *The Journal of Thoracic and Cardiovascular Surgery, 154*(1), 286–289.

Eshaq, R. S., Wright, W. S., & Harris, N. R. (2014). Oxygen delivery, consumption, and conversion to reactive oxygen species in experimental models of diabetic retinopathy. *Redox Biology, 2*, 661–666.

Ferey, J. L. A., et al. (2019). A maternal high-fat, high-sucrose diet induces transgenerational cardiac mitochondrial dysfunction independently of maternal mitochondrial inheritance. *American Journal of Physiology—Heart and Circulatory Physiology, 316*(5), H1202–h1210.

Fontecha-Barriuso, M., et al. (2020). The role of PGC-1α and mitochondrial biogenesis in kidney diseases. *Biomolecules, 10*(2).

Fu, A., et al. (2017). Mitotherapy for fatty liver by intravenous administration of exogenous mitochondria in male mice. *Frontiers in Pharmacology, 8*, 241.

Fukushima, A., & Lopaschuk, G. D. (2016). Cardiac fatty acid oxidation in heart failure associated with obesity and diabetes. *Biochimica et Biophysica Acta, 1861*(10), 1525–1534.

Galluzzi, L., et al. (2018). Molecular mechanisms of cell death: Recommendations of the nomenclature committee on cell death 2018. *Cell Death and Differentiation, 25*(3), 486–541.

Gan, B. (2021). Mitochondrial regulation of ferroptosis. *The Journal of Cell Biology, 220*(9).

Gruessner, A. C., Sutherland, D. E., & Gruessner, R. W. (2012). Long-term outcome after pancreas transplantation. *Current Opinion in Organ Transplantation, 17*(1), 100–105.

Guariento, A., et al. (2020). Preischemic autologous mitochondrial transplantation by intracoronary injection for myocardial protection. *The Journal of Thoracic and Cardiovascular Surgery, 160*(2), e15–e29.

Guariento, A., et al. (2021). Autologous mitochondrial transplantation for cardiogenic shock in pediatric patients following ischemia-reperfusion injury. *The Journal of Thoracic and Cardiovascular Surgery, 162*(3), 992–1001.

Holt, R. I. G., et al. (2021). The management of type 1 diabetes in adults. A consensus report by the American Diabetes Association (ADA) and the European Association for the Study of Diabetes (EASD). *Diabetes Care, 44*(11), 2589–2625.

Hoogeveen, E. K. (2022). The epidemiology of diabetic kidney disease. *Kidney and Dialysis, 2*(3), 433–442.

Hsu, M.-J., et al. (2020). Mitochondrial transfer by human mesenchymal stromal cells ameliorates hepatocyte lipid load in a mouse model of NASH. *Biomedicines, 8*(9), 350.

Jacobs, A. K., et al. (2021). Systems of care for ST-segment-elevation myocardial infarction: A policy statement from the American heart association. *Circulation, 144*(20), e310–e327.

Joseph, J. J., et al. (2022). Comprehensive management of cardiovascular risk factors for adults with type 2 diabetes: A scientific statement from the American heart association. *Circulation, 145*(9), e722–e759.

Jubaidi, F. F., et al. (2020). Mitochondrial dysfunction in diabetic cardiomyopathy: The possible therapeutic roles of phenolic acids. *International Journal of Molecular Sciences, 21*(17).

Jung, E. J., et al. (2011). Bone marrow-derived mesenchymal stromal cells support rat pancreatic islet survival and insulin secretory function in vitro. *Cytotherapy, 13*(1), 19–29.

Karakelides, H., et al. (2007). Effect of insulin deprivation on muscle mitochondrial ATP production and gene transcript levels in type 1 diabetic subjects. *Diabetes, 56*(11), 2683–2689.

Kaza, A. K., et al. (2017). Myocardial rescue with autologous mitochondrial transplantation in a porcine model of ischemia/reperfusion. *The Journal of Thoracic and Cardiovascular Surgery, 153*(4), 934–943.

Khan, M. A. B., et al. (2020). Epidemiology of type 2 diabetes—Global burden of disease and forecasted trends. *Journal of Epidemiology and Global Health, 10*(1), 107–111.

Kim, H., et al. (2022). Quality of life in patients with diabetic nephropathy: Findings from the KNOW-CKD (Korean cohort study for outcomes in patients with chronic kidney disease) cohort. *Kidney Research and Clinical Practice, 41*(1), 43–57.

Kim-Campbell, N., Gomez, H., & Bayir, H. (2019). Chapter 20—Cell death pathways: Apoptosis and regulated necrosis. In C. Ronco, et al. (Eds.), *Critical care nephrology* (3rd ed., pp. 113–121.e2). Philadelphia: Elsevier.

Kin, T., et al. (2008). Risk factors for islet loss during culture prior to transplantation. *Transplant International, 21*(11), 1029–1035.

Kodde, I. F., et al. (2007). Metabolic and genetic regulation of cardiac energy substrate preference. *Comparative Biochemistry and Physiology Part A: Molecular and Integrative Physiology, 146*(1), 26–39.

Konari, N., et al. (2019). Mitochondria transfer from mesenchymal stem cells structurally and functionally repairs renal proximal tubular epithelial cells in diabetic nephropathy in vivo. *Scientific Reports, 9*(1), 5184.

Kozák-Bárány, A., et al. (2004). Impaired left ventricular diastolic function in newborn infants of mothers with pregestational or gestational diabetes with good glycemic control. *Early Human Development, 77*(1–2), 13–22.

Kubli, D. A., & Gustafsson, B., Å (2012). Mitochondria and mitophagy: The yin and yang of cell death control. *Circulation Research, 111*(9), 1208–1221.

Larsen, T. D., et al. (2019). Diabetic pregnancy and maternal high-fat diet impair mitochondrial dynamism in the developing fetal rat heart by sex-specific mechanisms. *International Journal of Molecular Sciences, 20*(12).

Lejay, A., et al. (2016). Ischemia reperfusion injury, ischemic conditioning and diabetes mellitus. *Journal of Molecular and Cellular Cardiology, 91*, 11–22.

Li, C., et al. (2019a). Mesenchymal stem cells and their mitochondrial transfer: A double-edged sword. *Bioscience Reports, 39*(5).

Li, H., et al. (2019b). Mitochondrial transfer from bone marrow mesenchymal stem cells to motor neurons in spinal cord injury rats via gap junction. *Theranostics, 9*(7), 2017–2035.

Li, R., Toan, S., & Zhou, H. (2020). Role of mitochondrial quality control in the pathogenesis of nonalcoholic fatty liver disease. *Aging (Albany NY), 12*(7), 6467–6485.

Lindsey, M. L., et al. (2018). Guidelines for experimental models of myocardial ischemia and infarction. *American Journal of Physiology—Heart and Circulatory Physiology, 314*(4), H812–H838.

Livingston, G., et al. (2017). Dementia prevention, intervention, and care. *The Lancet, 390*(10113), 2673–2734.

Louwagie, E. J. (2023). *Doctoral thesis: Mitochondrial roles in developmentally programmed heart disease.* USD RED.

Louwagie, E. J., et al. (2020). Age and sex influence mitochondria and cardiac health in offspring exposed to maternal glucolipotoxicity. *iScience, 23*(11), 101746.

Louwagie, E. J., et al. (2021). Mitochondrial transfer improves cardiomyocyte bioenergetics and viability in male rats exposed to pregestational diabetes. *International Journal of Molecular Sciences, 22*(5).

Lu, Y., et al. (2010). Mesenchymal stem cells protect islets from hypoxia/reoxygenation-induced injury. *Cell Biochemistry and Function, 28*(8), 637–643.

Luo, J. S., et al. (2022). The role of mitochondrial quality control in cognitive dysfunction in diabetes. *Neurochemical Research, 47*(8), 2158–2172.

Luz-Crawford, P., et al. (2019). Mesenchymal stem cell repression of Th17 cells is triggered by mitochondrial transfer. *Stem Cell Research and Therapy, 10*(1), 232.

Ma, H., et al. (2020). Transplantation of platelet-derived mitochondria alleviates cognitive impairment and mitochondrial dysfunction in db/db mice. *Clinical Science, 134*(16), 2161–2175.

Maeda, A., & Fadeel, B. (2014). Mitochondria released by cells undergoing TNF-α-induced necroptosis act as danger signals. *Cell Death and Disease, 5*(7). e1312-e1312.

Mahrouf-Yorgov, M., et al. (2017). Mesenchymal stem cells sense mitochondria released from damaged cells as danger signals to activate their rescue properties. *Cell Death and Differentiation, 24*(7), 1224–1238.

Marin, W., et al. (2021). Mitochondria as a therapeutic target for cardiac ischemia-reperfusion injury (review). *International Journal of Molecular Medicine, 47*(2), 485–499.

Masuzawa, A., et al. (2013). Transplantation of autologously derived mitochondria protects the heart from ischemia-reperfusion injury. *American Journal of Physiology—Heart and Circulatory Physiology, 304*(7), H966–H982.

Matsuzaki, T., et al. (2010). Insulin resistance is associated with the pathology of Alzheimer disease: The Hisayama study. *Neurology, 75*(9), 764–770.

McCully, J. D., et al. (2009). Injection of isolated mitochondria during early reperfusion for cardioprotection. *American Journal of Physiology—Heart and Circulatory Physiology, 296*(1), H94–H105.

McCully, J. D., Emani, S. M., Del Nido, P. J., & Letter by McCully et al Regarding Article. (2020). Mitochondria do not survive calcium overload. *Circulation Research, 126*(8), e56–e57.

Mdaki, K. S., et al. (2016). Maternal high-fat diet impairs cardiac function in offspring of diabetic pregnancy through metabolic stress and mitochondrial dysfunction. *American Journal of Physiology—Heart and Circulatory Physiology, 310*(6), H681–H692.

Meyts, P. D. (2016). The insulin receptor and its signal transduction network. In K. R. Feingold, B. Anawalt, & M. R. Blackman (Eds.), *Endotext [internet].* NIH/NLM/NCBI Books: Online.

Mills, C. D. (2012). M1 and M2 macrophages: Oracles of health and disease. *Critical Reviews in Immunology, 32*(6), 463–488.

Minet, A. D., & Gaster, M. (2010). ATP synthesis is impaired in isolated mitochondria from myotubes established from type 2 diabetic subjects. *Biochemical and Biophysical Research Communications, 402*(1), 70–74.

Montaigne, D., et al. (2014). Myocardial contractile dysfunction is associated with impaired mitochondrial function and dynamics in type 2 diabetic but not in obese patients. *Circulation, 130*(7), 554–564.

Mootha, V. K., et al. (2003). PGC-1α-responsive genes involved in oxidative phosphorylation are coordinately downregulated in human diabetes. *Nature Genetics, 34*(3), 267–273.

Moskowitzova, K., et al. (2020). Mitochondrial transplantation enhances murine lung viability and recovery after ischemia-reperfusion injury. *American Journal of Physiology—Lung Cellular and Molecular Physiology, 318*(1), L78–L88.

Mostafa Tork, O., et al. (2019). Targeting altered mitochondrial biogenesis in the brain of diabetic rats: Potential effect of pioglitazone and exendin-4. *Reports of Biochemistry and Molecular Biology, 8*(3), 287–300.

Mottl, A. K., & Tuttle, K. R. (2022). Diabetic kidney disease: Pathogenesis and epidemiology. In J. P. Forman (Ed.), *UpToDate*. UpToDate: Online.

Mottl, A. K., Tuttle, K. R., & Bakris, G. L. (2022). Diabetic kidney disease: Manifestations, evaluation, and diagnosis. In J. P. Forman (Ed.), *UpToDate*. UpToDate: Online.

Murphy, E., et al. (2016). Mitochondrial function, biology, and role in disease: A scientific statement from the American heart association. *Circulation Research, 118*(12), 1960–1991.

Murphy, M. P. (2009). How mitochondria produce reactive oxygen species. *Biochemical Journal, 417*(1), 1–13.

Nah, J., Miyamoto, S., & Sadoshima, J. (2017). Mitophagy as a protective mechanism against myocardial stress. *Comprehensive Physiology, 7*(4), 1407–1424.

Nassir, F., & Ibdah, J. A. (2014). Role of mitochondria in nonalcoholic fatty liver disease. *International Journal of Molecular Sciences, 15*(5), 8713–8742.

Nishikawa, T., et al. (2000). Normalizing mitochondrial superoxide production blocks three pathways of hyperglycaemic damage. *Nature, 404*(6779), 787–790.

Pacak, C. A., et al. (2015). Actin-dependent mitochondrial internalization in cardiomyocytes: Evidence for rescue of mitochondrial function. *Biology Open, 4*(5), 622–626.

Pafili, K., & Roden, M. (2021). Nonalcoholic fatty liver disease (NAFLD) from pathogenesis to treatment concepts in humans. *Molecular Metabolism, 50*, 101122.

Page, E., & McCallister, L. P. (1973). Quantitative electron microscopic description of heart muscle cells. Application to normal, hypertrophied and thyroxin-stimulated hearts. *The American Journal of Cardiology, 31*(2), 172–181.

Papas, K. K., et al. (2007). Human islet oxygen consumption rate and DNA measurements predict diabetes reversal in nude mice. *American Journal of Transplantation, 7*(3), 707–713.

Patti, M. E., et al. (2003). Coordinated reduction of genes of oxidative metabolism in humans with insulin resistance and diabetes: Potential role of PGC1 and NRF1. *Proceedings of the National Academy of Sciences of the United States of America, 100*(14), 8466–8471.

Peng, X. R., et al. (2011). The role of diabetes mellitus as a risk factor of acute myocardial infarction. *Zhongguo Wei Zhong Bing Ji Jiu Yi Xue, 23*(6), 322–328.

Pourghasem, M., Shafi, H., & Babazadeh, Z. (2015). Histological changes of kidney in diabetic nephropathy. *Caspian Journal of Internal Medicine, 6*(3), 120–127.

Preble, J. M., et al. (2014). Rapid isolation and purification of mitochondria for transplantation by tissue dissociation and differential filtration. *Journal of Visualized Experiments: JoVE*, (91), e51682.

Preston, C. C., et al. (2020). Maternal high fat diet and diabetes disrupts transcriptomic pathways that regulate cardiac metabolism and cell fate in newborn rat hearts. *Frontiers in Endocrinology, 11*, 570846.

Rackham, C. L., et al. (2018). Mesenchymal stromal cell secretory factors induce sustained improvements in islet function pre- and post-transplantation. *Cytotherapy, 20*(12), 1427–1436.

Rackham, C. L., et al. (2020). Optimizing beta cell function through mesenchymal stromal cell-mediated mitochondria transfer. *Stem Cells, 38*(4), 574–584.

Rackham, C. L., et al. (2013). Pre-culturing islets with mesenchymal stromal cells using a direct contact configuration is beneficial for transplantation outcome in diabetic mice. *Cytotherapy, 15*(4), 449–459.

Rackham, C. L., et al. (2014). Preculturing islets with adipose-derived mesenchymal stromal cells is an effective strategy for improving transplantation efficiency at the clinically preferred intraportal site. *Cell Medicine, 7*(1), 37–47.

Ramirez-Barbieri, G., et al. (2019). Alloreactivity and allorecognition of syngeneic and allogeneic mitochondria. *Mitochondrion, 46*, 103–115.

Raza, H., John, A., & Howarth, F. C. (2015). Increased oxidative stress and mitochondrial dysfunction in zucker diabetic rat liver and brain. *Cellular Physiology and Biochemistry, 35*(3), 1241–1251.

Redza-Dutordoir, M., & Averill-Bates, D. A. (2016). Activation of apoptosis signalling pathways by reactive oxygen species. *Biochimica et Biophysica Acta (BBA)—Molecular Cell Research, 1863*(12), 2977–2992.

Ren, Y., et al. (2011). Characterization of fetal cardiac structure and function detected by echocardiography in women with normal pregnancy and gestational diabetes mellitus. *Prenatal Diagnosis, 31*(5), 459–465.

Reutens, A. T., & Atkins, R. C. (2011). Epidemiology of diabetic nephropathy. *Contributions to Nephrology, 170*, 1–7.

Rosengart, T., et al. (2022). L4. Mitochondrial transplantation following ischemia-reperfusion injury in a transgenic model of type 2 diabetes enhances upregulation of mitochondrial dependent pathways of cardiac and muscle metabolism. In *The American association for thoracic surgery 102nd annual meeting* (Boston, MA).

Rovira-Llopis, S., et al. (2018). Mitochondria, the NLRP3 inflammasome, and sirtuins in type 2 diabetes: New therapeutic targets. *Antioxidants and Redox Signaling, 29*(8), 749–791.

Ruegsegger, G. N., et al. (2018). Altered mitochondrial function in insulin-deficient and insulin-resistant states. *Journal of Clinical Investigation, 128*(9), 3671–3681.

Saben, J. L., et al. (2016). Maternal metabolic syndrome programs mitochondrial dysfunction via germline changes across three generations. *Cell Reports, 16*(1), 1–8.

Santa, F. D., et al. (2019). The role of metabolic remodeling in macrophage polarization and its effect on skeletal muscle regeneration. *Antioxidants and Redox Signaling, 30*(12), 1553–1598.

Schrauwen, P., & Hesselink, M. K. C. (2004). Oxidative capacity, lipotoxicity, and mitochondrial damage in type 2 diabetes. *Diabetes, 53*(6), 1412–1417.

Seifert, E. L., et al. (2010). Electron transport chain-dependent and -independent mechanisms of mitochondrial H2O2 emission during long-chain fatty acid oxidation. *Journal of Biological Chemistry, 285*(8), 5748–5758.

Sekhar, R. V. (2021). GlyNAC supplementation improves glutathione deficiency, oxidative stress, mitochondrial dysfunction, inflammation, aging hallmarks, metabolic defects, muscle strength, cognitive decline, and body composition: Implications for healthy aging. *Journal of Nutrition, 151*(12), 3606–3616.

Selby, N. M., & Taal, M. W. (2020). An updated overview of diabetic nephropathy: Diagnosis, prognosis, treatment goals and latest guidelines. *Diabetes, Obesity and Metabolism, 22*(S1), 3–15.

Shaham, O., et al. (2008). Metabolic profiling of the human response to a glucose challenge reveals distinct axes of insulin sensitivity. *Molecular Systems Biology, 4*, 214.

Shapiro, A. M., Pokrywczynska, M., & Ricordi, C. (2017). Clinical pancreatic islet transplantation. *Nature Reviews Endocrinology, 13*(5), 268–277.

Sherkhane, B., et al. (2021). Mitochondrial remodelling-a vicious cycle in diabetic complications. *Molecular Biology Reports, 48*(5), 4721–4731.

Shiota, M., & Printz, R. L. (2012). Diabetes in Zucker diabetic fatty rat. *Methods in Molecular Biology, 933*, 103–123.

Silva, D. F., et al. (2011). Mitochondria: The common upstream driver of amyloid-β and tau pathology in Alzheimer's disease. *Current Alzheimer Research, 8*(5), 563–572.

Speight, J., et al. (2010). Patient-reported outcomes following islet cell or pancreas transplantation (alone or after kidney) in type 1 diabetes: A systematic review. *Diabetic Medicine, 27*(7), 812–822.

Stuart, A., et al. (2013). Long-term cardiovascular risk in relation to birth weight and exposure to maternal diabetes mellitus. *International Journal of Cardiology, 168*(3), 2653–2657.

Sulaiman, M. K. (2019). Diabetic nephropathy: Recent advances in pathophysiology and challenges in dietary management. *Diabetology and Metabolic Syndrome, 11*(1), 7.

Suriano, F., et al. (2021). Novel insights into the genetically obese (ob/ob) and diabetic (db/db) mice: Two sides of the same coin. *Microbiome, 9*(1), 147.

Sweet, I. R., et al. (2008). Glucose-stimulated increment in oxygen consumption rate as a standardized test of human islet quality. *American Journal of Transplantation, 8*(1), 183–192.

Targher, G., et al. (2021). The complex link between NAFLD and type 2 diabetes mellitus—Mechanisms and treatments. *Nature Reviews Gastroenterology and Hepatology, 18*(9), 599–612.

Upadhyaya, B., et al. (2017). Prenatal exposure to a maternal high-fat diet affects histone modification of cardiometabolic genes in newborn rats. *Nutrients, 9*(4).

Vakifahmetoglu-Norberg, H., Ouchida, A. T., & Norberg, E. (2017). The role of mitochondria in metabolism and cell death. *Biochemical and Biophysical Research Communications, 482*(3), 426–431.

Wang, X., & Gerdes, H. H. (2015). Transfer of mitochondria via tunneling nanotubes rescues apoptotic PC12 cells. *Cell Death and Differentiation, 22*(7), 1181–1191.
Wang, Y., et al. (2017). Neuroprotective effect and mechanism of thiazolidinedione on dopaminergic neurons in vivo and in vitro in Parkinson's disease. *PPAR Research, 2017*, 4089214.
Wang, Z., et al. (2005). Regulation of renal lipid metabolism, lipid accumulation, and glomerulosclerosis in FVBdb/db mice with type 2 diabetes. *Diabetes, 54*(8), 2328–2335.
Wen, C. P., et al. (2017). Diabetes with early kidney involvement may shorten life expectancy by 16 years. *Kidney International, 92*(2), 388–396.
Wijesekara, N., et al. (2017). Amyloid-β and islet amyloid pathologies link Alzheimer's disease and type 2 diabetes in a transgenic model. *The FASEB Journal, 31*(12), 5409–5418.
Xu, J., et al. (2021). Mitochondrial targeting therapeutics: Promising role of natural products in non-alcoholic fatty liver disease. *Frontiers in Pharmacology, 12*.
Yasuda, K., et al. (2011). Tunneling nanotubes mediate rescue of prematurely senescent endothelial cells by endothelial progenitors: Exchange of lysosomal pool. *Aging, 3*(6), 597–608.
Younossi, Z. M., et al. (2016). Global epidemiology of nonalcoholic fatty liver disease-Meta-analytic assessment of prevalence, incidence, and outcomes. *Hepatology, 64*(1), 73–84.
Yuan, Y., et al. (2019). Mitochondrial ROS-induced lysosomal dysfunction impairs autophagic flux and contributes to M1 macrophage polarization in a diabetic condition. *Clinical Science, 133*(15), 1759–1777.
Yuan, Y., et al. (2021). Mitochondrial transfer from mesenchymal stem cells to macrophages restricts inflammation and alleviates kidney injury in diabetic nephropathy mice via PGC-1α activation. *Stem Cells, 39*(7), 913–928.
Zabielski, P., et al. (2016). Altered skeletal muscle mitochondrial proteome as the basis of disruption of mitochondrial function in diabetic mice. *Diabetes, 65*(3), 561–573.
Zabielski, P., et al. (2014). Impact of insulin deprivation and treatment on sphingolipid distribution in different muscle subcellular compartments of streptozotocin-diabetic C57Bl/6 mice. *American Journal of Physiology. Endocrinology and Metabolism, 306*(5), E529–E542.
Zablah, J. E., et al. (2017). Subclinical decrease in myocardial function in asymptomatic infants of diabetic mothers: A tissue Doppler study. *Pediatric Cardiology, 38*(4), 801–806.
Zhang, P. N., et al. (2021). Mitochondrial dysfunction and diabetic nephropathy: Nontraditional therapeutic opportunities. *Journal of Diabetes Research, 2021*, 1010268.
Zhang, Z., et al. (2022). Hippocampal mitochondrial transplantation alleviates age-associated cognitive decline via enhancing Wnt signaling and neurogenesis. *Computational Intelligence and Neuroscience*, 9325302, 2022.
Zheng, H., et al. (2017). Cognitive decline in type 2 diabetic db/db mice may be associated with brain region-specific metabolic disorders. *Biochimica et Biophysica Acta—Molecular Basis of Disease, 1863*(1), 266–273.
Zhu, L., et al. (2009). PGC-1α is a key regulator of glucose-induced proliferation and migration in vascular smooth muscle cells. *PLoS One, 4*(1), e4182.
Zhu, Y., et al. (2022). Administration of mesenchymal stem cells in diabetic kidney disease: Mechanisms, signaling pathways, and preclinical evidence. *Molecular and Cellular Biochemistry, 477*(8), 2073–2092.
Zimbudzi, E., et al. (2020). Health-related quality of life among patients with comorbid diabetes and kidney disease attending a codesigned integrated model of care: A longitudinal study. *BMJ Open Diabetes Research and Care, 8*(1), e000842.

CHAPTER

23

Aging and mitochondrial transplantation

Mustafa Baran[1], *Sevin Guney*[2], *Guler Ozturk*[1] *and Burcu Caykara Peran*[1]

[1]**Department of Physiology, Faculty of Medicine, Istanbul Medeniyet University, Istanbul, Turkey;** [2]**Department of Physiology, Faculty of Medicine, Gazi University, Ankara, Turkey**

Mitochondria and aging

Aging is defined as time dependent gradual disturbances in cellular and organ functions. Genetic and environmental factors are important determinants of this process. It is accepted that the aging process begins with deterioration at molecular level and then leads to cell, tissue, and organ dysfunction. As a result, the capacity of the organism to maintain homeostasis under stress conditions decreases and this can lead to many diseases (Rodríguez-Rodero et al., 2011). Cardiorespiratory diseases, stroke, cancer, type 2 diabetes, and dementia are among the most common diseases defined by age. The number of elderly people is expected to more than double over the next 30 years and exceed 1.5 billion by 2050. It is estimated that the elderly, who made up 6% of the world's population in the 1990s, will constitute about 16% by 2050. By the 2050s, it is expected that the average life expectancy will be 6.4 years longer in Australia and New Zealand and 14.2 years in Africa compared to present (Shinde et al., 2022). As a result, scientists have prioritized reducing the effects of aging and aging related diseases, providing physiological aging, and improving overall quality of life.

The mitochondrial theory of aging

Aging is a multifactorial complex process and there is no single theory to explain the mechanisms responsible for aging. Genomic instability, epigenetic changes, telomere shortening and loss of proteostasis, stem cell depletion, mitochondrial dysfunction, and alterations in intercellular communication are some of the factors that are hypothesized to contribute to

Mitochondrial Transplantation and Transfer
https://doi.org/10.1016/B978-0-443-18858-9.00023-8

aging (López-Otín et al., 2013; Payne & Chinnery, 2015). The mitochondrial theory of aging, which is one of the theories explaining aging, has been on the agenda for more than 60 years (Harman, 1992; Shigenaga et al., 1994). According to this theory, reactive oxygen species (ROS) released by aerobic metabolism of mitochondria cause oxidative damage via their high chemical reactivity to various cellular macromolecules (Schieber & Chandel, 2014).

When electrons are transported, the respiratory chain in the inner membrane of mitochondria serves as the primary site for the production of a lot of superoxide (Zhao et al., 2019). The enzyme superoxide dismutase converts the superoxide anion to hydrogen peroxide. Hydrogen peroxide is a weak nonfree-radical, but it can be converted to a highly reactive hydroxyl radical via the Fenton reaction in the presence of transition metals (Phaniendra et al., 2015). The hydroxyl radical is considered the most harmful form of ROS that causes oxidative damage to almost every molecule including proteins, lipids, and nucleic acids (Maurya, 2020). 8-hydroxy-2 deoxyguanosine, which accumulates in the mitochondrial DNA (mtDNA) of the human brain and heart with age, is used to detect oxidative damage.

Despite contradictory findings in the literature regarding the levels of mitochondrial antioxidant enzymes during the aging process, there is agreement that the ratio of oxidative stress to antioxidant enzymes in mitochondria changes in favor of oxidative stress and promotes more ROS production during aging (Bratic & Larsson, 2013; Shields et al., 2021). Guler et al. showed low mitochondrial SOD (superoxide dismutase)/GPx (glutathione peroxidase) ratio, which reveals the relative ratios of both the first and second phases of antioxidative enzyme defense, as well as elevated SOD activity following melatonin treatment in aged rats (Öztürk et al., 2012). Impaired cytochrome c oxidase function has been observed in cardiomyocytes of elderly human subjects (Lesnefsky et al., 2001).

Ultrastructural mitochondrial alterations and genetic mutations are also implicated in aging. Mitochondria enlarge and become fewer in number as a result of vacuolization, crystal breakage, and the formation of intramitochondrial paracrystalline aggregates. Somatic mtDNA mutations, like deletions, duplications, and point mutations, are observed in several tissues with aging (Zapico & Ubelaker, 2013; Zhang et al., 1993). For example, increased amount of mutant mtDNA has been shown in the skeletal muscles of elderly subjects (Bua et al., 2006). Several mtDNA deletions have been linked to aging in the human heart (Elorza & Soffia, 2021). These mutations accumulate with aging and cause oxidative damage to proteins, lipids, and DNA, leading to a further increase in ROS production as they impair the electron transport chain (ETC) function (Guo et al., 2003). The strengthening of the antioxidant system in mice prevents mtDNA damage and extends their average lifespan by 5.5 months, which is one of the remarkable evidences of the relationship between mitochondria and aging (Kara & Müdüroğlu, 2008) (Fig. 23.1).

The importance of mitochondrial mutations in aging

The mitochondria are responsible for more than 90% of the energy produced in the body by oxidative phosphorylation (OXPHOS) (Sato et al., 2006). Apart from ATP production, mitochondria have multiple functions such as the synthesis of various bioactive molecules, regulation of calcium, control of apoptosis, and signal center formation and play a role in the biosynthesis of amino acids, lipids, and heme biosynthesis, formation of iron–sulfur

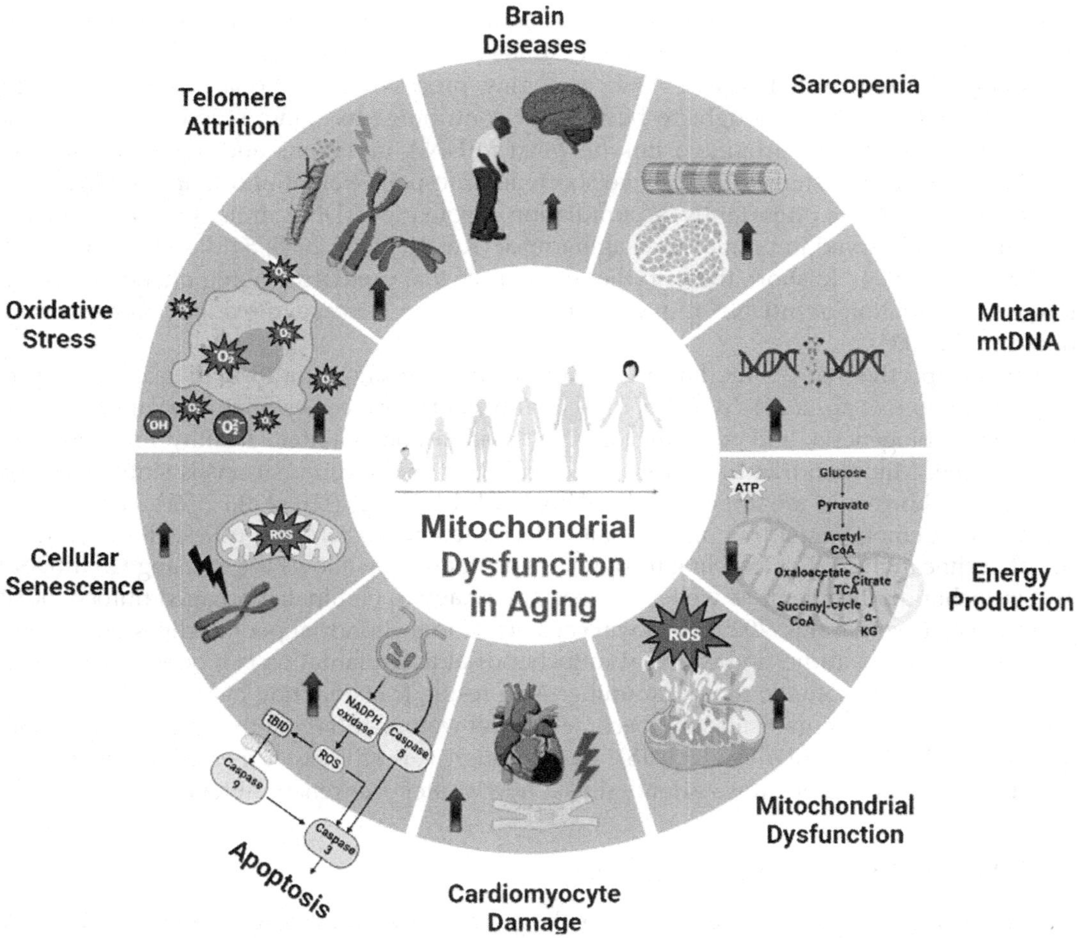

FIGURE 23.1 Mitochondrial dysfunction underlies the age related diseases.

complexes, and urea cycle. While mitochondria comprise about 1500 proteins, mtDNA encodes only ~1% of the mitochondrial proteome and 13 mitochondria encoded proteins which constitute critical components of OXPHOS complexes (Calvo & Mootha, 2010; Gammage & Frezza, 2019). A vast amount of the mitochondrial proteins are encoded by nuclear DNA and transferred to the mitochondria after being translated to proteins in the cytosol (Dewar et al., 2022). Because the majority of embryonic mitochondria (99.99%) are derived from oocytes, mitochondria are maternally inherited in humans and many other animals (Schwartz, 2021). Mutations in mtDNA can lead to mitochondrial diseases, cellular energy deficiency, and tissue dysfunction (Lax et al., 2011). mtDNA is more prone to mutations with the 10 times higher possibility than in the nuclear genome. mtDNA is more susceptible to chemical reagents or ROS because it does not have histone like protection as in the nuclear genome. Moreover, mitochondria do not have major repair systems. As a result, various somatic mtDNA mutations accumulate in tissues. Findings that the OXPHOS capacity of mitochondria decreases with aging in various organs give support to "the mitochondrial theory of aging" (Sato et al., 2006). These mutations also play an important role in the pathophysiology of age related diseases and the aging process itself (Hiona & Leeuwenburgh, 2008). Although ROS is thought to be a major contributor to aging, studies with mtDNA mutator mouse models provide a link between mtDNA mutations and aging. These mice manifest most of the signs of aging like shortened lifespan, weight and hair loss, kyphosis, and osteopenia. More importantly, accumulation of somatic mtDNA mutations and impaired ETC are also observed in prematurely aging mice (Kujoth et al., 2005; Safdar et al., 2021; Trifunovic et al., 2004). In short, these studies emphasize that premature aging phenotype is due to the accumulation of mtDNA mutations rather than the increase in oxidative stress (Trifunovic et al., 2005).

This perspective reveals the importance of healthy mitochondria formation in aging processes (Głombik et al., 2021). Mitochondrial quality control processes refer to balance between mitophagy, biogenesis, and recycling of mitochondria and maintain healthy mitochondrial pool for cell. Mitochondrial biogenesis is controlled at many different levels simultaneously and various hormones are important regulators of this process (Hu et al., 2021). For instance, it has been demonstrated that aging related decreased neurosteroid synthesis leads to functional decline in the ETC. Recycling through fission and fusion allows the exchange of mtDNA and mitochondrial content, as well as trigger mitophagy to eliminate damaged mitochondria (Markaki & Tavernarakis, 2020). Renewal capacity of mitochondrial pool declines with aging.

Recent research studies indicate that mitochondrial transplantation can boost cell proliferation and facilitate somatic reprogramming. As a result, transplanting healthy mitochondria into cells, tissues, or organisms may become a novel way of antiaging therapy (Adlimoghaddam et al., 2022; Babenko et al., 2018). Even though it is still experimental, mitochondrial transplantation/transfer draws attention as a new strategy to renew mitochondrial pool.

Mitochondrial transplantation in aging

mtDNA transfer studies revealed that pathogenic mtDNA mutations do not directly cause dysfunction. Instead, nuclear genome is thought to be responsible for the age-related decrease in OXPHOS capacity and accumulation of somatic mtDNA mutations during aging. One of the possible explanations of this phenomenon is mitochondrial complementation. The mitochondrial complementation idea suggests that mitochondria can exchange mtDNA via fission and fusion, dilute the damaged or mutant mtDNA, and thus compensate for each other's defects. Expanded reticular network in mitochondria and migration ability of the organelle allow the strengthening of complementation (Aryaman et al., 2019; Sato et al., 2006). During aging, fission and fusion capacities decrease and mitochondria undergo functional and morphological changes. Moreover, a loss of mitochondrial membrane potential and intact mtDNA can be observed with aging (Jendrach et al., 2005). When Human WI-38 fibroblasts cells were microinjected with mitochondria obtained from either young or old mice, their survival had changed, and mitochondria from old mice accelerated cellular degeneration in this cell line, according to an excellent work by Corbisier and Remacle (Corbisier & Remacle, 1993).

Increasing the capacity of mitochondria is of great importance for the prevention of aging-related effects and diseases. The most common strategies involve improving the ETC function, restoring antioxidant capacity and nitric oxide production, protecting cardiolipin, and enhancing mitochondrial biogenesis. For example, agents like coenzyme Q10, thiamine, riboflavin repair ETC function, cysteine donors, and arginine restore glutathione levels and nitric oxide production respectively. Although these strategies are mostly symptomatic, recent advances in biochemical and molecular methodologies have made it possible to understand mitochondrial damage and dysfunction, leading to develop more effective strategies. To overcome the limitations of these treatments, mitochondrial transfer/transplantation has emerged as a novel approach (El-Hattab et al., 2017).

Clark and Shay first applied mitochondrial transplantation, which means the transfer of functional exogenous mitochondria to the cell for recovering or preventing mitochondrial disease in 1982. They demonstrated that transplanting of mitochondria purified from efrapeptin and chloramphenicol resistant cells to antibiotic sensitive mammalian cells provides antibiotic resistance in these cells (Clark & Shay, 1982). Then King and Attardi showed that isolated human mitochondria injection into human tumor (osteosarcoma and fibrosarcoma) cell lines provided a rapid replacement of the recipient cell mtDNA with donor mtDNA (King & Attardi, 1988).

Various transfer or transplantation techniques have been used to deliver exogenous mitochondria to tissues or cells, including direct injection to tissue, systemic delivery, microinjection, and peptide-conjugated transplantation. However, these methods are required to be optimized to the cell or tissue type. Furthermore, several studies have shown that cells of high energy required organs like the heart and lung are more suitable donors for mitochondrial transfer/transplantation (Liu et al., 2022).

The development of new technologies for mitochondrial transfer/transplantation attracts attention, as promising results in the literature show mitochondrial diseases can be treated by mitochondrial transplantation.

Mitochondrial transplantation in age-related neurodegenerative diseases

It is not surprising that pathogenesis of most age-related neurodegenerative diseases is implicated with mitochondrial dysfunction. Because of their high energy consumption, neurons are more susceptible to mitochondrial dysfunction (Cui et al., 2012; Markaki & Tavernarakis, 2020). Apoptosis is one of the main causes underlying the symptoms of many age-related neurodegenerative disorders including Parkinson's disease (PD) and Alzheimer's disease (AD). Since it involves mitochondrial dysfunction and the production of proapoptotic substances like cytochrome c, which triggers caspase cascades, the intrinsic mechanism of apoptosis is also known as the mitochondrial pathway. Enhancing functions and the quantity of healthy mitochondria are crucial from this perspective. As previously indicated, most treatments that focus on improving mitochondrial functions are symptomatic.

The delivery of mitochondria from healthy tissue into the circulation or tissues afflicted by aging is referred to as mitochondrial transplantation and transfer. Adlimoghaddam et al. investigated whether healthy young mitochondria isolated from the liver may enhance mitochondrial function in older animals. They accomplished this by introducing mitochondria from young C57BL/6 mice into older C57BL/6 mice via IV injection. For the first time, their work indicated that mitochondrial transfusion improved mitochondrial activity in the hippocampus of elderly mice by upregulating the mitochondrial complex II protein component, SDHB. This discovery implied that mitochondrial complex II was involved in the aging process. As a result, mitochondrial complex II might be a promising antiaging target (Adlimoghaddam et al., 2022).

Mitochondrial dysfunction contributes to the onset and development of PD (Shults, 2004; Van Gurp et al., 2003). PD is a neurodegenerative disease characterized by the loss of dopaminergic neurons in the substantia nigra and the decrease of dopamine in the putamen emerging with aging. The role of oxidative stress in the etiology of PD is widely acknowledged. Patients with PD may have abnormally high quantities of iron in their brain tissue, which might serve as a source of ROS (Beal, 2005). Mutations in several genes related to mitochondrial quality control processes or oxidative stress have been found in patients with early PD. Leucine-rich repeat kinase 2 gene mutation, for instance, is linked to mitochondrial fission and more prevalent in familial PD (Niu et al., 2012), whereas DJ-1, parkin, and PINK-1 gene mutations are connected to autosomal recessive Parkinsonism. Ubiquitin proteasome system controls the mitochondrial proteome dynamics and mitophagy, thereby maintains the mitochondrial homeostasis. However, in the pathogenesis of PD, impairment in ubiquitin proteasome system is together with mitochondrial dysfunction (Rango & Bresolin, 2018).

Treatments that target mitochondrial function have the potential to decrease the course of PD (Prasuhn et al., 2021; Zambrano et al., 2022). Chang et al. tested the effectiveness of mitochondrial transplantation in PD treatment by using both in vitro (in pheochromocytoma cell line PC12) and in vivo (in rat) PD models. To mimic PD, they treated the cells and rats by 6-OHDA. By an in vitro model, they showed peptide-mediated delivery of mitochondria from allogeneic and xenogeneic sources effectively preserved mitochondrial function, cellular survival, and neurite outgrowth in PC12 cells induced with 6-OHDA. Pretreatment with mitochondria also provided a mitochondrial tolerance to 6-OHDA neurotoxicity in normal PC12 cells. Their results from in vivo experiments revealed that mitochondrial transplantation increased tyrosine hydroxylase expression in neurons of both substantia nigra pars compacta and striatum, and rectified mitochondrial hemostasis and turnover. They also emphasize the importance of reducing the level of inappropriate environment-induced decoupling of OXPHOS in isolated mitochondria before internalization (Chang et al., 2016). By another experimental PD model, Shi et al. first time showed that intravascular injection of mitochondria increased the activity of ETC, reduced ROS level, and restricted cell apoptosis and necrosis, and therefore dampened the course of PD (Shi et al., 2017).

AD is a neurodegenerative disorder, which causes age-related memory loss and impairment of many cognitive processes. Only 2% of cases of AD are inherited; genetic mutations are to blame for familial AD or early onset AD. Early-onset AD is related to genetic abnormalities in genes relating to beta-amyloid precursor protein or the presenilin 1 or presenilin 2 proteins. High presenilin expression causes DNA breakage and apoptosis which are both significant impacts of oxidative stress. The role of oxidative stress in the development of AD cannot be overstated (Fraser et al., 2000; Smith et al., 2000).

The brain's energy metabolism is disrupted as AD progresses with aging. Increased free radical generation disrupts brain mitochondrial OXPHOS and affects energy balance. Mitochondrial malfunction increases the pathogenicity of AD (Aliev et al., 2004; Beal, 2005).

Nitzan et al. showed that IV injection of freshly isolated human mitochondria into AD mice improved cognitive performance while effectively reducing neurodegeneration and gliosis in the hippocampus. This treatment also significantly elevated citrate synthase and cytochrome c oxidase activities in comparison to untreated AD mice, reaching non-AD mice activity levels. Another important result of the study was the absence of treatment related toxicity (Nitzan et al., 2019). Furthermore, even though the number of studies is limited, mitochondrial transplantation is suggested to reduce the formation and depositions of amyloid-β plaques, which is another hallmark of AD.

Cardiovascular aging and mitochondrial transplantation

The heart is a muscular organ that controls blood circulation. Cardiomyocytes, contractile cells that account for approximately 75% of the overall volume of a heart, are the most important component of this organ. The morphology of these contractile cells is determined by their location: ventricular cardiomyocytes are larger, longer, and narrower, whereas atrial cardiomyocytes are smaller and oval. Their quantity varies according to age. There are around six billion of these cells in the heart of a young person, but their quantity decreases significantly with age (Olivetti et al., 1991).

Mitochondria are extremely important for cardiomyocyte function. They account for around 30% of total cardiomyocyte volume and provide about 90% of cellular energy (Carvajal & Moreno-Sánchez, 2003). Cardiomyocytes are distinguished by their significant level of aerobic metabolism. Thus, the oxygen consumption of heart is quite high, and it is significantly higher when compared to other blood conditioning organs such as the liver, intestine, and kidney. The myocardium at rest requires around 10–12 mL of oxygen for each 100 g of tissue. Compared to most organs and tissues, skeletal muscle, and mainly the heart, gets the bulk of its energy from fatty acids (60%–100%) under physiological conditions. Glucose and lactate contribute only 0%–20% (Laloi-Michelin et al., 2006).

Polycarbonate fatty acids, as acyl-CoA derivatives, are transported through cell membranes to the mitochondrial matrix by special transporters such as carnitine, and then oxidized to acetyl-CoA during the oxidation cycle. Electrons from the reduced coenzyme are transported to the ETC of the inner mitochondrial membrane. The transport is strongly related to the ATP constituting mechanism via a procedure known as OXPHOS that creates the requisite number of ATP molecules for cardiac activity.

However, mitochondria are plastic and they are able to adapt bioenergetics, metabolism, and signaling to fulfill proper cell function and survival. By age, this plasticity decreases because of increased mtDNA mutations and ROS production. Increased mtDNA mutations and ROS production impair mitochondrial quality control processes and homeostasis in cardiac myocytes. More common age related cardiovascular diseases like heart failure and myocardial ischemia may cause inadequate perfusion of cardiomyocytes. Low oxygen delivery due to inadequate perfusion causes a shift from aerobic to anaerobic metabolism in the cell. Limited ATP production by glycolysis cannot support mitochondrial ATP consumption, which is required to maintain a stable membrane potential. This also triggers a vicious circle, increasing mitochondrial ATP consumption, impairing calcium handling in the cytoplasm, resulting in calcium overload. During reperfusion, mitochondria absorb calcium, which in turn leads to opening of mitochondrial permeability transition pore (mPTP) formation. Increased mitochondrial membrane permeability restricts proton gradient, which is required for ETC, and leads to enlargement of mitochondrial matrix. Secondary to mPTPs formation, cytochrome c and other proapoptotic mediators can leak into cytoplasm. Also high oxygen content of blood during reperfusion phase cannot concurrently restore the OXPHOS process because of shortcomings in ETC and ATP generating pathway enzymes (Elorza & Soffia, 2021; Lesnefsky et al., 2001; Liu et al., 2022).

By its high energy demand, heart is vulnerable to functional and morphological abnormalities of mitochondria, and abnormal mitochondria functions underline the pathogenesis of several cardiovascular diseases, such as heart failure, myocardial infarction, atherosclerosis, and hypertension. Therefore, treatment strategies involving mitochondrial health draw attention in recent years. Efficiency of many drugs or supplements that aim to increase mitochondrial capacity is not proved for cardiovascular diseases because these compounds aim to change one or more mitochondrial protein or pathway, which lay in some other biological pathways. For example, CoQ10, which targets the correction of mitochondrial ETC, is also an important regulator of the serine/threonine kinase AMP-activated protein kinase (AMPK) complex. Once activated AMPK diverts the metabolism to catabolic pathways via complex signaling pathways including mTOR (Herzig & Shaw, 2018). High doses of antioxidants may also have prooxidant effects, thereby inhibiting ETC (Zinovkin & Zamyatnin, 2019). Therefore, they may change other biological pathways other than targeted mitochondrial pathways. Even though it is still experimental, mitochondrial transplantation/transfer stands out a promising therapeutic strategy because it aims to increase mitochondrial quality.

Increased energy production induced by transplanted mitochondria increases the efficiency of function of many recipient tissues. Isolated mitochondria transplanted into an aging heart can diminish infarcted area, boost ATP generation, and enhance cardiac systolic performance (Masuzawa et al., 2013). Additionally, macrophages can increase their bioenergy by phagocytosing mitochondria in inflammatory and tissue damage conditions, enhancing their antiinflammatory and phagocytic capacities (Rousselle et al., 2020).

A plethora of therapies have been developed over the years to alleviate cardiac ischemia–reperfusion damage that progresses with age. The vast majority of these therapies have focused on a specific function or mechanism and have been demonstrated to provide a modest cardioprotection (Doulamis & McCully, 2021). The transplantation of healthy mitochondria extracted from a nonischemic area of the patient's body would help the heart recover from ischemia–reperfusion injury by restoring or enhancing damaged mitochondrial function.

With aging, there is a significant decline in vascular mitochondrial function, which is defined by elevated mtDNA breakage, increased mitochondrial ROS generation, and a loss in mitochondrial antioxidant enzyme resistance. Increasing mtDNA damage is directly related to age dependent rises in vascular resistance. Elevated mitochondrial ROS levels are an important contributor to age related vascular impairment. Another well known element of mitochondrial function is mitochondrial quality control mechanisms. Mitochondrial quality control mechanisms provide young, functional mitochondria and delay defective mitochondria formation by orchestrating proteostasis, biogenesis, dynamics, and mitophagy (LaRocca et al., 2014; Rossman et al., 2020). LaRocca et al. revealed that there is a correlation between impaired mitochondrial biogenesis and mitophagy mechanisms and vascular stiffness in old mice. Together with impaired mitochondrial dynamics, increased arterial mitochondrial ROS bioactivity as well as a reduced quantity of manganese superoxide dismutase (MnSOD), the main variant of SOD in mitochondria, have been reported by various studies (Gioscia-Ryan et al., 2014; Juan et al., 2008; LaRocca et al., 2014). These findings show that impaired mitochondrial quality control processes and functions may play an important role in cardiovascular aging, at least in part through increasing excess ROS (LaRocca et al., 2014).

A loss in mitochondrial stress resistance or decreased capacity to sustain function when confronted with external stresses is another symptom of mitochondrial malfunction that leads to age related cardiovascular dysfunction. Age related decreased vascular functions were demonstrated in studies using various mitochondrial stress models caused by exogenous stimuli like rotenone or the Western style diet. For instance, Gioscia-Ryan et al. found that whereas external mitochondrial stressors had no effect on the young mouse arteries, they further reduced endothelium function in older animals. They also suggested that mitochondria targeted antioxidant therapy recovers endothelium function by boosting NO bioavailability, lowering mitochondrial ROS, and enhancing mitochondrial stress resistance moderately. Moreover, these benefits were linked to increased MnSOD abundance and positive alterations in mitochondrial quantity and homeostasis indicators (Gioscia-Ryan et al., 2014).

Musculoskeletal aging and mitochondrial transplantation

As the elderly population grows, age related diseases become more prevalent. Especially, musculoskeletal diseases limit the physical performance and quality of individual life and put a burden on the health expenditures of states. Therefore, preventive and therapeutic strategies against these diseases are important for community health and preventive medicine.

Sarcopenia is frequently seen in the elderly population and is characterized by reduced muscle mass and muscle strength. Even though various factors may contribute to the pathogenesis of sarcopenia, impaired mitochondrial function in skeletal muscle is proposed as the major factor contributing to age dependent muscle degeneration. Age related declines in mitochondrial mass, activities of ETC enzymes, ATP synthesis, and OXPHOS are well known (Kim et al., 2017; Marzetti et al., 2013).

Although promising studies show that mitochondrial transplantation to skeletal muscle can improve muscle function and structure in various injury models, the effect of mitochondrial transplantation on age related muscle damage is limited. Sheu et al. showed that intramuscular transplantation of mitochondria prevented denervation induced muscle atrophy, restored muscle morphology and function, and augmented nerve regeneration (Sheu et al., 2021). In another study by Orfany et al., mitochondrial transplantation has been shown to reduce acute limb ischemia induced skeletal muscle injury, and to enhance hind limb function (Orfany et al., 2020). To maintain skeletal muscle health in the elderly, future studies focused on mitochondrial transfer/transplantation will be promising.

Osteoarthritis (OA), a degenerative joint disease, is common in the elderly and unfortunately; the only endpoint treatment available today is joint replacement surgery. Mitochondrial dysfunction of the chondrocyte is a risk factor for OA progression. An elegant study by Lee A.R. et al. suggests the therapeutic potential of mitochondrial transplantation for OA. They showed that mitochondria injection into the knee joint repaired cartilage destruction and bone loss, and improved mitochondria function in rats with experimentally induced OA (Lee et al., 2022).

Additionally, in inflammatory and tissue damage settings, macrophages can improve their bioenergy by phagocytosing mitochondria, improving their antiinflammatory and phagocytic capabilities (Rousselle et al., 2020). Delivering mitochondria into elderly tenocytes or rat models of tendinopathy reduced inflammation and apoptosis while improving collagen formation, which may be attributed to increased mitochondrial fusion (Lee et al., 2021).

Age related changes in the respiratory system and mitochondrial transplantation

Aging is a very complex process where functions gradually decrease in all structures from cells to organs, and it occurs at different rates in individuals due to reasons such as genetic characteristics, lifestyle, and diseases. While chest wall compliance and static elastic recoil decrease with aging, the alveolar canal and alveoli enlarge with age. Along with the aging process, alterations in the number and quality of progenitor cells like AT2, basal, or club cells in the lung impair self renewal and differentiation capacity. Proteostasis in the lung declines by age even without underlying disease; thence causing accumulation of misfolded, or damaged proteins (Schneider et al., 2021). Lung aging is accompanied by functional, mechanical, and structural changes involving mitochondria. Lung diseases such as idiopathic pulmonary fibrosis, chronic obstructive pulmonary disease, and acute respiratory distress syndrome occur frequently and severely in elderly populations. Mitochondrial changes have been demonstrated in the lung epithelium of the rat model of asthma and have been associated with a loss of mitochondrial ATP production (Cloonan et al., 2020). It has been reported that abnormal calcium homeostasis causes an increase in mitochondrial mass due to biogenesis in severe asthma patients (Trian et al., 2007). Tracheobronchial gland epithelial cells mutated with cystic fibrosis leading to chronic bacterial airway infection are associated with dysfunction in complex I (Valdivieso et al., 2007). Decreased complex I activity may initiate an increase in oxygen consumption and mitochondrial calcium, resulting in increased ROS in lung epithelial cells, and initiating a vicious circle (Luciani et al., 2010). Production and secretion of surfactant by alveolar type 2 cells are mitochondria dependent (Chung et al., 2019), and in response to injury, alveolar type 2 cells differentiate into type 1 cells, while mitochondria are reduced in number and size (Cloonan et al., 2020). When bone marrow stromal cells were injected into mouse lungs with lipopolysaccharide induced acute lung injury, it has been found that gap junctional channels are formed between these cells and there is a transfer of mitochondria containing microvesicles. It was found that alveolar ATP levels increased and animal mortality decreased after mitochondrial transfer/transplantation (Islam et al., 2012). Mitochondrial transplantation using pulmonary artery vascular delivery or tracheal aerosol delivery has also been reported to improve lung mechanics and reduce lung tissue damage (Moskowitzova et al., 2020). Lung adenocarcinoma cell line A549 ρ0 cells, which cannot proliferate due to a lack of mitochondria, can be induced to restore their proliferative ability by mitochondrial transfer when cocultured with human skin fibroblast cells or hMSCs (Spees et al., 2006). The success of lung tissue associated mitochondrial transplantation applications also gives hope in the prevention of aging related phenotype. To prevent lung aging, the effects of aging can be reduced or eliminated if the success of mitochondrial transplantation techniques is proven in vitro and in vivo.

Kidney aging and mitochondrial transplantation

There is a decrease in glomerular filtration rate (GFR) with the loss of cortex tissue and nephrons due to aging. Aging is accompanied by the loss of glomeruli due to cortex loss, thickening of the tubule cell membrane, tubular atrophy, and interstitial fibrosis. Intimal thickening of arterioles and great vessel walls and accompanying sclerosis are other aging related changes (Dede & Yazici Ozgur, 2017; Öztürk, 2021). Along with decreased GFR, interstitial fibrosis, impaired renal handling of sodium, and inadequate response to atrial natriuretic peptide in the kidney are also seen during aging (Dede & Yazici Ozgur, 2017). The kidney consumes a remarkable amount of energy to sustain the metabolism, acid–base, and fluid–electrolyte balance and to support hemodynamics. Therefore, after the heart, the kidney is the second organ which has the highest number of mitochondria and oxygen consumption (Duann & Lin, 2017). There is evidence that oxidative stress and mitochondrial damage may cause cell aging in the kidney (Perico et al., 2011). Significant deterioration in kidney function with mtDNA abnormalities and nuclear DNA damage and fibrosis accompanying mitochondrial dysfunction was observed in 24 month old proximal tubule specific autophagy deficient mice. Yamamoto et al. showed that long term autophagy deficiency in the kidney exaggerates age dependent mitochondrial dysfunction; therefore, they suggested that maintaining high basal autophagy activity is crucial to maintain kidney function through mitochondrial quality control processes in kidney aging (Yamamoto et al., 2016). According to a research deterioration of mitochondrial functions in aging rat kidney may cause improper VEGF regulation, which leads to disruption in angiogenesis and kidney damage (Satoh et al., 2011). It has been shown that the higher the mitochondrial copy number in blood samples from patients with reduced kidney function and healthy controls, the lower the risk of metabolic syndrome and type 2 diabetes (Fazzini et al., 2021). Mitochondrial transplantation has been accomplished by intraarterial injection or direct delivery into blood vessels to treat acute kidney injury (Doulamis et al., 2020) Mitochondrial transplantation has been reported to increase ETC activity, and decrease ROS levels, apoptosis, and necrosis in the kidney (Shi et al., 2017). In rats with diabetic nephropathy, mitochondrial transplantation under the renal capsule repairs renal proximal tubular epithelial cells both structurally and functionally while reducing ROS and apoptosis (Konari et al., 2019).

Liver aging and mitochondrial transplantation

Total liver mass and weight, regenerative capacity, hepatocyte and mitochondria count, enzyme levels, and hepatic blood flow decrease with aging (Öztürk, 2021; Palmer & Jensen, 2022). This results in a decrease in drug clearance and increases the incidence of polypharmacy, comorbidity, and drug-induced liver diseases. Moreover, altered hormone secretions by aging contribute to abnormalities in hepatic metabolism (Palmer & Jensen, 2022). Using flow cytometry, Sastre et al. showed that aging is associated with an increase in mitochondrial size and peroxide production and a decrease in mitochondrial membrane potential. They stated that mitochondrial damage plays an important role in aging, with the relationship between age related cell metabolism disorder and specific changes in mitochondrial function and morphology (Sastre et al., 1996). Based on the relationship between mitochondria and liver aging, we think that these pathological changes can be prevented by mitochondrial transplantation techniques. Mitochondria isolated from human hepatoma cells were transplanted to mouse hepatocytes with acetaminophen induced liver injury causing mitochondrial dysfunction. With this mitochondrial transplantation, ROS production was reduced and acetaminophen induced hepatotoxicity improved (Shi et al., 2018). When rabbit mitochondria were transplanted to ischemia–reperfusion damaged rat liver, a reduction in both the level of ROS production and apoptotic cells was observed, and the liver recovered (Lin & Lai, 2013). Injecting mitochondria from human hepatocytes into the liver of mice with fatty liver improved energy production, reduced lipid accumulation, and oxidative damage (Fu et al., 2017).

Total liver mass and weight, regenerative capacity, hepatocyte and mitochondria count, enzyme levels, and hepatic blood flow decrease with aging (Öztürk, 2021; Palmer & Jensen, 2022). This results in a decrease in drug clearance and increases the incidence of polypharmacy, comorbidity, and drug induced liver diseases. Moreover, altered hormone secretions by aging contribute abnormalities in hepatic metabolism (Palmer & Jensen, 2022). Using flow cytometry, Sastre et al. showed that aging is associated with an increase in mitochondrial size and peroxide production and a decrease in mitochondrial membrane potential. They stated that mitochondrial damage plays an important role in aging, with the relationship between age related cell metabolism disorder and specific changes in mitochondrial function and morphology (Sastre et al., 1996). Based on the relationship between mitochondria and liver aging, we think that these pathological changes can be prevented by mitochondrial transplantation techniques. Mitochondria isolated from human hepatoma cells were transferred to mouse hepatocytes with acetaminophen induced liver injury causing mitochondrial dysfunction. With this mitochondrial transplantation, ROS production was reduced and acetaminophen induced hepatotoxicity improved (Shi et al., 2018). When rabbit mitochondria were transferred to ischemia–reperfusion damaged rat liver, a reduction in both the level of ROS production and apoptotic cells was observed, and the liver recovered (Lin & Lai, 2013). Injecting mitochondria from human hepatocytes into the liver of mice with fatty liver improved energy production and reduced lipid accumulation and oxidative damage (Fu et al., 2017).

Conclusion

In conclusion, the mitochondrion is an important player in the healthy aging process, and most age-related diseases are related to mitochondrial dysfunction. Particularly, mitochondrial stress responses decline with aging, and cellular homeostasis is impaired. Hence, enhancing mitochondrial functions and quality control processes can be beneficial. Currently, various supplements and drugs in the market target mitochondrial ROS production and quality control processes. However, these treatment options are mostly symptomatic. Recent advances in mitochondrial transfer and transplantation studies are encouraging since they aimed to increase the healthy mitochondria population in target tissues. Even though these studies are still experimental, we foresee that mitochondrial transplantation will be among the treatment protocols for aging and age-related diseases in the near future.

References

Adlimoghaddam, A., Benson, T., & Albensi, B. C. (2022). Mitochondrial transfusion improves mitochondrial function through up-regulation of mitochondrial complex II protein subunit SDHB in the hippocampus of aged mice. *Molecular Neurobiology, 59*(10), 6009–6017. https://doi.org/10.1007/s12035-022-02937-w

Aliev, G., Smith, M. A., de la Torre, J. C., & Perry, G. (2004). Mitochondria as a primary target for vascular hypoperfusion and oxidative stress in Alzheimer's disease. *Mitochondrion, 4*(5–6), 649–663. https://doi.org/10.1016/j.mito.2004.07.018

Aryaman, J., Johnston, I. G., & Jones, N. S. (2019). Mitochondrial heterogeneity. *Frontiers in Genetics, 10*. https://doi.org/10.3389/fgene.2018.00718

Babenko, V. A., Silachev, D. N., Popkov, V. A., Zorova, L. D., Pevzner, I. B., Plotnikov, E. Y., Sukhikh, G. T., & Zorov, D. B. (2018). Miro1 enhances mitochondria transfer from multipotent mesenchymal stem cells (MMSC) to neural cells and improves the efficacy of cell recovery. *Molecules, 23*(3). https://doi.org/10.3390/molecules23030687

Beal, M. F. (2005). Mitochondria take center stage in aging and neurodegeneration. *Annals of Neurology, 58*(4), 495–505. https://doi.org/10.1002/ana.20624

Bratic, A., & Larsson, N. G. (2013). The role of mitochondria in aging. *Journal of Clinical Investigation, 123*(3), 951–957. https://doi.org/10.1172/JCI64125

Bua, E., Johnson, J., Herbst, A., Delong, B., McKenzie, D., Salamat, S., & Aiken, J. M. (2006). Mitochondrial DNA–deletion mutations accumulate intracellularly to detrimental levels in aged human skeletal muscle fibers. *The American Journal of Human Genetics, 79*(3), 469–480. https://doi.org/10.1086/507132

Calvo, S. E., & Mootha, V. K. (2010). The mitochondrial proteome and human disease. *Annual Review of Genomics and Human Genetics, 11*, 25–44. https://doi.org/10.1146/annurev-genom-082509-141720

Carvajal, K., & Moreno-Sánchez, R. (2003). Heart metabolic disturbances in cardiovascular diseases. *Archives of Medical Research, 34*(2), 89–99. https://doi.org/10.1016/S0188-4409(03)00004-3

Chang, J. C., Wu, S. L., Liu, K. H., Chen, Y. H., Chuang, C. S., Cheng, F. C., Su, H. L., Wei, Y. H., Kuo, S. J., & Liu, C. S. (2016). Allogeneic/xenogeneic transplantation of peptide-labeled mitochondria in Parkinson's disease: Restoration of mitochondria functions and attenuation of 6-hydroxydopamine-induced neurotoxicity. *Translational Research, 170*, 40–56.e3. https://doi.org/10.1016/j.trsl.2015.12.003

Chung, K. P., Hsu, C. L., Fan, L. C., Huang, Z., Bhatia, D., Chen, Y. J., Hisata, S., Cho, S. J., Nakahira, K., Imamura, M., Choi, M. E., Yu, C. J., Cloonan, S. M., & Choi, A. M. K. (2019). Mitofusins regulate lipid metabolism to mediate the development of lung fibrosis. *Nature Communications, 10*(1). https://doi.org/10.1038/s41467-019-11327-1

Clark, M. A., & Shay, J. W. (1982). Mitochondrial transformation of mammalian cells. *Nature, 295*(5850), 605–607. https://doi.org/10.1038/295605a0

Cloonan, S. M., Kim, K., Esteves, P., Trian, T., & Barnes, P. J. (2020). Mitochondrial dysfunction in lung ageing and disease. *European Respiratory Review, 29*(157). https://doi.org/10.1183/16000617.0165-2020

Corbisier, P., & Remacle, J. (1993). Influence of the energetic pattern of mitochondria in cell ageing. *Mechanisms of Ageing and Development, 71*(1–2), 47–58. https://doi.org/10.1016/0047-6374(93)90034-O

Cui, H., Kong, Y., & Zhang, H. (2012). Oxidative stress, mitochondrial dysfunction, and aging. *Journal of Signal Transduction, 2012*, 1–13. https://doi.org/10.1155/2012/646354

Dede, F., & Yazici Ozgur, C. (2017). Aging and the kidneys. *Istanbul Medical Journal, 18*(2), 53–57. https://doi.org/10.5152/imj.2017.26928

Dewar, C. E., Oeljeklaus, S., Mani, J., Mühlhäuser, W. W. D., von Känel, C., Zimmermann, J., Ochsenreiter, T., Warscheid, B., & Schneider, A. (2022). Mistargeting of aggregation prone mitochondrial proteins activates a nucleus-mediated posttranscriptional quality control pathway in trypanosomes. *Nature Communications, 13*(1). https://doi.org/10.1038/s41467-022-30748-z

Doulamis, I. P., Guariento, A., Duignan, T., Kido, T., Orfany, A., Saeed, M. Y., Weixler, V. H., Blitzer, D., Shin, B., Snay, E. R., Inkster, J. A., Packard, A. B., Zurakowski, D., Rousselle, T., Bajwa, A., Parikh, S. M., Stillman, I. E., del Nido, P. J., & McCully, J. D. (2020). Mitochondrial transplantation by intra-arterial injection for acute kidney injury. *American Journal of Physiology - Renal Physiology, 319*(3), F403–F413. https://doi.org/10.1152/AJPRENAL.00255.2020

Doulamis, I. P., & McCully, J. D. (2021). Mitochondrial transplantation for ischemia reperfusion injury. *Methods in Molecular Biology, 2277*, 15–37. https://doi.org/10.1007/978-1-0716-1270-5_2

Duann, P., & Lin, P. H. (2017). Mitochondria damage and kidney disease. *Advances in Experimental Medicine and Biology, 982*, 529–551. https://doi.org/10.1007/978-3-319-55330-6_27

El-Hattab, A. W., Zarante, A. M., Almannai, M., & Scaglia, F. (2017). Therapies for mitochondrial diseases and current clinical trials. *Molecular Genetics and Metabolism, 122*(3), 1–9. https://doi.org/10.1016/j.ymgme.2017.09.009

Elorza, A. A., & Soffia, J. P. (2021). mtDNA heteroplasmy at the core of aging-associated heart failure. An integrative view of OXPHOS and mitochondrial life cycle in cardiac mitochondrial Physiology. *Frontiers in Cell and Developmental Biology, 9*. https://doi.org/10.3389/fcell.2021.625020

Fazzini, F., Lamina, C., Raftopoulou, A., Koller, A., Fuchsberger, C., Pattaro, C., Del Greco, F. M., Döttelmayer, P., Fendt, L., Fritz, J., Meiselbach, H., Schönherr, S., Forer, L., Weissensteiner, H., Pramstaller, P. P., Eckardt, K. U., Hicks, A. A., & Kronenberg, F. (2021). Association of mitochondrial DNA copy number with metabolic syndrome and type 2 diabetes in 14 176 individuals. *Journal of Internal Medicine, 290*(1), 190–202. https://doi.org/10.1111/joim.13242

Fraser, P. E., Yang, D. S., Yu, G., Lévesque, L., Nishimura, M., Arawaka, S., Serpell, L. C., Rogaeva, E., & St George-Hyslop, P. (2000). Presenilin structure, function and role in Alzheimer disease. *Biochimica et Biophysica Acta - Molecular Basis of Disease, 1502*(1), 1–15. https://doi.org/10.1016/S0925-4439(00)00028-4

Fu, A., Shi, X., Zhang, H., & Fu, B. (2017). Mitotherapy for fatty liver by intravenous administration of exogenous mitochondria in male mice. *Frontiers in Pharmacology, 8*. https://doi.org/10.3389/fphar.2017.00241

Głombik, K., Detka, J., & Budziszewska, B. (2021). Hormonal regulation of oxidative phosphorylation in the brain in health and disease. *Cells, 10*(11), 2937. https://doi.org/10.3390/cells10112937

Gammage, P. A., & Frezza, C. (2019). Mitochondrial DNA: The overlooked oncogenome? *BMC Biology, 17*(1). https://doi.org/10.1186/s12915-019-0668-y

Gioscia-Ryan, R. A., LaRocca, T. J., Sindler, A. L., Zigler, M. C., Murphy, M. P., & Seals, D. R. (2014). Mitochondria-targeted antioxidant (MitoQ) ameliorates age-related arterial endothelial dysfunction in mice. *Journal of Physiology, 592*(12), 2549–2561. https://doi.org/10.1113/jphysiol.2013.268680

Guo, C., Sun, L., Chen, X., & Zhang, D. (2003). Oxidative stress, mitochondrial damage and neurodegenerative diseases. *Neural regeneration research, 8*.

Harman, D. (1992). Free radical theory of aging: History. *EXS, 62*, 1–10. https://doi.org/10.1007/978-3-0348-7460-1_1

Herzig, S., & Shaw, R. J. (2018). Ampk: Guardian of metabolism and mitochondrial homeostasis. *Nature Reviews Molecular Cell Biology, 19*(2), 121–135. https://doi.org/10.1038/nrm.2017.95

Hiona, A., & Leeuwenburgh, C. (2008). The role of mitochondrial DNA mutations in aging and sarcopenia: Implications for the mitochondrial vicious cycle theory of aging. *Experimental Gerontology, 43*(1), 24–33. https://doi.org/10.1016/j.exger.2007.10.001

Hu, B., Li, H., & Zhang, X. (2021). A balanced act: The effects of GH–GHR–IGF1 axis on mitochondrial function. *Frontiers in Cell and Developmental Biology, 9*. https://doi.org/10.3389/fcell.2021.630248

Islam, M. N., Das, S. R., Emin, M. T., Wei, M., Sun, L., Westphalen, K., Rowlands, D. J., Quadri, S. K., Bhattacharya, S., & Bhattacharya, J. (2012). Mitochondrial transfer from bone-marrow-derived stromal cells to pulmonary alveoli protects against acute lung injury. *Nature Medicine, 18*(5), 759–765. https://doi.org/10.1038/nm.2736

Jendrach, M., Pohl, S., Vöth, M., Kowald, A., Hammerstein, P., & Bereiter-Hahn, J. (2005). Morpho-dynamic changes of mitochondria during ageing of human endothelial cells. *Mechanisms of Ageing and Development, 126*(6–7), 813–821. https://doi.org/10.1016/j.mad.2005.03.002

Juan, M. E., Wenzel, U., Daniel, H., & Planas, J. M. (2008). Resveratrol induces apoptosis through ROS-dependent mitochondria pathway in HT-29 human colorectal carcinoma cells. *Journal of Agricultural and Food Chemistry, 56*(12), 4813–4818. https://doi.org/10.1021/jf800175a

Kara, I., & Müdüroğlu, A. (2008). Mitochondria and ageing. *Turkiye Klinikleri Cosmetic Dermatology - Special Topics,, 1*(2), 5–8.

Kim, Y., Triolo, M., & Hood, D. A. (2017). Impact of aging and exercise on mitochondrial quality control in skeletal muscle. *Oxidative Medicine and Cellular Longevity, 2017*. https://doi.org/10.1155/2017/3165396

King, M. P., & Attardi, G. (1988). Injection of mitochondria into human cells leads to a rapid replacement of the endogenous mitochondrial DNA. *Cell, 52*(6), 811–819. https://doi.org/10.1016/0092-8674(88)90423-0

Konari, N., Nagaishi, K., Kikuchi, S., & Fujimiya, M. (2019). Mitochondria transfer from mesenchymal stem cells structurally and functionally repairs renal proximal tubular epithelial cells in diabetic nephropathy in vivo. *Scientific Reports, 9*(1). https://doi.org/10.1038/s41598-019-40163-y

Kujoth, G. C., Hiona, A., Pugh, T. D., Someya, S., Panzer, K., Wohlgemuth, S. E., Hofer, T., Seo, A. Y., Sullivan, R., Jobling, W. A., Morrow, J. D., Van Remmen, H., Sedivy, J. M., Yamasoba, T., Tanokura, M., Weindruch, R., Leeuwenburgh, C., & Prolla, T. A. (2005). Mitochondrial DNA mutations, oxidative stress, and apoptosis in mammalian aging. *Science, 309*(5733), 481–484. https://doi.org/10.1126/science.1112125

López-Otín, C., Blasco, M. A., Partridge, L., Serrano, M., & Kroemer, G. (2013). The hallmarks of aging. *Cell, 153*(6), 1194. https://doi.org/10.1016/j.cell.2013.05.039

Laloi-Michelin, M., Virally, M., Jardel, C., Meas, T., Ingster-Moati, I., Lombès, A., Massin, P., Chabriat, H., Tielmans, A., Mikol, J., & Guillausseau, P. J. (2006). Kearns sayre syndrome: An unusual form of mitochondrial diabetes. *Diabetes and Metabolism, 32*(2), 182–186. https://doi.org/10.1016/s1262-3636(07)70267-7

LaRocca, T. J., Hearon, C. M., Henson, G. D., & Seals, D. R. (2014). Mitochondrial quality control and age-associated arterial stiffening. *Experimental Gerontology, 58*, 78–82. https://doi.org/10.1016/j.exger.2014.07.008

Lax, N. Z., Turnbull, D. M., & Reeve, A. K. (2011). Mitochondrial mutations: Newly discovered players in neuronal degeneration. *The Neuroscientist, 17*(6), 645–658. https://doi.org/10.1177/1073858411385469

Lee, A. R., Woo, J. S., Lee, S. Y., Na, H. S., Cho, K. H., Lee, Y. S., Lee, J. S., Kim, S. A., Park, S. H., Kim, S. J., & Cho, M. L. (2022). Mitochondrial transplantation ameliorates the development and progression of osteoarthritis. *Immune Network, 22*(2). https://doi.org/10.4110/in.2022.22.e14

Lee, J. M., Hwang, J. W., Kim, M. J., Jung, S. Y., Kim, K. S., Ahn, E. H., Min, K., & Choi, Y. S. (2021). Mitochondrial transplantation modulates inflammation and apoptosis, alleviating tendinopathy both in vivo and in vitro. *Antioxidants, 10*(5). https://doi.org/10.3390/antiox10050696

Lesnefsky, E. J., Moghaddas, S., Tandler, B., Kerner, J., & Hoppel, C. L. (2001). Mitochondrial dysfunction in cardiac disease: Ischemia - reperfusion, aging, and heart failure. *Journal of Molecular and Cellular Cardiology, 33*(6), 1065–1089. https://doi.org/10.1006/jmcc.2001.1378

Lin, H. C., & Lai, I. R. (2013). Isolated mitochondria infusion mitigates ischemia-reperfusion injury of the liver in rats: Reply. *Shock, 39*(6), 543.

Liu, Z., Sun, Y., Qi, Z., Cao, L., & Ding, S. (2022). Mitochondrial transfer/transplantation: An emerging therapeutic approach for multiple diseases. *Cell and Bioscience, 12*(1). https://doi.org/10.1186/s13578-022-00805-7

Luciani, A., Villella, V. R., Esposito, S., Brunetti-Pierri, N., Medina, D., Settembre, C., Gavina, M., Pulze, L., Giardino, I., Pettoello-Mantovani, M., D'Apolito, M., Guido, S., Masliah, E., Spencer, B., Quaratino, S., Raia, V., Ballabio, A., & Maiuri, L. (2010). Defective CFTR induces aggresome formation and lung inflammation in cystic fibrosis through ROS-mediated autophagy inhibition. *Nature Cell Biology, 12*(9), 863–875. https://doi.org/10.1038/ncb2090

Markaki, M., & Tavernarakis, N. (2020). Mitochondrial turnover and homeostasis in ageing and neurodegeneration. *FEBS Letters, 594*(15), 2370–2379. https://doi.org/10.1002/1873-3468.13802

Marzetti, E., Calvani, R., Cesari, M., Buford, T. W., Lorenzi, M., Behnke, B. J., & Leeuwenburgh, C. (2013). Mitochondrial dysfunction and sarcopenia of aging: From signaling pathways to clinical trials. *The International Journal of Biochemistry and Cell Biology, 45*(10), 2288–2301. https://doi.org/10.1016/j.biocel.2013.06.024

Masuzawa, A., Black, K. M., Pacak, C. A., Ericsson, M., Barnett, R. J., Drumm, C., Seth, P., Bloch, D. B., Levitsky, S., Cowan, D. B., & McCully, J. D. (2013). Transplantation of autologously derived mitochondria protects the heart from ischemia-reperfusion injury. *American Journal of Physiology - Heart and Circulatory Physiology, 304*(7), H966–H982. https://doi.org/10.1152/ajpheart.00883.2012

Maurya, P. K. (2020). Animal biotechnology as a tool to understand and fight aging. *Animal Biotechnology: Models in Discovery and Translation*, 235–250. https://doi.org/10.1016/B978-0-12-811710-1.00010-0

Moskowitzova, K., Orfany, A., Liu, K., Ramirez-Barbieri, G., Thedsanamoorthy, J. K., Yao, R., Guariento, A., Doulamis, I. P., Blitzer, D., Shin, B., Snay, E. R., Inkster, J. A. H., Iken, K., Packard, A. B., Cowan, D. B., Visner, G. A., del Nido, P. J., & McCully, J. D. (2020). Mitochondrial transplantation enhances murine lung viability and recovery after ischemia-reperfusion injury. *American Journal of Physiology - Lung Cellular and Molecular Physiology, 318*(1), L78–L88. https://doi.org/10.1152/ajplung.00221.2019

Nitzan, K., Benhamron, S., Valitsky, M., Kesner, E. E., Lichtenstein, M., Ben-Zvi, A., Ella, E., Segalstein, Y., Saada, A., Lorberboum-Galski, H., & Rosenmann, H. (2019). Mitochondrial transfer ameliorates cognitive deficits, neuronal loss, and gliosis in Alzheimer's disease mice. *Journal of Alzheimer's Disease: JAD, 72*(2), 587–604. https://doi.org/10.3233/JAD-190853

Niu, J., Yu, M., Wang, C., & Xu, Z. (2012). Leucine-rich repeat kinase 2 disturbs mitochondrial dynamics via dynamin-like protein. *Journal of Neurochemistry, 122*(3), 650–658. https://doi.org/10.1111/j.1471-4159.2012.07809.x

Öztürk, G. (2021). *Yaşlanma fizyolojisi 1. Baskı. İnsan fizyolojisi. Turkey: İstanbul medikal sağlık ve yayıncılık hiz. Tic. Ltd. Şti.\n.*

Öztürk, G., Akbulut, K. G., Güney, S., & Acuna-Castroviejo, D. (2012). Age-related changes in the rat brain mitochondrial antioxidative enzyme ratios: Modulation by melatonin. *Experimental Gerontology, 47*(9), 706–711. https://doi.org/10.1016/j.exger.2012.06.011

Olivetti, G., Melissari, M., Capasso, J. M., & Anversa, P. (1991). Cardiomyopathy of the aging human heart. Myocyte loss and reactive cellular hypertrophy. *Circulation Research, 68*(6), 1560–1568. https://doi.org/10.1161/01.RES.68.6.1560

Orfany, A., Arriola, C. G., Doulamis, I. P., Guariento, A., Ramirez-Barbieri, G., Moskowitzova, K., Shin, B., Blitzer, D., Rogers, C., del Nido, P. J., & McCully, J. D. (2020). Mitochondrial transplantation ameliorates acute limb ischemia. *Journal of Vascular Surgery, 71*(3), 1014–1026. https://doi.org/10.1016/j.jvs.2019.03.079

Palmer, A. K., & Jensen, M. D. (2022). Metabolic changes in aging humans: Current evidence and therapeutic strategies. *Journal of Clinical Investigation, 132*(16). https://doi.org/10.1172/JCI158451

Payne, B. A. I., & Chinnery, P. F. (2015). Mitochondrial dysfunction in aging: Much progress but many unresolved questions. *Biochimica et Biophysica Acta - Bioenergetics, 1847*(11), 1347–1353. https://doi.org/10.1016/j.bbabio.2015.05.022

Perico, N., Remuzzi, G., & Benigni, A. (2011). Aging and the kidney. *Current Opinion in Nephrology and Hypertension, 20*(3), 312–317. https://doi.org/10.1097/MNH.0b013e328344c327

Phaniendra, A., Jestadi, D. B., & Periyasamy, L. (2015). Free radicals: Properties, sources, targets, and their implication in various diseases. *Indian Journal of Clinical Biochemistry, 30*(1), 11–26. https://doi.org/10.1007/s12291-014-0446-0

Prasuhn, J., Davis, R. L., & Kumar, K. R. (2021). Targeting mitochondrial impairment in Parkinson's disease: Challenges and opportunities. *Frontiers in Cell and Developmental Biology, 8*. https://doi.org/10.3389/fcell.2020.615461

Rango, M., & Bresolin, N. (2018). Brain mitochondria, aging, and Parkinson's disease. *Genes, 9*(5), 250. https://doi.org/10.3390/genes9050250

Rodríguez-Rodero, S., Fernández-Morera, J. L., Menéndez-Torre, E., Calvanese, V., Fernández, A. F., & Fraga, M. F. (2011). Aging genetics and aging. *Aging and Disease, 2*(3), 186–195.

Rossman, M. J., Gioscia-Ryan, R. A., Clayton, Z. S., Murphy, M. P., & Seals, D. R. (2020). Targeting mitochondrial fitness as a strategy for healthy vascular aging. *Clinical Science, 134*(12), 1491–1519. https://doi.org/10.1042/CS20190559

Rousselle, T. V., Kuscu, C., Kuscu, C., Schlegel, K., Huang, L. P., Namwanje, M., Eason, J. D., Makowski, L., Maluf, D., Mas, V., & Bajwa, A. (2020). FTY720 regulates mitochondria biogenesis in dendritic cells to prevent kidney ischemic reperfusion injury. *Frontiers in Immunology, 11*. https://doi.org/10.3389/fimmu.2020.01278

Safdar, A., Khrapko, K., Flynn, J. M., Saleem, A., De Lisio, M., Johnston, A. P., Kratysberg, Y., Samjoo, I. A., Kitaoka, Y., & Ogborn, D. I. (2021). Retraction note to: Exercise-induced mitochondrial p53 repairs mtDNA mutations in mutator mice. *Skeletal Muscle, 11*.

Sastre, J., Pallardó, F. V., Plá, R., Pellín, A., Juan, G., O'Connor, J. E., Estrela, J. M., Miquel, J., & Viña, J. (1996). Aging of the liver: Age-associated mitochondrial damage in intact hepatocytes. *Hepatology, 24*(5), 1199–1205. https://doi.org/10.1002/hep.510240536

Sato, A., Nakada, K., & Hayashi, J. I. (2006). Mitochondrial dynamics and aging: Mitochondrial interaction preventing individuals from expression of respiratory deficiency caused by mutant mtDNA. *Biochimica et Biophysica Acta - Molecular Cell Research, 1763*(5–6), 473–481. https://doi.org/10.1016/j.bbamcr.2006.03.001

Satoh, M., Fujimoto, S., Horike, H., Ozeki, M., Nagasu, H., Tomita, N., Sasaki, T., & Kashihara, N. (2011). Mitochondrial damage-induced impairment of angiogenesis in the aging rat kidney. *Laboratory Investigation, 91*(2), 190–202. https://doi.org/10.1038/labinvest.2010.175

Schieber, M., & Chandel, N. S. (2014). ROS function in redox signaling and oxidative stress. *Current Biology, 24*(10), R453–R462. https://doi.org/10.1016/j.cub.2014.03.034

Schneider, J. L., Rowe, J. H., Garcia-de-Alba, C., Kim, C. F., Sharpe, A. H., & Haigis, M. C. (2021). The aging lung: Physiology, disease, and immunity. *Cell, 184*(8), 1990–2019. https://doi.org/10.1016/j.cell.2021.03.005

Schwartz, J. H. (2021). Evolution, systematics, and the unnatural history of mitochondrial DNA. *Mitochondrial DNA Part A, 32*(4), 126–151. https://doi.org/10.1080/24701394.2021.1899165

Sheu, M. L., Shen, C. C., Tsou, H. K., Yang, M. Y., Su, H. L., Sheehan, J., Chang, M. H., Chen, H. S., & Pan, H. C. (2021). Dual regeneration of muscle and nerve by intramuscular infusion of mitochondria in a nerve crush injury model. *Neurosurgery, 89*(1), E49–E59. https://doi.org/10.1093/neuros/nyab105

Shi, X., Bai, H., Zhao, M., Li, X., Sun, X., Jiang, H., & Fu, A. (2018). Treatment of acetaminophen-induced liver injury with exogenous mitochondria in mice. *Translational Research, 196*, 31–41. https://doi.org/10.1016/j.trsl.2018.02.003

Shi, X., Zhao, M., Fu, C., & Fu, A. (2017). Intravenous administration of mitochondria for treating experimental Parkinson's disease. *Mitochondrion, 34*, 91–100. https://doi.org/10.1016/j.mito.2017.02.005

Shields, H. J., Traa, A., & Van Raamsdonk, J. M. (2021). Beneficial and detrimental effects of reactive oxygen species on lifespan: A comprehensive review of comparative and experimental studies. *Frontiers in Cell and Developmental Biology, 9*. https://doi.org/10.3389/fcell.2021.628157

Shigenaga, M. K., Hagen, T. M., & Ames, B. N. (1994). Oxidative damageand mitochondrial decay in aging. *Proceedings of the National Academy of Sciences, 91*, 10771–11077.

Shinde, A., Deore, G., Navsariwala, K. P., Tabassum, H., & Wani, M. (2022). We are all aging, and here's why. *Aging Medicine, 5*(3), 211–231. https://doi.org/10.1002/agm2.12223

Shults, C. W. (2004). Mitochondrial dysfunction and possible treatments in Parkinson's disease - a review. *Mitochondrion, 4*(5–6), 641–648. https://doi.org/10.1016/j.mito.2004.07.028

Smith, M. A., Rottkamp, C. A., Nunomura, A., Raina, A. K., & Perry, G. (2000). Oxidative stress in Alzheimer's disease. *Biochimica et Biophysica Acta - Molecular Basis of Disease, 1502*(1), 139–144. https://doi.org/10.1016/S0925-4439(00)00040-5

Spees, J. L., Olson, S. D., Whitney, M. J., & Prockop, D. J. (2006). Mitochondrial transfer between cells can rescue aerobic respiration. *Proceedings of the National Academy of Sciences of the United States of America, 103*(5), 1283–1288. https://doi.org/10.1073/pnas.0510511103

Trian, T., Benard, G., Begueret, H., Rossignol, R., Girodet, P. O., Ghosh, D., Ousova, O., Vernejoux, J. M., Marthan, R., Tunon-De-Lara, J. M., & Berger, P. (2007). Bronchial smooth muscle remodeling involves calcium-dependent enhanced mitochondrial biogenesis in asthma. *Journal of Experimental Medicine, 204*(13), 3173–3181. https://doi.org/10.1084/jem.20070956

Trifunovic, A., Hansson, A., Wredenberg, A., Rovio, A. T., Dufour, E., Khvorostov, I., Spelbrink, J. N., Wibom, R., Jacobs, H. T., & Larsson, N.-G. (2005). Somatic mtDNA mutations cause aging phenotypes without affecting reactive oxygen species production. *Proceedings of the National Academy of Sciences, 102*(50), 17993–17998. https://doi.org/10.1073/pnas.0508886102

Trifunovic, A., Wredenberg, A., Falkenberg, M., Spelbrink, J. N., Rovio, A. T., Bruder, C. E., Bohlooly-Y, M., Gldlöf, S., Oldfors, A., Wibom, R., Törnell, J., Jacobs, H. T., & Larsson, N. G. (2004). Premature ageing in mice expressing defective mitochondrial DNA polymerase. *Nature, 429*(6990), 417–423. https://doi.org/10.1038/nature02517

Valdivieso, A. G., Marcucci, F., Taminelli, G., Guerrico, A. G., Álvarez, S., Teiber, M. L., Dankert, M. A., & Santa-Coloma, T. A. (2007). The expression of the mitochondrial gene MT-ND4 is downregulated in cystic fibrosis. *Biochemical and Biophysical Research Communications, 356*(3), 805–809. https://doi.org/10.1016/j.bbrc.2007.03.057

Van Gurp, M., Festjens, N., Van Loo, G., Saelens, X., & Vandenabeele, P. (2003). Mitochondrial intermembrane proteins in cell death. *Biochemical and Biophysical Research Communications, 304*(3), 487–497. https://doi.org/10.1016/S0006-291X(03)00621-1

Yamamoto, T., Takabatake, Y., Kimura, T., Takahashi, A., Namba, T., Matsuda, J., Minami, S., Kaimori, J. Y., Matsui, I., Kitamura, H., Matsusaka, T., Niimura, F., Yanagita, M., Isaka, Y., & Rakugi, H. (2016). Time-dependent dysregulation of autophagy: Implications in aging and mitochondrial homeostasis in the kidney proximal tubule. *Autophagy, 12*(5), 801–813. https://doi.org/10.1080/15548627.2016.1159376

Zambrano, K., Barba, D., Castillo, K., Noboa, L., Argueta-Zamora, D., Robayo, P., Arizaga, E., Caicedo, A., & Gavilanes, A. W. D. (2022). Fighting Parkinson's disease: The return of the mitochondria. *Mitochondrion, 64*, 34–44. https://doi.org/10.1016/j.mito.2022.02.003

Zapico, S. C., & Ubelaker, D. H. (2013). mtDNA mutations and their role in aging, diseases and forensic sciences. *Aging and Disease, 4*(6), 364–380. https://doi.org/10.14336/AD.2013.0400364

Zhang, C., Linnane, A. W., & Nagley, P. (1993). Occurrence of a particular base substitution (3243 A to G) in mitochondrial DNA of tissues of ageing humans. *Biochemical and Biophysical Research Communications, 195*(2), 1104–1110. https://doi.org/10.1006/bbrc.1993.2158

Zhao, R. Z., Jiang, S., Zhang, L., & Yu, Z. B. (2019). Mitochondrial electron transport chain, ROS generation and uncoupling (Review). *International Journal of Molecular Medicine, 44*(1), 3–15. https://doi.org/10.3892/ijmm.2019.4188

Zinovkin, R. A., & Zamyatnin, A. A. (2019). Mitochondria-targeted drugs. *Current Molecular Pharmacology, 12*(3), 202–214. https://doi.org/10.2174/1874467212666181127151059

CHAPTER

24

Mitochondrial transfer in the immune compartment

Colwyn Ansel Headley

Stanford Cardiovascular Institute, Stanford University School of Medicine, Palo Alto, CA, United States

Introduction

Mitochondria have been subjects of intense scientific scrutiny for many decades (Picard et al., 2016). Beyond their pivotal role in ATP synthesis, mitochondria also serve as intercellular signaling hubs (Bertero & Maack, 2018; Dan Dunn et al., 2015; Modesti et al., 2021; Peng et al., 2019; 2022; Rimessi et al., 2016; Shen et al., 2022; Stram & Payne, 2016), influencing a range of cellular processes including intracellular calcium modulation (Bertero & Maack, 2018; Modesti et al., 2021), redox balance (Bertero & Maack, 2018; Brand, 2016; Dan Dunn et al., 2015; Kang & Pervaiz, 2012; Rimessi et al., 2016), posttranslational protein modifications (Peng et al., 2022; Stram & Payne, 2016), and epigenetics (Bellizzi et al., 2012; Santos, 2021; Sun & St John, 2018). In terms of immunity, mitochondrial health is intrinsically linked to immune cell function (Cumming et al., 2018; Escoll et al., 2021; Gokhale et al., 2014; LaJeunesse et al., 2005; Li et al., 2005; Wang et al., 2019), and bacterial and viral pathogens have evolved mechanisms that induce mitochondrial dysfunction within immune cells (Cumming et al., 2018; Escoll et al., 2021; Gokhale et al., 2014; LaJeunesse et al., 2005; Li et al., 2005; Wang et al., 2019). For example, macrophages (Mθs) infected with *Mycobacterium tuberculosis* (*M.tb*), the causative agent of tuberculosis disease, undergo metabolic reprogramming that bias their cellular metabolism toward β-oxidation of fatty acids, which promotes *M.tb* growth and cellular quiescence (Cumming et al., 2018). *Legionella pneumophila*, the causative agent of Legionnaires' disease, inhibits apoptosis in Mθs by stalling the collapse of mitochondrial membrane potential ($\Delta\Psi m$) (Escoll et al., 2021), which effectively represses the cytochrome c-mediated apoptotic cascade (Escoll et al., 2021). Viruses like the Epstein–Barr virus (EBV) (LaJeunesse et al., 2005; Wang et al., 2019), and hepatitis C (Hep C) (Gokhale et al., 2014; Li et al., 2005) have also been shown to augment mitochondrial activity in immune cells.

Mitochondrial Transplantation and Transfer
https://doi.org/10.1016/B978-0-443-18858-9.00024-X

Hep C directly targets the mitochondrial antiviral-signaling protein (MAVS) response (Gokhale et al., 2014; Li et al., 2005), while EBV induces mitochondrial dysfunction to manipulate B cell metabolism to enhance its replication (LaJeunesse et al., 2005; Wang et al., 2019).

Independent of infections, chronic noncommunicable diseases including autoimmunity (multiple sclerosis (MS) and rheumatoid arthritis (RA)), cancer, and cardiovascular disease (CVD) are also characterized by mitochondrial dysfunction within immune cells, and this often accelerates disease pathogenesis and exacerbate disease pathology(Chen et al., 2016; De Biasi et al., 2019; Feng et al., 2021; Kumar et al., 2020; Li et al., 2019; Peng et al., 2019; Wu et al., 2021). In MS, compared to T cells from healthy individuals (controls), the mitochondria of T cells from primary progressive MS patients are hypopolarized (i.e., reduced $\Delta\Psi m$) with subpar mitochondrial respiration (OXPHOS) capacity (De Biasi et al., 2019). This biases T cells toward proinflammatory (Th1/Th17) differentiation (De Biasi et al., 2019). Similarly, In RA, T cells exhibit impaired mitochondrial respiratory function, that predisposes them to proinflammatory phenotypes (Li et al., 2019; Wu et al., 2021). In cancer, mitochondria-derived reactive oxygen species (mitoROS) in the tumor microenvironment suppress the activity of cytotoxic T cells (Chen et al., 2016; Kumar et al., 2020). In CVD like atherosclerosis, mitochondrial dysfunction impedes the transition of Mθs from proinflammatory (M1) toward antiinflammatory (M2) phenotypes (Feng et al., 2021; Peng et al., 2019).

In summary, interferences to mitochondrial fitness in immune cells compromise their function by dysregulating integral intracellular signaling cascades. The consequent immune dampening is permissive to the survival and replication of infectious pathogens within hosts, as well as chronic inflammation that can exacerbate the development of chronic noncommunicable diseases. Understandably, the examination of mitochondrial dysfunction, in contexts of disease pathogenesis and therapeutic opportunities, has also been intense areas of scientific investigation. In this regard, an intriguing and steadily expanding area of research is the delivery of healthy mitochondria to dysfunctional cells, termed mitochondrial transfer or transplantation (mito-transfer), as a basis for rescuing mitochondrial associated cellular dysregulations (Chang et al., 2019; Lightowlers et al., 2020; Liu, Sun, et al., 2022; Masuzawa et al., 2013; Plotnikov et al., 2015; Roushandeh et al., 2019; Sercel et al., 2021; Zhang et al., 2021). Currently, a myriad of disease models including heart failure (Ikeda et al., 2021; Masuzawa et al., 2013), spinal cord injury (Gollihue et al., 2018), diabetic nephropathy (Konari et al., 2019), and osteoarthritis (Wang et al., 2020) have examined and demonstrated the restorative potentials of mito-transfer therapy.

Regarding immune cells, research also suggests that immune cells can naturally exchange mitochondria in certain contexts (Akhter et al., 2023; Borcherding et al., 2022; Brestoff et al., 2021; Court et al., 2020; Kidwell et al., 2023; Kim et al., 2022; Levoux et al., 2021; Liu, Wu, et al., 2022; Piekarska et al., 2022; Rousselle et al., 2020; Saha et al., 2022; Scozzi et al., 2019; van der Vlist et al., 2022). Although the precise triggers for immune cell mito-transfer remain unclear, a deeper understanding of these dynamics, especially in the context of regenerative mito-transfer, may significantly shift current perspectives on immune signaling, and the development of immune based therapies. Moreover, I posit that intercellular mito-transfer is likely an intrinsic method by which the immune system modulates inflammatory and reparative processes. In this chapter, I catalog some of the evidence of immune cell associated mito-transfer and explore the potential therapeutic applications of this phenomenon.

Broad overview of the immune system

The immune system is subdivided into innate and adaptive components (Castelo-Branco & Soveral, 2014; Chaplin, 2010; Nicholson, 2016). The innate branch provides immediate defense via physical barriers and nonspecific cellular responses via myeloid derived immune cells (neutrophils, Mθ, natural killer (NK) cells, eosinophils, mast cells, dendritic cells (DCs), and platelets) (Koenderman et al., 2014; Uthaisangsook et al., 2002). The main role of the innate immune response is to inhibit the growth of pathogens and facilitate clearance from the host (Koenderman et al., 2014). Typically, this requires a proinflammatory milieu consisting of complement factors (Rus et al., 2005), cytokines (IL-1β, IL-6, IL-12, TNF-α, IFNγ) (Striz et al., 2014; Zhang & An, 2007), and chemokines (IL-8, MCP-1α/β, CCL5/RANTES) (Sokol & Luster, 2015). On the other hand, the adaptive branch of the immune system provides highly specific responses that are complementary to innate immunity (Bonilla & Oettgen, 2010; Dempsey et al., 2003). Cognate antigen presentation via Mθs and DCs initiate effector T and B cell reprogramming (activation), differentiation, and proliferation (Bonilla & Oettgen, 2010; Dempsey et al., 2003). Activated $CD4^+$ T cells produce cytokines that further mobilize innate and adaptive immune cells (Bonilla & Oettgen, 2010; Dempsey et al., 2003; Luckheeram et al., 2012). Activated $CD8^+$ T cells directly eliminate infected cells via cytotoxicity (Reiser & Banerjee, 2016; Verdon et al., 2020), while B cells produce antibodies (IgM, IgD, IgG, IgA, and IgE) aimed at neutralizing pathogens (Castaldi & Mira, 1967; Hoffman et al., 2016). Importantly, the adaptive system retains a memory of these antigens, enabling a more efficient response to subsequent encounters (Bonilla & Oettgen, 2010; Castaldi & Mira, 1967; Dempsey et al., 2003; Hoffman et al., 2016; Luckheeram et al., 2012).

Metabolic reprogramming during immune cell activation

Immune cells undergo significant metabolic reprogramming upon their activation (Ganeshan & Chawla, 2014; Kominsky et al., 2010; Soto-Heredero et al., 2020; Sun et al., 2020). In general, quiescent or resting immune cells preferentially employ OXPHOS for ATP production (Kominsky et al., 2010; Sun et al., 2020). Activating stimuli (pathogens and/or inflammatory cytokines) rewire immune cells toward aerobic glycolysis or Warburg metabolism (Chapman & Chi, 2022; Kominsky et al., 2010; Sun et al., 2020). Warburg metabolic reprogramming enables accelerated synthesis of ATP, at the expense of efficiency (Ganeshan & Chawla, 2014; Kornberg, 2020; Soto-Heredero et al., 2020). The accelerated generation of ATP also allows for concurrent production and accumulation of molecular intermediates required for the biosynthesis of macromolecules such as nucleotides, amino acids, and lipids, and cofactors such as nicotinamide adenine dinucleotide phosphate hydrogen (NADPH) and ribose 5-phosphate, that are essential for cell growth and function. Overall, these metabolic activities support the upsurge in immune cell proliferation, trafficking, and secretion of effector cytokines that steer inflammation and immunity (Ganeshan & Chawla, 2014; Kornberg, 2020; Soto-Heredero et al., 2020).

Mitochondrial dynamics and immune function

Concurrent with metabolic rewiring, there are shifts in mitochondrial morphology during the course of immune activation, with mitochondrial morphology typically transitioning from fused to fragmented intracellular networks (Cervantes-Silva et al., 2021). Mitochondrial fragmentation increases intracellular mitoROS levels (Jezek et al., 2018; Park et al., 2013). These augmented levels of mitoROS influence intracellular redox signaling pathways, which modulate the activation of transcription factors like nuclear factor kappa B (NF-κB) and hypoxia-inducible factor 1-alpha (HIF-1α), which are central to the expression of inflammatory genes (Andreyev et al., 2020; Sena & Chandel, 2012). Mitochondrial metabolism and mitoROS additionally support other aspects of the immune responses (Elesela & Lukacs, 2021), including the induction of antiviral signaling (MAVS) and NLR family pyrin domain containing 3 (NLRP3) signaling pathways. During viral infections, mitoROS propagate the MAVS signaling cascade (Elesela & Lukacs, 2021), which initiates the production of antiviral cytokines type I interferons (IFNs), TNFα, IL-6, and IL-12 (Mohanty et al., 2019). Similarly, mitoROS and oxidized mtDNA are key to NLRP3 inflammasome activation and downstream production of inflammatory cytokines IL-1β and IL-18, as well as the induction of cell death (pyroptosis) (Kelley et al., 2019; Mohanty et al., 2019).

Other immune related cellular processes mediated by mitochondria include calcium buffering (Giorgi et al., 2012; Modesti et al., 2021; Osellame et al., 2012), and immunological synapse formation (Sena & Chandel, 2012). Variations in mitochondrial ultrastructure impact cytosolic calcium concentrations (Modesti et al., 2021), and this affects immune cell functions like migration (Paupe & Prudent, 2018) and T cell activation (Trebak & Kinet, 2019). Further, the migration of mitochondria (within cells) is essential for the immunological synapse formation between antigen-presenting cells such as MΘs and DCs, with T cells (Quintana et al., 2007). Collectively, these and other roles underscore mitochondria's centrality in modulating both innate and adaptive immune responses. For a deeper understanding of the role of mitochondria in immune responses, a curated selection of primary studies and reviews on this topic is provided in Table 24.1.

Methods of intercellular mito-transfer

Several mechanisms have been described in facilitating the process of mito-transfer:

1. **Tunneling Nanotubes (TNTs):** These are slender filamentous structures acting as cellular bridges, enabling the direct exchange of cellular constituents, including mitochondria (Liu, Sun, et al., 2022).
2. **Gap Junctions:** Serving as intercellular channels, these junctions facilitate cytoplasmic crosstalk between cells, which might accommodate mitochondrial transfers (Norris, 2021).

TABLE 24.1 Summary of the role of mitochondria in different immune cells.

Immune cell	Role of mitochondria in cell functionality
MΘs	MitoROS and morphology (fusion/fission) impact inflammatory signaling (Afroz et al., 2023); fatty acid β-oxidation supports the antiinflammatory (M2- MΘ) responses (Wang et al., 2021), mitochondrial dysfunction promotes irreversible M1 polarization (Afroz et al., 2023; Cai et al., 2023; Van den Bossche et al., 2016)
DCs	DC maturation (Moller et al., 2022); chemotaxis/migration (Guak et al., 2018); phagocytosis (Chougnet et al., 2015); antigen presentation to T cells (Del Prete et al., 2008; Du et al., 2018)
Neutrophils	Extracellular trap (NET) formation (Douda et al., 2015; Vorobjeva & Chernyak, 2020), apoptosis (Maianski et al., 2004), respiratory burst (Vorobjeva et al., 2017); migration and adhesion (Bao et al., 2015; Cao et al., 2022; Zhou et al., 2018)
Eosinophils	Eosinophil activation and survival (Bonjour et al., 2022; Ilmarinen et al., 2014a, 2014b).
Mast cells	Mast cell degranulation and the release of histamine (Chelombitko et al., 2020; Cuong et al., 2016; Erlich et al., 2014)
Platelets	Platelet aggregation (Kulkarni et al., 2022; Melchinger et al., 2019), inflammatory response (Boudreau et al., 2014; Garcia-Souza & Oliveira, 2014; Wang et al., 2017)
NK cells	Modulates NK cell cytotoxicity (Weng et al., 2021).
B cells	B Cell development and maturation (Iperi et al., 2021; Sandoval et al., 2018; Yazicioglu et al., 2023); B cell activation (Urbanczyk et al., 2022)
T cells	T Cell activation (Ron-Harel et al., 2015), differentiation (Buck et al., 2016; Dimeloe et al., 2017; Patel & Powell, 2017), memory formation (Buck et al., 2016; Lanna & Dustin, 2016; Yanes et al., 2019), T cell exhaustion (Alrubayyi et al., 2022; Fisicaro et al., 2017; Soto-Heredero et al., 2021; Vardhana et al., 2020)

3. **Cell Fusion:** In specific scenarios, two cells might merge, pooling their intracellular contents, which naturally include mitochondria, into a unified cytoplasm (Liu, Sun, et al., 2022).
4. **Active Capture:** Cells can actively uptake mitochondria, potentially guided by molecular signals reminiscent of phagocytosis cues (Liu, Sun, et al., 2022).

Mito-transfer in innate immune cells

Studies by Brestoff et al. and Borcherding et al. reported that innate cells including monocytes, DCs, MΘs, and neutrophils acquired extracellular mitochondria from the surrounding environment, specifically in adipose tissue (Borcherding et al., 2022; Brestoff et al., 2021). This was done by using a bone marrow transplantation model: bone marrow from wild type (WT) mice (CD45.1) was transplanted into genetically modified mice (CD45.2) designed to express a fluorescent protein, Dendra2, tethered to a mitochondrial targeting sequence (mtD2). This allowed mtD2-tagged mitochondria to be detected in potential recipient cells. Approximately 40% of donor derived immune cells in the epididymal white adipose tissue (eWAT) were mtD2+, indicating mito-transfer from host adipocytes, with the majority of immune cells being identified as Mθs (Brestoff et al., 2021).

Borcherding et al. further validated these results using methods similar to those of Brestoff and team, but additionally stratified mito-transfer by type of adipose tissue (Borcherding et al., 2022). Specifically, Borcherding and team investigated the diversity of intercellular mito-transfer from adipocytes to other cell types in epididymal and inguinal white adipose tissues (eWAT and iWAT, respectively), and interscapular brown adipose tissue (BAT), in adipocyte-specific mitochondria reporter (mtD2+) mice (referred to as MitoFat mice). Monocytes, Mθs, DCs, neutrophils, and B cells were identified as the recipients of adipocyte derived mtD2+ mitochondria (Borcherding et al., 2022). In summary, both studies highlighted that myeloid cells, including monocytes, Mθs, DCs, eosinophils, and neutrophils, participated in mito-transfer in vivo.

Mito-transfer in macrophages (Mθs)

Specific to Mθs, it was discovered that the uptake of mitochondria by Mθs from adipocytes was mediated through heparan sulfate (HS) interactions at the surface (Borcherding et al., 2022). The acquisition of adipocyte derived mitochondria also impacted the metabolic health of Mθs in vivo. It was demonstrated that metabolically compromised Mθs, that either lacked NDUFS4, a subunit of complex 1 of the electron transport chain, or that were exposed to mitochondrial electron chain uncouplers (rotenone and antimycin A), had their aerobic respiration restored after mito-transfer from adipocytes (Borcherding et al., 2022). This implies that metabolically injured Mθs can mobilize healthy mitochondria from other cells to maintain cellular bioenergetics. Borcherding et al.'s study principally focused on Mθs; whether this compensatory mechanism is consistent across other immune cells that retained mtD2+ mitochondria, such as DCs, neutrophils, and B cells, remains unclear (Borcherding et al., 2022).

Conversely, Mθs have also been reported to donate mitochondria to rescue dysfunction. Liu et al. demonstrated this through both, cell cocultures and in vivo tracking; the author's overarching goal was to assess whether the adoptive transfer of antiinflammatory polarized M2-Mθs could mitigate doxorubicin (DOX) induced cardiac dysfunction in mice (Liu, Wu, et al., 2022). Compared to "control" DOX exposed C57BL/6 mice, the adoptive transfer of M2-Mθs significantly reduced left ventricular dysfunction, cardiac remodeling, cardiomyocyte apoptosis, increased the levels of the Th2 cytokine IL-4 and Th2 polarized $CD4^+$ T cells in DOX exposed mice (Liu, Wu, et al., 2022). Additionally, at 24 h post adoptive transfer, MitoTracker Red tagged mitochondria from M2-Mθs were detected in cardiomyocytes of DOX treated mice. The authors additionally confirmed mito-transfer by coculturing cardiomyocytes with either M2-Mθs or naked mitochondria isolated from M2-Mθs; with both experimental conditions resulting in the uptake of mitochondria (Liu, Wu, et al., 2022). It was also shown that the retention of donor mitochondria depended on the extent of DOX induced damage in the cardiomyocytes. Although the oxidative metabolic activities of cardiomyocytes after mito-transfer were not assessed in this study, the improved cardiac functionality of DOX-mice and decreased fibrosis after mito-transfer, suggest that OXPHOS capacities were likely improved.

A study by van der Vlist et al. further corroborated the ability of M2-Mθs to donate their mitochondria (van der Vlist et al., 2022). While investigating the molecular mechanisms involved in the resolution of neuronal pain, van der Vlist and team discovered that M2-Mθs donate mitochondria to injured neurons. A key finding of this study was that the interaction between the Mθ receptor, CD200R, and its ligand iSec1, located on sensory neurons, is necessary for the mito-transfer to occur (van der Vlist et al., 2022). Moreover, the absence of CD200R on Mθ or iSec1 on neurons hindered the alleviation of inflammatory pain (van der Vlist et al., 2022). While the studies by van der Vlist and Liu et al. provided valuable insights, they did not elucidate why M2-Mθ donate their mitochondria to distressed cells. The concept of Mθ polarization is increasingly viewed as a spectrum, rather than simply being at opposing ends of inflammation and resolution (Yuan et al., 2020; Zhu et al., 2015). M1-Mθs, linked to proinflammatory responses, lean toward glycolysis/Warburg metabolism. While, M2-Mθs, associated with tissue repair and antiinflammatory actions, are more dependent on OXPHOS (Yuan et al., 2020; Zhu et al., 2015). Although the metabolic differences defining these polarizations have been extensively studied, the distinction in mitochondrial volume/density between M1-Mθs and M2-Mθs is still unclear. Further, existing research on mito-transfer involving Mθs primarily demonstrates their capacity to ameliorate cellular dysfunction; whether pathogenic mito-transfer is also prevalent in disease pathogenesis is largely unexplored.

Research by Kidwell et al. examined both of these aspects (mitochondrial density in polarized Mθs, and pathogenic mito-transfer); first by performing confocal analysis on the mitochondrial networks of M1-Mθs and M2-Mθs, and then assessing the impact of mito-transfer between healthy Mθs and tumor cells (Kidwell et al., 2023). Regarding mitochondrial density, compared to M1-Mθs, M2-Mθs had a higher prevalence of fragmented mitochondria. This is surprising, given the fact that fused mitochondrial networks are more efficient producers of ATP and tricarboxylic acid cycle (TCA) intermediates. Additionally, in the comparative assessment of mito-transfer rates to 231 human epithelial breast cancer cells (231 cells), mito-transfer was more pronounced between M2-Mθs and 231 cells, as opposed to unpolarized Mθs and M1-Mθs(Kidwell et al., 2023). These data suggest fragmented mitochondria in M2-Mθs enhance their propensity for mito-transfer. In further regards to pathogenic mito-transfer, it was also revealed that cancer cells primarily sequestered damaged, depolarized, and prooxidant mitochondria from Mθs. Such sequestration was associated with an elevated tumorigenic potential, as patient-derived tumor cells inheriting mitochondria from Mθs, exhibited enhanced proliferation relative to controls (nonmito-transferred cancer cells) (Kidwell et al., 2023).

Mito-transfer in dendritic cells

DCs are considered sentinels of the immune system; they help bridge innate and adaptive immunity via antigen presentation to T cells (Banchereau et al., 2000; Mellman, 2013). DC dysregulation or dysfunction can contribute to conditions such as autoimmune disorders (Blanco et al., 2008), allergies (Morianos & Semitekolou, 2020), and transplant rejection (Ingulli, 2010).

Research by Rousselle and colleagues reported on the protective role of mito-transfer between DCs and Mθs, in the context of kidney ischemia–reperfusion injury (IRI) (Rousselle et al., 2020). The researchers initially aimed to investigate the protective mechanisms of FTY720 treatment of bone marrow–derived dendritic cells (BMDCs), in a preclinical mouse model of kidney IRI. FTY720 is a sphingosine 1-phosphate (S1P) receptor agonist known to modulate immunity (Bajwa et al., 2010, 2015), and it was hypothesized that treating BMDCs with FTY720 contributed to improvements in kidney IRI. FTY720 treatment led to metabolic reprogramming in DCs, evidenced by elevated basal oxygen consumption rates (OCRs) (i.e., basal OXPHOS) and OCR consumption related to mitochondria ATP synthesis, both measured by extracellular flux analysis. Compared to untreated (control) BMDCs, FTY720-primed BMDCs (FTY720-BMDCs) had significantly higher amounts of mitochondria with MitoTracker CMXRos Red (MTCR) staining. MTCR is a $\Delta\Psi m$ sensitive probe; increased MTCR staining (via confocal) indicated both the increased amount and polarization of mitochondria in FTY720-BMDCs. Additionally, compared to controls, FTY720-BMDCs exposed to lipopolysaccharide (LPS) had significantly higher mean fluorescence intensity for the mitochondrial dyes MitoSox and MitoTracker Green (MTG). MitoSox is a mitoROS reporter, while MTG is a mitochondrial mass reporter. These data collectively demonstrated that FTY720-BMDCs contained more active mitochondria than controls. Increased mRNA levels for peroxisome proliferator–activated receptor gamma coactivator 1-alpha, a transcriptional coactivator of mitochondrial biogenesis, were also reported and further corroborated the increased mitochondrial mass and activity of FTY720-BMDCs (Rousselle et al., 2020). When FTY720-BMDCs interacted with splenic Mθs in vivo, mito-transfer occurred. The authors compared the inflammatory gene expression of control Mθs (treated with nonFTY720-BMDCs) with that of Mθs which received donor mitochondria from FTY720-BMDCs, after stimulation with LPS. Their results showed that Mθs with mitochondria from FTY720-BMDCs exhibited repressed expression of inflammatory genes, including Nos2, TNFa, IL1b, and IL-16. In essence, this mito-transfer rewired Mθs to adopt a less reactive phenotype when faced with inflammatory triggers (Rousselle et al., 2020). Unfortunately, the precise mechanism of mito-transfer (i.e., surface level interactions, or exosome-mediated transfer) was not elucidated. Similar to M2-Mθs having increased amounts of mitochondria that were fragmented, perhaps a similar mechanism is occurring in mito-transfer from FTY720-BMDCs to splenic Mθs.

Mito-transfer in neutrophils

Neutrophils are the most abundant type of leukocytes in the human body and play a crucial role in the immune response against bacterial infections (Burn et al., 2021; Rosales, 2018). Their involvement extends to various pathological conditions, including chronic inflammatory diseases (Herrero-Cervera et al., 2022), cancer (Uribe-Querol & Rosales, 2015), and complications related to organ transplantation (Burn et al., 2021; Rosales, 2018). While studies by Brestoff et al. and Borcherding et al. detected adipocyte derived mitochondria in neutrophils, the impact of these mito-transfers on neutrophil functionality was not examined (Borcherding et al., 2022; Brestoff et al., 2021).

Unrelated work by Scozzi and team identified pathogenic mito-transfer-induced neutrophil dysfunction (Scozzi et al., 2019). The researchers' original aim was to uncover molecular mechanisms that contributed to the development of primary graft dysfunction (PGD), a severe inflammatory complication seen in lung transplant recipients, that occurs within a 72 hour postoperative window (Scozzi et al., 2019). In this context, an important finding of this study was the association between mitochondria damage and PGD severity. Utilizing real-time PCR, the researchers quantified circulating plasma mtDNA (measured as MT-CYB levels by PCR) in patients; patients with moderate or severe PGD exhibited significantly higher levels of circulating mtDNA compared to patients without PGD or those with mild PGD. In other words, high perioperative circulating levels of mtDNA were associated with severe PGD (Scozzi et al., 2019). The authors further validated this finding in a mouse orthotopic lung transplant model. In this model, lungs are preserved at 4°C in a low potassium dextran glucose solution for 18 hours and then orthotopically transplanted into mice (Scozzi et al., 2019). The bronchoalveolar lavage fluids from these grafts were then analyzed for extracellular mitochondria, by using antibodies specific against TOM22, a mitochondria-outer membrane protein. Compared to sham-operated controls, significantly higher amounts of cell free mitochondria were detected in transplanted grafts (Scozzi et al., 2019). Their experiments also established a critical role for formylated peptide receptor 1 (FPR1), a chemotaxis sensor for N-formylated peptides that are released by damaged mitochondria. By utilizing donor lungs from mice that express mtD2+, the researchers showed that FPR1 mediated neutrophil trafficking is coupled with the engulfment of damaged mitochondria, which triggered ROS induced pulmonary edema (Scozzi et al., 2019). Another important finding was that neutrophils lacking FPR1 have a reduced capacity to engulf damaged mitochondria and decreased ROS levels, when in the presence of low amounts of damaged mitochondria. However, at higher concentrations of damaged mitochondria, FPR1 deficient neutrophils mirrored FPR1 carrying counterparts (Scozzi et al., 2019). This indicated that although FPR1 mediates mito-transfer in neutrophils, compensatory mechanisms for mitochondrial engulfment likely exist that can be exploited for reparative therapy.

Mito-transfer in platelets

Beyond their widely recognized function in blood clotting, platelets play a role in the immune response and act as cellular communicators (George, 2000; van der Meijden & Heemskerk, 2019). Their capabilities encompass pathogen recognition, leukocyte interaction, and the release of immune modulators (George, 2000; van der Meijden & Heemskerk, 2019). Platelets can also transfer functional mitochondria to other cells (MSCs and fibroblasts), affecting their metabolism and function (Kim et al., 2022; Levoux et al., 2021).

Research by Levoux et al. (2021) specifically showed that activated platelets can transfer both functional and dysfunctional (rotenone and antimycin A exposed) mitochondria to MSCs. This mito-transfer was found to predominantly occur through dynamin dependent clathrin mediated endocytosis. The authors also showed that mito-transfer from activated platelets to MSCs accelerated the healing of cutaneous wounds in mice. While MSCs can acquire functional or dysfunctional platelet mitochondria at similar rates, their acquisition does not affect MSC differentiation into adipocytes or osteoblasts. Further, compared to control MSCs, mito-transfer to MSCs did not enhance the ability of MSCs to elicit the antiinflammatory activity of immune cells (T cells and MΘs) during wound healing, nor did the proliferative capacity of MSCs increase after mito-transfer (Levoux et al., 2021). However, the mito-transfer of intact mitochondria from platelets to MSCs induced metabolic rewiring; increasing both the OCR and extracellular acidification rate of MSCs, and intracellular levels of ATP. Unbiased metabolomics also showed that the mito-transfer of intact mitochondria upregulated TCA cycle activity. There was also a notable activation of the de novo fatty acid synthesis pathway in the MSCs after mito-transfer, evidenced by increased expression of enzymes involved in fatty acid synthesis (ACLY, ACC, and FAS) at both mRNA and protein levels (Levoux et al., 2021). This upregulation was associated with the release of proangiogenic factors Vascular Endothelial Growth Factor and Hepatocyte Growth Factor (Levoux et al., 2021), which promote wound healing. Research by Kim et al. further substantiated that platelet derived mitochondria accelerated wound closure. Albeit this mito-transfer involved human dermal fibroblast (hDFs) instead of MSCs (Kim et al., 2022). Nevertheless, comparable to the results of mito-transfer from platelets to MSCs, hDFs showed enhanced bioenergetics after mito-transfer. The mito-transfer also suppressed ROS production and significantly increased the proliferative and migratory capacity of hDFs in vitro (Kim et al., 2022).

Mito-transfer in adaptive immune cells

Adaptive immune cells, B and T cells, also have been reported to participate in mito-transfer. B cell associated mito-transfer remains largely uncharted territory, with Borcherding et al. demonstrating B cells acquiring mitochondria from surrounding adipocytes (Borcherding et al., 2022), and Court et al. reporting mito-transfer from MSC to B cells in coculture systems (Court et al., 2020). Conversely, T cell associated mito-transfer has been more thoroughly explored (Akhter et al., 2023; Court et al., 2020; Piekarska et al., 2022).

Mito-transfer in CD4$^+$ T cells

CD4$^+$ T cells, commonly referred to as helper T cells, orchestrate the activities of other immune cells, including MΘs, CD8$^+$ T cells, and B cells (Lefebvre & Haynes, 2012; Luckheeram et al., 2012). Hough et al. originally showed that myeloid derived regulatory cells in the airway produce mitochondria rich exosomes, and these exosomes are captured by CD4$^+$ T cells (Hough et al., 2018). Most other CD4$^+$ T cell associated intracellular mito-transfer research is in relation to mito-transfer from MSCs. Unrelated studies by (Akhter et al., 2023; Court et al., 2020; Piekarska et al., 2022) have collectively illustrated that mito-transfer from MSCs to CD4$^+$ T cells occurs through direct and indirect contact with allogeneic MSCs. These studies also reported significant alterations in the function and phenotype of T cells postmito-transfer, with the acquisition of MSC mitochondria preferentially bolstering the differentiation of CD4$^+$ T cells into T regulatory cells (Tregs), that have augmented suppressive capacities (Akhter et al., 2023; Court et al., 2020; Piekarska et al., 2022).

Court and colleagues specifically showed that mito-transfer from MSCs to human CD4$^+$ T cells led to the upregulation of mRNA transcripts associated with T cell activation and Treg differentiation (Court et al., 2020). MSCs were also reported to donate mitochondria to B cells and NK cells, but the impact on these cells was not further investigated. The authors also showed that this mito-transfer was unidirectional, as CD4$^+$ T cells were not captured donating mitochondria to MSCs, and neither were MSCs seen donating mitochondria to other MSCs. The resulting mito-transferred T cells showed enhanced glycolysis and migratory capabilities, but no changes in proliferative potential, compared to nonmito-transferred T cells (Court et al., 2020). The differentiated Tregs of mito-transferred T cells were also notably suppressive, marked by increased CD25 and FoxP3 coexpression, and these Tregs significantly repressed peripheral blood mononuclear cell (PBMC) proliferation in coculture conditions. In vivo examination of mito-transferred T cells, in PBMC adoptive transfers to graft versus host disease (GVHD) mouse models, showed that mito-transferred T cells (and PBMCs) significantly decreased GVHD associated tissue inflammation, and increased the survival rates of GVHD mice.

Piekarska et al. further supported these findings related to Tregs, by reporting an augmented frequency of highly suppressive Tregs postMSC interaction (Piekarska et al., 2022). In this study, human Tregs were cocultured with MSCs derived from adipose tissue (MSC-Tregs), *and* compared against nonexposed Tregs (controls). Compared to controls, MSC-Tregs exhibited higher frequencies of FoxP3+CD25$^+$ coexpression, indicative of increased Treg differentiation. MSC-Tregs also exhibited a higher frequency and intensity of CD69$^+$ and CD73$^+$ surface markers, and were significantly better at suppressing the proliferation of activated T cells, as compared to controls (Piekarska et al., 2022). Similar to Court et al. this study uncovered that mito-transfer mediated this enhanced Treg activity. However unlike Court et al., Piekarska et al. reported that mito-transfer required direct cell contact between MSC and Tregs. Interestingly, the mechanism of this mito-transfer was not dependent on TNT formation as the presence of gap junction blockers (18-β-glycyrrhetinic acid), or actin filament inhibitors (latrunculin A) had no impact on mito-transfer. It was instead dependent on human leukocyte antigens (HLA) compatibility. This was specifically demonstrated when no detectable mito-transfer was observed in coculture experiments between MSCs and K562 lymphoblast cells, which are immature lymphocytes that lack HLA. Furthermore, significant reductions in mito-transfer to Tregs were also reported when HLA class II molecules were blocked on MSCs (Piekarska et al., 2022).

Akhter et al. also reported the ability of MSCs to repress CD4$^+$ T cell growth and inflammatory action, through mito-transfer (Akhter et al., 2023). Their study utilized CD4$^+$ T cells from HNT TCR transgenic mice, known for their Th1 skewed responses upon antigenic challenges. MSCs, especially those derived from bone marrow, dampened the diabetogenic potential of these HNT CD4$^+$ T cells in a mouse model, and this was correlated with the ability of T cells to incorporate mitochondria from MSCs (Akhter et al., 2023). This mito-transfer led to curtailed T cell proliferation and reduced IFNγ production. Furthermore, both the coculture setup and isolated MSC mito-transfers prevented the upregulation of T-bet, the master Th1 transcription factor, in activated CD4$^+$ T cells (Akhter et al., 2023). Finally, Luz-Crawford and colleagues showed that mito-transfer from MSC to proinflammatory Th17 CD4$^+$ cells decreased the production of the proinflammatory cytokine IL-17 and increased Th17-to-Treg conversion (Luz-Crawford et al., 2019). Importantly, this mitochondrial relay was disrupted in synovial MSCs from RA patients, implying a potential role in sustained inflammation. MSCs were also shown to dampen the expression of T-bet in T cells, corroborating their influence in pivoting away from Th1 CD4$^+$ T cell differentiation (Luz-Crawford et al., 2019).

Mito-transfer in $CD8^+$ T cells

$CD8^+$ T cells, also referred to as cytotoxic T cells, are instrumental in the immune response due to their ability to directly neutralize infected or malignant cells (Verdon et al., 2020; Zhang & Bevan, 2011). Research spearheaded by Saha et al. provided insights into the mito-transfer between cancer cells and $CD8^+$ T cells (Saha et al., 2022). The team employed field emission scanning electron microscopy to visualize nanotubes (TNT) between $CD8^+$ T cells and tumors to monitor the transfer of MitoTracker Green-tagged mitochondria. The authors found that the mito-transfer process between $CD8^+$ T cells and tumor cells required direct contact between cells, as minimal transfer was observed in cells separated by a porous membrane (Saha et al., 2022). Importantly, this mito-transfer was found to enhance the metabolic activity of cancer cells while simultaneously decreasing it in $CD8^+$ T cells, suggesting that mito-transfer might play an underappreciated role in cancer progression by providing cancer cells a metabolic advantage (Saha et al., 2022). This mito-transfer process was found to be reliant on exocyst proteins and mitochondrial Rho GTPase-1, as their partial knockdown led to a decrease in mito-transfer. Evidence of this process was also observed in vivo. Notably, the study found that L-778,123, a drug that inhibits nanotube formation, can disrupt this intercellular interaction, offering a potential therapeutic approach to mitigate the metabolic advantage gained by cancer cells (Saha et al., 2022).

Discussion

Potential implications of mito-transfer in Immunomodulation

Traditionally, our understanding of immune regulation was framed around cytokines and surface receptor interactions. This model has been fundamental in unraveling myriad immune processes, spanning from the identification of pathogenic invaders (Casasnovas, 2013; Mogensen & Paludan, 2001), to the modulation of inflammatory cascades (Nathan, 2002). This foundational knowledge has also paved the way for innovative therapies like monoclonal antibodies (Weiner et al., 2009) and checkpoint inhibitors (Sharma & Allison, 2015), advancements that have been transformative in treating cancer (Posner et al., 2019; Shiravand et al., 2022; Topalian et al., 2015), autoimmune disorders (Hafeez et al., 2018; Hansel et al., 2010), infectious diseases (Boonyaratanakornkit et al., 2022; Motley et al., 2019), and facilitating organ transplantation (Mahmud et al., 2010).

The burgeoning field of immunometabolism continues to uncover novel perspectives related to the metabolic interplays that underpin immune responses (O'Neill et al., 2016). Within this metabolic paradigm, the phenomenon of intercellular mito-transfer emerges as a potential yet largely unexplored mechanism of intercellular communication in the immune system. Mito-transfer has been observed in a diverse range of innate immune cells, including Mθs, DCs, neutrophils, platelets, and adaptive immune cells such as B and T cells Table 24.2. Importantly, some instances of mito-transfer are not merely a mechanism of resource exchange; they actively shape the function and fate of recipient cells. This brings forth the intriguing concept of the "mitochondrial biome" or "mitobiome" of immune cells, a term referring to the collective mitochondrial content and its functional implications within

TABLE 24.2 Summary of mito-transfer in different immune cells.

Immune cell	Mito-transfer Direction	Mechanism of mito-transfer	Impact on recipient cells	Observations
Mθ	Adipocytes to Mθ	Heparan sulfate (HS) interactions at the surface	↑ metabolic health in vivo; restored aerobic respiration	Specific to adipose tissue
M2-Mθ	M2-Mθ to cardiomyocytes	Not Specified	↓ left ventricular dysfunction, cardiac remodeling, cardiomyocyte apoptosis.↑ IL-4 and Th2+ $CD4^+$ T cells	Retention of donor mitochondria depends on doxorubicin-induced damage
M2-Mθ	M2-Mθ to neurons	Mθ receptor CD200R and ligand iSec1 on neurons	↓ inflammatory pain in injured neurons	The presence of CD200R on Mθ and iSec1 on neurons necessary for mito-transfer
M1-Mθ/M2 Mθs	Mθ to cancer cells	Not Specified	↑ cancer cell proliferation	M2-Mθ → more fragmented mitochondria; more mito-transfer to cancer cells
DCs	BMDCs to macrophages	FTY720 treatment	↓ inflammatory genes (Nos2, TNFa, IL1b, IL-16)	Less inflammatory phenotype
Neutrophils	Damaged mitochondria to neutrophils	FPR1 receptor	↑ ROS-induced pulmonary edema	Role in primary graft dysfunction (PGD) after lung transplantation
Platelets	Platelets to MSCs and hDFs	Dynamin-dependent clathrin-mediated endocytosis	↑ fatty acid synthesis in MSCs; ↑ proangiogenic factors and angiogenesis; ↑ proliferation and migration in hDFs	Enhances therapeutic capabilities of MSCs and fibroblasts
B cells	Adipocytes and MSCs to B cells	Not Specified	Functional consequences not extensively studied	Largely uncharted mito-transfer
$CD4^+$ T cells	MSCs to $CD4^+$ T cells,	Direct cell contact and engulfment of naked mitochondria	↑ Treg differentiation and suppressive abilities,↓ IFNγ production.	Mito-transfer between MSCs and $CD4^+$ T cells leads to the upregulation of vital mRNA transcripts for T cell activation and Treg differentiation
$CD8^+$ T cells	$CD8^+$ T cells to cancer cells	Requires direct cell contact	↑ cancer cell metabolism,↓ $CD8^+$ T cell metabolism	Mito-transfer may give cancer cells a metabolic advantage

immune cells. This is particularly exemplified in the studies by (Rousselle et al., 2020), where mito-transfer between primed DCs and splenic Mθs leads to antiinflammatory adaptations in recipient Mθs. Furthermore, Court et al. (2020) and others (Akhter et al., 2023; Luz-Crawford et al., 2019; Piekarska et al., 2022) collectively show that mito-transfer from MSC to T cells primarily primed T cells for Treg differentiation. However, the induction of Treg phenotypes was not observed with mito-transfer occurring between T cells and other cells like fibroblasts (Court et al., 2020). A possible explanation for the observed induction of these antiinflammatory, immunosuppressive reprogrammings may be related to changes in conserved redox signaling pathways (NF-κB, Hif-1α, mTOR, and Nrf2) that are mediated by mitochondrial activity in respective immune cell subsets (Perez & Rius-Perez, 2022). For example, mito-transfer might foster a metabolic shift in Mθs toward increased OXPHOS and reduced glycolysis, a hallmark of M2-Mθs. Changes in OXPHOS dependence also influence mitoROS release from mitochondria within Mθ, as M1-Mθ requires higher levels of intracellular ROS for the induction of inflammatory genes (Brune et al., 2013; Gostner et al., 2013; Perez & Rius-Perez, 2022). However, all of this is still rather speculative; future studies will be needed to better interrogate which pathways are predominantly modulated through mito-transfer.

Understanding the precise triggers, conditions, and molecular signals under which mito-transfer occurs is of paramount importance. Of the studies discussed, the most common method of mito-transfer involved direct cell to cell contact, with some coculture studies also demonstrating that naked mitochondria can be engulfed by the surrounding environment. Unfortunately, not much is understood regarding the surface level interactions that facilitate mito-transfer in immune cells, and this knowledge could facilitate the manipulation of this process in diseases, where immune responses are either insufficient or hyperactive. The studies by Scozzi et al. (2019) showed that neutrophils selectively uptake damaged mitochondria, and this was related to FPR1 expression, which contributed to PGD in pulmonary transplant patients. These findings provide the basis for testing whether healthy mitochondria modified to express FPR1, would lower the neutrophil mediated damage in the lungs of PGD patients. Similarly, the work by Saha et al. (2022) establishes a basis for investigating whether the delivery of proapoptotic via mito-transfer between $CD8^+$ T cells and tumor cells can increase the tumor sensitivity to chemotherapeutics.

Another critical area for future investigation lies in understanding how mito-transfer impacts immune cells within different tissue microenvironments. Both Borcherding et al. (2022) and Brestoff et al. (2021) demonstrated that mito-transfer from adipocytes to immune cells, in WAT and BAT, is a common occurrence. A unique feature of BAT is its specialized mitochondria, which express uncoupling protein 1 (UCP1) (Azzu & Brand, 2010; Brand et al., 2004; Jastroch et al., 2010). UCP1 is critical for the thermogenic capability of BAT, as it allows mitochondria to uncouple fatty acid oxidation from ATP production, thereby generating heat (Azzu & Brand, 2010; Brand et al., 2004; Jastroch et al., 2010). Given this unique metabolic profile, a key question that arises is whether the acquisition of BAT derived mitochondria by Mθs might lead to a functional shift in these immune cells. M2-Mθs, which are often associated with tissue repair and antiinflammatory functions, exhibit enhanced oxidative metabolism (Brune et al., 2013; Zhu et al., 2015). Therefore, UCP1 rich mitochondria may conceivably bias an antiinflammatory phenotype.

Another key avenue for exploration is the role of mito-transfer in understudied immune cell types. B cells remain largely uncharted in the context of mito-transfer, but initial studies suggest that they may acquire mitochondria from surrounding adipocytes. Unraveling the implications of this transfer on B cell function, especially regarding antigen presentation and antibody production, could yield significant insights. NK cells, key players in innate immunity with pivotal roles in antiviral and antitumor responses, represent another unexplored domain. Assessing mito-transfer in NK cells could have far-reaching implications for the treatment of infectious diseases and cancer.

Lastly, the translational aspect of mito-transfer in immune cells is deserving of focused attention. As has been alluded to, the manipulation of mito-transfer could have significant therapeutic implications, particularly for diseases involving immune dysregulation such as autoimmunity, cancer, infectious diseases, and aging. Regarding aging, Headley et al. examined the ex vivo and in vivo functional consequences of mito-transfer on $CD4^+$ T cells isolated from old mice (Headley et al., 2024). Mitochondria isolated from mouse embryonic fibroblast were transferred by centrifugation into purified $CD4^+$ T cells from old mice. Compared to unmanipulated $CD4^+$ T cells from old mice, mito-transferred $CD4^+$ T cells displayed lower basal levels of mitoROS and oxidative stress, and improved OXPHOS metabolism. Mito-transferred $CD4^+$ T cells also showed improved activation, proliferation, and cytokine production, particularly IL-2 (Headley et al., 2024). The researchers then compared the ability of unmanipulated and mito-transferred $CD4^+$ T cells from old mice to protect against pathogens (influenza and *M.tb*) in vivo. Compared to unmanipulated naïve $CD4^+$ T cells from old mice, in both disease models, mito-transferred $CD4^+$ T cells conferred significant protection (Headley et al., 2024), exemplifying the broad applicability of extracellular mito-transfer. Thus, future research should be oriented toward developing targeted therapies that modulate mito-transfer from extracellular sources.

Conclusion

In conclusion, mitochondria mediate a significant amount of signaling cascades in immune cells. Emerging research suggests that mito-transfer is an intrinsic process by which immune cells communicate with other cells of the body. The mechanisms that initiate mito-transfer in immune cells are largely unclear. Unraveling the molecular cues that initiate mito-transfer in immune cells will likely revolutionize approaches to designing immunotherapies in the future.

Funding

This work was supported by the Stanford University Propel Post-Doctoral Fellowship, The Burroughs Wellcome Fund Postdoctoral Diversity Enrichment Program Award (1069497), and the National Heart, Lung, and Blood Institute (1T32HL098049).

References

Afroz, S. F., Raven, K. D., Lawrence, G., Kapetanovic, R., Schroder, K., & Sweet, M. J. (2023). Mitochondrial dynamics in macrophages: Divide to conquer or unite to survive? *Biochemical Society Transactions, 51*(1), 41–56. https://doi.org/10.1042/BST20220014

Akhter, W., Nakhle, J., Vaillant, L., Garcin, G., Le Saout, C., Simon, M., Crozet, C., Djouad, F., Jorgensen, C., Vignais, M. L., & Hernandez, J. (2023). Transfer of mesenchymal stem cell mitochondria to CD4(+) T cells contributes to repress Th1 differentiation by downregulating T-bet expression. *Stem Cell Research and Therapy, 14*(1), 12. https://doi.org/10.1186/s13287-022-03219-x

Alrubayyi, A., Moreno-Cubero, E., Hameiri-Bowen, D., Matthews, R., Rowland-Jones, S., Schurich, A., & Peppa, D. (2022). Functional restoration of exhausted CD8 T cells in chronic HIV-1 infection by targeting mitochondrial dysfunction. *Frontiers in Immunology, 13*, 908697. https://doi.org/10.3389/fimmu.2022.908697

Andreyev, A. Y., Kushnareva, Y. E., Starkova, N. N., & Starkov, A. A. (2020). Metabolic ROS signaling: To immunity and beyond. *Biochemistry, 85*(12), 1650–1667. https://doi.org/10.1134/S0006297920120160

Azzu, V., & Brand, M. D. (2010). The on-off switches of the mitochondrial uncoupling proteins. *Trends Biochem Sci, 35*(5), 298–307. https://doi.org/10.1016/j.tibs.2009.11.001

Bajwa, A., Jo, S. K., Ye, H., Huang, L., Dondeti, K. R., Rosin, D. L., Haase, V. H., Macdonald, T. L., Lynch, K. R., & Okusa, M. D. (2010). Activation of sphingosine-1-phosphate 1 receptor in the proximal tubule protects against ischemia-reperfusion injury. *Journal of the American Society of Nephrology, 21*(6), 955–965. https://doi.org/10.1681/ASN.2009060662

Bajwa, A., Rosin, D. L., Chroscicki, P., Lee, S., Dondeti, K., Ye, H., Kinsey, G. R., Stevens, B. K., Jobin, K., Kenwood, B. M., Hoehn, K. L., Lynch, K. R., & Okusa, M. D. (2015). Sphingosine 1-phosphate receptor-1 enhances mitochondrial function and reduces cisplatin-induced tubule injury. *Journal of the American Society of Nephrology, 26*(4), 908–925. https://doi.org/10.1681/ASN.2013121351

Banchereau, J., Briere, F., Caux, C., Davoust, J., Lebecque, S., Liu, Y. J., Pulendran, B., & Palucka, K. (2000). Immunobiology of dendritic cells. *Annual Review of Immunology, 18*, 767–811. https://doi.org/10.1146/annurev.immunol.18.1.767

Bao, Y., Ledderose, C., Graf, A. F., Brix, B., Birsak, T., Lee, A., Zhang, J., & Junger, W. G. (2015). mTOR and differential activation of mitochondria orchestrate neutrophil chemotaxis. *The Journal of Cell Biology, 210*(7), 1153–1164. https://doi.org/10.1083/jcb.201503066

Bellizzi, D., D'Aquila, P., Giordano, M., Montesanto, A., & Passarino, G. (2012). Global DNA methylation levels are modulated by mitochondrial DNA variants. *Epigenomics, 4*(1), 17–27. https://doi.org/10.2217/epi.11.109

Bertero, E., & Maack, C. (2018). Calcium signaling and reactive oxygen species in mitochondria. *Circulation Research, 122*(10), 1460–1478. https://doi.org/10.1161/CIRCRESAHA.118.310082

Blanco, P., Palucka, A. K., Pascual, V., & Banchereau, J. (2008). Dendritic cells and cytokines in human inflammatory and autoimmune diseases. *Cytokine Growth Factor Review, 19*(1), 41–52. https://doi.org/10.1016/j.cytogfr.2007.10.004

Bonilla, F. A., & Oettgen, H. C. (2010). Adaptive immunity. *The Journal of Allergy and Clinical Immunology, 125*(Suppl. 2), S33–S40. https://doi.org/10.1016/j.jaci.2009.09.017

Bonjour, K., Palazzi, C., Silva, T. P., Malta, K. K., Neves, V. H., Oliveira-Barros, E. G., Neves, I., Kersten, V. A., Fortuna, B. T., Samarasinghe, A. E., Weller, P. F., Bandeira-Melo, C., & Melo, R. C. N. (2022). Mitochondrial population in mouse eosinophils: Ultrastructural dynamics in cell differentiation and inflammatory diseases. *Frontiers in Cell and Developmental Biology, 10*, Article 836755. https://doi.org/10.3389/fcell.2022.836755

Boonyaratanakornkit, J., Boeckh, M., & Waghmare, A. (2022). Monoclonal antibodies for prophylaxis and treatment of respiratory viral infections. *Current Opinion in Infectious Diseases, 35*(4), 280–287. https://doi.org/10.1097/QCO.0000000000000846

Borcherding, N., Jia, W., Giwa, R., Field, R. L., Moley, J. R., Kopecky, B. J., Chan, M. M., Yang, B. Q., Sabio, J. M., Walker, E. C., Osorio, O., Bredemeyer, A. L., Pietka, T., Alexander-Brett, J., Morley, S. C., Artyomov, M. N., Abumrad, N. A., Schilling, J., Lavine, K., Crewe, C., & Brestoff, J. R. (2022). Dietary lipids inhibit mitochondria transfer to macrophages to divert adipocyte-derived mitochondria into the blood. *Cell Metabolism, 34*(10), 1499–1513 e8. https://doi.org/10.1016/j.cmet.2022.08.010

Boudreau, L. H., Duchez, A. C., Cloutier, N., Soulet, D., Martin, N., Bollinger, J., Pare, A., Rousseau, M., Naika, G. S., Levesque, T., Laflamme, C., Marcoux, G., Lambeau, G., Farndale, R. W., Pouliot, M., Hamzeh-Cognasse, H., Cognasse, F., Garraud, O., Nigrovic, P. A., … Boilard, E. (2014). Platelets release mitochondria serving as substrate for bactericidal group IIA-secreted phospholipase A2 to promote inflammation. *Blood, 124*(14), 2173–2183. https://doi.org/10.1182/blood-2014-05-573543

Brand, M. D. (2016). Mitochondrial generation of superoxide and hydrogen peroxide as the source of mitochondrial redox signaling. *Free Radical Biology and Medicine, 100*, 14–31. https://doi.org/10.1016/j.freeradbiomed.2016.04.001

Brand, M. D., Affourtit, C., Esteves, T. C., Green, K., Lambert, A. J., Miwa, S., Pakay, J. L., & Parker, N. (2004). Mitochondrial superoxide: Production, biological effects, and activation of uncoupling proteins. *Free Radical Biology and Medicine, 37*(6), 755–767. https://doi.org/10.1016/j.freeradbiomed.2004.05.034

Brestoff, J. R., Wilen, C. B., Moley, J. R., Li, Y., Zou, W., Malvin, N. P., Rowen, M. N., Saunders, B. T., Ma, H., Mack, M. R., Hykes, B. L., Jr., Balce, D. R., Orvedahl, A., Williams, J. W., Rohatgi, N., Wang, X., McAllaster, M. R., Handley, S. A., Kim, B. S., … Teitelbaum, S. L. (2021). Intercellular mitochondria transfer to macrophages regulates white adipose tissue homeostasis and is impaired in obesity. *Cell Metabolism, 33*(2), 270–282 e8. https://doi.org/10.1016/j.cmet.2020.11.008

Brune, B., Dehne, N., Grossmann, N., Jung, M., Namgaladze, D., Schmid, T., von Knethen, A., & Weigert, A. (2013). Redox control of inflammation in macrophages. *Antioxidants and Redox Signaling, 19*(6), 595–637. https://doi.org/10.1089/ars.2012.4785

Buck, M. D., O'Sullivan, D., Klein Geltink, R. I., Curtis, J. D., Chang, C. H., Sanin, D. E., Qiu, J., Kretz, O., Braas, D., van der Windt, G. J., Chen, Q., Huang, S. C., O'Neill, C. M., Edelson, B. T., Pearce, E. J., Sesaki, H., Huber, T. B., Rambold, A. S., & Pearce, E. L. (2016). Mitochondrial dynamics controls T cell fate through metabolic programming. *Cell, 166*(1), 63–76. https://doi.org/10.1016/j.cell.2016.05.035

Burn, G. L., Foti, A., Marsman, G., Patel, D. F., & Zychlinsky, A. (2021). The neutrophil. *Immunity, 54*(7), 1377–1391. https://doi.org/10.1016/j.immuni.2021.06.006

Cai, S., Zhao, M., Zhou, B., Yoshii, A., Bugg, D., Villet, O., Sahu, A., Olson, G. S., Davis, J., & Tian, R. (2023). Mitochondrial dysfunction in macrophages promotes inflammation and suppresses repair after myocardial infarction. *Journal of Clinical Investigation, 133*(4). https://doi.org/10.1172/JCI159498

Cao, Z., Zhao, M., Sun, H., Hu, L., Chen, Y., & Fan, Z. (2022). Roles of mitochondria in neutrophils. *Frontiers in Immunology, 13*, 934444. https://doi.org/10.3389/fimmu.2022.934444

Casasnovas, J. M. (2013). Virus-receptor interactions and receptor-mediated virus entry into host cells. *Subcellular Biochemistry, 68*, 441–466. https://doi.org/10.1007/978-94-007-6552-8_15

Castaldi, G., & Mira, L. (1967). [Critical considerations on the evaluation of psychomotor and language development of children in the first 3 years of life]. *Rivista Sperimentale di Freniatria e Medicina Legale Delle Alienazioni Mentali, 91*(6), 1536–1547.

Castelo-Branco, C., & Soveral, I. (2014). The immune system and aging: A review. *Gynecological Endocrinology, 30*(1), 16–22. https://doi.org/10.3109/09513590.2013.852531

Cervantes-Silva, M. P., Cox, S. L., & Curtis, A. M. (2021). Alterations in mitochondrial morphology as a key driver of immunity and host defence. *EMBO Reports, 22*(9), Article e53086. https://doi.org/10.15252/embr.202153086

Chang, J. C., Chang, H. S., Wu, Y. C., Cheng, W. L., Lin, T. T., Chang, H. J., Kuo, S. J., Chen, S. T., & Liu, C. S. (2019). Mitochondrial transplantation regulates antitumour activity, chemoresistance and mitochondrial dynamics in breast cancer. *Journal of Experimental and Clinical Cancer Research, 38*(1), 30. https://doi.org/10.1186/s13046-019-1028-z

Chaplin, D. D. (2010). Overview of the immune response. *The Journal of Allergy and Clinical Immunology, 125*(Suppl. 2), S3–S23. https://doi.org/10.1016/j.jaci.2009.12.980

Chapman, N. M., & Chi, H. (2022). Metabolic adaptation of lymphocytes in immunity and disease. *Immunity, 55*(1), 14–30. https://doi.org/10.1016/j.immuni.2021.12.012

Chelombitko, M. A., Chernyak, B. V., Fedorov, A. V., Zinovkin, R. A., Razin, E., & Paruchuru, L. B. (2020). The role played by mitochondria in FcepsilonRI-dependent mast cell activation. *Frontiers in Immunology, 11*, 584210. https://doi.org/10.3389/fimmu.2020.584210

Chen, X., Song, M., Zhang, B., & Zhang, Y. (2016). Reactive oxygen species regulate T cell immune response in the tumor microenvironment. *Oxidative Medicine and Cellular Longevity, 2016*, Article 1580967. https://doi.org/10.1155/2016/1580967

Chougnet, C. A., Thacker, R. I., Shehata, H. M., Hennies, C. M., Lehn, M. A., Lages, C. S., & Janssen, E. M. (2015). Loss of phagocytic and antigen cross-presenting capacity in aging dendritic cells is associated with mitochondrial dysfunction. *Journal of Immunology, 195*(6), 2624–2632. https://doi.org/10.4049/jimmunol.1501006

Court, A. C., Le-Gatt, A., Luz-Crawford, P., Parra, E., Aliaga-Tobar, V., Batiz, L. F., Contreras, R. A., Ortuzar, M. I., Kurte, M., Elizondo-Vega, R., Maracaja-Coutinho, V., Pino-Lagos, K., Figueroa, F. E., & Khoury, M. (2020). Mitochondrial transfer from MSCs to T cells induces Treg differentiation and restricts inflammatory response. *EMBO Reports, 21*(2), Article e48052. https://doi.org/10.15252/embr.201948052

Cumming, B. M., Addicott, K. W., Adamson, J. H., & Steyn, A. J. (2018). Mycobacterium tuberculosis induces decelerated bioenergetic metabolism in human macrophages. *Elife, 7*. https://doi.org/10.7554/eLife.39169

Cuong, D. V., Kim, H. K., Marquez, J., Kim, N., Ko, K. S., Rhee, B. D., & Han, J. (2016). Mitochondrial calcium uniporter inhibition attenuates mouse bone marrow-derived mast cell degranulation induced by beta-1,3-glucan. *Korean Journal of Physiology and Pharmacology, 20*(2), 213–220. https://doi.org/10.4196/kjpp.2016.20.2.213

Dan Dunn, J., Alvarez, L. A., Zhang, X., & Soldati, T. (2015). Reactive oxygen species and mitochondria: A nexus of cellular homeostasis. *Redox Biology, 6*, 472–485. https://doi.org/10.1016/j.redox.2015.09.005

De Biasi, S., Simone, A. M., Bianchini, E., Lo Tartaro, D., Pecorini, S., Nasi, M., Patergnani, S., Carnevale, G., Gibellini, L., Ferraro, D., Vitetta, F., Pinton, P., Sola, P., Cossarizza, A., & Pinti, M. (2019). Mitochondrial functionality and metabolism in T cells from progressive multiple sclerosis patients. *European Journal of Immunology, 49*(12), 2204–2221. https://doi.org/10.1002/eji.201948223

Del Prete, A., Zaccagnino, P., Di Paola, M., Saltarella, M., Oliveros Celis, C., Nico, B., Santoro, G., & Lorusso, M. (2008). Role of mitochondria and reactive oxygen species in dendritic cell differentiation and functions. *Free Radical Biology and Medicine, 44*(7), 1443–1451. https://doi.org/10.1016/j.freeradbiomed.2007.12.037

Dempsey, P. W., Vaidya, S. A., & Cheng, G. (2003). The art of war: Innate and adaptive immune responses. *Cellular and Molecular Life Sciences, 60*(12), 2604–2621. https://doi.org/10.1007/s00018-003-3180-y

Dimeloe, S., Burgener, A. V., Grahlert, J., & Hess, C. (2017). T-cell metabolism governing activation, proliferation and differentiation; a modular view. *Immunology, 150*(1), 35–44. https://doi.org/10.1111/imm.12655

Douda, D. N., Khan, M. A., Grasemann, H., & Palaniyar, N. (2015). SK3 channel and mitochondrial ROS mediate NADPH oxidase-independent NETosis induced by calcium influx. *Proceedings of the National Academy of Sciences of the United States of America, 112*(9), 2817–2822. https://doi.org/10.1073/pnas.1414055112

Du, X., Wen, J., Wang, Y., Karmaus, P. W. F., Khatamian, A., Tan, H., Li, Y., Guy, C., Nguyen, T. M., Dhungana, Y., Neale, G., Peng, J., Yu, J., & Chi, H. (2018). Hippo/Mst signalling couples metabolic state and immune function of CD8alpha(+) dendritic cells. *Nature, 558*(7708), 141–145. https://doi.org/10.1038/s41586-018-0177-0

Elesela, S., & Lukacs, N. W. (2021). Role of mitochondria in viral infections. *Life (Basel), 11*(3). https://doi.org/10.3390/life11030232

Erlich, T. H., Yagil, Z., Kay, G., Peretz, A., Migalovich-Sheikhet, H., Tshori, S., Nechushtan, H., Levi-Schaffer, F., Saada, A., & Razin, E. (2014). Mitochondrial STAT3 plays a major role in IgE-antigen-mediated mast cell exocytosis. *The Journal of Allergy and Clinical Immunology, 134*(2), 460–469. https://doi.org/10.1016/j.jaci.2013.12.1075

Escoll, P., Platon, L., Drame, M., Sahr, T., Schmidt, S., Rusniok, C., & Buchrieser, C. (2021). Reverting the mode of action of the mitochondrial F(O)F(1)-ATPase by Legionella pneumophila preserves its replication niche. *Elife, 10*. https://doi.org/10.7554/eLife.71978

Feng, X., Chen, W., Ni, X., Little, P. J., Xu, S., Tang, L., & Weng, J. (2021). Metformin, macrophage dysfunction and atherosclerosis. *Frontiers in Immunology, 12*, Article 682853. https://doi.org/10.3389/fimmu.2021.682853

Fisicaro, P., Barili, V., Montanini, B., Acerbi, G., Ferracin, M., Guerrieri, F., Salerno, D., Boni, C., Massari, M., Cavallo, M. C., Grossi, G., Giuberti, T., Lampertico, P., Missale, G., Levrero, M., Ottonello, S., & Ferrari, C. (2017). Targeting mitochondrial dysfunction can restore antiviral activity of exhausted HBV-specific CD8 T cells in chronic hepatitis B. *Nature Medicine, 23*(3), 327–336. https://doi.org/10.1038/nm.4275

Ganeshan, K., & Chawla, A. (2014). Metabolic regulation of immune responses. *Annual Review of Immunology, 32*, 609–634. https://doi.org/10.1146/annurev-immunol-032713-120236

Garcia-Souza, L. F., & Oliveira, M. F. (2014). Mitochondria: Biological roles in platelet physiology and pathology. *International Journal of Biochemistry and Cell Biology, 50*, 156–160. https://doi.org/10.1016/j.biocel.2014.02.015

George, J. N. (2000). Platelets. *Lancet, 355*(9214), 1531–1539. https://doi.org/10.1016/S0140-6736(00)02175-9

Giorgi, C., Baldassari, F., Bononi, A., Bonora, M., De Marchi, E., Marchi, S., Missiroli, S., Patergnani, S., Rimessi, A., Suski, J. M., Wieckowski, M. R., & Pinton, P. (2012). Mitochondrial Ca^{2+} and apoptosis. *Cell Calcium, 52*(1), 36–43. https://doi.org/10.1016/j.ceca.2012.02.008

Gokhale, N. S., Vazquez, C., & Horner, S. M. (2014). Hepatitis C virus. Strategies to evade antiviral responses. *Future Virology, 9*(12), 1061–1075. https://doi.org/10.2217/fvl.14.89

Gollihue, J. L., Patel, S. P., Eldahan, K. C., Cox, D. H., Donahue, R. R., Taylor, B. K., Sullivan, P. G., & Rabchevsky, A. G. (2018). Effects of mitochondrial transplantation on bioenergetics, cellular incorporation, and functional recovery after spinal cord injury. *Journal of Neurotrauma, 35*(15), 1800–1818. https://doi.org/10.1089/neu.2017.5605

Gostner, J. M., Becker, K., Fuchs, D., & Sucher, R. (2013). Redox regulation of the immune response. *Redox Report, 18*(3), 88–94. https://doi.org/10.1179/1351000213Y.0000000044

Guak, H., Al Habyan, S., Ma, E. H., Aldossary, H., Al-Masri, M., Won, S. Y., Ying, T., Fixman, E. D., Jones, R. G., McCaffrey, L. M., & Krawczyk, C. M. (2018). Glycolytic metabolism is essential for CCR7 oligomerization and dendritic cell migration. *Nature Communications, 9*(1), 2463. https://doi.org/10.1038/s41467-018-04804-6

Hafeez, U., Gan, H. K., & Scott, A. M. (2018). Monoclonal antibodies as immunomodulatory therapy against cancer and autoimmune diseases. *Current Opinion in Pharmacology, 41*, 114–121. https://doi.org/10.1016/j.coph.2018.05.010

Hansel, T. T., Kropshofer, H., Singer, T., Mitchell, J. A., & George, A. J. (2010). The safety and side effects of monoclonal antibodies. *Nature Reviews Drug Discovery, 9*(4), 325–338. https://doi.org/10.1038/nrd3003

Headley, C. A., Gautam, S., Olmo-Fontanez, A., Akhter, A., Schami, A., Ault, R., Zhang, H., Cai, H., Whigham, A., Delgado, J., Hicks, A., Tsao, P. S., Gelfond, J., Martinez-Sobrido, L., Wang, Y., Torrelles, J. B., Turner, J., et al. (2024). Extracellular delivery of functional mitochondria rescues the dysfunction of CD4+ T cells in aging. *Advanced Science, 11*. https://doi.org/10.1002/advs.202303664. In press.

Herrero-Cervera, A., Soehnlein, O., & Kenne, E. (2022). Neutrophils in chronic inflammatory diseases. *Cell Molecular Immunology, 19*(2), 177–191. https://doi.org/10.1038/s41423-021-00832-3

Hoffman, W., Lakkis, F. G., & Chalasani, G. (2016). B cells, antibodies, and more. *Clinical Journal of the American Society of Nephrology, 11*(1), 137–154. https://doi.org/10.2215/CJN.09430915

Hough, K. P., Trevor, J. L., Strenkowski, J. G., Wang, Y., Chacko, B. K., Tousif, S., Chanda, D., Steele, C., Antony, V. B., Dokland, T., Ouyang, X., Zhang, J., Duncan, S. R., Thannickal, V. J., Darley-Usmar, V. M., & Deshane, J. S. (2018). Exosomal transfer of mitochondria from airway myeloid-derived regulatory cells to T cells. *Redox Biology, 18*, 54–64. https://doi.org/10.1016/j.redox.2018.06.009

Ikeda, G., Santoso, M. R., Tada, Y., Li, A. M., Vaskova, E., Jung, J. H., O'Brien, C., Egan, E., Ye, J., & Yang, P. C. (2021). Mitochondria-rich extracellular vesicles from autologous stem cell-derived cardiomyocytes restore energetics of ischemic myocardium. *Journal of the American College of Cardiology, 77*(8), 1073–1088. https://doi.org/10.1016/j.jacc.2020.12.060

Ilmarinen, P., Moilanen, E., & Kankaanranta, H. (2014a). Mitochondria in the center of human eosinophil apoptosis and survival. *International Journal of Molecular Sciences, 15*(3), 3952–3969. https://doi.org/10.3390/ijms15033952

Ilmarinen, P., Moilanen, E., & Kankaanranta, H. (2014b). Regulation of spontaneous eosinophil apoptosis-a neglected area of importance. *Journal of Cell Death, 7*, 1–9. https://doi.org/10.4137/JCD.S13588

Ingulli, E. (2010). Mechanism of cellular rejection in transplantation. *Pediatric Nephrology, 25*(1), 61–74. https://doi.org/10.1007/s00467-008-1020-x

Iperi, C., Bordron, A., Dueymes, M., Pers, J. O., & Jamin, C. (2021). Metabolic Program of regulatory B lymphocytes and influence in the control of malignant and autoimmune situations. *Frontiers in Immunology, 12*, Article 735463. https://doi.org/10.3389/fimmu.2021.735463

Jastroch, M., Divakaruni, A. S., Mookerjee, S., Treberg, J. R., & Brand, M. D. (2010). Mitochondrial proton and electron leaks. *Essays in Biochemistry, 47*, 53–67. https://doi.org/10.1042/bse0470053

Jezek, J., Cooper, K. F., & Strich, R. (2018). Reactive oxygen species and mitochondrial dynamics: The yin and yang of mitochondrial dysfunction and cancer progression. *Antioxidants, 7*(1). https://doi.org/10.3390/antiox7010013

Kang, J., & Pervaiz, S. (2012). Mitochondria: Redox metabolism and dysfunction. *Biochemistry Research International, 2012*, Article 896751. https://doi.org/10.1155/2012/896751

Kelley, N., Jeltema, D., Duan, Y., & He, Y. (2019). The NLRP3 inflammasome: An overview of mechanisms of activation and regulation. *International Journal of Molecular Sciences, 20*(13). https://doi.org/10.3390/ijms20133328

Kidwell, C. U., Casalini, J. R., Pradeep, S., Scherer, S. D., Greiner, D., Bayik, D., Watson, D. C., Olson, G. S., Lathia, J. D., Johnson, J. S., Rutter, J., Welm, A. L., Zangle, T. A., & Roh-Johnson, M. (2023). Transferred mitochondria accumulate reactive oxygen species, promoting proliferation. *Elife, 12*. https://doi.org/10.7554/eLife.85494

Kim, S., Kim, Y., Yu, S. H., Lee, S. E., Park, J. H., Cho, G., Choi, C., Han, K., Kim, C. H., & Kang, Y. C. (2022). Platelet-derived mitochondria transfer facilitates wound-closure by modulating ROS levels in dermal fibroblasts. *Platelets, 34*(1). https://doi.org/10.1080/09537104.2022.2151996

Koenderman, L., Buurman, W., & Daha, M. R. (2014). The innate immune response. *Immunology Letters, 162*(2), 95–102. https://doi.org/10.1016/j.imlet.2014.10.010

Kominsky, D. J., Campbell, E. L., & Colgan, S. P. (2010). Metabolic shifts in immunity and inflammation. *Journal of Immunology, 184*(8), 4062. https://doi.org/10.4049/jimmunol.0903002

Konari, N., Nagaishi, K., Kikuchi, S., & Fujimiya, M. (2019). Mitochondria transfer from mesenchymal stem cells structurally and functionally repairs renal proximal tubular epithelial cells in diabetic nephropathy in vivo. *Scientific Reports, 9*(1), 5184. https://doi.org/10.1038/s41598-019-40163-y

Kornberg, M. D. (2020). The immunologic Warburg effect: Evidence and therapeutic opportunities in autoimmunity. *Wiley Interdisciplinary Reviews: Systems Biology and Medicine, 12*(5), e1486. https://doi.org/10.1002/wsbm.1486

Kulkarni, P. P., Ekhlak, M., Sonkar, V. K., & Dash, D. (2022). Mitochondrial ATP generation in stimulated platelets is essential for granule secretion but dispensable for aggregation and procoagulant activity. *Haematologica, 107*(5), 1209–1213. https://doi.org/10.3324/haematol.2021.279847

Kumar, A., Chamoto, K., Chowdhury, P. S., & Honjo, T. (2020). Tumors attenuating the mitochondrial activity in T cells escape from PD-1 blockade therapy. *Elife, 9*. https://doi.org/10.7554/eLife.52330

LaJeunesse, D. R., Brooks, K., & Adamson, A. L. (2005). Epstein-Barr virus immediate-early proteins BZLF1 and BRLF1 alter mitochondrial morphology during lytic replication. *Biochemical and Biophysical Research Communications, 333*(2), 438–442. https://doi.org/10.1016/j.bbrc.2005.05.120

Lanna, A., & Dustin, M. L. (2016). Mitochondrial fusion fuels T cell memory. *Cell Research, 26*(9), 969–970. https://doi.org/10.1038/cr.2016.94

Lefebvre, J. S., & Haynes, L. (2012). Aging of the CD4 T cell compartment. *Open Longevity Science, 6*, 83–91. https://doi.org/10.2174/1876326X01206010083

Levoux, J., Prola, A., Lafuste, P., Gervais, M., Chevallier, N., Koumaiha, Z., Kefi, K., Braud, L., Schmitt, A., Yacia, A., Schirmann, A., Hersant, B., Sid-Ahmed, M., Ben Larbi, S., Komrskova, K., Rohlena, J., Relaix, F., Neuzil, J., & Rodriguez, A. M. (2021). Platelets facilitate the wound-healing capability of mesenchymal stem cells by mitochondrial transfer and metabolic reprogramming. *Cell Metabolism, 33*(2), 283–299 e9. https://doi.org/10.1016/j.cmet.2020.12.006

Li, X. D., Sun, L., Seth, R. B., Pineda, G., & Chen, Z. J. (2005). Hepatitis C virus protease NS3/4A cleaves mitochondrial antiviral signaling protein off the mitochondria to evade innate immunity. *Proceedings of the National Academy of Sciences of the United States of America, 102*(49), 17717–17722. https://doi.org/10.1073/pnas.0508531102

Li, Y., Shen, Y., Jin, K., Wen, Z., Cao, W., Wu, B., Wen, R., Tian, L., Berry, G. J., Goronzy, J. J., & Weyand, C. M. (2019). The DNA repair nuclease MRE11A functions as a mitochondrial protector and prevents T cell pyroptosis and tissue inflammation. *Cell Metabolism, 30*(3), 477–492 e6. https://doi.org/10.1016/j.cmet.2019.06.016

Lightowlers, R. N., Chrzanowska-Lightowlers, Z. M., & Russell, O. M. (2020). Mitochondrial transplantation-a possible therapeutic for mitochondrial dysfunction? Mitochondrial transfer is a potential cure for many diseases but proof of efficacy and safety is still lacking. *EMBO Reports, 21*(9), Article e50964. https://doi.org/10.15252/embr.202050964

Liu, Y., Wu, M., Zhong, C., Xu, B., & Kang, L. (2022a). M2-like macrophages transplantation protects against the doxorubicin-induced heart failure via mitochondrial transfer. *Biomaterials Research, 26*(1), 14. https://doi.org/10.1186/s40824-022-00260-y

Liu, Z., Sun, Y., Qi, Z., Cao, L., & Ding, S. (2022b). Mitochondrial transfer/transplantation: An emerging therapeutic approach for multiple diseases. *Cell and Bioscience, 12*(1), 66. https://doi.org/10.1186/s13578-022-00805-7

Luckheeram, R. V., Zhou, R., Verma, A. D., & Xia, B. (2012). CD4(+)T cells: Differentiation and functions. *Clinical and Developmental Immunology, 2012*, Article 925135. https://doi.org/10.1155/2012/925135

Luz-Crawford, P., Hernandez, J., Djouad, F., Luque-Campos, N., Caicedo, A., Carrere-Kremer, S., Brondello, J. M., Vignais, M. L., Pene, J., & Jorgensen, C. (2019). Mesenchymal stem cell repression of Th17 cells is triggered by mitochondrial transfer. *Stem Cell Research and Therapy, 10*(1), 232. https://doi.org/10.1186/s13287-019-1307-9

Mahmud, N., Klipa, D., & Ahsan, N. (2010). Antibody immunosuppressive therapy in solid-organ transplant: Part I. *MAbs, 2*(2), 148–156. https://doi.org/10.4161/mabs.2.2.11159

Maianski, N. A., Geissler, J., Srinivasula, S. M., Alnemri, E. S., Roos, D., & Kuijpers, T. W. (2004). Functional characterization of mitochondria in neutrophils: A role restricted to apoptosis. *Cell Death and Differentiation, 11*(2), 143–153. https://doi.org/10.1038/sj.cdd.4401320

Masuzawa, A., Black, K. M., Pacak, C. A., Ericsson, M., Barnett, R. J., Drumm, C., Seth, P., Bloch, D. B., Levitsky, S., Cowan, D. B., & McCully, J. D. (2013). Transplantation of autologously derived mitochondria protects the heart from ischemia-reperfusion injury. *American Journal of Physiology - Heart and Circulatory Physiology, 304*(7), H966–H982. https://doi.org/10.1152/ajpheart.00883.2012

Melchinger, H., Jain, K., Tyagi, T., & Hwa, J. (2019). Role of platelet mitochondria: Life in a nucleus-free zone. *Frontiers in Cardiovascular Medicine, 6*, 153. https://doi.org/10.3389/fcvm.2019.00153

Mellman, I. (2013). Dendritic cells: Master regulators of the immune response. *Cancer Immunology Research, 1*(3), 145–149. https://doi.org/10.1158/2326-6066.CIR-13-0102

Modesti, L., Danese, A., Angela Maria Vitto, V., Ramaccini, D., Aguiari, G., Gafa, R., Lanza, G., Giorgi, C., & Pinton, P. (2021). Mitochondrial Ca^{2+} signaling in health, disease and therapy. *Cells, 10*(6). https://doi.org/10.3390/cells10061317

Mogensen, T. H., & Paludan, S. R. (2001). Molecular pathways in virus-induced cytokine production. *Microbiology and Molecular Biology Reviews, 65*(1), 131–150. https://doi.org/10.1128/MMBR.65.1.131-150.2001

Mohanty, A., Tiwari-Pandey, R., & Pandey, N. R. (2019). Mitochondria: The indispensable players in innate immunity and guardians of the inflammatory response. *Journal of Cell Communication and Signaling, 13*(3), 303–318. https://doi.org/10.1007/s12079-019-00507-9

Moller, S. H., Wang, L., & Ho, P. C. (2022). Metabolic programming in dendritic cells tailors immune responses and homeostasis. *Cell Molecular Immunology, 19*(3), 370–383. https://doi.org/10.1038/s41423-021-00753-1

Morianos, I., & Semitekolou, M. (2020). Dendritic cells: Critical regulators of allergic asthma. *International Journal of Molecular Sciences, 21*(21). https://doi.org/10.3390/ijms21217930

Motley, M. P., Banerjee, K., & Fries, B. C. (2019). Monoclonal antibody-based therapies for bacterial infections. *Current Opinion in Infectious Diseases, 32*(3), 210–216. https://doi.org/10.1097/QCO.0000000000000539

Nathan, C. (2002). Points of control in inflammation. *Nature, 420*(6917), 846–852. https://doi.org/10.1038/nature01320

Nicholson, L. B. (2016). The immune system. *Essays in Biochemistry, 60*(3), 275–301. https://doi.org/10.1042/EBC20160017

Norris, R. P. (2021). Transfer of mitochondria and endosomes between cells by gap junction internalization. *Traffic, 22*(6), 174–179. https://doi.org/10.1111/tra.12786

O'Neill, L. A., Kishton, R. J., & Rathmell, J. (2016). A guide to immunometabolism for immunologists. *Nature Reviews Immunology, 16*(9), 553–565. https://doi.org/10.1038/nri.2016.70

Osellame, L. D., Blacker, T. S., & Duchen, M. R. (2012). Cellular and molecular mechanisms of mitochondrial function. *Best Practice and Research Clinical Endocrinology and Metabolism, 26*(6), 711–723. https://doi.org/10.1016/j.beem.2012.05.003

Park, J., Choi, H., Min, J. S., Park, S. J., Kim, J. H., Park, H. J., Kim, B., Chae, J. I., Yim, M., & Lee, D. S. (2013). Mitochondrial dynamics modulate the expression of pro-inflammatory mediators in microglial cells. *Journal of Neurochemistry, 127*(2), 221–232. https://doi.org/10.1111/jnc.12361

Patel, C. H., & Powell, J. D. (2017). Targeting T cell metabolism to regulate T cell activation, differentiation and function in disease. *Current Opinion in Immunology, 46*, 82–88. https://doi.org/10.1016/j.coi.2017.04.006

Paupe, V., & Prudent, J. (2018). New insights into the role of mitochondrial calcium homeostasis in cell migration. *Biochemical and Biophysical Research Communications, 500*(1), 75–86. https://doi.org/10.1016/j.bbrc.2017.05.039

Peng, W., Cai, G., Xia, Y., Chen, J., Wu, P., Wang, Z., Li, G., & Wei, D. (2019). Mitochondrial dysfunction in atherosclerosis. *DNA and Cell Biology, 38*(7), 597–606. https://doi.org/10.1089/dna.2018.4552

Peng, Y., Liu, H., Liu, J., & Long, J. (2022). Post-translational modifications on mitochondrial metabolic enzymes in cancer. *Free Radical Biology and Medicine, 179*, 11–23. https://doi.org/10.1016/j.freeradbiomed.2021.12.264

Perez, S., & Rius-Perez, S. (2022). Macrophage polarization and reprogramming in acute inflammation: A redox perspective. *Antioxidants, 11*(7). https://doi.org/10.3390/antiox11071394

Picard, M., Wallace, D. C., & Burelle, Y. (2016). The rise of mitochondria in medicine. *Mitochondrion, 30*, 105–116. https://doi.org/10.1016/j.mito.2016.07.003

Piekarska, K., Urban-Wojciuk, Z., Kurkowiak, M., Pelikant-Malecka, I., Schumacher, A., Sakowska, J., Spodnik, J. H., Arcimowicz, L., Zielinska, H., Tymoniuk, B., Renkielska, A., Siebert, J., Slominska, E., Trzonkowski, P., Hupp, T., & Marek-Trzonkowska, N. M. (2022). Mesenchymal stem cells transfer mitochondria to allogeneic Tregs in an HLA-dependent manner improving their immunosuppressive activity. *Nature Communications, 13*(1), 856. https://doi.org/10.1038/s41467-022-28338-0

Plotnikov, E. Y., Babenko, V. A., Silachev, D. N., Zorova, L. D., Khryapenkova, T. G., Savchenko, E. S., Pevzner, I. B., & Zorov, D. B. (2015). Intercellular transfer of mitochondria. *Biochemistry, 80*(5), 542–548. https://doi.org/10.1134/S0006297915050041

Posner, J., Barrington, P., Brier, T., & Datta-Mannan, A. (2019). Monoclonal antibodies: Past, present and future, *260*. *Handbook of Experimental Pharmacology* (pp. 81–141). https://doi.org/10.1007/164_2019_323

Quintana, A., Schwindling, C., Wenning, A. S., Becherer, U., Rettig, J., Schwarz, E. C., & Hoth, M. (2007). T cell activation requires mitochondrial translocation to the immunological synapse. *Proceedings of the National Academy of Sciences of the United States of America, 104*(36), 14418–14423. https://doi.org/10.1073/pnas.0703126104

Reiser, J., & Banerjee, A. (2016). Effector, memory, and dysfunctional CD8(+) T cell fates in the antitumor immune response. *Journal of Immunology Research, 2016*, Article 8941260. https://doi.org/10.1155/2016/8941260

Rimessi, A., Previati, M., Nigro, F., Wieckowski, M. R., & Pinton, P. (2016). Mitochondrial reactive oxygen species and inflammation: Molecular mechanisms, diseases and promising therapies. *International Journal of Biochemistry and Cell Biology, 81*, 281–293. https://doi.org/10.1016/j.biocel.2016.06.015

Ron-Harel, N., Sharpe, A. H., & Haigis, M. C. (2015). Mitochondrial metabolism in T cell activation and senescence: A mini-review. *Gerontology, 61*(2), 131–138. https://doi.org/10.1159/000362502

Rosales, C. (2018). Neutrophil: A cell with many roles in inflammation or several cell types? *Frontiers in Physiology, 9*, 113. https://doi.org/10.3389/fphys.2018.00113

Roushandeh, A. M., Kuwahara, Y., & Roudkenar, M. H. (2019). Mitochondrial transplantation as a potential and novel master key for treatment of various incurable diseases. *Cytotechnology, 71*(2), 647–663. https://doi.org/10.1007/s10616-019-00302-9

Rousselle, T. V., Kuscu, C., Kuscu, C., Schlegel, K., Huang, L., Namwanje, M., Eason, J. D., Makowski, L., Maluf, D., Mas, V., & Bajwa, A. (2020). FTY720 regulates mitochondria biogenesis in dendritic cells to prevent kidney ischemic reperfusion injury. *Frontiers in Immunology, 11*, 1278. https://doi.org/10.3389/fimmu.2020.01278

Rus, H., Cudrici, C., & Niculescu, F. (2005). The role of the complement system in innate immunity. *Immunologic Research, 33*(2), 103–112. https://doi.org/10.1385/IR:33:2:103

Saha, T., Dash, C., Jayabalan, R., Khiste, S., Kulkarni, A., Kurmi, K., Mondal, J., Majumder, P. K., Bardia, A., Jang, H. L., & Sengupta, S. (2022). Intercellular nanotubes mediate mitochondrial trafficking between cancer and immune cells. *Nature Nanotechnology, 17*(1), 98–106. https://doi.org/10.1038/s41565-021-01000-4

Sandoval, H., Kodali, S., & Wang, J. (2018). Regulation of B cell fate, survival, and function by mitochondria and autophagy. *Mitochondrion, 41*, 58–65. https://doi.org/10.1016/j.mito.2017.11.005

Santos, J. H. (2021). Mitochondria signaling to the epigenome: A novel role for an old organelle. *Free Radical Biology and Medicine, 170*, 59–69. https://doi.org/10.1016/j.freeradbiomed.2020.11.016

Scozzi, D., Ibrahim, M., Liao, F., Lin, X., Hsiao, H. M., Hachem, R., Tague, L. K., Ricci, A., Kulkarni, H. S., Huang, H. J., Sugimoto, S., Krupnick, A. S., Kreisel, D., & Gelman, A. E. (2019). Mitochondrial damage-associated molecular patterns released by lung transplants are associated with primary graft dysfunction. *American Journal of Transplantation, 19*(5), 1464–1477. https://doi.org/10.1111/ajt.15232

Sena, L. A., & Chandel, N. S. (2012). Physiological roles of mitochondrial reactive oxygen species. *Molecular Cell, 48*(2), 158–167. https://doi.org/10.1016/j.molcel.2012.09.025

Sercel, A. J., Patananan, A. N., Man, T., Wu, T. H., Yu, A. K., Guyot, G. W., Rabizadeh, S., Niazi, K. R., Chiou, P. Y., & Teitell, M. A. (2021). Stable transplantation of human mitochondrial DNA by high-throughput, pressurized isolated mitochondrial delivery. *Elife, 10*. https://doi.org/10.7554/eLife.63102

Sharma, P., & Allison, J. P. (2015). Immune checkpoint targeting in cancer therapy: Toward combination strategies with curative potential. *Cell, 161*(2), 205–214. https://doi.org/10.1016/j.cell.2015.03.030

Shen, K., Pender, C. L., Bar-Ziv, R., Zhang, H., Wickham, K., Willey, E., Durieux, J., Ahmad, Q., & Dillin, A. (2022). Mitochondria as cellular and organismal signaling hubs. *Annual Review of Cell and Developmental Biology, 38*, 179–218. https://doi.org/10.1146/annurev-cellbio-120420-015303

Shiravand, Y., Khodadadi, F., Kashani, S. M. A., Hosseini-Fard, S. R., Hosseini, S., Sadeghirad, H., Ladwa, R., O'Byrne, K., & Kulasinghe, A. (2022). Immune checkpoint inhibitors in cancer therapy. *Current Oncology, 29*(5), 3044–3060. https://doi.org/10.3390/curroncol29050247

Sokol, C. L., & Luster, A. D. (2015). The chemokine system in innate immunity. *Cold Spring Harbor Perspectives in Biology, 7*(5). https://doi.org/10.1101/cshperspect.a016303

Soto-Heredero, G., Desdin-Mico, G., & Mittelbrunn, M. (2021). Mitochondrial dysfunction defines T cell exhaustion. *Cell Metabolism, 33*(3), 470–472. https://doi.org/10.1016/j.cmet.2021.02.010

Soto-Heredero, G., Gomez de Las Heras, M. M., Gabande-Rodriguez, E., Oller, J., & Mittelbrunn, M. (2020). Glycolysis - a key player in the inflammatory response. *FEBS Journal, 287*(16), 3350–3369. https://doi.org/10.1111/febs.15327

Stram, A. R., & Payne, R. M. (2016). Post-translational modifications in mitochondria: Protein signaling in the powerhouse. *Cellular and Molecular Life Sciences, 73*(21), 4063–4073. https://doi.org/10.1007/s00018-016-2280-4

Striz, I., Brabcova, E., Kolesar, L., & Sekerkova, A. (2014). Cytokine networking of innate immunity cells: A potential target of therapy. *Clinical Science, 126*(9), 593–612. https://doi.org/10.1042/CS20130497

Sun, L., Yang, X., Yuan, Z., & Wang, H. (2020). Metabolic reprogramming in immune response and tissue inflammation. *Arteriosclerosis, Thrombosis, and Vascular Biology, 40*(9), 1990–2001. https://doi.org/10.1161/ATVBAHA.120.314037

Sun, X., & St John, J. C. (2018). Modulation of mitochondrial DNA copy number in a model of glioblastoma induces changes to DNA methylation and gene expression of the nuclear genome in tumours. *Epigenetics and Chromatin, 11*(1), 53. https://doi.org/10.1186/s13072-018-0223-z

Topalian, S. L., Drake, C. G., & Pardoll, D. M. (2015). Immune checkpoint blockade: A common denominator approach to cancer therapy. *Cancer Cell, 27*(4), 450–461. https://doi.org/10.1016/j.ccell.2015.03.001

Trebak, M., & Kinet, J. P. (2019). Calcium signalling in T cells. *Nature Reviews Immunology, 19*(3), 154–169. https://doi.org/10.1038/s41577-018-0110-7

Urbanczyk, S., Baris, O. R., Hofmann, J., Taudte, R. V., Guegen, N., Golombek, F., Castiglione, K., Meng, X., Bozec, A., Thomas, J., Weckwerth, L., Mougiakakos, D., Schulz, S. R., Schuh, W., Schlotzer-Schrehardt, U., Steinmetz, T. D., Brodesser, S., Wiesner, R. J., & Mielenz, D. (2022). Mitochondrial respiration in B lymphocytes is essential for humoral immunity by controlling the flux of the TCA cycle. *Cell Reports, 39*(10), Article 110912. https://doi.org/10.1016/j.celrep.2022.110912

Uribe-Querol, E., & Rosales, C. (2015). Neutrophils in cancer: Two sides of the same coin. *Journal of Immunology Research, 2015*, Article 983698. https://doi.org/10.1155/2015/983698

Uthaisangsook, S., Day, N. K., Bahna, S. L., Good, R. A., & Haraguchi, S. (2002). Innate immunity and its role against infections. *Annals of Allergy, Asthma, and Immunology, 88*(3), 253–264. https://doi.org/10.1016/S1081-1206(10)62005-4

Van den Bossche, J., Baardman, J., Otto, N. A., van der Velden, S., Neele, A. E., van den Berg, S. M., Luque-Martin, R., Chen, H. J., Boshuizen, M. C., Ahmed, M., Hoeksema, M. A., de Vos, A. F., & de Winther, M. P. (2016). Mitochondrial dysfunction prevents repolarization of inflammatory macrophages. *Cell Reports, 17*(3), 684–696. https://doi.org/10.1016/j.celrep.2016.09.008

van der Meijden, P. E. J., & Heemskerk, J. W. M. (2019). Platelet biology and functions: New concepts and clinical perspectives. *Nature Reviews Cardiology, 16*(3), 166–179. https://doi.org/10.1038/s41569-018-0110-0

van der Vlist, M., Raoof, R., Willemen, H., Prado, J., Versteeg, S., Martin Gil, C., Vos, M., Lokhorst, R. E., Pasterkamp, R. J., Kojima, T., Karasuyama, H., Khoury-Hanold, W., Meyaard, L., & Eijkelkamp, N. (2022). Macrophages transfer mitochondria to sensory neurons to resolve inflammatory pain. *Neuron, 110*(4), 613–626 e9. https://doi.org/10.1016/j.neuron.2021.11.020

Vardhana, S. A., Hwee, M. A., Berisa, M., Wells, D. K., Yost, K. E., King, B., Smith, M., Herrera, P. S., Chang, H. Y., Satpathy, A. T., van den Brink, M. R. M., Cross, J. R., & Thompson, C. B. (2020). Impaired mitochondrial oxidative phosphorylation limits the self-renewal of T cells exposed to persistent antigen. *Nature Immunology, 21*(9), 1022–1033. https://doi.org/10.1038/s41590-020-0725-2

Verdon, D. J., Mulazzani, M., & Jenkins, M. R. (2020). Cellular and molecular mechanisms of CD8(+) T cell differentiation, dysfunction and exhaustion. *International Journal of Molecular Sciences, 21*(19). https://doi.org/10.3390/ijms21197357

Vorobjeva, N., Prikhodko, A., Galkin, I., Pletjushkina, O., Zinovkin, R., Sud'ina, G., Chernyak, B., & Pinegin, B. (2017). Mitochondrial reactive oxygen species are involved in chemoattractant-induced oxidative burst and degranulation of human neutrophils in vitro. *European Journal of Cell Biology, 96*(3), 254–265. https://doi.org/10.1016/j.ejcb.2017.03.003

Vorobjeva, N. V., & Chernyak, B. V. (2020). NETosis: Molecular mechanisms, role in physiology and pathology. *Biochemistry, 85*(10), 1178–1190. https://doi.org/10.1134/S0006297920100065

Wang, L. W., Shen, H., Nobre, L., Ersing, I., Paulo, J. A., Trudeau, S., Wang, Z., Smith, N. A., Ma, Y., Reinstadler, B., Nomburg, J., Sommermann, T., Cahir-McFarland, E., Gygi, S. P., Mootha, V. K., Weekes, M. P., & Gewurz, B. E. (2019). Epstein-barr-virus-induced one-carbon metabolism drives B cell transformation. *Cell Metabolism, 30*(3), 539–555 e11. https://doi.org/10.1016/j.cmet.2019.06.003

Wang, L., Wu, Q., Fan, Z., Xie, R., Wang, Z., & Lu, Y. (2017). Platelet mitochondrial dysfunction and the correlation with human diseases. *Biochemical Society Transactions, 45*(6), 1213–1223. https://doi.org/10.1042/BST20170291

Wang, R., Maimaitijuma, T., Ma, Y. Y., Jiao, Y., & Cao, Y. P. (2020). Mitochondrial transfer from bone-marrow-derived mesenchymal stromal cells to chondrocytes protects against cartilage degenerative mitochondrial dysfunction in rats chondrocytes. *Chinese Medical Journal, 134*(2), 212–218. https://doi.org/10.1097/CM9.0000000000001057

Wang, Y., Li, N., Zhang, X., & Horng, T. (2021). Mitochondrial metabolism regulates macrophage biology. *Journal of Biological Chemistry, 297*(1), Article 100904. https://doi.org/10.1016/j.jbc.2021.100904

Weiner, L. M., Dhodapkar, M. V., & Ferrone, S. (2009). Monoclonal antibodies for cancer immunotherapy. *Lancet, 373*(9668), 1033–1040. https://doi.org/10.1016/S0140-6736(09)60251-8

Weng, X., Kumar, A., Cao, L., He, Y., Morgun, E., Visvabharathy, L., Zhao, J., Sena, L. A., Weinberg, S. E., Chandel, N. S., & Wang, C. R. (2021). Mitochondrial metabolism is essential for invariant natural killer T cell development and function. *Proceedings of the National Academy of Sciences of the United States of America, 118*(13). https://doi.org/10.1073/pnas.2021385118

Wu, B., Zhao, T. V., Jin, K., Hu, Z., Abdel, M. P., Warrington, K. J., Goronzy, J. J., & Weyand, C. M. (2021). Mitochondrial aspartate regulates TNF biogenesis and autoimmune tissue inflammation. *Nature Immunology, 22*(12), 1551–1562. https://doi.org/10.1038/s41590-021-01065-2

Yanes, R. E., Zhang, H., Shen, Y., Weyand, C. M., & Goronzy, J. J. (2019). Metabolic reprogramming in memory CD4 T cell responses of old adults. *Clinical Immunology, 207*, 58–67. https://doi.org/10.1016/j.clim.2019.07.003

Yazicioglu, Y. F., Marin, E., Sandhu, C., Galiani, S., Raza, I. G. A., Ali, M., Kronsteiner, B., Compeer, E. B., Attar, M., Dunachie, S. J., Dustin, M. L., & Clarke, A. J. (2023). Dynamic mitochondrial transcription and translation in B cells control germinal center entry and lymphomagenesis. *Nature Immunology, 24*(6), 991–1006. https://doi.org/10.1038/s41590-023-01484-3

Yuan, Z., Lu, Y., Wei, J., Wu, J., Yang, J., & Cai, Z. (2020). Abdominal aortic aneurysm: Roles of inflammatory cells. *Frontiers in Immunology, 11*. https://doi.org/10.3389/fimmu.2020.609161

Zhang, J. M., & An, J. (2007). Cytokines, inflammation, and pain. *International Anesthesiology Clinics, 45*(2), 27–37. https://doi.org/10.1097/AIA.0b013e318034194e

Zhang, N., & Bevan, M. J. (2011). CD8(+) T cells: Foot soldiers of the immune system. *Immunity, 35*(2), 161. https://doi.org/10.1016/j.immuni.2011.07.010

Zhang, Z., Yan, C., Miao, J., Pu, K., Ma, H., & Wang, Q. (2021). Muscle-derived mitochondrial transplantation reduces inflammation, enhances bacterial clearance, and improves survival in sepsis. *Shock, 56*(1), 108–118. https://doi.org/10.1097/SHK.0000000000001681

Zhou, W., Cao, L., Jeffries, J., Zhu, X., Staiger, C. J., & Deng, Q. (2018). Neutrophil-specific knockout demonstrates a role for mitochondria in regulating neutrophil motility in zebrafish. *Disease Models and Mechanisms, 11*(3). https://doi.org/10.1242/dmm.033027

Zhu, L., Zhao, Q., Yang, T., Ding, W., & Zhao, Y. (2015). Cellular metabolism and macrophage functional polarization. *International Reviews of Immunology, 34*(1), 82–100. https://doi.org/10.3109/08830185.2014.969421

CHAPTER

25

The past, present, and close future of mitochondria as a therapeutic agent

Romina Maya[1,2,3], *Annalia Valdivieso*[1,2,3], *Paola Robayo*[1,2,3] *and Andrés Caicedo*[1,2,3,4,5]

[1]Universidad San Francisco de Quito USFQ, Colegio de Ciencias de la Salud, Escuela de Medicina, Quito, Ecuador; [2]Universidad San Francisco de Quito USFQ, Instituto de Investigaciones en Biomedicina iBioMed, Quito, Ecuador; [3]Mito-Act Research Consortium, Quito, Ecuador; [4]Sistemas Médicos SIME, Universidad San Francisco de Quito USFQ, Quito, Ecuador; [5]Universidad San Francisco de Quito USFQ, USFQ Space Front, Quito, Ecuador

Introduction

Mitochondria are unique organelles that are part of eukaryotic cells, thanks to an endosymbiotic relationship that happened approximately two billion years ago. This relationship gave the cell a tremendous evolutionary advantage that led to the diversification of life on earth (Nobs et al., 2022). Mitochondria are known as the "powerhouse of the cell," as they play a fundamental role in ATP production and in several cellular processes, including cell differentiation, proliferation, apoptosis, and autophagy, among others (Chinnery & Schon, 2003; Hamanaka & Chandel, 2010; Lill & Mühlenhoff, 2008; McBride et al., 2006). Furthermore, the newly discovered functions performed by mitochondria are not only limited to what happens inside the cell but also how they participate in cell to cell communication and repair. Mitochondria can be transferred through tunneling nanotubes (TNTs), exist extracellularly inside microvesicles (MVs), or free of a membrane cover, referred as FreeMitos (Caicedo & Singh, 2023; Caicedo et al., 2024; Caicedo, Zambrano, Sanon, & Gavilanes, 2021; Caicedo, Zambrano, Sanon, Luis Vélez, et al., 2021; Nunnari & Suomalainen, 2012; Zambrano, Barba, Castillo, Noboa, et al., 2022).

The transfer of mitochondria from one cell to another, especially from mesenchymal stem/stromal cells (MSCs) to damaged cells and tissues, restores their loss of functionality and viability (Babenko et al., 2018; Islam et al., 2012). This evidence, together with the symbiotic and extracellular origin of this organelle, inspired scientists and physicians to explore the possibility of using artificial mitochondrial transfer/transplant (AMT/T) as a method to

Mitochondrial Transplantation and Transfer
https://doi.org/10.1016/B978-0-443-18858-9.00025-1

regenerate harmed cells and tissues affected by disease or lesions (Caicedo & Singh, 2023). The use of AMT/T implies different techniques based on the isolation of mitochondria and their transfer to cells and tissues. These different techniques include the "ex vivo transfer to cells" (mitochondrial augmentation therapy MAT, MitoCeption, among others), the in situ administration at the wounded site by injecting a mitochondrial rich solution, aerosolized compositions, and the mix of mitochondria with a hydrogel matrix. These techniques have shown positive and promising results from in vitro to clinical studies, opening the possibility of using mitochondria as a therapeutic agent and to treat different conditions (Cabrera et al., 2022; Jacoby et al., 2018, 2021, 2022; McCully and Del Nido, 2020; Patel et al., 2022).

Worldwide interest in the use of AMT/T was reflected in the financial support of approximately 40 million USD from international investors in start-ups and companies (based on available and public information). Mitochondrial companies (MitoCompanies) aim to use AMT/T to treat genetic mitochondrial related disorders, chronic diseases, and wounds. Numerous patents have been granted or are patent pending to scientists belonging to these companies that are at the cutting edge of this new therapeutic option. The scope of these patents includes mitochondrial compositions, products and methods for isolation and transfer, the enrichment of mammalian cells with functional mitochondria, mitochondrial enriched genetically modified cells, aerosolized compositions with mitochondria, prophylactic, and then therapeutic use and combined mitochondrial agents, among other inventions (CO2022015278A2 - Genetically modified cells enriched with mitochondria and their uses - Google Patents, n.d.; Schmitt et al., 2013; US11441124B2 - mammalian cells enriched with functional mitochondria - Google Patents, n.d.; Yivgi-Ohana & Halavee, 2019). In this work, we provide key information about how mitochondria is naturally transferred between cells, the exciting and recent evidence of the clinical applications of AMT/T, and their therapeutic application in the close future.

How mitochondria are exchanged from one cell to another

Mitochondrial integrity and functionality are key to maintaining cellular and tissue health. Different processes that support mitochondrial fitness can be highlighted, such as germline quality control during ovogenesis, selective DNA replication, asymmetric division during mitosis, and mitophagy (Palozzi et al., 2022; Torralba et al., 2016). Interestingly, recent studies have shown that cells can maintain their mitochondrial fitness by uptaking them horizontally from donor cells (horizontal mitochondrial transfer). This transfer can occur between cells of the same or different lineages or from healthy to damaged cells. The transfer and incorporation process of mitochondria generate changes in the metabolism of the recipient cell, altering or repairing certain functional properties (Velarde et al., 2022). It has even been suggested that this process can induce stem cell differentiation, reprogram differentiated cells, and activate proinflammatory or immune regulatory signaling pathways (Caicedo, Zambrano, Sanon, & Gavilanes, 2021; Torralba et al., 2016). Some described ways of mitochondrial transport include TNTs formation, MVs, gap junctions, or cell fusion (Paliwal et al., 2018). Although this process has been widely reported, the mechanisms of mitochondrial trafficking, internalization, and effects inside the recipient cell are still under investigation.

Cell to cell communication is necessary to induce the mitochondrial transfer process (Liu et al., 2021). It has been reported that stress factors can induce the formation of TNTs leading to tubular shaped membrane structures whose dimensions vary between 50 and 1500 nm in diameter and between 10 and 100 μm in length (Liu et al., 2021; Vignais et al., 2017). TNTs allow the trafficking of different intracellular elements thanks to the continuity of the cytoplasmic space between the linked cells. The formation of these connections can occur by two means: first, through the extension of protrusions similar to filopodia that cross and fuse with the recipient cell; and second, through the maintenance of a membrane link when contiguous cells separate from one another (Vignais et al., 2017). The presence of actin filaments and microtubules are fundamental in the process of TNT development and organelle transport. TNTs are structures not attached to the extracellular matrix space, and therefore they depend on a cytoskeleton that allows membrane and cytoplasm continuity between cells. Additionally, their presence, in conjunction with various motor associated adaptor proteins, is responsible for the long distance movement of mitochondria (Melkov & Abdu, 2018).

Another important process of mitochondrial trafficking depends on extracellular vesicles (EVs). EVs correspond to a diverse group of spherical shaped lipid bilayer membrane structures that usually have different origins, sizes, and functions (Velarde et al., 2022). In general, the dimensions of EVs are highly variable, ranging from 50 to 500 nm or 1–10 μm. EVs usually transport various biomolecules, metabolites, and even iRNA and mtDNA (Caicedo et al., 2017). These have received various classifications dividing EVs into exosomes, MVs, and apoptotic bodies (Doyle & Wang, 2019). Exosomes are produced through an endosomal pathway, and MVs from the plasma membrane (van Niel et al., 2018). MVs may contain whole or partial loads of mitochondria, whereas exosomes usually possess only mtDNA (Velarde et al., 2022). The transport of mitochondria through MVs has been observed in different cell types as a response to stress conditions not only for the restoration of cellular health but also for the transcellular degradation process in which organelles are recycled (Crewe et al., 2021; D'Souza et al., 2021).

TNTs and MVs are the most commonly observed mechanisms of natural mitochondria transfer between cells. This can be attributed to the fact that most of the research methods to estimate mitochondrial transfer are more easily performed in vitro and by coculture systems. However, gap junctions, which are membrane protein microdomains of intercellular communication, are also involved in mitochondria mobilization between cells in vivo (Evans & Martin, 2002; Islam et al., 2012). These structures are formed by proteins known as connexins in vertebrate organisms. Interestingly, it has been observed that MSCs formed gap junctions containing connexin 43 with alveolar cells releasing MVs with mitochondria that the epithelia absorbed (Nielsen et al., 2012). Additionally, another means of exchange of organelles and cytosolic compounds is cell fusion, where different cells join their cytoplasm transiently or permanently to share mitochondria. It is believed that this process is triggered by injury or inflammation and that it can influence the stemness of cells and oncogenesis (Qin et al., 2021).

In 2020, Al Amir Dache and colleagues detected cell free functional mitochondria free of wrapping membranes in the blood of healthy individuals (Al Amir Dache et al., 2020). The evidence of the physiological presence of extracellular mitochondria opened the possibility to investigate the health implications of FreeMitos and their origin (Caicedo et al., 2024; Caicedo, Zambrano, Sanon, Luis Vélez, et al., 2021). This same year, Yong Zhao's team provided further data of the presence of circulating mitochondria in human and animal blood. These mitochondria displayed the immune tolerance associated molecules CD270 and PD-L1 and the capacity to contribute to T and B cell activation and reduce the production of inflammatory cytokines (Song et al., 2020). Interestingly, Crewe and colleagues detected that adipocytes respond to stress by releasing small EVs (sEVs) with respiratory competent but damaged mitochondria. These sEVs mitochondria were taken up by cardiomyocytes increasing reactive oxygen species (ROS) production and a compensatory antioxidant mechanism. This effect led to a decreased cardiac ischemia/reperfusion injury in mice (Crewe et al., 2021). Brestoff's team determined that the majority of the extracellular mitochondria from adipocytes were negative for EV markers but positive for TOM22. This suggested the presence of FreeMitos mostly derived from white adipocytes, confirming a heterogeneous diversity of mitochondrial types (Borcherding et al., 2022; Caicedo et al., 2024). Furthermore, it was observed that white adipocytes transfer mitochondria to macrophages in order to maintain metabolic homeostasis. However, this process is inhibited in obese individuals releasing a greater quantity of FreeMitos in blood (Borcherding et al., 2022). Understanding the origin, diversity, and coexistence of extracellular mitochondria in circulation and in the cerebrospinal fluid may lead to better therapeutic cues for systemic diseases (Caicedo, Zambrano, Sanon, & Gavilanes, 2021).

Evidence of the existence of physiologically competent extracellular mitochondria, their interorgan and cell transfer, has inspired an entire field to work on how to replicate this process artificially for further clinical applications. Key questions still need to be answered in order to develop new therapeutic approaches and use AMT/T in circulation or in situ lesions (Balcázar et al., 2020). The future of the physiological implications of the presence of extracellular mitochondria will provide new knowledge on their origin and dynamics. The possible presence of distinct types of extracellular mitochondria has been suggested, varying in their concentration as we age and face disease, and even between people on earth and astronauts in space (Bezdan et al., 2020; Bisserier et al., 2021; Gómez et al., 2021).

History and present of AMT/T

In 1982, Clark and Shay where one of the first to describe the transfer process of functional isolated mitochondria from antibiotic resistant cells to antibiotic sensitive cells. This process resulted in passing the resistance, coded in the mitochondrial DNA (mtDNA) and belonging to the extracellular mitochondria that were being internalized by recipient cells. Clark and Shay's assays are now considered to be part of the firsts AMT/T experiments (Caicedo et al., 2017; Clark & Shay, 1982). It is known that certain limitations may restrict the use of AMT/T. Among these is the large amount of starting tissue needed in order to obtain a significant amount of transferable mitochondria and the time the process would take to complete, requiring 90–120 minutes. Hence, a rapid mitochondrial isolation methodology has been developed, requiring only a small amount of tissue and less than 30 min (McCully et al., 2022a, 2022b). By using this procedure, isolated mitochondria can be delivered by direct injection or by vascular infusion (McCully et al., 2022a, 2022b).

Different studies have shown that AMT/T recovers cell function in damaged cells. Both direct and intraarterial delivery methods have shown that the transplanted mitochondria remain in the treated tissue (McCully et al., 2022a, 2022b). Nevertheless, there are also different methods that may aid the mitochondrial uptake from cells: first, the use of cell-penetrating peptide (Pep-1); second, the use of centrifugation (MitoCeption); and last, the use of isolated mitochondria linked to a carrier that allows the injected organelle to be directed to the liver. This last linker would lead to the use of systemic delivery of mitochondria (Cabrera et al., 2019, 2022; Caicedo et al., 2015; Chang et al., 2017; McCully et al., 2022a, 2022b; Nzigou Mombo et al., 2017).

Mitochondria showing outer membrane integrity and viable respiration seem to be more effectively transferred between cells or by AMT/T for organ function recovery. Studies have shown that AMT/T increases ATP content and promotes myocardial viability after 2 hours of reperfusion and 28 days of recovery after ischemia and reperfusion injury in myocardiocytes. In other words, it increases energy and oxygen uptake as well as cellular respiration in cardiac cells after damage. Lastly, AMT/T also plays an important role in cardioprotection through the release of different cytokines for angiogenesis, immunomodulation, apoptosis prevention, and progenitor cell migration in the lesion's microenvironment (McCully et al., 2022a, 2022b).

It has been established that the concentration of mitochondria needed for an effective therapeutic effect may depend on the organ to be treated. For example, it has been proposed that when treating the myocardium, concentrations of 1×10^5–1×10^6 per gram wet heart weight of mitochondria might be efficacious (McCully et al., 2009; McCully et al., 2022a, 2022b; Shin et al., 2019). In other in vivo studies, mitochondrial concentrations might need to be increased when modeling pulmonary or skeletal muscle damage (Moskowitzova et al., 2020). On the other hand, the use of multiple mitochondrial injections in pre- and post-ischemic scenarios showed not to be statistically different regarding cardioprotection (Guariento, Blitzer, et al., 2020; Guariento, Doulamis, et al., 2020; McCully et al., 2022a, 2022b; Shin et al., 2019).

Nowadays, the AMT/T therapeutic effects continue to be studied in different organ lesions modeled in animals. Some of these studies are focused on the repair of neurodegenerative disorders. It has been suggested that AMT/T could be a good therapeutic candidate for prion disease, Alzheimer's, and Parkinson's (Zambrano, Barba, Castillo, Noboa, et al., 2022; Zambrano, Barba, Castillo, Robayo, Argueta-Zamora, et al., 2022; Zambrano, Barba, Castillo, Robayo, Arizaga, et al., 2022). As an example, the use of AMT/T led to a recovery of the mitochondrial function and reduced brain degeneration in a Parkinson's disease model (Chang et al., 2016). Additionally, AMT/T reduced cellular oxidative stress, apoptosis, reactive astrogliosis, and brain infarct volume after a middle cerebral artery stroke in rats (Zhang et al., 2019). Despite the improvements observed when using AMT/T in treating neurodegenerative diseases and spinal cord injuries, this evidence still needs to be further analyzed to extrapolate its use in clinical scenarios (Patel et al., 2022).

In addition, AMT/T has been proven to reduce ROS stress and increase the expression of neurotrophic factors in injured nerves and denervated muscles in rats (Kuo et al., 2017). On the other hand, AMT/T was found to reduce neutrophil infiltration, apoptosis, and interstitial edema either by direct delivery to the pulmonary artery or by aerosol through the trachea in a lung ischemic scenario in a murine model (Moskowitzova et al., 2020). Moreover, AMT/T decreased serum enzyme markers of liver injury in a liver ischemia/reperfusion injury in rats (Lin et al., 2013). Lastly, AMT/T enhanced histological and functional restoration even more than the administration of N-acetylcysteine in a rat model of acetaminophen induced liver toxicity (Ulger et al., 2021).

Furthermore, AMT/T restored renal function and viability in both rat and pig models of acute kidney injury (AKI) (Doulamis et al., 2020; Jabbari et al., 2020). In another study, the transplantation of mitochondria derived from MSCs reduced cellular oxidative stress and protein accumulation and encouraged tubular cell regeneration in a rat model of AKI secondary to doxorubicin nephrotoxicity (Kubat et al., 2021). Last but not least, after an acute limb ischemia in a murine model, AMT/T reduced apoptosis and the size of the infarct (McCully et al., 2022a, 2022b; Orfany et al., 2020). This key evidence of the therapeutic use of AMT/T in animal models supported the development of clinical trials.

To further assess the effectiveness and safety of AMT/T, current clinical trials could be expanded. For instance, the study conducted at Boston Children's Hospital provides an example of such a trial, where the impact of AMT/T on patients requiring postcardiotomy extracorporeal membrane oxygenation (ECMO) for refractory cardiogenic shock after myocardial ischemia–reperfusion injury was examined (Guariento et al., 2021). The results of this trial showed that AMT/T was associated with superior ventricular strain magnitude, shorter recovery time, and durable time separation from ECMO (Guariento et al., 2021).

So far, the use of AMT/T in treating different modeled diseases and in clinical trials has not been shown to induce an observable inflammatory response as those associated with damage-associated molecular patterns. These observations have led to further use of allogenic cells or tissue as a source of mitochondria. At present, AMT/T is being studied and investigated, mostly directed to cardiac cells. Yet further investigations to improve knowledge on mitochondria availability, viability, delivery method, dosage, and timing are being done for future potential therapeutic opportunities (Caìcedo et al., 2017).

A transition from the present to the close future of AMT/T

Mitochondria fulfill several functions inside and outside the cell, and its deregulation is associated with premature aging, cardiomyopathies, metabolic syndrome, neurodegenerative diseases, immune system disorders, and cancer development (Ahlqvist et al., 2012; Friedland-Leuner et al., 2014; Liu et al., 2022; Nunnari & Suomalainen, 2012). Moreover, since mitochondrial diseases are incurable, AMT/T has great potential to be used as a therapeutic option. On the other hand, it has been seen that AMT/T is not only useful in patients with diseases related to mitochondrial dysfunction but can also serve as a revitalization treatment for any damaged organ. Several companies and entrepreneurial initiatives have been founded to exploit these new therapies based on AMT/T. The leading companies in this field include Minovia Therapeutics, Mitobridge, Cellvie, Dragon Biomed, and LUCA Science Inc., along with university spin-offs such as Dragon Biomed. The following information is intended to provide an overview of companies that specialize in AMT/T. This overview is based on publicly available information that may be subject to change. The details presented were collected from the companies' websites and patent data were obtained from the Espacenet and PatentScope websites.

To start with, Minovia Therapeutics aims to treat patients with single large scale mtDNA deletion syndromes which cause multisystemic disease, such as Pearson or Kearns–Sayre syndromes, in childhood or adulthood, respectively. Pearson syndrome negatively affects hematopoietic stem cells (HSCs) and pancreas function. For this purpose, the safety and therapeutic effects of transplantation of autologous $CD34^+$ cells enriched with healthy blood derived mitochondria through MAT were evaluated in interventional phase I and II clinical studies. In a recently published article, $CD34^+$ cells treated with MAT resulted in a decrease of heteroplasmy, the presence of healthy mtDNA, and improvement of ATP production in peripheral blood mononuclear cells, leading to better muscle quality, endurance, strength, and weight (Jacoby et al., 2022). In summary, MAT is applied for the enrichment of patients' HSCs with healthy mitochondria derived from a donor cell (A study to evaluate the safety and therapeutic effects of transplantation of MNV-BM-BLD in pediatric patients with pearson syndrome - full text view - ClinicalTrials.gov, n.d.). In addition, Minovia has been responsible for an observational study focused on the collection of mitochondria from placenta tissue for the use of MAT. This study was conducted in 2018 with 20 participants between 18 and 40 years old. The only factors taken into account were the weight and health of a placenta belonging to a mother who had not received any treatment before or after the placenta donation (Collection of placenta for mitochondria isolation - full text view - ClinicalTrials.gov, n.d.).

Regarding Minovia's patents, in accordance with the Espacenet and PatentScope websites, these are focused on the protection of MAT for renal, liver, pancreatic, brain, muscle, ocular, and primary mitochondrial diseases (Espacenet - advanced search, n.d.). Key processes like the use of intact mitochondria for skin and hair loss management, the production of stem cells augmented with functional mitochondria, and the induction of angiogenesis were additionally granted (Espacenet - advanced search, n.d.).

Mitobridge is another company dedicated to the development of new therapeutic options for diseases with nonfunctional mitochondria. They are also involved in clinical trials for the study of mitophagy, dynamics malfunction, oxidative phosphorylation, oxidative stress, vascular inflammation, and many other processes. Furthermore, Mitobridge has also been engaged in studying the role that dysfunctional mitochondria play over the progression of age related muscle decline and frailty. This observational study started in 2015 and ended in 2016 with no conclusive results in the muscle biopsy due to the lack of biomarkers at that time. However, a reduced autophagic flux and decreased elimination of dysfunctional elements were demonstrated (Mitophagy and autophagy in elderly subjects - full text view - ClinicalTrials.gov, n.d.).

Concerning Mitobridge's patents, some of them are related to modulate PPARδ (Peroxisome Proliferator-Activated Receptor Delta) to improve muscle strength and mitochondrial fatty acid utilization. The application of these treatments aims to heal patients with mitochondrial vascular, muscular, and demyelinating diseases. Additionally, Mitobridge has other patents focused on the intercellular transfer of isolated mitochondria and on treatments of mitochondrial diseases (Espacenet - advanced search, n.d.).

Cellvie is another MitoCompany dedicated to the development of therapies based on AMT/T with the main goal of reducing the impact of ischemia/reperfusion injury. Their focus of research and development is on kidney transplantation, given the high demand for these organs and the high number of patient deaths. It is expected that AMT/T will translate into a higher probability of survival. In addition, the company plans to expand its services and technologies to solid organ transplantation, long term cardiac surgery, and even the treatment of degenerative conditions associated with aging with AMT/T (Home - cellvie Inc., n.d.).

Thus, a clinical trial is currently underway to evaluate the transplantation of autologous mitochondria to cardiac tissue affected by ischemia/reperfusion injury in patients with ECMO. The interventional clinical trial has 16 participants enrolled. It began on August 2nd, 2017. Mitochondria are obtained during the reoperation or catheterization, by removing biopsies of approximately 6 mm from skeletal chest muscle, and processed immediately (Transplantation of autologously derived mitochondria following ischemia - full text view - ClinicalTrials.gov, n.d.).

The application of isolated mitochondria, regarding their type and concentration, may depend on the type of medical procedure and disease to be treated. Therefore, if the patient is subjected to a reoperation, the mitochondria are transplanted via injection into the proximal aorta to ensure adequate distribution through the coronary arteries. Once the injection is completed, the cross-clamp can be removed. If a cross-clamp is not required, the injection is made into the myocardium through the ventricle. Similarly, if there is tissue damage, mitochondria will be applied by a series of five or 10 injections in the affected area. On the other hand, if a catheterization takes place, the transplantation method relies on intracoronary infusion of the mitochondria suspended in a 5 mL buffer followed by a 5 mL normal saline flush (Transplantation of autologously derived mitochondria following ischemia - full text view - ClinicalTrials.gov, n.d.).

In addition, the company and their founders have developed different procedures, treatments, and patented products including a kit and method for mitochondria isolation (WO2015192020 products and methods to isolate mitochondria, n.d.), reactants for mitochondrial extraction (WO2017095946 methods and compositions relating to chondrisomes, n.d.), compositions for cultured cells, mitochondria and associated agents used in therapeutic treatments for cancer and heart failure (WO2017124037 therapeutic use of mitochondria and combined mitochondrial agents, n.d.), automated methods of mitochondrial isolation, and aerosolized mitochondria compositions (Espacenet — search results, n.d.).

Another example is the entrepreneurial initiative of the Universidad San Francisco de Quito (USFQ) Dragon Biomed. Dragon Biomed's main focus is to understand cellular and tissue processes that are involved in regeneration, illness development, and aging. In addition, it works on prognosis biomarkers as well as AMT/T and its effects (Dragon BioMed Universidad San Francisco de Quito, n.d.). The research done by previous collaborations of the inventors in this entrepreneurial initiative has resulted in one patent which involves a procedure of mitochondrial isolation for further enrichment of AMT/T to recipient cells (WO2016008937 methods for the intercellular transfer of isolated mitochondria in recipient cells, n.d.).

Finally, LUCA Science Inc. main purpose is to apply isolated mitochondria as a biopharmaceutical product for several therapeutic areas that include oncology, neurology, obstetrics, and cardiology, among others (LUCA science Inc., n.d.). The company remains on the preclinical stage but has generated several patents as well, by the company itself and its researchers. These include procedures for isolation and AMT/T to lymphoid tissue (WO2020054829 transplantation of mitochondria into lymphoid organ and composition therefor, n.d.), methods for respiration activity of mitochondrial complex (WO2020054824 method for measuring activity of mitochondrial respiratory complex, n.d.), lipid nanoparticle manufacturing process that can act as MITO-Porters (WO2020203961 lipid membrane structure and manufacturing method therefor, n.d.), procedures to isolate and encapsulate mitochondria (WO2021015298 method of obtaining mitochondria from cells and obtained mitochondria, n.d.), lipid membranes to deliver substances to mitochondria and their respective recombinant vectors of expression (WO2021132735 isolated mitochondria with smaller size and lipid membrane-based vesicles encapsulating isolated mitochondria, n.d.), nucleic acids for mitochondrial delivery and protein expression, kits for mitochondrial ribosomal RNA mutation and specific tDNA mutation at position 3243 in mtDNA (WO2020230601 Kit including primer DNA set for detecting mitochondrial ribosomal RNA mutation, nucleic acid for expressing mitochondrial ribosomal RNA, lipid membrane structure obtained by encapsulating the nucleic acid, and uses of these, n.d.), lipid nanoparticles for the transportation of soluble compounds, mitochondrial respirometry to evaluate mitochondrial function in healthy and sick tissues, and compounds known to stimulate the activity of mitochondria (Espacenet – search results, n.d.).

Concluding remarks

In spite of the high expectations for the future of mitochondrial transplantation, there are still questions to be answered, such as which source is best for isolating therapeutic mitochondria and how patents would affect the further development of AMT/T by scientists and physicians in public and private institutions. One key factor that may be limiting the exponential growth of the use of AMT/T is the source of cells and tissues to isolate mitochondria. Isolating mitochondria from cultured cells could represent the most plausible option, in order to obtain a high concentration, purity, and replicability among different assays. As more knowledge is generated about the origin and physiological role of extracellular mitochondria in blood, this could bring important cues about how to use them as a therapeutic agent (Borcherding et al., 2022). This could suggest that the use of AMT/T in the form of intravascular infusions may lead to protecting harmed organs and achieving positive effects in those suffering from disease or obesity.

The options for the therapeutic delivery of extracellular mitochondria are becoming more diverse, including in situ injection, aerosolized compositions, mitochondria mixed with hydrogels, and the ex vivo transfer to cells that are reinjected in the bodies of patients. Different companies are developing their own application of AMT/T, ranging from in situ cardiac delivery to MAT to treat mitochondrial hereditary diseases. In the close future, based on the evidence published in recent clinical trials, mitochondria could become a therapeutic option for those in need. It is possible to suggest that MAT could represent a new therapy for mitochondrial diseases and to regenerate ischemic tissue. Nevertheless, AMT/T could become an expensive therapeutic option similar to available gene therapies (Hendricks et al., 2022).

In the far future, we expect AMT/T to become part of the field of synthetic biology. We suggest that synthetic biology may develop artificial mitochondria, contributing to the manufacturing of an organelle resembling their natural counterparts and performing the same functions. The procedure commonly used for these studies is called membrane protein reconstitution, in which protein complexes are isolated and purified to be inserted in an artificial membranous environment. Future investigations aim to assemble a functional respiratory chain putting together at least the following proteins: Complex I, III, IV, and ATP synthase (Biner et al., 2018).

The future of mitochondrial transplantation, or AMT/T, holds great promise for the treatment of mitochondrial diseases and the regeneration of ischemic tissue. Despite the remaining challenges, the future of AMT/T is promising, with the potential to greatly benefit those suffering from mitochondrial diseases and other conditions. Mitochondria companies are dedicated to the development of AMT/T as a therapeutic option or "Living Drug" for mitochondrial dysfunction and even aging (Caicedo et al., 2023; Caicedo & Singh, 2023). These companies hold patents focused on the protection and application of AMT/T for which further research and clinical trials are underway to evaluate the transplantation of autologous or allogeneic mitochondria in various organs and conditions.

Acknowledgments

We thank the School of Medicine at the Universidad San Francisco de Quito, USFQ, the "Instituto de Investigaciones en Biomedicina, USFQ," and the Mito-Act Research Consortium in Quito, Ecuador, for their constant support of our work and initiatives. Romina Maya wishes to express her profound gratitude to her parents for their unconditional love and support, and for being her steadfast source of strength. Andrés Caicedo expresses his heartfelt gratitude to his wife, Katherine Ramírez, for her unconditional love, patience, and support.

Author contributions

RM, AV, and AC wrote the manuscript; RM, AV, PR, and AC reviewed the manuscript and cured the information; RM, AV, AC critically reviewed and commented on the manuscript to its final form and approval. AC supervised and mentored the authors team. AC conceptualization.

Funding

This work was supported by the Air Force Office of Scientific Research (AFOSR) grant (FA9550-19-S-0003; FA9550-20-1-0407, Project Leader Andrés Caicedo), and also partially supported by the Escuela de Medicina, Colegio de Ciencias de la Salud COCSA, Universidad San Francisco de Quito USFQ.

Declaration of competing interest

Andrés Caicedo is the scientific founder and advisor of Dragon Biomed, an entrepreneurial initiative at the Universidad San Francisco de Quito (USFQ). He also serves as a scientific advisor in the Research and Development department of Luvigix. In these roles, he provides scientific guidance and expertise but does not participate in the decision-making processes or operational activities of either company. Other authors declare that they have no known competing financial interests or personal relationships that could have appeared to influence the work reported in this paper.

References

A Study to evaluate the safety and therapeutic effects of transplantation of MNV-BM-BLD in pediatric patients with pearson syndrome - full text view - ClinicalTrials.gov (n.d.). Retrieved from https://clinicaltrials.gov/ct2/show/NCT03384420?term=minovia&draw=2&rank=3. (Accessed 9 December 2022).

Ahlqvist, K. J., Hämäläinen, R. H., Yatsuga, S., Uutela, M., Terzioglu, M., Götz, A., Forsström, S., et al. (2012). Somatic progenitor cell vulnerability to mitochondrial DNA mutagenesis underlies progeroid phenotypes in polg mutator mice. *Cell Metabolism, 15*(1), 100–109.

Al Amir Dache, Z., Otandault, A., Tanos, R., Pastor, B., Meddeb, R., Sanchez, C., Arena, G., et al. (2020). Blood contains circulating cell-free respiratory competent mitochondria. *The FASEB Journal, 34*(3), 3616–3630.

Babenko, V. A., Silachev, D. N., Popkov, V. A., Zorova, L. D., Pevzner, I. B., Plotnikov, E. Y., Sukhikh, G. T., et al. (2018). Miro1 enhances mitochondria transfer from multipotent mesenchymal stem cells (MMSC) to neural cells and improves the efficacy of cell recovery. *Molecules, 23*(3).

Balcázar, M., Cañizares, S., Borja, T., Pontón, P., Bisiou, S., Carabasse, E., Bacilieri, A., et al. (2020). Bases for treating skin aging with artificial mitochondrial transfer/transplant (AMT/T). *Frontiers in Bioengineering and Biotechnology, 8*, 919.

Bezdan, D., Grigorev, K., Meydan, C., Vatter, F. A. P., Cioffi, M., Rao, V., MacKay, M., Nakahira, K., Burnham, P., Afshinnekoo, E., & Westover, C. (2020). Cell-free DNA (cfDNA) and exosome profiling from a year-long human spaceflight reveals circulating biomarkers. *Iscience, 23*(12).

Biner, O., Schick, T., Ganguin, A. A., & von Ballmoos, C. (2018). Towards a synthetic mitochondrion. *Chimia, 72*(5), 291–296.

Bisserier, M., Shanmughapriya, S., Rai, A. K., Gonzalez, C., Brojakowska, A., Garikipati, V. N. S., Madesh, M., Mills, P. J., Walsh, K., Arakelyan, A., & Kishore, R. (2021). Cell-free mitochondrial DNA as a potential biomarker for astronauts' health. *Journal of the American Heart Association, 10*(21), e022055.

Borcherding, N., Jia, W., Giwa, R., Field, R. L., Moley, J. R., Kopecky, B. J., Chan, M. M., et al. (2022). Dietary lipids inhibit mitochondria transfer to macrophages to divert adipocyte-derived mitochondria into the blood. *Cell Metabolism, 34*(10), 1499–1513.e8.

Cabrera, F., Castañeda, V., Morales, E., Velarde, F., Ortega, M., Leon-Sosa, A., Jorgensen, C., et al. (2022). Early evidence of the artificial transfer/transplant of mitochondria to oocytes and zygotes by MitoCeption. *Mitochondrion, 65*, 102–112.

Cabrera, F., Ortega, M., Velarde, F., Parra, E., Gallardo, S., Barba, D., Soto, L., et al. (2019). Primary allogeneic mitochondrial mix (PAMM) transfer/transplant by MitoCeption to address damage in PBMCs caused by ultraviolet radiation. *BMC Biotechnology, 19*(1), 42.

Caicedo, A., Aponte, P. M., Cabrera, F., Hidalgo, C., & Khoury, M. (2017). Artificial mitochondria transfer: Current challenges, advances, and future applications. *Stem Cells International, 2017*, 7610414.

Caicedo, A., Benavides-Almeida, A., Haro-Vinueza, A., Peña-Cisneros, J., Pérez-Meza, Á. A., Michelson, J., Peñaherrera, S., & Picard, M. (2024). Decoding the nature and complexity of extracellular mtDNA: Types and implications for health and disease. *Mitochondrion, 75*, 101848.

Caicedo, A., Fritz, V., Brondello, J.-M., Ayala, M., Dennemont, I., Abdellaoui, N., de Fraipont, F., et al. (2015). MitoCeption as a new tool to assess the effects of mesenchymal stem/stromal cell mitochondria on cancer cell metabolism and function. *Scientific Reports, 5*, 9073.

Caicedo, A., Morales, E., Moyano, A., Peñaherrera, S., Peña-Cisneros, J., Benavides-Almeida, A., Pérez-Meza, Á. A., Haro-Vinueza, A., Ruiz, C., Robayo, P., & Tenesaca, D. (2023). Powering Prescription: Mitochondria as "Living Drugs"-Definition. *Clinical Applications, and Industry Advancements. Pharmacological Research*, 107018.

Caicedo, A., & Singh, K. K. (2023). Advancing mitochondria as a therapeutic agent. *Mitochondrion, 69*, 33–35.

Caicedo, A., & Singh, K. K. (2023). Mitochondria makeover: unlocking the path to healthy longevity. *Expert Opinion on Therapeutic Targets*, 1–4.

Caicedo, A., Zambrano, K., Sanon, S., & Gavilanes, A. W. D. (2021). Extracellular mitochondria in the cerebrospinal fluid (CSF): Potential types and key roles in central nervous system (CNS) physiology and pathogenesis. *Mitochondrion, 58*, 255–269.

Caicedo, A., Zambrano, K., Sanon, S., Luis Vélez, J., Montalvo, M., Jara, F., Moscoso, S. A., et al. (2021). The diversity and coexistence of extracellular mitochondria in circulation: A friend or foe of the immune system. *Mitochondrion, 58*, 270–284.

Chang, J.-C., Hoel, F., Liu, K.-H., Wei, Y.-H., Cheng, F.-C., Kuo, S.-J., Tronstad, K. J., et al. (2017). Peptide-mediated delivery of donor mitochondria improves mitochondrial function and cell viability in human cybrid cells with the MELAS A3243G mutation. *Scientific Reports, 7*(1), 10710.

Chang, J.-C., Wu, S.-L., Liu, K.-H., Chen, Y.-H., Chuang, C.-S., Cheng, F.-C., Su, H.-L., et al. (2016). Allogeneic/xenogeneic transplantation of peptide-labeled mitochondria in Parkinson's disease: Restoration of mitochondria functions and attenuation of 6-hydroxydopamine-induced neurotoxicity. *Translational Research, 170*, 40–56.e3.

Chinnery, P. F., & Schon, E. A. (2003). Mitochondria. *Journal of Neurology, Neurosurgery, and Psychiatry, 74*(9), 1188–1199.

Clark, M. A., & Shay, J. W. (1982). Mitochondrial transformation of mammalian cells. *Nature, 295*(5850), 605–607.

CO2022015278A2 - genetically modified cells enriched with mitochondria and their uses - Google Patents (n.d.). Retrieved from https://patents.google.com/patent/CO2022015278A2/en?inventor=NATALIE+YIVGI-OHANA&oq=NATALIE+YIVGI-OHANA. (Accessed 26 January 2023).

Collection of Placenta for mitochondria isolation - full text view - ClinicalTrials.gov (n.d.). Retrieved from https://clinicaltrials.gov/ct2/show/NCT03762512?term=minovia&draw=2&rank=2. (Accessed 9 December 2022).

Crewe, C., Funcke, J.-B., Li, S., Joffin, N., Gliniak, C. M., Ghaben, A. L., An, Y. A., et al. (2021). Extracellular vesicle-based interorgan transport of mitochondria from energetically stressed adipocytes. *Cell Metabolism, 33*(9), 1853–1868.e11.

D'Souza, A., Burch, A., Dave, K. M., Sreeram, A., Reynolds, M. J., Dobbins, D. X., Kamte, Y. S., et al. (2021). Microvesicles transfer mitochondria and increase mitochondrial function in brain endothelial cells. *Journal of Controlled Release, 338*, 505–526.

Doulamis, I. P., Guariento, A., Duignan, T., Kido, T., Orfany, A., Saeed, M. Y., Weixler, V. H., et al. (2020). Mitochondrial transplantation by intra-arterial injection for acute kidney injury. *American Journal of Physiology - Renal Physiology, 319*(3), F403–F413.

Doyle, L. M., & Wang, M. Z. (2019). Overview of extracellular vesicles, their origin, composition, purpose, and methods for exosome isolation and analysis. *Cells, 8*(7).

Dragon BioMed Universidad San Francisco de Quito (n.d.). Retrieved from https://www.usfq.edu.ec/es/dragon-biomed. (Accessed 31 January 2023).

Espacenet - advanced search (n.d.). Retrieved from https://worldwide.espacenet.com/advancedSearch?locale=en_EP. (Accessed 8 February 2023).

Espacenet – search results (n.d.). Retrieved from https://worldwide.espacenet.com/patent/search/family/069180899/publication/WO2020021534A1?q=pn%3DWO2020021534A1. (Accessed 7 February 2023).

Evans, W. H., & Martin, P. E. M. (2002). Gap junctions: Structure and function (review). *Molecular Membrane Biology, 19*(2), 121–136.

Friedland-Leuner, K., Stockburger, C., Denzer, I., Eckert, G. P., & Müller, W. E. (2014). Mitochondrial dysfunction: Cause and consequence of Alzheimer's disease. *Progress in Molecular Biology and Translational Science, 127*, 183–210.

Gómez, X., Sanon, S., Zambrano, K., Asquel, S., Bassantes, M., Morales, J. E., Otáñez, G., et al. (2021). Key points for the development of antioxidant cocktails to prevent cellular stress and damage caused by reactive oxygen species (ROS) during manned space missions. *NPJ Microgravity, 7*(1), 35.

Guariento, A., Blitzer, D., Doulamis, I., Shin, B., Moskowitzova, K., Orfany, A., Ramirez-Barbieri, G., et al. (2020). Preischemic autologous mitochondrial transplantation by intracoronary injection for myocardial protection. *The Journal of Thoracic and Cardiovascular Surgery, 160*(2), e15–e29.

Guariento, A., Doulamis, I. P., Duignan, T., Kido, T., Regan, W. L., Saeed, M. Y., Hoganson, D. M., et al. (2020). Mitochondrial transplantation for myocardial protection in ex-situ–perfused hearts donated after circulatory death. *The Journal of Heart and Lung Transplantation, 39*(11), 1279–1288.

Guariento, A., Piekarski, B. L., Doulamis, I. P., Blitzer, D., Ferraro, A. M., Harrild, D. M., Zurakowski, D., et al. (2021). Autologous mitochondrial transplantation for cardiogenic shock in pediatric patients following ischemia-reperfusion injury. *The Journal of Thoracic and Cardiovascular Surgery, 162*(3), 992–1001.

Hamanaka, R. B., & Chandel, N. S. (2010). Mitochondrial reactive oxygen species regulate cellular signaling and dictate biological outcomes. *Trends in Biochemical Sciences, 35*(9), 505–513.

Hendricks, C. L., Alessandrini, M., & Pepper, M. S. (2022). Equitable access to cell and gene therapies in South Africa: Opportunities and hurdles. *Gene Therapy, 30*.

Home - cellvie Inc. (n.d.). Retrieved from https://cellvie.bio/. (Accessed 30 January 2023).

Islam, M. N., Das, S. R., Emin, M. T., Wei, M., Sun, L., Westphalen, K., Rowlands, D. J., et al. (2012). Mitochondrial transfer from bone-marrow-derived stromal cells to pulmonary alveoli protects against acute lung injury. *Nature Medicine, 18*(5), 759–765.

Jabbari, H., Roushandeh, A. M., Rostami, M. K., Razavi-Toosi, M. T., Shokrgozar, M. A., Jahanian-Najafabadi, A., Kuwahara, Y., et al. (2020). Mitochondrial transplantation ameliorates ischemia/reperfusion-induced kidney injury in rat. *Biochimica et biophysica acta. Molecular Basis of Disease, 1866*(8), 165809.

Jacoby, E., Bar-Yosef, O., Gruber, N., Lahav, E., Varda-Bloom, N., Bolkier, Y., Bar, D., et al. (2022). Mitochondrial augmentation of hematopoietic stem cells in children with single large-scale mitochondrial DNA deletion syndromes. *Science Translational Medicine, 14*(676), eabo3724.

Jacoby, E., Ben Yakir-Blumkin, M., Blumenfeld-Kan, S., Brody, Y., Meir, A., Melamed-Book, N., Napso, T., et al. (2021). Mitochondrial augmentation of CD34+ cells from healthy donors and patients with mitochondrial DNA disorders confers functional benefit. *Npj Regenerative Medicine, 6*(1), 58.

Jacoby, E., Blumkin, M., Anikster, Y., Varda-Bloom, N., Pansheen, J., Bar Yoseph, O., Gruber, N., et al. (2018). First-in-Human mitochondrial augmentation of hematopoietic stem cells in Pearson syndrome. *Blood, 132*(Suppl. 1), 1024.

Kubat, G. B., Ozler, M., Ulger, O., Ekinci, O., Atalay, O., Celik, E., Safali, M., et al. (2021). The effects of mesenchymal stem cell mitochondrial transplantation on doxorubicin-mediated nephrotoxicity in rats. *Journal of Biochemical and Molecular Toxicology, 35*(1), Article e22612.

Kuo, C.-C., Su, H.-L., Chang, T.-L., Chiang, C.-Y., Sheu, M.-L., Cheng, F.-C., Chen, C.-J., et al. (2017). Prevention of axonal degeneration by perineurium injection of mitochondria in a sciatic nerve crush injury model. *Neurosurgery, 80*(3), 475–488.

Lill, R., & Mühlenhoff, U. (2008). Maturation of iron-sulfur proteins in eukaryotes: Mechanisms, connected processes, and diseases. *Annual Review of Biochemistry, 77*, 669–700.

Lin, H.-C., Liu, S.-Y., Lai, H.-S., & Lai, I.-R. (2013). Isolated mitochondria infusion mitigates ischemia-reperfusion injury of the liver in rats. *Shock, 39*(3), 304–310.

Liu, D., Gao, Y., Liu, J., Huang, Y., Yin, J., Feng, Y., Shi, L., et al. (2021). Intercellular mitochondrial transfer as a means of tissue revitalization. *Signal Transduction and Targeted Therapy, 6*(1), 65.

Liu, Y., Chen, C., Wang, X., Sun, Y., Zhang, J., Chen, J., & Shi, Y. (2022). An epigenetic role of mitochondria in cancer. *Cells, 11*(16).

LUCA science Inc (n.d.). Retrieved from https://www.luca-science.com/. (Accessed 31 January 2023).

McBride, H.\M., Neuspiel, M., & Wasiak, S. (2006). Mitochondria: More than just a powerhouse. *Current Biology, 16*(14), R551–R560.

McCully, J., & Del Nido, P. (April 15, 2020). *WO2020214644A1 aerosolized compositions comprising mitochondria and methods of use thereof. Espacenet*. from https://worldwide.espacenet.com/patent/search/family/072837565/publication/WO2020214644A1?q=pn%3DWO2020214644A1. (Accessed 19 July 2022).

McCully, J. D., Cowan, D. B., Pacak, C. A., Toumpoulis, I. K., Dayalan, H., & Levitsky, S. (2009). Injection of isolated mitochondria during early reperfusion for cardioprotection. *American Journal of Physiology - Heart and Circulatory Physiology, 296*(1), H94–H105.

McCully, J. D., del Nido, P. J., & Emani, S. M. (2022a). Mitochondrial transplantation for organ rescue. *Mitochondrion, 64*, 27–33.

McCully, J. D., del Nido, P. J., & Emani, S. M. (2022b). Therapeutic mitochondrial transplantation. *Current Opinion in Physiology, 27*, 100558.

Melkov, A., & Abdu, U. (2018). Regulation of long-distance transport of mitochondria along microtubules. *Cellular and Molecular Life Sciences, 75*(2), 163–176.

Mitophagy and autophagy in elderly subjects - full text view - ClinicalTrials.gov (n.d.). Retrieved from https://clinicaltrials.gov/ct2/show/NCT02472340?cond=NCT02472340&draw=2&rank=1. (Accessed 9 December 2022).

Moskowitzova, K., Orfany, A., Liu, K., Ramirez-Barbieri, G., Thedsanamoorthy, J. K., Yao, R., Guariento, A., et al. (2020). Mitochondrial transplantation enhances murine lung viability and recovery after ischemia-reperfusion injury. *American Journal of Physiology - Lung Cellular and Molecular Physiology, 318*(1), L78–L88.

Nielsen, M. S., Axelsen, L. N., Sorgen, P. L., Verma, V., Delmar, M., & Holstein-Rathlou, N.-H. (2012). Gap junctions. *Comprehensive Physiology, 2*(3), 1981–2035.

Nobs, S.-J., MacLeod, F. I., Wong, H. L., & Burns, B. P. (2022). Eukarya the chimera: Eukaryotes, a secondary innovation of the two domains of life? *Trends in Microbiology, 30*(5), 421–431.

Nunnari, J., & Suomalainen, A. (2012). Mitochondria: In sickness and in health. *Cell, 148*(6), 1145–1159.

Nzigou Mombo, B., Gerbal-Chaloin, S., Bokus, A., Daujat-Chavanieu, M., Jorgensen, C., Hugnot, J.-P., & Vignais, M.-L. (2017). Mitoception: Transferring isolated human MSC mitochondria to glioblastoma stem cells. *Journal of Visualized Experiments, 120*.

Orfany, A., Arriola, C. G., Doulamis, I. P., Guariento, A., Ramirez-Barbieri, G., Moskowitzova, K., Shin, B., et al. (2020). Mitochondrial transplantation ameliorates acute limb ischemia. *Journal of Vascular Surgery, 71*(3), 1014–1026.

Paliwal, S., Chaudhuri, R., Agrawal, A., & Mohanty, S. (2018). Regenerative abilities of mesenchymal stem cells through mitochondrial transfer. *Journal of Biomedical Science, 25*(1), 31.

Palozzi, J. M., Jeedigunta, S. P., Minenkova, A. V., Monteiro, V. L., Thompson, Z. S., Lieber, T., & Hurd, T. R. (2022). Mitochondrial DNA quality control in the female germline requires a unique programmed mitophagy. *Cell Metabolism, 34*(11), 1809–1823.e6.

Patel, S. P., Michael, F. M., Arif Khan, M., Duggan, B., Wyse, S., Darby, D. R., Chaudhuri, K., et al. (2022). Erodible thermogelling hydrogels for localized mitochondrial transplantation to the spinal cord. *Mitochondrion, 64*, 145–155.

Qin, Y., Jiang, X., Yang, Q., Zhao, J., Zhou, Q., & Zhou, Y. (2021). The functions, methods, and mobility of mitochondrial transfer between cells. *Frontiers in Oncology, 11*, 672781.

Schmitt, S., Saathoff, F., Meissner, L., Schropp, E.-M., Lichtmannegger, J., Schulz, S., Eberhagen, C., et al. (2013). A semi-automated method for isolating functionally intact mitochondria from cultured cells and tissue biopsies. *Analytical Biochemistry, 443*(1), 66–74.

Shin, B., Saeed, M. Y., Esch, J. J., Guariento, A., Blitzer, D., Moskowitzova, K., Ramirez-Barbieri, G., et al. (2019). A novel biological strategy for myocardial protection by intracoronary delivery of mitochondria: Safety and efficacy. *JACC. Basic to Translational Science, 4*(8), 871–888.

Song, X., Hu, W., Yu, H., Wang, H., Zhao, Y., Korngold, R., & Zhao, Y. (2020). Existence of circulating mitochondria in human and animal peripheral blood. *International Journal of Molecular Sciences, 21*(6).

Torralba, D., Baixauli, F., & Sánchez-Madrid, F. (2016). Mitochondria know no boundaries: Mechanisms and functions of intercellular mitochondrial transfer. *Frontiers in Cell and Developmental Biology, 4*, 107.

Transplantation of autologously derived mitochondria following ischemia - full text view - ClinicalTrials.gov (n.d.). Retrieved from https://clinicaltrials.gov/ct2/show/NCT02851758?term=mitochondria&draw=2&rank=1. (Accessed 30 January 2023).

Ulger, O., Kubat, G. B., Cicek, Z., Celik, E., Atalay, O., Suvay, S., & Ozler, M. (2021). The effects of mitochondrial transplantation in acetaminophen-induced liver toxicity in rats. *Life Sciences, 279*, 119669.

US11441124B2 - Mammalian cells enriched with functional mitochondria - Google Patents (n.d.). Retrieved from https://patents.google.com/patent/US11441124B2/en?inventor=NATALIE+YIVGI-OHANA&oq=NATALIE+YIVGI-OHANA. (Accessed 26 January 2023).

van Niel, G., D'Angelo, G., & Raposo, G. (2018). Shedding light on the cell biology of extracellular vesicles. *Nature Reviews Molecular Cell Biology, 19*(4), 213–228.

Velarde, F., Ezquerra, S., Delbruyere, X., Caicedo, A., Hidalgo, Y., & Khoury, M. (2022). Mesenchymal stem cell-mediated transfer of mitochondria: Mechanisms and functional impact. *Cellular and Molecular Life Sciences, 79*(3), 177.

Vignais, M.-L., Caicedo, A., Brondello, J.-M., & Jorgensen, C. (2017). Cell connections by tunneling nanotubes: Effects of mitochondrial trafficking on target cell metabolism, homeostasis, and response to therapy. *Stem Cells International, 2017*, Article 6917941.

WO2015192020 products and methods to isolate mitochondria (n.d.). Retrieved from https://patentscope.wipo.int/search/en/detail.jsf?docId=WO2015192020. (Accessed 8 February 2023).

WO2016008937 methods for the intercellular transfer of isolated mitochondria in recipient cells (n.d.). Retrieved from https://patentscope.wipo.int/search/en/detail.jsf?docId=WO2016008937. (Accessed 8 February 2023).

WO2017095946 methods and compositions relating to chondrisomes (n.d.). Retrieved from https://patentscope.wipo.int/search/en/detail.jsf?docId=WO2017095946. (Accessed 8 February 2023).

WO2017124037 therapeutic use of mitochondria and combined mitochondrial agents (n.d.). Retrieved from https://patentscope.wipo.int/search/en/detail.jsf?docId=WO2017124037. (Accessed 8 February 2023).

WO2020054824 method for measuring activity of mitochondrial respiratory complex (n.d.). Retrieved from https://patentscope.wipo.int/search/en/detail.jsf?docId=WO2020054824. (Accessed 8 February 2023).

WO2020054829 transplantation of mitochondria into lymphoid organ and composition therefor (n.d.). Retrieved from https://patentscope.wipo.int/search/en/detail.jsf?docId=WO2020054829. (Accessed 8 February 2023).

WO2020203961 lipid membrane structure and manufacturing method therefor (n.d.). Retrieved from https://patentscope.wipo.int/search/en/detail.jsf?docId=WO2020203961. (Accessed 8 February 2023).

WO2020230601 kit including primer DNA set for detecting mitochondrial ribosomal RNA mutation, nucleic acid for expressing mitochondrial ribosomal RNA, lipid membrane structure obtained by encapsulating the nucleic acid, and uses of these (n.d.). Retrieved from https://patentscope.wipo.int/search/en/detail.jsf?docId=WO2020230601. (Accessed 8 February 2023).

WO2021015298 method of obtaining mitochondria from cells and obtained mitochondria (n.d.). Retrieved from https://patentscope.wipo.int/search/en/detail.jsf?docId=WO2021015298. (Accessed 8 February 2023).

WO2021132735 isolated mitochondria with smaller size and lipid membrane-based vesicles encapsulating isolated mitochondria (n.d.). Retrieved from https://patentscope.wipo.int/search/en/detail.jsf?docId=WO2021132735. (Accessed 8 February 2023).

Yivgi-Ohana, N., & Halavee, U. (2019). *Compositions of functional mitochondria and uses thereof.* U.S. Patent No. 10,213,459, (10).

Zambrano, K., Barba, D., Castillo, K., Noboa, L., Argueta-Zamora, D., Robayo, P., Arizaga, E., et al. (2022). Fighting Parkinson's disease: The return of the mitochondria. *Mitochondrion, 64*, 34–44.

Zambrano, K., Barba, D., Castillo, K., Robayo, P., Argueta-Zamora, D., Sanon, S., Arizaga, E., et al. (2022). The war against Alzheimer, the mitochondrion strikes back. *Mitochondrion, 64*, 125–135.

Zambrano, K., Barba, D., Castillo, K., Robayo, P., Arizaga, E., Caicedo, A., & Gavilanes, A. W. D. (2022). A new hope: Mitochondria, a critical factor in the war against prions. *Mitochondrion, 65*, 113–123.

Zhang, Z., Ma, Z., Yan, C., Pu, K., Wu, M., Bai, J., Li, Y., et al. (2019). Muscle-derived autologous mitochondrial transplantation: A novel strategy for treating cerebral ischemic injury. *Behavioural Brain Research, 356*, 322–331.

Index

Note: 'Page numbers followed by *f* indicate figures and *t* indicate tables.'

Made in the USA
Monee, IL
14 September 2026

63198303R10308